Metal Oxides for Biomedical and Biosensor Applications

The Metal Oxides Book Series Edited by Ghenadii Korotcenkov

Forthcoming Titles

- *Palladium Oxides Material Properties, Synthesis and Processing Methods, and Applications*, Alexander M. Samoylov, Vasily N. Popov, 9780128192238
- *Metal Oxides for Non-volatile Memory*, Panagiotis Dimitrakis, Ilia Valov, Stefan Tappertzhofen, 9780128146293
- *Metal Oxide Nanostructured Phosphors*, H. Nagabhushana, Daruka Prasad, S.C. Sharma, 9780128118528
- *Nanostructured Zinc Oxide*, Kamlendra Awasthi, 9780128189009
- *Multifunctional Piezoelectric Oxide Nanostructures*, Sang-Jae Kim, Nagamalleswara Rao Alluri, Yuvasree Purusothaman, 9780128193327
- *Transparent Conductive Oxides*, Mirela Petruta Suchea, Petronela Pascariu, Emmanouel Koudoumas, 9780128206317
- *Metal oxide-based nanofibers and their applications*, Vincenzo Esposito, Debora Marani, 9780128206294
- *Metal-oxides for Biomedical and Biosensor Applications*, Kunal Mondal, 9780128230336
- *Metal Oxide-Carbon Hybrid Materials*, Muhammad Akram, Rafaqat Hussain, Faheem K Butt, 9780128226940
- *Metal Oxide-based heterostructures*, Naveen Kumar, Bernabe Mari Soucase, 9780323852418
- *Metal Oxides and Related Solids for Electrocatalytic Water Splitting*, Junlei Qi, 9780323857352
- *Advances in Metal Oxides and Their Composites for Emerging Applications*, Sagar Delekar, 9780323857055
- *Metallic Glasses and Their Oxidation*, Xinyun Wang, Mao Zhang, 9780323909976
- *Solution Methods for Metal Oxide Nanostructures*, Rajaram S. Mane, Vijaykumar Jadhav, Abdullah M. Al-Enizi, 9780128243534
- *Metal Oxide Defects*, Vijay Kumar, Sudipta Som, Vishal Sharma, Hendrik Swart, 9780323855884
- *Renewable Polymers and Polymer-Metal Oxide Composites*, Sajjad Haider, Adnan Haider, 9780323851558
- *Metal Oxides for Optoelectronics and Optics-based Medical Applications*, Suresh Sagadevan, Jiban Podder, Faruq Mohammad, 9780323858243
- *Graphene Oxide-Metal Oxide and Other Graphene Oxide-Based Composites in Photocatalysis and Electrocatalysis*, Jiaguo Yu, Liuyang Zhang, Panyong Kuang, 9780128245262

Published Titles

- *Metal Oxides in Nanocomposite-Based Electrochemical Sensors for Toxic Chemicals*, Alagarsamy Pandikumar, Perumal Rameshkumar, 9780128207277
- *Metal Oxide-Based Nanostructured Electrocatalysts for Fuel Cells, Electrolyzers, and Metal-Air Batteries*, Teko Napporn, Yaovi Holade, 9780128184967
- *Titanium Dioxide (TiO2) and its applications*, Leonardo Palmisano, Francesco Parrino, 9780128199602
- *Solution Processed Metal Oxide Thin Films for Electronic Applications*, Zheng Cui, 9780128149300
- *Metal Oxide Powder Technologies*, Yarub Al-Douri, 9780128175057
- *Colloidal Metal Oxide Nanoparticles*, Sabu Thomas, Anu Tresa Sunny, Prajitha V, 9780128133576
- *Cerium Oxide*, Salvatore Scire, Leonardo Palmisano, 9780128156612
- *Tin Oxide Materials*, Marcelo Ornaghi Orlandi, 9780128159248
- *Metal Oxide Glass Nanocomposites*, Sanjib Bhattacharya, 9780128174586
- *Gas Sensors Based on Conducting Metal Oxides*, Nicolae Barsan, Klaus Schierbaum, 9780128112243
- *Metal Oxides in Energy Technologies*, Yuping Wu, 9780128111673
- *Metal Oxide Nanostructures*, Daniela Nunes, Lidia Santos, Ana Pimentel, Pedro Barquinha, Luis Pereira, Elvira Fortunato, Rodrigo Martins, 9780128115121
- *Gallium Oxide*, Stephen Pearton, Fan Ren, Michael Mastro, 9780128145210
- *Metal Oxide-Based Photocatalysis*, Adriana Zaleska-Medynska, 9780128116340
- *Metal Oxides in Heterogeneous Catalysis*, Jacques C. Vedrine, 9780128116319
- *Magnetic, Ferroelectric, and Multiferroic Metal Oxides*, Biljana Stojanovic, 9780128111802
- *Iron Oxide Nanoparticles for Biomedical Applications*, Sophie Laurent, Morteza Mahmoudi, 9780081019252
- *The Future of Semiconductor Oxides in Next-Generation Solar Cells*, Monica Lira-Cantu, 9780128111659
- *Metal Oxide-Based Thin Film Structures, Nini Pryds*, Vincenzo Esposito, 9780128111666
- *Metal Oxides in Supercapacitors*, Deepak Dubal, Pedro Gomez-Romero, 9780128111697
- *Transition Metal Oxide Thin Film-Based Chromogenics and Devices*, Pandurang Ashrit, 9780081018996

Metal Oxides

Metal Oxides for Biomedical and Biosensor Applications

Edited by

Kunal Mondal

Materials Science and Engineering Department, Idaho National Laboratory, Idaho Falls, ID, United States

Series Editor

Ghenadii Korotcenkov

Department of Physics and Engineering, Moldova State University, Chisinau, Republic of Moldova; Proprietor—Battelle Energy Alliance, LLC (BEA), 2525 North Fremont Avenue, P.O. Box 1625, Idaho Falls, ID 83415-3899, United States

ELSEVIER

Elsevier
Radarweg 29, PO Box 211, 1000 AE Amsterdam, Netherlands
The Boulevard, Langford Lane, Kidlington, Oxford OX5 1GB, United Kingdom
50 Hampshire Street, 5th Floor, Cambridge, MA 02139, United States

British Library Cataloguing-in-Publication Data
A catalogue record for this book is available from the British Library

Library of Congress Cataloging-in-Publication Data
A catalog record for this book is available from the Library of Congress

ISBN: 978-0-12-823033-6

Publisher: Matthew Deans
Acquisitions Editor: Kayla Dos Santos
Editorial Project Manager: Ruby Smith
Production Project Manager: Debasish Ghosh
Cover Designer: Miles Hitchen

Typeset by MPS Limited, Chennai, India

Contents

List of contributors

Keshaw Ram Aadil Center for Basic Sciences, Pt. Ravishankar Shukla University, Raipur, India

Md. Azahar Ali Mechanical Engineering, Carnegie Mellon University, Pittsburgh, PA, United States

Bhuvaneshwari Balasubramaniam Department of Chemical Engineering, Indian Institute of Technology Kanpur, Kanpur, India

Tandrima Banerjee Department of Chemical Sciences, Indian Institute of Science Education and Research (IISER), Kolkata, India

Souravi Bardhan Department of Physics, Jadavpur University, Kolkata, India

Jayesh Bellare Department of Chemical Engineering, Indian Institute of Technology Bombay, Mumbai, India; Wadhwani Research Centre for Bioengineering, Indian Institute of Technology Bombay, Mumbai, India

Snehasis Biswas Department of Chemical Engineering, Indian Institute of Technology Bombay, Mumbai, India

Naveen Bunekar Department of Chemistry, Center for Nanotechnology, Chung Yuan Christian University, Chung Li, Taiwan (ROC)

Richa Chaturvedi Department of Chemical Engineering, Indian Institute of Technology Kanpur, Kanpur, India

Abhishek Chaudhary Department of Biotechnology and Bioinformatics, Jaypee University of Information Technology, Solan, India

Gaurav Chauhan School of Engineering and Sciences, Tecnologico de Monterrey, Monterrey, Mexico

Mohit Chawla Gyan Sankul, Gyan Sankul Complex, Kullu, India

Antonella V. Dan Córdoba Universidad Nacional de Córdoba, Facultad de Ciencias Químicas, Departamento de Ciencias Farmacéuticas, Ciudad Universitaria, Córdoba, Argentina; Consejo Nacional de Investigaciones Científicas y Técnicas, CONICET, Unidad de Investigación y Desarrollo en Tecnología Farmacéutica, UNITEFA, Córdoba, Argentina

Sukhen Das Department of Physics, Jadavpur University, Kolkata, India

Alokmay Datta CSIR-Central Glass and Ceramic Research Institute and University of Calcutta, Kolkata, India

Anca Dinischiotu Department of Biochemistry and Molecular Biology, Faculty of Biology, University of Bucharest, Bucharest, Romania

Sunil Dutt Department of Chemistry, Government Post Graduate College Una, Una, India

Suparna Dutta-Sinha Living Systems Institute, University of Exeter, Exeter, United Kingdom

Nirmal G. R. Graduate Institute of Biomedical Scienece, Chang Gung University, Taoyuan City, Taiwan

Mónica C. García Universidad Nacional de Córdoba, Facultad de Ciencias Químicas, Departamento de Ciencias Farmacéuticas, Ciudad Universitaria, Córdoba, Argentina; Consejo Nacional de Investigaciones Científicas y Técnicas, CONICET, Unidad de Investigación y Desarrollo en Tecnología Farmacéutica, UNITEFA, Córdoba, Argentina

Bidipta Ghosh Department of Chemical Engineering, National Institute of Technology Durgapur, Durgapur, India

Agnivo Gosai Corning Inc., Science & Technology, Painted Post, NY, United States

Kishan Gugulothu Department of chemistry, Osmania University, Hyderabad, India

Abhishek Gupta Department of Desalination and Water Treatment, Zuckerberg Institute for Water Research, The Jacob Blaustein Institutes for Desert Research, Ben-Gurion University of the Negev, Beer Sheva, Israel

Abhishek Kumar Gupta Organic Semiconductor Centre, EaStCHEM School of Chemistry, University of St. Andrews, Fife, United Kingdom; Organic Semiconductor Centre, SUPA, School of Physics and Astronomy, University of St. Andrews, Fife, United Kingdom

Ankur Gupta Department of Mechanical Engineering, Indian Institute of Technology, Jodhpur, India

Raju Kumar Gupta Department of Chemical Engineering, Indian Institute of Technology Kanpur, Kanpur, India

Uttam Gupta Max Planck Institute for Chemical Physics of Solids, Dresden, Germany

James Hartmann Florida Atlantic University, Boca Raton, FL, United States

Monsur Islam Institute of Microstructure Technology, Karlsruhe Institute of Technology, Karlsruhe, Germany

Vinay Kishnani Department of Mechanical Engineering, Indian Institute of Technology, Jodhpur, India

Jan G. Korvink Institute of Microstructure Technology, Karlsruhe Institute of Technology, Karlsruhe, Germany

Ashwani Kumar Department of Chemistry, Government College Anni, Kullu, India

Raj Kumar Department of Pharmaceutical Sciences, University of Michigan, Ann Arbor, MI, United States; School of Basic Sciences and Advanced Material Research Center, Indian Institute of Technology Mandi, Mandi, India

Rudra Kumar School of Engineering and Sciences, Tecnologico de Monterrey, Monterrey, Mexico

Yogeenth Kumaresan Electronics & Nanoscale Engineering, School of Engineering, University of Glasgow, Glasgow, United Kingdom

Lingeshwar Reddy Kumbam Department of Pharmaceutical Sciences, University of Michigan, Ann Arbor, MI, United States; Department of Chemistry, Indian Institute of Science Education and Research Tirupati, Tirupati, India

Genevieve M. Liddle Florida Atlantic University, Boca Raton, FL, United States

Marcela Longhi Universidad Nacional de Córdoba, Facultad de Ciencias Químicas, Departamento de Ciencias Farmacéuticas, Ciudad Universitaria, Córdoba, Argentina; Consejo Nacional de Investigaciones Científicas y Técnicas, CONICET, Unidad de Investigación y Desarrollo en Tecnología Farmacéutica, UNITEFA, Córdoba, Argentina

Dario Mager Institute of Microstructure Technology, Karlsruhe Institute of Technology, Karlsruhe, Germany

Shweta J. Malode Department of Chemistry, School of Advanced Sciences, KLE Technological University, Hubballi, India

Venkateswarulu Mangili Department of Pharmaceutical Sciences, University of Michigan, Ann Arbor, MI, United States; Department of Inorganic and Physical Chemistry, Indian Institute of Science, Bangalore, India

Miguel Manso Silvan Departamento de Física Aplicada and Instituto de Ciencia de Materiales Nicolás Cabrera, Universidad Autónoma de Madrid, Madrid, Spain

Sergio O. Martinez-Chapa School of Engineering and Sciences, Tecnologico de Monterrey, Monterrey, Mexico

Vivek K. Mishra Independent Researcher, Groningen, The Netherlands

Jaba Mitra Material Science & Engineering, University of Illinois, Urbana Champaign, IL, United States; Thermo Fisher Scientific, South San Francisco, CA, United States

Joyee Mitra Inorganic Materials & Catalysis Division, CSIR-CSMCRI, Bhavnagar, India; Academy of Scientific and Innovative Research (AcSIR), Ghaziabad, India

Mousumi Mitra Department of Physics, University of Virginia, Charlottesville, VA, United States

Kunal Mondal Idaho National University, Pocatello, ID, United States; Materials Science and Engineering Department, Idaho National Laboratory, Idaho Falls, ID, United States

Nagaraju Nakka Department of Pharmaceutical Sciences, University of Michigan, Ann Arbor, MI, United States

Ionela Cristina Nica Research Institute of the University of Bucharest–ICUB, University of Buchares, Bucharest, Romania

Yegor Piskarev School of Engineering, Swiss Federal Institute of Technology Lausanne, Lausanne, Switzerland

Praveen Kumar Poola Electronics and Communication Enhineering, K.L. University Hyderabad, Hyderabad, India

Guruprasad Reddy Pulikanti School of Basic Sciences and Advanced Material Research Center, Indian Institute of Technology Mandi, Mandi, India; Department of Chemistry, Indian Institute of Science Education and Research Tirupati, Tirupati, India

Nitin Puri Delhi Technological University, Delhi, India

Yuhao Qiang Florida Atlantic University, FL, United States

Darsi Rambabu Department of Pharmaceutical Sciences, University of Michigan, Ann Arbor, MI, United States; Molecular Chemistry, Materials and Catalysis, Institute of Condensed Matter and Nanosciences, Universite Catholique de Louvain, Brussels, Belgium

Shubham Roy Department of Physics, Jadavpur University, Kolkata, India

Ahsana Sadaf Institute of Microstructure Technology, Karlsruhe Institute of Technology, Karlsruhe, Germany

Prateep Singh Sagara Department of Pharmaceutical Sciences, University of Michigan, Ann Arbor, MI, United States

Abhijit Samanta School of Science and Technology, The Neotia University, Sarisha, India

Deepika Sandil Bhagwan Parshuram Institute of Technology, Delhi, India

Konathala Ravi Shankar Department of Pharmaceutical Sciences, University of Michigan, Ann Arbor, MI, United States; School of Nano Sciences, Central University of Gujrat, Gandhinagar, India

Nagaraj P. Shetti Department of Chemistry, School of Advanced Sciences, KLE Technological University, Hubballi, India

Jun Shintake School of Informatics and Engineering, The University of Electro-Communications, Tokyo, Japan

Miruna Silvia Stan Research Institute of the University of Bucharest–ICUB, University of Buchares, Bucharest, Romania

Suchitra Max Planck Institute for Chemical Physics of Solids, Dresden, Germany

Jazmín Torres Universidad Nacional de Córdoba, Facultad de Ciencias Químicas, Departamento de Ciencias Farmacéuticas, Ciudad Universitaria, Córdoba, Argentina; Consejo Nacional de Investigaciones Científicas y Técnicas, CONICET, Unidad de Investigación y Desarrollo en Tecnología Farmacéutica, UNITEFA, Córdoba, Argentina

Paula M. Uberman Universidad Nacional de Córdoba, Facultad de Ciencias Químicas, Departamento de Química Orgánica, Ciudad Universitaria, Córdoba, Argentina; Consejo Nacional de Investigaciones Científicas y Técnicas, CONICET, Instituto de Investigaciones en Fisico-Química de Córdoba, INFIQC, Córdoba, Argentina

Jianning Wei Florida Atlantic University, Boca Raton, FL, United States

Anshul Yadav Membrane Science and Separation Technology, CSIR-Central Salt & Marine Chemicals Research Institute, Bhavnagar, India

Midathala Yogesh Department of Pharmaceutical Sciences, University of Michigan, Ann Arbor, MI, United States

Section 1

Biomedical applications of metal oxides

Bioactivity, biocompatibility, and toxicity of metal oxides

1

Snehasis Biswas[1] *and Jayesh Bellare*[1,2]
[1]Department of Chemical Engineering, Indian Institute of Technology Bombay, Mumbai, India, [2]Wadhwani Research Centre for Bioengineering, Indian Institute of Technology Bombay, Mumbai, India

1.1 Introduction

Metal oxides (MOs) are gaining research and commercial interest due to their diversity of physicochemical properties (Rastogi et al., 2017). The ability to adopt numbers of structural geometry, vacant lattice structure, ease of inducing dopant, photocatalytic activity, and supermagnetism make them a potential technological application (Estelrich, Escribano, Queralt, & Busquets, 2015; Fernández-García & A., 2011; Srigurunathan et al., 2019). Moreover, in the nano-dimension, they show several unique properties as compared to the bulk MOs, which enhance their applicability in the biomedical and industrial fields. MOs are used in a wide range of applications such as catalysts, fuel cells, semiconductors, insulator sensors, piezoelectric devices, biomaterials, and medicines (Andreescu, Maryna, S, Ana, & C, 2011; McNamara & Tofail, 2017). Some MOs are extremely bioactive and biocompatible, which makes them a potential candidate for numerous biomedical applications. MOs such as Fe_3O_4, SiO_2, ZnO, TiO_2, CuO, and CeO_2 are being used in drug delivery, biosensing, bioimaging, separation of biomacromolecules and antimicrobial agents (Andreescu et al., 2011; Modi & Bellare, 2019; Singh & Kumar, 2020; Sudtha et al., 2020). Historical evidence shows that MOs were used as oral medicines in traditional medicinal systems such as Ayurveda for centuries. For instance, ZnO particles were traditionally used in Ayurveda as a therapeutic agent against diabetes (Bhowmick, Suresh, Kane, Joshi, & Bellare, 2009).

MOs show different bioactivity mechanisms depending on their physicochemical characteristics. Properties such as redox potential, dissolution rate, photocatalytic activity, and surface defects are the crucial factors that influence the bioactivity directly or indirectly. Depending upon their properties, MOs can be tailored for different biomedical applications. For example, TiO_2 has photocatalytic activities that can be used for photodynamic therapy to kill cancer cells (Yin, Wu, Gui Yang, & Hua Su, 2013). TiO_2 has excellent UV absorption properties, which are useful for sunscreen applications. ZnO and CuO possess a high dissolution rate, for which they show antimicrobial properties (Zhang et al., 2012). Fe_2O_3, on the other hand, can induce magnetic hyperthermia in tumors (Grüttner, Müller, Teller, & Westphal, 2013): as cancer cells are more susceptible to temperature with respect to healthy

Metal Oxides for Biomedical and Biosensor Applications. DOI: https://doi.org/10.1016/B978-0-12-823033-6.00001-6

cells, iron oxide nanoparticles are used to kill cancer cells without affecting the healthy cells using magnetic hyperthermia. Likewise, ZrO_2 is used in implants and dental crown applications as it aids osteoblast proliferation and differentiation (Chen, Roohani-Esfahani, Lu, Zreiqat, & Dunstan, 2015).

In the last few decades, nanoparticles of MOs have become an important class of new materials in biomedical applications. The development of engineered MO nanoparticles enables us to overcome technological challenges, which was impossible previously. Engineered nanoparticles are designed to have one or more unique properties that are absent in the bulk sample. The higher surface/volume ratio of nanosized particles enables them to exhibit exceptional physical and chemical functionality that is useful for biomedical applications (McNamara & Tofail, 2017).

On the other hand, in recent times, engineered MO nanoparticles have been produced industrially on a large scale. Due to the massive production of MO nanoparticles, the possibility of environmental exposure has increased (Cai et al., 2017). Also, improper disposal and management of nanoparticle waste could contaminate the environment. With the rise in applications, concern has been also raised about the toxicity of MO nanoparticles. In the nanoscale, not only the bulk properties affect the toxicity but also several inherent properties of nanosize such as size, shape, surface coating, and surface charge are responsible for possible adverse effects. For instance, SiO_2 is considered as a safe material, although several reports have indicated its toxicity in the nanoscale (Chen et al., 2018). Due to the larger surface to volume ratio, nanoparticles cause higher chemical reactivity toward cells. Indeed, as surfaces of nanoparticles are the site of interaction with the biological system, the surface area of nanoparticles plays a crucial role in their toxicity. The toxicity of MO nanoparticles has been comprehensively studied in the literature. Some MOs show significant in vitro and in vivo toxicity. Most studies have indicated enhanced production of reactive oxygen species (ROS) as one of the main reason for this toxicity (Horie & Fujita, 2011).

This chapter focuses on a brief description of the bioactivity, biocompatibility, and toxicity of MOs. Their potential biomedical applications also are described. In the last section of the chapter, a review is provided of the toxicity of MOs, with particular emphasis on MO nanoparticles.

1.2 Metals oxides and their biomedical applications

Metal oxides are those compounds having metal as a cation and oxygen as an anion, where the oxidation number of oxygen is −2. Almost all metals naturally form oxides and are found in the Earth's crust (with exceptions such as Au and Pt). The primary source of metals is metal oxide ores. Most interestingly, due to multiple oxidation numbers of some metals, a particular metal can have a number of oxides. Furthermore, polymorphic variation in the crystalline structure is also possible in MOs. In modern society, almost every household uses MO-derived consumer or industrial products. In addition, its applications in the biomedical field have enhanced its importance to modern civilizations. The metals whose oxides are widely used in biomedical fields are shown in Fig. 1.1 and listed in Table 1.1.

MOs have been used as medicines from the historical era in Indian traditional medicine Ayurveda (Pandey & Chaudhary, 2018). Rasa-Shastra (Vedic-chemistry) is one of the disciplines of Ayurveda, which deals with metal/metal oxide preparations called Bhasma. Bhasmas are the class of medicine which mainly consists of nano- and micron-size particles of metals or metal oxides (Biswas et al., 2020). Ayurveda, which is the oldest traditional medicinal system rooted since 3000 BCE,

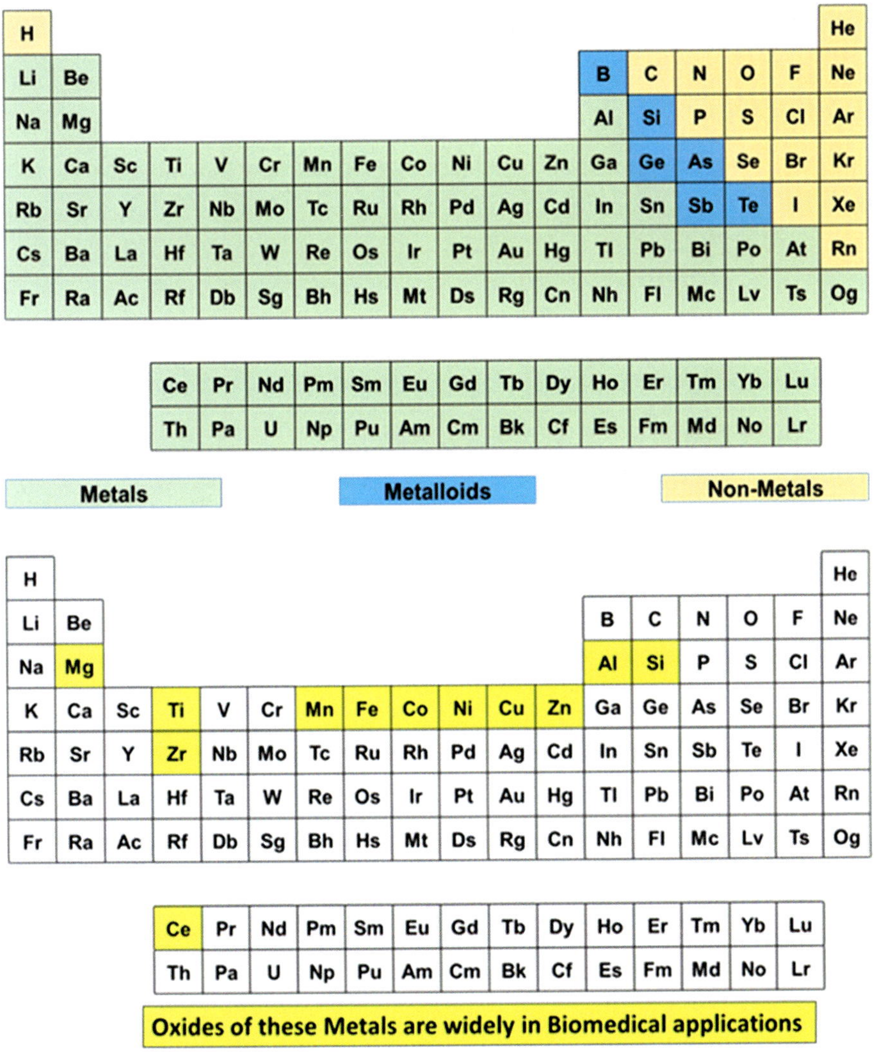

Figure 1.1

Table 1.1 Metal oxides and their applications.

Elements	Oxides formulae	Possible biomedical applications	References
Ti	TiO, $\mathbf{TiO_2}$, Ti_2O_3	Cosmetics, drug delivery, photodynamic therapy, dental applications	Santiago-Medina et al. (2015), Yin et al. (2013)
V	VO, V_2O_3, V_2O_5, $\mathbf{VO_2}$	Antibacterial and osteogenic activity	Guo et al. (2018)
Mn	MnO, $\mathbf{MnO_2}$	Gene therapy, biosensors	Chen et al. (2020)
Fe	FeO, Fe_2O_3, $\mathbf{Fe_3O_4}$	Drug delivery, hyperthermia, MRI image	Sangaiya and Jayaprakash (2018)
Co	CoO, Co_3O_4	Enzyme inhibition properties	Iravani and Varma (2020)
Ni	**NiO**, Ni_2O_3	Antimicrobial agent, radiotherapy	Marzban et al. (2020)
Cu	**CuO**, Cu_2O	Antimicrobial agent, radiotherapy	Verma and Kumar (2019)
Zr	$\mathbf{ZrO_2}$	Dental applications	(Murthy, Effiong, & Fei, 2020)
Mg	**MgO**	Alleviation of heartburn, stomach pains, and for bone regeneration	Mazaheri et al. (2019)
Zn	**ZnO**	Antimicrobial agent, diabetes	(Sruthi et al., 2018)
Ce	$\mathbf{CeO_2}$	Antioxidant, biocatalyst, neuroprotective agent	Singh, Nayak, Sarkar and Singh (2020)
Si	$\mathbf{SiO_2}$	Drug delivery	Croissant, Fatieiev, Almalik and Khashab (2018)

MRI, magnetic resonance imaging. Bold entries are widely used chemical formula of a particular MO developed a critical and rigorous procedure to manufacture colloidal MOs from raw metals with numerous unit operation steps such as purification, drying, and heat treatment. Bhasmikaran is a special treatment, in which metals are oxidized at a higher temperature, that induces some unique characteristics, such as nonstoichiometric MOs with oxygen deficiency and smaller crystallite size (Bhowmick et al., 2009; Shilpa Chakra & Venkateswara Rao, 2014). Yasad Bhasma (ZnO), Tamra Bhasma (CuO), and Louha Bhasma (iron oxides) are some examples of ancient metal oxide formulations used in therapeutic applications such as diabetes, liver diseases, anemia, and chronic fever (Bhowmick et al., 2009; Pandey & Chaudhary, 2018). These ancient medicinal formulations are still used in India on a large scale to treat a wide range of diseases from jaundice to cancer. These metal oxide medicines are usually given orally, although, in some cases, they are applied topically due to their antimicrobial activity. For example, Yasad Bhasma (ZnO) is one of the main component in commercially used Ayurvedic antimicrobial ointments.

In modern times, MOs are becoming valuable materials for the drug and health-related industry. The applications of MOs in the bulk material are limited, whereas, due to their cellular penetration capability, nanoparticles of MOs have made remarkable breakthroughs in targeted drug delivery, cell imaging, biomolecule sensors, and antimicrobial activity. Due to their small size, they can interact with a number of cellular and biochemical components. Metal oxide nanoparticles such as SiO_2, Fe_3O_4, TiO_2, ZrO_2, and CeO_2 show excellent biocompatibility and are of great interest to researchers. In the following section, a brief description is given for the potential biomedical applications of MOs.

1.2.1 Iron oxide

Iron oxides consist of several polymorphic phases, such as α, β, γ, δ, ε-Fe_2O_3, and γ-Fe_2O_3 (maghemite) (McNamara & Tofail, 2017). Maghemite and magnetite are commonly employed in biomedical applications. In the nanoscale, a supermagnetic property is obtained from Fe_3O_4 (magnetite) or Fe_2O_3 (maghemite) nanoparticles that enhance the performance in biomedical applications such as for targeted drug delivery. However, magnetite (Fe_3O_4) is preferable over maghemite due to the saturation in magnetization and magnetic susceptibility.

Iron oxides are usually used as nano- or colloidal particles for biomedical applications. Iron oxide nanoparticles (IONPs) have potential uses in cell imaging, magnetic resonance imaging (MRI), targeted drug delivery, iron supplements, and magnetic separation of cells (Estelrich et al., 2015; Remya, Syama, Sabareeswaran, & Mohanan, 2016; Singh et al., 2012). Supermagnetism of iron oxide nanoparticles is the key property for targeted drug delivery. IONPs are very efficient as a contrasting agent for MRI images; they are used for imaging tumors, the gastrointestinal system, and nervous system. As the magnetic field can influence IONPs' pharmacokinetics, they are already being used clinically for targeted drug delivery (Laffon et al., 2018). The surface modifications of IONPs with core–shell structure (IONP core) allow it to carry a wide range of drugs. Furthermore, with their magnetic properties, they have become an excellent tool for guided drug delivery

(Singh & Kumar, 2020). Various therapeutic agents such as chemotherapeutics, antibiotics, and radiotherapeutic can be delivered to targeted organs by tagging them with IONPs.

IONPs can have another remarkable potential application in magnetic hyperthermia, as they can generate heat in a variable magnetic field. Magnetic IONPs can be circulated through the blood circulatory system to a specific target organ or tissue. Further, with the induction of a variable magnetic field, hyperthermia can be induced at the targeted cells. Hyperthermia is one of the alternative treatment procedures to kill cancer cells, as these cells are more vulnerable to temperature than healthy cells (Singh & Kumar, 2020).

Presently, several IONP formulations have been approved by the FDA for biomedical applications, including Endorem®, GastroMARK™, Umirem®, Feraheme®, Feridex®, and Gastromark® (Vakili-Ghartavol et al., 2020).

1.2.2 ZnO

ZnO nanoparticles are extremely attractive as they can form some unique nanostructures, such as nanohelices and nanorings (Sruthi, Ashtami, & Mohanan, 2018). ZnO has an excellent UV-blocking property and so it is used in sunscreen. Furthermore, ZnO nanoparticles have tremendous antimicrobial and antifungal properties. For these advantages, ZnO is commercially used in cosmetics, ointments, rash-preventing agents, antidandruff shampoos, and food packaging (Sruthi et al., 2018). The antimicrobial effects of ZnO are useful for wound healing and skin inflammation. It is also very common as a part of nutritional products and diet supplements (Kolodziejczak-Radzimska & Jesionowski, 2014). ZnO nanoparticles are a reliable alternative when considering a convenient and inexpensive tool for bioimaging, due to their inherent fluorescence properties with excitonic emission at near-UV, blue, and green regions. In addition, ZnO nanoparticles demonstrate high stability and reduced photobleaching properties (Sruthi et al., 2018).

1.2.3 TiO_2

TiO_2 is poorly soluble and found in two crystal structures: anatase and rutile. Anatase is chemically more reactive than rutile. TiO_2-coated titanium alloys are very common in dental applications because they aid in the differentiation of osteoblasts (Tsukimura, Kojima, Att, Kameyama, & Maeda, 2008). The higher water wettability and oxygen concentration of the TiO_2 surface increase ALP activity, which is an important biomarker for the mineralization of bone (Santiago-Medina, Sundaram, & Diffoot-Carlo, 2015). TiO_2 nanoparticles are one of the most investigated and well-characterized nanoparticles in the literature. TiO_2 nanoparticles are commercially used in sunscreens due to their ability to block UV light. They have strong bone-bonding and protein adsorption characteristics. TiO_2 can also be used as a biosensor for biological assay and genetic engineering (Yin et al., 2013). Most interestingly, TiO_2 can be used in photodynamic therapy of tumor cells, which involves direct illumination by UV or visible light (Yin et al., 2013). Photodynamic therapy is a promising alternative and noninvasive treatment approach for cancer.

1.2.4 SiO_2

Silica (SiO_2) is used in biomedical applications due to its excellent chemical stability and ability to bond covalently between atoms. The two main applications of silica-based materials are bone regeneration and controlled drug delivery (María & Francisco, 2008). The hydrated silica gel formation on the surface makes SiO_2 favorable for bone cells and tissues (Priyadarshini, Rama, Chetan, & Vijayalakshmi, 2019). One of the greatest potential applications of SiO_2 nanoparticles is delivery of drugs by the oral route. Silica has two basic forms: amorphous and crystalline. The lattices of crystalline silica are arranged regularly, while lattices are arranged irregularly for amorphous silica. Crystalline silica possesses a high pore to volume ratio, and tunable porosity can be achieved in silica nanoparticles. Mesoporous silica nanoparticles having 2–100 nm pore size are ideal candidates for controlled drug release. Amorphous silica also has numerous biomedical applications in gene carriers and diagnostic probes (Chen et al., 2018; Tang & Cheng, 2013).

1.2.5 CuO

CuO nanoparticles can restrict the growth of bacteria, and are highly active against human lung cancer as an antitumor agent. Thus, CuO can be incorporated into medical devices, ointments, and bandages for its antimicrobial properties. Furthermore, CuO plays a key role as an antiangiogenic agent (angiogenesis is the transformation of tumor from benign to malignant) (Barui, Nethi, Haque, Basuthakur, & Patra, 2019). CuO nanoparticles are reported to have induced apoptosis in tumor cells. They can be fabricated as a biosensor for a number of biomolecules such as glucose, cholesterol, and DNA (Verma & Kumar, 2019).

1.2.6 CeO_2

CeO_2 nanoparticles can be used as a biological antioxidant (McNamara & Tofail, 2017). The adaptable redox properties of CeO_2 nanoparticles can modulate cellular ROS for multiple pathological conditions such as wounds, cancer, and diabetes. CeO_2 and doped CeO_2 have exceptional catalytic and free radical scavenging properties (Abuid, Gattás-Asfura, LaShoto, Poulos, & Stabler, 2020). They can protect nerves and other cells from ROS-induced stress. Ceria nanoparticles extend the life span of the brain. Furthermore, CeO_2 nanoparticles are capable of regenerating their redox activity and act as catalysts for numerous bioreactions for an unlimited number of times (Popov et al., 2018).

1.2.7 MgO

Currently, MgO is used to treat diseases such as stomach indigestion (antacid), heartburn, and is an important material in toxic waste remediation (Kannan, Radhika, Sadasivuni, Reddy, & Raghu, 2020). In the nano size, MgO has tremendous bioactivity toward microorganisms. The toxicity of MgO nanoparticles on microbes can be justified based on the ROS formation theory. The electrostatic interaction between MgO nanoparticles and the microbial surface is the main reason for cell death. The formation of ROS during the interaction of MgO nanoparticles is mostly executed outside the cell by destroying the cell membranes (Suresh, Pradheesh, Alexramani, Sundrarajan, & Hong, 2018).

1.2.8 ZrO_2

ZrO_2 can exist in three polymorphs: monoclinic, tetragonal, and cubic (Srigurunathan et al., 2019). ZrO_2 is generally used to coat orthopedic implants as it has greater strength, toughness, high wear resistance, and chemical inertness. Additionally, due to enhanced anticorrosion and esthetic properties, it is widely used in dental implantations.

1.2.9 MnO_2

MnO_2 nanoparticles possess excellent catalytic activity. MnO_2 nanosheets can be utilized to build biosensors (Chen, Cong, Shen, & Yu, 2020). Also, they have potential applications in gene therapy and bioimaging.

1.2.10 NiO

NiO nanoparticles have antimicrobial activities (Imran Din & Rani, 2016) and are used to study the cytotoxic effect against human cancers cells.

1.3 Bioactivity of metal oxides

The bioactivity of a compound is the capability to respond to any kind of activity toward the biological system by exerting either beneficial or adverse effects (Guaadaoui, Benaicha, Elmajdoub, Bellaoui, & Hamal, 2014). As mentioned earlier, factors including redox potential, oxidative state, surface charge, and surface defects can lead to the bioactivity of MOs (Fig. 1.2A). The dissolution rate of nanoparticles in the biological medium is another decisive factor for their bioactivity. Furthermore, some MOs possess magnetic or photocatalytic activity, which influences their bio-interactions with cells and tissues. In contrast to molecular chemicals, MOs are in solid state, and their three-dimensional structure interacts with biological systems. MOs, either in bulk or nanosize, can attach to cells or biomolecules on the surface only. Hence, surface properties play a significant role in their

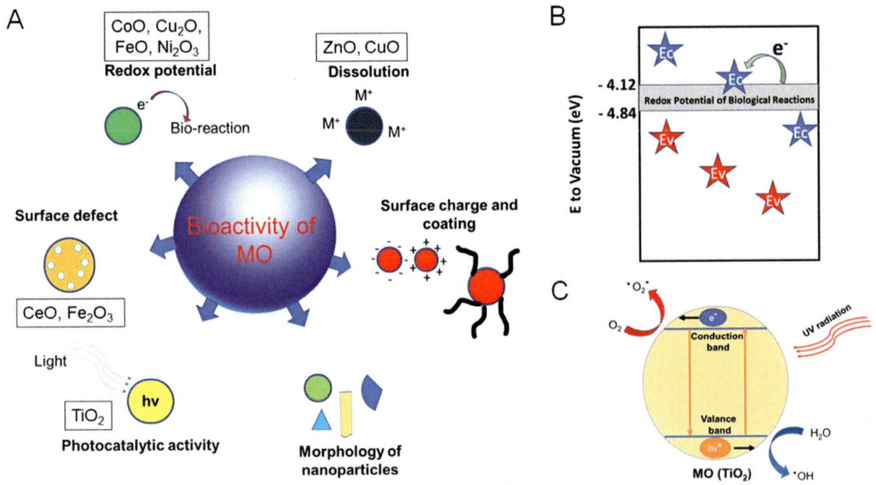

Figure 1.2 Schematic diagram of (A) various bioactivity mechanism of metal oxide nanoparticles, (B) redox potential of MO and biointeraction, and (C) photocatalytic activity of TiO_2 nanoparticles.

bioactivity. Moreover, as the surface of MOs is the site of interaction with the biological system, the bulk properties such as redox potential, photocatalytic effects, or structural defects are reflected on its surface only. Thus, the intensity of bioactivity of any MO is largely determined by its surface area. As nanoparticles have a very high surface area as compared to the bulk materials, nanosize MOs are reported to be highly bioactive, though their bulk material is relatively inactive.

The dissolution rate of MOs in the biological medium enhances the bioavailability of the metal ions. The degree of bioactivity also depends on the dissolution rates for some MOs. The dissolution rate is not a fundamental physicochemical property of MOs, rather it is an experimental evaluation (Kaweeteerawat et al., 2015). Most metal oxides have a lower dissolution rate in a biological medium. However, CuO and ZnO are the two most commonly used MO nanoparticles in biomedical applications, which show a higher dissolution rate as compared to other MOs. Experimental studies on nanoparticles show that the rate of dissolution of an MO increases with decreasing oxidation numbers (Simeone & Costa, 2019). An estimated $>10\%$ dissolution is reported for MOs having an oxidation number ≤ 2, whereas a drastic decrease in dissolution ($< 1\%$) has been reported for MOs having an oxidation number > 3. Several other factors that increase the rate of dissolution of MOs include elevated metal solubility, higher surface/volume ratios, surface roughness, and low pH of the medium (Teske & Detweiler, 2015).

MOs show semiconductive properties. It has been widely accepted that the redox potential of MOs determines its interaction with the cell's redox system. The bandgap hypothesis proposes that, depending upon its redox potential, MOs could catalyze bioreactions in the cellular system by accepting or donating electrons (Zhang et al., 2012). MOs like Cr_2O_3, CoO, and ZrO_2 could interfere with the cell's redox system and eventually affect the cell functionality. In a cell, a number of redox systems exist. When the energy level between the valence (E_v) and conduction (E_c) bands of semiconductive MOs are similar to the energy level of cellular redox potential (E°), the interaction could take place due to electron transfer between two systems (B). The relative positions of E_v or E_c with respect to E° dictate the direction of electron transfer between the oxide and cellular redox system. If the energy level (E°) of the biological redox system is slightly higher than the E_c of the MO, then electrons can be transferred from the biological redox system to the vacant conduction band of MO (oxidation of biomolecules) (Teske & Detweiler, 2015). On the other hand, if the redox potential of biomolecules is closer to the MO's valance band (E_v), then the electron will be accepted by the biomolecule (reduction of biomolecules). It is reported that the range of biological redox potential is −4.12 to −4.84 eV (Zhang et al., 2012). MOs having E_c and E_v in that range show higher bioactivity Fig. 1.2B.

Another mechanism of bioactivity can be seen in some MOs, especially in TiO_2, called photocatalytic activity Fig. 1.2C that are susceptible to UV activation (Yin et al., 2013). The large bandgap of TiO_2 (~3.1 eV) is activated under ultraviolet (UV) irradiation (Yin et al., 2013). However, with the induction of a suitable dopant, the bandgap of TiO_2 can be reduced so that it can be activated by visible light irradiation. TiO_2 can absorb a photon with energy greater than the bandgap (Fig. 1.2C). The excitation of TiO_2 by a photon shifts one electron from a valance band to the conduction band, generating one excited electron in it. Simultaneously, the electron–hole is created in the valance band. The excited electrons and holes separate and migrate to the TiO_2 surface, which can lead to chemical reactions with biomolecules. The surface reaction can generate free radicals (hydroxyl radicals and superoxide radical anions) when reacting with a free redox system of cells.

The surface defects of several MOs can also contribute a significant role to their bioactivity. The oxygen vacancy at the surface of CeO_2 nanoparticles could be the reason for its antioxidant properties. For CeO_2, the common oxidative state of Ce is +4. However, a lower oxidation state of Ce (+3) is also possible, resulting oxygen vacancies in the cryastal structure (Abuid et al., 2020). Due to the oxygen vacancies at the surface of the CeO_2 NP, it can act as a Lewis base. A Lewis base can provide a negative charge to electron-deficient species (e.g., reactive oxygen species) (Simeone & Costa, 2019). This behavior of CeO_2 nanoparticles is responsible for its antioxidant properties. Moreover, several researchers have reported that due to the oxygen vacancy, CeO_2 nanoparticles show superoxide-dismutase (SOD) and catalase mimetic activity to neutralize ROS (Abuid et al., 2020).

In addition, the bioactivity of MO nanoparticles is largely governed by the inherent properties of nanosize, especially size, shape, and surface charge (zeta potential). It has been reported that the internalization of positively charged nanoparticles in cells more easily occurs as compared to negatively charged or neutral nanoparticles (Hyun, Hyoung, Geol, & Hun, 2015)

1.4 Biocompatibility of metal oxide

The earlier concept of biocompatibility for any material was intended to remain within an individual for a long duration without causing any side effects. Earlier, the least reactive materials were considered as biomaterials for bioimplants, such as vanadium steels or stainless steel. Biomaterials were selected based on that they would be nontoxic, nonthrombogenic, nonimmunogenic, nonirritant, and noncarcinogenic to the host (Williams, 2008). However, with modern scientific advances, the idea of biocompatibility has been reevaluated based on the following three factors. First, the material characteristics are not the only decisive factor—the site of interaction is also important for its biocompatibility. Second, in several applications, it is required that the material should explicitly interact with the tissues rather than be ignored by them, which is not possible in the case of an inert material. Finally, in some applications, biodegradation or clearance of material (especially for nonbiodegradable MO nanoparticles) from the body is required. Thus, with the development of numerous new classes of biomaterials (biodegradable polymers, or nanoparticles), the definition of biocompatibility was redefined. The definition given by Williams (2008) is one of the most accepted ones, as follows:

> *"Biocompatibility refers to the ability of a biomaterial to perform its desired function with respect to medical therapy, without eliciting any undesirable local or systemic effects in the recipient or beneficiary of that therapy, but generating the most appropriate beneficial cellular or tissue response in that specific situation, and optimizing the clinically relevant performance of that therapy."*

However, biocompatibility for any material is not universal. Instead, it is a relative subject: the ratio of benefits to risk determines its feasibility in biomedical applications (Naahidi et al., 2013).

As reported in the literature, the biocompatibility of several MOs is a great advantage for a wide range of biomedical applications. MOs can be tailored to instigate bioreactions that could be beneficial for the therapy. MOs are mainly used for medical diagnostic and therapeutic interventions such as an implants, drug delivery, magnetic hyperthermia, and biomolecule detection. Most MOs used in the bulk state for biomedical applications are nontoxic and relatively biocompatible. However, MO nanoparticles exhibit different biocompatibility profiles as compared to their bulk material. For biomedical applications, nanoparticles should enter the body and can contact directly with various types of cells and tissues. Hence, it is essential for MO nanoparticles to be biocompatible. It is a well-established fact that the physicochemical properties of nanoparticles can influence the extent of biocompatibility.

Unfortunately, the use of nanoparticles in biomedical applications has several limitations. The interception of nanoparticles by the immune system is one of the barriers to drug delivery by nanoparticles at a targeted site. When nanoparticles enter the blood circulatory system, an immune response cascade is initiated upon the adsorption of proteins like immunoglobulins or complement proteins on the nanoparticle surface. The adsorption of such proteins can result in nanoparticles being treated as foreign substances, which enhances their uptake by the phagocytic cells of the reticuloendothelial system (RES) (Hillaireau & Couvreur, 2009). Most studies have suggested that surface properties such as surface charge and hydrophobicity significantly determine the immune response.

Furthermore, the route of administration is another decisive factor for the biocompatibility of nanoparticles. Oral, parenteral, pulmonary, and dermal delivery are the possible routes of administration for nanoparticles. The toxicokinetics or excretion of nanoparticles largely depends on the drug administration route. For example, administration by oral or intravenous intake of nanoparticles could determine their fate differently in the body.

The biocompatibility of MO nanoparticles does not depend on the nanomaterial alone, but also on the concentration and duration of time the nanoparticle spends in the cells and tissues. Generally, clearance of nanoparticles from the body is essential after their desired performance. An indefinite stay in the body could lead to the accumulation of nanoparticles in tissues such as kidneys, liver and brain, which is undesirable. MO nanoparticles are nonbiodegradable. The clearance of nanoparticles, which is largely size dependent, is possible by a different mechanism. Nanoparticles having a size 20–30 nm could be rapidly cleared by renal excretion (Adabi et al., 2017). Particles having ~200 nm dimension are more efficiently taken up by being cleared by the reticuloendothelial system. At the same time, particles of sizes 200–400 nm undergo prompt hepatic clearance. The clearance of nanoparticles is assisted by the opsonization of complement proteins and blood components on the particle surface (Owens & Peppas, 2006). On the other hand, rapid clearance of nanoparticles before any therapeutic action has taken place is also undesirable. Nanoparticles are reportedly removed from the bloodstream within seconds by macrophages if the surface of nanoparticles are not modified to prevent adsorption of opsonins (Dobrovolskaia & McNeil, 2007). Without surface modifications, nanoparticles mostly are taken up by phagocytic cells, causing an unfavorable interaction with the immune system, which causes toxicity to the host. For the biomedical applicability of MO nanoparticles, suitable surface modifications are necessary strategies to overcome this problem. Researchers have suggested that surface modification of MO nanoparticles could enhance their biocompatibility and facilitate smart drug delivery and other biomedical applications. A high degree of biocompatibility can be achieved for MO nanoparticles by capping them with various surface modification approaches.

Moreover, nanoparticles usually tend to agglomerate in the aqueous system due to their large surface to volume ratio. The agglomerated nanoparticles result in fast detection by the immune system. To control this problem, an appropriate surface coating is usually used on the MO nanoparticles, depending on the biomedical application. This surface coating improves compatibility with the blood, allows for functionalization of the surface, averts nonspecific adsorption of plasma proteins, and can prevent degradation by macrophages. Coating with polymeric compounds, such as polyvinyl alcohol (PVA), polyethylene glycol (PEG), dextran, and chitosan are the most common surface modifiers for MO nanoparticles (Umut, 2013). Recently,

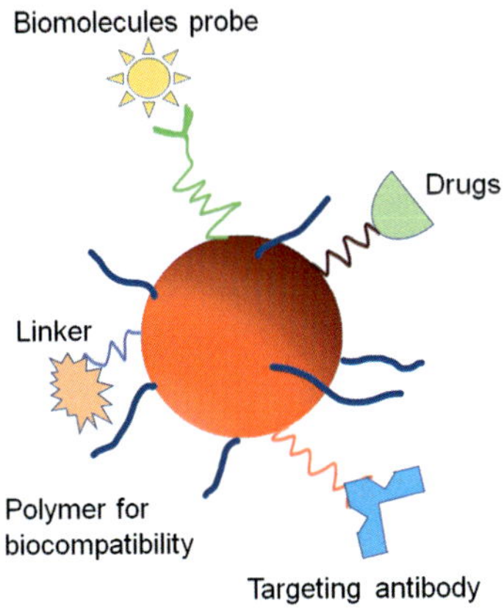

Figure 1.3 Functionalization of MO nanoparticles with the inclusion of various surface coating strategies (core–shell structure with a MO nanoparticle core). No permission required.

nonpolymeric materials, such as carboxylates, oleic acid, and alkyl phosphonates, have provided excellent stability to nanoparticles. Gold and silica are also used on MO nanoparticles to improve their biocompatibility (Keshtkar, Shahbazi-Gahrouei, Mehrgardi, Aghaei, & Khoshfetrat, 2018; Pham et al., 2020). Besides the stability, the core–shell engineered structure enables MO nanoparticles to have multiple functions such as anchoring for biomolecules (receptors, antibodies, proteins) or imaging agents (radionuclides and fluorophores), and enhance intracellular behavior Fig. 1.3.

1.5 Toxicity of metal oxides

MOs interacts with human cells and tissues in various ways. In the bulk state, the interaction of MOs with the biological system is limited, and most do not show considerable in vitro or in vivo toxicity. However, with the decreasing size of MO particles, the intensity of the interaction is enhanced exponentially. It has been widely reported that several nanosize MOs show significant toxicity. With the increasing industrial and medicinal applications, the risk of ecotoxicity, as well as human interactions with these MO nanoparticles, raises a serious health concern. Especially for biomedical applications, cellular internalization offers the basis for therapeutic applications and diagnostics of nanoparticles (Boyles et al., 2017). Thus, emerging MO nanostructures require rigorous evaluations for their biological safety. Therefore, to ensure the safety of MO nanoparticles, it is essential to develop a precise understanding of the biological effects. It is worth noting that, sometimes, the characteristic toxicity of MO nanoparticles can be advantageous in the medicinal application such as in antimicrobial or anticancer agents. In this section, we will discuss the toxicity mechanisms of various MO nanoparticles.

The interaction between nanoparticles and the biological system is not yet fully understood. Cell culture studies show that the accumulation of nanoparticles occurs via passive or active mechanisms. However, the mechanism and pathways are difficult to understand. A slight change in properties could lead to quite different biointeractions. Unlike classical toxicology, the toxicity study of nanoparticles, such as dose metrics, is not straightforward. At present, almost no biocompatibility assessment standards have been established for nanoparticles used in drug delivery systems. Further, the complexity increases greatly in animal models as compared to cell culture studies. Toxicological data on nanoparticles are challenging to replicate because a huge number of parameters influence the interactions, and several of these parameters are not easy to control (Mahmoudi, Hofmann, Rothen-Rutishauser, & Petri-Fink, 2012). In this context, nanoparticle toxicity refers to "the ability of the particles to adversely affect the normal physiology as well as to directly interrupt the normal structure of organs and tissues of humans and animals" (Li, Wang, Fan, Feng, & Cui, 2012).

Researchers have struggled to address the toxicity issues related to nanoparticles. This is perhaps due to the lack of a comprehensive assessment of: (1) the size, surface charge, and composition of NPs; (2) thickness density and stability of the surface coating under physiological environments; (3) a detailed comparison of (1) and (2) across numerous in vivo models; (4) differences in manufacturing procedures and variability in raw materials used to produce these particles; and (5) systematic toxicological characterization (Dobrovolskaia & McNeil, 2007). Hence protocolization for bioassay for nanoparticles is yet to be formalized for authenticating uniform toxicological data.

Internalization of MO nanoparticles in cells can be possible using various pathways, such as endocytic internalization, and receptor-mediated endocytosis depending on their size, shape, and surface charge. The MO nanoparticles, after internalization, can affect a number of biomolecules in cells according to their bioactivity mechanisms, as discussed earlier. A study on 24 MO nanoparticles in mammalian cell lines and bacteria by Zhang et al. (2012) and Kaweeteerawat et al. (2015) showed a strong correlation between ROS-related toxicity and redox potential of MOs (Fig. 1.4A). The bandgap hypothesis to predict the toxicity of semiconductive MO nanoparticles has been theoretically and experimentally explored. According to this hypothesis, prediction of ROS generation by any MO nanoparticles can be possible by comparing the E_v and E_c levels to the cellular redox potential. Additionally, the occurrence of electron transfer depends on resemblances in the energetic states of the nanomaterials and ambient redox-active aqueous substances. It was demonstrated that the overlap of the redox potential of cells (from −4.2 to −4.8 eV) and conduction band of MOs could lead to significant toxicity. The toxicity associated with the redox potential of MO nanoparticles dovetails with hierarchical oxidative stress hypothesis as shown in Fig. 1.4B. This hierarchical oxidative stress concept, which is based on the perception that the tier-1 response, attempts to maintain and restore the cellular redox equilibrium by increased expression of several redox-active substances, including glutathione (Meng, Xia, George, & Nel, 2009). This is followed by the tier-2 response to the activation of the Nrf2- pathway. If, however, this defense is overwhelmed by an escalating amount of ROS, the cellular response shifts to the activation of proinflammatory signaling cascades such as MAP or NF-κB (tier-3), and ultimately to mitochondrial-mediated cell death.

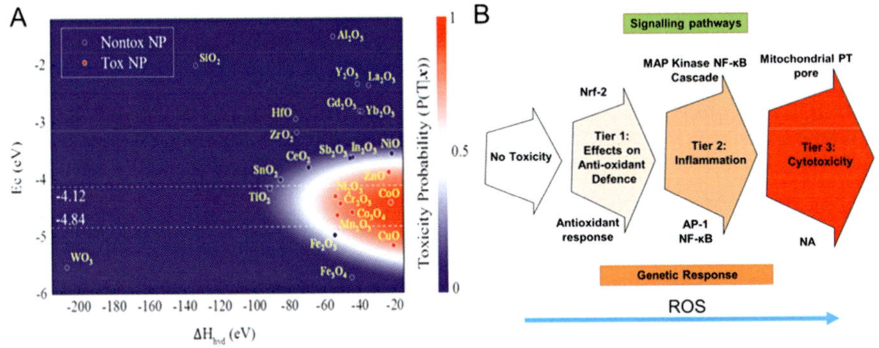

Figure 1.4 (A) Toxicity prediction from bandgap hypothesis of MO nanoparticles [image source Kaweeteerawat et al. (2015)], and (B) hierarchical oxidative stress model. From Kaweeteerawat, C., Ivask, A., Liu, R., Zhang, H., Chang, C. H., Low-Kam, C., Fischer, H., Ji, Z., Pokhrel, S., Cohen, Y., Telesca, D., Zink, J., Mädler, L., Holden, P. A., Nel, A., & Godwin, H. (2015). Toxicity of metal oxide nanoparticles in Escherichia coli correlates with conduction band and hydration energies. Environmental Science and Technology, 49(2), 1105–1112. https://doi.org/10.1021/es504259s

Furthermore, the metal ion release from the surface of MOs can lead to serious cytotoxicity. The dissolution of any MO nanoparticles depends on the physicochemical properties, biological environment, and solubility. ZnO and CuO exhibit high solubility, leading to a major toxicological impact by the released metal ions. Thus, for MOs the dissolution factor and catalytic surface are the foremost contributors to toxicity and ROS generation. The surface charge was also demonstrated to play a crucial role in membrane damage and particle internalization in cells (Baek et al., 2011; Stankic, Suman, Haque, & Vidic, 2016). The electrostatic interaction with the cellular membrane and MO nanoparticles leads to morphological changes to cells. For instance, upon binding with the bacterial cell surface, ZnO nanoparticles cause leakage in the membrane. Likewise, MO nanoparticles may specifically bind with various proteins, or lipids situated at the cell membrane, causing death (several bactericidal effects of MO nanoparticles act through this mechanism). It can be observed that not only nanoparticle characteristics, but also the cell type (in vitro) or administration route (in vivo) largely determine the fate and toxicity of MO nanoparticles (Cai et al., 2017; Kumari et al., 2013).

Both in vitro and in vivo toxicity follow different mechanisms for different MO nanoparticles. In the following section, the toxicity of some of the most important MO nanoparticles are discussed.

1.5.1 Iron oxide

Since the biomedical applications require the introduction of iron oxide nanoparticles (IONPs) directly into the human body, the biocompatibility, safety, and risk assessments are the top priority in investigating their advantages in medicinal uses.

Despite diverse biomedical applications, the toxicity of IONPs in biological systems has been reported. To understand the potential risk of IONPs, cellular, molecular, and organism-level toxicity of IONPs have been investigated in the literature. Cell culture studies show time- and concentration-dependent decreases in viability due to IONP exposure (Szalay, Tátrai, Nyíro, Vezér, & Dura, 2012). Several studies have shown an increase in ROS activity due to IONP exposure toward a wide ranges of cell lines, such as fibroblasts, human lung A549 cells, and brain microglia cells (Petters, Thiel, & Dringen, 2016; Radu et al., 2015; Rafieepour et al., 2019). Additionally, IONPs show suppressive activity toward antioxidative markers such as glutathione (GSH) and SOD (Alarifi, Ali, Alkahtani, & Alhader, 2014). IONPs can cause mitochondrial damage by producing altered membrane potential in mitochondria. Several studies have suggested that IONPs cause DNA damage, including strand breaks and micronucleus (MN) formation (Alarifi et al., 2014). However, some studies did not find any genotoxic alterations at concentrations lower than 500 ppm (Hong et al., 2011).

The evaluation of cell viability after IONP exposure demonstrated no conclusive results. Most of the cell viability study reported it as noncytotoxic at $<100\ \mu g/mL$ concentration (Mahmoudi et al., 2012). However, the toxicity of IONPs is cell-specific. Furthermore, IONPs are applied acutely to the brain for analytical and therapeutic purposes, such as MRI or cancer treatment via hyperthermia, or they remain at the site of administration and are taken up by phagocytic cells and astrocytes. Geppert, Hohnholt, Nürnberger, and Dringen (2012) demonstrated that long-term exposure did not compromise cell viability during prolonged incubation on astrocytes. Even an exposure of up to $200\ \mu g/mL$ of IONPs (25 nm size) had no measurable effect on Neuro-2A cells (Jeng & James, 2007).

IONPs with proper surface coating are often considered as biocompatible and safe. Injection of IONPs through the tail vein at a dose of 10 mg Fe/kg bw in the rat causes no severe toxicity (Jain, Reddy, Morales, Leslie-Pelecky, & Labhasetwar, 2008). However, many studies have shown that IONPs can alter biochemical and histopathological parameters. Hepatic injury and inflammation in the lung due to IONPs have also been reported for in vivo IONP exposure (Sadeghi, Yousefi Babadi, & Espanani, 2015). It was noticed that a repeated dose of IONPs causes serious in vivo toxicity; on the other hand, a single dose showed lower toxicity in most animal studies. A repeating dose shows the accumulation of IONPs in various tissues including the brain, heart, liver, lungs, kidneys, and spleen, causing considerable toxicity. It is crucial to consider the surface coating on IONPs in their toxicity. For instance, Feng et al. reported no toxicity for PEGylated IONPs in mice, whereas PEI-coated IONPs showed dose-dependent lethal toxicity (Feng et al., 2018).

1.5.2 ZnO

Zn is an essential component of many enzymes and transcription factors; thus, disruption of Zn homeostasis in cells causes oxidative stress, loss of viability, and mitochondrial dysfunction. Cell culture studies found a small range of ZnO nanoparticle concentrations, where cell viability decreases sharply from 100% to zero (Vandebriel & Jong, 2012). The toxicity of ZnO nanoparticles has been tested on a wide range of cell lines. ZnO nanoparticles of size ~20 nm induce time-dependent cytotoxicity in BEAS-2B cells, along with alteration of the genes associated with oxidative stress and apoptosis (Vandebriel & Jong, 2012). Additionally, IL-8 messenger production was stimulated by 24–70 nm ZnO nanoparticles (2 – 8 μg/mL) (Wu, Samet, Peden, & Bromberg, 2010). On the other hand, its exposure to A549 cells induced ROS generation and activated caspase-3 and caspase-9 in a time- and dose-dependent manner (Ahamed et al., 2011). Additionally, an antiapoptotic bcl-2 protein was downregulated, and proapoptotic proteins were upregulated, due to ZnO exposure.

ZnO is typically used in sunscreens, therefore its absorption through the skin has been extensively studied in the literature. The effect of ZnO NPs on various dermal cell lines such as human dermal fibroblasts, HaCaT human keratinocytes, and BJ human neonatal foreskin cells has been investigated (Vandebriel & Jong, 2012). Most of these studies suggested that ZnO NPs activate the p53 pathway, which leads to cell apoptosis. It is interesting to note that rod-shaped ZnO is more toxic than the spherical one. Most clinical studies have shown that ZnO could not penetrate beyond the stratum corneum. Animal studies support that ZnO nanoparticles cannot penetrate to a large extent through skin. Clinical research with isotopic ZnO nanoparticles (with ^{68}Zn) by Gulson et al. (2010) showed that only 0.1% of total Zn was of ^{68}Zn in the blood after a 5-day application on the skin. Therefore it can be concluded that ZnO nanoparticles can penetrate only to a limited extent through the skin.

Inhalation exposure in rats showed that exposure to both micron- (size: 3 μm, and concentration 25 and 50 mg/m^3) and nanosized ZnO (size: 300 nm, and concentration: 1 and 5 mg/kg body weight) cause inflammation in bronchoalveolar lavage (BAL). ZnO NPs (<10 nm) induced fibrotic and granulomatous inflammation (Cho et al., 2010). The mechanism of inflammation is due to the release of Zn^{2+} ions in the acidic environment of the lung-lining fluid. The release of ionic Zn increases lactate dehydrogenase (LDH), neutrophil content, and protein content. The pH-dependent solubility of ZnO nanoparticles is the main reason behind the toxicity of ZnO NPs.

1.5.3 TiO_2

TiO_2 particles (> 100 nm) are classified as biologically inert in both animals and humans. However, TiO_2 nanoparticles show enhanced cellular penetration capability that could make TiO_2 toxic. The opacity of TiO_2 is used to enhance the whiteness of products such as medicines, toothpaste, lotions, and cottage cheese. The concentration of TiO_2 can reach >100 μg/mL in toothpaste and sunscreens. El-Said, Ali, Kanehira, and Taniguchi (2014) concluded that exposure to TiO_2 nanoparticles causes oxidative stress, leading to p53 activation and DNA damage. It induces apoptosis by releasing cytochrome c into the cytoplasm and activating caspase-3. In addition to this, the overproduction of toll-like receptors (TLR4) enhances the ROS activity mediated by TiO_2 nanoparticles. They showed that expressions of 17 DNA damage marker genes, such as IP6K3 and ATM genes, were altered by TiO_2 NP exposure (El-Said et al., 2014). The Ames test, Comet assay micronucleus assay, sister chromatid exchange assay, wing somatic mutation, and recombination assay suggested that TiO_2 can induce genotoxicity which is mainly stress-driven (Chen, Yan, & Li, 2014). Evidence shows that the genotoxicity of TiO_2 nanoparticles was largely observed for in vitro systems, whereas results were mostly negative for in vivo systems. Additionally, the result for the mutagenicity of TiO_2 was also negative (Chen et al., 2014).

As skin contact is one of the major routes of exposure to TiO_2 nanoparticles, several studies have been carried out in skin cell lines. Tucci et al. (2013) examined the toxic effect of TiO_2 nanoparticles in the human keratinocyte HaCaT cell line. They found activation of cellular stress and reduced metabolic capacity due to 100 μg/mL TiO_2 exposure. Another great risk of TiO_2 NP exposure is to the respiratory system, as it has been used in cosmetic grade powder. TiO_2 exhibits toxicity to A549 cells (Wang et al., 2015). It has been demonstrated that exposure to 100 and 200 μg/mL TiO_2 nanoparticles by A549 cells increases the proportion of apoptotic cells. Additionally, TiO_2 nanoparticles also interrupted the mitochondrial membrane potential. In support of this, Oberdorster, Ferin, Gelein, Soderholm, and Finkelstein (1992) reported that TiO_2 exhibits chronic pulmonary inflammation in rats. The intravenous exposure of TiO_2 nanoparticles could induce pathological lesions of the brain, kidneys, liver, and spleen at a very high dose (Shi, Magaye, Castranova, & Zhao, 2013).

1.5.4 CeO_2

CeO_2 nanoparticles have been found to have both antioxidant and pro-oxidant impacts on various cellular systems. Studies have reported protective effects of nanoceria against ROS-mediated degeneration in certain neurological disorders (Abuid et al., 2020). Additionally, it shows other advantages, such as protection against radiation damage and inflammation. It has also been reported as a ROS generator in a number of in vitro studies. Pulido-Reyes et al. (2015) have demonstrated that the per cent of Ce^{3+} ions at the surface is the determining toxicity factor for ceria nanoparticles. They showed that the higher the percentage of Ce^{3+} at the surface site, the greater the toxicity. In fact, the in vitro toxicity of nanoceria is controversial as contradictory outcomes have been described in the literature. CeO_2 nanoparticles are commonly reported as having a low toxicity profile (Urner et al., 2014). Several studies have reported that CeO_2 nanoparticles do not produce inflammation or cytotoxicity (Fisichella et al., 2014; Xia et al., 2008). Another cell culture study showed that ceria nanoparticles have low toxicity in short-term (24 h) exposure, but over the long term (10 days), they showed significant toxicity toward A549, CaCo2, and HepG2 cell lines over a range of nanoparticle concentrations (0.5–5000 μg/mL) (De Marzi et al., 2013).

The toxic effect of nanoceria was observed in blood monocytes after 20–40 hours of exposure (concentration $\geq$ 5 μg/mL), which involved the induction of cell death by apoptosis and autophagy (Hussain et al., 2012). Mitochondrial damage and p53-modulated cytotoxicity have also been reported for CeO_2 NP-induced apoptosis (Hussain et al., 2012).

1.5.5 SiO_2

SiO_2 is a "generally regarded as safe" (GRAS) component by the United States Food and Drug Administration (US-FDA) (Rubio et al., 2019). Although SiO_2 is highly biocompatible, some reports have shown the toxicity of SiO_2 nanoparticles such as immunotoxicity, neurotoxicity, and genotoxicity (Murugadoss et al., 2017; Yazdimamaghani, Moos, Dobrovolskaia, & Ghandehari, 2019). Furthermore, being the most-produced nanoparticles, they have several ecotoxicological effects on aquatic life. The environmental exposure of SiO_2 NPs leads to oxidative damage, DNA impairment, and hematological and histological changes in fish (Athif et al., 2020). SiO_2 also induces behavior alterations and neurotoxicity in fish.

Animal studies of SiO_2 nanoparticles represent it as nontoxic after 90 days of oral dose in rats (Liang et al., 2018). It did not affect histological, hematological parameters, or mortality rate up to 1500 mg/kg bw/day. Another 90-day oral toxicity study of SiO_2 nanoparticles showed a no-observed-adverse-effect-level (NOAEL) up to 2000 mg/kg (Kim et al., 2014). However, an intranasal injection of SiO_2 nanoparticles caused toxicity including blood coagulation and platelet count effects indicating a coagulation disorder (Yoshida et al., 2013).

1.5.6 CuO

In the pharmaceutical industry, CuO is used in the nanoparticle form in the production of antimicrobial fabrics for treatment or prevention of infections. Humans could be exposed to CuO nanoparticles by two main routes: inhalation and skin exposure. CuO nanoparticles induce cell death to human lung cells (A549, H1650, and CNE-2Z cell line) (Sun, Yan, Zhao, Guo, & Jiang, 2012). They cause autophagic cell death to A549 cells (Sun et al., 2012). Exposure of CuO nanoparticles to kidney epithelial cells (A6 cell line) showed ROS-induced DNA damage and cell death via apoptosis (Thit, Selck, & Bjerregaard, 2015).

An oral dose of CuO causes alterations to hematological and clinical parameters in rats. Alterations to alkaline phosphatase (ALP) and aspartate aminotransferase (AST) enzymes indicated the presence of liver toxicity (De Jong et al., 2019). Histopathological changes have been reported in liver, indicating liver damage due to a CuO oral dose.

1.5.7 MgO

Recent results have uncovered the possible utility of MgO nanoparticles in the treatment of cancer, including nano-cryosurgery and hyperthermia (Krishnamoorthy, Moon, Hyun, Cho, & Kim, 2012). MgO nanoparticles have extremely high bactericidal property. The formation of a water layer outside the MgO surface is inferred as the main cause of this antibacterial activity. The pH of the water layer outside MgO nanoparticles increased compared to the equilibrium value under solution sets. The high pH level of the water surrounded by the nanoparticles causes membrane damage to bacteria when they come into close contact (Mazaheri, Naghsh, Karimi, & Salavati, 2019).

An in vivo study by Mazaheri et al. (2019) showed hematological (such as increased white blood cells, red blood cells, and hemoglobin) and histopathological (bile ductules, congestion, and apoptotic cells in the liver) changes in high-dose groups (250 and 500 μg/mL) after intraperitoneal injection of MgO nanoparticles. However, they reported no adverse toxicity when the concentration of MgO nanoparticles was low (62.5 and 125 μg/mL) in the rats over a 28-day study period. An oral dose of MgO nanoparticles in rats illustrated genotoxicity in a dose-dependent manner and significant toxicity was observed after 28 days of treatment (Mangalampalli, Dumala, Perumalla Venkata, & Grover, 2018).

1.5.8 ZrO_2

ZrO_2 is used as a high-end product, especially in dentistry, due to its biocompatibility, strong corrosion resistance, and pure white color. ZrO_2 provides good stability and mechanical strength to orthopedic implants. On the other hand, some indication of the toxicity of ZrO_2 nanoparticles has been reported, but notably, the exposure concentration of ZrO_2 nanoparticles is extremely high. ZrO_2 nanoparticles are capable of drug delivery applications. ZrO_2 nanoparticles (size 25.4 ± 2.8 nm) exhibit good biocompatibility toward osteoblast-like 3T3-E1 cells (10 μg/mL), although high-concentration exposure showed elevated ROS leading to cytotoxicity at high concentrations (Ye & Shi, 2018). Kozelskaya et al. (2016) found that ZrO_2 nanoparticles induced an increase in membrane microviscosity and surface cracks on the red blood cells due to enhanced oxidative stress.

Most in vivo studies have shown low toxicity of ZrO_2. Intravenous administration of ZrO_2 nanoparticles on mice showed alterations to serum ALT levels, and reduced levels of SOD and white blood cell counts significantly at very high concentration (500 mg/kg mice) (Yang et al., 2019). Another study on rats showed increasing MDA level and decreased GP_X, CAT, and SOD activities after intraperitoneal administration of 100 ppm ZrO_2 (1 mL) (Arefian, Pishbin, Negahdary, & Ajdary, 2015). The lower dose (50 ppm or less) did not show toxicity (Arefian et al., 2015).

1.5.9 MnO_2

Cell culture studies have shown that MnO_2 nanoparticles have cytotoxic effects on human breast cancer epithelial (MCF-7), HeLa, and human fibrosarcoma epithelial (HT1080) cells (Li et al., 2010). MnO_2 nanoparticles induce oxidative stress and apoptosis mediated by the P53 pathway (Alhadlaq, Akhtar, & Ahamed, 2019).

Oral administration of MnO_2 nanoparticles (300, 1000 mg/kg) and microparticles (1000 mg/kg) in rats showed a significant increase in DNA damage in bone marrow cells (Singh et al., 2013). Significant dose-dependent changes were observed in lactate dehydrogenase (LDH), alanine aminotransferase (ALT), and aspartate aminotransferase (AST), in the serum liver and kidney (Singh et al., 2013). Subcutaneous injection of MnO_2 nanoparticles to males caused a significant reduction in the number of sperm, spermatocytes, diameter of seminiferous tubes, spermatogonia, and in the motility of sperm (Yousefalizadegan, Mousavi, Rastegar, Razavi, & Najafizadeh, 2019).

1.5.10 NiO

NiO nanoparticles showed a decrease in cell viability in carcinogenic cell lines such as HEp-2, HT-29, SW620, and A549 (AlSalhi et al., 2019; Khan et al., 2019; Siddiqui et al., 2012). Oxidative stress-mediated apoptosis or necrosis due to the destructive inflammatory response caused impairments to DNA/mitochondria in cancer cells (AlSalhi et al., 2019). Khan et al. (2019) reported downregulation of Bcl-2 and Bcl-xL protein levels in HT-29 cells, suggesting that NiO nanoparticles can induce cell death through alterations to the mitochondrial pathway of apoptosis.

A 28-day repeated oral dose of NiO nanoparticles with an average size of 12.9 ($\pm$ 3.4) nm in rats showed a significant increase in catalase activity and a decrease in superoxide dismutase activity (Dumala, Mangalampalli, Kalyan Kamal, & Grover, 2019). Further, a decrease in GSH activity was also reported, indicating enhanced reactive oxygen species in a dose-dependent manner (Dumala et al., 2019). NiO had impacts on the vital organs of the rats, such as the kidneys and liver. The genotoxicity of NiO nanoparticles (size 15.62 $\pm$ 2.59 nm) in rats after a single oral dose (at 250 mg/ kg bw) was reported by Dumala et al. (2017). Comet assay revealed significant DNA damage in kidney and liver cells. A 50 mg/kg intraperitoneal dose of NiO for 8 days showed neurotoxicological effects in rats (Marzban, Seyedalipour, Mianabady, Taravati, & Hoseini, 2020). The GSH level decreased significantly due to NiO exposure. Additionally, histopathological alterations, such as necrosis, hyperemia, gliosis, and spongy changes in brain tissue were also reported (Marzban et al., 2020).

1.6 Summary

This chapter was intended to provide an overview of the recent advancements in MOs in the biomedical field. It is inevitable that MOs will become a high-potential and efficient candidate for biomedical applications. However, several toxic effects, especially for nanoparticles, hinder its acceptance in this field. Several MO nanoparticles are toxic even at low concentrations, while others have low toxicity and toxicity that is marked only after exposure to high doses. Increases of ROS, inflammatory responses to DNA damage, and cell death have been reported for MO nanoparticles throughout the literature. Despite its vast potency, concern regarding its toxicity is yet to be answered. Encouragingly, recent developments in engineered

MO nanoparticles with surface modifications can reduce the inherent toxicity of nanoparticles and enhance their biocompatibility. There is no doubt that MOs are gaining increased presence in the biomedical field. In future, they could deliver promising breakthroughs in clinical applications for biosensors, drug delivery, cancer research, and cell imaging techniques.

References

Abuid, N. J., Gattás-Asfura, K. M., LaShoto, D. J., Poulos, A. M., & Stabler, C. L. (2020). *Biomedical applications of cerium oxide nanoparticles: A potent redox modulator and drug delivery agent. Nanoparticles for biomedical applications* (pp. 283–301). Available from https://doi.org/10.1016/b978-0-12-816662-8.00017-5.

Adabi, M., Naghibzadeh, M., Adabi, M., Zarrinfard, M. A., Esnaashari, S. S., Seifalian, A. M., ... Ghanbari, H. (2017). Biocompatibility and nanostructured materials: Applications in nanomedicine. *Artificial Cells, Nanomedicine and Biotechnology*, *45*(4), 833–842. Available from https://doi.org/10.1080/21691401.2016.1178134.

Ahamed, M., Akhtar, M. J., Raja, M., Ahmad, I., Siddiqui, M. K. J., AlSalhi, M. S., & Alrokayan, S. A. (2011). ZnO nanorod-induced apoptosis in human alveolar adenocarcinoma cells via p53, survivin and bax/bcl-2 pathways: Role of oxidative stress. *Nanomedicine: Nanotechnology, Biology, and Medicine*, *7*(6), 904–913. Available from https://doi.org/10.1016/j.nano.2011.04.011.

Alarifi, S., Ali, D., Alkahtani, S., & Alhader, M. S. (2014). Iron oxide nanoparticles induce oxidative stress, DNA damage, and caspase activation in the human breast cancer cell line. *Biological Trace Element Research*, *159*(1–3), 416–424. Available from https://doi.org/10.1007/s12011-014-9972-0.

Alhadlaq, H. A., Akhtar, M. J., & Ahamed, M. (2019). Different cytotoxic and apoptotic responses of MCF-7 and HT1080 cells to MnO2 nanoparticles are based on similar mode of action. *Toxicology*, *411*, 71–80. Available from https://doi.org/10.1016/j.tox.2018.10.023.

AlSalhi, M. S., Aziz, M. H., Atif, M., Fatima, M., Shaheen, F., Devanesan, S., & Aslam Farooq, W. (2019). Synthesis of NiO nanoparticles and their evaluation for photodynamic therapy against HeLa cancer cells. *Journal of King Saud University – Science*, *32*, 1395–1402. Available from https://doi.org/10.1016/j.jksus.2019.11.033.

Andreescu, S., Maryna, O., S, E. J., Ana, E., & C, L. J. (2011). *Biomedical applications of metal oxide nanoparticles. Fine particles in medicine and pharmacy* (pp. 57–100). Springer Science and Business Media LLC, https://doi.org/10.1007/978-1-4614-0379-1_3.

Arefian, Z., Pishbin, F., Negahdary, M., & Ajdary, M. (2015). Potential toxic effects of zirconia oxide nanoparticles on liver and kidney factors. *Biomedical Research (India)*, *26*(1), 89–97. Available from http://biomedres.info/yahoo_site_admin/assets/docs/89-97.336195302.pdf.

Athif, P., Suganthi, P., Murali, M., Sadiq Bukhari, A., Syed Mohamed, H. E., Basu, H., & Singhal, R. K. (2020). Hepatic toxicological responses of SiO2 nanoparticle on Oreochromis mossambicus. *Environmental Toxicology and Pharmacology*, *78*, 103398. Available from https://doi.org/10.1016/j.etap.2020.103398.

Baek, M., Kim, M. K., Cho, H. J., Lee, J. A., Yu, J., Chung, E. H., & Choi, J. S. (2011). Factors influencing the cytotoxicity of zinc oxide nanoparticles: particle size and surface charge. *Journal of Physics: Conference Series*, *304*, 012044. Available from https://doi.org/10.1088/1742-6596/304/1/012044.

Barui, A. K., Nethi, S. K., Haque, S., Basuthakur, P., & Patra, C. R. (2019). Recent development of metal nanoparticles for angiogenesis study and their therapeutic applications. *ACS Applied Bio Materials*, *2*(12), 5492–5511. Available from https://doi.org/10.1021/acsabm.9b00587.

Bhowmick, T. K., Suresh, A. K., Kane, S. G., Joshi, A. C., & Bellare, J. R. (2009). Physicochemical characterization of an Indian traditional medicine, Jasada Bhasma: Detection of nanoparticles containing non-stoichiometric zinc oxide. *Journal of Nanoparticle Research*, *11*(3), 655–664. Available from https://doi.org/10.1007/s11051-008-9414-z.

Biswas, S., Dhumal, R., Selkar, N., Bhagat, S., Chawda, M., Thakur, K., ... Bellare, J. (2020). Physicochemical characterization of Suvarna Bhasma, its toxicity profiling in rat and behavioural assessment in zebrafish model. *Journal of Ethnopharmacology*, *249*, 112388. Available from https://doi.org/10.1016/j.jep.2019.112388.

Boyles, M. S. P., Powell, L. G., Kermanizadeh, A., Johnston, H. J., Stone, V., & Clift, M. J. D. (2017). *An overview of nanoparticle biocompatibility for their use in nanomedicine. Pharmaceutical nanotechnology: Innovation and production* (pp. 443–467). .

Cai, X., Lee, A., Ji, Z., Huang, C., Chang, C. H., Wang, X., ... Li, R. (2017). Reduction of pulmonary toxicity of metal oxide nanoparticles by phosphonate-based surface passivation. *Particle and Fibre Toxicology*, *14*(1). Available from https://doi.org/10.1186/s12989-017-0193-5.

Chen, L., Liu, J., Zhang, Y., Zhang, G., Kang, Y., Chen, A., ... Shao, L. (2018). The toxicity of silica nanoparticles to the immune system. *Nanomedicine: Nanotechnology, Biology, and Medicine*, *13*(15), 1939–1962. Available from https://doi.org/10.2217/nnm-2018-0076.

Chen, T., Yan, J., & Li, Y. (2014). Genotoxicity of titanium dioxide nanoparticles. *Journal of Food and Drug Analysis*, *22*(1), 95–104. Available from https://doi.org/10.1016/j.jfda.2014.01.008.

Chen, Y., Cong, H., Shen, Y., & Yu, B. (2020). Biomedical application of manganese dioxide nanomaterials. *Nanotechnology*, *31*(20). Available from https://doi.org/10.1088/1361-6528/ab6fe1.

Chen, Y., Roohani-Esfahani, S. I., Lu, Z. F., Zreiqat, H., & Dunstan, C. R. (2015). Zirconium ions up-regulate the BMP/SMAD signaling pathway and promote the proliferation and differentiation of human osteoblasts. *PLoS One*, *10*(1). Available from https://doi.org/10.1371/journal.pone.0113426.

Cho, W. S., Duffn, R., Poland, C. A., Howie, S. E. M., Macnee, W., Bradley, M., ... Donaldson, K. (2010). Metal oxide nanoparticles induce unique infammatory footprints in the lung: Important implications for nanoparticle testing. *Environmental Health Perspectives*, *118*(12), 1699–1706. Available from https://doi.org/10.1289/ehp.1002201.

Croissant, J. G., Fatieiev, Y., Almalik, A., & Khashab, N. M. (2018). Mesoporous silica and organosilica nanoparticles: Physical chemistry, biosafety, delivery strategies, and biomedical applications. *Advanced Healthcare Materials*, *7*(4), 1700831. Available from https://doi.org/10.1002/adhm.201700831.

De Jong, W. H., De Rijk, E., Bonetto, A., Wohlleben, W., Stone, V., Brunelli, A., ... Cassee, F. R. (2019). Toxicity of copper oxide and basic copper carbonate nanoparticles after short-term oral exposure in rats. *Nanotoxicology*, *13*(1), 50–72. Available from https://doi.org/10.1080/17435390.2018.1530390.

De Marzi, L., Monaco, A., De Lapuente, J., Ramos, D., Borras, M., Di Gioacchino, M., ... Poma, A. (2013). Cytotoxicity and genotoxicity of ceria nanoparticles on different cell lines in vitro. *International Journal of Molecular Sciences*, *14*(2), 3065–3077. Available from https://doi.org/10.3390/ijms14023065.

Dobrovolskaia, M. A., & McNeil, S. E. (2007). Immunological properties of engineered nanomaterials. *Nature Nanotechnology*, *2*(8), 469–478. Available from https://doi.org/10.1038/nnano.2007.223.

Dumala, N., Mangalampalli, B., Chinde, S., Kumari, S. I., Mahoob, M., Rahman, M. F., & Grover, P. (2017). Genotoxicity study of nickel oxide nanoparticles in female Wistar rats after acute oral exposure. *Mutagenesis*, *32*(4), 417–427. Available from https://doi.org/10.1093/mutage/gex007.

Dumala, N., Mangalampalli, B., Kalyan Kamal, S. S., & Grover, P. (2019). Repeated oral dose toxicity study of nickel oxide nanoparticles in Wistar rats: A histological and biochemical perspective. *Journal of Applied Toxicology*, *39*(7), 1012–1029. Available from https://doi.org/10.1002/jat.3790.

El-Said, K. S., Ali, E. M., Kanehira, K., & Taniguchi, A. (2014). Molecular mechanism of DNA damage induced by titanium dioxide nanoparticles in toll-like receptor 3 or 4 expressing human hepatocarcinoma cell lines. *Journal of Nanobiotechnology*, *12*(1), 48. Available from https://doi.org/10.1186/s12951-014-0048-2.

Estelrich, J., Escribano, E., Queralt, J., & Busquets, M. A. (2015). Iron oxide nanoparticles for magnetically-guided and magnetically-responsive drug delivery. *International Journal of Molecular Sciences*, *16*(4), 8070–8101. Available from https://doi.org/10.3390/ijms16048070.

Feng, Q., Liu, Y., Huang, J., Chen, K., Huang, J., & Xiao, K. (2018). Uptake, distribution, clearance, and toxicity of iron oxide nanoparticles with different sizes and coatings. *Scientific Reports*, *8*(1). Available from https://doi.org/10.1038/s41598-018-19628-z.

Fernández-García, M., & A, R. J. (2011). *Metal oxide nanoparticles. Encyclopedia of inorganic and bioinorganic chemistry*. Wiley, https://doi.org/10.1002/9781119951438.eibc0331.

Fisichella, M., Berenguer, F., Steinmetz, G., Auffan, M., Rose, J., & Prat, O. (2014). Toxicity evaluation of manufactured CeO2 nanoparticles before and after alteration: Combined physicochemical and whole-genome expression analysis in Caco-2 cells. *BMC Genomics*, *15*(1), 700. Available from https://doi.org/10.1186/1471-2164-15-700.

Geppert, M., Hohnholt, M. C., Nürnberger, S., & Dringen, R. (2012). Ferritin up-regulation and transient ROS production in cultured brain astrocytes after loading with iron oxide nanoparticles. *Acta Biomaterialia*, *8*(10), 3832–3839. Available from https://doi.org/10.1016/j.actbio.2012.06.029.

Grüttner, C., Müller, K., Teller, J., & Westphal, F. (2013). Synthesis and functionalisation of magnetic nanoparticles for hyperthermia applications. *International Journal of Hyperthermia*, *29*(8), 777–789. Available from https://doi.org/10.3109/02656736.2013.835876.

Guaadaoui, A., Benaicha, S., Elmajdoub, N., Bellaoui, M., & Hamal, A. (2014). What is a bioactive compound? A combined definition for a preliminary consensus. *International Journal of Food Sciences and Nutrition*, *3*(3), 17–179. Available from https://doi.org/10.11648/j.ijnfs.20140303.16.

Gulson, B., Mccall, M., Korsch, M., Gomez, L., Casey, P., Oytam, Y., ... Greenoak, G. (2010). Small amounts of zinc from zinc oxide particles in sunscreens applied outdoors are absorbed through human skin. *Toxicological Sciences*, *118*(1), 140–149. Available from https://doi.org/10.1093/toxsci/kfq243.

Guo, J., Zhou, H., Wang, J., Liu, W., Cheng, M., Peng, X., ... Zhang, X. (2018). Nano vanadium dioxide films deposited on biomedical titanium: A novel approach for simultaneously enhanced osteogenic and antibacterial effects. *Artificial Cells, Nanomedicine and Biotechnology*, *46*(2), 58–74. Available from https://doi.org/10.1080/21691401.2018.1452020.

Hillaireau, H., & Couvreur, P. (2009). Nanocarriers' entry into the cell: Relevance to drug delivery. *Cellular and Molecular Life Sciences*, *66*(17), 2873–2896. Available from https://doi.org/10.1007/s00018-009-0053-z.

Hong, S. C., Lee, J. H., Lee, J., Kim, H. Y., Park, J. Y., Cho, J., ... Han, D. W. (2011). Subtle cytotoxicity and genotoxicity differences in superparamagnetic iron oxide nanoparticles coated with various functional groups. *International Journal of Nanomedicine*, *6*, 3219–3231.

Horie, M., & Fujita, K. (2011). *Toxicity of metal oxides nanoparticles*, . *Advances in molecular toxicology* (Vol. 5, pp. 145–178). Elsevier B.V, https://doi.org/10.1016/B978-0-444-53864-2.00004-9.

Hussain, S., Al-Nsour, F., Rice, A. B., Marshburn, J., Yingling, B., Ji, Z., ... Garantziotis, S. (2012). Cerium dioxide nanoparticles induce apoptosis and autophagy in human peripheral blood monocytes. *ACS Nano*, *6*(7), 5820–5829. Available from https://doi.org/10.1021/nn302235u.

Hyun, J. D., Hyoung, K. J., Geol, L. T., & Hun, K. J. (2015). Size, surface charge, and shape determine therapeutic effects of nanoparticles on brain and retinal diseases. *Nanomedicine: Nanotechnology, Biology and Medicine*, *11*, 1603–1611. Available from https://doi.org/10.1016/j.nano.2015.04.015.

Imran Din, M., & Rani, A. (2016). Recent advances in the synthesis and stabilization of nickel and nickel oxide nanoparticles: A green adeptness. *International Journal of Analytical Chemistry*, *2016*. Available from https://doi.org/10.1155/2016/3512145.

Iravani, S., & Varma, R. S. (2020). Sustainable synthesis of cobalt and cobalt oxide nanoparticles and their catalytic and biomedical applications. *Green Chemistry*, *22*(9), 2643–2661. Available from https://doi.org/10.1039/d0gc00885k.

Jain, T. K., Reddy, M. K., Morales, M. A., Leslie-Pelecky, D. L., & Labhasetwar, V. (2008). Biodistribution, clearance, and biocompatibility of iron oxide magnetic nanoparticles in rats. *Molecular Pharmaceutics*, *5*(2), 316–327. Available from https://doi.org/10.1021/mp7001285.

Jeng, A. H., & James, S. (2007). Toxicity of metal oxide nanoparticles in mammalian cells. *Journal of Environmental Science and Health, Part A*, *41*, 2699–2711. Available from https://doi.org/10.1080/10934520600966177.

Kannan, K., Radhika, D., Sadasivuni, K. K., Reddy, K. R., & Raghu, A. V. (2020). Nanostructured metal oxides and its hybrids for photocatalytic and biomedical applications. *Advances in Colloid and Interface Science*, *281*, 102178. Available from https://doi.org/10.1016/j.cis.2020.102178.

Kaweeteerawat, C., Ivask, A., Liu, R., Zhang, H., Chang, C. H., Low-Kam, C., ... Godwin, H. (2015). Toxicity of metal oxide nanoparticles in *Escherichia coli* correlates with conduction band and hydration energies. *Environmental Science and Technology*, *49*(2), 1105–1112. Available from https://doi.org/10.1021/es504259s.

Keshtkar, M., Shahbazi-Gahrouei, D., Mehrgardi, M. A., Aghaei, M., & Khoshfetrat, S. M. (2018). Synthesis and cytotoxicity assessment of gold-coated magnetic iron oxide nanoparticles. *Journal of Biomedical Physics and Engineering*, *8*(3), 357–364. Available from http://www.jbpe.org/Journal_OJS/JBPE/index.php/jbpe/article/download/588/336.

Khan, S., Ansari, A. A., Malik, A., Chaudhary, A. A., Syed, J. B., & Khan, A. A. (2019). Preparation, characterizations and in vitro cytotoxic activity of nickel oxide nanoparticles on HT-29 and SW620 colon cancer cell lines. *Journal of Trace Elements in Medicine and Biology*, *52*, 12–17. Available from https://doi.org/10.1016/j.jtemb.2018.11.003.

Kim, Y. R., Lee, S. Y., Lee, E. J., Park, S. H., Seong, N. W., Seo, H. S., ... Kim, M. K. (2014). Toxicity of colloidal silica nanoparticles administered orally for 90 days in rats. *International Journal of Nanomedicine*, *9*, 67–78. Available from https://doi.org/10.2147/IJN.S57925.

Kolodziejczak-Radzimska, A., & Jesionowski, T. (2014). Zinc oxide-from synthesis to application: A review. *Materials*, *7*(4), 2833–2881. Available from https://doi.org/10.3390/ma7042833.

Kozelskaya, A. I., Panin, A. V., Khlusov, I. A., Mokrushnikov, P. V., Zaitsev, B. N., Kuzmenko, D. I., & Vasyukov, G. Y. (2016). Morphological changes of the red blood cells treated with metal oxide nanoparticles. *Toxicology In Vitro*, *37*, 34–40. Available from https://doi.org/10.1016/j.tiv.2016.08.012.

Krishnamoorthy, K., Moon, J. Y., Hyun, H. B., Cho, S. K., & Kim, S. J. (2012). Mechanistic investigation on the toxicity of MgO nanoparticles toward cancer cells. *Journal of Materials Chemistry*, *22*(47), 24610–24617. Available from https://doi.org/10.1039/c2jm35087d.

Kumari, M., Rajak, S., Singh, S. P., Murty, U. S. N., Mahboob, M., Grover, P., & Rahman, M. F. (2013). Biochemical alterations induced by acute oral doses of iron oxide nanoparticles in Wistar rats. *Drug and Chemical Toxicology*, *36*(3), 296–305. Available from https://doi.org/10.3109/01480545.2012.720988.

Laffon, B., Fernández-Bertólez, N., Costa, C., Brandão, F., Teixeira, J. P., Pásaro, E., & Valdiglesias, V. (2018). *Cellular and molecular toxicity of iron oxide nanoparticles*, . *Advances in experimental medicine and biology* (Vol. 1048, pp. 199–213). Springer New York LLC, https://doi.org/10.1007/978-3-319-72041-8_12.

Li, X., Wang, L., Fan, Y., Feng, Q., & Cui, F. Z. (2012). Biocompatibility and toxicity of nanoparticles and nanotubes. *Journal of Nanomaterials*, *2012*. Available from https://doi.org/10.1155/2012/548389.

Liang, C. L., Xiang, Q., Cui, W. M., Fang, J., Sun, N. N., Zhang, X. P., ... Jia, X. D. (2018). Subchronic Oral Toxicity of Silica Nanoparticles and Silica Microparticles in Rats. *Biomedical and Environmental Sciences*, *31*(3), 197–207. Available from https://doi.org/10.3967/bes2018.025.

Mahmoudi, M., Hofmann, H., Rothen-Rutishauser, B., & Petri-Fink, A. (2012). Assessing the in vitro and in vivo toxicity of superparamagnetic iron oxide nanoparticles. *Chemical Reviews*, *112*(4), 2323–2338. Available from https://doi.org/10.1021/cr2002596.

Mangalampalli, B., Dumala, N., Perumalla Venkata, R., & Grover, P. (2018). Genotoxicity, biochemical, and biodistribution studies of magnesium oxide nano and microparticles in albino wistar rats after 28-day repeated oral exposure. *Environmental Toxicology*, *33*(4), 396–410. Available from https://doi.org/10.1002/tox.22526.

María, V.-R., & Francisco, B. (2008). Silica Materials for Medical Applications. *The Open Biomedical Engineering Journal,*, *2*, 1–9. Available from https://doi.org/10.2174/1874120700802010001.

Marzban, A., Seyedalipour, B., Mianabady, M., Taravati, A., & Hoseini, S. M. (2020). Biochemical, toxicological, and histopathological outcome in rat brain following treatment with NiO and NiO nanoparticles. *Biological Trace Element Research*, *196*(2), 528–536. Available from https://doi.org/10.1007/s12011-019-01941-x.

Mazaheri, N., Naghsh, N., Karimi, A., & Salavati, H. (2019). In vivo toxicity investigation of magnesium oxide nanoparticles in rat for environmental and biomedical applications. *Iranian Journal of Biotechnology*, *17*(1), 1–9. Available from https://doi.org/10.21859/ijb.1543.

McNamara, K., & Tofail, S. A. M. (2017). Nanoparticles in biomedical applications. *Advances in Physics: X*, *2*(1), 54–88. Available from https://doi.org/10.1080/23746149.2016.1254570.

Meng, H., Xia, T., George, S., & Nel, A. E. (2009). A predictive toxicological paradigm for the safety assessment of nanomaterials. *ACS Nano*, *3*(7), 1620–1627. Available from https://doi.org/10.1021/nn9005973.

Modi, A., & Bellare, J. (2019). Efficient separation of biological macromolecular proteins by polyethersulfone hollow fiber ultrafiltration membranes modified with Fe3O4 nanoparticles-decorated carboxylated graphene oxide nanosheets. *International Journal of Biological Macromolecules*, *135*, 798–807. Available from https://doi.org/10.1016/j.ijbiomac.2019.05.200.

Murthy, Sudtha, Effiong, Paul, & Fei, Chee Chin (2020). *Metal oxide nanoparticles in biomedical applications. Metal Oxide Powder Technologies*. Elsevier. Available from https://doi.org/10.1016/b978-0-12-817505-7.00011-7.

Murugadoss, S., Lison, D., Godderis, L., Van Den Brule, S., Mast, J., Brassinne, F., . . . Hoet, P. H. (2017). Toxicology of silica nanoparticles: An update. *Archives of Toxicology*, *91* (9), 2967–3010. Available from https://doi.org/10.1007/s00204-017-1993-y.

Naahidi, S., Jafari, M., Edalat, F., Raymond, K., Khademhosseini, A., & Chen, P. (2013). Biocompatibility of engineered nanoparticles for drug delivery. *Journal of Controlled Release*, *166*(2), 182–194. Available from https://doi.org/10.1016/j.jconrel.2012.12.013.

Oberdorster, G., Ferin, J., Gelein, R., Soderholm, S. C., & Finkelstein, J. (1992). Role of the alveolar macrophage in lung injury: Studies with ultrafine particles. *Environmental Health Perspectives*, *97*, 193–199. Available from https://doi.org/10.1289/ehp.97-1519541.

Owens, D. E., & Peppas, N. A. (2006). Opsonization, biodistribution, and pharmacokinetics of polymeric nanoparticles. *International Journal of Pharmaceutics*, *307*(1), 93–102. Available from https://doi.org/10.1016/j.ijpharm.2005.10.010.

Pandey, S., & Chaudhary, A. (2018). *Toxicity of Bhasmas and chelating agents used in Ayurveda. Biomedical applications of metals* (pp. 237–255). Springer International Publishing, https://doi.org/10.1007/978-3-319-74814-6_11.

Petters, C., Thiel, K., & Dringen, R. (2016). Lysosomal iron liberation is responsible for the vulnerability of brain microglial cells to iron oxide nanoparticles: Comparison with neurons and astrocytes. *Nanotoxicology*, *10*(3), 332–342. Available from https://doi.org/10.3109/17435390.2015.1071445.

Pham, X. H., Hahm, E., Kim, H. M., Son, B. S., Jo, A., An, J., . . . Jun, B. H. (2020). Silica-coated magnetic iron oxide nanoparticles grafted onto graphene oxide for protein isolation. *Nanomaterials*, *10*(1), 117. Available from https://doi.org/10.3390/nano10010117.

Popov, A. L., Popova, N. R., Tarakina, N. V., Ivanova, O. S., Ermakov, A. M., Ivanov, V. K., & Sukhorukov, G. B. (2018). Intracellular delivery of antioxidant CeO2 nanoparticles via polyelectrolyte microcapsules. *ACS Biomaterials Science and Engineering*, *4* (7), 2453–2462. Available from https://doi.org/10.1021/acsbiomaterials.8b00489.

Priyadarshini, B., Rama, M., Chetan., & Vijayalakshmi, U. (2019). Bioactive coating as a surface modification technique for biocompatible metallic implants: A review. *Journal of Asian Ceramic Societies*, *7*(4), 397–406. Available from https://doi.org/10.1080/21870764.2019.1669861.

Pulido-Reyes, G., Rodea-Palomares, I., Das, S., Sakthivel, T. S., Leganes, F., Rosal, R., ... Fernández-Pinãs, F. (2015). Untangling the biological effects of cerium oxide nanoparticles: The role of surface valence states. *Scientific Reports*, *5*, 15613. Available from https://doi.org/10.1038/srep15613.

Radu, M., Dinu, D., Sima, C., Burlacu, R., Hermenean, A., Ardelean, A., & Dinischiotu, A. (2015). Magnetite nanoparticles induced adaptive mechanisms counteract cell death in human pulmonary fibroblasts. *Toxicology In Vitro*, *29*(7), 1492–1502. Available from https://doi.org/10.1016/j.tiv.2015.06.002.

Rafieepour, A., Azari, M. R., Peirovi, H., Khodagholi, F., Jaktaji, J. P., Mehrabi, Y., ... Mohammadian, Y. (2019). Investigation of the effect of magnetite iron oxide particles size on cytotoxicity in A549 cell line. *Toxicology and Industrial Health*, *35*(11–12), 703–713. Available from https://doi.org/10.1177/0748233719888077.

Rastogi, A., Zivcak, M., Sytar, O., Kalaji, H. M., He, X., Mbarki, S., & Brestic, M. (2017). Impact of metal and metal oxide nanoparticles on plant: A critical review. *Frontiers in Chemistry*, *5*, 78. Available from https://doi.org/10.3389/fchem.2017.00078.

Remya, N. S., Syama, S., Sabareeswaran, A., & Mohanan, P. V. (2016). Toxicity, toxicokinetics and biodistribution of dextran stabilized Iron oxide Nanoparticles for biomedical applications. *International Journal of Pharmaceutics*, *511*(1), 586–598. Available from https://doi.org/10.1016/j.ijpharm.2016.06.119.

Rubio, L., Pyrgiotakis, G., Beltran-Huarac, J., Zhang, Y., Gaurav, J., Deloid, G., ... Demokritou, P. (2019). Safer-by-design flame-sprayed silicon dioxide nanoparticles: The role of silanol content on ROS generation, surface activity and cytotoxicity. *Particle and Fibre Toxicology*, *16*(1), 40. Available from https://doi.org/10.1186/s12989-019-0325-1.

Sadeghi, L., Yousefi Babadi, V., & Espanani, H. R. (2015). Toxic effects of the Fe2O3 nanoparticles on the liver and lung tissue. *Bratislava Medical Journal*, *116*(6), 373–378. Available from https://doi.org/10.4149/BLL_2015_071.

Sangaiya, P., & Jayaprakash, R. (2018). A review on iron oxide nanoparticles and their biomedical applications. *Journal of Superconductivity and Novel Magnetism*, *31*(11), 3397–3413. Available from https://doi.org/10.1007/s10948-018-4841-2.

Santiago-Medina, P., Sundaram, P. A., & Diffoot-Carlo, N. (2015). Titanium oxide: A bioactive factor in osteoblast differentiation. *International Journal of Dentistry*, *2015*, 357653. Available from https://doi.org/10.1155/2015/357653.

Shi, H., Magaye, R., Castranova, V., & Zhao, J. (2013). Titanium dioxide nanoparticles: A review of current toxicological data. *Particle and Fibre Toxicology*, *10*(1), 15. Available from https://doi.org/10.1186/1743-8977-10-15.

Shilpa Chakra, C., & Venkateswara Rao, K. (2014). Bioprocess variables of magnetite nanoparticles using modified modern bhasmikaran method. *International Journal of Research in Ayurveda and Pharmacy*, *5*(2), 205–208. Available from https://doi.org/10.7897/2277-4343.05241.

Siddiqui, M. A., Ahamed, M., Ahmad, J., Majeed Khan, M. A., Musarrat, J., Al-Khedhairy, A. A., & Alrokayan, S. A. (2012). Nickel oxide nanoparticles induce cytotoxicity, oxidative stress and apoptosis in cultured human cells that is abrogated by the dietary antioxidant curcumin. *Food and Chemical Toxicology*, *50*(3–4), 641–647. Available from https://doi.org/10.1016/j.fct.2012.01.017.

Simeone, F. C., & Costa, A. L. (2019). Assessment of cytotoxicity of metal oxide nanoparticles on the basis of fundamental physical-chemical parameters: A robust approach to grouping. *Environmental Science: Nano*, *6*(10), 3102–3112. Available from https://doi.org/10.1039/c9en00785g.

Singh, A., & Kumar, V. (2020). *Iron oxide nanoparticles in biosensors, imaging and drug delivery applications—a complete tool, . Intelligent systems reference library* (Vol. 180, pp. 243–252). Springer, https://doi.org/10.1007/978-3-030-39119-5_20.

Singh, K. R. B., Nayak, V., Sarkar, T., & Singh, R. P. (2020). Cerium oxide nanoparticles: Properties, biosynthesis and biomedical application. *RSC Advances*, *10*(45), 27194–27214. Available from https://doi.org/10.1039/d0ra04736h.

Singh, S. P., Kumari, M., Kumari, S. I., Rahman, M. F., Mahboob, M., & Grover, P. (2013). Toxicity assessment of manganese oxide micro and nanoparticles in Wistar rats after 28days of repeated oral exposure. *Journal of Applied Toxicology*, *33*(10), 1165–1179. Available from https://doi.org/10.1002/jat.2887.

Srigurunathan, K., Meenambal, R., Guleria, A., Kumar, D., Ferreira, J. M. D. F., & Kannan, S. (2019). Unveiling the effects of rare-earth substitutions on the structure, mechanical, optical, and imaging features of ZrO2 for biomedical applications. *ACS Biomaterials Science and Engineering*, *5*(4), 1725–1743. Available from https://doi.org/10.1021/acsbiomaterials.8b01570.

Sruthi, S., Ashtami, J., & Mohanan, P. V. (2018). Biomedical application and hidden toxicity of Zinc oxide nanoparticles. *Materials Today Chemistry*, *10*, 175–186. Available from https://doi.org/10.1016/j.mtchem.2018.09.008.

Stankic, S., Suman, S., Haque, F., & Vidic, J. (2016). Pure and multi metal oxide nanoparticles: Synthesis, antibacterial and cytotoxic properties. *Journal of Nanobiotechnology*, *14* (1), 73. Available from https://doi.org/10.1186/s12951-016-0225-6.

Sun, T., Yan, Y., Zhao, Y., Guo, F., & Jiang, C. (2012). Copper oxide nanoparticles induce autophagic cell death in a549 cells. *PLOS ONE*, *7*(8), e43442. Available from https://doi.org/10.1371/journal.pone.0043442.

Suresh, J., Pradheesh, G., Alexramani, V., Sundrarajan, M., & Hong, S. I. (2018). Green synthesis and characterization of hexagonal shaped MgO nanoparticles using insulin plant (Costus pictus D. Don) leave extract and its antimicrobial as well as anticancer activity. *Advanced Powder Technology*, *29*(7), 1685–1694. Available from https://doi.org/10.1016/j.apt.2018.04.003.

Szalay, B., Tátrai, E., Nyíro, G., Vezér, T., & Dura, G. (2012). Potential toxic effects of iron oxide nanoparticles in in vivo and in vitro experiments. *Journal of Applied Toxicology*, *32*(6), 446–453. Available from https://doi.org/10.1002/jat.1779.

Tang, L., & Cheng, J. (2013). Nonporous silica nanoparticles for nanomedicine application. *Nano Today*, *8*(3), 290–312. Available from https://doi.org/10.1016/j.nantod.2013.04.007.

Teske, S. S., & Detweiler, C. S. (2015). The biomechanisms of metal and metal-oxide nanoparticles' interactions with cells. *International Journal of Environmental Research and Public Health*, *12*(2), 1112–1134. Available from https://doi.org/10.3390/ijerph120201112.

Thit, A., Selck, H., & Bjerregaard, H. F. (2015). Toxic mechanisms of copper oxide nanoparticles in epithelial kidney cells. *Toxicology In Vitro*, *29*(5), 1053–1059. Available from https://doi.org/10.1016/j.tiv.2015.03.020.

Tsukimura, N., Kojima, K., Att, K. T., Kameyama, Y., & Maeda, H. (2008). The effect of superficial chemistry of titanium on osteoblastic function & T J. *Biomed. Mater. Res. Part A*, *84*, 108–116. Available from https://doi.org/10.1002/jbm.a.31422.

Tucci, P., Porta, G., Agostini, M., Dinsdale, D., Iavicoli, I., Cain, K., ... Willis, A. (2013). Metabolic effects of TIO2 nanoparticles, a common component of sunscreens and cosmetics, on human keratinocytes. *Cell Death and Disease*, *4*(3), e549. Available from https://doi.org/10.1038/cddis.2013.76.

Umut, E. (2013). *Surface modification of nanoparticles used in biomedical applications. Modern surface engineering treatments* (pp. 185–208). Available from https://doi.org/10.5772/55746.

Urner, M., Schlicker, A., Z'Graggen, B. R., Stepuk, A., Booy, C., Buehler, K. P., ... Beck-Schimmer, B. (2014). Inflammatory response of lung macrophages and epithelial cells after exposure to redox active nanoparticles: Effect of solubility and antioxidant treatment. *Environmental Science and Technology*, *48*(23), 13960–13968. Available from https://doi.org/10.1021/es504011m.

Vakili-Ghartavol, R., Momtazi-Borojeni, A. A., Vakili-Ghartavol, Z., Aiyelabegan, H. T., Jaafari, M. R., Rezayat, S. M., & Arbabi Bidgoli, S. (2020). Toxicity assessment of superparamagnetic iron oxide nanoparticles in different tissues. *Artificial Cells, Nanomedicine and Biotechnology*, *48*(1), 443–451. Available from https://doi.org/10.1080/21691401.2019.1709855.

Vandebriel, J. R., & Jong, H. W. (2012). A review of mammalian toxicity of ZnO nanoparticles. *Nanotechnology, Science and Applications*, *5*, 61–71.

Verma, N., & Kumar, N. (2019). Synthesis and biomedical applications of copper oxide nanoparticles: An expanding horizon. *ACS Biomaterials Science and Engineering*, *5*(3), 1170–1188. Available from https://doi.org/10.1021/acsbiomaterials.8b01092.

Wang, Y., Cui, H., Zhou, J., Li, F., Wang, J., Chen, M., & Liu, Q. (2015). Cytotoxicity, DNA damage, and apoptosis induced by titanium dioxide nanoparticles in human non-small cell lung cancer A549 cells. *Environmental Science and Pollution Research*, *22* (7), 5519–5530. Available from https://doi.org/10.1007/s11356-014-3717-7.

Williams, D. F. (2008). On the mechanisms of biocompatibility. *Biomaterials*, *29*(20), 2941–2953. Available from https://doi.org/10.1016/j.biomaterials.2008.04.023.

Wu, W., Samet, J. M., Peden, D. B., & Bromberg, P. A. (2010). Phosphorylation of p65 is required for zinc oxide nanoparticle-induced interleukin 8 expression in human bronchial epithelial cells. *Environmental Health Perspectives*, *118*(7), 982–987. Available from https://doi.org/10.1289/ehp.0901635.

Xia, T., Kovochich, M., Liong, M., Mädler, L., Gilbert, B., Shi, H., ... Nel, A. E. (2008). Comparison of the mechanism of toxicity of zinc oxide and cerium oxide nanoparticles based on dissolution and oxidative stress properties. *ACS Nano*, *2*(10), 2121–2134. Available from https://doi.org/10.1021/nn800511k.

Yang, Y., Bao, H., Chai, Q., Wang, Z., Sun, Z., Fu, C., ... Liu, T. (2019). Toxicity, biodistribution and oxidative damage caused by zirconia nanoparticles after intravenous injection. *International Journal of Nanomedicine*, *14*, 5175–5186. Available from https://doi.org/10.2147/IJN.S197565.

Yazdimamaghani, M., Moos, P. J., Dobrovolskaia, M. A., & Ghandehari, H. (2019). Genotoxicity of amorphous silica nanoparticles: Status and prospects. *Nanomedicine: Nanotechnology, Biology, and Medicine*, *16*, 106–125. Available from https://doi.org/10.1016/j.nano.2018.11.013.

Ye, M., & Shi, B. (2018). Zirconia nanoparticles-induced toxic effects in osteoblast-like 3T3-E1 cells. *Nanoscale Research Letters*, *13*, 353. Available from https://doi.org/10.1186/s11671-018-2747-3.

Yin, Z. F., Wu, L., Gui Yang, H., & Hua Su, Y. (2013). Recent progress in biomedical applications of titanium dioxide. *Physical Chemistry Chemical Physics*, *15*(14), 4844–4858. Available from https://doi.org/10.1039/c3cp43938k.

Yoshida, T., Yoshioka, Y., Tochigi, S., Hirai, T., Uji, M., Ichihashi, Ki, ... Tsutsumi, Y. (2013). Intranasal exposure to amorphous nanosilica particles could activate intrinsic coagulation cascade and platelets in mice. *Particle and Fibre Toxicology*, *10*(1), 41. Available from https://doi.org/10.1186/1743-8977-10-41.

Yousefalizadegan, N., Mousavi, Z., Rastegar, T., Razavi, Y., & Najafizadeh, P. (2019). Reproductive toxicity of manganese dioxide in forms of micro-and nanoparticles in male rats. *International Journal of Reproductive BioMedicine*, *17*(5), 361–370. Available from https://doi.org/10.18502/ijrm.v17i5.4603.

Zhang, H., Ji, Z., Xia, T., Meng, H., Low-Kam, C., Liu, R., ... Nel, A. E. (2012). Use of metal oxide nanoparticle band gap to develop a predictive paradigm for oxidative stress and acute pulmonary inflammation. *ACS Nano*, *6*(5), 4349–4368. Available from https://doi.org/10.1021/nn3010087.

Further reading

Li, Y., Tian, X., Lu, Z., Yang, C., Yang, G., Zhou, X., ... Yang, X. (2010). Mechanism for α-MnO 2 nanowire-Induced cytotoxicity in hela cells. *Journal of Nanoscience and Nanotechnology*, *10*(1), 397–404. Available from https://doi.org/10.1166/jnn.2010.1719.

Singh, N., Jenkins, G. J. S., Nelson, B. C., Marquis, B. J., Maffeis, T. G. G., Brown, A. P., ... Doak, S. H. (2012). The role of iron redox state in the genotoxicity of ultrafine superparamagnetic iron oxide nanoparticles. *Biomaterials*, *33*(1), 163–170. Available from https://doi.org/10.1016/j.biomaterials.2011.09.087.

Sudtha, M., Paul, E., & Chin, F. C. (2020). *Metal oxide nanoparticles in biomedical applications* (pp. 233–251). Elsevier BV, https://doi.org/10.1016/b978-0-12-817505-7.00011-7.

Drug delivery using metal oxide nanoparticles

2

Mónica C. García[1,2], Jazmín Torres[1,2], Antonella V. Dan Córdoba[1,2], Marcela Longhi[1,2] and Paula M. Uberman[3,4]

[1]Universidad Nacional de Córdoba, Facultad de Ciencias Químicas, Departamento de Ciencias Farmacéuticas, Ciudad Universitaria, Córdoba, Argentina, [2]Consejo Nacional de Investigaciones Científicas y Técnicas, CONICET, Unidad de Investigación y Desarrollo en Tecnología Farmacéutica, UNITEFA, Córdoba, Argentina, [3]Universidad Nacional de Córdoba, Facultad de Ciencias Químicas, Departamento de Química Orgánica, Ciudad Universitaria, Córdoba, Argentina, [4]Consejo Nacional de Investigaciones Científicas y Técnicas, CONICET, Instituto de Investigaciones en Fisico-Química de Córdoba, INFIQC, Córdoba, Argentina

2.1 Introduction

During the last decades, the use of nanotechnology in order to reach purposes from biomedical and pharmaceutical fields has had a progressively increasing impact, even on preclinical developments, shaping the emerging scientific field of nanomedicine. Currently, many of these developments are entering to the clinical area and most of them are based on nanosystems for diagnosis and/or therapy within the body, which are commonly composed by a carrier nanoplatform and a payload for imaging, sensing, and/or therapy; and optionally they can contain targeting ligands (Lehner, Wang, Marsch, & Hunziker, 2013). Even though the innocuity of nanoderivative-containing medicines raises questions among the general society, as well as in the health professionals (Vance et al., 2015), their intrinsic chemical, physical, and biological properties render these nanomaterials particularly interesting in the biomedical field.

In this sense, different types of nanomaterials have been studied and reported. The development of nanoparticles (NPs) as drug carriers began in the late 1960s and early 1970s. Since then, there have been numerous reports and studies conducted every year and their number has increased exponentially. NPs for pharmaceutical and medical application are around now for over 50 years (Kreuter, 2007).

The last few decades have been witness of the active exploration from scientists with respect to the synthesis of inorganic NPs for applications in several fields, including biomedicine (Hassan et al., 2017). In fact, inorganic NPs with tunable and diverse properties have promising potential in the field of nanomedicine; however, they have experienced a restricted clinical translation to date due to their non-negligible toxicity in healthy tissues and organs (Yang, Phua, Bindra, & Zhao,

Metal Oxides for Biomedical and Biosensor Applications. DOI: https://doi.org/10.1016/B978-0-12-823033-6.00029-6

2019). The wide research regarding inorganic nanomaterials has provided plenty of understanding to control the properties at the nano–bio interface, which has led to some successful clinical trials and translations (Hassan et al., 2017). These types of NPs have also been exploited as effective imaging and contrast agents due to their optical properties because of their quantum size effect (Nam et al., 2013). The size of inorganic NPs can be modulated. Moreover, a variety of chemical functionalization can be applied on their surface, such as PEGylation, charge modulation, conjugation with ligands, and incorporation of stimuli-responsive moieties and small-molecule probes for improving therapies (Hassan et al., 2017).

Among the different types of inorganic NPs, metal oxide NPs (MONPs) have captivated the attention of several scientists and pharmaceutical industries. They exhibited several advantages, including high stability, tunable shape, porosity, easy engineering to the desired size, simple preparation processes, no swelling variations, simple incorporation into hydrophobic/hydrophilic systems, and ability for further functionalization by different molecules due to their negative surface charge (Fig. 2.1). Also, the interaction of MONPs with *in vivo* systems is different, depending on their properties (viz., size, shape, purity, stability, and surface properties) (Nikolova & Chavali, 2020). Therefore their morphology and interfacial properties need to be comprehensively studied.

The use of MONPs for diagnostics and/or therapy, including drug delivery offers many advantages for biomedical applications. MONPs have been explored for bioimaging, drug and gene delivery, hyperthermia treatments, antioxidant therapy, photodynamic therapy (PDT), antimicrobial therapy, and so on (Fig. 2.1).

This chapter aims to cover the main aspects regarding MONPs, including iron oxide, cerium oxide, titanium dioxide, zinc oxide, copper oxides, silver oxide, magnesium oxide, calcium oxide, nickel oxide, and aluminum oxide. Their biomedical applications are highlighted, mainly focusing on their preclinical and clinical evaluation, and some representative examples on their use for drug delivery are described.

2.2 Metal oxide-based nanoparticles for drug delivery and other biomedical applications

In the last 20 years, new and unimaginable technologies emerged thanks to the outstanding advance in nanotechnology. This progress has been directly associated with the large amount of research dedicated to improving the synthesis and characterization of novel nanomaterials. In the same direction, the research focused into the synthesis and modification of MONPs with biologically relevant molecules have offered to the scientific community deep insights into the interaction with small pharmaceutical drugs and biomacromolecules such as antibodies and genes (Kwon et al., 2018; Yadavalli & Shukla, 2017). Due to the unique optical, mechanical, and magnetic properties exhibited by metal oxide-based nanomaterials, they have been successfully used in the fabrication of nanodevices for drug delivery, hyperthermia treatments, imaging, and sensing (Lee, Choi, Hyeon, & Nano-sized, 2013; Lee et al., 2015; Rajh, Dimitrijevic, Bissonnette,

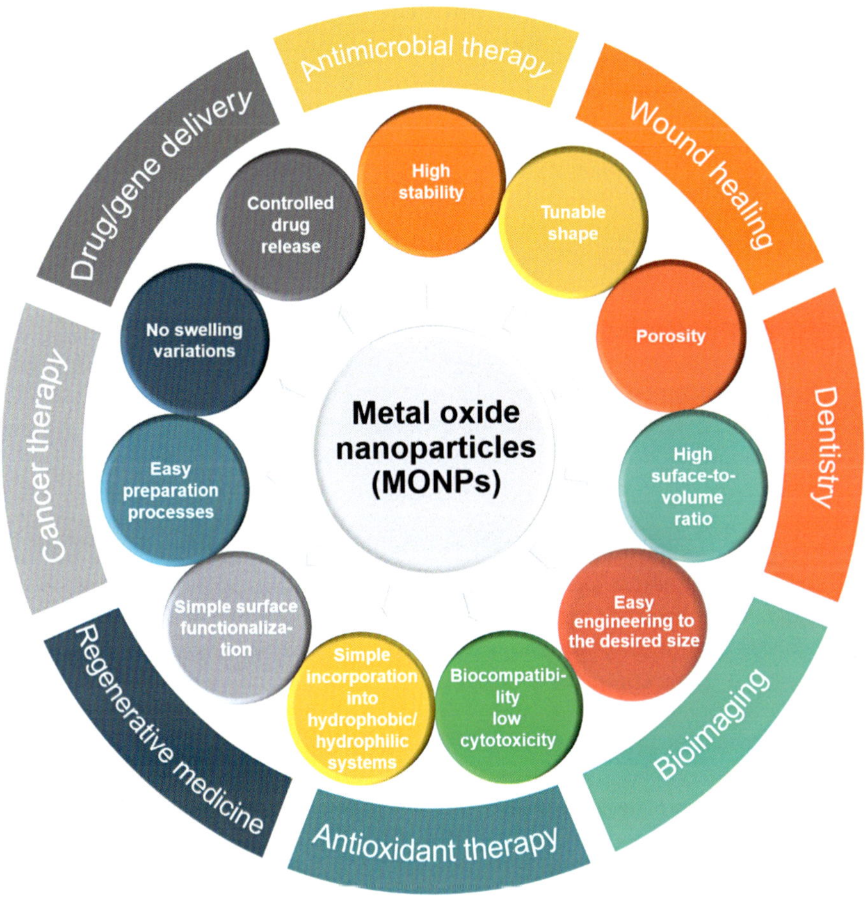

Figure 2.1 Advantages and properties of MONPs and their main biomedical applications discussed in this chapter.

Koritarov, & Konda, 2014; Xu & Qu, 2014). Moreover, the high level of control achieved in size and shape of MONPs has been used to modulate their chemical and physical properties as well as their circulation in the bloodstream and biodistribution (Choi et al., 2007; Duan & Li, 2013; Yadavalli & Shukla, 2017). Furthermore, the high stability and relatively low production costs make them fascinating materials to be employed in nanomedicine. In the next sections, some selected applications for Fe, Ce, Ti, Zn, Cu, Ag, Mg, Ca, Ni, and Al oxides-based nanomaterials are introduced and discussed.

2.2.1 Iron oxide

The high biocompatibility of iron oxide-based nanomaterials enables them to be widely used for biomedical applications. Both maghemite (γ-Fe_2O_3) and magnetite

(Fe_3O_4) phases are the most studied Fe-oxides NPs in nanomedicine. These oxides present superparamagnetic behavior, which allows manipulating them by an external source, magnetic field, as well as generating thermal energy when they are exposed to an alternating magnetic field (AMF). Furthermore, from the superparamagnetic behavior of these materials arise important biomedical applications such as magnetic resonance imaging (MRI) and magnetic separation (Iv et al., 2015), drug delivery (Castro, Gatti, Martín, Uberman, & García, 2021; Mody et al., 2013; Wahajuddin & Arora, 2012) as well as cell targeting and magnetic hyperthermia (Banobre-Lopez, Teijeiro, & Rivas, 2013; Laurent, Dutz, Hafeli, & Mahmoudi, 2011).

However, the use of magnetic NPs *in vivo* requires of surface modification with biocompatible ligands such as polyethylene glycol (PEG) or natural biopolymers [i.e., dextran, hyaluronic acid (HA), chitosan, etc.] in order to protect magnetic NPs from mononuclear phagocyte system; avoiding their aggregation and increasing their stability (Mody et al., 2013; Wahajuddin & Arora, 2012).

2.2.1.1 Bioimaging applications

The use as MRI contrast agents is perhaps the biomedical application most intensively studied for Fe-oxide superparamagnetic NPs. MRI is the most powerful noninvasive and nonionizing technique that provides excellent anatomical images, enhancing soft tissue contrast with high spatial resolution and without depth limitation in the organism (Lee et al., 2015; Soltis & Penn, 2016). In this technique, the ^1H nuclei located in the specific tissue of interest are detected to construct a cross-sectional MR image. However, it is not always possible to generate sufficient contrast to detect some pathologies. To overcome this issue, contrast agents are usually administered. It is important to notice that the contrast agent does not generate a signal by itself, but it affects the ^1H relaxation rate. Hence, while the proton density of a tissue is fixed, the relaxation time of ^1H can be varied by the contrast agent, altering the magnetic characteristics of nearby water protons. Due to their smaller size and longer blood half-life, small superparamagnetic NPs are suited for MRI. Fe-oxide contrast agents are characterized by high longitudinal ($r_1 = 1/T_1$) and transverse ($r_2 = 1/T_2$) relaxivities. These "T_1- and T_2-reducing" agents can be used to represent the degree of contrast agent uptake within a tissue (Beckmann et al., 2009; Corot, Robert, Idee, & Port, 2006; Liu et al., 2021): higher relaxivity agents improve the quality of imaging and lesion depiction (Giesel, Mehndiratta, & Essig, 2010).

Advanced technique of MRI currently faces several challenges, for example, detecting and characterizing tumor margins for surgical resection, or differentiate benign postoperative changes from potential residual tumor following surgery (Nie et al., 2012; Spickler et al., 1990). In this context, development of new and strongest contrast agents is of great importance for improving the use of MRI in clinical diagnosis.

Among therapeutic paramagnetic agents, Gd^{3+} ion gives the strongest contrast effect; however, Gd^{3+} itself presents some toxicity and can induce undesirable side effects, such as nephrogenic systemic fibrosis in patients with renal dysfunction (Penfield & Reilly, 2007). Hence, ultrasmall Fe-oxide NPs (mean hydrodynamic diameter <50 nm) are the best choice as contrast agents for MRI (Iv et al., 2015; Jeon, Halbert, Stephen, & Zhang, 2021; Liu et al., 2021). For instance, Liu et al. (2021) synthesized ultrasmall Fe@Fe_3O_4 NPs for T_1-T_2 dual-mode MRI. The authors observed that the particle size affected the MRI performance. Among the prepared NPs, those modified with 3-(3,4-dihydroxyphenyl)propionic acid (DHCA) and with mean diameter of 8 nm exhibited the optimal T_1-T_2 dual-mode MRI performance. The NPs were conjugated with F56 peptide, which targets the vascular endothelial growth factor receptor, for enhanced tumor accumulation for targeted imaging. *In vitro* and *in vivo* studies demonstrated that the resulting F56-DHCA-Fe@Fe_3O_4 NPs exhibited good T_1-T_2 dual-mode imaging and tumor targeting as shown in human umbilical vein endothelial cells (HUVECs) and human colorectal cancer (HCT-116) tumor-bearing BALB/c nude mice, respectively (Fig. 2.2).

Up to date, some Fe-oxide NPs have been studied in humans for clinical imaging applications as MRI contrast agent, such as ferumoxtran-10 (dextran) (Clement & Luciani, 2004; Harisinghani et al., 2003; Simon et al., 2006), ferumoxytol (carboxymethyl dextran) (Daldrup-Link et al., 2011), VSOP (citrate) (Wagner, Schnorr, Pilgrimm, Hamm, & Taupitz, 2002), feruglose (PEGylated starch) (Taylor et al., 1999), and SHU-555C (carboxydextran) (Tombach et al., 2004) (Table 2.1). These NPs are of similar design: a Fe-oxide NP covered with a biocompatible polymer. However, they differ in the nature of the coating, particle size, and particle charge. Multiple factors determine the efficacy of these contrast agents. Their chemical composition had a notable impact over their efficacy for MRI. The nature of magnetic core and coating, the hydrodynamic size of the NP, effective charge, etc. also modified their physicochemical characteristics as well as their stability, biodistribution, and metabolism.

For example, Ferumoxtran-10 (Sinerem; Guerbet, Paris, France, and Combidex, AMI-227; Advanced Magnetics, MA, USA) consists of Fe-oxide NPs coated with dextran, which present magnetic cores of ~ 6 nm, and mean hydrodynamic diameter of 35 nm. It has the longest blood half-life for small Fe-oxide-based contrast agents (Table 2.1), and they have been studied in Phase I–III clinical trials (Anzai et al., 2003; Harisinghani et al., 2003; Heesakkers et al., 2008). It must be administered as a slow infusion to avoid hypotension side effects.

Ferumoxytol (Feraheme, AMI-7228, Advanced Magnetics, MA, USA) possesses a carboxymethyl dextran coating, with magnetic core size of 6.8 nm and mean hydrodynamic diameter of 30 nm. The formulation exhibited high stability and it could be administered as a rapid intravenous bolus injection (Neuwelt et al., 2007; Pai & Garba, 2012). Ferumoxytol has the fastest r_1 and r_2 relaxivity, allowing the improvement of lesion detection compared with the other contrast agents. Ferumoxytol has received the approval from US Food and Drug Administration (FDA), Canada and several countries of Europe for the treatment of iron deficiency anemia in adults with chronic kidney disease (Lu, Cohen, Rieves, & Pazdur, 2010). Thus Ferumoxytol is now available for clinical imaging.

These small Fe-oxide NPs have proved to be important tools for detecting brain tumors and monitoring treatment efficacy. Their long circulation time in the

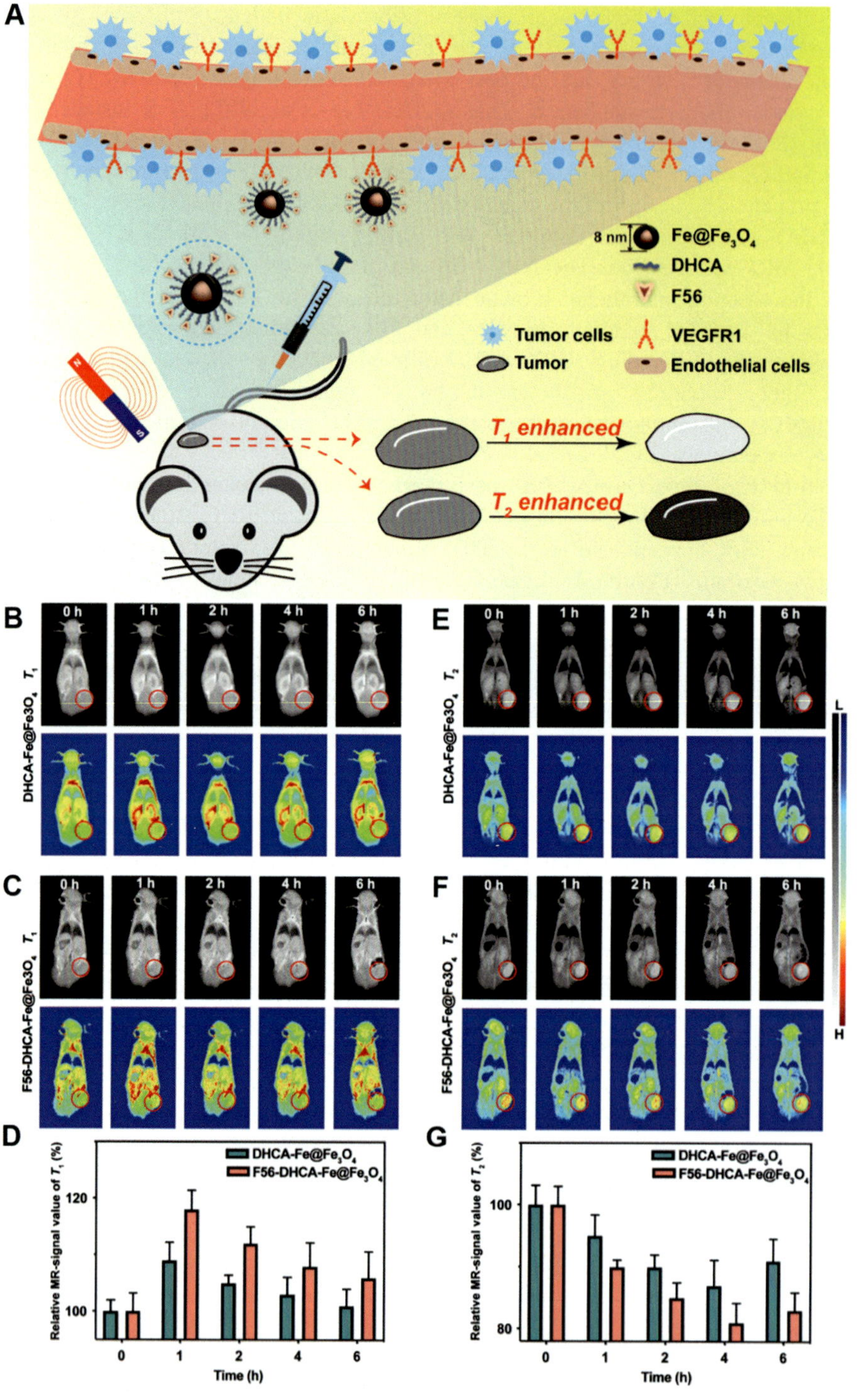
A
8 nm
$Fe@Fe_3O_4$
DHCA
F56
Tumor cells
Tumor
VEGFR1
Endothelial cells
T_1 enhanced
T_2 enhanced
B
E
C
F
D
G
0 h
1 h
2 h
4 h
6 h
DHCA-Fe@Fe3O4 T_1
F56-DHCA-Fe@Fe3O4 T_1
DHCA-Fe@Fe3O4 T_2
F56-DHCA-Fe@Fe3O4 T_2
L
H
$DHCA-Fe@Fe_3O_4$
$F56-DHCA-Fe@Fe_3O_4$
Relative MR-signal value of T_1 (%)
Relative MR-signal value of T_2 (%)
Time (h)

(*Continued*)

bloodstream allows their use on short-term follow-up imaging without requiring repeat injections. Also, their capacity for early retention in the intravascular space also improves the assessment of tumor perfusion and determination of relative cerebral blood volume. Due to their phagocytosis by macrophages, they can be used to quantify the intrinsic immune response to malignant gliomas, which is associated with tumor aggressiveness and prognosis, and to monitor new cancer immunotherapies. Thus far, Ferumoxytol has been well tolerated in patients, and its relative kidney-independent pharmacokinetics make it a suitable contrast agent, alternative to Ga^{3+} chelates, in patients with chronic kidney disease.

2.2.1.2 Drug and gene delivery applications

The suitable results obtained in the clinical translation of biocompatible Fe-oxide NPs as contrast agents for bioimaging encouraged the scientific community to investigate their potential application in other fields of nanomedicine such as drug and gene delivery nanocarriers (El-Boubbou, 2018; Neuberger, Schöpf, Hofmann, Hofmann, & von Rechenberg, 2005). The capacity of magnetic NPs to be conjugated with organic surface modifiers, such as antibodies, peptides, nucleic acids, polymers, and sugar derivatives, as well as with inorganic components, allowed preparing nanomaterials with specific targeting properties, which led to enhance the treatment efficacy, providing a high precision drug or gene delivery, minimizing, for example, the conventional drug-originated systemic toxic effect.

There are plenty of examples on the use of Fe-oxides NPs for controlled and/or targeted drug delivery (Ayyanaar et al., 2020; Gholami et al., 2020; Vangijzegem, Stanicki, & Laurent, 2019) and gene delivery (Bi et al., 2020; Dowaidar, 2021; Huang, Liu, Weng, & Chang, 2021). A recent example is the work reported by Garcia's research group (Castro et al., 2021). The authors synthesized biocompatible hybrid magnetic nanoplatforms based on Fe_3O_4 NPs, which were conjugated with L-cysteine (*L*-Cys) and/or HA. Tamoxifen (TMX) was loaded by electrostatic interaction with high loading efficiency ($>60\%$). The nanoplatforms exhibited

Figure 2.2 (A) Schematic depiction of F56-DHCA-Fe@Fe_3O_4 nanoparticles (NPs) designed and synthesized as a T_1-T_2 dual-mode targeted MRI contrast agent. DHCA-Fe@Fe_3O_4 NPs (8 nm) performed good dual-mode MRI effect. The F56 peptide attached on the surface of DHCA-Fe@Fe_3O_4 NPs could bind to vascular endothelial growth factor receptor 1 (VEGFR1), which is highly expressed in tumor vascular endothelial cells. *In vivo* MRI performance of DHCA-Fe@Fe_3O_4 and F56-DHCA-Fe@Fe_3O_4. (B) T_1-weighted imaging of DHCA-Fe@Fe_3O_4 after injection. (C) T_1-weighted imaging of F56-DHCA-Fe@Fe_3O_4 after injection. (D) MRI signal intensities for the images shown in B and C. (E) T_2-weighted imaging of DHCA-Fe@Fe_3O_4 after injection. (F) T_2-weighted imaging of F56-DHCA-Fe@Fe_3O_4 after injection. (G) MRI signal intensities for the images shown in E and F. *Source*: Reproduced with permission from Liu, D., Li, J., Wang, C., An, L., Lin, J., Tian, Q., et al. (2021). Ultrasmall Fe@Fe_3O_4 nanoparticles as T_1-T_2 dual-mode MRI contrast agents for targeted tumor imaging. *Nanomedicine: Nanotechnology, Biology and Medicine*, *32*, 1–11. Copyright 2021 Elsevier.

Table 2.1 Summary of some Fe-oxide-based nanomaterials under clinical studies for magnetic resonance imaging applications.

Name	Magnetic core (size)	Hydrodynamic size (nm)	Coating	R1 relaxivity	R2 relaxivity	Blood half-life	Type of studying	Application	References
Ferumoxtran-10 (Sinerem)	Iron oxide (6 nm)	35	Dextran T-10	23	53	>24 h (human)	Clinical	– Metastatic lymph node imaging – Macrophage image – Blood pool agent – Cellular labeling	McLachlan et al. (1994)
Ferumoxytol (Feraheme)	Iron oxide (6.8 nm)	30	Carboxymethyl dextran	38	83	10–14 h (human)	Clinical	– Macrophage imaging – Blood pool agent – Cellular labeling	Li et al. (2005)
VSOP-CI84	Iron oxide (8 nm)	19	Citrate	14	33	1 h	Clinical	– Blood pool agent – Cellular labeling	Taupitz and Wagner (2004)
Feruglose (Clariscan)	Iron oxide (20 nm)	n. a.	PEGylated starch	n. a.	n. a.	6 h	Abandoned	– Blood pool agent	Daldrup-Link et al. (2003)
SHU-555C (Supravist)	Iron oxide (4 nm)	21 nm	Carboxydextran	11	38	6 h	Clinical	– Blood pool agent – Cellular labeling	Simon et al. (2006)

n. a., not available.

superparamagnetic behavior, nanometric size (11 and 14 nm for Fe_3O_4-*L*-Cys-HA and Fe_3O_4-HA, respectively), hydrophilic behavior, and aqueous dispersibility, and they were hemocompatible. The amino acid improved the aqueous dispersibility and colloidal stability of the Fe_3O_4-*L*-Cys-HA. TMX was released in a controlled from the hybrid magnetic nanoplatforms. Both nanoplatforms exhibited low cytotoxicity against breast normal (MCF-10A) cells and showed enhanced drug efficacy against breast cancer (MCF-7) cells compared to the free drug (Fig. 2.3).

Active delivery of chemotherapeutics from nanoformulations has been extensively studied and great efforts have been made in translation of nanomedicine into the clinic. While magnetic NPs and targeting molecules, such as antibodies, have been approved for clinical use, systems combining those materials do not have approved products yet (Kamaly, Xiao, Valencia, Radovic-Moreno, & Farokhzad, 2012; Polakis, 2016). Moreover, as far as we know, until now, no anticancer drug-loaded magnetic NPs have been clinically approved for clinical setting (El-Boubbou, 2018).

Among the Fe-oxide magnetic NPs used as drug delivery nanocarriers that are under clinical trials, it can be mentioned the nanosystem based on a mixture of Fe and carbon noncovalently loaded with doxorubicin (DOX) denoted as magnetic-targeted carriers (DOX-MTC) (Goodwin, Bittner, Peterson, & Wong, 2001; Goodwin, Peterson, Hoh, & Bittner, 1999), which was tested in the clinic for the treatment of hepatocellular carcinoma. Goodwin et al. (2001) evaluated the hepatic toxicity of DOX adsorbed into the

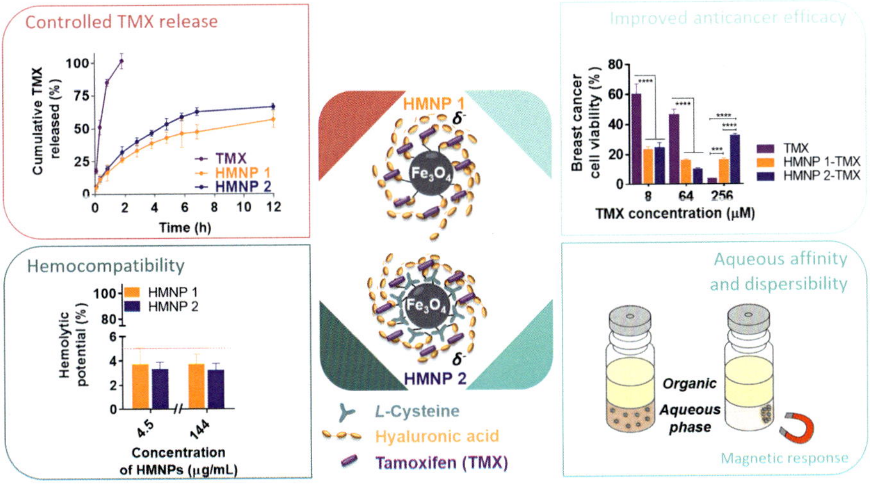

Figure 2.3 Schematic depiction of tamoxifen-loaded hybrid magnetic nanoplatforms (HMNPs) with and without L-cysteine and their main properties, including hydrophilic behavior, aqueous dispersibility and magnetic response, controlled drug release, hemocompatibility, enhanced anticancer activity compared to the free drug.
Source: Conceptualized from Castro, D. C., Gatti, G., Martín, S. E., Uberman, P. M., & García, M. C. (2021). Promising tamoxifen-loaded biocompatible hybrid magnetic nanoplatforms against breast cancer cells: Synthesis, characterization and biological evaluation. *New Journal of Chemistry, 45*(8), 4032–4045.

microparticles (~0.5–5 μm) administered by intraarterial infusion. DOX-MTC was localized and retained by liver tissue. The no-adverse-effect level (NOAEL) was determined to be the MTC-DOX low-dose group, and high doses of DOX-MTC led to liver necrosis in testing animals (>75 mg MTC). Further studies of DOX-MTC regarding Phase I/II clinical trials in patients with hepatocellular carcinoma proved that DOX-MTC was localized in the tumors in 94% (30/32) of all the patients, and no clinically toxicity was observed (Koda, Venook, Walser, & Goodwin, 2002). A multinational clinical study of DOX-MTC Phase II/III (with 240 patients with hepatocellular carcinoma) was stopped since no significant increase in median survival time was observed for DOX-MTC-treated patients relative to patients treated with free DOX. Thus even preclinical studies demonstrate a complete and permanent tumor remission; the Phase I–III clinical trials resulted in no clinically significant efficacies and had a relatively poor tumor response. This lack of applicability of nanomaterials in medicine, is related to the difference observed between the study's results in animal models and the translation to human use (Park, 2013; Venditto & Szoka, 2013). In most cases, the nanoformulations do not provide considerable improvements in the drug performance, which encourages the pharmaceutical industries to invest in their developments.

As it was already mentioned, there is a magnetic NPs-based nanomaterial approved for the treatment of iron deficiency: Feraheme (Fe-oxide coated with polyglucose sorbitol carboxymethylether; ~30 nm, also known as Ferumoxytol-AMAG Pharmaceuticals). It has received FDA-cleared status for use in humans (Bullivant et al., 2013). Feraheme was designed to minimize immunological sensitivity, and it has been employed with success for the treatment of iron deficiency anemia in adult patients with chronic kidney disease (Pai & Garba, 2012). Feraheme has an excellent blood half-life (~15 h) and minimal appearance of free iron in blood, allowing the formulation to be given in relatively large doses with a rapid rate of administration (Provenzano et al., 2009).

2.2.1.3 Hyperthermia treatments

Fe-oxide nanomaterials have received good acceptance to be used for developing magnetoresponsive devices for clinical applications, and few nanoformulations can be found that have been approved by the FDA. For example, NanoTherm (15 nm aminosilane-coated magnetic NPs; #BU48; Magforce, Inc.) is a commercial formulation used for hyperthermia treatments in patients with brain tumors (Jordan et al., 1999). The therapy consists of the intratumoral injection of the magnetic NPs and then heating-induced under AMF (thermoablation ~44.6°C). In addition, Magnablate I (clinicaltrials.gov, NCT02033447) has been under clinical trials (ClinicalTrials, 2014). Magnablate I has been used to thermally ablate solid prostate tumors using AMF.

Due to the excellent biocompatibility of Fe-oxide NPs, they present great potential for future clinical use for imaging or drug delivery. The promising results observed for magnetic-drug delivery and magnetic hyperthermia treatments prompted the use of these nanomaterials for cancer treatment. However, advanced preclinical studies and more clinical trials are still needed to better define the biological performance before effective translation to clinical use.

2.2.2 Cerium oxide

Ce-oxide NPs, also known as nanoceria, have been used for decades in several industrial applications such as glass polishing, catalytic convertors, solid oxide fuel cells, sensors, and catalysts (Sun, Li, & Chen, 2012; Xu & Qu, 2014). In the middle of the 2000s, it was proved that nanoceria exhibited interesting antioxidant character in cell culture models (Chen, Patil, Seal, & McGinnis, 2006; Korsvik, Patil, Seal, & Self, 2007; Rzigalinski et al., 2005; Schubert, Dargusch, Raitano, & Chan, 2006). The outstanding results obtained in those studies encouraged the scientific community to further study nanoceria for biomedical applications, as well as its safe and effective use in nanomedicine, nanobiology, and regenerative medicine.

The extraordinary antioxidant properties of nanoceria are mainly due to its ability of being a regenerative antioxidant. Ce presents two possible oxidation states: Ce^{3+} and Ce^{4+}, being the state Ce^{4+} the most stable electronic configuration; however, both CeO_2 and Ce_2O_3 are naturally occurring oxides. At the nanoscale, Ce-oxide structure corresponds to CeO_2. Nevertheless, several studies indicated that at the surface of the nanostructure, both Ce^{3+} and Ce^{4+} states can coexist (Deshpande & Patil, 2005). Furthermore, the relative amount of Ce^{3+}/Ce^{4+} is a function of the particle size, increasing the Ce^{3+} portion as the particle size decreases (Conesa, 1995; Suzuki, Kosacki, Anderson, & Colomban, 2001). An important feature of Ce-oxide nanomaterials is their low reduction potential (~ 1.52 V) for the Ce^{3+}/Ce^{4+} couple, which allows easy switching between them. This important property makes this couple interchangeable and regenerative (Das et al., 2007; Karakoti, Singh, Dowding, Seal, & Self, 2010). In addition, the presence of oxygen defects into the nanostructure together with the availability of both oxidation states, Ce^{3+} and Ce^{4+}, makes nanoceria prone to act as redox catalysts.

Due to the redox switching ability of the couple Ce^{3+}/ Ce^{4+}, nanoceria is capable of react mimicking two essential antioxidant enzymes: superoxide dismutase (SOD) and catalase (CAT) (Heckert, Karakoti, Seal, & Self, 2008; Pirmohamed et al., 2010). The intrinsic activities of nanoceria have been also highlighted as effective scavenge of reactive oxygen species (ROS) and reactive nitrogen species (RNS) (Celardo, Pedersen, Traversa, & Ghibelli, 2011). Since the redox properties of nanoceria can be tuned by controlling the Ce^{3+}/Ce^{4+} ratio (associated to the size of the nanostructure), and taking into account that Ce^{3+} ions are responsible for scavenge the deleterious ROS such as superoxide anions (O_2^-) and hydroxyl radicals (•OH), maximizing its presence in the structure will lead to an increase in the antioxidant activities of nanomaterials (Korsvik et al., 2007).

2.2.2.1 Applications as antioxidant, anticancer, and regenerative nanomaterials

Considering that ROS and RNS are associated with inflammation and cell death, nanoceria has been investigated as a therapeutic agent in several pathologies associated with excessive oxidative stress and inflammation. Nanoceria was able to protect biological tissues, including neurons or photoreceptor cells, against oxidative stress induced by different agents or radiations. It can also act as a modulator of proangiogenesis and improve wound healing, control or reduce the growth and proliferation of tumors, reduce brain damage in ischemic stroke, and reduce chronic inflammation (Das et al., 2013).

Chronic inflammation, due to high levels of ROS or RNS, can initiate complex immunological disorders, resulting in irreversible organ damage and promoting the development and progression of several pathologies such as retinal degenerative diseases, sepsis, cancers, vascular and neurodegenerative disorders, among others. The intrinsic antioxidant activity of nanoceria can reduce such inflammation, protecting cells or tissues from damage. Antioxidant properties of nanoceria have been tested using *in vitro* cell culture models, applying two predominant forms of ROS, namely O_2^- and H_2O_2. To protect cells, nanoceria acts via the SOD- and CAT-mimetic activities, for O_2^- and H_2O_2, respectively (Celardo et al., 2011; Das et al., 2013). For example, oxidative stress was induced in mouse leukemic monocyte macrophage (RAW 264.7) and human bronchial epithelial (BEAS-2B) cells, by treating them with pro-oxidative organic diesel exhaust particle extract (Xia et al., 2008). Cells pretreated with nanoceria showed a significant decrease in intracellular ROS levels in comparison with stimulated nonprotected cells. Nanoceria-pretreated murine macrophage cells (J774A.1) showed a twofold decrease in the ROS generation, being the nanoceria activity concentration-dependent (Hirst et al., 2009).

There are some reports on *in vivo* studies that tested the antioxidant activity of nanoceria in murine models. For instance, Hirst et al. (2013) evaluated the biomimetic activity of nanoceria as ROS scavenger both *in vitro* and *in vivo*. The *in vitro* studies performed in J774A.1 macrophages demonstrated that nanoceria could effectively quench ROS production. However, *in vivo* studies conducted in an adult mice model were less conclusive. In this *in vivo* study, nanoceria was intraperitoneally administered in a CCl_4 dispersion. A number of factors could modify the nanoceria activity in biological media. Mostly, the aggregation of the particles and formation of protein corona may lead to immune recognition. Nevertheless, this study suggested that nanoceria may reduce ROS levels during periods of oxidative stress and may exert its antioxidant effects for a week or even more following administration. This supports the theory that nanoceria is regenerative so its administration does not need to be repeated. Furthermore, the same study suggested that nanoceria may increase ROS under physiologically normal conditions. This problem could be circumvented, by targeting nanoceria to specific sites of inflammation.

Recently, Nourmohammadi, Khoshdel-Sarkarizi, Nedaeinia, Darroudi, and Oskuee (2020) evaluated the anticancer effects of CeO_2 NPs in a mouse model of fibrosarcoma. Mouse fibrosarcoma (WEHI164) cells were subcutaneously injected into the flank of female adult BALB/c mice. Nanoceria was intraperitoneally administered (0.5 mg/kg), twice a week for 4 weeks. The results showed that nanoceria preferably accumulated in the tumor and significantly decreased tumor growth and volume in tumor-bearing mice (Fig. 2.4).

Nanoceria has been also effective to reduce the radiation toxicity to normal cell under radiotherapy. Radiotherapy is the primary method for treatment of a wide range of human diseases. However, radiation exposure does not differentiate between normal and cancer cells. Hence, there is an actual demand for new agents able to protect the normal tissues surrounding the tumor from the radiation toxicity. Several studies were conducted employing nanoceria as cell protector (Das et al., 2013). For example, nanoceria improved the cell viability after irradiation of cultured normal mammary epithelial (CRL8798) cells, whereas the tumor counterparts (MCF-7 cells) irradiated under identical conditions were affected (Tarnuzzer, Colon, Patil, & Seal, 2005).

Nanoceria was also exploited as a protector of retinal neurons. Chen et al. (2006) demonstrated by *in vivo* studies that nanoceria can effectively protect mammalian retinal neurons, by injecting less than 0.5 ng of nanoceria into the eyes of albino rats that were exposed to 2700 lux of light for 6 h. Almost complete preservation of retinal morphology was achieved with a dose of 344 ng, even under conditions of light damage that may cause blindness in noninjected or saline-injected rodents.

Regarding the biocompatibility of nanoceria, some *in vivo* studies showed that this nanomaterial is well tolerated by experimental animals and no histological

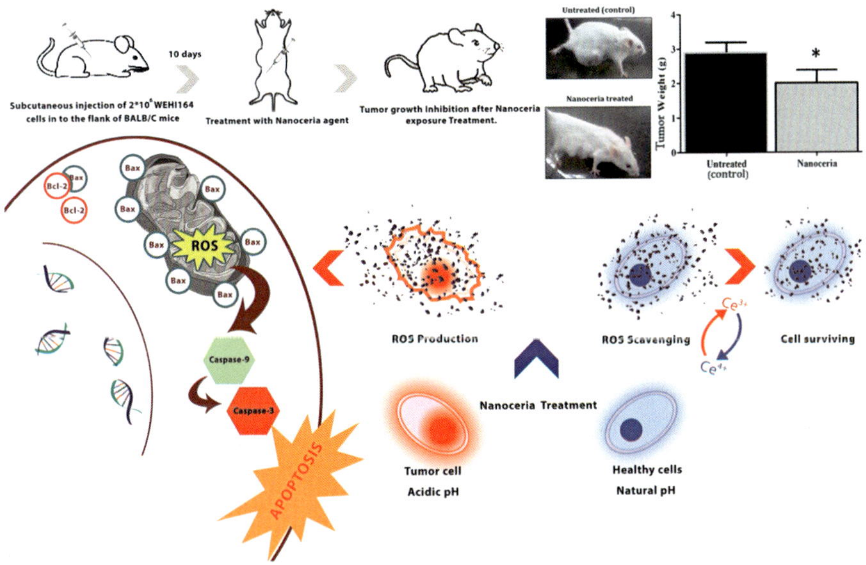

Figure 2.4 Ce-oxide nanoparticles (nanoceria) exhibited anticancer effects in a mouse model of fibrosarcoma. Nanoceria exerted double properties regarding environmental pH. At acidic pH (tumor microenvironment), it could generate and stabilize reactive oxygen species (ROS), whereas at neutral pH (normal cells), it could act as an antioxidant enzyme and attenuate free radicals. After nanoceria treatment the tumor volume and weight significantly decreased compared to the untreated group.
Source: Reproduced with permission from Nourmohammadi, E., Khoshdel-Sarkarizi, H., Nedaeinia, R., Darroudi, M., & Oskuee, R. K. (2020). Cerium oxide nanoparticles: A promising tool for the treatment of fibrosarcoma in-vivo. *Materials Science and Engineering: C, 109*, 1–10. Copyright 2020 Elsevier.

alterations have been reported up to 30 days after injection (Hirst et al., 2009); more important, no increase in the animal mortality rate has been observed after 9 weeks of treatment (Xia et al., 2008). Nanoceria can be administered to experimental animals by intraperitoneal or intravenous injection or as a food additive, with notable effects observed in the organ specifically examined, that is, heart, eye, lungs, etc. (Chen et al., 2006; Niu, Azfer, Rogers, Wang, & Kolattukudy, 2007). Thus nanoceria is delivered efficiently to the target organs after injection. However, it is important to notice that high doses of nanoceria can promote the expression of inflammatory cytokines in cultured human endothelial cells (Gojova et al., 2009). There are a few studies that showed deleterious effects over laboratory animals treated with nanoceria; for example, some animals exhibited lung inflammation, cytotoxicity, lung injury, alveolar macrophage functional changes, and induction of phospholipidosis and release of proinflammatory and fibrotic cytokines after the administration of nanoceria (Ma et al., 2011). It has been also found that nanoceria can produce heart fibrosis, inducing myocardial fibroblast proliferation and collagen deposition in rats (Ma et al., 2012). Finally, Ce is not found in the human body and there are no clearance mechanisms for it. By all the exposure, it is evident that further systematic investigations are required, regarding the cytotoxicity of nanoceria in order to effectively translate it to clinical applications.

2.2.3 *Titanium dioxide*

The use of TiO_2 as a white pigment has been well-known from ancient times (Rajh et al., 2014). TiO_2 has been extensively used to prepare from cosmetics to paints, due to its high chemical stability, low toxicity, efficient photoactivity, and low cost. However, the application of TiO_2 in the biomedical field is relatively new. From the 1990s, an increasing number of researches have been published regarding the use of TiO_2 NPs in nanomedicine (Yin, Wu, Yang, & Su, 2013). Currently, TiO_2-based nanomaterials present important applications in the design of implants for bone regeneration and antimicrobial activity for dental implants (Parnia, Yazdani, Javaherzadeh, & Dizaj, 2017), PDT for cancer treatment (Çeşmeli & Biray Avci, 2018), drug delivery and cell imaging (Rajh et al., 2014), and also they have been used as biosensors and for genetic engineering (Yin et al., 2013).

2.2.3.1 *Photodynamic therapy*

TiO_2-based nanomaterials have been extensively studied in several photocatalytic processes (Fujishima, Zhang, & Tryk, 2008; Lun Pang, Lindsay, & Thornton, 2008), and it has been demonstrated that the photocatalytic activity of TiO_2 is size-dependent, in which higher surface led to a greater photocatalytic property. Therefore the photocatalytic activity of TiO_2 NPs has a special place among their biomedical applications, being extensively used for PDT, especially in the treatment of cancer (Çeşmeli & Biray Avci, 2018). Recent works showed that TiO_2-based nanomaterials killed cancer cells under UV irradiation (Cai et al., 1992; Kubota et al., 1994; Paunesku et al., 2003). Since then, TiO_2 NPs have been employed as a photosensitizing agent in PDT, producing ROS by light activation at a specific wavelength. Compared to others conventional treatments for cancer such us surgical, radiological, and chemotherapeutic, PDT is an alternative noninvasive treatment (Allison, Bagnato, Cuenca, Downie, & Sibata, 2006; Allison & Sibata, 2010; Dougherty et al., 1998; Triesscheijn, Baas, Schellens, & Stewart, 2006).

When TiO_2 NPs were irradiated by UV light (<385 nm), an electron from the valence band was excited to the conduction band, leaving a hole in the valence band. The electrons and holes accommodated in the surface of the NPs can further react with water or hydroxyl ions to form ROS (Linsebigler, Lu, & Yates, 1995), which are capable of destroying bacteria, fungi, tumor cells, and so on. *In vitro* studies have proved the ability of TiO_2 NPs under photoinduction of UVA irradiation to kill a series of human cancer cells, such as cervical cancer (HeLa) (Cai et al., 1992), bladder cancer (T24) (Kubota et al., 1994), monocytic leukemia (U937) (Huang, Min-hua, Yuan, & Rui-rong, 1997), adenocarcinoma (SPC-A1) (Xu, Huang, Xiao, & Lu, 1998), colon carcinoma (Ls-174-t) cells (Zhang & Sun, 2004), among others. The anticancer effect of TiO_2 NPs has been also demonstrated *in vivo* (Cai et al., 1992; Fujishima, Rao, & Tryk, 2000). The pioneer study performed by Cai et al. (1992) proved the effectiveness of TiO_2 for PDT *in vivo*. Nude mice were preinjected with HeLa cells under the skin to produce tumors, and the tumors were left to grow until 0.5 cm. Then, a dispersion of TiO_2 NPs was injected into the tumors. After 2 or 3 days, the skin covering the tumor was cut open to be exposed and irradiated by UVA, and the procedure was repeated for 13 days after. The treatment was able to inhibit the tumor growth and a remarkable antineoplastic effect was observed (Yin et al., 2013) (Fig. 2.5).

The major drawback in the use of Ti-oxide NPs for PDT is that pristine TiO_2 NPs can only be activated by UV light, due to their high-energy band gap (3.0–3.2 eV). In addition, UV light can result in harm to the human body. Therefore extending the optical absorption of TiO_2 toward the visible light region is highly required. This can be achieved by incorporating different dyes over the TiO_2 NPs. Some dyes that have been used for this purpose are hypocrellin B (Xu, Shen, Chen, Zhang, & Shen, 2002), chlorine E6 (Tokuoka, Yamada, Kawashima, & Miyasaka, 2006), and zinc phthalocyanine (Yurt et al., 2018), all of them are well-known PDT sensitizers. Another approach to modify the absorption band of Ti-based nanomaterials consist in the doping method, in which pristine TiO_2 NPs are combined with metals (i.e., Pt, Fe, etc.) or nonmetal elements (i.e., C, N, etc.) (Asahi, Morikawa, Ohwaki, Aoki, & Taga, 2001; Hu et al., 2012; Pan et al., 2017), which led to a bathochromic shift into the UV-visible absorption spectra of the nanomaterial. For instance, Li, Mi, Wang, and Chen (2011) synthesized nitrogen-doped TiO_2 NPs (N-TiO_2) by calcining the anatase TiO_2 NPs under ammonia atmosphere. The N-TiO_2 not only exhibited higher absorbance in the visible region but also, they showed a higher antitumor activity than pristine TiO_2 NPs.

Since TiO_2 NPs are typically insoluble in aqueous medium, different researches have been focused on the improvement of their water solubility. For instance, Seo et al. (2007) prepared water-soluble and biocompatible TiO_2 nanorods surface

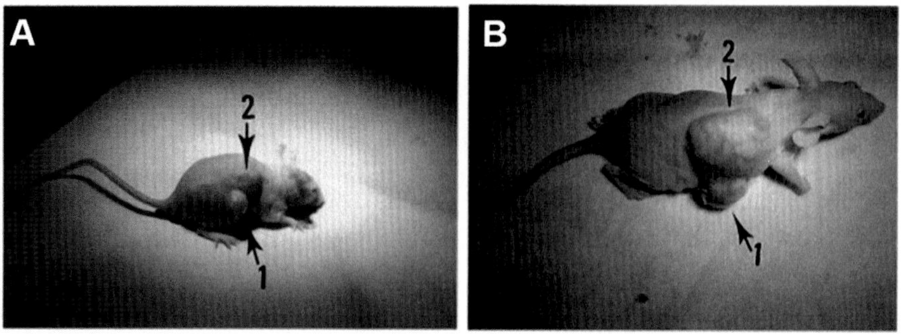

Figure 2.5 (A) Tumors induced in nude mouse before treatment and (B) 4 weeks after treatment. TiO_2 nanoparticles (0.4 mg) were injected into tumor 1, and tumor 2 was not injected with TiO_2 particles. Both tumors were opened surgically and irradiated by an Hg lamp for 1 h.
Source: Reproduced with permission from Yin, Z. F., Wu, L., Yang, H. G., & Su, Y. H. (2013). Recent progress in biomedical applications of titanium dioxide. *Physical Chemistry Chemical Physics, 15*(14), 4844–4858. Copyright 2013 Royal Society of Chemistry.

modified with 2,3-dimercaptosuccinic acid, by a high-temperature nonhydrolytic method. The nanorods exhibited high toxicity on human melanoma (A375) cells when they were irradiated under UV irradiation.

In order to improve the efficiency and selectivity of PDT with TiO_2 NPs, it is the key to increase their selectivity to tumor cells, reducing the normal cell death (Yin et al., 2013). A common strategy is the incorporation of monoclonal antibody proteins such as CEA (Xu et al., 2007), pre-S1/S2 or IL13a2R (Ogino et al., 2010) onto the surface of TiO_2 NPs. Since these proteins are overexpressed on certain cancer cells, the immobilization of antibodies on the TiO_2 NPs surface gives specific direction over targeted cancer cells. Recently, it has been prepared a hybrid-conjugate nano-biocomplex TiO_2–DLDHRGD, based on commercial TiO_2 NPs (Degussa P-25) and bioengineering modified human protein dihydrolipoamide dehydrogenase (DLDH) (Dayan, Fleminger, & Ashur-Fabian, 2018). The obtained nanobiocomplex showed high selectivity toward cancer cells and was effective in producing cell death under UVA irradiation.

2.2.3.2 *Drug delivery*

Ti-based nanomaterials have received considerable attention for preparing drug delivery systems. Several TiO_2 NPs with different shapes have been proposed as drug delivery platforms, mostly due to their low toxicity, and also because numerous drugs can be adsorbed over the surface of TiO_2 or incorporated inside a hollow structure such as TiO_2 nanotubes. In addition, TiO_2-based nanomaterials are suitable to build stimuli-response systems to be activated by light for delivering toxic drugs in a safety way to the targeting tissue (Sortino, 2012). For example, TiO_2 has been studied as a carrier material for sodium valproic acid (Uddin et al., 2011), temozolomide (López, Sotelo, Navarrete, & Ascencio, 2006), and DOX (Wu et al., 2011).

The incorporation of a drug into a TiO_2-based nanoplatform can be performed in two ways, by covalently or noncovalently interaction between the drug and the platform. The effectiveness of both attachments has been studied by Qin et al. (2011). The authors built a water-dispersible TiO_2 NPs for incorporating DOX, as model anticancer drug, by noncovalent complexation (TiO_2/DOX) or by covalent conjugation (TiO_2-DOX). Both nanoplatforms, TiO_2/DOX and TiO_2-DOX, exhibited important differences in their cytotoxicity against C6 glioma cells. It was observed that from TiO_2/DOX, the main fraction of DOX was released inside the nuclei, demonstrating a pH-responsive release of the drug. Conversely, TiO_2-DOX nanoplatform was mainly distributed in the cytoplasm, leading to a significant decrease in the antitumor activity of DOX, due to limited release of the drug. In the same way, TiO_2/DOX exhibited a significantly higher cytotoxicity toward C6 glioma cells than free DOX (Fig. 2.6).

Likewise, TiO_2 nanotubes have also proved to be ideal carriers for several drugs (Wang et al., 2016; Wang, Huang, et al. 2017). Schematic depiction of four methods for drug loading into TiO_2 nanotubes are shown in the Fig. 2.7.

The controlled-release behavior of TiO_2-based nanomaterials has also been exploited for orthopedic implants. Since both bacterial infection and tissue inflammation are the major causes of early failure of Ti-based orthopedic implants, surgical implants with drug releasing properties are promising materials for addressing this issue. TiO_2 nanotubes were used as ibuprofen-drug release platforms into surgical implants (Wang, Weng, et al., 2017). A nanohybrid system composed of poly(lactic-co-glycolic acid) (PLGA) and TiO_2 nanotubes was loaded with ibuprofen. Drug release profile from this nanohybrid material was dependent on the size of TiO_2 nanotubes and the thickness of the polymer film. Highest drug loading was achieved with TiO_2 nanotube with an average size of 80 nm. The maximum time in drug release was 40 days, when 10 layers of polymers were used as a cup of the nanotube. In addition, PLGA layers may favor the proliferation and osteogenesis of MC3T3-E1 mouse cells at an earlier stage. Therefore this nanohybrid material could be employed as an effective nanoplatform for improving both self-antibacterial performance and biocompatibility of Ti-based biomaterials.

2.2.3.3 Applications in dentistry

Currently, TiO_2 NPs are under clinical trials to be employed in dentistry as coating of dental implants. The benefits of using TiO_2 NPs as bioinert dental materials are high flexural strength, biocompatibility, bone generative and regenerative potential and osteogenic potential (Besinis, De Peralta, Tredwin, & Handy, 2015; Parnia et al., 2017). TiO_2 NPs have proved to be able to increase the density of osteoblast cells on the implant and, therefore, lead to a better implant stability (Parnia et al., 2017). Shokuhfar et al. (2014) studied the interaction of osteoblast (MC3T3) cells with surfaces modified with different TiO_2 nanotubes. This study demonstrated

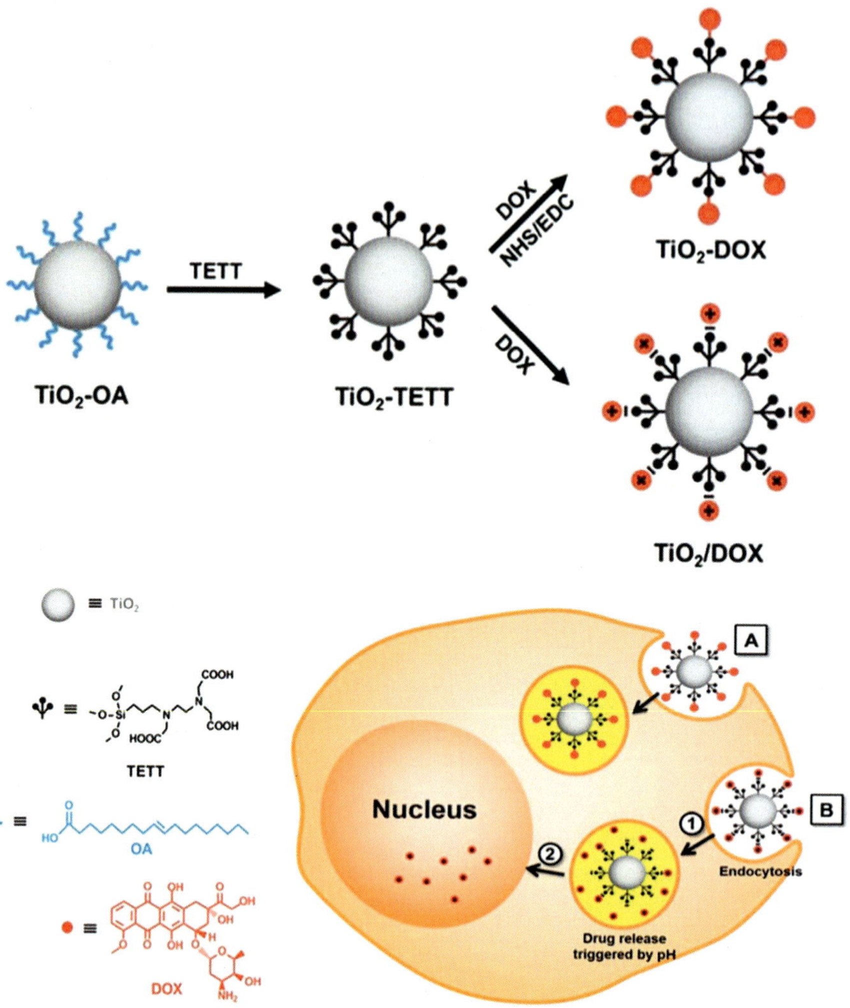

Figure 2.6 Schematic depiction of the synthesis of highly water-dispersible TiO_2 nanoparticles and loading of doxorubicin (DOX) via noncovalent complexation (TiO_2/DOX) or covalent conjugation (TiO_2-DOX). Effect of loading mode on the intracellular location of DOX. (A) Most of TiO_2-DOX was distributed in the cytoplasm. (B) The main fraction of released DOX from TiO_2/DOX was found inside the nuclei.
Source: Reproduced with permission from Qin, Y., Sun, L., Li, X., Cao, Q., Wang, H., Tang, X., et al. (2011). Highly water-dispersible TiO_2 nanoparticles for doxorubicin delivery: Effect of loading mode on therapeutic efficacy. *Journal of Materials Chemistry, 21*(44), 18003. Copyright 2011 Royal Society of Chemistry.

that crystalline TiO_2 nanotubes could increase the density of osteoblast cells over the surface in comparison with surfaces without nanotubes. It was also observed that the crystallinity of the material improved the surface wettability, increasing

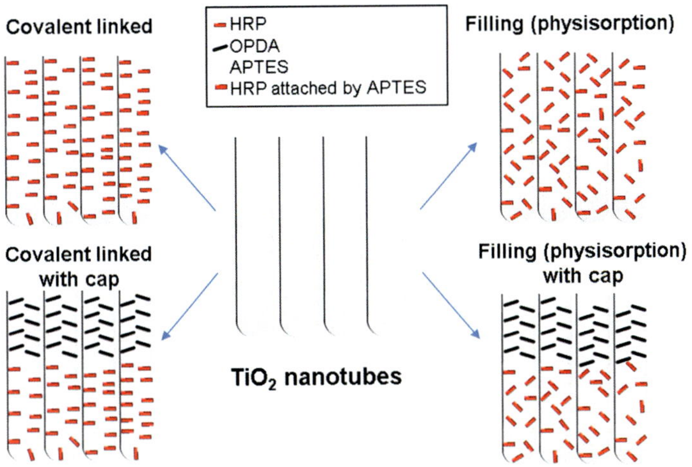

Figure 2.7 Schematic representation of four methods for drug loading over TiO_2 nanoparticles.
Source: Conceptualized from Wang, Q., Huang, J. Y., Li, H. Q., Chen, Z., Zhao, A. Z., Wang, Y., et al. (2016). TiO_2 nanotube platforms for smart drug delivery: A review. *International Journal of Nanomedicine*, *11*, 4819–4834.

cell spreading and extending the cell filopodia into the hollow space of nanotubes. This can be explained considering the super hydrophilicity properties of crystalline nanotubc's surfaces.

It is also important to highlight the antimicrobial properties of TiO_2 nanomaterials (Gomez-Florit et al., 2016; Lin et al., 2014). For instance, Zhang et al. (2018) prepared a TiO_2 nanocomposite for a blue-light-activation of nondestructive teeth whitening system (Zhang et al., 2018). The nanocomposite was obtained by incorporation of polydopamine polymerized over TiO_2 NPs (TiO_2@PDA), producing a nanocomposite capable of absorbing light in the visible light region. A dispersion of TiO_2@PDA was placed on the surface of enamel evenly and the device was irradiated by blue light-emitting diode (LED). The final color of the teeth was compared with a dental professional color card. After 4 h of treatment, a notable whitening change was achieved, similar to the effect of clinic H_2O_2-based whitening agents (Fig. 2.8). In this study was also evaluated the antibacterial properties of TiO_2@PDA. The teeth were cocultured with bacterial medium before being treated with TiO_2@PDA, and compared with an on-sale antibacterial toothpaste. TiO_2@PDA presented an extraordinary antibacterial activity against *Staphylococcus aureus*, a Gram-positive bacterium, even better than the on-sell antibacterial toothpaste.

TiO_2-based nanomaterials present exceptional properties for their use in biomedical applications, among which their high stability, extraordinary photoreactivity, low cost, and biocompatibility are the most outstanding. Thus TiO_2-based nanomaterials have a special place in the future of the biomedical field.

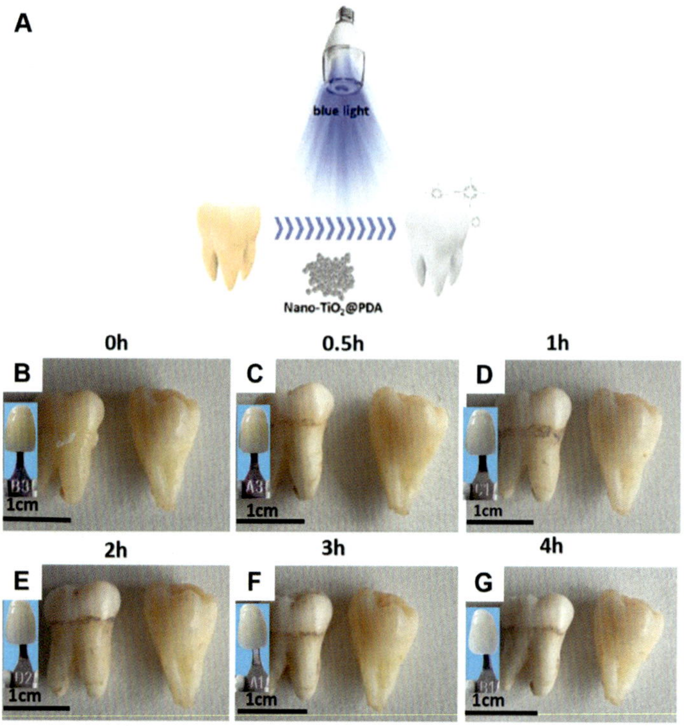

Figure 2.8 (A) Schematic depiction the nanocomposite obtained, based on the incorporation of polydopamine polymerized over TiO_2 nanoparticles (TiO_2@PDA), which is capable of absorbing light in the visible light region and act as a tooth whitening system. (B–G) Left tooth was whitened by nano-TiO_2@PDA for 0.5, 1, 1.5, 2, 3, and 4 h, respectively. The right tooth was used as the control.
Source: Reproduced with permission from Zhang, F., Wu, C., Zhou, Z., Wang, J., Bao, W., Dong, L., et al. (2018). Blue-light-activated nano-TiO_2@PDA for highly effective and nondestructive tooth whitening. *ACS Biomaterials Science & Engineering,4*(8), 3072–3077. Copyright 2018 American Chemical Society.

2.2.4 Zinc oxide

Zn-oxide (ZnO) NPs have been a focus of attention in a broad range of research fields. Several new approaches, including physical, chemical, and biological methods have been used for the synthesis of ZnO-based nanomaterials with different morphologies (Król, Pomastowski, Rafińska, Railean-Plugaru, & Buszewski, 2017). ZnO, an emerging wide band-gap semiconductor, has been used in electronic and optoelectronic devices, photocatalysis (Singh, Das, & Sil, 2020), biosensors and bioimaging (Zhu et al., 2016), and water disinfection (Dimapilis, Hsu, Mendoza, & Lu, 2018). In addition, ZnO NPs have excellent UV-absorbing properties and transparency for visible light, making them excellent sunscreen agents (Mishra, Mishra, Ekielski, Talegaonkar, & Vaidya, 2017).

Currently, ZnO NPs receive great interest in a wide variety of biomedical applications since they exhibit cytostatic activity against cancer cells (Hassan, Mansour, Abo-Youssef, Elsadek, & Messiha, 2017), antifungal activity (Miri, Mahdinejad, Ebrahimy, Khatami, & Sarani, 2019), antiinflammatory activity (Agarwal & Shanmugam, 2019; Rangeela, Rajeshkumar, Lakshmi, & Roy, 2019), ability to accelerate wound healing (Batool et al., 2021), and antidiabetic properties (Bayrami, Haghgooie, Pouran, Arvanag, & Habibi-Yangjeh, 2020; Gudkov et al., 2021). Also, ZnO NPs can be considered as a potential therapeutic for the treatment of mast cell-mediated allergic diseases (Kim, Seo, Kim, & Jeong, 2014). Recent reports have shown that ZnO NPs exhibited antibacterial and antibiofilm activity against a broad spectrum of microbes such as *Pseudomona aeruginosa*, *Streptococcus pneumoniae*, *Listeria monocytogens, Salmonella enteritidis*, and *Escherichia coli* (Bai et al., 2015; Mahamuni-Badiger et al., 2020), and ZnO NPs inhibited the growth and biofilm formation of vancomycin-resistant *S. aureus* (Jasim et al., 2020). The FDA has recognized ZnO (bulk material) as a generally recognized as safe (GRAS) substance. Also, ZnO NPs larger than 100 nm have demonstrated to be relatively biocompatible, supporting their use for drug delivery and other biomedical applications (Mishra et al., 2017).

The use of stabilizers and biocompatible polymers as coating agents for ZnO NPs prevents their aggregation and avoids nonspecific protein adsorption onto their surface in biological media. Also, coated ZnO NPs have shown less toxicity (Singh et al., 2020).

2.2.4.1 Anticancer activity and drug delivery

Among several biomedical applications, the use of ZnO NPs in cancer has been well explored (Mishra et al., 2017). Some evidence on the cytotoxicity of ZnO NPs has demonstrated their anticancer activity against cervical, breast, lung, prostate, liver, ovarian, colon, tongue/oral, blood, skin, brain, and pancreatic cancer cells (Singh et al., 2020). ZnO NPs have been tested in combination with Ag, Fe, Fe_2O_3, WO_3, chitosan, L-asparaginase, PEG, apotransferrin and albumin, Triton X-100, etc. This surface modification allowed, in some cases, decreasing the release of Zn^{2+} ions, and these hybrid systems showed less toxicity in normal cells. Also, these reports evidenced that the cytotoxicity could significantly increase when ZnO NPs were combined with these compounds (Mishra et al., 2017; Singh et al., 2020). Recently, AbuMousa et al. (2020) studied WO_3/ZnO nanostructured semiconducting materials against HeLa cells in different doses (0.200–200 μg/mL) and duration of light irradiation (5–60 min). The results showed a remarkable reduction of cell survival viability for WO_3/ZnO. In addition, at 200 μg/mL dose of WO_3/ZnO, after 60 min of photocatalytic reaction, the cancer cell viability was reduced to 15% as compared to 65% cell viability observed after treatment with the separate components, WO_3 or ZnO, under the same concentration and irradiation time. In this study, the composite formation between WO_3 and ZnO increased the visible light absorption and reduced charge recombination, which favorably improved the photocatalytic killing of HeLa cancer cells.

Another interesting report studied ZnO NPs obtained by green methods, which were explored as anticancer agents. This approach included synthesis through plants, bacteria, fungi, algae, etc. Particularly, plant extract secretes some phytochemicals that act both as a reducing agent and as a capping or stabilization agent (Agarwal, Kumar, & Rajeshkumar, 2017). Some studies evidenced that green synthesis of ZnO NPs provide higher reproducibility, stability, and cytotoxic effect (Singh et al., 2020). For instance, Shamasi, Es-haghi, Taghavizadeh Yazdi, Amiri, and Homayouni-Tabrizi (2021) evaluated ZnO NPs synthesized by aqueous extract of *Rubia tinctorum* against the MCF-7 cancer cells. The results revealed that ZnO NPs induced the apoptosis of MCF-7 cells after 24 h, with dose-dependent behavior.

Given that ZnO NPs exhibit inherent anticancer properties, researchers have utilized them for loading active drugs (Mishra et al., 2017). For example, Pairoj et al. (2021) formulated drug-NPs complexes between ZnO, DOX, and carboplatin (CP). Their anticancer activity was evaluated against different human cancer cells, namely, colon adenocarcinoma (HT-29), oral cavity carcinoma (KB), HeLa, hepatocarcinoma (HepG-2), and MCF-7 cell lines. Among human cancer cell lines, MCF-7 was remarkably sensitive to CP/DOX/ZnO complex under UV exposition, after 24 h of incubation (half-maximal inhibitory concentration, IC_{50} CP/DOX/ZnO = 0.137 μg/mL), indicating the highest cytotoxicity compared with previous studies. This report suggested that cancer cells endocytosed and selectively internalized the CP/DOX/ZnO NPs; thus the free chemotherapeutic drug, DOX, was released in the cytoplasm, inducing acute apoptosis.

It has been demonstrated that ZnO NPs could target some drugs to various cells and tissues (Mirzaei & Darroudi, 2017). In a recent study reported by Sathishkumar et al. (2021), a reasonable amount of the hydrophobic drug quercetin was attached onto the surface of ZnO NPs. The attachment was achieved without any copolymers or functional materials. The anticancer activity of free quercetin, ZnO NPs, and ZnO/quercetin nanocomposite was evaluated against MCF-7 cancer cells, after 72 h of treatment. A maximum cell death, reaching almost 99%, was found at 0.5 μg/mL of ZnO/quercetin nanocomposite. This study has shown that the rate of quercetin release was found to be faster in acidic conditions (tumor microenvironment pH) than in physiological conditions (neutral pH), suggesting that quercetin could be released rapidly from the constructed nanocomposite to the targeted cancer sites. Li, Zhang, Gong, Li, and Zhao (2019) studied the surface functionalization of ZnO NPs by *N*-acetyl-L-cysteine (NAC) for anticancer camptothecin (CPT) delivery. The obtained NPs, namely ZnO/NAC/CPT NPs, exhibited near-spherical shape and uniform dispersion with an average diameter of ~70 nm. The conjugated NPs have almost no hemolytic activity. The IC_{50} values against lung cancer (A549) cells were 1.17 μg/mL and 0.66 μg/mL for free CPT and ZnO/NAC/CPT NPs, respectively. This result showed an augmented cancer-inhibitory effect in the nanoconjugate complex (Fig. 2.9).

In addition, the use of ZnO-based nanomaterials for gene delivery is associated with different advantages. For instance, the expression of plasmid-containing gene onto ZnO NPs surface may ensure safe and efficient gene targeting to the target tissues (Mirzaei & Darroudi, 2017).

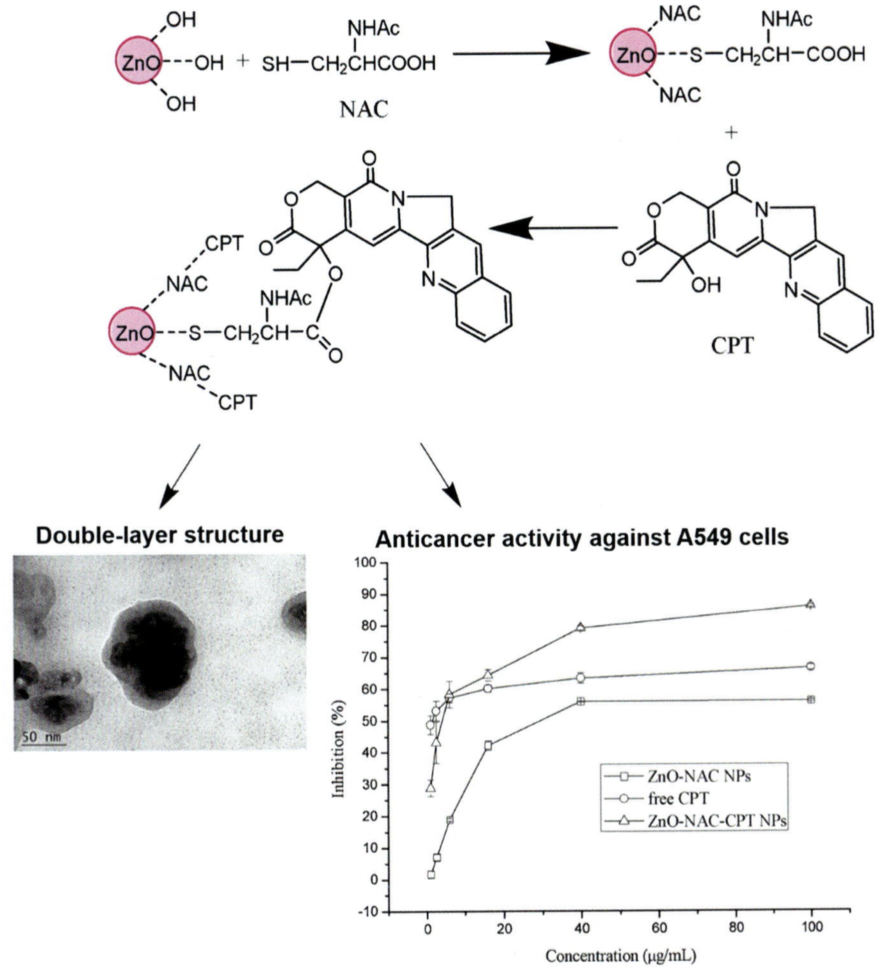

Figure 2.9 Schematic diagram of *N*-acetyl-L-cysteine (NAC)-capped ZnO nanoparticles (NPs) loaded with camptothecin (CPT), TEM image showing their double-layer structure and percentage of inhibition of free CPT, ZnO-NAC NPs, and ZnO-NAC-CPT NPs on the viability of lung cancer A549 cells after 24 h exposure.
Source: Reproduced with permission from Li, C., Zhang, H., Gong, X., Li, Q., & Zhao, X. (2019). Synthesis, characterization, and cytotoxicity assessment of *N*-acetyl-L-cysteine capped ZnO nanoparticles as camptothecin delivery system. *Colloids and Surfaces B: Biointerfaces, 174*, 476–482. Copyright 2019 Elsevier.

2.2.4.2 Antidiabetic activity

Different inorganic NPs such as ZnO, Se, MgO, Cu-oxide, and Ce-oxide NPs play an important role in controlling diabetes (Alkazazz & Taher, 2021). ZnO NPs are currently under investigation to treat diabetes mellitus and diabetic complications because of their ability to deliver Zn^{2+} ions. The pharmacological mechanisms by which ZnO NPs alleviate these conditions are antihyperglycemic activity, enhancement of pancreatic functions, upregulation of glucose transporters and biosensors, weight maintenance, antidyslipidemic activity, antiinflammatory activity, inhibition of α-amylase and α-glucosidase, improvement of insulin sensitivity, and antioxidative activity. A recent study evaluated the acute effects of ZnO NPs on glycemia during 6 h, after oral or intraperitoneal administration in healthy and diabetic rats. Male Wistar rats intraperitoneally administered with streptozotocin-nicotinamide were used as a diabetes model. At the lowest dosage tested (10 mg/kg), ZnO NPs did not change the baseline glucose in any group. Nevertheless, a dose of 100 mg/kg induced a short-term hyperglycemic response depending on the dose, route of administration, and health status (Virgen-Ortiz et al., 2020).

Another preclinical study compared the antidiabetic activity and toxic effects of ZnO NPs in diabetic rats compared to $ZnSO_4$. ZnO (1−10 mg/kg) was orally administered for 56 days. ZnO NPs showed higher antidiabetic activity in comparison to $ZnSO_4$ as evidenced by enhanced glucose disposal, insulin concentration, and zinc status. In addition, it was shown the altered activities of antioxidant enzymes in erythrocytes as well as increased levels of lipid peroxidation and a noticeable reduction of total antioxidant capacity. Also, ZnO NPs severely elicited oxidative stress particularly at higher doses (Nazarizadeh & Asri-Rezaie, 2016).

Bayrami et al. (2020) evaluated the effectiveness of ZnO NPs (10 mg/dL). *Urtica diocia* leaf extract was used for synthesizing them and ZnO NPs were analyzed for treating alloxan-caused diabetic rats. The results were compared with those obtained for plant extract (150 mg/dL) and ZnO/extract (8 mg/dL), after daily intraperitoneal injections for 16 days. Among all the tested treatments, ZnO/extract performed the best performance for controlling the common complications accompanying diabetes compared to the only extract or pure ZnO NPs. These results confirmed the synergistic relationship between ZnO and *Urtica diocia* leaf extract.

ZnO-based nanomaterials exhibit interesting properties and advantages for biomedical applications, and different *in vitro* and preclinical studies have demonstrated their intrinsic as well as combined activity with other compounds, including drugs for controlled delivery. Despite that, advanced assessments are still required to better understand their interaction with more complex biological environments and to define their potential translation to clinical use.

2.2.5 Copper oxides

Cu-oxide is a well-known semiconductor compound with a monoclinic structure and CuO is the simplest member of the family of copper compounds. It has a variety of useful properties such as high-temperature superconductivity, electron correlation effects, and dynamic rotation (Mandal, 2017; Sánchez-Sanhueza, Fuentes-Rodríguez, & Bello-Toledo, 2016).

Copper is a heavy metal that is toxic to mammalian cells. However, with the advancement of nanotechnology, Cu-oxide NPs can target specific cells with lesser side effects (Nagajyothi, Muthuraman, Sreekanth, Kim, & Shim, 2017).

For the synthesis of Cu and CuO NPs, several procedures are available, including physical, chemical, and biological synthesis (Mandal, 2017; Sankar, Maheswari, Karthik, Shivashangari, & Ravikumar, 2014). Among these methods, biological synthesis has had the greatest upgrade since there is an incipient need to develop an environmentally benign metallic NP synthesis method that does not use toxic chemicals in synthesis protocols to avoid adverse effects in biomedical applications (Sankar et al., 2014). In recent years, different researches have been carried out for obtaining Cu-oxide NPs mainly on the basis of the synthesis by a biological method, also known as green synthesis or photosynthesis. NPs obtained by biological synthesis are relatively safe for bioapplications, due to the use of nontoxic components of biological origin as reducing and stabilizing agents. In addition, this synthesis is simple and it allows obtaining NPs with different sizes and shapes (Augustine & Hasan, 2020; Nagajyothi et al., 2017).

Cu-oxide NPs and Cu NPs have evolved in the field of nanomedicine. They are recommended for various applications, such as antimicrobial agents, anticancer agents, image contrast agents, and drug delivery systems.

2.2.5.1 Antimicrobial applications

With regard to the biomedical use of Cu-oxide nanomaterials, their use as antimicrobial agents is one of the most studied applications, since, in recent decades, a greater number of multiresistant bacteria (MDR) have been recorded (Ermini & Voliani, 2021). It has been believed that the microbial toxicity of Cu-oxide NPs has been mainly due to the fact that Cu-oxide NPs have certain properties such as high surface-to-volume ratio, small size, and high dispersion, which allowed them to interact with microbial surfaces (Hassanien, Husein, & Al-Hakkani, 2018) and thus they could penetrate the membrane of bacteria-releasing Cu^{2+} ions inside the cells (Karlsson et al., 2013). Later, the released Cu^{2+} ions can bind with DNA molecules and cause a disorder of the helical structure (Abboud et al., 2014).

Currently, there are a variety of investigations that are based on the use of plants and their derivatives for the synthesis of Cu-based NPs for antimicrobial applications. For example, one study was based on the biosynthesis of Cu NPs, using Tilia extract. Cu NPs with hemispherical shape and different sizes were obtained, in the range of 4.7–17.4 nm (Hassanien et al., 2018). The antimicrobial activity of the biosynthesized Cu NPs was determined by the diffusion method. The bacteria studied were *P. aeruginosa*, *E. coli* (Gram-negative bacteria) and *Bacillus subtilis*, *S. aureus* (Gram-positive bacteria). All bacteria (Gram-negative and Gram-positive) had a good response to Cu NPs in the concentration range studied (25–200 μg/mL). Furthermore, it was observed that there was a direct proportionality between the concentration and the antibacterial activity (Sankar et al., 2014).

Another example is the use of a brown seaweed (*Bifurcaria bifurcata*) in the biosynthesis of Cu-oxide NPs. Abboud et al. (2014) reported that the NPs obtained exhibited dimensions of 5–45 nm. Through different tests it was found that these Cu-oxide NPs exhibited high antibacterial activity against two different strains of bacteria, namely *Enterobacter aerogenes* (Gram negative) and *S. aureus* (Gram positive). Recently, Karthikeyan et al. (2021) use an ecofriendly synthesis method for obtaining a chitosan-Cu-oxide (CCuO) nanomaterial. Cu salt was nucleated with *Psidium Guajava* leaves extract to form the nanomaterial in the chitosan network, which exhibited an average size of ~52.49 nm. The antibacterial study showed that CCuO exhibited higher activity than the commercial amoxicillin against both Gram-positive (*S. aureus, B. subtilis*) and Gram-negative (*Klebsiella pneumonia, E. coli*) bacteria. CCuO displayed a green emission (oxygen vacancies) at ~516 nm in the photoluminescence spectrum, which was attributed to the generation of ROS and may explain the observed biocidal effect (Fig. 2.10).

Applerot et al. (2012) observed that Cu-oxide NPs can produce ROS. Bacteria cells were exposed to Cu-oxide NPs and it was noticed that *E. coli* was more affected than *S. aureus*. From these results it was proposed that the differences in the membranes of Gram-negative and Gram-positive bacteria may influence the resistance to ROS. Moreover, a correlation between the number of killed bacteria and the size of Cu-oxide NPs was observed. Smaller NPs (2 nm) were more effective against bacteria compared to bigger NPs (30 nm) since they disrupted and penetrated the bacterial membrane more efficiently than 30 nm Cu-oxide NPs. Furthermore, smaller NPs were associated with an increased amount of O_2^- anions, generating more intense oxidative stress.

2.2.5.2 Applications in cancer therapy and drug delivery

In recent years, multiple researches have been developed around cancer drug discovery in combination with a smart drug delivery system for targeting cancer cells. Different studies have proven that *in vitro* metal NP systems are more efficient and biocompatible than the traditional use of chemotherapeutic agents (Sankar et al., 2014). Cu-oxide NPs may facilitate the selective cytotoxicity at the cancer site, thus minimizing side effects and improving the therapeutic index of conventional chemotherapeutic treatment (Hossen et al., 2019).

Cu-oxide oxide NPs have been synthesized using *Trichoderma asperellum* cell-free extract (TA-CuO NPs) and their phototherapy-induced *in vitro* anticancer activity was evaluated against A549 cancer cells (Saravanakumar et al., 2019). The obtained NPs were found to have a crystalline and spherical shape with an average size of 110 nm. TA-CuO NPs induced the photothermolysis of A549 cancer cells in a concentration-dependent manner (Saravanakumar et al., 2019).

Surface modifications of Cu-oxide NPs have also been evaluated. For instance, Cu-oxide NPs were synthesized using *Helianthus tuberosus* extracts and folic acid and starch were anchored onto their surface to improve the anticancer effects against triple negative breast cancer (MDA-MB-231) cells. Folic acid acted by directing the Cu-oxide NPs to the folate-receptor overexpressed in cancer cells

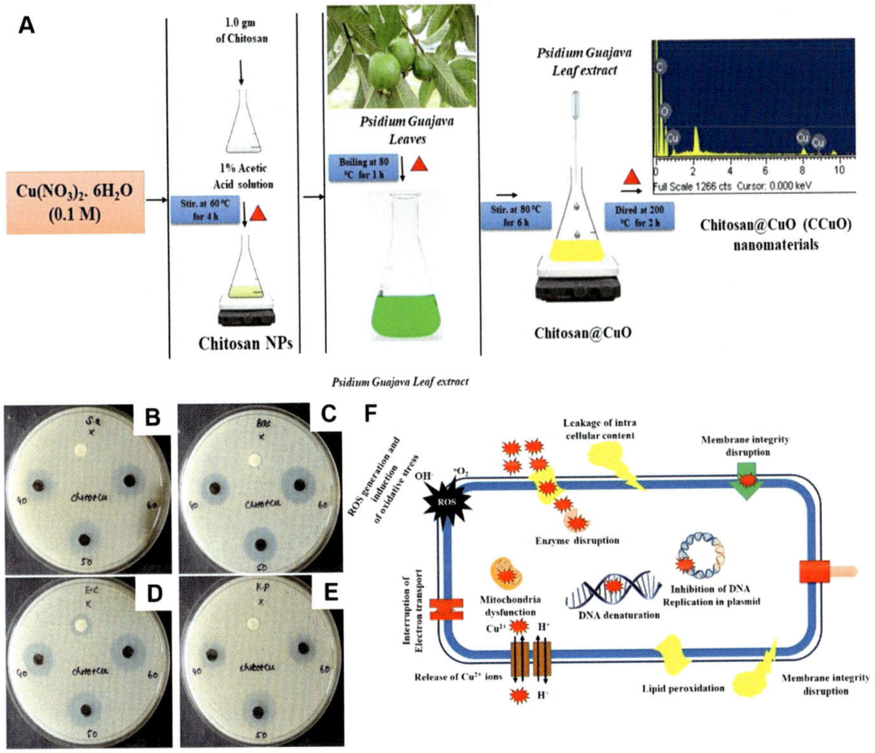

Figure 2.10 (A) Schematic diagram of the green synthesis of chitosan-copper oxide (CCuO) nanomaterial. (B–E) Antibacterial activities of CCuO, tested against *Staphylococcus aureus*, *Bacillus subtilis*, *Escherichia coli*, and *Klebsiella pneumonia*, respectively. (F) Schematic depictions of the possible antimicrobial mechanisms.
Source: Reproduced with permission from Karthikeyan, C., Varaprasad, K., Venugopal, S. K., Shakila, S., Venkatraman, B. R., & Sadiku, R. (2021). Biocidal (bacterial and cancer cells) activities of chitosan/CuO nanomaterial, synthesized *via* a green process. *Carbohydrate Polymers*, *259*, 117762. Copyright 2021 Elsevier.

while the starch facilitated the pH-dependent release of the payload. Hexagonal and oval-shaped surface-modified Cu-oxide NPs with an average size of 108.83 nm were obtained, which exerted enhanced anticancer activity and retarded the cancer growth via apoptosis (Mariadoss, Saravanakumar, Sathiyaseelan, Venkatachalam, & Wang, 2020). Leaf extract of *Ficus religious*, acting as a reducing and protective agent, has been also used for green synthesis of Cu-oxide NPs. Spherical-shaped NPs with an average particle size of 577 nm were obtained and their anticancer activity against A549 cancer cells demonstrated cytotoxic effects in dose-dependent concentration (Sankar et al., 2014).

Is has been revealed that Cu-oxide NPs exhibit their anticancer activity by activating the apoptosis pathway through the generation of ROS, which alters the membrane potential of cancer cells, leading to cell death (Mariadoss et al., 2020; Sankar et al., 2014; Saravanakumar et al., 2019).

Cu-oxide NPs are promising nanomaterials in the management of infectious diseases and treatment of cancer, and their advantageous properties among which their excellent biocompatibility is highlighted, especially if they are synthesized by biological methods, emphasize their biomedical use. However, more comprehensive studies are still required to obtain reliable data on local toxicity and long-term effects on the body. That is, advanced preclinical and more clinical trials are needed before an effective translation into clinical use can be achieved.

2.2.6 Other metal oxides and their main biomedical applications

Metals have been used for centuries as bactericidal agents. There are reports on the use of Ag, Cu, Au, Ti, and Zn as bactericides since 1500 BP in Egypt (Alexander, 2009; Fasiku, Owonubi, Malima, Hassan, & Revaprasadu, 2020). Recently, MONPs have attracted particular attention in the development of antimicrobial nanomaterials due to their various properties, including potency as well as a broad spectrum of antimicrobial activities against bacteria, viruses, and protozoans (Abo-zeid & Williams, 2020; Cavalieri, Tortora, Stringaro, Colone, & Baldassarri, 2014).

At present, the development of microbial resistance to the currently approved antimicrobial drugs is one of the major global healthcare concerns. It has been estimated that by 2050, the antibacterial resistance will cause 10 million deaths per year (Sugden, Kelly, & Davies, 2016). Consequently, the development of newer and more effective antibiotics is of great importance in medicine. In this regard, MONPs have shown interesting properties against several strains of MDR bacteria (Dizaj, Lotfipour, Barzegar-Jalali, Zarrintan, & Adibkia, 2014; Malarkodi et al., 2014; Vithiya, Kumar, & Sen, 2014), in addition with their targeted and extended antibacterial activity at low doses (Martinez-Gutierrez et al., 2010). Furthermore, the use of MONPs as antibacterial agents reduces the possibility of resistance developing, mainly due to their ability to participate in multiple mechanisms of action that can simultaneously attack many sites in the microorganism (Fig. 2.11). It has been generally accepted that the main action mechanism of MONPs involves the ROS production or the release of metal ions. Moreover, they also are prone to participate in alteration of cell components, mostly in the disruption of membrane potential and integrity of the cell wall, and DNA/RNA damage (Fasiku et al., 2020; Nikolova & Chavali, 2020). For example, it was determined that the photocatalytic toxicity of TiO_2 NPs in *E. coli* was associated with lipid peroxidation, which eventually led to respiratory dysfunction and death of bacteria (Beyth, Houri-Haddad, Domb, Khan, & Hazan, 2015). ZnO NPs bonded strongly to bacterial walls and destroy the components of membrane (lipids and proteins), increasing membrane permeability which can cause cell death (Pelgrift & Friedman, 2013). Ag NPs can modify metal ion uptake into the cells, followed by the decreased cellular ATP levels and depletion of DNA replication (Lok et al., 2006).

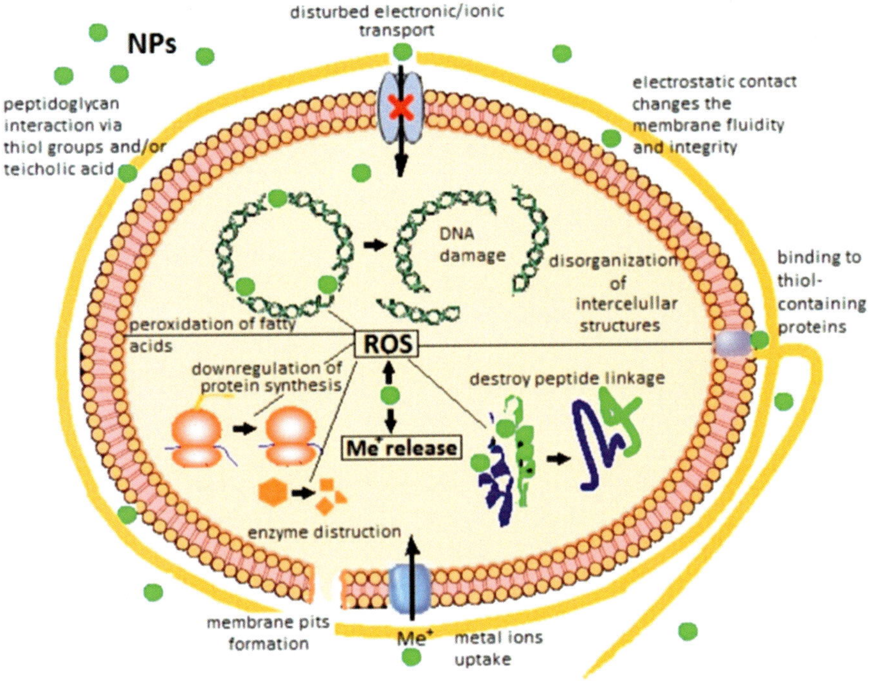

Figure 2.11 Mechanisms of bacteria cell damage by metal oxide nanoparticles (MONPs). Possible interactions between MONP and bacterial cell wall, membrane components, DNA, enzymes, and other proteins are schematized and the influence of external reactive oxygen species (ROS) on membrane integrity, protein synthesis, and functioning are depicted. Different MONPs may cause toxicity via one or more of the described mechanisms. *Source*: Reproduced with permission from Nikolova, M. P., & Chavali, M. S. (2020). Metal oxide nanoparticles as biomedical materials. *Biomimetics, 5*(2), 27. Copyright 2020 MDPI. Creative Common CC BY license.

Despite the interesting features that these nanomaterials have presented as antimicrobial agents, there are to date no MONPs clinically approved for antimicrobial therapy (Fasiku et al., 2020; Malarkodi et al., 2014).

Among MONPs, Ag-oxide NPs have been widely explored as antimicrobial agents. Since ancient times, Ag has been used in medicine for the treatment of wounds, burns, and several infectious diseases, in addition with water purification and disinfection of medical devices (Beyth et al., 2015). Ag and Ag-oxide NPs, such as Ag_2O, AgO, and Ag_2O_3 oxides, have received great attention for their use in nanomedicine. They have exhibited high bactericidal activity against various microorganisms, even at low concentrations (Aisida et al., 2021; Allahverdiyev, Abamor, Bagirova, & Rafailovich, 2011; Dizaj et al., 2014; Manikandan et al., 2017; Nikolova & Chavali, 2020; Siddiqi, Husen, & Rao, 2018). The major improvement of the use of Ag-oxide NPs over other NPs, is their ability to act as an antibacterial agent, even in the solid state (Siddiqi et al., 2018).

The antimicrobial properties of Ag and Ag-oxide NPs have been associated to their ability of ROS generation, the tendency of Ag atoms to interact with the thiol or disulfide groups of proteins that lead to disruption of metabolic processes, and the production of Ag^+ ions, which can lyse or kill the bacteria cells, by the induction of pits and gaps in the bacterial membrane, and then fragmenting the cells (Jung et al., 2008; Nikolova & Chavali, 2020; Yun, Kim, Choi, & Lee, 2013).

The antimicrobial activity of Ag-oxide NPs, in a similar way to Ag NPs, depends on the size and shape, and surface chemistry of the NPs. For example, hemolysis efficiency of several Ag and Ag-oxide NPs chemically and phytochemically prepared were evaluated (Ashokraja, Sakar, & Balakumar, 2017). In this study, the chemically synthesized Ag_2O NPs exhibited higher hemolytic potential than that the chemically prepared Ag NPs; and the biosynthesized Ag_2O and Ag NPs by reduction of Ag^+ with Fenugreek leaf and Papaya extracts showed similar hemolytic potential, which was lower than their chemically obtained counterparts. This behavior was associated with the size, surface chemistry, and physicochemical properties of each group of NPs. The higher lysis properties of Ag_2O NPs were explained considering that Ag_2O NPs released more Ag^+ than that of the more stable Ag NPs.

The bactericidal activity of biosynthesized Ag_2O NPs using *Ficus benghalensis* prop root extract was evaluated against two-dental bacteria, namely *Streptococcus mutans* and *Lactobacilli* sp. (Fernandez, Thomas, & Shailaja Raj, 2016). Ag_2O NPs with particle size of 43 nm were obtained by reduction with the root extract. The combination Ag_2O NPs with the root extract presented remarkable antimicrobial activity, followed by those of commercial Ag NPs and Ag^+ ions. The proposed blend of Ag_2O NPs with the root extract could be useful in the preparation of toothpaste as a germicidal agent.

Li et al. (2019) presented the biosynthesis of Ag_2O NPs of 20 nm by using leaf extract of *Lippia citriodora* powder. The Ag_2O NPs exhibited outstanding antibacterial against *S. aureus* and antifungal activity against *Aspergillus aureus*. Also, wound healing studies were performed in excision skin wound model developed in albino Wistar rats. The Ag_2O NPs were incorporated in a hydrogel (3% *w/v* of hydroxypropyl methylcellulose) in order to apply the formulation in the skin wound model as a dressing material. It was found that the Ag_2O NPs-containing hydrogel exhibited an accelerated wound healing effect compared to the untreated control and the treatment with the plant extract (Fig. 2.12). The promising properties of the obtained materials could be exploited in the development of dressings for treatment of diabetic wounds and for nursing care. In the same way, Kim et al. (2018) incorporated Ag-oxide NPs into a thermosensitive methylcellulose (MC) hydrogel, in a one-pot synthesis. The resulting Ag_2O-MC hydrogel exhibited an excellent antimicrobial activity against *S. aureus, K. pneumonia*, and *E. coli*, probably associated with the release of the Ag^+ from the MC hydrogel. Burn wound healing effect was examined through the burn wound test in rats. The treatment with the Ag_2O-MC hydrogel prevented necrosis and inflammation induced by burn damage in skin. The control group showed extensive necrosis in the epithelium and the dermis showed bulla, necrosis, and infiltrated inflammation. After 7 days, the group treated

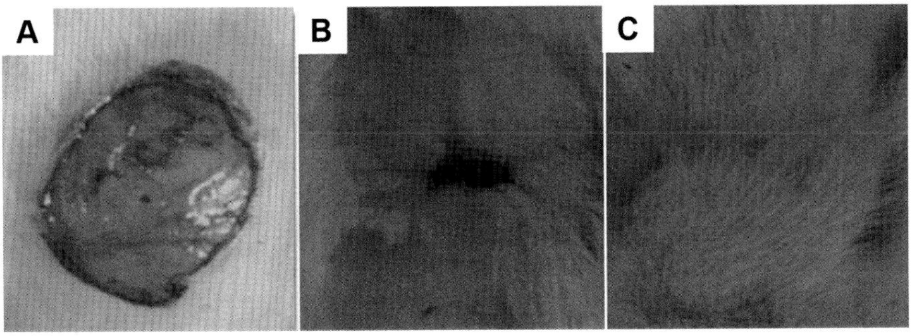

Figure 2.12 Effect of different treatments in the development of cutaneous wound healing in rat groups. (A) Untreated control, (B) *Lippia citriodora* extract (100 mg/mL), and (C) biosynthesized Ag_2O nanoparticles (50 μg/mL).
Source: Reproduced with permission from Li, R., Chen, Z., Ren, N., Wang, Y., Wang, Y., & Yu, F. (2019). Biosynthesis of silver oxide nanoparticles and their photocatalytic and antimicrobial activity evaluation for wound healing applications in nursing care. *Journal of Photochemistry and Photobiology B: Biology*, *199*, 111593. Copyright 2019 Elsevier.

with the Ag_2O-MC hydrogel showed a little infiltrated inflammation and the cure process. Therefore the thermoresponsive Ag_2O-MC hydrogel has great potential as an injectable hydrogel for wound regeneration.

The combination of Ag-oxide NPs with other nanooxides of relevance in nanomedicine has been also explored. For example, tantalum oxides have been of recent interest in the development of medical devices, mostly due to their excellent biocompatibility and bioactivity. They have been evaluated for chemothermal cancer therapy as well as for the fabrication of dental and orthopedic implants (Balla, Banerjee, Bose, & Bandyopadhyay, 2010). Sarraf et al. (2018) proposed the combination of Ta_2O_5 nanotubes coated on their surface with medical grade Ti alloy for osseointegration and Ag_2O NPs in order to improve the antibacterial characteristic of the final material. Ta_2O_5 nanotubes were obtained via anodization of a Ta layer, and then covered with Ti_6Al_4V. The edges and walls of nanotubes were then decorated with Ag_2O NPs through physical vapor deposition. The Ag_2O-Ta_2O_5 nanotubes promoted the formation of bone-like apatite layer, with a notable reducing viability of *E. coli* cells. The authors suggested that the decoration of Ta_2O_5 nanotubes with Ag_2O NPs could improve antibacterial activity and the osseointegration of Ti_6Al_4V implants.

Mg-oxide NPs also exhibited interesting properties for their use in nanomedicine. MgO NPs have shown effective antimicrobial activity against both Gram-positive and Gram-negative bacteria (Imani & Safaei, 2019; Mirhosseini & Afzali, 2016; Pugazhendhi, Prabhu, Muruganantham, & Shanmuganathan, 2019) and fungi (Sierra-Fernandez et al., 2017). The strong antibacterial activity of MgO NPs has been related to two main factors: their alkalinity and the ability to generate active oxygen species. Like other MONPs, it has been proposed that the antibacterial mechanism is triggered by the generation of O_2^- ions on the surface of MgO NPs

(Huang et al., 2005). MgO NPs can damage the cell membrane, causing the leakage of intracellular contents, which eventually lead to the bacterial cell death (Jin & He, 2011). Leung et al. (2014) observed that MgO NPs produced cell membrane damage on *E. coli* bacteria, but no evidence of oxidative stress or lipid peroxidation was found. The robust antibacterial activity observed for MgO NPs was associated with the membrane damage produced by the combination of attachment of the NPs to the bacteria membrane, together with the change in the local pH and Mg^{2+} release. The interaction of MgO NPs with bacteria cells was favored by electrostatic attraction since MgO NPs exhibit a positive charge in pH range from 4 to 8 (Cai et al., 2018). MgO NPs were able to react with water to produce $Mg(OH)_2$, which could release OH^- and Mg^{2+} ions into the solution. In addition, O_2^- ions exhibited higher stability in the alkaline environment (Yao et al., 2018). All these aspects could contribute to the enhanced antibacterial effect of MgO NPs (Wetteland, Nguyen, & Liu, 2016).

It has been demonstrated that the antibacterial activity of MgO NPs decreased with the increasing particle size (Sawai, Igarashi, Hashimoto, Kokugan, & Shimizu, 1996). Moreover, they have shown an improved inhibition for rod-shaped bacteria (*E. coli*) in comparison to spherical-shaped bacteria. MgO NPs displayed high antibacterial activities even at low doses (Jin & He, 2011). They also possess the advantage of being noncytotoxic to human cells at low concentrations (Ge et al., 2011; Wetteland et al., 2016). However, the bactericidal activity increases with raising the concentration of MgO NPs.

MgO NPs have been also used in nanomedicine as relief of heartburns, in bone regeneration and as antitumor agents (Krishnamoorthy, Moon, Hyun, Cho, & Kim, 2012; Pugazhendhi et al., 2019). In cancer therapy, MgO NPs exhibited preferential cytotoxicity against HeLa, gastric adenocarcinoma (AGS), and gastric carcinoma (SNU-16) cells. The proposed mechanism involves ROS generation, which induces lipid peroxidation (Krishnamoorthy et al., 2012). In bone repair, a composite of MgO NPs loaded into polycaprolactone polymer (PLCP) was prepared as a bone—soft tissue engineering scaffold (Suryavanshi, Khanna, Sindhu, Bellare, & Srivastava, 2017). The MgO-PLCP composite showed significant improvement in tensile strength and elastic modulus, together with an enhanced *in vitro* performance, improving the adhesion and proliferation of osteoblast-like (MG-63) cells. *In vivo* studies in Sprague Dawley rats showed initial moderate inflammatory tissue response near the implant site that subsided later, with no side effect on vital organ functionalities, which demonstrated the potential application of MgO NPs as an efficient scaffold material for bone—soft tissue engineering applications.

Ca-oxide NPs, in similar fashion to MgO NPs, present antimicrobial activity through the formation of ROS on the NPs surface and because of their alkalinity (Sawai, 2003). Therefore an important factor for their antibacterial activity is the contact between NPs and bacterial cells. CaO NPs together with CaOH NPs were evaluated against *Enterococcus faecalis* in a dentinal block model (Louwakul, Saelo, & Khemaleelakul, 2017). The results indicated that CaOH NPs were more efficient than CaO NPs for the elimination of bacterial content in the dentinal tubules. Furthermore, CaO NPs were significantly more active against *E. faecalis* than bulk CaO and $Ca(OH)_2$. The authors suggested the potential use of CaO and CaOH NPs as an intracanal medicament in root canal therapy.

The antifungal activity of CaO was evaluated against *Candida albicans, Saccharomyces cerevisiae, Rhizopus stolonifer*, and *Aspergillus niger*. The application of CaO powder as slurries in physiological saline showed antimicrobial activities against all fungi evaluated. The same behavior was observed for MgO. They were able to kill spores of *B. subtilis*, a more robust microbial. Even more, they exhibited higher activity than ZnO NPs against fungal cells. These results suggest that the alkali behavior of CaO and MgO NPs improved their antifungal activity, since the yeasts were killed by a combination between ROS generation and the high pH of the alkali oxides (Sawai & Yoshikawa, 2004).

Ni-oxide NPs also present bactericidal activity against several bacterial strains such as *P. aeruginosa, S. aureus, E. coli, K. pneumoniae, B. subtilis, Serratia marcescens, Staphylococcus epidermidis*, and *Micrococcus luteus*; however, their antibacterial activity depends on the bacterial species and concentration (Behera et al., 2019; Ezhilarasi et al., 2018; Helan et al., 2016; Ilbeigi, Kariminik, & Moshafi, 2019). NiO NPs have exhibited a higher antibacterial activity against Gram-positive compared to Gram-negative bacteria. The action mechanism involves the ROS production on the bacterial membrane, which results in membrane damage and the eventual bacterial cell death (Ezhilarasi et al., 2018). Moreover, it has been proposed that Ni^{2+} ions can interfere with intracellular Ca^{2+} metabolism. NiO NPs have been also able to interact with the bacteria DNA and proteins, which in combination might led to bacteria death.

Al-oxide NPs have been evaluated as antimicrobial agents, with disparate results (Beyth et al., 2015; Pelgrift & Friedman, 2013). Even though Al_2O_3 NPs have presented antibacterial activity against *E. coli, S. aureus, Salmonella typhimurium*, and *S. mutans*, it was found that they worked at high concentrations (Ansari et al., 2014; Ansari, Khan, Khan, Pal, & Cameotra, 2013; Balasubramanyam et al., 2010). The antibacterial mechanism observed in *E. coli* includes diffusion and accumulation of Al_2O_3 NPs inside cells, causing the formation of pits and membrane disorganization, leading to cell death (Ansari et al., 2014). It has been proposed that the interaction between the lipopolysaccharides from the cell wall and Al_2O_3 NPs may disturb the membrane integrity, being the main factor for their bactericidal activity. Unlike other MONPs, there is no conclusive evidence to support that the bactericidal activity was associated with ion release (Jiang, Mashayekhi, & Xing, 2009). In addition, Al_2O_3 NPs showed higher toxicity than their bulk counterparts. For that reason, it is not clear if these MONPs are suitable for antibacterial therapy. Besides their mild bactericidal effect, it has been proposed that Al_2O_3 NPs can promote horizontal transfer of antibiotic-resistance genes mediated by plasmids across genera (Qiu et al., 2012). Therefore Al_2O_3 NPs might increase the likelihood of development of drug resistance. Qiu et al. (2012) demonstrated that Al_2O_3 NPs are prone to promote horizontal transfer of antibiotic-resistance genes by conjugation from *E. coli* to *Salmonella*. The authors suggested that the oxidative damage of bacterial membranes caused by Al_2O_3 NPs triggered an increase of the expression of conjugative genes and a decrease in expression of global regulatory factor genes for RP4 plasmid conjugation. As a result, an increase in the likelihood of acquisition by bacteria of one or multiple drug resistance genes was induced.

Al_2O_3 NPs have been also evaluated in tissue engineering, cell growth, and drug delivery (Davoodi, Zhianmanesh, Montazerian, Milani, & Hoorfar, 2020; Jani, Losic, & Voelcker, 2013). With regard to their tissue engineering and cell growth applications, nanoporous Al_2O_3 were used in bone regeneration (La Flamme et al., 2007; Popat et al., 2005; Swan, Popat, Grimes, & Desai, 2005). Osteoblasts were seeded on nanoporous Al_2O_3 membranes to investigate short-term adhesion, proliferation as well as long-term functionality and matrix production (Popat et al., 2005). The obtained results indicated an improved osteoblast adhesion, proliferation, and increased matrix production after 4 weeks. The nanopores of Al_2O_3 membranes produced an active matrix that contained calcium and phosphorus, and osteoblasts adhered on Al_2O_3 membranes laid more matrix compared to cells adhered on other surfaces.

In addition, porous Al_2O_3 coated with hydroxyapatite (HAP) have been used as a ceramic bioinert scaffold for cell attachment of rat pituitary (PR1) tumor cells and osteosarcoma (SAOS) cells (Bose et al., 2002). Hulbert, Matthews, Klawitter, Sauer, and Leonard (1974) explored the use of Al_2O_3 implants in rabbits to examine endosteal bone ingrowth into Al_2O_3 cone-shaped plugs in amputated tibiae. By radiographic, histologic, and microradiographic analyses of the *in vivo* implants a little mineralized bone ingrowth into the cone-shaped Al_2O_3 implants in the amputated tibiae was observed; but the pellet implants in the tibiae and femora of the nonamputated limbs demonstrated outstanding mineralized bone ingrowth into the surface and internal pores. By comparing these results, it was determined that the damage into the musculature and vascularity were interfering with the normal bone activity, and therefore bone ingrowth. The Al_2O_3 implants showed to be nontoxic and elicited no inflammatory responses. Ohgushi et al. (2005) introduced tissue engineering into bone-prosthesis for the treatment of osteoarthritis. The authors collected fresh bone marrow cells from the patient's iliac crest; then applied the extended mesenchymal cells into a composite of Al_2O_3-HAP ankle prosthesis and cultured them to form an osteoblasts/bone matrix on the prosthesis. Later, the prosthesis was chirurgical introduced on three patients suffering from ankle arthritis, and followed their progress for at least 2 years. X-ray analysis revealed a radiodense appearance around the cell-seeded areas of the prostheses about 2 months after the operation. All patients did not exhibit inflammatory reactions. These preliminary results indicated that this approach could be used to prevent aseptic loosening of the total ankle arthroplasty.

Furthermore, Al_2O_3 porous NPs have been used to develop drug delivery systems (Aw, Simovic, Addai-Mensah, & Losic, 2011). For example, Simovic, Losic, and Vasilev (2010) prepared a nanoporous Al_2O_3 membrane for controlled release of vancomycin as a model drug by applying a thin polymer film on the top of Al_2O_3 membrane after drug loading. The material showed an extended drug release behavior that can be exploited for various types of drugs.

The aforementioned biomedical applications of different MONPs demonstrate their versatility to be applied for various diseases. These MONPs have exhibited outstanding properties, which are, in some cases, even better than reference drugs or counterparts bulk materials.

2.3 Concluding remarks and perspectives

MONPs exhibit great potential as antimicrobial agents, which can be used for fighting MDR microorganisms. The application of MONPs as bioimaging agents, drug and gene carriers, scaffolds, and anticancer therapeutics is attractive in the field of nanomedicine; however, there are many challenges ahead. Even though several action mechanisms are related to ROS generation, there are also certain unknown mechanisms, unrelated to ROS production, that need to be successfully established.

Because of their interesting properties and advantages, including high stability, tunable shape, porosity, easy preparation processes, simple surface functionalization, easy engineering to the desired size, no swelling variations, simple incorporation into hydrophobic/hydrophilic systems, etc., MONPs have been evaluated for a broad range of biomedical applications, namely, bioimaging, drug and gene delivery, hyperthermia treatments, antioxidant therapy, PDT, dentistry, wound healing, antidiabetic treatment, bone regeneration, and, specially, for antimicrobial therapy, etc. This chapter covered the main aspects regarding MONPs, highlighting the main biomedical applications of different metal oxide-based nanomaterials, emphasizing their preclinical and clinical evaluation.

MONPs have exhibited outstanding results both *in vitro* and *in vivo* in small animals. Even though those results were encouraging, they do not guarantee their further success *in vivo* in bigger animal models as well as in humans when thinking in clinical use. Despite the promising and interesting properties of MONPs, more toxicity studies at all levels are needed to clearly define their advanced delivery prospects. Even though some clinical trials demonstrated their use for different biomedical applications and also there are some MONPs-containing approved products commercially available, there are still many gaps related to their biomedical use. Their performance in more complex biological scenarios must be evaluated. Therefore more and advanced *in vivo* studies are critical to better understand the interactions between MONPs and living organisms to really define their use and potential translation to clinical use.

References

Abboud, Y., Saffaj, T., Chagraoui, A., El Bouari, A., Brouzi, K., Tanane, O., et al. (2014). Biosynthesis, characterization and antimicrobial activity of copper oxide nanoparticles (CONPs) produced using brown alga extract (*Bifurcaria bifurcata*). *Applied Nanoscience*, *4*(5), 571–576.

Abo-zeid, Y., & Williams, G. R. (2020). The potential anti-infective applications of metal oxide nanoparticles: A systematic review. *Wiley Interdisciplinary Reviews: Nanomedicine and Nanobiotechnology*, *12*(2), e1592.

AbuMousa, R. A., Baig, U., Gondal, M. A., Dastageer, M. A., AlSalhi, M. S., Moftah, B., et al. (2020). Investigation of the survival viability of cervical cancer cells (HeLa) under visible light induced photo-catalysis with facile synthesized WO_3/ZnO nanocomposite. *Saudi Journal of Biological Sciences*, *27*(7), 1743–1752.

Agarwal, H., Kumar, S. V., & Rajeshkumar, S. (2017). A review on green synthesis of zinc oxide nanoparticles—An eco-friendly approach. *Resource-Efficient Technologies*, *3*(4), 406–413.

Agarwal, H., & Shanmugam, V. K. (2019). Synthesis and optimization of zinc oxide nanoparticles using *Kalanchoe pinnata* towards the evaluation of its anti-inflammatory activity. *Journal of Drug Delivery Science and Technology*, *54*, 101291.

Aisida, S. O., Ugwu, K., Nwanya, A. C., Bashir, A. K. H., Nwankwo, N. U., Ahmed, I., et al. (2021). Biosynthesis of silver oxide nanoparticles using leave extract of *Telfairia occidentalis* and its antibacterial activity. *Materials Today: Proceedings*, *36*, 208–213.

Alexander, J. W. (2009). History of the medical use of silver. *Surgical Infections*, *10*(3), 289–292.

Alkazazz, F. F., & Taher, Z. A. (Eds.), (2021). A review on nanoparticles as a promising approach to improving diabetes mellitus. *Journal of Physics: Conference Series*, *1853*.

Allahverdiyev, A. M., Abamor, E. S., Bagirova, M., & Rafailovich, M. (2011). Antimicrobial effects of TiO_2 and Ag_2O nanoparticles against drug-resistant bacteria and leishmania parasites. *Future Microbiology*, *6*(8), 933–940.

Allison, R. R., Bagnato, V. S., Cuenca, R., Downie, G. H., & Sibata, C. H. (2006). The future of photodynamic therapy in oncology. *Fiuture Oncology*, *2*(1), 53–71.

Allison, R. R., & Sibata, C. H. (2010). Oncologic photodynamic therapy photosensitizers: A clinical review. *Photodiagnosis and Photodynamic Therapy*, *7*(2), 61–75.

Ansari, M. A., Khan, H. M., Khan, A. A., Cameotra, S. S., Saquib, Q., & Musarrat, J. (2014). Interaction of Al_2O_3 nanoparticles with *Escherichia coli* and their cell envelope biomolecules. *Journal of Applied Microbiology*, *116*(4), 772–783.

Ansari, M. A., Khan, H. M., Khan, A. A., Pal, R., & Cameotra, S. S. (2013). Antibacterial potential of Al_2O_3 nanoparticles against multidrug resistance strains of *Staphylococcus aureus* isolated from skin exudates. *Journal of Nanoparticle Research*, *15*(10), 1–12.

Anzai, Y., Piccoli, C. W., Outwater, E. K., Stanford, W., Bluemke, D. A., Nurenberg, P., et al. (2003). Evaluation of neck and body metastases to nodes with ferumoxtran 10-enhanced MR imaging phase III safety and efficacy study. *Radiology*, *228*(3), 777–788.

Applerot, G., Lellouche, J., Lipovsky, A., Nitzan, Y., Lubart, R., Gedanken, A., et al. (2012). Understanding the antibacterial mechanism of CuO nanoparticles: Revealing the route of induced oxidative stress. *Small (Weinheim an der Bergstrasse, Germany)*, *8*(21), 3326–3337.

Asahi, R., Morikawa, T., Ohwaki, T., Aoki, K., & Taga, Y. (2001). Visible-light photocatalysis in nitrogen-doped titanium oxides. *Science (New York, NY)*, *293*(5528), 269–271.

Ashokraja, C., Sakar, M., & Balakumar, S. (2017). A perspective on the hemolytic activity of chemical and green-synthesized silver and silver oxide nanoparticles. *Materials Research Express*, *4*(10), 105406.

Augustine, R., & Hasan, A. (2020). Emerging applications of biocompatible phytosynthesized metal/metal oxide nanoparticles in healthcare. *Journal of Drug Delivery Science and Technology*, *56*, 101516.

Aw, M. S., Simovic, S., Addai-Mensah, J., & Losic, D. (2011). Polymeric micelles in porous and nanotubular implants as a new system for extended delivery of poorly soluble drugs. *Journal of Materials Chemistry*, *21*(20), 7082–7089.

Ayyanaar, S., Kesavan, M. P., Balachandran, C., Rasala, S., Rameshkumar, P., Aoki, S., et al. (2020). Iron oxide nanoparticle core-shell magnetic microspheres: Applications toward targeted drug delivery. *Nanomedicine: Nanotechnology, Biology and Medicine, 24*, 102134.

Bai, X., Li, L., Liu, H., Tan, L., Liu, T., & Meng, X. (2015). Solvothermal synthesis of ZnO nanoparticles and anti-infection application *in vivo*. *ACS Applied Materials & Interfaces*, *7*(2), 1308–1317.

Balasubramanyam, A., Sailaja, N., Mahboob, M., Rahman, M. F., Hussain, S. M., & Grover, P. (2010). *In vitro* mutagenicity assessment of aluminium oxide nanomaterials using the Salmonella/microsome assay. *Toxicology In Vitro*, *24*(6), 1871–1876.

Balla, V. K., Banerjee, S., Bose, S., & Bandyopadhyay, A. (2010). Direct laser processing of a tantalum coating on titanium for bone replacement structures. *Acta Biomaterialia*, *6* (6), 2329–2334.

Banobre-Lopez, M., Teijeiro, A., & Rivas, J. (2013). Magnetic nanoparticle-based hyperthermia for cancer treatment. *Reports of Practical Oncology and Radiotherapy: Journal of Greatpoland Cancer Center in Poznan and Polish Society of Radiation Oncology*, *18*(6), 397–400.

Batool, M., Khurshid, S., Qureshi, Z., & Daoush, W. M. (2021). Adsorption, antimicrobial and wound healing activities of biosynthesised zinc oxide nanoparticles. *Chemical Papers*, *75*(3), 893–907.

Bayrami, A., Haghgooie, S., Pouran, S. R., Arvanag, F. M., & Habibi-Yangjeh, A. (2020). Synergistic antidiabetic activity of ZnO nanoparticles encompassed by *Urtica dioica* extract. *Advanced Powder Technology*, *31*(5), 2110–2118.

Beckmann, N., Cannet, C., Babin, A. L., Blé, F. X., Zurbruegg, S., Kneuer, R., et al. (2009). *In vivo* visualization of macrophage infiltration and activity in inflammation using magnetic resonance imaging. *Wiley Interdisciplinary Reviews: Nanomedicine and Nanobiotechnology*, *1*(3), 272–298.

Behera, N., Arakha, M., Priyadarshinee, M., Pattanayak, B. S., Soren, S., Jha, S., et al. (2019). Oxidative stress generated at nickel oxide nanoparticle interface results in bacterial membrane damage leading to cell death. *RSC Advances*, *9*(43), 24888–24894.

Besinis, A., De Peralta, T., Tredwin, C. J., & Handy, R. D. (2015). Review of nanomaterials in dentistry: Interactions with the oral microenvironment, clinical applications, hazards, and benefits. *ACS Nano*, *9*(3), 2255–2289.

Beyth, N., Houri-Haddad, Y., Domb, A., Khan, W., & Hazan, R. (2015). Alternative antimicrobial approach: Nano-antimicrobial materials. *Evidence-Based Complementary and Alternative Medicine, 2015*.

Bi, Q., Song, X., Hu, A., Luo, T., Jin, R., Ai, H., et al. (2020). Magnetofection: Magic magnetic nanoparticles for efficient gene delivery. *Chinese Chemical Letters*, *31*(12), 3041–3046.

Bose, S., Darsell, J., Hosick, H. L., Yang, L., Sarkar, D. K., & Bandyopadhyay, A. (2002). Processing and characterization of porous alumina scaffolds. *Journal of Materials Science: Materials in Medicine*, *13*(1), 23–28.

Bullivant, J. P., Zhao, S., Willenberg, B. J., Kozissnik, B., Batich, C. D., & Dobson, J. (2013). Materials characterization of feraheme/ferumoxytol and preliminary evaluation of its potential for magnetic fluid hyperthermia. *International Journal of Molecular Sciences*, *14*(9), 17501–17510.

Cai, L., Chen, J., Liu, Z., Wang, H., Yang, H., & Ding, W. (2018). Magnesium oxide nanoparticles: Effective agricultural antibacterial agent against *Ralstonia solanacearum*. *Frontiers in Microbiology*, *9*, 790.

Cai, R., Kubota, Y., Shuin, T., Sakai, H., Hashimoto, K., & Fujishima, A. (1992). Induction of cytotoxicity by photoexcited TiO_2 particles. *Cancer Research*, *52*(8), 2346–2348.

Castro, D. C., Gatti, G., Martín, S. E., Uberman, P. M., & García, M. C. (2021). Promising tamoxifen-loaded biocompatible hybrid magnetic nanoplatforms against breast cancer cells: Synthesis, characterization and biological evaluation. *New Journal of Chemistry*, *45*(8), 4032–4045.

Cavalieri, F., Tortora, M., Stringaro, A., Colone, M., & Baldassarri, L. (2014). Nanomedicines for antimicrobial interventions. *Journal of Hospital Infection*, *88*(4), 183–190.

Celardo, I., Pedersen, J. Z., Traversa, E., & Ghibelli, L. (2011). Pharmacological potential of cerium oxide nanoparticles. *Nanoscale*, *3*(4), 1411–1420.

Çeşmeli, S., & Biray Avci, C. (2018). Application of titanium dioxide (TiO_2) nanoparticles in cancer therapies. *Journal of Drug Targeting*, *27*(7), 762–766.

Chen, J., Patil, S., Seal, S., & McGinnis, J. F. (2006). Rare earth nanoparticles prevent retinal degeneration induced by intracellular peroxides. *Nature Nanotechnology*, *1*(2), 142–150.

Choi, H. S., Liu, W., Misra, P., Tanaka, E., Zimmer, J. P., Itty Ipe, B., et al. (2007). Renal clearance of quantum dots. *Nature Biotechnology*, *25*(10), 1165–1170.

Clement, O., & Luciani, A. (2004). Imaging the lymphatic system: Possibilities and clinical applications. *European Radiology*, *14*(8), 1498–1507.

ClinicalTrials. (2014). *Magnetic nanoparticle thermoablation-retention and maintenance in the prostate: A Phase 0 study in men*. Retrieved from https://clinicaltrials.gov/ct2/show/NCT02033447.

Conesa, J. (1995). Computer modeling of surfaces and defects on cerium dioxide. *Surface Science*, *339*(3), 337–352.

Corot, C., Robert, P., Idee, J. M., & Port, M. (2006). Recent advances in iron oxide nanocrystal technology for medical imaging. *Advanced Drug Delivery Reviews*, *58*(14), 1471–1504.

Daldrup-Link, H. E., Golovko, D., Ruffell, B., DeNardo, D. G., Castaneda, R., Ansari, C., et al. (2011). MRI of tumor-associated macrophages with clinically applicable iron oxide nanoparticles. *Clinical Cancer Research*, *17*(17), 5695–5704.

Daldrup-Link, H. E., Kaiser, A., Helbich, T., Werner, M., Bjørnerud, A., Link, T. M., et al. (2003). Macromolecular contrast medium (feruglose) vs small molecular contrast medium (gadopentetate) enhanced magnetic resonance imaging. *Academic Radiology*, *10*(11), 1237–1246.

Das, S., Dowding, J. M., Klump, K. E., McGinnis, J. F., Self, W., & Seal, S. (2013). Cerium oxide nanoparticles applications and prospect in nanomedicine. *Nanomedicine: Nanotechnology, Biology, and Medicine*, *8*(9), 1483–1508.

Das, M., Patil, S., Bhargava, N., Kang, J. F., Riedel, L. M., Seal, S., et al. (2007). Auto-catalytic ceria nanoparticles offer neuroprotection to adult rat spinal cord neurons. *Biomaterials*, *28*(10), 1918–1925.

Davoodi, E., Zhianmanesh, M., Montazerian, H., Milani, A. S., & Hoorfar, M. (2020). Nano-porous anodic alumina: Fundamentals and applications in tissue engineering. *Journal of Materials Science: Materials in Medicine*, *31*(7), 1–16.

Dayan, A., Fleminger, G., & Ashur-Fabian, O. (2018). RGD-modified dihydrolipoamide dehydrogenase conjugated to titanium dioxide nanoparticles switchable integrin-targeted photodynamic treatment of melanoma cells. *RSC Advances*, *8*(17), 9112–9119.

Deshpande, S., Patil, S., Kuchibhatla, S. V. N. T., & Seal, S. (2005). Size dependency variation in lattice parameter and valency states in nanocrystalline cerium oxide. *Applied Physics Letters*, *87*(13), 133113.

Dimapilis, E. A. S., Hsu, C.-S., Mendoza, R. M. O., & Lu, M.-C. (2018). Zinc oxide nanoparticles for water disinfection. *Sustainable Environment Research*, *28*(2), 47–56.

Dizaj, S. M., Lotfipour, F., Barzegar-Jalali, M., Zarrintan, M. H., & Adibkia, K. (2014). Antimicrobial activity of the metals and metal oxide nanoparticles. *Materials Science and Engineering: C.*, *44*, 278–284.

Dougherty, T. J., Gomer, C. J., Henderson, B. W., Jori, G., Kessel, D., Korbelik, M., et al. (1998). Photodynamic therapy. *JNCI: Journal of the National Cancer Institute*, *90*(12), 889–905.

Dowaidar, M. (2021). *Magnetic iron oxide nanoparticles have potential on gene therapy effectiveness and biocompatibility*.

Duan, X., & Li, Y. (2013). Physicochemical characteristics of nanoparticles affect circulation, biodistribution, cellular internalization, and trafficking. *Small (Weinheim an der Bergstrasse, Germany)*, *9*(9–10), 1521–1532.

El-Boubbou, K. (2018). Magnetic iron oxide nanoparticles as drug carriers clinical relevance. *Nanomedicine: Nanotechnology, Biology, and Medicine*, *13*(8), 953–971.

Ermini, M. L., & Voliani, V. (2021). Antimicrobial nano-agents: The copper age. *ACS Nano*, *15*(4), 6008–6029.

Ezhilarasi, A. A., Vijaya, J. J., Kaviyarasu, K., Kennedy, L. J., Ramalingam, R. J., & Al-Lohedan, H. A. (2018). Green synthesis of NiO nanoparticles using *Aegle marmelos* leaf extract for the evaluation of in-vitro cytotoxicity, antibacterial and photocatalytic properties. *Journal of Photochemistry and Photobiology B: Biology*, *180*, 39–50.

Fasiku, V. O., Owonubi, S. J., Malima, N. M., Hassan, D., & Revaprasadu, N. (2020). *Metal oxide nanoparticles: A welcome development for targeting bacteria. Antibiotic materials in healthcare* (pp. 261–286). Elsevier.

Fernandez, C., Thomas, A., & Shailaja Raj, M. (2016). Green synthesis of silver oxide nanoparticle and its antimicrobial activity against organisms causing dental plaques. *International Journal of Pharma and Bio Sciences*, *7*, 14–19.

Fujishima, A., Rao, T. N., & Tryk, D. A. (2000). Titanium dioxide photocatalysis. *Journal of Photochemistry and Photobiology C: Photochemistry Reviews*, *1*(1), 1–21.

Fujishima, A., Zhang, X., & Tryk, D. (2008). TiO_2 photocatalysis and related surface phenomena. *Surface Science Reports*, *63*(12), 515–582.

Ge, S., Wang, G., Shen, Y., Zhang, Q., Jia, D., Wang, H., et al. (2011). Cytotoxic effects of MgO nanoparticles on human umbilical vein endothelial cells *in vitro*. *IET Nanobiotechnology*, *5*(2), 36–40.

Gholami, A., Mousavi, S. M., Hashemi, S. A., Ghasemi, Y., Chiang, W.-H., & Parvin, N. (2020). Current trends in chemical modifications of magnetic nanoparticles for targeted drug delivery in cancer chemotherapy. *Drug Metabolism Reviews*, *52*(1), 205–224.

Giesel, F. L., Mehndiratta, A., & Essig, M. (2010). High-relaxivity contrast-enhanced magnetic resonance neuroimaging: A review. *European Radiology*, *20*(10), 2461–2474.

Gojova, A., Lee, J. T., Jung, H. S., Guo, B., Barakat, A. I., & Kennedy, I. M. (2009). Effect of cerium oxide nanoparticles on inflammation in vascular endothelial cells. *Inhalation Toxicology*, *21*(Suppl. 1), 123–130.

Gomez-Florit, M., Pacha-Olivenza, M. A., Fernandez-Calderon, M. C., Cordoba, A., Gonzalez-Martin, M. L., Monjo, M., et al. (2016). Quercitrin-nanocoated titanium surfaces favour gingival cells against oral bacteria. *Scientific Reports*, *6*, 22444.

Goodwin, S. C., Bittner, C. A., Peterson, C. L., & Wong, G. (2001). Single-dose toxicity study of hepatic intra-arterial infusion of doxorubicin coupled to a novel magnetically targeted drug carrier. *Toxicological Sciences: An Official Journal of the Society of Toxicology*, *60*(1), 177–183.

Goodwin, S., Peterson, C., Hoh, C., & Bittner, C. (1999). Targeting and retention of magnetic targeted carriers (MTCs) enhancing intra-arterial chemotherapy. *Journal of Magnetism and Magnetic Materials*, *194*(1–3), 132–139.

Gudkov, S. V., Burmistrov, D. E., Serov, D. A., Rebezov, M. B., Semenova, A. A., & Lisitsyn, A. B. (2021). A mini review of antibacterial properties of ZnO nanoparticles. *Frontiers in Physics*, *9*, 641481. Available from https://doi.org/10.3389/fphy.2021.641481.

Harisinghani, M. G., Barentsz, J., Hahn, P. F., Deserno, W. M., Tabatabaei, S., van de Kaa, C. H., et al. (2003). Noninvasive detection of clinically occult lymph-node metastases in prostate cancer. *The New England Journal of Medicine*, *348*(25), 2491–2499.

Hassan, H. F. H., Mansour, A. M., Abo-Youssef, A. M. H., Elsadek, B. E. M., & Messiha, B. A. S. (2017). Zinc oxide nanoparticles as a novel anticancer approach; *in vitro* and *in vivo* evidence. *Clinical and Experimental Pharmacology and Physiology*, *44*(2), 235–243.

Hassan, S., Prakash, G., Ozturk, A. B., Saghazadeh, S., Sohail, M. F., Seo, J., et al. (2017). Evolution and clinical translation of drug delivery nanomaterials. *Nano Today*, *15*, 91–106.

Hassanien, R., Husein, D. Z., & Al-Hakkani, M. F. (2018). Biosynthesis of copper nanoparticles using aqueous Tilia extract: Antimicrobial and anticancer activities. *Heliyon.*, *4*(12), e01077.

Heckert, E. G., Karakoti, A. S., Seal, S., & Self, W. T. (2008). The role of cerium redox state in the SOD mimetic activity of nanoceria. *Biomaterials*, *29*(18), 2705–2709.

Heesakkers, R. A., Hövels, A. M., Jager, G. J., van den Bosch, H. C., Witjes, J. A., Raat, H. P., et al. (2008). MRI with a lymph-node-specific contrast agent as an alternative to CT scan and lymph-node dissection in patients with prostate cancer a prospective multicohort study. *The Lancet Oncology*, *9*(9), 850–856.

Helan, V., Prince, J. J., Al-Dhabi, N. A., Arasu, M. V., Ayeshamariam, A., Madhumitha, G., et al. (2016). Neem leaves mediated preparation of NiO nanoparticles and its magnetization, coercivity and antibacterial analysis. *Results in Physics*, *6*, 712–718.

Hirst, S. M., Karakoti, A., Singh, S., Self, W., Tyler, R., Seal, S., et al. (2013). Biodistribution and *in vivo* antioxidant effects of cerium oxide nanoparticles in mice. *Environmental Toxicology*, *28*(2), 107–118.

Hirst, S. M., Karakoti, A. S., Tyler, R. D., Sriranganathan, N., Seal, S., & Reilly, C. M. (2009). Anti-inflammatory properties of cerium oxide nanoparticles. *Small (Weinheim an der Bergstrasse, Germany)*, *5*(24), 2848–2856.

Hossen, S., Hossain, M. K., Basher, M. K., Mia, M. N. H., Rahman, M. T., & Uddin, M. J. (2019). Smart nanocarrier-based drug delivery systems for cancer therapy and toxicity studies: A review. *Journal of Advanced Research*, *15*, 1–18.

Hu, Z., Huang, Y., Sun, S., Guan, W., Yao, Y., Tang, P., et al. (2012). Visible light driven photodynamic anticancer activity of graphene oxide/TiO_2 hybrid. *Carbon*, *50*(3), 994–1004.

Huang, L., Li, D.-Q., Lin, Y.-J., Wei, M., Evans, D. G., & Duan, X. (2005). Controllable preparation of Nano-MgO and investigation of its bactericidal properties. *Journal of Inorganic Biochemistry*, *99*(5), 986–993.

Huang, R.-Y., Liu, Z.-H., Weng, W.-H., & Chang, C.-W. (2021). Magnetic nanocomplexes for gene delivery applications. *Journal of Materials Chemistry B*.

Huang, N. P., Min-hua, X., Yuan, C. W., & Rui-rong, Y. (1997). The study of the photokilling effect and mechanism of ultrafine TiO_2 particles on U937 cells. *Journal of Photochemistry and Photobiology A: Chemistry*, *108*(2–3), 229–233.

Hulbert, S. F., Matthews, J. R., Klawitter, J. J., Sauer, B. W., & Leonard, R. B. (1974). Effect of stress on tissue ingrowth into porous aluminum oxide. *Journal of Biomedical Materials Research*, *8*(3), 85–97.

Ilbeigi, G., Kariminik, A., & Moshafi, M. (2019). The antibacterial activities of NiO nanoparticles against some Gram-positive and Gram-negative bacterial strains. *International Journal of Basic Science in Medicine, [online]*, *4*(2), 69–74.

Imani, M. M., & Safaei, M. (2019). Optimized synthesis of magnesium oxide nanoparticles as bactericidal agents. *Journal of Nanotechnology*, *2019*.

Iv, M., Telischak, N., Feng, D., Holdsworth, S. J., Yeom, K. W., & Daldrup-Link, H. E. (2015). Clinical applications of iron oxide nanoparticles for magnetic resonance imaging of brain tumors. *Nanomedicine: Nanotechnology, Biology, and Medicine*, *10*(6), 993–1018.

Jani, A. M. M., Losic, D., & Voelcker, N. H. (2013). Nanoporous anodic aluminium oxide: Advances in surface engineering and emerging applications. *Progress in Materials Science*, *58*(5), 636–704.

Jasim, N. A., Al-Gasha'a, F. A., Al-Marjani, M. F., Al-Rahal, A. H., Abid, H. A., Al-Kadhmi, N. A., et al. (2020). ZnO nanoparticles inhibit growth and biofilm formation of vancomycin-resistant *S. aureus* (VRSA). *Biocatalysis and Agricultural Biotechnology.*, *29*, 101745.

Jeon, M., Halbert, M. V., Stephen, Z. R., & Zhang, M. (2021). Iron oxide nanoparticles as T_1 contrast agents for magnetic resonance imaging: Fundamentals, challenges, applications, and prospectives. *Advanced Materials*, *33*(23), 1906539.

Jiang, W., Mashayekhi, H., & Xing, B. (2009). Bacterial toxicity comparison between nano- and micro-scaled oxide particles. *Environmental Pollution*, *157*(5), 1619–1625.

Jin, T., & He, Y. (2011). Antibacterial activities of magnesium oxide (MgO) nanoparticles against foodborne pathogens. *Journal of Nanoparticle Research*, *13*(12), 6877–6885.

Jordan, A., Scholz, R., Wust, P., Schirra, H., Schiestel, T., Schmidt, H., et al. (1999). Endocytosis of dextran and silan-coated magnetite nanoparticles and the effect of intracellular hyperthermia on human mammary carcinoma cells *in vitro*. *Journal of Magnetism and Magnetic Materials*, *194*(1–3), 185–196.

Jung, W. K., Koo, H. C., Kim, K. W., Shin, S., Kim, S. H., & Park, Y. H. (2008). Antibacterial activity and mechanism of action of the silver ion in *Staphylococcus aureus* and *Escherichia coli*. *Applied and Environmental Microbiology*, *74*(7), 2171–2178.

Kamaly, N., Xiao, Z., Valencia, P. M., Radovic-Moreno, A. F., & Farokhzad, O. C. (2012). Targeted polymeric therapeutic nanoparticles design. *Chemical Society Reviews*, *41*(7), 2971–3010.

Karakoti, A., Singh, S., Dowding, J. M., Seal, S., & Self, W. T. (2010). Redox-active radical scavenging nanomaterials. *Chemical Society Reviews*, *39*(11), 4422–4432.

Karlsson, H. L., Cronholm, P., Hedberg, Y., Tornberg, M., De Battice, L., Svedhem, S., et al. (2013). Cell membrane damage and protein interaction induced by copper containing nanoparticles—Importance of the metal release process. *Toxicology*, *313*(1), 59–69.

Karthikeyan, C., Varaprasad, K., Venugopal, S. K., Shakila, S., Venkatraman, B. R., & Sadiku, R. (2021). Biocidal (bacterial and cancer cells) activities of chitosan/CuO nanomaterial, synthesized via a green process. *Carbohydrate Polymers*, *259*, 117762.

Kim, M. H., Park, H., Nam, H. C., Park, S. R., Jung, J.-Y., & Park, W. H. (2018). Injectable methylcellulose hydrogel containing silver oxide nanoparticles for burn wound healing. *Carbohydrate Polymers*, *181*, 579–586.

Kim, M.-H., Seo, J.-H., Kim, H.-M., & Jeong, H.-J. (2014). Zinc oxide nanoparticles, a novel candidate for the treatment of allergic inflammatory diseases. *European Journal of Pharmacology*, *738*, 31–39.

Koda, J., Venook, A., Walser, E., & Goodwin, S. (2002). *A multicenter, phase I/II trial of hepatic intra-arterial delivery of doxorubicin hydrochloride adsorbed to magnetic targeted carriers in patients with hepatocellular carcinoma.*

Korsvik, C., Patil, S., Seal, S., & Self, W. T. (2007). Superoxide dismutase mimetic properties exhibited by vacancy engineered ceria nanoparticles. *Chemical Commununications (Cambridge, UK)* (10), 1056–1058.

Kreuter, J. (2007). Nanoparticles—A historical perspective. *International Journal of Pharmaceutics*, *331*(1), 1–10.

Krishnamoorthy, K., Moon, J. Y., Hyun, H. B., Cho, S. K., & Kim, S.-J. (2012). Mechanistic investigation on the toxicity of MgO nanoparticles toward cancer cells. *Journal of Materials Chemistry*, *22*(47), 24610–24617.

Król, A., Pomastowski, P., Rafińska, K., Railean-Plugaru, V., & Buszewski, B. (2017). Zinc oxide nanoparticles: Synthesis, antiseptic activity and toxicity mechanism. *Advances in Colloid and Interface Science*, *249*, 37–52.

Kubota, Y., Shuin, T., Kawasaki, C., Hosaka, M., Kitamura, H., Cai, R., et al. (1994). Photokilling of T-24 human bladder cancer cells with titanium dioxide. *British Journal of Cancer*, *70*(6), 1107.

Kwon, H. J., Shin, K., Soh, M., Chang, H., Kim, J., Lee, J., et al. (2018). Large-scale synthesis and medical applications of uniform-sized metal oxide nanoparticles. *Advanced Materials*, *30*(42), e1704290.

La Flamme, K. E., Popat, K. C., Leoni, L., Markiewicz, E., La Tempa, T. J., Roman, B. B., et al. (2007). Biocompatibility of nanoporous alumina membranes for immunoisolation. *Biomaterials*, *28*(16), 2638–2645.

Laurent, S., Dutz, S., Hafeli, U. O., & Mahmoudi, M. (2011). Magnetic fluid hyperthermia: Focus on superparamagnetic iron oxide nanoparticles. *Advances in Colloid and Interface Science*, *166*(1–2), 8–23.

Lee, N., Choi, S. H., & Hyeon, T. (2013). Nano-sized CT contrast agents. *Advanced Materials*, *25*(19), 2641–2660.

Lee, N., Yoo, D., Ling, D., Cho, M. H., Hyeon, T., & Cheon, J. (2015). Iron oxide based nanoparticles for multimodal imaging and magnetoresponsive therapy. *Chemical Reviews*, *115*(19), 10637–10689.

Lehner, R., Wang, X., Marsch, S., & Hunziker, P. (2013). Intelligent nanomaterials for medicine: Carrier platforms and targeting strategies in the context of clinical application. *Nanomedicine: Nanotechnology, Biology and Medicine*, *9*(6), 742–757.

Leung, Y. H., Ng, A. M. C., Xu, X., Shen, Z., Gethings, L. A., Wong, M. T., et al. (2014). Mechanisms of antibacterial activity of MgO: Non-ROS mediated toxicity of MgO nanoparticles towards *Escherichia coli*. *Small (Weinheim an der Bergstrasse, Germany)*, *10*(6), 1171–1183.

Li, R., Chen, Z., Ren, N., Wang, Y., Wang, Y., & Yu, F. (2019). Biosynthesis of silver oxide nanoparticles and their photocatalytic and antimicrobial activity evaluation for wound healing applications in nursing care. *Journal of Photochemistry and Photobiology B: Biology*, *199*, 111593.

Li, Z., Mi, L., Wang, P. N., & Chen, J. Y. (2011). Study on the visible-light-induced photo-killing effect of nitrogen-doped TiO_2 nanoparticles on cancer cells. *Nanoscale Research Letters*, *6*(1), 356.

Li, W., Tutton, S., Vu, A. T., Pierchala, L., Li, B. S., Lewis, J. M., et al. (2005). First-pass contrast-enhanced magnetic resonance angiography in humans using ferumoxytol, a novel ultrasmall superparamagnetic iron oxide (USPIO)-based blood pool agent. *Journal of Magnetic Resonance Imaging: JMRI*, *21*(1), 46–52.

Li, C., Zhang, H., Gong, X., Li, Q., & Zhao, X. (2019). Synthesis, characterization, and cytotoxicity assessment of *N*-acetyl-L-cysteine capped ZnO nanoparticles as camptothecin delivery system. *Colloids and Surfaces B: Biointerfaces*, *174*, 476–482.

Lin, W. T., Tan, H. L., Duan, Z. L., Yue, B., Ma, R., He, G., et al. (2014). Inhibited bacterial biofilm formation and improved osteogenic activity on gentamicin-loaded titania nanotubes with various diameters. *International Journal of Nanomedicine*, *9*, 1215–1230.

Linsebigler, A. L., Lu, G., & Yates, J. T., Jr. (1995). Photocatalysis on TiO_2 surfaces: Principles, mechanisms, and selected results. *Chemical Reviews (Washington, DC, USA)*, *95*(3), 735–758.

Liu, D., Li, J., Wang, C., An, L., Lin, J., Tian, Q., et al. (2021). Ultrasmall Fe@ Fe_3O_4 nanoparticles as T1–T2 dual-mode MRI contrast agents for targeted tumor imaging. *Nanomedicine: Nanotechnology, Biology and Medicine*, *32*, 1–11.

Lok, C.-N., Ho, C.-M., Chen, R., He, Q.-Y., Yu, W.-Y., Sun, H., et al. (2006). Proteomic analysis of the mode of antibacterial action of silver nanoparticles. *Journal of Proteome Research*, *5*(4), 916–924.

López, T., Sotelo, J., Navarrete, J., & Ascencio, J. A. (2006). Synthesis of TiO_2 nanostructured reservoir with temozolomide: Structural evolution of the occluded drug. *Optical Materials*, *29*(1), 88–94.

Louwakul, P., Saelo, A., & Khemaleelakul, S. (2017). Efficacy of calcium oxide and calcium hydroxide nanoparticles on the elimination of *Enterococcus faecalis* in human root dentin. *Clinical Oral Investigations*, *21*(3), 865–871.

Lu, M., Cohen, M. H., Rieves, D., & Pazdur, R. (2010). FDA report: Ferumoxytol for intravenous iron therapy in adult patients with chronic kidney disease. *American Journal of Hematology*, *85*(5), 315–319.

Lun Pang, C., Lindsay, R., & Thornton, G. (2008). Chemical reactions on rutile $TiO_2(110)$. *Chemical Society Reviews*, *37*(10), 2328–2353.

Ma, J. Y., Mercer, R. R., Barger, M., Schwegler-Berry, D., Scabilloni, J., Ma, J. K., et al. (2012). Induction of pulmonary fibrosis by cerium oxide nanoparticles. *Toxicology and Applied Pharmacology*, *262*(3), 255–264.

Ma, J. Y., Zhao, H., Mercer, R. R., Barger, M., Rao, M., Meighan, T., et al. (2011). Cerium oxide nanoparticle-induced pulmonary inflammation and alveolar macrophage functional change in rats. *Nanotoxicology.*, *5*(3), 312–325.

Mahamuni-Badiger, P. P., Patil, P. M., Badiger, M. V., Patel, P. R., Thorat-Gadgil, B. S., Pandit, A., et al. (2020). Biofilm formation to inhibition: Role of zinc oxide-based nanoparticles. *Materials Science and Engineering: C.*, *108*, 110319.

Malarkodi, C., Rajeshkumar, S., Paulkumar, K., Vanaja, M., Gnanajobitha, G., & Annadurai, G. (2014). Biosynthesis and antimicrobial activity of semiconductor nanoparticles against oral pathogens. *Bioinorganic Chemistry and Applications*, *2014*.

Mandal, A. K. (2017). Copper nanomaterials as drug delivery system against infectious agents and cancerous cells. *Journal of Applied Life Sciences International*, 1–8.

Manikandan, V., Velmurugan, P., Park, J.-H., Chang, W.-S., Park, Y.-J., Jayanthi, P., et al. (2017). Green synthesis of silver oxide nanoparticles and its antibacterial activity against dental pathogens. *3 Biotech*, *7*(1), 72.

Mariadoss, A. V. A., Saravanakumar, K., Sathiyaseelan, A., Venkatachalam, K., & Wang, M.-H. (2020). Folic acid functionalized starch encapsulated green synthesized copper oxide nanoparticles for targeted drug delivery in breast cancer therapy. *International Journal of Biological Macromolecules*, *164*, 2073–2084.

Martinez-Gutierrez, F., Olive, P. L., Banuelos, A., Orrantia, E., Nino, N., Sanchez, E. M., et al. (2010). Synthesis, characterization, and evaluation of antimicrobial and cytotoxic effect of silver and titanium nanoparticles. *Nanomedicine: Nanotechnology, Biology and Medicine*, *6*(5), 681–688.

McLachlan, S. J., Morris, M. R., Lucas, M. A., Fisco, R. A., Eakins, M. N., Fowler, D. R., et al. (1994). Phase I clinical evaluation of a new iron oxide MR contrast agent. *Journal of Magnetic Resonance Imaging*, *4*(3), 301–307.

Mirhosseini, M., & Afzali, M. (2016). Investigation into the antibacterial behavior of suspensions of magnesium oxide nanoparticles in combination with nisin and heat against *Escherichia coli* and *Staphylococcus aureus* in milk. *Food Control*, *68*, 208–215.

Miri, A., Mahdinejad, N., Ebrahimy, O., Khatami, M., & Sarani, M. (2019). Zinc oxide nanoparticles: Biosynthesis, characterization, antifungal and cytotoxic activity. *Materials Science and Engineering: C*, *104*, 109981.

Mirzaei, H., & Darroudi, M. (2017). Zinc oxide nanoparticles: Biological synthesis and biomedical applications. *Ceramics International*, *43*(1), 907–914.

Mishra, P. K., Mishra, H., Ekielski, A., Talegaonkar, S., & Vaidya, B. (2017). Zinc oxide nanoparticles: A promising nanomaterial for biomedical applications. *Drug Discovery Today*, *22*(12), 1825–1834.

Mody, V. V., Cox, A., Shah, S., Singh, A., Bevins, W., & Parihar, H. (2013). Magnetic nanoparticle drug delivery systems for targeting tumor. *Applied Nanoscience*, *4*(4), 385–392.

Nagajyothi, P. C., Muthuraman, P., Sreekanth, T. V. M., Kim, D. H., & Shim, J. (2017). Green synthesis: In-vitro anticancer activity of copper oxide nanoparticles against human cervical carcinoma cells. *Arabian Journal of Chemistry*, *10*(2), 215–225.

Nam, J., Won, N., Bang, J., Jin, H., Park, J., Jung, S., et al. (2013). Surface engineering of inorganic nanoparticles for imaging and therapy. *Advanced Drug Delivery Reviews*, *65*(5), 622–648.

Nazarizadeh, A., & Asri-Rezaie, S. (2016). Comparative study of antidiabetic activity and oxidative stress induced by zinc oxide nanoparticles and zinc sulfate in diabetic rats. *AAPS PharmSciTech*, *17*(4), 834–843.

Neuberger, T., Schöpf, B., Hofmann, H., Hofmann, M., & von Rechenberg, B. (2005). Superparamagnetic nanoparticles for biomedical applications: Possibilities and limitations of a new drug delivery system. *Journal of Magnetism and Magnetic Materials*, *293*(1), 483–496.

Neuwelt, E. A., Várallyay, C. G., Manninger, S., Solymosi, D., Haluska, M., Hunt, M. A., et al. (2007). The potential of ferumoxytol nanoparticle magnetic resonance imaging perfusion and angiography in central nervous system malignancy a pilot study. *Neurosurgery*, *60*(4), 601–612.

Nie, G., Hah, H. J., Kim, G., Lee, Y. E., Qin, M., Ratani, T. S., et al. (2012). Hydrogel nanoparticles with covalently linked coomassie blue for brain tumor delineation visible to the surgeon. *Small (Weinheim an der Bergstrasse, Germany)*, *8*(6), 884–891.

Nikolova, M. P., & Chavali, M. S. (2020). Metal oxide nanoparticles as biomedical materials. *Biomimetics*, *5*(2), 27.

Niu, J., Azfer, A., Rogers, L. M., Wang, X., & Kolattukudy, P. E. (2007). Cardioprotective effects of cerium oxide nanoparticles in a transgenic murine model of cardiomyopathy. *Cardiovascular Research*, *73*(3), 549–559.
Nourmohammadi, E., Khoshdel-Sarkarizi, H., Nedaeinia, R., Darroudi, M., & Oskuee, R. K. (2020). Cerium oxide nanoparticles: A promising tool for the treatment of fibrosarcoma in-vivo. *Materials Science and Engineering: C.*, *109*, 1–10.
Ogino, C., Shibata, N., Sasai, R., Takaki, K., Miyachi, Y., Kuroda, S., et al. (2010). Construction of protein-modified TiO_2 nanoparticles for use with ultrasound irradiation in a novel cell injuring method. *Bioorganic & Medicinal Chemistry Letters*, *20*(17), 5320–5325.
Ohgushi, H., Kotobuki, N., Funaoka, H., Machida, H., Hirose, M., Tanaka, Y., et al. (2005). Tissue engineered ceramic artificial joint—Ex vivo osteogenic differentiation of patient mesenchymal cells on total ankle joints for treatment of osteoarthritis. *Biomaterials*, *26*(22), 4654–4661.
Pai, A. B., & Garba, A. O. (2012). Ferumoxytol: A silver lining in the treatment of anemia of chronic kidney disease or another dark cloud? *Journal of Blood Medicine*, *3*, 77–85.
Pairoj, S., Damrongsak, P., Damrongsak, B., Jinawath, N., Kaewkhaw, R., Ruttanasirawit, C., et al. (2021). Antitumor activities of carboplatin–doxorubicin–ZnO complexes in different human cancer cell lines (breast, cervix uteri, colon, liver and oral) under UV exposition. *Artificial Cells, Nanomedicine, and Biotechnology*, *49*(1), 120–135.
Pan, X., Liang, X., Yao, L., Wang, X., Jing, Y., Ma, J., et al. (2017). Study of the photodynamic activity of N-doped $TiO(_2)$ nanoparticles conjugated with aluminum phthalocyanine. *Nanomaterials (Basel)*, *7*(10).
Park, K. (2013). Facing the truth about nanotechnology in drug delivery. *ACS Nano*, *7*(9), 7442–7447.
Parnia, F., Yazdani, J., Javaherzadeh, V., & Dizaj, S. M. (2017). Overview of nanoparticle coating of dental implants for enhanced osseointegration and antimicrobial purposes. *Journal of Pharmacy & Pharmaceutical Sciences*, *20*, 148–160.
Paunesku, T., Rajh, T., Wiederrecht, G., Maser, J., Vogt, S., Stojicevic, N., et al. (2003). Biology of TiO_2-oligonucleotide nanocomposites. *Nature Materials*, *2*(5), 343–346.
Pelgrift, R. Y., & Friedman, A. J. (2013). Nanotechnology as a therapeutic tool to combat microbial resistance. *Advanced Drug Delivery Reviews*, *65*(13–14), 1803–1815.
Penfield, J. G., & Reilly, R. F., Jr. (2007). What nephrologists need to know about gadolinium. *Nature Reviews Nephrology.*, *3*(12), 654.
Pirmohamed, T., Dowding, J. M., Singh, S., Wasserman, B., Heckert, E., Karakoti, A. S., et al. (2010). Nanoceria exhibit redox state-dependent catalase mimetic activity. *Chemical Communications (Cambridge, England)*, *46*(16), 2736–2738.
Polakis, P. (2016). Antibody drug conjugates for cancer therapy. *Pharmacological Reviews*, *68*(1), 3–19.
Popat, K. C., Swan, E. E. L., Mukhatyar, V., Chatvanichkul, K.-I., Mor, G. K., Grimes, C. A., et al. (2005). Influence of nanoporous alumina membranes on long-term osteoblast response. *Biomaterials*, *26*(22), 4516–4522.
Provenzano, R., Schiller, B., Rao, M., Coyne, D., Brenner, L., & Pereira, B. J. (2009). Ferumoxytol as an intravenous iron replacement therapy in hemodialysis patients. *Clinical Journal of the American Society of Nephrology: CJASN*, *4*(2), 386–393.
Pugazhendhi, A., Prabhu, R., Muruganantham, K., Shanmuganathan, R., & Natarajan S. (2019). Anticancer, antimicrobial and photocatalytic activities of green synthesized magnesium oxide nanoparticles (MgONPs) using aqueous extract of *Sargassum wightii*. *Journal of Photochemistry and Photobiology B: Biology*, *190*, 86–97.

Qin, Y., Sun, L., Li, X., Cao, Q., Wang, H., Tang, X., et al. (2011). Highly water-dispersible TiO_2 nanoparticles for doxorubicin delivery: Effect of loading mode on therapeutic efficacy. *Journal of Materials Chemistry*, *21*(44), 18003.

Qiu, Z., Yu, Y., Chen, Z., Jin, M., Yang, D., Zhao, Z., et al. (2012). Nanoalumina promotes the horizontal transfer of multiresistance genes mediated by plasmids across genera. *Proceedings of the National Academy of Sciences*, *109*(13), 4944–4949.

Rajh, T., Dimitrijevic, N. M., Bissonnette, M., Koritarov, T., & Konda, V. (2014). Titanium dioxide in the service of the biomedical revolution. *Chemical Reviews*, *114*(19), 10177–10216.

Rangeela, M., Rajeshkumar, S., Lakshmi, T., & Roy, A. (2019). Anti-inflammatory activity of zinc oxide nanoparticles prepared using amla fruits. *Drug Invention Today*, *11*(10).

Rzigalinski, B. A., Meehan, K., Davis, R. M., Xu, Y., Miles, W. C., & Cohen, C. A. (2005). Radical nanomedicine. *Nanomedicine: Nanotechnology, Biology, and Medicine*, *1*(4), 399–412.

Sánchez-Sanhueza, G., Fuentes-Rodríguez, D., & Bello-Toledo, H. (2016). Copper nanoparticles as potential antimicrobial agent in disinfecting root canals. A systematic review. *International Journal of Odontostomatology*, *10*(3), 547–554.

Sankar, R., Maheswari, R., Karthik, S., Shivashangari, K. S., & Ravikumar, V. (2014). Anticancer activity of *Ficus religiosa* engineered copper oxide nanoparticles. *Materials Science and Engineering: C.*, *44*, 234–239.

Saravanakumar, K., Shanmugam, S., Varukattu, N. B., MubarakAli, D., Kathiresan, K., & Wang, M.-H. (2019). Biosynthesis and characterization of copper oxide nanoparticles from indigenous fungi and its effect of photothermolysis on human lung carcinoma. *Journal of Photochemistry and Photobiology B: Biology*, *190*, 103–109.

Sarraf, M., Dabbagh, A., Razak, B. A., Nasiri-Tabrizi, B., Hosseini, H. R. M., Saber-Samandari, S., et al. (2018). Silver oxide nanoparticles-decorated tantala nanotubes for enhanced antibacterial activity and osseointegration of Ti6Al4V. *Materials & Design*, *154*, 28–40.

Sathishkumar, P., Li, Z., Govindan, R., Jayakumar, R., Wang, C., & Gu, F. L. (2021). Zinc oxide-quercetin nanocomposite as a smart nano-drug delivery system: Molecular-level interaction studies. *Applied Surface Science*, *536*, 147741.

Sawai, J. (2003). Quantitative evaluation of antibacterial activities of metallic oxide powders (ZnO, MgO and CaO) by conductimetric assay. *Journal of Microbiological Methods*, *54* (2), 177–182.

Sawai, J., Igarashi, H., Hashimoto, A., Kokugan, T., & Shimizu, M. (1996). Effect of particle size and heating temperature of ceramic powders on antibacterial activity of their slurries. *Journal of Chemical Engineering of Japan*, *29*(2), 251–256.

Sawai, J., & Yoshikawa, T. (2004). Quantitative evaluation of antifungal activity of metallic oxide powders (MgO, CaO and ZnO) by an indirect conductimetric assay. *Journal of Applied Microbiology*, *96*(4), 803–809.

Schubert, D., Dargusch, R., Raitano, J., & Chan, S. W. (2006). Cerium and yttrium oxide nanoparticles are neuroprotective. *Biochemical and Biophysical Research Communications*, *342*(1), 86–91.

Seo, J. W., Chung, H., Kim, M. Y., Lee, J., Choi, I. H., & Cheon, J. (2007). Development of water-soluble single-crystalline TiO_2 nanoparticles for photocatalytic cancer-cell treatment. *Small (Weinheim an der Bergstrasse, Germany)*, *3*(5), 850–853.

Shamasi, Z., Es-haghi, A., Taghavizadeh Yazdi, M. E., Amiri, M. S., & Homayouni-Tabrizi, M. (2021). Role of *Rubia tinctorum* in the synthesis of zinc oxide nanoparticles and apoptosis induction in breast cancer cell line. *Nanomedicine Journal.*, *8*(1), 65–72.

Shokuhfar, T., Hamlekhan, A., Chang, J. Y., Choi, C. K., Sukotjo, C., & Friedrich, C. (2014). Biophysical evaluation of cells on nanotubular surfaces: The effects of atomic ordering and chemistry. *International Journal of Nanomedicine*, *9*, 3737–3748.

Siddiqi, K. S., Husen, A., & Rao, R. A. K. (2018). A review on biosynthesis of silver nanoparticles and their biocidal properties. *Journal of Nanobiotechnology*, *16*(1), 1–28.

Sierra-Fernandez, A., De la Rosa-García, S. C., Gomez-Villalba, L. S., Gómez-Cornelio, S., Rabanal, M. E., Fort, R., et al. (2017). Synthesis, photocatalytic, and antifungal properties of MgO, ZnO and Zn/Mg oxide nanoparticles for the protection of calcareous stone heritage. *ACS Applied Materials & Interfaces*, *9*(29), 24873–24886.

Simon, G. H., von Vopelius-Feldt, J., Fu, Y., Schlegel, J., Pinotek, G. M. S., Wendland, M. F., et al. (2006). Ultrasmall supraparamagnetic iron oxide-enhanced magnetic resonance imaging of antigen-induced arthritis a comparative study between SHU 555 C ferumoxtran-10 and ferumoxytol. *Investigative Radiology*, *41*(1), 45–51.

Simovic, S., Losic, D., & Vasilev, K. (2010). Controlled drug release from porous materials by plasma polymer deposition. *Chemical Communications*, *46*(8), 1317–1319.

Singh, T. A., Das, J., & Sil, P. C. (2020). Zinc oxide nanoparticles: A comprehensive review on its synthesis, anticancer and drug delivery applications as well as health risks. *Advances in Colloid and Interface Science*, 102317.

Soltis, J. A., & Penn, R. L. (2016). Medical applications of iron oxide nanoparticles. In D. Faivre (Ed.), *Iron oxides from nature to applications*. Weinheim, Germany: Wiley-VCH.

Sortino, S. (2012). Photoactivated nanomaterials for biomedical release applications. *Journal of Materials Chemistry*, *22*(2), 301–318.

Spickler, E. M., Dion, J. E., Lufkin, R. B., Rachner, T., Vinuela, F., Duckwiler, G., et al. (1990). The MR appearance of endovascular embolic agents *in vitro* with clinical correlation. *Computerized Medical Imaging and Graphics*, *14*(6), 415–423.

Sugden, R., Kelly, R., & Davies, S. (2016). Combatting antimicrobial resistance globally. *Nature Microbiology*, *1*(10), 1–2.

Sun, C., Li, H., & Chen, L. (2012). Nanostructured ceria-based materials: Synthesis, properties, and applications. *Energy & Environmental Science*, *5*(9), 8475.

Suryavanshi, A., Khanna, K., Sindhu, K. R., Bellare, J., & Srivastava, R. (2017). Magnesium oxide nanoparticle-loaded polycaprolactone composite electrospun fiber scaffolds for bone–soft tissue engineering applications: In-vitro and in-vivo evaluation. *Biomedical Materials*, *12*(5), 055011.

Suzuki, T., Kosacki, I., Anderson, H. U., & Colomban, P. (2001). Electrical conductivity and lattice defects in nanocrystalline cerium oxide thin films. *Journal of the American Ceramic Society*, *84*(9), 2007–2014.

Swan, E. E. L., Popat, K. C., Grimes, C. A., & Desai, T. A. (2005). Fabrication and evaluation of nanoporous alumina membranes for osteoblast culture. *Journal of Biomedical Materials Research Part A: An Official Journal of The Society for Biomaterials, The Japanese Society for Biomaterials, and The Australian Society for Biomaterials and the Korean Society for Biomaterials*, *72*(3), 288–295.

Tarnuzzer, R. W., Colon, J., Patil, S., & Seal, S. (2005). Vacancy engineered ceria nanostructures for protection from radiation-induced cellular damage. *Nano Letters*, *5*(12), 2573–2577.

Taupitz, M., Wagner, S., Schnorr, J., Kravec, I., Pilgrimm, H., Bergmann-Fritsch, H., et al. (2004). Phase I clinical evaluation of citrate-coated monocrystalline very small superparamagnetic iron oxide particles as a new contrast medium for magnetic resonance imaging. *Investigative Radiology*, *39*(7), 394–405.

Taylor, A. M., Panting, J. R., Keegan, J., Gatehouse, P. D., Amin, D., Jhooti, P., et al. (1999). Safety and preliminary findings with the intravascular contrast agent NC100150 injection for MR coronary angiography. *Journal of Magnetic Resonance Imaging*, *9*(2), 220–227.

Tokuoka, Y., Yamada, M., Kawashima, N., & Miyasaka, T. (2006). Anticancer effect of dye-sensitized TiO_2 nanocrystals by polychromatic visible light irradiation. *Chemical Letters*, *35*(5), 496–497.

Tombach, B., Reimer, P., Bremer, C., Allkemper, T., Engelhardt, M., Mahler, M., et al. (2004). First-pass and equilibrium-MRA of the aortoiliac region with a superparamagnetic iron oxide blood pool MR contrast agent (SH U 555 C): Results of a human pilot study. *NMR in Biomedicine*, *17*(7), 500–506.

Triesscheijn, M., Baas, P., Schellens, J. H., & Stewart, F. A. (2006). Photodynamic therapy in oncology. *The Oncologist*, *11*(9), 1034–1044.

Uddin, M. J., Mondal, D., Morris, C. A., Lopez, T., Diebold, U., & Gonzalez, R. D. (2011). An *in vitro* controlled release study of valproic acid encapsulated in a titania ceramic matrix. *Applied Surface Science*, *257*(18), 7920–7927.

Vance, M. E., Kuiken, T., Vejerano, E. P., McGinnis, S. P., Hochella, M. F., Jr., Rejeski, D., et al. (2015). Nanotechnology in the real world: Redeveloping the nanomaterial consumer products inventory. *Beilstein Journal of Nanotechnology*, *6*(1), 1769–1780.

Vangijzegem, T., Stanicki, D., & Laurent, S. (2019). Magnetic iron oxide nanoparticles for drug delivery: Applications and characteristics. *Expert Opinion on Drug Delivery*, *16*(1), 69–78.

Venditto, V. J., & Szoka, F. C., Jr. (2013). Cancer nanomedicines:So many papers and so few drugs!. *Advanced Drug Delivery Reviews*, *65*(1), 80–88.

Virgen-Ortiz, A., Apolinar-Iribe, A., Díaz-Reval, I., Parra-Delgado, H., Limón-Miranda, S., Sánchez-Pastor, E. A., et al. (2020). Zinc oxide nanoparticles induce an adverse effect on blood glucose levels depending on the dose and route of administration in healthy and diabetic rats. *Nanomaterials.*, *10*(10).

Vithiya, K., Kumar, R., & Sen, S. (2014). Antimicrobial activity of biosynthesized silver oxide nanoparticles. *Journal of Pure and Applied Microbiology*, *4*, 3263–3268.

Wagner, S., Schnorr, J., Pilgrimm, H., Hamm, B., & Taupitz, M. (2002). Monomer-coated very small superparamagnetic iron oxide particles as contrast medium for magnetic resonance imaging preclinical *in vivo* characterization. *Investigative Radiology*, *37*(4), 167–177.

Wahajuddin., & Arora, S. (2012). Superparamagnetic iron oxide nanoparticles: Magnetic nanoplatforms as drug carriers. *International Journal of Nanomedicine*, *7*, 3445–3471.

Wang, Q., Huang, J. Y., Li, H. Q., Chen, Z., Zhao, A. Z., Wang, Y., et al. (2016). TiO_2 nanotube platforms for smart drug delivery: A review. *International Journal of Nanomedicine*, *11*, 4819–4834.

Wang, Q., Huang, J. Y., Li, H. Q., Zhao, A. Z., Wang, Y., Zhang, K. Q., et al. (2017). Recent advances on smart TiO_2 nanotube platforms for sustainable drug delivery applications. *International Journal of Nanomedicine*, *12*, 151–165.

Wang, T., Weng, Z., Liu, X., Yeung, K. W. K., Pan, H., & Wu, S. (2017). Controlled release and biocompatibility of polymer/titania nanotube array system on titanium implants. *Bioactive materials*, *2*(1), 44–50.

Wetteland, C. L., Nguyen, N.-Y. T., & Liu, H. (2016). Concentration-dependent behaviors of bone marrow derived mesenchymal stem cells and infectious bacteria toward magnesium oxide nanoparticles. *Acta Biomaterialia*, *35*, 341–356.

Wu, K. C., Yamauchi, Y., Hong, C. Y., Yang, Y. H., Liang, Y. H., Funatsu, T., et al. (2011). Biocompatible, surface functionalized mesoporous titania nanoparticles for intracellular imaging and anticancer drug delivery. *Chemical Communications (Cambridge, England)*, *47*(18), 5232–5234.

Xia, T., Kovochich, M., Liong, M., Madler, L., Gilbert, B., Shi, H., et al. (2008). Comparison of the mechanism of toxicity of zinc oxide and cerium oxide nanoparticles based on dissolution and oxidative stress properties. *ACS Nano*, *2*(10), 2121–2134.

Xu, M., Huang, N., Xiao, Z., & Lu, Z. (1998). Photoexcited TiO_2 nanoparticles through OH-radicals induced malignant cells to necrosis. *Supramolecular Science*, *5*(5–6), 449–451.

Xu, S., Shen, J., Chen, S., Zhang, M., & Shen, T. (2002). Active oxygen species ($^1O_2$, $O_2\cdot -$) generation in the system of TiO_2 colloid sensitized by hypocrellin B. *Journal of Photochemistry and Photobiology B: Biology*, *67*(1), 64–70.

Xu, J., Sun, Y., Huang, J., Chen, C., Liu, G., Jiang, Y., et al. (2007). Photokilling cancer cells using highly cell-specific antibody-TiO_2 bioconjugates and electroporation. *Bioelectrochemistry (Amsterdam, Netherlands)*, *71*(2), 217–222.

Xu, C., & Qu, X. (2014). Cerium oxide nanoparticle: A remarkably versatile rare earth nanomaterial for biological applications. *NPG Asia Materials*, *6*(3), e90.

Yadavalli, T., & Shukla, D. (2017). Role of metal and metal oxide nanoparticles as diagnostic and therapeutic tools for highly prevalent viral infections. *Nanomedicine: Nanotechnology, Biology, and Medicine*, *13*(1), 219–230.

Yang, G., Phua, S. Z. F., Bindra, A. K., & Zhao, Y. (2019). Degradability and clearance of inorganic nanoparticles for biomedical applications. *Advanced Materials (Weinheim, Ger)*, *31*(10), 1805730, 1–23.

Yao, C., Wang, W., Wang, P., Zhao, M., Li, X., & Zhang, F. (2018). Near-infrared upconversion mesoporous cerium oxide hollow biophotocatalyst for concurrent pH-/H_2O_2-responsive O_2-evolving synergetic cancer therapy. *Advanced Materials*, *30*(7), 1704833.

Yin, Z. F., Wu, L., Yang, H. G., & Su, Y. H. (2013). Recent progress in biomedical applications of titanium dioxide. *Physical Chemistry Chemical Physics: PCCP*, *15*(14), 4844–4858.

Yun, H., Kim, J. D., Choi, H. C., & Lee, C. W. (2013). Antibacterial activity of CNT-Ag and GO-Ag nanocomposites against Gram-negative and Gram-positive bacteria. *Bulletin of the Korean Chemical Society*, *34*(11), 3261–3264.

Yurt, F., Ocakoglu, K., Ince, M., Colak, S. G., Er, O., Soylu, H. M., et al. (2018). Photodynamic therapy and nuclear imaging activities of zinc phthalocyanine-integrated TiO_2 nanoparticles in breast and cervical tumors. *Chemical Biology & Drug Design*, *91*(3), 789–796.

Zhang, A. P., & Sun, Y. P. (2004). Photocatalytic killing effect of TiO_2 nanoparticles on Ls-174-t human colon carcinoma cells. *World Journal of Gastroenterology*, *10*(21), 3191.

Zhang, F., Wu, C., Zhou, Z., Wang, J., Bao, W., Dong, L., et al. (2018). Blue-light-activated nano-TiO_2@PDA for highly effective and nondestructive tooth whitening. *ACS Biomaterials Science & Engineering*, *4*(8), 3072–3077.

Zhu, P., Weng, Z., Li, X., Liu, X., Wu, S., Yeung, K. W. K., et al. (2016). Biomedical applications of functionalized ZnO nanomaterials: From biosensors to bioimaging. *Advanced Materials Interfaces*, *3*(1), 1500494.

Cancer therapy, immunotherapy, photothermal therapy

3

Genevieve M. Liddle, Jianning Wei and James Hartmann
Florida Atlantic University, Boca Raton, FL, United States

3.1 Introduction

The term "nanotechnology" was created in 1974 by the Japanese researcher Norio Taniguchi. Nanotechnology, a technological study that deals with matter between the scale of 1–100 nm, is a rapidly developing multidisciplinary field that integrates biology, chemistry, physics, and mechanical engineering (Taniguchi, 1974). The versatility of nanotechnology in the scientific field provides advancements in nanomedicine as seen in cancer therapy, immunotherapy, and photothermal therapy. These advancements are due in part to specific types of heterogeneous nanocatalysts called metal oxides. Metal oxides are compounds that have at least one oxygen atom and an element such as iron, titanium, zinc, cerium, copper, silicon, graphene, etc (Vinardell & Mitjans, 2015). In addition, metal oxides can be composed of transitional metals such as tungsten, molybdenum, niobium, antimony, vanadium, etc. (Ogata et al., 2005).

Key physicochemical properties of nanoparticles (NPs) include size, surface energy, physicochemical stability, chemical composition, and shape (Beberwyck, Surendranath, & Paul Alivisatos, 2013; Xu, Liang, Yang, & Yu, 2018). The smaller the NP's diameter, the larger its surface area to volume ratio. A high value of surface area to volume ratio allows for a greater number of atoms to be located on the surface making NPs highly reactive (Xu et al., 2018). NPs exist far from an equilibrium state because of their high surface energy (Andrievski, 2003). This causes both positive and negative characteristics that give NPs the dual nature of poor stability and high reactivity (Andrievski, 2003). Due to their high reactivity, NPs can be processed via redox reactions (Akhavan et al., 2012; Stankovich et al., 2007), galvanic exchanges (Anderson & Hill, 2002; Xia, Wang, Ruditskiy, & Xia, 2013), ion exchanges (Anderson & Tracy 2014; Beberwyck et al., 2013), etc. As a result of these processes, NPs are more applicable to nanomedical applications since their surfaces can be functionalized, thus modifying the NPs to have improved stability (Zhang et al., 2008), to be highly sensitive (Chen et al., 2013; Peng, Liang, & Calderon, 2019), and to have rapid dynamic responses (Xu et al., 2018). In many therapeutic applications, functionalized NPs have their surface conjugated with ligands that are tumor specific thus enabling targeted drug delivery to cancer cells (Ediriwickrema & Saltzman, 2015). NPs can be conjugated with monoclonal antibodies to prevent unwanted immunological responses (Moyano, Liu, Peer, & Rotello, 2016) and for the use as immunotherapy agents to treat primary tumors and secondary metastasis

Metal Oxides for Biomedical and Biosensor Applications. DOI: https://doi.org/10.1016/B978-0-12-823033-6.00028-4

(Surendran, Moon, Park, & Jeong, 2018). Moreover, NPs can be implemented in gene therapy as carries of foreign nucleic acids that encode tumor-killing proteins (Mokhtarzadeh et al., 2016). Shape has a significant impact on NP performance since it plays an important role in therapeutic delivery processes such as particle adhesion, particle dispersal, and cell internalization (Salata, 2004). Most NPs have a spherical shape. However, advancements in nanofabrication techniques have yielded NPs with unique geometrical properties such as nanorods (Huang, El-Sayed, Qian, & El-Sayed, 2006), nanochains (Sardar & Shumaker-Parry, 2008), nanoworms (Park et al., 2009), nanonecklaces (Dai, Worden, Trullinger, & Huo, 2005), filomicelles (Oltra et al., 2013), etc.

Currently, the clinical application of NPs has greatly impacted the detection, treatment, and prevention of cancer. This is possible through the ability to manipulate and utilize the surface structure of NPs to design therapeutic agents. The capabilities of metal oxide NPs as therapeutic agents are promising since their unique structure can undergo various reaction which include redox reactions (Grimaud et al., 2017), dehydration (Kostestkyy, Yu, Gorte, & Mpourmpakis, 2014), dehydrogenation (Heracleous & Lemonidou, 2010), and isomerization (Anderson & Tracy, 2014). Another advantage of metal oxide NPs is the ability to tailor their surfaces with specific ligands and antibody conjugates to target cancerous cells (Huang et al., 2006; Vinardell & Mitjans, 2015). In addition, metal oxide NPs have photoreactive properties which have encouraged the development of phototherapy. This therapy employs near-infrared irradiation techniques that can noninvasively target and ablate cancer cells (Estelrich & Busquets, 2018).

3.2 Cancer biology and nanoparticles

Engineering NPs for cancer therapy remains a challenge since biological barriers need to be seriously considered. Critical issues such as kidney filtration and immune clearance via the reticuloendothelial system affect the efficacy and mode of drug administration (inhalation, oral, intravenous, etc.) at the organ and cellular level (Gonçalves et al., 2013). To go to the cellular level, NPs need to be transported from the bloodstream and be able to cross tissue and cell membranes. Many advancements have been made to improve the knowledge of cancer biology including the tumor environment, tumor signaling pathways, and metastatic growth (Munn & Mellor, 2016). However, the tumor environment is a complexity of heterogeneous tumor cells, immune cells, and neovacularization. A variety of drug-resistant mechanisms within cancer cells makes it difficult to optimize current cancer therapies across multiple cancer types. In the tumor environment, the development of new blood vessels, or angiogenesis, is essential for tumor growth and progression (Forster, Harriss-Phillips, Douglass, & Bezak, 2017). When tumor cells do not have access to blood vessels, the oxygen and nutrient content supplied to the tumor decreases. Decreased oxygen levels are called hypoxia. Hypoxic environments trigger tumor cells and nearby cells to start producing growth factors that stimulate blood vessel formation (Al Tameemi, Dale, Al-Jumaily, & Forsyth, 2019). By increasing the surrounding vasculature of the tumor environment, the tumor can overcome nutrient diffusion limitations. In addition, vascularization increases internal pressure which leads to an outward interstitial flow of fluid (Munson & Shieh, 2014). This results in decreased drug diffusion into the tumor environment. However, drugs and NPs that do gain interstitial access to the tumor environment have higher retention than seen in normal tissues (Bogdan, Pławińska-Czarnak, & Zarzyńska, 2017). This phenomenon is known as enhanced permeability and retention (EPR) effect and allows for selective concentrations of NPs (>7 nm) to accumulate in the tumor tissues (Golombek et al., 2018). From the biological standpoint, the instability of tumors can be attributed to the accumulation of tumor growth, increased angiogenesis, increased internal pressure, and a dense interstitial environment. These instabilities lead to transport barriers for NPs. Thus, NP internalization in the tumor remains a challenge and is not consistent to all areas of the tumor environment. As such, alternative strategies for targeting cancer cells with NPs need to be considered.

3.3 Metal oxides for cancer therapy

Current cancer treatments involve alkylating agents, antimetabolites, biological agents, etc (Rasmussen, Martinez, Louka, & Wingett, 2010). However, a principle issue of these cancer treatments is systematic toxicity which is a side effect due to difficulties in differentiating between tumor cells and healthy cells. Metal oxide NPs have been explored as drug delivery vehicles for the inhibition of tumor formation, development, and progression (Felice, Prabhakaran, Rodríguez, & Ramakrishna, 2014). Their anticancer effects are either related to intrinsic features such as antioxidant activity or dependent on external stimuli such as hyperthermia in response to infrared rays or magnetic fields as seen with phototherapy (Vinardell & Mitjans, 2015). In general, NPs can treat cancer either passively or actively. Passive treatment can include taking advantage of EPR such as leaky vasculature in the tumor environment which enables diffusion of NPs into cancer tissues at limited amounts (Attia et al., 2019). However, passive treatment can have side effects since inflamed tissues have leaky vasculature similar to the tumor environment resulting in faulty targeted drug delivery (Attia et al., 2019). Active treatments with NPs can improve tumor targeting via functionalization of the NP's surface with biomolecules and ligand receptors that significantly increase targeted tumor killing and drug delivery (Vinardell & Mitjans, 2015). Currently, metal oxide NPs have been implemented in the direct killing of tumor cells in vitro and in vivo. The results of a variety of metal oxide NPs treatments for cancer therapy are presented below and their mechanism of action is considered.

3.4 Iron oxides for cancer therapy

Supermagnetic iron oxide NPs have dual imaging and therapeutic capabilities for noninvasive detection and management of tumors (Veiseh, Gunn, & Zhang, 2010). In addition, iron oxide NPs that are conjugated with tumor-specific targeting ligands have multifunctional treatment modalities for the specific delivery of anticancer drugs to tumors while reducing the severe side effects associated with the accumulation of NPs in healthy tissues (Nasongkla et al., 2006). Cross-linking polymeric coatings to functionalize NPs with amine or carboxyl surface groups are used to conjugate fluorescent dyes and drugs. These polymeric coatings allow for imaging of iron oxide NPs in a noninvasive way as well as improve drug homing and efficacy (Sosnovik, Nahrendorf, & Weissleder, 2008). However, iron oxide NPs are compromised in terms of solubility in aqueous media due to having multimodality in magnetic and fluorescent capabilities (Santra, Kaittanis, Grimm, & Manuel Perez, 2009) as well as multifunctionality for imaging and therapeutic aspects (Yu et al., 2008). In addition, a reduced number of available functional ligand groups used for attaching ligands are seen (Santra et al., 2009). To solve these issues, coencapsulation with a lipophilic near-infrared dye along with hydrophobic anticancer drugs pocketed in the polymeric matrix of poly(acrylic) acid is used to coat iron oxide NPs (Santra et al., 2009). A green chemistry approach is used to synthesize these NPs and includes five components: (1) Taxol, an encapsulated chemotherapeutic agent used for cancer therapy, (2) a folic acid ligand for surface functionality and for specific targeting of cancer cells, (3) the use of click chemistry-based conjugation for targeting ligands, (4) fluorescent imaging capabilities via encapsulation of near-infrared dye, and (5) a superparamagnetic iron oxide core that is compatible with MRI (Santra et al., 2009) (Fig. 3.1).

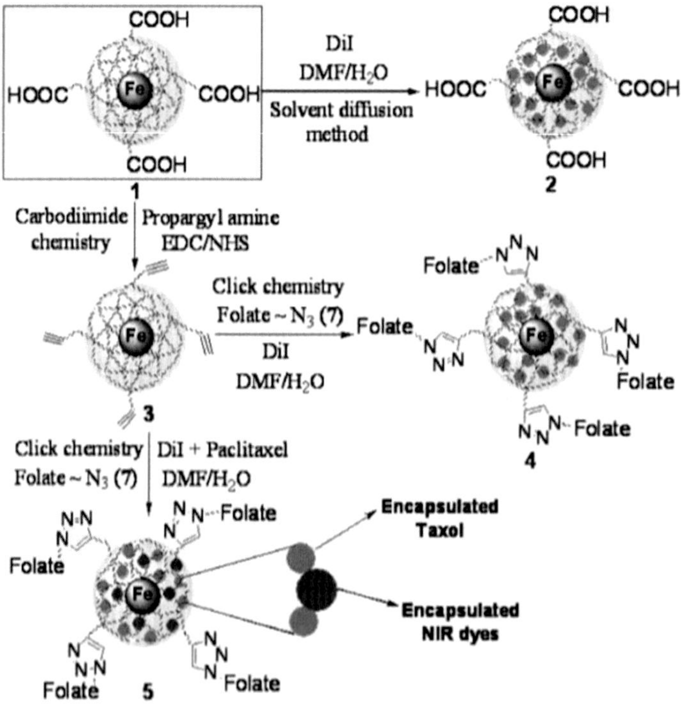

Figure 3.1 Schematic representation of multimodal and multifactorial iron oxide nanoparticles (NPs). Click chemistry and carbodiimide chemistry are used for the synthesis of functional iron oxide NPs. Near-infrared dyes and panclitaxel coencapsulation iron oxide NPs are prepared in water using the modified solvent diffusion method (Santra et al., 2009).

Iron oxide NPs are advantageous for cancer treatment since they have low toxicity, are biocompatible, and are biodegradable (Gobbo et al., 2015; Martinkova, Brtnicky, Kynicky, & Pohanka, 2018). Iron oxide NPs are typically injected intravenously since the most effective way to reach targeted organs is via the blood stream (Chertok et al., 2008; Elgqvist, 2017). Thus the distribution of iron oxide NPs in vivo is strongly related to particle size and shape. For example, particles larger than 50 nm passively target the reticuloendothelial system whereas particles smaller than 50 nm have low uptake by phagocytic immune cells and no degradation nor accumulation in the reticuloendothelial system (Elgqvist, 2017; Wahajuddin & Arora, 2012). Targeted distribution of iron oxide NPs can be achieved using three different approaches: passive, active, or magnetic (Yigit, Moore, & Medarova, 2012). Passive targeting does not require any further surface functionalization nor modification and depends on particle size for passive diffusion into leaky tumor capillaries (Elgqvist, 2017). With active targeting, particles are functionalized with ligands that bind to receptors at the targeted tumor site (Danhier, Le Breton, & Préat, 2012). For magnetic targeting, an external magnetic field is applied over the targeted tumor causing a magnetic field gradient suitable for iron oxide NP accumulation within the target tissue (He et al., 2013). Thus magnetic targeting of iron oxide NPs is an ideal tool for *in vivo* cancer diagnosis and cancer therapy (Wahajuddin & Arora, 2012).

3.5 Titanium dioxides for cancer therapy

Titanium dioxide NPs are typically used in photo and sonodynamic processes due to their unique physical and electrochemical properties (Kim et al., 2011). The NPs have a large surface area capable of absorbing UV radiation, resulting in greater photocatalytic activity. Thus titanium dioxide NPs can be utilized as radio sensitizing agents in radiotherapy where ionizing radiation is used to target tumor cells (Youkhana, Feltis, Blencowe, & Geso, 2017). Radio sensitizing agents have the potential to target cancer cells since they have a high Z or atomic number. Tumors incorporated with high Z agents have different absorbance properties compared to healthy surrounding tissues which may partially spare healthy tissues from damage during radiation (Youkhana et al., 2017).

In addition to radiotherapy, titanium dioxide NPs are utilized in sonodynamic therapy, a method in which sonosensitizers and ultrasound are used. Ultrasound is more effective than UV radiation since it can target deeper into the tumor that has incorporated NP sonosensitizers (Ninomiya, Fukuda, Ogino, & Shimizu, 2014). Sonosensitizers activated with ultrasound induce reactive oxygen species production in targeted tumors cells leading to cytotoxicity (Ninomiya et al., 2014). To prevent degradation of titanium dioxide NPs by reactive oxygen species produced by sonodynamic therapy, the NPs are hydrophilized with a carboxymethyl dextran polymers (You et al., 2016). Carboxymethyl dextran polymers enhance the stability of titanium dioxide NPs, leading to increased the titanium dioxide NPs time in blood circulation, allowing for titanium dioxide NPs extravasation into the tumor, and increasing reactive oxygen species levels with ultrasound treatment (You et al. 2016). Polyethylene glycol (PEG), another polymer tcommonly used in medical applications, is used to stabilize titanium dioxide NPs encapsulated inside micelles, while improving biocompatibility (Harada, Ono, Yuba, & Kono, 2013). Thus, titanium dioxide NPs encapsulated inside micelles is another method used with ultrasound energy to generate reactive oxygen species which leads to tumor cytotoxicity (Harada et al., 2013).

3.6 Zinc oxides for cancer therapy

Zinc oxide NPs are known to be widely used in cosmetic products and dental-related materials (Abdalla et al., 2018; Guimarães et al., 2018). Recently, zinc oxide NPs are shown to target multiple cancer types such as prostate, lung (Hassan et al., 2017), head and neck squamous cell carcinoma (Gehrke et al., 2017), colorectal adenocarcinoma, and lymphoblastoid cancer (Yin, Casey, Mccall, & Fenech, 2015). The anticancer effects of zinc oxide NPs have multiple modes of action which include inhibition of cancer proliferation, increasing tumor sensitivity to anticancer drugs, preventing metastasis, and restoring immunosurveillance against cancer cells (Wang, Lee, Kim, & Zhu, 2017).

Mitophagy is a physiological process that maintains the overall health of the mitochondrial organelles, the providers of energy to cells (Van Der Bliek, Sedensky, & Morgan, 2017). In addition, mitochondrial organelles respond to and provide adaptation to cellular stress during human development and throughout life (Van Der Bliek et al., 2017). Cancer cells exhibit a disparity in mitophagy from which neoplastic progression of tumor cells as well as drug resistance arises (Qian, Chao, & Ding, 2018; Yang & Suda, 2018). Zinc oxide NPs reverse this disparity in mitophagy by significantly inducing autophagy or "self-eating" of cancer cells via the production of reactive oxygen species (Bai et al., 2017). In addition, zinc oxide NPs maintain the activity of the tumor suppressor gene p53 which is a gene responsible for regulating apoptosis activity (Ng et al., 2011). It has also been reported that the size of zinc oxide NPs, which is around 50 nm, can exert toxic effects on cancer cells by inhibiting autophagic flux, leading to cancer cell death (B. Wang et al., 2018).

Zinc oxide NPs have been shown to have enhanced permeation and retention within tumor cells. This effect is due to their size and surface properties allowing diffusion through blood vessels and localization in tumor cells (Akhtar et al., 2012). Zinc oxide NPs have a positively charged surface in physiological conditions seen in fluid and blood whereas cancer cells have a negatively charged phospholipid outer membrane leading to an electrostatic attraction between the NPs and the cancer cells (Thurber et al., 2012).

Zinc oxide NPs are known to have immunomodulatory effects which are beneficial for the treatment of cancer. Zinc oxide NPs increase antigen-specific immune reactions via enhancement of antigen-specific antibodies found in blood serum (Roy et al., 2014). These antibodies include immunoglobulin G and immunoglobulin E which are both involved in enhancing the functionality of certain immune cells in reversing the immunosuppressed tumor environment. In addition, zinc oxide

NPs have an adjuvant effect toward the T helper 2 cell response by affecting the production and activation of specific cytokines involved in immune activation and suppression (Roy et al., 2014).

3.7 Transition metal oxides for cancer therapy

Among the various NPs that have proven anticancerous effects, the properties of transition metal oxides are slowly gaining credibility as antitumor therapeutics. In the EU, spherical iron (III) oxide NPs are utilized for magnetic tumor hyperthermia therapy for brain and prostate cancers (Gamarra et al., 2011; Johannsen, Thiesen, Wust, & Jordan, 2010). Cerium oxide NPs are reported to increase oxidative stress and apoptosis in cancer cells without harming normal tissues (Neri & Supuran, 2011; Tarnuzzer, Colon, Patil, & Seal, 2005). Copper (II) oxide NPs synthesized from plant extracts have apoptotic and cytotoxic effects on lung and breast cancer cell lines (Sankar et al., 2014; Sivaraj et al., 2014). In another study, three transition metal oxide NPs, copper oxide, nickel oxide, and ferric oxide were synthesized and characterized for their anticancerous activity against lung cancer cell lines in hypoxic and normoxic conditions (Pandey, Dhiman, Srivastava, & Majumder, 2016). Hypoxia is a relevant feature of tumor environments that drives metabolic and epigenetic modifications in tumor cells. Thus mimicking the tumor environment expands the understanding of the therapeutic potentials that transition metals have in cancer therapy (Pandey et al., 2016). It was seen that copper oxide contributed to cytotoxicity of cancer cells via reactive oxygen species under normoxic and hypoxic conditions (Pandey et al., 2016). Nickel oxide and ferric oxide did not induce reactive oxygen species production but did negatively affect cancer cell viability under normal and hypoxic conditions (Pandey et al., 2016).

Polyoxometalates are negatively charged metal oxide cluster compounds that have early transitional metal ions such as tungsten, molybdenum, niobium, antimony, and vanadium (Ogata et al., 2005). Several polyoxometalates are reported to inhibit the replication of human immunodeficiency virus (Kim, Judd, Hill, & Schinazi, 1994), herpes simplex virus (Dan et al., 2003), and methicillin-resistant *Staphylococcus aureus* (Fukuda, Yamase, & Tajima, 1999). One polyoxometalate that has been recognized as significantly antitumoral is PM-8. This transitional metal oxide inhibits tumor growth through DNA ladder formation and DNA fragmentation. Thus the antitumor activity of PM-8 is the activation of apoptotic pathways in tumor cells (Ogata et al., 2005).

3.8 Metal oxides for immunotherapy

3.8.1 Involvement of innate and adaptive immune responses in immunotherapy

Immunotherapy is an essential branch of medicine that involves the induction, enhancement, suppression, or modulation of the immune response. The immune system is composed of lymphocytes [T cells, B cells, dendritic cells (DCs), natural killer (NK) cells, etc.] along with the corresponding organs and tissues of the lymph system. This system is proficient in recognizing and neutralizing foreign antigens or exogenous molecules through an array of mechanisms that involve the innate and adaptive immune subsystems (Chaplin, 2010). Both subsystems work closely together when triggered by foreign antigens, exogenous molecules, or cancer cells. The innate immune system provides a general, nonspecific defense against harmful antigens and is composed of NK cells and phagocytes or "eating cells" such as macrophages, DCs, etc. The adaptive immune system delivers a "specific" immune response via T cells and B cells. T cells consist of $CD8^+$ killer cells and $CD4^+$ helper T cells. $CD8^+$ killer cells react to a foreign peptides presented on the surface of an altered host cell thereby leading to the host cell's death. In contrast, $CD4^+$ helper T cells indirectly kill by producing signal molecules to activate phagocytic cells to engulf and destroy invading pathogens to stimulating other lymphocytes such as $CD8^+$ cells and B cells. B cells secrete antibodies that specifically bind to the antigens present on invading pathogens. Antigens that are identified via antibody binding result in antigen–antibody complexes. The antigen–antibody complexes are recognized by Fc receptors present on the surface of multiple types of lymphocytes which lead to activation of the lymphocyte and elimination of the antigen–antibody complex (Chaplin, 2010).

3.9 Nanoparticles elicit immune responses by delivering targeted antigens

Immunotherapy is used as an alternative strategy for cancer treatments since conventional chemotherapy has deleterious effects on both cancer cells and healthy cells. NPs make suitable immunotherapeutic agents since they have a tendency to be internalized by a variety of cell types (Kostarelos et al., 2007; Kovacsovics-Bankowski & Rock, 1995). The ability to be internalized makes NPs suitable for delivering antigens into antigen-presenting cells including DCs, macrophages, and B cells (Fig. 3.2). Antigen-presenting cells mediate the immune response via internalizing antigens and presentation of peptides derived from the antigen to T cells resulting in an adaptive immune response. Antigens encapsulated into NPs are effective in cancer immunotherapy for multiple reasons: (1) NPs prevent antigen degradation by proteolytic enzymes (Shen, Reznikoff, Dranoff, & Rock, 1997), (2) the surface area and functionality of NPs increase the efficacy of antigen loading (Nakaoka, Tabata, & Ikada, 1996; Waeckerle-Men & Groettrup, 2005), (3) NPs prolong the release of the antigen, enhancing antigen presentation to the T cells via MHC-peptide presentation (Audran et al., 2003; Prasad et al., 2011), and (4) NPs can reduce the need for adjuvants (Mutis et al., 1999; Prasad et al., 2011) (Fig. 3.2). A variety of metal oxide NPs used for immunotherapy will be discussed below.

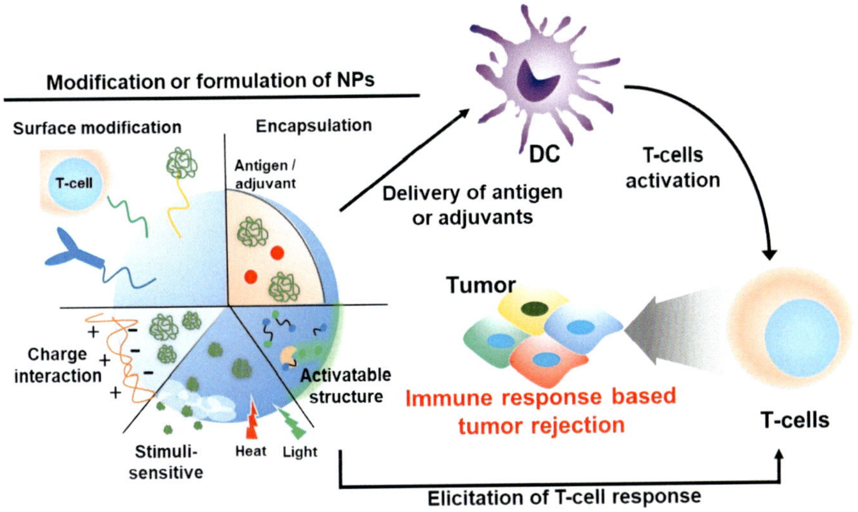

Figure 3.2 Illustration of the multiple surface modifications and formulations that metal oxide nanoparticles (NPs) can undergo to be used as immunotherapeutic agents. NPs delivery to dendritic cells (DCs) is a well-known method to activate a T-cell response against tumor cells in immunotherapy (Yoon et al., 2018).

3.10 Cobalt oxides as immunotherapeutic agents

Multiple studies have shown that vaccines containing antigen-encapsulated NPs have antitumor effects via antigen presentation to immune cells in mouse and human cancer models (Livingston et al., 1994; Mutis et al., 1999; Vujanovic & Butterfield, 2007; Waeckerle-Men & Groettrup, 2005). One example is cobalt, a transition metal that can easily enter cells (Papis et al., 2009) since it plays a physiological role as a cofactor of vitamin B_{12}. Phosphonomethyliminodiacetic acid (PMIDA) conjugated to cobalt oxide NPs can bind cancer lysates to stimulate anticancer immune responses by activating macrophages, T cells, and NK cells to produce tumor necrosis factor-α, a cytokine that suppresses tumor growth (Chattopadhyay et al., 2013). PMIDA-cobalt oxide NPs activate macrophages are as evidenced by increased levels of both IFN-γ and tumor necrosis factor-α (Chattopadhyay et al., 2016). Mice immunized with PMIDA-cobalt oxide NPs loaded with cancer cell lysates have induced anticancer IgG antibody production via B cells. The resulting antibody-dependent cellular cytotoxicity responses using the cancer lysate was greater than in the control groups (Chattopadhyay et al., 2016). Whole tumor lysates are often used in cancer immunotherapy since the presence of multiple antigens reduces the risk of immune cell evasion. Additionally, identification of effective tumor-specific antigens is not required with whole tumor lysates. Thus NPs loaded with cancer cell lysates is a feasible choice of treatment for cancers with either tumor-specific and/or tumor-associated antigens (Chattopadhyay et al., 2013; Mutis et al., 1999).

3.11 Hydroxides as immunotherapeutic agents

Indole amine 2,3-dioxidase (IDO) is an intracellular enzyme that metabolizes tryptophan into kynurenine, a molecule that suppresses T and NK cells. Thus, IDO is an immunosuppressive protein that allows tumor growth and metastasis (Munn & Mellor, 2016). IDO is upregulated in the tumor environment, where high levels enhance the differentiation of $CD4^+$ T cells into tumor-promoting T cells (Jenkins, Barbie, & Flaherty, 2018). Activated T cells secrete interferon γ which leads to higher expression of the immunosuppressive programmed death-ligand 1 on tumor cells and macrophages as well as upregulated IDO activity (Sharma, Hu-Lieskovan, Wargo, & Ribas, 2017). Layered double hydroxide NPs have been used to codeliver IDO inhibitors along with a cisplatin prodrug, an immune chemotherapeutic drug (N. Wang et al., 2018). Layered, double hydroxide NPs are hexagonal and brucite-like flakes that consist of multilayers made up of magnesium hydroxides and aluminum hydroxides (Wang et al., 2015). Each layer is positively charged which allows the NPs to make electrostatic complexes with anionic molecules (Wang et al., 2015). Thus, a negatively charged cisplatin prodrug and IDO inhibitors can be loaded into hydroxide NPs by intercalating into the gaps of each layer. An *in vitro* study of IDO inhibitor-loaded hydroxide NPs confirmed the downregulation of cellular kynurenine levels in human cervical cancer HeLa cells (Wang et al., 2015). In addition, HeLa cells treated with the NPs and cocultured with blood monocytes increased T cell proliferation and reversed tumor-induced immune suppression (Wang et al., 2015).

3.12 Iron oxides as immunotherapeutic agents

DCs are involved in initiating and regulating adaptive immune responses (Seong & Matzinger, 2004) and have been successfully used as therapeutic vaccines against human cancers (Bae, Cho, & Seong, 2009; Palucka & Banchereau, 2012; Palucka, Karolina, Fay, & Banchereau, 2007). In order for dendritic cells to induce an antitumor response, they must phagocytose the target antigen, travel to the lymph nodes, and present the peptides to T cells (De Vries et al., 2005). Their migration can be tracked using magnetic resonance imaging (MRI) by employing superparamagnetic iron oxide NPs (De Vries et al., 2005). Iron oxide NPs have exceptional image contrast in comparison to lymphoid tissue and good signal-to-noise ratios in vivo (De Vries et al., 2005). In clinical applications, iron oxide NPs must be surface modified with a water-soluble polymer such as dextran so that they are biocompatible (De Vries et al., 2005). Phagocytosis of antigens via DCs require incubation times greater than 6 hours with the use of transfection agents and surface modification (Kunzmann et al., 2011; Martin et al., 2008; Rogers, Meyer, & Kramer, 2006). However, transfection agents are in general cytotoxic and have a narrow range of nontoxic concentrations that need to be optimized for clinical settings (Rogers et al., 2006). To overcome cytotoxicity issues, disulfonated indocyanine green optical probes and iron oxide NPs can be coated with poly(lactide-co-glycolide) which improves biocompatibility without impeding optical imaging (Lim et al., 2008). NPs consisting of a superparamagnetic iron oxide core can be covered with a photonic zinc oxide shell to eliminate the need for transfection agents (Cho et al., 2011). These zinc-shelled iron oxide NPs are effectively taken up by DCs and imaged using confocal microscopy or MRI techniques. The NPs also have zinc oxide-binded peptides that are effective for stabilizing the loaded antigens (Cho et al., 2011). Developing a clinically relevant DC-based immunotherapeutic can be challenging since a sufficient amount of antigens need to be phagocytosed by the DC to generate potent $CD4^+$ T-cell and cytotoxic $CD8^+$ T-cell responses which collaborate to kill tumor cells (Gilboa, 2007). Thus NPs make for optimal candidates for delivering antigens to DCs since their large surface area allows for the delivery of multiple therapeutic agents through surface modification (Klippstein & Pozo, 2010; Sun, Lee, & Zhang, 2008).

3.13 Graphene oxides as immunotherapeutic agents

Graphene is a one-atom-thick two-dimensional layer of sp^2-bonded carbon that was discovered in 2004 (Chung et al., 2013; Feng, Wu, & Qu, 2013). Graphene oxide NPs are highly oxidized, chemically modified graphene that are structured with a single-atom carbon sheet composed of carboxylate groups on the periphery (Ni et al., 2012). Thus the location of the carboxylate groups provides a pH-dependent, negatively charged surfacc as well as colloidal stability (Ni et al., 2012). Graphene and its derivatives, graphene oxide and reduced graphene oxide have been used in biomedical applications as nanocarriers for molecular delivery of anticancer drugs and immune stimulators since their large surface area can be functionalized with multiple molecular variants (Sun et al., 2008). In terms of the immunotherapeutic aspects of graphene, it has been reported to activate macrophages and trigger proinflammatory cytokine production, making graphene oxide NPs a good immune adjuvant candidate (Chen et al., 2012; Zhou et al., 2012). Graphene oxide NPs are relatively safe therapeutic agents, up to 250 ug when administered in vivo in mice (Sanchez, Jachak, Hurt, & Kane, 2012). Commercial chemotherapeutics agents such as 5-fluorouracil and doxorubicin are typically slowly released in the tumor environment (Sanchez et al., 2012). These drugs benefit from local administration of graphene oxide NPs since they can ensure that chemotherapeutic agents and immune stimulating cytokines remain in high concentration for longer periods of time in the tumor environment (Zhang et al., 2010). Graphene oxide NPs can also bind to single-stranded DNA to protect it from degradative enzymes (Sanchez et al., 2012) which is beneficial for conjugating adjuvant such as lipopolysaccharides and CpG oligonucleotides in order to reverse the tumor suppressive environment and elicit a robust $CD8^+$ T-cell response (Jarnicki et al., 2008; Vicari et al., 2002).

Key immune cells such as T cells, DCs, macrophages, and NK cells recognize targeted cancer cells and communicate with other immune cells through an organization of membrane proteins called immunological synapses (Davis & Dustin, 2004; Dustin, Chakraborty, & Shaw, 2010). These immunological synapses involve nanoscale membrane molecules of 10–400 nm which correlate with cell signaling and nanostructural changes that drive immune cell activation (Delcassian et al., 2013; Oszmiana et al., 2016). Acellular soluble reagents that exploit membrane receptor nanoclustering have been created to activate immune cells (Loftus, Mezida Saeed, Davis, & Dunlop, 2018) and consist of artificial immunoreceptor nanoclusters that mimic leukocyte-stimulating ligands in terms of size and molecular number (Loftus et al., 2018). These artificial ligand molecules are mounted on planar graphene oxide NPs that have an increasing size (>50 nm) and no sharp curvatures due to their planar shape to help prevent immunogenic reactions (Choudhuri et al., 2005). One of the best characterized NK-activating receptors is CD16 which recognizes and binds to the Fc portion of antibodies that target and bind to targeted cancer cells (Mandelboim et al., 1999). As a result of CD16 and Fc binding, NK cells induce antibody-dependent cellular cytotoxicity as well as cytokine secretion (Mandelboim et al., 1999). In the same way, graphene oxide NPs can be functionalized with artificial nanoclusters that specifically bind with human NK cells via CD16 (Loftus et al., 2018). Thus these artificial nanoclusters mounted to graphene oxide NPs function as activating reagents to enhance NK cell's cytotoxic catalytic response to targeted cancer cells (Loftus et al., 2018).

IL-10 is an antiinflammatory cytokine that regulates the immune response to self or foreign antigens (Saraiva & O'Garra, 2010). IL-10 messenger RNA and proteins are found in a variety of excised tumors. Upregulation of IL-10 by the tumor cells leads to a negative prognosis of the disease (Saraiva & O'Garra, 2010). In addition to tumor cells, immune cells such as T cells and macrophages present in the tumor environment produce IL-10, leading to further tumor progression (Sato et al., 2011). Graphene oxide NPs absorbed with anti-IL-10 are bioreactive *in vitro* and *in vivo* and are more efficient than free anti-IL-10 receptor antibodies for eliciting a tumor-killing $CD8^+$ T-cell response (Ni et al., 2012). In addition, graphene oxide NPs are able to absorb anti-IL-10 production which can be slowly released into the tumor environment due to the pH-dependent, negatively charged surface of the graphene oxide NP (Ni et al., 2012).

In the past few decades, CpG oligonucleotide therapy has become a well-investigated treatment for cancer (Stahel & Zangemeister-Wittke, 2003). Specifically, therapeutic nucleic acids called unmethylated CpG motifs are known to promote strong immune stimulatory activities (Meng, Yamazaki, Nishida, & Hanagata, 2011). The mammalian immune system recognizes CpG oligonucleotides via toll-like receptor 9, causing production of proinflammatory cytokines such as tumor necrosis factor and interleukin 6 (Klinman et al., 1996). These proinflammatory cytokines then stimulate a cascade of innate and adaptive immune responses which are necessary for targeting cancer cells (Klinman et al., 1996). Although, synthetic CpG oligonucleotides are a favorable tool for immunotherapeutic applications, the delivery of these synthetic nucleic acids to the tumor environment remains a challenge (Cutler, Auyeung, & Mirkin, 2012). Since nucleic acids are negatively charged, they cannot effectively cross the electronegative cell membrane of cancer cells (Cutler et al., 2012). Also, nucleic acids can be degraded by nucleases (Campolongo, Tan, Xu, & Luo, 2010). Thus it is important to develop CpG delivery vehicles that can not only transfer CpG into the targeted tumor cells but also prevent its degradation (Campolongo et al., 2010). Nanobiotechnology has provided effective opportunities for the development of vectors to transport CpG oligonucleotides into targeted cells. These include nanoliposomes (Erikçi, Gursel, & Gürsel, 2011), gold NPs (Tao, Ju, Li, et al. 2014), and carbon nanotubes (Bianco et al., 2005). Although great progress has been made concerning these nanomaterials, nanocomplex materials such as graphene oxide-PEG-polyethyleneimine conjugated to CpG can significantly improve proinflammatory cytokine production and enhance the immunostimulatory response (Tao, Ju, Ren, & Qu, 2014). Graphene oxide can also be thermally heated using near-infrared reflectance (NIR) which enhances the intracellular delivery of CpG oligonucleotides (Tao, Ju, Ren, et al., 2014). Thus graphene oxide nanocomplexes can also be synergistically combined with photothermal therapy *in vivo* to improve tumor reduction (Tao, Ju, Ren, et al., 2014).

3.14 Metal oxides for photothermal therapy

3.14.1 Photothermal therapy for targeting primary tumors and secondary tumor metastasis

One of the greatest challenges in treating cancer is targeting metastase. Cancer metastasis is responsible for 90% of cancer-related deaths across multiple cancer types (Eckhardt, Francis, Parker, & Anderson, 2012; Owonikoko et al., 2014; Schroeder et al., 2012). Metastasis refers to the dissemination of cancer cells from the primary tumor that colonize at distant secondary sites (Schroeder et al., 2012). Unfortunately, metastases that occur in distant organs likely result in incurable and fatal outcomes (Schroeder et al., 2012). Currently, therapies including aggressive surgery, radiotherapy, and chemotherapy are utilized for cancer treatment but have limited effects in targeting metastatic cancers (Okamoto et al., 2011; Park, Zhang, Vykhodtseva, & McDannold, 2012; Ramanlal Chaudhari et al., 2012; Zardavas, Baselga, & Piccart, 2013). Even with the recent progress in therapeutic agents, targeted therapies, and combinational therapies, their therapeutic efficacy on cancer metastasis remains pessimistic with improvements in survivals increasing by only a few months (Devulapally et al., 2015; Fernández et al., 2010). In fact, patients undergoing multimodal treatments for metastatic cancer only have a 5-year survival rate of about 20% (Fernández et al., 2010; Peiris et al., 2012).

Photothermal therapy is a recent and promising strategy that utilizes NIR photothermal agents that generate heat for thermal ablation of cancer cells targeted with NIR laser irradiation (Alkilany et al., 2012; Shanmugam, Selvakumar, & Yeh, 2014; Zhang, Wang, & Chen, 2013). In comparison to conventional therapeutic strategies, photothermal therapy has the potential to provide unique advantages including increased targeted specificity, minimal invasiveness, and precise spatial–temporal selectivity (Hu et al., 2014; Ma et al., 2013; Shanmugam et al., 2014; Yan et al., 2014; Zhang et al., 2013). Photothermal therapy can directly eradicate the primary cancer and may combat the initial stages of nearby secondary metastasis when used in combination with therapeutic modalities that treat metastatic sites (Chen, Liang, Wang, & Liu, 2015; He et al., 2015; Wang et al., 2014). The efficacy of photothermal therapy significantly depends on the transformation of light to sufficient heat using photothermal agents in the nanoscale size (Wang et al., 2014). Presently, photothermal nanotherapeutics including noble metal nanostructures, nanocarbons, transition metal sulfides/oxides nanomaterials, and organic nanoagents have been explored (Okuno et al., 2013; Orecchioni, Cabizza, Bianco, & Delogu, 2015; Wang et al., 2014; Zhang et al., 2013).

Photothermal therapy causes hyperthermia when targeted tissues that have incorperated photothermal agents are exposed to high temperatures compared to healthy, surrounding tissues (Beik et al., 2016). Cancer cells can be directly destroyed via temperatures above 47°C while temperatures 41°C–45°C allows susceptibility to drug treatment via increased blood flow, counteracting the hypoxic environment created by the cancer cells (Lim et al., 2018). Thus photothermal therapy, along with a photothermal agent that can be excited via vibrational energy, leads to light-induced heat release in the targeted cancer tissue. When NPs are used in photothermal therapy, light induces resonance of free electrons through surface plasmon resonance where electrons on the metal surface are excited by photons of incident light at a certain angle of incidence that causes an excitation propagation parallel to the metal surface (Lim et al., 2018). The absorption wavelength of nanomaterials can be tuned to the optical window of NIR wavelengths of 650–900 nm to ensure deep penetration into the cancer tissues (Weissleder, 2001). Thus the involvement of metal oxides as photothermal agents for the treatment of primary tumors and metastasis will be presented below.

3.15 Graphene oxides as photothermal agents

NIR phototherapy based on NPs offers an encouraging therapeutic strategy that ablates tumors with minor injury to healthy surrounding tissue. When combined with chemotherapy, there is improved tumor killing and reduced drug resistance within the targeted tumor (Banu et al., 2015; Lee, Park, & Yoo, 2010; Tang et al., 2010). Nanographene oxide is a two-dimensional material containing a variety of reactive oxygen functional groups including epoxy and hydroxyl on the basal plane whereas carboxylic acid groups occupy the sheet edges (Cai et al., 2008; Lerf, He, Forster, & Klinowski, 1998). Much work has been done using combinational targeted therapy with chemotherapy and graphene oxide acting as a nanocarrier (Shen et al., 2012; Yang et al., 2012; Zhang et al., 2010). In addition, multifunctional nanographene oxide sheets have been implemented for targeted chemo-photothermal therapy (Qin et al., 2013). An example is polyvinylpyrrolidone (PVP), a nonionic, biocompatible polymer surfactant that is typically used as a stabilizing agent and dispersant during the synthesis of metal nanostructures (Tan et al., 2009). PVP-functionalized nanographene oxide can be fabricated in a noncovalent – stacking interaction (Tan et al., 2009). Folic acid molecules can be covalently bonded with the PVP nanographene oxide for targeting folate receptors overexpressed on the surface of numerous cancer cells (Kaaki et al., 2012; Zhou et al., 2009). Folic acid–PVP nanographene oxides exhibit an ultrahigh loading ratio for the anticancer drug doxorubicin which allows delivery of both heat and drugs to the targeted tumor regions (Zhang et al., 2010). Graphene oxide can be functionalized with PEG for loading the photothermal agent chlorin e6 via – stacking (Tian et al., 2011). Graphene-based photothermal delivery significantly increases delivery of chlorin e6 which enhances the cancer killing efficacy in combination with NIR (Tian et al., 2011). In addition, graphene has a high NIR absorption that can "cook" cancer cells at temperatures above 50°C (Robinson et al., 2010). Graphene NPs with high NIR absorption can also produce local heating under mild NIR laser irradiation of around 43°C which improves PEG-coated graphene oxide uptake into the tumor by twofold in combination with chlorin e6 (Tian et al., 2011).

In order to scale up production of graphene, a chemical exfoliation method using a strong reluctant such as hydrazine was applied for the synthesis of graphene oxide suspensions (Liao, Lin, MacOsko, & Haynes, 2011). However, for photothermal therapy purposes, hydrazine is highly toxic and is not biocompatible (Liao et al., 2011). As an alternative to prevent aggregation of reduced graphene oxide while utilizing a biocompatible reduction process at a neutral pH, glucose has been used for green reduction as well as the functionalization of graphene oxide (Akhavan et al., 2012). Graphene oxide reduction and functionalization by glucose in the presence of an iron catalyst has an improved NIR photothermal killing effect on prostate cancer cells compared to hydrazine-reduced graphene oxide (Akhavan et al., 2012). In addition, the functionalization of graphene oxide by gluconate ions prevents suspension aggregation and has reduced cytotoxic effects on healthy cells compared to hydrazine-reduced graphene oxide (Akhavan et al., 2012). Thus glucose-reduced graphene sheets can be used as a biocompatible alternative for graphene oxide functionalization instead of PEG polymers (Akhavan et al., 2012).

3.16 Titanium dioxide as photothermal agents

Titanium dioxide is a promising photocatalyst that is stable in various solvents under photoirradiation and is an inexpensive material that is simple to synthesize in a laboratory setting (Ivanković, Gotić, Jurin, & Music, 2003). Photocatalytic studies with cancer cells have been done with commercially available P25 titanium dioxide NPs which are nonporous 75:25 anatase-to-rutile mixture with crystallite sizes of around 30 nm in 200–400 aggregates (Ivanković et al., 2003). A main disadvantage of using pure titanium dioxide for photochemical applications is its low absorption to visible light (Ivanković et al., 2003). Thus doping titanium dioxide with iron extends the absorption threshold to visible light (Ivanković et al., 2003). In contrast, substitutional iron doping of titanium dioxide further increases the charge-carrier lifetime from nanoseconds for pure titanium dioxide to minutes or hours for iron-doped, nonpure titanium dioxide (Huang, Xu, Yuan, & Yu, 1997). It was found that the synthesis procedure strongly correlates with the distribution of the iron in the titanium dioxide matrix. Samples with increased uniformity in iron distribution in titanium dioxide had enhanced photocatalytic activity (Wang, Böttcher, Bahnemann, & Dohrmann, 2003). Thus iron-doped titanium dioxide in the presence of UV irradiation can produce hydroxyl radials that are toxic to tumor cells (Ivanković et al., 2003).

3.17 Iron oxides as photothermal agents

There is significant interest in utilizing physical forces such as magnetic fields as a novel tumor targeting approach (He et al., 2013; Mikhaylov et al., 2011). Magnetic fields and magnetic NPs may deliver therapeutic agents and improve therapeutic efficacy toward targeted tumors (Ghosh et al., 2012; Ma, Cheng, & Chen, 2015; Mathieu & Martel, 2009). Magnetic targeting strategies have been developed to enhance the accumulation of NPs in the sentinel lymph nodes that host metastatic tumor cells coming from a nearby primary tumor (Veronesi et al., 1996). One strategy involves gold-shelled iron oxide nanoclusters coated with PEG which provides stable molecular structure in physiological solutions (Liang et al., 2015) (Fig. 3.3). Following intratumoral injection delivery, modified iron oxide NPs can be monitored by magnetic resonance to follow their active traveling through the lymphatic system to the sentinel lymph nodes where they accumulate (Liang et al., 2015). After accumulation, an 808 nm NIR laser is used to photothermally ablate the primary tumor and the sentinel lymph nodes. The result is inhibition of lymphatic cancer metastasis and increased patient survival time (Liang et al., 2015).

3.18 Manganese oxides as photothermal agents

Manganese oxide NPs attached to carbon nanotubes have been used for photothermal therapy and as a contrast agent for MRI applications (Wang et al., 2016). In addition, manganese oxide NPs are used for alleviating hypoxic tumor environments by decomposing hydrogen peroxide into oxygen (Song et al., 2016). Although NPs are encouraging carriers for antitumor compounds and are capable of controlled release, one side effect that is seen with antitumor drugs used for photothermal therapy is their unwanted accumulation in healthy tissues (Gulfam & Chung, 2014; Seo et al., 2015). To combat this side effect, more selective targeting with active, anticancer compounds have been developed. For example, arginine-glycine-aspartic acid peptides are used to target cancer cells via $\alpha_v\beta_3$ ligand recognition (Danhier et al., 2012). In another example, manganese oxide NPs covered with graft terpolymers improved their passive accumulation in cancer cells by enhancing the cancer cells permeability and retention of the NPs (Gordijo et al., 2015). Covering NPs with graft terpolymers allows for the hydrophobicity of the NPs to be tailored which reduces the decomposition of hydrogen peroxide in the tumor environment to create a more controlled tumor-killing effect (Gordijo et al., 2015). Alternatively, linking NPs with folic acid is used since many cancer cells overexpress folic acid receptors (Low & Kularatne, 2009). In the case of graphene oxide nanosheets, folic acid and manganese oxide NPs are used to functionalize the

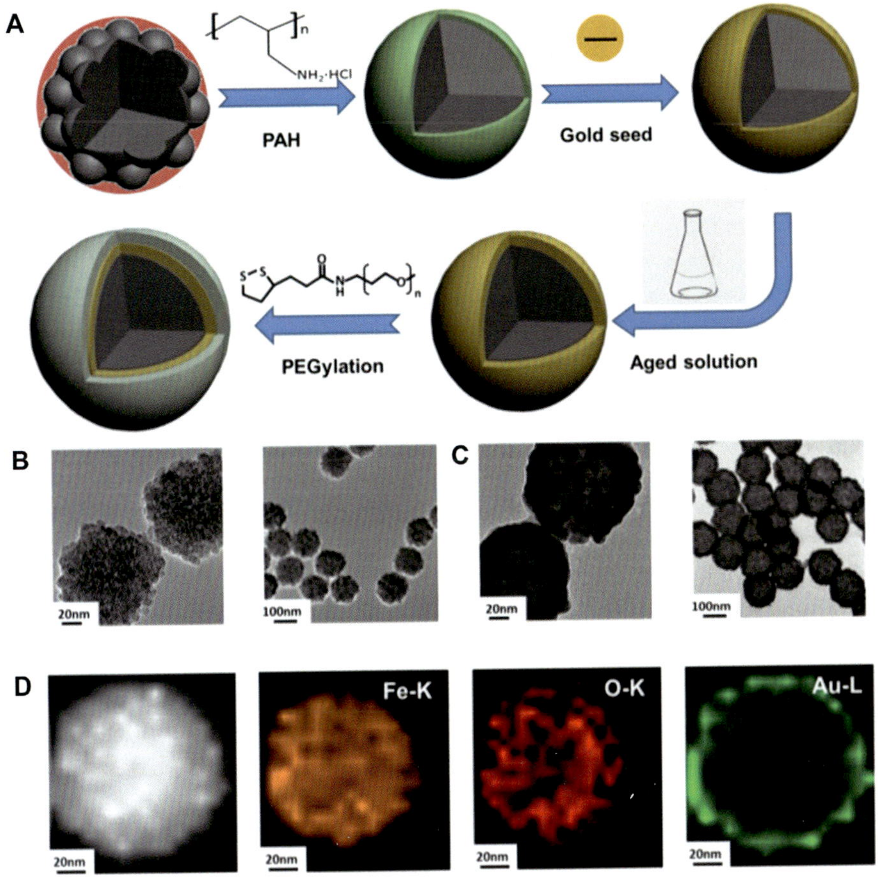

Figure 3.3 Fabrication and characterization of a PEG-functionalized gold-shelled iron oxide nanoclusters. (A) Iron oxide nanoclusters coated with poly(acrylic acid) are synthesized with a one-step hydrothermal process that does not require any surfactants or templates. (B) As clusters of small nanocrystals, the iron oxide nanoclusters have a uniform spherical morphology with an average diameter of 100 nm. (C) and (D) With the addition of $HAuCl_4$ as the growth solution and the reducing agent formaldehyde, a dense gold shell forms on the iron oxide nanoclusters (Liang et al., 2015).

nanosheet's surface for photothermal therapy with absorption in the NIR range (Lim et al., 2018). With this therapeutic strategy, the manganese oxide NPs alleviate the hypoxic tumor environment while reducing the tumors immunity to chemotherapeutic agents and photothermal therapy. Manganese oxide NPs can also be used as a contrast agent for MRI to monitor the treatment. Furthermore, folic acid conjunction reduces the side effects of NPs accumulating in healthy tissues. A more recent study has utilized hollow manganese oxide NPs to codeliver the photosensitizer chlorin e6 and chemotherapeutic doxorubicin (Yang et al., 2017). Since hollow manganese oxide NPs have a negative surface charge, a cationic poly(allylamine hydrochloride) polymer along with an anionic poly (acrylic acid) polymer can be used to coat the NPs using a layer-by-layer process. Photothermal therapy using 660 nm light irradiation along with the codelivery of the chlorin e6 and doxorubicin enhanced the hollow manganese oxide NPs antitumor effects as well as induced a systematic immune response when cotreated with a programmed death-ligand 1 antibody (Yang et al., 2017). In fact, with the 4T1 breast cancer model, combining hollow manganese oxide NPs with programmed death-ligand 1 antibodies inhibited tumor growth in the primary tumor irradiated with light as well as distant metastasis not treated with irradiation.

3.19 Conclusions

Multiple strategies using metal oxides as anticancer therapeutic agents have been reviewed in this chapter. Metal oxides can undergo a multitude of surface modifications with ligands and be functionalized to improve cancer targeting and cytotoxicity for cancer therapy. As immunotherapeutic agents, metal oxides can be modified with antitumor drugs along with immune stimulatory agents to elicit innate and adaptive immune responses to both reverse the immunosuppressive tumor environment and kill tumor cells. In photothermal therapy, various therapeutic modalities may be utilized in conjunction with photothermal metal oxide agents to target and thermally ablate tumor cells along with secondary metastasis. Overall, metal oxides are a valuable therapeutic that can be used by themselves or in combinational therapies for the treatment of primary and secondary cancers.

References

Abdalla, A. M. E., Xiao, L., Ullah, M. W., Yu, M., Ouyang, C., & Yang, G. (2018). Current challenges of cancer anti-angiogenic therapy and the promise of nanotherapeutics. *Theranostics*, *8*(2), 533–549, Ivyspring International Publisher. Available from https://doi.org/10.7150/thno.21674.

Akhavan, O., Ghaderi, E., Aghayee, S., Fereydooni, Y., & Talebi, A. (2012). The use of a glucose-reduced graphene oxide suspension for photothermal cancer therapy. *Journal of Materials Chemistry*, *22*(27), 13773–13781. Available from https://doi.org/10.1039/c2jm31396k.

Akhtar, M. J., Ahamed, M., Kumar, S., Majeed Khan, M. A., Ahmad, J., & Alrokayan, S. A. (2012). Zinc oxide nanoparticles selectively induce apoptosis in human cancer cells through reactive oxygen species. *International Journal of Nanomedicine*, *7*, 845–857. Available from https://doi.org/10.2147/IJN.S29129.

Al Tameemi, W., Dale, T. P., Al-Jumaily, R. M. K., & Forsyth, N. R. (2019). Hypoxia-Modified Cancer Cell Metabolism. *Frontiers in Cell and Developmental Biology*, *7*, 4. Available from https://doi.org/10.3389/fcell.2019.00004.

Alkilany, A. M., Thompson, L. B., Boulos, S. P., Sisco, P. N., & Murphy, C. J. (2012). Gold nanorods: Their potential for photothermal therapeutics and drug delivery, tempered by

the complexity of their biological interactions. *Advanced Drug Delivery Reviews*, *64*(2), 190–199. Available from https://doi.org/10.1016/j.addr.2011.03.005.
Anderson, B. D., & Tracy, J. B. (2014). Nanoparticle conversion chemistry: Kirkendall effect, galvanic exchange, and anion exchange. *Nanoscale*, *6*(21), 12195–12216, Royal Society of Chemistry. Available from https://doi.org/10.1039/c4nr02025a.
Anderson, T. M., & Hill, C. L. (2002). Modeling reactive metal oxides. Kinetics, thermodynamics, and mechanism of M3 cap isomerization in polyoxometalates. *Inorganic Chemistry*, *41*(16), 4252–4258. Available from https://doi.org/10.1021/ic020224d.
Andrievski, R. A. (2003). Stability of nanostructured materials. *Journal of Materials Science*, *38*(7), 1367–1375, Springer. Available from https://doi.org/10.1023/A:1022988706296.
Attia, M. F., Anton, N., Wallyn, J., Omran, Z., & Vandamme, T. F. (2019). An overview of active and passive targeting strategies to improve the nanocarriers efficiency to tumour sites. *Journal of Pharmacy and Pharmacology*, *71*(8), 1185–1198. Available from https://doi.org/10.1111/jphp.13098.
Audran, R., Peter, K., Dannull, J., Men, Y., Scandella, E., Groettrup, M., Gander, B., & Corradin, G. (2003). Encapsulation of peptides in biodegradable microspheres prolongs their MHC class-I presentation by dendritic cells and macrophages in vitro. *Vaccine*, *21* (11–12), 1250–1255. Available from https://doi.org/10.1016/S0264-410X(02)00521-2.
Bae, M. Y., Cho, N. H., & Seong, S. Y. (2009). Protective anti-tumour immune responses by murine dendritic cells pulsed with recombinant Tat-carcinoembryonic antigen derived from Escherichia coli. *Clinical and Experimental Immunology*, *157*(1), 128–138. Available from https://doi.org/10.1111/j.1365-2249.2009.03943.x.
Bai, D. P., Zhang, X. F., Zhang, G. L., Huang, Y. F., & Gurunathan, S. (2017). Zinc oxide nanoparticles induce apoptosis and autophagy in human ovarian cancer cells. *International Journal of Nanomedicine*, *12*, 6521–6535. Available from https://doi.org/10.2147/IJN.S140071.
Banu, H., Sethi, D. K., Edgar, A., Sheriff, A., Rayees, N., Renuka, N., Faheem, S. M., Premkumar, K., & Vasanthakumar, G. (2015). Doxorubicin loaded polymeric gold nanoparticles targeted to human folate receptor upon laser photothermal therapy potentiates chemotherapy in breast cancer cell lines. *Journal of Photochemistry and Photobiology B: Biology*, *149*, 116–128. Available from https://doi.org/10.1016/j.jphotobiol.2015.05.008.
Beberwyck, B. J., Surendranath, Y., & Alivisatos, A. P. (2013). Cation exchange: A versatile tool for nanomaterials synthesis. *Journal of Physical Chemistry C*, *117*(39), 19759–19770. Available from https://doi.org/10.1021/jp405989z.
Beik, J., Abed, Z., Ghoreishi, F. S., Hosseini-Nami, S., Mehrzadi, S., Shakeri-Zadeh, A., & Kamrava, S. K. (2016). Nanotechnology in hyperthermia cancer therapy: From fundamental principles to advanced applications. *Journal of Controlled Release*, *235*, 205–221, Elsevier B.V.. Available from https://doi.org/10.1016/j.jconrel.2016.05.062.
Bianco, A., Hoebeke, J., Godefroy, S., Chaloin, O., Pantarotto, D., Briand, J. P., Muller, S., Prato, M., & Partidos, C. D. (2005). Cationic carbon nanotubes bind to CpG oligodeoxynucleotides and enhance their immunostimulatory properties. *Journal of the American Chemical Society*, *127*(1), 58–59. Available from https://doi.org/10.1021/ja044293y.
Bogdan, J., Pławińska-Czarnak, J., & Zarzyńska, J. (2017). Nanoparticles of Titanium and Zinc Oxides as Novel Agents in Tumor Treatment: a Review. *Nanoscale Research Letters*, *12*(1), 1–15. Available from https://doi.org/10.1186/s11671-017-2007-y.
Cai, W., Piner, R. D., Stadermann, F. J., Park, S., Shaibat, M. A., Ishii, Y., Yang, D., Velamakanni, A., Sung, J. A., Stoller, M., An, J., Chen, D., & Ruoff, R. S. (2008). Synthesis and solid-state NMR structural characterization of 13C-labeled graphite oxide.

Science, *321*(5897), 1815–1817. Available from https://doi.org/10.1126/science.1162369.

Campolongo, M. J., Tan, S. J., Xu, J., & Luo, D. (2010). DNA nanomedicine: Engineering DNA as a polymer for therapeutic and diagnostic applications. *Advanced Drug Delivery Reviews*, *62*(6), 606–616. Available from https://doi.org/10.1016/j.addr.2010.03.004, Elsevier.

Chaplin, D. D. (2010). Overview of the immune response. *Journal of Allergy and Clinical Immunology*, *125*(2), S3. Available from https://doi.org/10.1016/j.jaci.2009.12.980, SUPPL. 2.

Chattopadhyay, S., Dash, S. K., Ghosh, T., Das, S., Tripathy, S., Mandal, D., Das, D., Pramanik, P., & Roy, S. (2013). Anticancer and immunostimulatory role of encapsulated tumor antigen containing cobalt oxide nanoparticles. *Journal of Biological Inorganic Chemistry*, *18*(8), 957–973. Available from https://doi.org/10.1007/s00775-013-1044-y.

Chattopadhyay, S., Dash, S. K., Mandal, D., Das, B., Tripathy, S., Dey, A., Pramanik, P., & Roy, S. (2016). Metal based nanoparticles as cancer antigen delivery vehicles for macrophage based antitumor vaccine. *Vaccine*, *34*(7), 957–967. Available from https://doi.org/10.1016/j.vaccine.2015.12.053.

Chen, G. Y., Yang, H. J., Lu, C. H., Chao, Y. C., Hwang, S. M., Chen, C. L., Lo, K. W., Sung, L. Y., Luo, W. Y., Tuan, H. Y., & Hu, Y. C. (2012). Simultaneous induction of autophagy and toll-like receptor signaling pathways by graphene oxide. *Biomaterials*, *33* (27), 6559–6569. Available from https://doi.org/10.1016/j.biomaterials.2012.05.064.

Chen, H., Zhen, Z., Todd, T., Chu, P. K., & Xie, J. (2013). Nanoparticles for improving cancer diagnosis. In Materials Science and Engineering R: Reports, *74*(3), 35–69, Elsevier Ltd. Available from https://doi.org/10.1016/j.mser.2013.03.001.

Chen, Q., Liang, C., Wang, C., & Liu, Z. (2015). An Imagable and Photothermal "Abraxane-Like" Nanodrug for Combination Cancer Therapy to Treat Subcutaneous and Metastatic Breast Tumors. *Advanced Materials*, *27*(5), 903–910. Available from https://doi.org/10.1002/adma.201404308.

Chertok, B., Moffat, B. A., David, A. E., Yu, F., Bergemann, C., Ross, B. D., & Yang, V. C. (2008). Iron oxide nanoparticles as a drug delivery vehicle for MRI monitored magnetic targeting of brain tumors. *Biomaterials*, *29*(4), 487–496. Available from https://doi.org/10.1016/j.biomaterials.2007.08.050.

Cho, N. H., Cheong, T. C., Min, J. H., Wu, J. H., Lee, S. J., Kim, D., Yang, J. S., Kim, S., Kim, Y. K., & Seong, S. Y. (2011). A multifunctional core-shell nanoparticle for dendritic cell-based cancer immunotherapy. *Nature Nanotechnology*, *6*(10), 675–682. Available from https://doi.org/10.1038/nnano.2011.149.

Choudhuri, K., Wiseman, D., Brown, M. H., Gould, K., & Van Der Merwe, P. A. (2005). T-cell receptor triggering is critically dependent on the dimensions of its peptide-MHC ligand. *Nature*, *436*(7050), 578–582. Available from https://doi.org/10.1038/nature03843.

Chung, C., Kim, Y. K., Shin, D., Ryoo, S. R., Hong, B. H., & Min, D. H. (2013). Biomedical applications of graphene and graphene oxide. *Accounts of Chemical Research*, *46*(10), 2211–2224. Available from https://doi.org/10.1021/ar300159f.

Cutler, J. I., Auyeung, E., & Mirkin, C. A. (2012). Spherical nucleic acids. *Journal of the American Chemical Society*, *134*(3), 1376–1391. Available from https://doi.org/10.1021/ja209351u, American Chemical Society.

Dai, Q., Worden, J. G., Trullinger, J., & Huo, Q. (2005). A "nanonecklace" synthesized from monofunctionalized gold nanoparticles. *Journal of the American Chemical Society*, *127*(22), 8008–8009. Available from https://doi.org/10.1021/ja042610v.

Dan, K., Miyashita, K., Seto, Y., Fujita, H., & Yamase, T. (2003). Mechanism of the Protective Effect of Heteropolyoxotungstate against Herpes Simplex Virus Type 2. *Pharmacology*, *67*(2), 83–89. Available from https://doi.org/10.1159/000067738.

Danhier, F., Breton, A., Le., & Préat, V. (2012). RGD-based strategies to target alpha(v) beta (3) integrin in cancer therapy and diagnosis. *Molecular Pharmaceutics*, *9*(11), 2961–2973. Available from https://doi.org/10.1021/mp3002733.

Davis, D. M., & Dustin, M. L. (2004). What is the importance of the immunological synapse? *Trends in Immunology*, *25*(6), 323–327, Trends Immunol. Available from https://doi.org/10.1016/j.it.2004.03.007.

De Vries, I. J. M., Lesterhuis, W. J., Barentsz, J. O., Verdijk, P., Van Krieken, J. H., Boerman, O. C., Oyen, W. J. G., Bonenkamp, J. J., Boezeman, J. B., Adema, G. J., Bulte, J. W. M., Scheenen, T. W. J., Punt, C. J. A., Heerschap, A., & Figdor, C. G. (2005). Magnetic resonance tracking of dendritic cells in melanoma patients for monitoring of cellular therapy. *Nature Biotechnology*, *23*(11), 1407–1413. Available from https://doi.org/10.1038/nbt1154.

Delcassian, D., Depoil, D., Rudnicka, D., Liu, M., Davis, D. M., Dustin, M. L., & Dunlop, I. E. (2013). Nanoscale ligand spacing influences receptor triggering in T cells and NK cells. *Nano Letters*, *13*(11), 5608–5614. Available from https://doi.org/10.1021/nl403252x.

Devulapally, R., Sekar, N. M., Sekar, T. V., Foygel, K., Massoud, T. F., Willmann, J. K., & Paulmurugan, R. (2015). Polymer nanoparticles mediated codelivery of AntimiR-10b and AntimiR-21 for achieving triple negative breast cancer therapy. *ACS Nano*, *9*(3), 2290–2302. Available from https://doi.org/10.1021/nn507465d.

Dustin, M. L., Chakraborty, A. K., & Shaw, A. S. (2010). Understanding the structure and function of the immunological synapse. *Cold Spring Harbor perspectives in biology*, *2*(1), Cold Spring Harbor Laboratory Press. Available from https://doi.org/10.1101/cshperspect.a002311.

Eckhardt, B. L., Francis, P. A., Parker, B. S., & Anderson, R. L. (2012). Strategies for the discovery and development of therapies for metastatic breast cancer. *Nature Reviews Drug Discovery*, *11*(6), 479–497. Available from https://doi.org/10.1038/nrd2372.

Ediriwickrema, A., & Saltzman, W. M. (2015). Nanotherapy for Cancer: Targeting and Multifunctionality in the Future of Cancer Therapies. *ACS Biomaterials Science and Engineering*, *1*(2), 64–78, American Chemical Society. Available from https://doi.org/10.1021/ab500084g.

Elgqvist, J. (2017). Nanoparticles as theranostic vehicles in experimental and clinical applications-focus on prostate and breast cancer. *International Journal of Molecular Sciences*, *18*(5), MDPI AG. Available from https://doi.org/10.3390/ijms18051102.

Erikçi, E., Gursel, M., & Gürsel, I. (2011). Differential immune activation following encapsulation of immunostimulatory CpG oligodeoxynucleotide in nanoliposomes. *Biomaterials*, *32*(6), 1715–1723. Available from https://doi.org/10.1016/j.biomaterials.2010.10.054.

Estelrich, J., & Antònia Busquets, M. (2018). Iron oxide nanoparticles in photothermal therapy. *Molecules*, *23*(7), MDPI AG. Available from https://doi.org/10.3390/molecules23071567.

Felice, B., Prabhakaran, M. P., Rodríguez, A. P., & Ramakrishna, S. (2014). Drug delivery vehicles on a nano-engineering perspective. *Materials Science and Engineering C*, *41*, 178–195, Elsevier BV. Available from https://doi.org/10.1016/j.msec.2014.04.049.

Feng, L., Wu, L., & Qu, X. (2013). New horizons for diagnostics and therapeutic applications of graphene and graphene oxide. *Advanced Materials*, *25*(2), 168–186. Available from https://doi.org/10.1002/adma.201203229.

Fernández, Y., Cueva, J., Palomo, A. G., Ramos, M., de Juan, A., Calvo, L., García-Mata, J., García-Teijido, P., Peláez, I., & García-Estévez, L. (2010). Novel therapeutic approaches to the treatment of metastatic breast cancer. *Cancer Treatment Reviews*, *36*(1), 33–42. Available from https://doi.org/10.1016/j.ctrv.2009.10.001.

Forster, J., Harriss-Phillips, W., Douglass, M., & Bezak, E. (2017). A review of the development of tumor vasculature and its effects on the tumor microenvironment. *Hypoxia*, *5*, 21–32. Available from https://doi.org/10.2147/hp.s133231.

Fukuda, N., Yamase, T., & Tajima, Y. (1999). Inhibitory effect of polyoxotungstates on the production of penicillin- binding proteins and β-lactamase against methicillin-resistant Staphylococcus aureus. *Biological and Pharmaceutical Bulletin*, *22*(5), 463–470. Available from https://doi.org/10.1248/bpb.22.463.

Gamarra, L., Silva, A. C., Oliveira, T. R., Mamani, J. B., Malheiros, S. M. F., Malavolta, L., Pavon, L. F., Sibov, T. T., Amaro, E., Jr, & Gamarra, L. (2011). Application of hyperthermia induced by superparamagnetic iron oxide nanoparticles in glioma treatment. *International Journal of Nanomedicine*, *6*, 591. Available from https://doi.org/10.2147/ijn.s14737.

Gehrke, T., Scherzad, A., Ickrath, P., Schendzielorz, P., Hagen, R., Kleinsasser, N., & Hackenberg, S. (2017). Zinc oxide nanoparticles antagonize the effect of Cetuximab on head and neck squamous cell carcinoma in vitro. *Cancer Biology and Therapy*, *18*(7), 513–518. Available from https://doi.org/10.1080/15384047.2017.1323598.

Ghosh, D., Lee, Y., Thomas, S., Kohli, A. G., Yun, D. S., Belcher, A. M., & Kelly, K. A. (2012). M13-templated magnetic nanoparticles for targeted in vivo imaging of prostate cancer. *Nature Nanotechnology*, *7*(10), 677–682. Available from https://doi.org/10.1038/nnano.2012.146.

Gilboa, E. (2007). DC-based cancer vaccines. *Journal of Clinical Investigation*, *117*(5), 1195–1203, American Society for Clinical Investigation. Available from https://doi.org/10.1172/JCI31205.

Gobbo, O. L., Sjaastad, K., Radomski, M. W., Volkov, Y., & Prina-Mello, A. (2015). Magnetic nanoparticles in cancer theranostics. *Theranostics*, *5*(11), 1249–1263, Ivyspring International Publisher. Available from https://doi.org/10.7150/thno.11544.

Golombek, S. K., May, J. N., Theek, B., Appold, L., Drude, N., Kiessling, F., & Lammers, T. (2018). Tumor targeting via EPR: Strategies to enhance patient responses. *Advanced Drug Delivery Reviews*, *130*, 17–38, Elsevier B.V.. Available from https://doi.org/10.1016/j.addr.2018.07.007.

Gonçalves, G., Vila, M., Portolés, M. T., Vallet-Regi, M., Gracio, J., & Marques, P. A. A. P. (2013). Nano-graphene oxide: A potential multifunctional platform for cancer therapy. *Advanced Healthcare Materials*, *2*(8), 1072–1090. Available from https://doi.org/10.1002/adhm.201300023.

Gordijo, C. R., Abbasi, A. Z., Amini, M. A., Lip, H. Y., Maeda, A., Cai, P., O'Brien, P. J., Dacosta, R. S., Rauth, A. M., & Wu, X. Y. (2015). Design of hybrid MnO2-polymer-lipid nanoparticles with tunable oxygen generation rates and tumor accumulation for cancer treatment. *Advanced Functional Materials*, *25*(12), 1858–1872. Available from https://doi.org/10.1002/adfm.201404511.

Grimaud, A., Diaz-Morales, O., Han, B., Hong, W. T., Lee, Y. L., Giordano, L., Stoerzinger, K. A., Koper, M. T. M., & Shao-Horn, Y. (2017). Activating lattice oxygen redox reactions in metal oxides to catalyse oxygen evolution. *Nature Chemistry*, *9*(5), 457–465. Available from https://doi.org/10.1038/nchem.2695.

Guimarães, P. P. G., Gaglione, S., Sewastianik, T., Carrasco, R. D., Langer, R., & Mitchell, M. J. (2018). Nanoparticles for Immune Cytokine TRAIL-Based Cancer Therapy. *ACS*

Nano, *12*(2), 912–931, American Chemical Society. Available from https://doi.org/10.1021/acsnano.7b05876.

Gulfam, M., & Chung, B. G. (2014). Development of pH-responsive chitosan-coated mesoporous silica nanoparticles. *Macromolecular Research*, *22*(4), 412–417. Available from https://doi.org/10.1007/s13233-014-2063-4.

Harada, A., Ono, M., Yuba, E., & Kono, K. (2013). Titanium dioxide nanoparticle-entrapped polyion complex micelles generate singlet oxygen in the cells by ultrasound irradiation for sonodynamic therapy. *Biomaterials Science*, *1*(1), 65–73. Available from https://doi.org/10.1039/c2bm00066k.

Hassan, H. F. H., Mansour, A. M., Abo-Youssef, A. M. H., Elsadek, B. E. M., & Messiha, B. A. S. (2017). Zinc oxide nanoparticles as a novel anticancer approach; in vitro and in vivo evidence. *Clinical and Experimental Pharmacology and Physiology*, *44*(2), 235–243. Available from https://doi.org/10.1111/1440-1681.12681.

He, H., David, A., Chertok, B., Cole, A., Lee, K., Zhang, J., Wang, J., Huang, Y., & Yang, V. C. (2013). Magnetic nanoparticles for tumor imaging and therapy: A so-called theranostic system. *Pharmaceutical Research*, *30*(10), 2445–2458, NIH Public Access. Available from https://doi.org/10.1007/s11095-013-0982-y.

He, X., Bao, X., Cao, H., Zhang, Z., Yin, Q., Gu, W., Chen, L., Yu, H., & Li, Y. (2015). Tumor-Penetrating Nanotherapeutics Loading a Near-Infrared Probe Inhibit Growth and Metastasis of Breast Cancer. *Advanced Functional Materials*, *25*(19), 2831–2839. Available from https://doi.org/10.1002/adfm.201500772.

Heracleous, E., & Lemonidou, A. A. (2010). Ni-Me-O mixed metal oxides for the effective oxidative dehydrogenation of ethane to ethylene - Effect of promoting metal Me. *Journal of Catalysis*, *270*(1), 67–75. Available from https://doi.org/10.1016/j.jcat.2009.12.004.

Hu, S.-H., Fang, R.-H., Chen, Y.-W., Liao, B.-J., Chen, I.-W., & Chen, S.-Y. (2014). Photoresponsive Protein-Graphene-Protein Hybrid Capsules with Dual Targeted Heat-Triggered Drug Delivery Approach for Enhanced Tumor Therapy. *Advanced Functional Materials*, *24*(26), 4144–4155. Available from https://doi.org/10.1002/adfm.201400080.

Huang, N. P., Xu, M. H., Yuan, C. W., & Yu, R. R. (1997). The study of the photokilling effect and mechanism of ultrafine TiO2 particles on U937 cells. *Journal of Photochemistry and Photobiology A: Chemistry*, *108*(2–3), 229–233. Available from https://doi.org/10.1016/S1010-6030(97)00093-2.

Huang, X., El-Sayed, I. H., Qian, W., & El-Sayed, M. A. (2006). Cancer cell imaging and photothermal therapy in the near-infrared region by using gold nanorods. *Journal of the American Chemical Society*, *128*(6), 2115–2120. Available from https://doi.org/10.1021/ja057254a.

ICPE, N. T.-P. of the, & 1974, undefined. (n.d.). On the basic concept of nanotechnology. *Ci.Nii.Ac.Jp*. Retrieved September 29, 2020, from https://ci.nii.ac.jp/naid/10008480916/.

Ivanković, S., Gotić, M., Jurin, M., & Music, S. (2003). Photokilling squamous carcinoma cells SCCVII with ultrafine particles of selected metal oxides. *Journal of Sol-Gel Science and Technology*, *27*(2), 225–233. Available from https://doi.org/10.1023/A:1023715004575.

Jarnicki, A. G., Conroy, H., Brereton, C., Donnelly, G., Toomey, D., Walsh, K., Sweeney, C., Leavy, O., Fletcher, J., Lavelle, E. C., Dunne, P., & Mills, K. H. G. (2008). Attenuating Regulatory T Cell Induction by TLR Agonists through Inhibition of p38 MAPK Signaling in Dendritic Cells Enhances Their Efficacy as Vaccine Adjuvants and Cancer Immunotherapeutics. *The Journal of Immunology*, *180*(6), 3797–3806. Available from https://doi.org/10.4049/jimmunol.180.6.3797.

Jenkins, R. W., Barbie, D. A., & Flaherty, K. T. (2018). Mechanisms of resistance to immune checkpoint inhibitors. *British Journal of Cancer*, *118*(1), 9–16. Available from https://doi.org/10.1038/bjc.2017.434.

Johannsen, M., Thiesen, B., Wust, P., & Jordan, A. (2010). Magnetic nanoparticle hyperthermia for prostate cancer. *International Journal of Hyperthermia*, *26*(8), 790–795. Available from https://doi.org/10.3109/02656731003745740.

Kaaki, K., Hervé-Aubert, K., Chiper, M., Shkilnyy, A., Soucé, M., Benoit, R., Paillard, A., Dubois, P., Saboungi, M. L., & Chourpa, I. (2012). Magnetic nanocarriers of doxorubicin coated with poly(ethylene glycol) and folic acid: Relation between coating structure, surface properties, colloidal stability, and cancer cell targeting. *Langmuir*, *28*(2), 1496–1505. Available from https://doi.org/10.1021/la2037845.

Kim, D., Yu, M. K., Lee, T. S., Park, J. J., Jeong, Y. Y., & Jon, S. (2011). Amphiphilic polymer-coated hybrid nanoparticles as CT/MRI dual contrast agents. *Nanotechnology*, *22*(15). Available from https://doi.org/10.1088/0957-4484/22/15/155101.

Kim, G. S., Judd, D. A., Hill, C. L., & Schinazi, R. F. (1994). Synthesis, Characterization, and Biological Activity of a New Potent Class of Anti-HIV Agents, the Peroxoniobium-Substituted Heteropolytungstates. *Journal of Medicinal Chemistry*, *37*(6), 816–820. Available from https://doi.org/10.1021/jm00032a016.

Klinman, D. M., Yi, A. K., Beaucage, S. L., Conover, J., & Krieg, A. M. (1996). CpG motifs present in bacterial DNA rapidly induce lymphocytes to secrete interleukin 6, interleukin 12, and interferon γ. *Proceedings of the National Academy of Sciences of the United States of America*, *93*(7), 2879–2883. Available from https://doi.org/10.1073/pnas.93.7.2879.

Klippstein, R., & Pozo, D. (2010). Nanotechnology-based manipulation of dendritic cells for enhanced immunotherapy strategies. *Nanomedicine: Nanotechnology, Biology, and Medicine*, *6*(4), 523–529. Available from https://doi.org/10.1016/j.nano.2010.01.001, Nanomedicine.

Kostarelos, K., Lacerda, L., Pastorin, G., Wu, W., Wieckowski, S., Luangsivilay, J., Godefroy, S., Pantarotto, D., Briand, J. P., Muller, S., Prato, M., & Bianco, A. (2007). Cellular uptake of functionalized carbon nanotubes is independent of functional group and cell type. *Nature Nanotechnology*, *2*(2), 108–113. Available from https://doi.org/10.1038/nnano.2006.209.

Kostestkyy, P., Yu, J., Gorte, R. J., & Mpourmpakis, G. (2014). Structure-activity relationships on metal-oxides: Alcohol dehydration. *Catalysis Science and Technology*, *4*(11), 3861–3869. Available from https://doi.org/10.1039/c4cy00632a.

Kovacsovics-Bankowski, M., & Rock, K. L. (1995). A phagosome-to-cytosol pathway for exogenous antigens presented on MHC class I molecules. *Science*, *267*(5195), 243–246. Available from https://doi.org/10.1126/science.7809629.

Kunzmann, A., Andersson, B., Vogt, C., Feliu, N., Ye, F., Gabrielsson, S., Toprak, M. S., Buerki-Thurnherr, T., Laurent, S., Vahter, M., Krug, H., Muhammed, M., Scheynius, A., & Fadeel, B. (2011). Efficient internalization of silica-coated iron oxide nanoparticles of different sizes by primary human macrophages and dendritic cells. *Toxicology and Applied Pharmacology*, *253*(2), 81–93. Available from https://doi.org/10.1016/j.taap.2011.03.011.

Lee, S. M., Park, H., & Yoo, K. H. (2010). Synergistic cancer therapeutic effects of locally delivered drug and heat using multifunctional nanoparticles. *Advanced Materials*, *22*(36), 4049–4053. Available from https://doi.org/10.1002/adma.201001040.

Lerf, A., He, H., Forster, M., & Klinowski, J. (1998). Structure of graphite oxide revisited. *Journal of Physical Chemistry B*, *102*(23), 4477–4482. Available from https://doi.org/10.1021/jp9731821.

Liang, C., Song, X., Chen, Q., Liu, T., Song, G., Peng, R., & Liu, Z. (2015). Magnetic Field-Enhanced Photothermal Ablation of Tumor Sentinel Lymph Nodes to Inhibit Cancer Metastasis. *Small*, *11*(37), 4856–4863. Available from https://doi.org/10.1002/smll.201501197.

Liao, K. H., Lin, Y. S., MacOsko, C. W., & Haynes, C. L. (2011). Cytotoxicity of graphene oxide and graphene in human erythrocytes and skin fibroblasts. *ACS Applied Materials and Interfaces*, *3*(7), 2607–2615. Available from https://doi.org/10.1021/am200428v.

Lim, J. H., Kim, D. E., Kim, E. J., Ahrberg, C. D., & Chung, B. G. (2018). Functional Graphene Oxide-Based Nanosheets for Photothermal Therapy. *Macromolecular Research*, *26*(6), 557–565. Available from https://doi.org/10.1007/s13233-018-6067-3.

Lim, Y. T., Noh, Y. W., Han, J. H., Cai, Q. Y., Yoon, K. H., & Chung, B. H. (2008). Biocompatible polymer-nanoparticle-based bimodal imaging contrast agents for the labeling and tracking of dendritic cells. *Small*, *4*(10), 1640–1645. Available from https://doi.org/10.1002/smll.200800582.

Livingston, P. O., Wong, G. Y. C., Adluri, S., Tao, Y., Padavan, M., Parente, R., Hanlon, C., Calves, M. J., Helling, F., Ritter, G., Oettgen, H. F., & Old, L. J. (1994). Improved survival in stage III melanoma patients with GM2 antibodies: A randomized trial of adjuvant vaccination with GM2 ganglioside. *Journal of Clinical Oncology*, *12*(5), 1036–1044. Available from https://doi.org/10.1200/JCO.1994.12.5.1036.

Loftus, C., Saeed, M., Davis, D. M., & Dunlop, I. E. (2018). Activation of Human Natural Killer Cells by Graphene Oxide-Templated Antibody Nanoclusters. *Nano Letters*, *18*(5), 3282–3289. Available from https://doi.org/10.1021/acs.nanolett.8b01089.

Low, P. S., & Kularatne, S. A. (2009). Folate-targeted therapeutic and imaging agents for cancer. *Current Opinion in Chemical Biology*, *13*(3), 256–262. Available from https://doi.org/10.1016/j.cbpa.2009.03.022.

Ma, J., Cheng, L., & Chen, K. (2015). Impact of dispersants on relaxivities of magnetite contrast agents. *Journal of Applied Physics*, *117*(15), 154701. Available from https://doi.org/10.1063/1.4918553.

Ma, Y., Liang, X., Tong, S., Bao, G., Ren, Q., & Dai, Z. (2013). Gold nanoshell nanomicelles for potential magnetic resonance imaging, light-triggered drug release, and photothermal therapy. *Advanced Functional Materials*, *23*(7), 815–822. Available from https://doi.org/10.1002/adfm.201201663.

Mandelboim, O., Malik, P., Davis, D. M., Jo, C. H., Boyson, J. E., & Strominger, J. L. (1999). Human CD16 as a lysis receptor mediating direct natural killer cell cytotoxicity. *Proceedings of the National Academy of Sciences of the United States of America*, *96*(10), 5640–5644. Available from https://doi.org/10.1073/pnas.96.10.5640.

Martin, A. L., Bernas, L. M., Rutt, B. K., Foster, P. J., & Gillies, E. R. (2008). Enhanced cell uptake of superparamagnetic iron oxide nanoparticles functionalized with dendritic guanidines. *Bioconjugate Chemistry*, *19*(12), 2375–2384. Available from https://doi.org/10.1021/bc800209u.

Martinkova, P., Brtnicky, M., Kynicky, J., & Pohanka, M. (2018). Iron Oxide Nanoparticles: Innovative Tool in Cancer Diagnosis and Therapy. *Advanced Healthcare Materials*, *7*(5). Available from https://doi.org/10.1002/adhm.201700932, Wiley-VCH Verlag.

Mathieu, J. B., & Martel, S. (2009). Aggregation of magnetic microparticles in the context of targeted therapies actuated by a magnetic resonance imaging system. *Journal of Applied Physics*, *106*(4), 044904. Available from https://doi.org/10.1063/1.3159645.

Meng, W., Yamazaki, T., Nishida, Y., & Hanagata, N. (2011). Nuclease-resistant immunostimulatory phosphodiester CpG oligodeoxynucleotides as human Toll-like receptor 9 agonists. *BMC Biotechnology*, *11*(1), 88. Available from https://doi.org/10.1186/1472-6750-11-88.

Mikhaylov, G., Mikac, U., Magaeva, A. A., Itin, V. I., Naiden, E. P., Psakhye, I., Babes, L., Reinheckel, T., Peters, C., Zeiser, R., Bogyo, M., Turk, V., Psakhye, S. G., Turk, B., & Vasiljeva, O. (2011). Ferri-liposomes as an MRI-visible drug-delivery system for targeting tumours and their microenvironment. *Nature Nanotechnology*, *6*(9), 594–602. Available from https://doi.org/10.1038/nnano.2011.112.

Mokhtarzadeh, A., Alibakhshi, A., Yaghoobi, H., Hashemi, M., Hejazi, M., & Ramezani, M. (2016). Recent advances on biocompatible and biodegradable nanoparticles as gene carriers. *Expert Opinion on Biological Therapy*, *16*(6), 771–785, Taylor and Francis Ltd. Available from https://doi.org/10.1517/14712598.2016.1169269.

Moyano, D. F., Liu, Y., Peer, D., & Rotello, V. M. (2016). Modulation of Immune Response Using Engineered Nanoparticle Surfaces. *Small*, *12*(1), 76–82, Wiley-VCH Verlag. Available from https://doi.org/10.1002/smll.201502273.

Munn, D. H., & Mellor, A. L. (2016). IDO in the Tumor Microenvironment: Inflammation, Counter-Regulation, and Tolerance. *Trends in Immunology*, *37*(3), 193–207, Elsevier Ltd. Available from https://doi.org/10.1016/j.it.2016.01.002.

Munson, J. M., & Shieh, A. C. (2014). Interstitial fluid flow in cancer: Implications for disease progression and treatment. *Cancer Management and Research*, *6*(1), 317–318, Dove Press. Available from https://doi.org/10.2147/CMAR.S65444.

Mutis, T., Verdijk, R., Schrama, E., Esendam, B., Brand, A., & Goulmy, E. (1999). Feasibility of immunotherapy of relapsed leukemia with ex vivo-generated cytotoxic T lymphocytes specific for hematopoietic system-restricted minor histocompatibility antigens. *Blood*, *93*(7), 2336–2341. Available from https://doi.org/10.1182/blood.v93.7.2336.407k26_2336_2341.

Nakaoka, R., Tabata, Y., & Ikada, Y. (1996). Adjuvant effect of biodegradable poly(DL-lactic acid) granules capable for antigen release following intraperitoneal injection. *Vaccine*, *14*(17–18), 1671–1676. Available from https://doi.org/10.1016/S0264-410X(96)00098-9.

Nasongkla, N., Bey, E., Ren, J., Ai, H., Khemtong, C., Guthi, J. S., Chin, S. F., Sherry, A. D., Boothman, D. A., & Gao, J. (2006). Multifunctional polymeric micelles as cancer-targeted, MRI-ultrasensitive drug delivery systems. *Nano Letters*, *6*(11), 2427–2430. Available from https://doi.org/10.1021/nl061412u.

Neri, D., & Supuran, C. T. (2011). Interfering with pH regulation in tumours as a therapeutic strategy. *Nature Reviews Drug Discovery*, *10*(10), 767–777. Available from https://doi.org/10.1038/nrd3554.

Ng, K. W., Khoo, S. P. K., Heng, B. C., Setyawati, M. I., Tan, E. C., Zhao, X., Xiong, S., Fang, W., Leong, D. T., & Loo, J. S. C. (2011). The role of the tumor suppressor p53 pathway in the cellular DNA damage response to zinc oxide nanoparticles. *Biomaterials*, *32*(32), 8218–8225. Available from https://doi.org/10.1016/j.biomaterials.2011.07.036.

Ni, G., Wang, Y., Wu, X., Wang, X., Chen, S., & Liu, X. (2012). Graphene oxide absorbed anti-IL10R antibodies enhance LPS induced immune responses in vitro and in vivo. *Immunology Letters*, *148*(2), 126–132. Available from https://doi.org/10.1016/j.imlet.2012.10.001.

Ninomiya, K., Fukuda, A., Ogino, C., & Shimizu, N. (2014). Targeted sonocatalytic cancer cell injury using avidin-conjugated titanium dioxide nanoparticles. *Ultrasonics Sonochemistry*, *21*(5), 1624–1628. Available from https://doi.org/10.1016/j.ultsonch.2014.03.010.

Ogata, A., Mitsui, S., Yanagie, H., Kasano, H., Hisa, T., Yamase, T., & Eriguchi, M. (2005). A novel anti-tumor agent, polyoxomolybdate induces apoptotic cell death in AsPC-1 human pancreatic cancer cells. *Biomedicine and Pharmacotherapy*, *59*(5), 240–244. Available from https://doi.org/10.1016/j.biopha.2004.11.008.

Okamoto, H., Shiraki, K., Yasuda, R., Danjo, K., & Watanabe, Y. (2011). Chitosan-interferon-β gene complex powder for inhalation treatment of lung metastasis in mice. *Journal of Controlled Release*, *150*(2), 187–195. Available from https://doi.org/10.1016/j.jconrel.2010.12.006.

Okuno, T., Kato, S., Hatakeyama, Y., Okajima, J., Maruyama, S., Sakamoto, M., Mori, S., & Kodama, T. (2013). Photothermal therapy of tumors in lymph nodes using gold nanorods and near-infrared laser light. *Journal of Controlled Release*, *172*(3), 879–884. Available from https://doi.org/10.1016/j.jconrel.2013.10.014.

Oltra, N. S., Swift, J., Mahmud, A., Rajagopal, K., Loverde, S. M., & Discher, D. E. (2013). Filomicelles in nanomedicine-from flexible, fragmentable, and ligand-targetable drug carrier designs to combination therapy for brain tumors. *Journal of Materials Chemistry B*, *1* (39), 5177–5185. Available from https://doi.org/10.1039/c3tb20431f.

Orecchioni, M., Cabizza, R., Bianco, A., & Delogu, L. G. (2015). Graphene as cancer theranostic tool: Progress and future challenges. *Theranostics*, *5*(7), 710–723, Ivyspring International Publisher. Available from https://doi.org/10.7150/thno.11387.

Oszmiana, A., Williamson, D. J., Cordoba, S. P., Morgan, D. J., Kennedy, P. R., Stacey, K., & Davis, D. M. (2016). The Size of Activating and Inhibitory Killer Ig-like Receptor Nanoclusters Is Controlled by the Transmembrane Sequence and Affects Signaling. *Cell Reports*, *15*(9), 1957–1972. Available from https://doi.org/10.1016/j.celrep.2016.04.075.

Owonikoko, T. K., Arbiser, J., Zelnak, A., Shu, H. K. G., Shim, H., Robin, A. M., Kalkanis, S. N., Whitsett, T. G., Salhia, B., Tran, N. L., Ryken, T., Moore, M. K., Egan, K. M., & Olson, J. J. (2014). Current approaches to the treatment of metastatic brain tumours. *Nature Reviews Clinical Oncology*, *11*(4), 203–222, Nature Publishing Group. Available from https://doi.org/10.1038/nrclinonc.2014.25.

Palucka, A. K., Ueno, H., Fay, J. W., & Banchereau, J. (2007). Taming cancer by inducing immunity via dendritic cells. *Immunological Reviews*, *220*(1), 129–150, John Wiley & Sons, Ltd. Available from https://doi.org/10.1111/j.1600-065X.2007.00575.x.

Palucka, K., & Banchereau, J. (2012). Cancer immunotherapy via dendritic cells. *Nature Reviews Cancer*, *12*(4), 265–277, NIH Public Access. Available from https://doi.org/10.1038/nrc3258.

Pandey, N., Dhiman, S., Srivastava, T., & Majumder, S. (2016). Transition metal oxide nanoparticles are effective in inhibiting lung cancer cell survival in the hypoxic tumor microenvironment. *Chemico-Biological Interactions*, *254*, 221–230. Available from https://doi.org/10.1016/j.cbi.2016.06.006.

Papis, E., Rossi, F., Raspanti, M., Dalle-Donne, I., Colombo, G., Milzani, A., Bernardini, G., & Gornati, R. (2009). Engineered cobalt oxide nanoparticles readily enter cells. *Toxicology Letters*, *189*(3), 253–259. Available from https://doi.org/10.1016/j.toxlet.2009.06.851.

Park, E. J., Zhang, Y. Z., Vykhodtseva, N., & McDannold, N. (2012). Ultrasound-mediated blood-brain/blood-tumor barrier disruption improves outcomes with trastuzumab in a breast cancer brain metastasis model. *Journal of Controlled Release*, *163*(3), 277–284. Available from https://doi.org/10.1016/j.jconrel.2012.09.007.

Park, J. H., Von Maltzahn, G., Zhang, L., Derfus, A. M., Simberg, D., Harris, T. J., Ruoslahti, E., Bhatia, S. N., & Sailor, M. J. (2009). Systematic surface engineering of magnetic nanoworms for in vivo tumor targeting. *Small*, *5*(6), 694–700. Available from https://doi.org/10.1002/smll.200801789.

Peiris, P. M., Toy, R., Doolittle, E., Pansky, J., Abramowski, A., Tam, M., Vicente, P., Tran, E., Hayden, E., Camann, A., Mayer, A., Erokwu, B. O., Berman, Z., Wilson, D., Baskaran, H., Flask, C. A., Keri, R. A., & Karathanasis, E. (2012). Imaging metastasis using an integrin-targeting chain-shaped nanoparticle. *ACS Nano*, *6*(10), 8783–8795. Available from https://doi.org/10.1021/nn303833p.

Peng, J., Liang, X., & Calderon, L. (2019). Progress in research on gold nanoparticles in cancer management. *Medicine (United States)*, *98*(18). Available from https://doi.org/10.1097/MD.0000000000015311.

Prasad, S., Cody, V., Saucier-Sawyer, J. K., Saltzman, W. M., Sasaki, C. T., Edelson, R. L., Birchall, M. A., & Hanlon, D. J. (2011). Polymer nanoparticles containing tumor lysates as antigen delivery vehicles for dendritic cell-based antitumor immunotherapy. *Nanomedicine: Nanotechnology, Biology, and Medicine*, *7*(1), 1–10. Available from https://doi.org/10.1016/j.nano.2010.07.002.

Qian, H., Chao, X., & Ding, W. X. (2018). A PINK1-mediated mitophagy pathway decides the fate of tumors—to be benign or malignant? *Autophagy*, *14*(4), 563–566, Taylor and Francis Inc. Available from https://doi.org/10.1080/15548627.2018.1425057.

Qin, X. C., Guo, Z. Y., Liu, Z. M., Zhang, W., Wan, M. M., & Yang, B. W. (2013). Folic acid-conjugated graphene oxide for cancer targeted chemo-photothermal therapy. *Journal of Photochemistry and Photobiology B: Biology*, *120*, 156–162. Available from https://doi.org/10.1016/j.jphotobiol.2012.12.005.

Ramanlal Chaudhari, K., Kumar, A., Megraj Khandelwal, V. K., Ukawala, M., Manjappa, A. S., Mishra, A. K., Monkkonen, J., & Ramachandra Murthy, R. S. (2012). Bone metastasis targeting: A novel approach to reach bone using Zoledronate anchored PLGA nanoparticle as carrier system loaded with Docetaxel. *Journal of Controlled Release*, *158*(3), 470–478. Available from https://doi.org/10.1016/j.jconrel.2011.11.020.

Rasmussen, J. W., Martinez, E., Louka, P., & Wingett, D. G. (2010). Zinc oxide nanoparticles for selective destruction of tumor cells and potential for drug delivery applications. *Expert Opinion on Drug Delivery*, *7*(9), 1063–1077. Available from https://doi.org/10.1517/17425247.2010.502560.

Robinson, J. T., Welsher, K., Tabakman, S. M., Sherlock, S. P., Wang, H., Luong, R., & Dai, H. (2010). High performance in vivo near-IR (>1 μm) imaging and photothermal cancer therapy with carbon nanotubes. *Nano Research*, *3*(11), 779–793. Available from https://doi.org/10.1007/s12274-010-0045-1.

Rogers, W. J., Meyer, C. H., & Kramer, C. M. (2006). Technology Insight: In vivo cell tracking by use of MRI. *Nature Clinical Practice Cardiovascular Medicine*, *3*(10), 554–562, Nature Publishing Group. Available from https://doi.org/10.1038/ncpcardio0659.

Roy, R., Kumar, S., Verma, A. K., Sharma, A., Chaudhari, B. P., Tripathi, A., Das, M., & Dwivedi, P. D. (2014). Zinc oxide nanoparticles provide an adjuvant effect to ovalbumin via a th2 response in Balb/c mice. *International Immunology*, *26*(3), 159–172. Available from https://doi.org/10.1093/intimm/dxt053.

Salata, O. V. (2004). Applications of nanoparticles in biology and medicine. *Journal of Nanobiotechnology*, *2*(1), 3, BioMed Central. Available from https://doi.org/10.1186/1477-3155-2-3.

Sanchez, V. C., Jachak, A., Hurt, R. H., & Kane, A. B. (2012). Biological interactions of graphene-family nanomaterials: An interdisciplinary review. *Chemical Research in Toxicology*, *25*(1), 15–34. Available from https://doi.org/10.1021/tx200339h.

Sankar, R., Maheswari, R., Karthik, S., Shivashangari, K. S., & Ravikumar, V. (2014). Anticancer activity of Ficus religiosa engineered copper oxide nanoparticles. *Materials*

Science and Engineering C, *44*, 234–239. Available from https://doi.org/10.1016/j.msec.2014.08.030.

Santra, S., Kaittanis, C., Grimm, J., & Perez, J. M. (2009). Drug/dye-loaded, multifunctional iron oxide nanoparticles for combined targeted cancer therapy and dual optical/magnetic resonance imaging. *Small*, *5*(16), 1862–1868. Available from https://doi.org/10.1002/smll.200900389.

Saraiva, M., & O'Garra, A. (2010). The regulation of IL-10 production by immune cells. *Nature Reviews Immunology*, *10*(3), 170–181, Nature Publishing Group. Available from https://doi.org/10.1038/nri2711.

Sardar, R., & Shumaker-Parry, J. S. (2008). Asymmetrically functionalized gold nanoparticles organized in one-dimensional chains. *Nano Letters*, *8*(2), 731–736. Available from https://doi.org/10.1021/nl073154m.

Sato, T., Terai, M., Tamura, Y., Alexeev, V., Mastrangelo, M. J., & Selvan, S. R. (2011). Interleukin 10 in the tumor microenvironment: A target for anticancer immunotherapy. *Immunologic Research*, *51*(2–3), 170–182. Available from https://doi.org/10.1007/s12026-011-8262-6.

Schroeder, A., Heller, D. A., Winslow, M. M., Dahlman, J. E., Pratt, G. W., Langer, R., Jacks, T., & Anderson, D. G. (2012). Treating metastatic cancer with nanotechnology. *Nature Reviews Cancer*, *12*(1), 39–50, Nature Publishing Group. Available from https://doi.org/10.1038/nrc3180.

Seo, H. I., Cho, A. N., Jang, J., Kim, D. W., Cho, S. W., & Chung, B. G. (2015). Thermo-responsive polymeric nanoparticles for enhancing neuronal differentiation of human induced pluripotent stem cells. *Nanomedicine: Nanotechnology, Biology, and Medicine*, *11*(7), 1861–1869. Available from https://doi.org/10.1016/j.nano.2015.05.008.

Seong, S. Y., & Matzinger, P. (2004). Hydrophobicity: An ancient damage-associated molecular pattern that initiates innate immune responses. *Nature Reviews Immunology*, *4*(6), 469–478, Nature Publishing Group. Available from https://doi.org/10.1038/nri1372.

Shanmugam, V., Selvakumar, S., & Yeh, C. S. (2014). Near-infrared light-responsive nanomaterials in cancer therapeutics. *Chemical Society Reviews*, *43*(17), 6254–6287, Royal Society of Chemistry. Available from https://doi.org/10.1039/c4cs00011k.

Sharma, P., Hu-Lieskovan, S., Wargo, J. A., & Ribas, A. (2017). Primary, Adaptive, and Acquired Resistance to Cancer Immunotherapy. *Cell*, *168*(4), 707–723, Cell Press. Available from https://doi.org/10.1016/j.cell.2017.01.017.

Shen, A. J., Li, D. L., Cai, X. J., Dong, C. Y., Dong, H. Q., Wen, H. Y., Dai, G. H., Wang, P. J., & Li, Y. Y. (2012). Multifunctional nanocomposite based on graphene oxide for in vitro hepatocarcinoma diagnosis and treatment. *Journal of Biomedical Materials Research - Part A*, *100 A* (9), 2499–2506. Available from https://doi.org/10.1002/jbm.a.34148.

Shen, Z., Reznikoff, G., Dranoff, G., & Rock2, K. (1997). Cloned Dendritic Cells Can Present Exogenous Antigens on Both M H C Class I and Class II Molecules'. *The journal of Immunology* (158). Available from http://www.jimmunol.org/.

Sivaraj, R., Rahman, P. K. S. M., Rajiv, P., Narendhran, S., & Venckatesh, R. (2014). Biosynthesis and characterization of Acalypha indica mediated copper oxide nanoparticles and evaluation of its antimicrobial and anticancer activity. *Spectrochimica Acta - Part A: Molecular and Biomolecular Spectroscopy*, *129*, 255–258. Available from https://doi.org/10.1016/j.saa.2014.03.027.

Song, M., Liu, T., Shi, C., Zhang, X., & Chen, X. (2016). Bioconjugated manganese dioxide nanoparticles enhance chemotherapy response by priming tumor-Associated macrophages toward m1-like phenotype and attenuating tumor hypoxia. *ACS Nano*, *10*(1), 633–647. Available from https://doi.org/10.1021/acsnano.5b06779.

Sosnovik, D. E., Nahrendorf, M., & Weissleder, R. (2008). Magnetic nanoparticles for MR imaging: Agents, techniques and cardiovascular applications. *Basic Research in Cardiology*, *103*(2), 122–130, NIH Public Access. Available from https://doi.org/10.1007/s00395-008-0710-7.

Stahel, R. A., & Zangemeister-Wittke, U. (2003). Antisense oligonucleotides for cancer therapy - An overview. *Lung Cancer*, *41*(SUPPL. 1), 81–88. Available from https://doi.org/10.1016/S0169-5002(03)00147-8.

Stankovich, S., Dikin, D. A., Piner, R. D., Kohlhaas, K. A., Kleinhammes, A., Jia, Y., Wu, Y., Nguyen, S. B. T., & Ruoff, R. S. (2007). Synthesis of graphene-based nanosheets via chemical reduction of exfoliated graphite oxide. *Carbon*, *45*(7), 1558–1565. Available from https://doi.org/10.1016/j.carbon.2007.02.034.

Sun, C., Lee, J. S. H., & Zhang, M. (2008). Magnetic nanoparticles in MR imaging and drug delivery. *Advanced Drug Delivery Reviews*, *60*(11), 1252–1265. Available from https://doi.org/10.1016/j.addr.2008.03.018, Elsevier.

Sun, X., Liu, Z., Welsher, K., Robinson, J. T., Goodwin, A., Zaric, S., & Dai, H. (2008). Nano-graphene oxide for cellular imaging and drug delivery. *Nano Research*, *1*(3), 203–212. Available from https://doi.org/10.1007/s12274-008-8021-8.

Surendran, S. P., Moon, M. J., Park, R., & Jeong, Y. Y. (2018). Bioactive nanoparticles for cancer immunotherapy. *International Journal of Molecular Sciences*, *19*(12), MDPI AG. Available from https://doi.org/10.3390/ijms19123877.

Tan, X., Wang, Z., Yang, J., Song, C., Zhang, R., & Cui, Y. (2009). Polyvinylpyrrolidone-(PVP-) coated silver aggregates for high performance surface-enhanced Raman scattering in living cells. *Nanotechnology*, *20*(44). Available from https://doi.org/10.1088/0957-4484/20/44/445102.

Tang, Y., Lei, T., Manchanda, R., Nagesetti, A., Fernandez-Fernandez, A., Srinivasan, S., & McGoron, A. J. (2010). Simultaneous delivery of chemotherapeutic and thermal-optical agents to cancer cells by a polymeric (PLGA) nanocarrier: An in vitro study. *Pharmaceutical Research, 27(10), 2242–2253*. Available from https://doi.org/10.1007/s11095-010-0231-6.

Tao, Y., Ju, E., Li, Z., Ren, J., & Qu, X. (2014). Engineered CpG-Antigen Conjugates Protected Gold Nanoclusters as Smart Self-Vaccines for Enhanced Immune Response and Cell Imaging. *Advanced Functional Materials*, *24*(7), 1004–1010. Available from https://doi.org/10.1002/adfm.201302347.

Tao, Y., Ju, E., Ren, J., & Qu, X. (2014). Immunostimulatory oligonucleotides-loaded cationic graphene oxide with photothermally enhanced immunogenicity for photothermal/immune cancer therapy. *Biomaterials*, *35*(37), 9963–9971. Available from https://doi.org/10.1016/j.biomaterials.2014.08.036.

Tarnuzzer, R. W., Colon, J., Patil, S., & Seal, S. (2005). Vacancy engineered ceria nanostructures for protection from radiation-induced cellular damage. *Nano Letters*, *5*(12), 2573–2577. Available from https://doi.org/10.1021/nl052024f.

Thurber, A., Wingett, D. G., Rasmussen, J. W., Layne, J., Johnson, L., Tenne, D. A., Zhang, J., Hanna, C. B., & Punnoose, A. (2012). Improving the selective cancer killing ability of ZnO nanoparticles using Fe doping. *Nanotoxicology*, *6*(4), 440–452. Available from https://doi.org/10.3109/17435390.2011.587031.

Tian, B., Wang, C., Zhang, S., Feng, L., & Liu, Z. (2011). Photothermally enhanced photodynamic therapy delivered by nano-graphene oxide. *ACS Nano*, *5*(9), 7000–7009. Available from https://doi.org/10.1021/nn201560b.

Van Der Bliek, A. M., Sedensky, M. M., & Morgan, P. G. (2017). Cell biology of the mitochondrion. *Genetics*, *207*(3), 843–871. Available from https://doi.org/10.1534/genetics.117.300262.

Veiseh, O., Gunn, J. W., & Zhang, M. (2010). Design and fabrication of magnetic nanoparticles for targeted drug delivery and imaging. *Advanced Drug Delivery Reviews*, *62*(3), 284–304, NIH Public Access. Available from https://doi.org/10.1016/j.addr.2009.11.002.

Veronesi, U., Paganelli, G., Galimberti, V., Viale, G., Zurrida, S., Bedoni, M., Costa, A., De Cicco, C., Geraghty, J. G., Luini, A., Sacchini, V., & Veronesi, P. (1996). Sentinel-node biopsy to avoid axillary dissection in breast cancer with clinically negative lymph-nodes. *Lancet*, *349*(9069), 1864–1867. Available from https://doi.org/10.1016/S0140-6736(97)01004-0.

Vicari, A. P., Chiodoni, C., Vaure, C., Aït-Yahia, S., Dercamp, C., Matsos, F., Reynard, O., Taverne, C., Merle, P., Colombo, M. P., O'Garra, A., Trinchieri, G., & Caux, C. (2002). Reversal of tumor-induced dendritic cell paralysis by CpG immunostimulatory oligonucleotide and anti-interleukin 10 receptor antibody. *Journal of Experimental Medicine*, *196*(4), 541–549. Available from https://doi.org/10.1084/jem.20020732.

Vinardell, M. P., & Mitjans, M. (2015). Antitumor activities of metal oxide nanoparticles. *Nanomaterials*, *5*(2), 1004–1021, MDPI AG. Available from https://doi.org/10.3390/nano5021004.

Vujanovic, L., & Butterfield, L. H. (2007). Melanoma cancer vaccines and anti-tumor T cell responses. *Journal of Cellular Biochemistry*, *102*(2), 301–310. Available from https://doi.org/10.1002/jcb.21473.

Waeckerle-Men, Y., & Groettrup, M. (2005). PLGA microspheres for improved antigen delivery to dendritic cells as cellular vaccines. *Advanced Drug Delivery Reviews*, *57*(3 SPEC. ISS), 475–482, Elsevier B.V. Available from https://doi.org/10.1016/j.addr.2004.09.007.

Wahajuddin., & Arora, S. (2012). Superparamagnetic iron oxide nanoparticles: Magnetic nanoplatforms as drug carriers. *International Journal of Nanomedicine*, *7*, 3445–3471. Available from https://doi.org/10.2147/IJN.S30320.

Wang, B., Zhang, J., Chen, C., Xu, G., Qin, X., Hong, Y., Bosc, D. D., Qiu, F., & Zou, Z. (2018). The size of zinc oxide nanoparticles controls its toxicity through impairing autophagic flux in A549 lung epithelial cells. *Toxicology Letters*, *285*, 51–59. Available from https://doi.org/10.1016/j.toxlet.2017.12.025.

Wang, C. Y., Böttcher, C., Bahnemann, D. W., & Dohrmann, J. K. (2003). A comparative study of nanometer sized Fe(III)-doped TiO2 photocatalysts: Synthesis, characterization and activity. *Journal of Materials Chemistry*, *13*(9), 2322–2329. Available from https://doi.org/10.1039/b303716a.

Wang, D., Xu, Z., Yu, H., Chen, X., Feng, B., Cui, Z., Lin, B., Yin, Q., Zhang, Z., Chen, C., Wang, J., Zhang, W., & Li, Y. (2014). Treatment of metastatic breast cancer by combination of chemotherapy and photothermal ablation using doxorubicin-loaded DNA wrapped gold nanorods. *Biomaterials*, *35*(29), 8374–8384. Available from https://doi.org/10.1016/j.biomaterials.2014.05.094.

Wang, J., Lee, J. S., Kim, D., & Zhu, L. (2017). Exploration of Zinc Oxide Nanoparticles as a Multitarget and Multifunctional Anticancer Nanomedicine. *ACS Applied Materials and Interfaces*, *9*(46), 39971–39984. Available from https://doi.org/10.1021/acsami.7b11219.

Wang, N., Wang, Z., Xu, Z., Chen, X., & Zhu, G. (2018). A Cisplatin-Loaded Immunochemotherapeutic Nanohybrid Bearing Immune Checkpoint Inhibitors for

Enhanced Cervical Cancer Therapy. *Angewandte Chemie - International Edition*, *57*(13), 3426–3430. Available from https://doi.org/10.1002/anie.201800422.

Wang, S., Zhang, Q., Yang, P., Yu, X., Huang, L. Y., Shen, S., & Cai, S. (2016). Manganese Oxide-Coated Carbon Nanotubes As Dual-Modality Lymph Mapping Agents for Photothermal Therapy of Tumor Metastasis. *ACS Applied Materials and Interfaces*, *8*(6), 3736–3743. Available from https://doi.org/10.1021/acsami.5b08087.

Wang, Z., Ma, R., Yan, L., Chen, X., & Zhu, G. (2015). Combined chemotherapy and photodynamic therapy using a nanohybrid based on layered double hydroxides to conquer cisplatin resistance. *Chemical Communications*, *51*(58), 11587–11590. Available from https://doi.org/10.1039/c5cc04376j.

Weissleder, R. (2001). A clearer vision for in vivo imaging: Progress continues in the development of smaller, more penetrable probes for biological imaging. *Nature Biotechnology*, *19*(4), 316–317. Available from https://doi.org/10.1038/86684.

Xia, X., Wang, Y., Ruditskiy, A., & Xia, Y. (2013). 25th anniversary article: Galvanic replacement: A simple and versatile route to hollow nanostructures with tunable and well-controlled properties. *Advanced Materials*, *25*(44), 6313–6333. Available from https://doi.org/10.1002/adma.201302820.

Xu, L., Liang, H. W., Yang, Y., & Yu, S. H. (2018). Stability and Reactivity: Positive and Negative Aspects for Nanoparticle Processing. *Chemical Reviews*, *118*(7), 3209–3250, American Chemical Society. Available from https://doi.org/10.1021/acs.chemrev.7b00208.

Yan, X., Liu, J., Huang, J., Huang, M., He, F., Ye, Z., Xiao, W., Hu, X., & Luo, Z. (2014). Electrical stimulation induces calcium-dependent neurite outgrowth and immediate early genes expressions of dorsal root ganglion neurons. *Neurochemical Research*, *39*(1), 129–141. Available from https://doi.org/10.1007/s11064-013-1197-7.

Yang, C., & Suda, T. (2018). Hyperactivated mitophagy in hematopoietic stem cells news-and-views. *Nature Immunology*, *19*(1), 2–3, Nature Publishing Group. Available from https://doi.org/10.1038/s41590-017-0008-8.

Yang, G., Xu, L., Chao, Y., Xu, J., Sun, X., Wu, Y., Peng, R., & Liu, Z. (n.d.). *Hollow MnO 2 as a tumor-microenvironment-responsive biodegradable nano-platform for combination therapy favoring antitumor immune responses*. https://doi.org/10.1038/s41467-017-01050-0

Yang, Y., Zhang, Y. M., Chen, Y., Zhao, D., Chen, J. T., & Liu, Y. (2012). Construction of a graphene oxide based noncovalent multiple nanosupramolecular assembly as a scaffold for drug delivery. *Chemistry - A European Journal*, *18*(14), 4208–4215. Available from https://doi.org/10.1002/chem.201103445.

Yigit, M. V., Moore, A., & Medarova, Z. (2012). Magnetic nanoparticles for cancer diagnosis and therapy. *Pharmaceutical Research*, *29*(5), 1180–1188. Available from https://doi.org/10.1007/s11095-012-0679-7.

Yin, H., Casey, P. S., Mccall, M. J., & Fenech, M. (2015). Size-dependent cytotoxicity and genotoxicity of ZnO particles to human lymphoblastoid (WIL2-NS) cells. *Environmental and Molecular Mutagenesis*, *56*(9), 767–776. Available from https://doi.org/10.1002/em.21962.

Yoon, H. Y., Selvan, S. T., Yang, Y., Kim, M. J., Yi, D. K., Kwon, I. C., & Kim, K. (2018). Engineering nanoparticle strategies for effective cancer immunotherapy. *Biomaterials*, *178*, 597–607. Available from https://doi.org/10.1016/j.biomaterials.2018.03.036.

You, D. G., Deepagan, V. G., Um, W., Jeon, S., Son, S., Chang, H., Yoon, H. I., Cho, Y. W., Swierczewska, M., Lee, S., Pomper, M. G., Kwon, I. C., Kim, K., & Park, J. H. (2016).

ROS-generating TiO2 nanoparticles for non-invasive sonodynamic therapy of cancer. *Scientific Reports*, *6*(1), 1−12. Available from https://doi.org/10.1038/srep23200.

Youkhana, E. Q., Feltis, B., Blencowe, A., & Geso, M. (2017). Titanium dioxide nanoparticles as radiosensitisers: An in vitro and phantom-based study. *International Journal of Medical Sciences*, *14*(6), 602−614. Available from https://doi.org/10.7150/ijms.19058.

Yu, M. K., Jeong, Y. Y., Park, J., Park, S., Kim, J. W., Min, J. J., Kim, K., & Jon, S. (2008). Drug-loaded superparamagnetic iron oxide nanoparticles for combined cancer imaging and therapy in vivo. *Angewandte Chemie - International Edition*, *47*(29), 5362−5365. Available from https://doi.org/10.1002/anie.200800857.

Zardavas, D., Baselga, J., & Piccart, M. (2013). Emerging targeted agents in metastatic breast cancer. *Nature Reviews Clinical Oncology*, *10*(4), 191−210. Available from https://doi.org/10.1038/nrclinonc.2013.29.

Zhang, L., Xia, J., Zhao, Q., Liu, L., & Zhang, Z. (2010). Functional graphene oxide as a nanocarrier for controlled loading and targeted delivery of mixed anticancer drugs. *Small*, *6*(4), 537−544. Available from https://doi.org/10.1002/smll.200901680.

Zhang, Y., Chen, Y., Westerhoff, P., Hristovski, K., & Crittenden, J. C. (2008). Stability of commercial metal oxide nanoparticles in water. *Water Research*, *42*(8−9), 2204−2212. Available from https://doi.org/10.1016/j.watres.2007.11.036.

Zhang, Z., Wang, J., & Chen, C. (2013). Near-infrared light-mediated nanoplatforms for cancer thermo-chemotherapy and optical imaging. *Advanced Materials*, *25*(28), 3869−3880. Available from https://doi.org/10.1002/adma.201301890.

Zhou, F., Xing, D., Ou, Z., Wu, B., Resasco, D. E., & Chen, W. R. (2009). Cancer photothermal therapy in the near-infrared region by using single-walled carbon nanotubes. *Journal of Biomedical Optics*, *14*(2), 021009. Available from https://doi.org/10.1117/1.3078803.

Zhou, H., Zhao, K., Li, W., Yang, N., Liu, Y., Chen, C., & Wei, T. (2012). The interactions between pristine graphene and macrophages and the production of cytokines/chemokines via TLR- and NF-κB-related signaling pathways. *Biomaterials*, *33*(29), 6933−6942. Available from https://doi.org/10.1016/j.biomaterials.2012.06.064.

Metal oxides for cosmetics and sunscreens

4

Tandrima Banerjee[1] *and Abhijit Samanta*[2]
[1]Department of Chemical Sciences, Indian Institute of Science Education and Research (IISER), Kolkata, India, [2]School of Science and Technology, The Neotia University, Sarisha, India

4.1 Introduction

Throughout the ages, humans have used physical barriers such as hats, umbrellas, and cloths wrapped over the head and face (Olson, 2009). Tars and oils were also used to enhance their beauty and for the purposes of cleaning, perfuming, protection, changing the appearance, disguising body odors, and keeping the skin in good condition. These endeavors have been mostly directed at avoiding the darkening effects of ultraviolet (UV) radiation from sunlight as, in some cultures, a fairer skin tone, especially for women's beauty as perceived by society, was expected (Benjatikul, Mahamongkol, & Wongtrakul, 2020; Cole, Shyr, & Ou-Yang, 2016; Hanneman, Cooper, & Baron, 2006; Jou, Feldman, & Tomecki, 2012; Neale, Williams, & Green, 2002). The effects of prolonged sunlight exposure, that include its carcinogenicity and harmfulness to human skin cells, were not studied until the 1800s (Urbach, 2001). Hence it has become very important to produce materials that can prevent this damage. There has also been a significant boost to beauty industries in the last few decades through manufacturing a variety of cosmetics for different parts of the human body, such as the skin, hair, nails, lips, external genital organs, and teeth.

"Cosmetics" is a well-known term in maintaining good looks and attractiveness and is also identified as helping with mental health. These cosmetics are mainly required for skin, hair, nail, and teeth care for cleansing and imparting attractive properties. The major properties of cosmetics are cleansing, beautifying and altering the appearance, adding fragrance, stopping the development of bad odors, and not having any adverse side effects. Sunscreens are also important cosmetics that can protect the skin from inflammation, immunosuppression, early photo-aging, DNA damage, and photocarcinogenesis caused by dangerous levels of UV radiation from sunlight (Marrot et al., 2004). Sunscreens prevent direct cellular damage by reducing reactive oxygen species (ROS) in skin cells which can lead to an impairment of the endogenous antioxidant defense systems.

The "chemical" and "physical" ingredients are the two most important active constituents used in modern cosmetics. Different aromatic compounds with a carbonyl group such as benzophenones, cinnamates, dibenzoyl methanes or camphor

Metal Oxides for Biomedical and Biosensor Applications. DOI: https://doi.org/10.1016/B978-0-12-823033-6.00002-8

derivatives, para-amino benzoates (PABA), and salicylates, are the main active "chemical" ingredients present in cosmetics, and different metal oxides particles such as ZnO, TiO_2, Fe_2O_3, CeO_2, and ZrO_2 are used as the "physical" constituents (Huber, 2005; Husen & Siddiqi, 2014; Larese, Gianpietro, Venier, Maina, & Renzi, 2007; Midander, Wallinder, & Leygraf, 2007). These components are usually both electron donors as well as electron acceptors which can easily absorb the UV radiation of sunlight (Maier & Korting, 2005; Wolf, Matz, Orion, & Lipozenčić, 2003).

There are many different compositions of cosmetics currently available from a number of manufacturers. However, the development of cosmetics brings many technical and toxicological issues, which stem from the requirement of having contrasting ingredients with opposite needs in the same formula. They are made of chemicals that do not leave heavy metals that have a toxic effect on the body, but metal oxides are one of the major components in cosmetics as well as sunscreens. The development of good-quality cosmetics is required to be maintained by the producers and controlled by the regulatory bodies. Different metal oxides such as TiO_2, ZnO, Fe_2O_3, and Al_2O_3 are also used as ingredients in several cosmetics (Larese et al., 2007; Midander et al., 2007). Typically the pigment size ranges between 20–150 nm for TiO_2 and 40–100 nm for ZnO when used in sunscreens to absorb and scatter UV radiation, which make sunscreens appear transparent on the skin, which is a desirable quality. The growing applications of nanotechnology has led to them being widely employed in the fields of nanoelectronics, nanomaterials, biology, and medicine (Hofmann-Amtenbrink, Hofmann, Hool, & Roubert, 2014; Roco, 2011). The nanoparticles (NPs) of different metal oxides have been utilized in the manufacture of cosmetics and sunscreens globally as they can easily penetrate through biological barriers and cell membranes (Biener et al., 2009; Cristina, I, & Kevin, 2007; De, Ghosh, & Rotello, 2008; Draelos, 2004; Kulinowski, 2004; Morabito, Shapley, Steeley, & Tripathi, 2011; Raj, Jose, Sumod, & Sabitha, 2012). Globally, the major nanometal oxides used in cosmetics and sunscreens are TiO_2NPs, ZnONPs, SiO_2NPs, γ-Fe_2O_3NPs, Fe_3O_4NPs, and Al_2O_3NPs (Pepić, Vujičić, Lovrić, & Filipović-Grčić, 2012; Schmid & Riediker, 2008). It has been reported that the market for NP-containing cosmetics and sunscreens has grown remarkably in recent years (Ngoc, 2019) and a survey report has shown that the supply of cosmetics products includes approximately 70%–80% of TiO_2NPs and 70% of ZnONPs (Fabiano, Fadri, Stefan, & Bernd, 2012; Niska, Zielinska, Radomski, & Inkielewicz-Stepniak, 2018). This study discusses the applications and effects of different metal oxides and their NPs in cosmetics and sunscreens.

4.2 Applications of metal oxides in cosmetics and sunscreens

Various metal oxides and their NPs have attracted attention from the cosmetics industries largely due to the broad spectrum of their biological activities and their unique physicochemical properties. The Nanodermatology Society was established in 2010 to permit and control the regulations of applications of nanotechnology in cosmetics and sunscreens with respect to their scientific and medical aspects (Nasir, Wang, & Friedman, 2011). Metal oxides have attracted enormous interest from cosmetologists with advanced knowledge of chemistry, toxicology, dermatology, rheology, and even marketing. The applications of different metal oxides as one of the active ingredients in cosmetics and sunscreens are discussed below.

4.2.1 Titanium dioxide (TiO_2) and zinc oxide (ZnO)

Inorganic metal oxides, titanium dioxide (TiO_2) and zinc oxide (ZnO), are utilized as the main active ingredients in many cosmetics and sunscreen compositions to prevent skin damage from UV radiation. TiO_2 and ZnO are popular because they are insoluble and do not undergo any chemical decomposition, creating a layer on the top of the skin and they are not absorbed systemically. Hence, these particles represent alternative and safer chemicals in cosmetics and sunscreens. Moreover, they are widely recommended for protection from skin damage in preference to other organic compounds.

TiO_2 appears as a white powder and titanium dioxide nanoparticles (TiO_2NPs) are preferred in cosmetics manufacturing as it avoids white coloration of the skin after application. The study of percutaneous absorption of microfine TiO_2 from sunscreens has been carried out in skin surgery for skin biopsies and it has been observed that the titanium concentration after absorption through the skin is higher than in controls (Tan, Commens, Burnett, & Snitch, 1996). These results were contradicted by other researchers who found that penetration of TiO_2 was not found in viable layers of the epidermis. Lademann and coworkers found that a small amount of metal oxide was detected in the hair follicles of volunteers (Lademann et al., 1999). An optical and electron microscopy study for surface and particle size characterizations of micronized TiO_2 did not observed any dermal absorption of these pigments (Schulz et al., 2002).

The particles of micronized TiO_2 were exclusively set down on the outermost layer in the human epidermis, but not in deeper surfaces of the dermis. A sunscreen containing TiO_2 nanoparticles has been formulated and experimented on some volunteers in vivo by Mavon and coresearchers (Mavon, Miquel, Lejeune, Payre, & Moretto, 2007). Transmission electron microscopy (TEM) and particle-induced X-ray emission (PIXE) techniques were used to locate the TiO_2 in the skin in the same experiments. There were no traces of TiO_2 found in the hair follicles, viable epidermis, or dermis in in vivo or in vitro permeation studies. It was found that more than 90% of the applied sunscreen was recovered and remaining 10% was confined to a small area in the furrows and the opened infundibulum. Menzel, Reinert, Vogt and Butz (2004) noticed the presence of TiO_2 in the deeper viable epidermal layers of freeze-dried cross-sections of pig skin biopsies using spatially resolved ion beam analysis (PIXE, RBS, STIM, and secondary electron imaging), while other authors observed that TiO_2NPs do not enter through the stratum corneum of human foreskin transplants (Kiss et al., 2008; Zs. et al., 2005). The

NANODERM project was carried out for the formulation of TiO_2NP-containing sunscreens and confirmed no indication of NP transcutaneous penetration [(NANODERM (2007) Quality of Skin as a Barrier to Ultra-Fine Particles. Final Report, 2007)]. The European project EDETOX specified the protocols on intact and damaged human skin using in vitro skin absorption of TiO_2NPs (Williams, 2004) which correlated with the studies of other types of NPs (Larese et al., 2007; Larese Filon et al., 2011, 2013).

However, the penetration of microfine TiO_2 in different formulations reached deeper into the epidermis or dermis of human skin from an oily diffusion than from an aqueous one. Sometimes, greater penetration was also caused by encapsulation of the pigments into liposomes, as observed by the same authors. Additionally, greater penetration was reported for hairy skin due to surface penetration through hair follicles or pores. An experimental study has shown that the average concentration of Ti after 24 h of skin interaction was below the level of detection (LOD), which indicated damage to the epidermal layer of intact skin (Crosera et al., 2015) (Fig. 4.1).

In vitro absorption of TiO_2 and ZnO was investigated in cosmetic formulations via porcine skin by Gamer and coworkers (Gamer, Leibold, & Van Ravenzwaay, 2006). Modified Franz static dermal penetration cells were tested using one ZnO and two TiO_2 formulations of cosmetics by the same authors who concluded that neither microfine ZnO nor microfine TiO_2 was capable of passing through porcine skin. Cross et al. (2007) investigated skin absorption of ZnO using Franz-type diffusion cells. They compared two different compositions of cosmetics of 26–30 nm ZnO particles and one cream formulation with no ZnONPs. The experimental

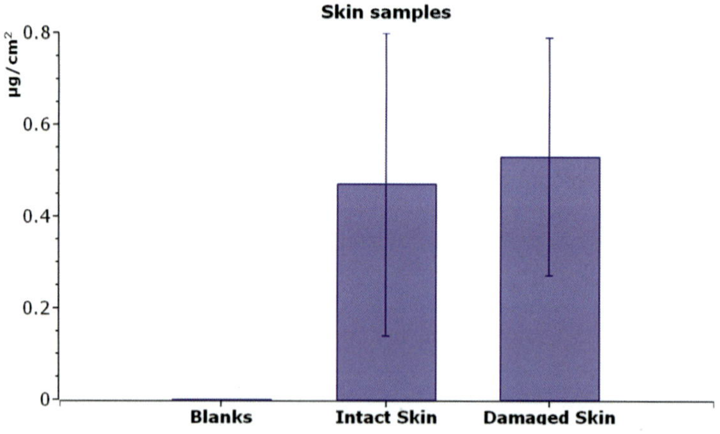

Figure 4.1 Ti content inside the skin of blank cells, intact skin, and damaged skin (by TiO_2NPs). From Crosera, M, Prodi, A, Mauro, M. Pelin, M. Florio, C, Bellomo, F,....Filon, F.L. (2015). Titanium dioxide nanoparticle penetration into the skin and effects on HaCaT cells. *International Journal of Environmental Research and Public Health*. 12(8),9282–97 https://doi.org/10.3390/ijerph120809282.

results showed that <0.03% of the applied zinc was detected in the receptor phase and no zinc content was found in the viable epidermis by electron microscopic study, indicating that minimal nanoparticle penetration occurs through the human epidermis. Numerous research studies have recommended that metal oxide NPs can show photoelectric activities in skin cells by forming free radicals which can damage DNA (Dunford et al., 1997; Fujishima, Shuin, Hashimoto, & Fujishima, 1992; Serpone, Salinaro, & Emeline, 2001; Wamer, Yin, & Wei, 1997). This effect may disrupt the normal dermal layers, their functions, and cell feasibility (Sayes et al., 2006); hence discussion is still open to evaluate their safety and percutaneous penetration for use in human skin.

4.2.2 Iron oxides (Fe_2O_3 or FeO)

Fe_2O_3 is used in a wide array of cosmetic products from eye shadow to talcum powder —even products that are marketed as natural or organic. They are naturally occurring mineral deposits and utilized as color shades in cosmetics, however the iron oxides in cosmetics and personal care products are prepared synthetically. Synthetically produced Fe_2O_3 is applied in formulations for a wide variety of cosmetics, including makeup and skin care preparations. The different shades of cosmetics vary from orange, red, yellow, to black depending on the specific Fe_2O_3 or mixtures of iron oxide pigments used. Iron oxides are gentle and nontoxic in cosmetic products placed on the skin surface; usually they do not irritate the skin and they are not known to be allergenic. Iron oxides typically do not cause problems even for people with sensitive skin. Contamination levels with heavy metals in iron oxides of cosmetics formulations are controlled by the Food and Drug Administration (FDA) due to safety and risks to human health. Arlettaz, Christe, Surai and Pape Møller (2002) proposed that those iron oxides with pro-oxidant effects to kill bacteria and increase the mobility of vitamin A and also functioned as scavengers of free radicals for antioxidants (Surai, Speake, & Sparks, 2001). Berry, Charles, Wells, Dalby and Curtis (2004) observed that underivatized Fe_2O_3NPs caused rapid disruption to the cell cytoskeleton and reduced proliferation. They also stimulated cell proliferation and were not internalized, but became visible when connecting to the external parts of the cell membranes. Some smaller FeONPs (size <10 nm) were able to penetrate through the viable epidermis and hair follicle orifices and reached the deepest layers of the stratum granulosum and hair follicles (Baroli et al., 2007).

Yu and Li (2011) investigated the activities of ZnONPs, TiO_2NPs, and FeONPs using human colon adenocarcinoma HT29 cells by expressing fluorescence resonance energy transfer (FRET)-based reporters for caspase-3 and caspase-9 indicators, SCAT3 and SCAT9, respectively. Fig. 4.2 displays no caspase activation was observed after treatment for 18 h. Staurosporine (STS), the positive control, induced caspase activation and showed loss of FRET, which indicated that the assay was working. This result varies from related investigations, where it was stated that ZnONPs and TiO_2NPs would result in apoptosis (Deng et al., 2009; Hussain et al., 2010).

Some authors have reported the morphologies of cells after 18 h treatment and found that cells were normal with TiO_2NPs and FeONPs (up to 100 μg/mL), but rounded up and died with ZnONPs with 5–20 μg/mL, which could not be inhibited

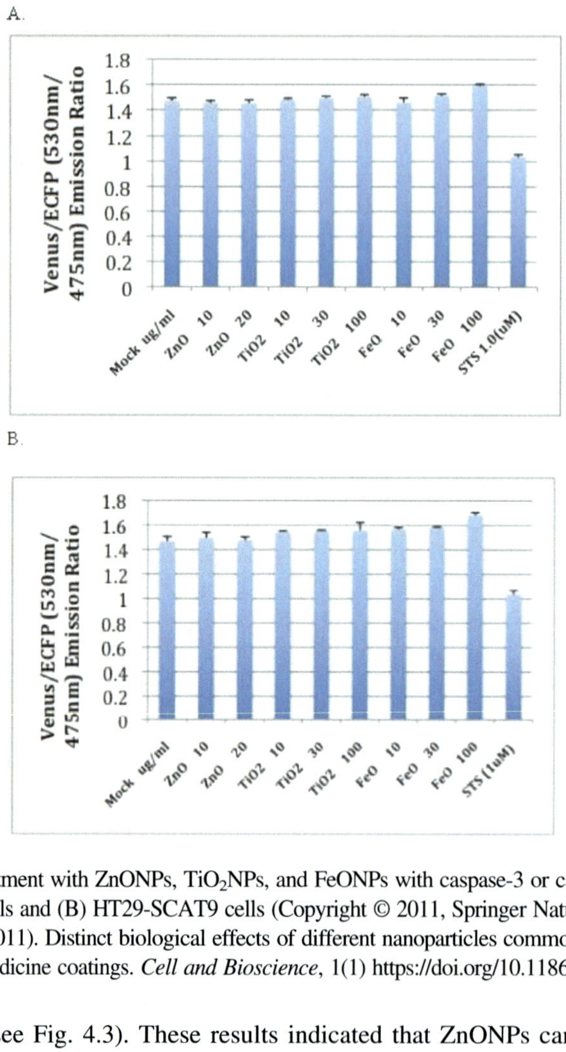

Figure 4.2 Treatment with ZnONPs, TiO_2NPs, and FeONPs with caspase-3 or caspase-9 for (A) HT29-SCAT3 cells and (B) HT29-SCAT9 cells (Copyright © 2011, Springer Nature). From Yu, J. X., & Li, T.H. (2011). Distinct biological effects of different nanoparticles commonly used in cosmetics and medicine coatings. *Cell and Bioscience*, 1(1) https://doi.org/10.1186/2045-3701-1-19.

by IDN6556 (see Fig. 4.3). These results indicated that ZnONPs can result in caspase activation via a nonapoptosis path which matched with preexisting facts (Linton et al., 2005; Yuan, Lipinski, & Degterev, 2003).

4.2.3 Chromium oxides (Cr_2O_3)

Another metal oxide, chromium dioxide or chromium oxide green, is a pigment that ranges from dull olive green, to a blue green, or bright green, that is used in various cosmetics formulations. It is extensively used in the production of hair-coloring products, makeup, soap, nail polish, and other skin care products. It provides superior color in purity and saturation, excellent dispersibility, and no aggregate formation. Sometimes it can be dyed with TiO_2 to create different green shades. Although the EU

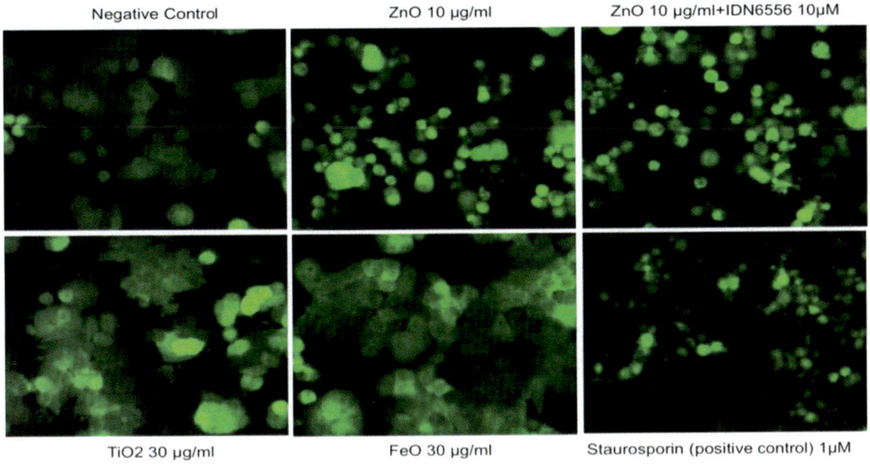

Figure 4.3 Morphological study of HT29-SCAT3 cells treated with ZnONPs, TiO_2NPs, and FeONPs (Copyright 2011, Springer Nature). From Yu, J.X., & Li, T.H. (2011). Distinct biological effects of different nanoparticles commonly used in cosmetics and medicine coatings. *Cell and Bioscience*, 1(1) https://doi.org/10.1186/2045-3701-1-19.

has approved the use of these kinds of different green shades for lips, eyes, and face products, the FDA has prohibited it in the United States. However, its use in other cosmetics like foundations, blushes, bronzers, and similar products was confirm after meeting FDA specifications.

4.2.4 Other metal oxides

Besides the above-mentioned metal oxides, researchers have developed other metal oxides and their NPs for applications in cosmetics and sunscreen formulations. Metal oxides can release metal ions in biological systems which can subsequently penetrate through the skin, but NPs were much less able to pass through. Cobalt oxide NPs (Co_3O_4NPs) and micro-sized particles were applied to human fibroblasts and the cytotoxic and genotoxic effects compared (Papageorgiou et al., 2007). Metal oxide NPs can be used as anticancer agents. Mauro et al. (2015) investigated the absorption and toxicity of keratinocytes of Co_3O_4NPs in human skin and observed that these showed in vitro membrane damage and genotoxicity, but not cytotoxicity to keratinocytes.

Steckiewicz and Inkielewicz-Stepniak (2020) reported that dextran-coated cerium oxide nanoparticles (CeO_2NPs) were noncancerous and more effectual in an acid medium against osteosarcoma cells. The cytotoxicity of CeO_2NPs was minimal to noncancerous bone cells in the same acidic conditions and the effects of ZnONPs also had the similar results (Alpaslan, Yazici, Golshan, Ziemer, & Webster, 2015; Sisubalan et al., 2018). The anticancer activity of TiO_2NPs was induced by aluminum oxide nanoparticles (Al_2O_3NPs) (Di Virgilio, Reigosa, & Lorenzo De Mele, 2010). In some sunscreens, trace amounts of SiO_2, aluminum silicates, iron oxide, and MgO are used as minor components with other metal oxides with $CaCO_3$ as the major component (Benjatikul et al., 2020; Davies, Scully, & Preston, 2010; Olson, 2009).

4.3 Mechanism of dermal absorption and toxicity of metal oxides in human skin

Exposure to UV radiation can cause ROS in human skin cells and damage to DNA in cells, leading to skin cancer (Kvam & Tyrrell, 1997). Hence, effective chemical or physical filters are required to protect the skin against both UVA and UVB radiation and they act as micromirrors against cancer development. Metal oxides and metal oxide NPs may be used as skin shields (Raj et al., 2012). However, these metal oxides and their NPs may also have some toxic effects on human skin as UV light can increase the phototoxic and photoallergic effects of NPs. ZnONPs and TiO_2NPs can enhance ROS generation systems (Dayem et al., 2017; Tran & Salmon, 2011). These effects can cause photocarcinogenicity and premature aging of the skin. The mechanism of ROS generation by NPs primarily depends on three major factors: (1) prooxidant functional groups of NPs, (2) active redox cycling of NPs due to transition metals, and (3) interactions between cells and particles (Manke, Wang, & Rojanasakul, 2013). The electrons of nanostructured ZnO or TiO_2 become photoexcited due to the absorption of UV radiation and then these excited electrons or electron–hole pairs can react with oxygen or water and form free radicals (Fu, Xia, Hwang, Ray, & Yu, 2014). However, these two metal oxides can easily generate highly oxidizing radicals such as hydroxyl radical (·OH) and superoxide anion ($O_2^{-\cdot}$ as well as other cytotoxic and genotoxic species singlet oxygen ($^1O_2$) and hydrogen peroxide (H_2O_2) (Fu et al., 2014; He et al., 2016) (see (Fig. 4.4)). In this reaction,

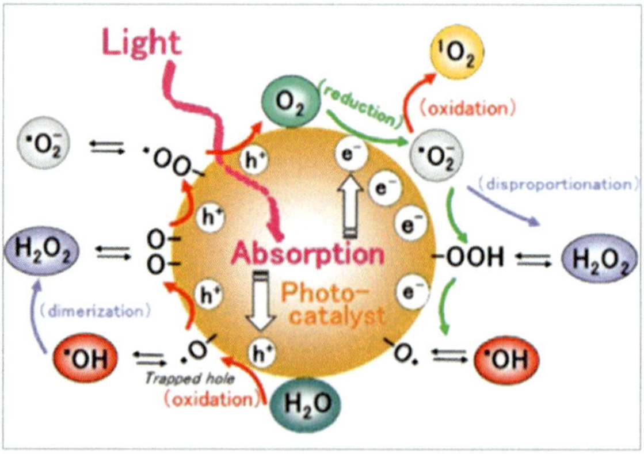

Figure 4.4 ROS generatation in the photocatalytic reduction and oxidation steps of oxygen and water (Copyright 2017, American Chemical Society). From Nosaka, Y., & Nosaka, A. Y. (2017). Generation and Detection of Reactive Oxygen Species in Photocatalysis. *Chemical Reviews*, 117(17), 11302–11336 https://doi.org/10.1021/acs.chemrev.7b00161.

firstly the superoxide anion radical ($O_2^{-\cdot}$) is produced by transfer of an electron from photo-excited TiO_2 to molecular oxygen leads (see Equation 6.1) and then formation of hydroxyl radicals can be done by electron release from water (see Eq. 6.2).

$$TiO_2 + O_2 \xrightarrow{h\nu} TiO_2^{+\cdot} + O_2^{-\cdot} \quad (6.1)$$

$$H_2O \xrightarrow{TiO_2\ h\nu} H^+ + e + {}^{\cdot}OH \quad (6.2)$$

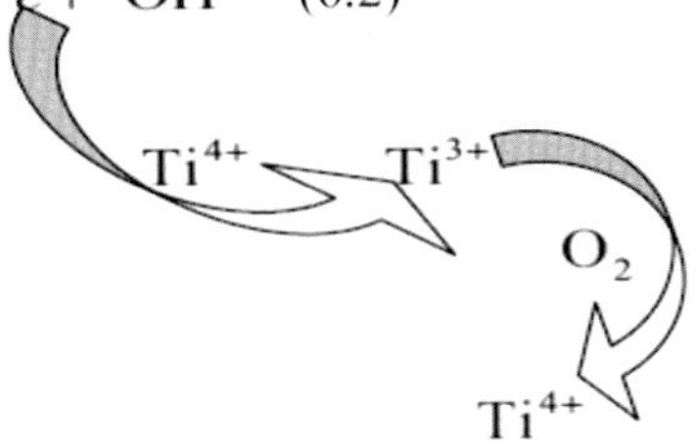

The Ti^{4+} ions are produced again from Ti^{3+} ions by a reoxidation process and ZnO also shows similar mechanisms. Hence, Al_2O_3 or SiO_2 coatings are used with TiO_2 for sun protection cream in order to prevent oxygen radical formation. Several researchers have quantified and compared dynamic ROS generation using some common types of metal oxide NPs ($nTiO_2$, nZnO, $nCeO_2$, nV_2O_5, nFe_2O_3, and nAl_2O_3) by continuous-flow chemiluminescence (CFCL) methods (Dabin et al., 2015; Wang, Zhao, Ma, Zhang, & Guo, 2017; Wang, Zhao, Guo, & Zhang, 2014). It has been reported that initially the production of $O_2\cdot^-$ increased sharply and continued to rise until it ultimately flattened out. The average concentration of generated $O_2\cdot^-$ followed the order $nTiO_2 > nZnO > nV_2O_5 > nFe_2O_3 > nCeO_2 > nAl_2O_3$ and $bTiO_2 > bZnO$ (see Fig. 4.5).

The penetration of NPs mainly depends on their size, shape, chemical composition, surface area, and charge. In addition, skin textures and state are also key factors for skin absorption of NPs. Several studies have recommended the ranges of sizes of NPs and their action to skin (Baroli et al., 2007; Dayem et al., 2017; Nosaka & Nosaka, 2017) showing that a larger size of NPs ($>$45 nm) can neither permeate nor penetrate the skin, NPs of 21–45 nm in size could permeate and enter only damaged skin, and on the other hand smaller sizes ($\leq$4 nm) can penetrate and leak into intact skin, with NPs of 4–20 nm potentially going through and saturating intact and damaged skin (see Fig. 4.6). Sometimes, intracellular ROS by NPs can cause mitochondrial respiratory chain disruption, oxidative stress on DNA cleavage, lipid peroxidation, and NADPH-like enzyme system activation (Khan, Naqvi, & Ahmad, 2015; Larese Filon, Mauro, Adami, Bovenzi, & Crosera, 2015). Hence more research is required to elucidate the effects of NPs under UV radiation in human skin. Most other heavy metals that are present in cosmetics and sunscreens lead to toxic effects in human skin. Various cosmetics such as lipstick, eye shadow, and face-whitening cream contain heavy metals like cadmium and lead. Various scientific reports have revealed that several eminent leading cosmetic brands containing metal oxides and their NPs show genotoxicity and cytotoxicity toward

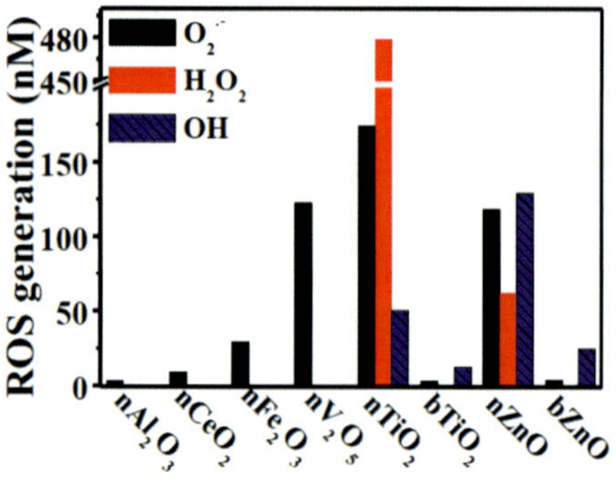

Figure 4.5 Generation of ROS using some common metal oxides NPs (Copyright 2017, American Chemical Society). From Wang, D., Zhao, L., Ma, H., Zhang, H., & Guo, L.H. (2017). Quantitative analysis of reactive oxygen species photogenerated on metal oxide nanoparticles and their bacteria toxicity: The role of superoxide radicals. *Environmental Science and Technology*, 51(17), 10137−10145 https://doi.org/10.1021/acs.est.7b00473.

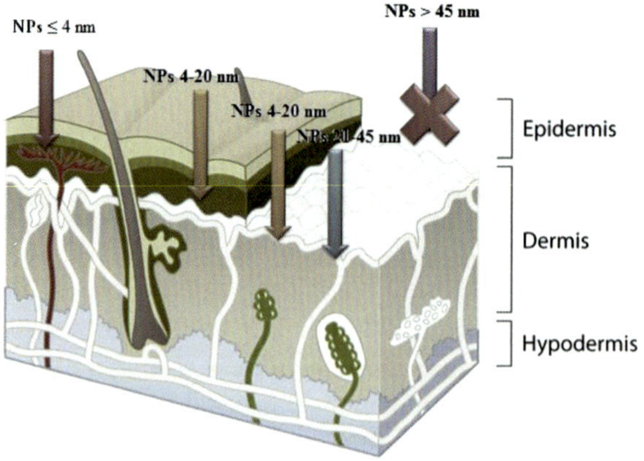

Figure 4.6 Penetration of nanoparticles through human skin (© 2017 Elsevier B.V. All rights reserved). From Niska, K., Zielinska, E., Radomski, M.W., & Inkielewicz-Stepniak, I. (2018). Metal nanoparticles in dermatology and cosmetology: Interactions with human skin cells. Chemico-Biological Interactions, 295, 38−51 https://doi.org/10.1016/j.cbi.2017.06.018.

human skin (Ivask et al., 2014; Yin et al., 2012). Metal oxide nanoparticles are more cytotoxic compared with metal oxides as the particle size decreases (Oberdörster, Oberdörster, & Oberdörster, 2005). For example, with a lower TiO_2 concentration, the toxicity is controlled by particle nature, dimension, and exposure time, whereas at a higher TiO_2 concentration, toxicity is influenced by nanoparticle aggregation (Nasterlack, Zober, & Oberlinner, 2008; Progress toward Safe Nanotechnology in the Workplace—a Report from the NIOSH Nanotechnology Research Center. DHHS (NIOSH) Publication No, 2007; Schulte, Geraci, Zumwalde, Hoover, & Kuempe, 2008). The threat of exposure of dermal contact by respiration and ingestion has been elevated with daily use of ZnONP- and TiO_2NP-based cosmetics (Lee, Lin, Hou, Li, & Chang, 2020). Recently, long-term exposure to such products has been shown to pose undesirable health hazards and several consequences, and also an enormous adverse impact on the environment.

4.4 Discussion and conclusions

Despite the recent advances in the production of cosmetics and sunscreens, different metal oxides and their NPs are among the major ingredients used by leading producers due to their unique physicochemical properties. These can be directly cytotoxic to cancer cells and more effective in protection from sunlight UV radiation. Several factors including the type, size and shape, concentration, pH of the environment, and capping agents of metal oxide NPs affect their biological properties. They also act as antimicrobial agents. Sometimes, metal NPs can penetrate intact or damaged skin, through the dermal layer. The consequences can be clarified by the dynamic ROS generation using some common types of metal oxide NPs in agreement with numerous studies in the literature. The toxicity, health hazards, and environmental impact of heavy metals from metal oxides are discussed with evidence from several literature sources.

In future studies, a more insightful mechanism of metal oxides and their NPs in biological systems is required. Nanotoxicological studies should be perused to illustrate the positive characteristics of NPs in cosmetics and sunscreens and for a better understanding of the harmful effects of these materials. Further research and safety evaluations of metal oxides in cosmetics and sunscreens is required to avoid any side effects. Detailed information about the interactions between metal oxides and biological cells in terms of cytotoxicity, anticancer, and antimicrobial characteristics will help in the design of nanoparticle-based cosmetics formulations with extremely constructive dermatological/toxicological profiles. This study will be a powerful tool for improving the usage of metal oxides and their NPs in cosmetics; however there remains a lot to be done to understand the real-world scenarios of long-term applications.

References

Alpaslan, E., Yazici, H., Golshan, N. H., Ziemer, K. S., & Webster, T. J. (2015). PH-dependent activity of dextran-coated cerium oxide nanoparticles on prohibiting osteosarcoma cell proliferation. *ACS Biomaterials Science and Engineering*, *1*(11), 1096–1103. Available from https://doi.org/10.1021/acsbiomaterials.5b00194.

Arlettaz, R., Christe, P., Surai, P. F., & Pape Møller, A. (2002). Deliberate rusty staining of plumage in the bearded vulture: Does function precede art? *Animal Behaviour*, *64*(3), F1–F3. Available from https://doi.org/10.1006/anbe.2002.3097.

Baroli, B., Ennas, M. G., Loffredo, F., Isola, M., Pinna, R., & López-Quintela, M. A. (2007). Penetration of metallic nanoparticles in human full-thickness skin. *Journal of*

Investigative Dermatology, *127*(7), 1701–1712. Available from https://doi.org/10.1038/sj.jid.5700733.

Benjatikul, K., Mahamongkol, H., & Wongtrakul, P. (2020). Sunscreen properties of marl. *Journal of Oleo Science*, *69*(1), 73–82. Available from https://doi.org/10.5650/jos.ess19232.

Berry, C. C., Charles, S., Wells, S., Dalby, M. J., & Curtis, A. S. G. (2004). The influence of transferrin stabilised magnetic nanoparticles on human dermal fibroblasts in culture. *International Journal of Pharmaceutics*, *269*(1), 211–225. Available from https://doi.org/10.1016/j.ijpharm.2003.09.042.

Biener, J., Wittstock, A., Baumann, T. F., Weissmüller, J., Bäumer, M., & Hamza, A. V. (2009). Surface chemistry in nanoscale materials. *Materials*, *2*(4), 2404–2428. Available from https://doi.org/10.3390/ma2042404.

Cole, C., Shyr, T., & Ou-Yang, H. (2016). Metal oxide sunscreens protect skin by absorption, not by reflection or scattering. *Photodermatology Photoimmunology and Photomedicine*, *32*(1), 5–10. Available from https://doi.org/10.1111/phpp.12214.

Cristina, B., I, P. I., & Kevin, R. (2007). Nanomaterials and nanoparticles: Sources and toxicity. *Biointerphases*, MR17–MR71. Available from https://doi.org/10.1116/1.2815690.

Crosera, M., Prodi, A., Mauro, M., Pelin, M., Florio, C., Bellomo, F., ... Filon, F. L. (2015). Titanium dioxide nanoparticle penetration into the skin and effects on HaCaT cells. *International Journal of Environmental Research and Public Health*, *12*(8), 9282–9297. Available from https://doi.org/10.3390/ijerph120809282.

Cross, S. E., Innes, B., Roberts, M. S., Tsuzuki, T., Robertson, T. A., & McCormick, P. (2007). Human skin penetration of sunscreen nanoparticles: In-vitro assessment of a novel micronized zinc oxide formulation. *Skin Pharmacology and Physiology*, *20*(3), 148–154. Available from https://doi.org/10.1159/000098701.

Dabin, W., Lixia, Z., Lianghong, G., Hui, Z., Bin, W., & Yu, Y. (2015). Online quantification of O2•- (center dot-) and H_2O_2 and their formation kinetics in ultraviolet (UV)-irradiated nano-TiO_2 suspensions by continuous flow chemiluminescence. *Acta Chimia Sinica*, *73*, 388–394.

Davies, R., Scully, C., & Preston, A. J. (2010). Dentifrices-an update. *Medicina Oral, Patologia Oral y Cirugia Bucal*, *15*, 976–982.

Dayem, D. A., Hossain, M. K., Lee, S. B., Kim, K., Saha, S. K., Yang, G. M., ... Cho, S. G. (2017). The role of reactive oxygen species (ROS) in the biological activities of metallic nanoparticles. *International Journal of Molecular Sciences*, 18.

De, M., Ghosh, P. S., & Rotello, V. M. (2008). Applications of nanoparticles in biology. *Advanced Materials*, *20*(22), 4225–4241. Available from https://doi.org/10.1002/adma.200703183.

Deng, X., Luan, Q., Chen, W., Wang, Y., Wu, M., Zhang, H., & Jiao, Z. (2009). Nanosized zinc oxide particles induce neural stem cell apoptosis. *Nanotechnology*, *20*(11). Available from https://doi.org/10.1088/0957-4484/20/11/115101.

Di Virgilio, A. L., Reigosa, M., & Lorenzo De Mele, M. F. (2010). Response of UMR 106 cells exposed to titanium oxide and aluminum oxide nanoparticles. *Journal of Biomedical Materials Research - Part A*, *92*(1), 80–86. Available from https://doi.org/10.1002/jbm.a.32339.

Draelos, Z. D. (2004). What are nanoparticles and where do they go? *Journal of Cosmetic Dermatology*, *24*, 306–308.

Dunford, R., Salinaro, A., Cai, L., Serpone, N., Horikoshi, S., Hidaka, H., & Knowland, J. (1997). Chemical oxidation and DNA damage catalysed by inorganic sunscreen ingredients. *FEBS Letters*, *418*(1–2), 87–90. Available from https://doi.org/10.1016/S0014-5793(97)01356-2.

Fabiano, P., Fadri, G., Stefan, S., & Bernd, N. (2012). Industrial production quantities and uses of ten engineered nanomaterials in Europe and the world. *Journal of Nanoparticle Research*. Available from https://doi.org/10.1007/s11051-012-1109-9.

Fu, P. P., Xia, Q., Hwang, H. M., Ray, P. C., & Yu, H. (2014). Mechanisms of nanotoxicity: Generation of reactive oxygen species. *Journal of Food and Drug Analysis*, *22*(1), 64–75. Available from https://doi.org/10.1016/j.jfda.2014.01.005.

Fujishima, A., Shuin, T., Hashimoto, K., & Fujishima, A. (1992). Induction of cytotoxicity by photoexcited TiO_2 particles. *Cancer Research*, *52*(8), 2346–2348.

Gamer, A. O., Leibold, E., & Van Ravenzwaay, B. (2006). The in vitro absorption of microfine zinc oxide and titanium dioxide through porcine skin. *Toxicology In Vitro*, *20*(3), 301–307. Available from https://doi.org/10.1016/j.tiv.2005.08.008.

Hanneman, K. K., Cooper, K. D., & Baron, E. D. (2006). Ultraviolet immunosuppression: Mechanisms and consequences. *Dermatologic Clinics*, *24*(1), 19–25. Available from https://doi.org/10.1016/j.det.2005.08.003.

He, W., Jia, H., Cai, J., Han, X., Zheng, Z., Wamer, W. G., & Yin, J. J. (2016). Production of reactive oxygen species and electrons from photoexcited ZnO and ZnS nanoparticles: A comparative study for unraveling their distinct photocatalytic activities. *Journal of Physical Chemistry C*, *120*(6), 3187–3195. Available from https://doi.org/10.1021/acs.jpcc.5b11456.

Hofmann-Amtenbrink, M., Hofmann, H., Hool, A., & Roubert, F. (2014). Nanotechnology in medicine: European research and its implications. *Swiss Medical Weekly*, 144. Available from https://doi.org/10.4414/smw.2014.14044.

Huber, D. L. (2005). Synthesis, properties, and applications of iron nanoparticles. *Small (Weinheim an der Bergstrasse, Germany)*, *1*(5), 482–501. Available from https://doi.org/10.1002/smll.200500006.

Husen, A., & Siddiqi, K. S. (2014). Phytosynthesis of nanoparticles: Concept, controversy and application. *Nanoscale Research Letters*, *9*(1), 1–24. Available from https://doi.org/10.1186/1556-276X-9-229.

Hussain, S., Thomassen, L. C. J., Ferecatu, I., Borot, M. C., Andreau, K., Martens, J. A., ... Boland, S. (2010). Carbon black and titanium dioxide nanoparticles elicit distinct apoptotic pathways in bronchial epithelial cells. *Particle and Fibre Toxicology*, *7*. Available from https://doi.org/10.1186/1743-8977-7-10.

Ivask, A., Kurvet, I., Kasemets, K., Blinova, I., Aruoja, V., Suppi, S., ... Kahru, A. (2014). Size-dependent toxicity of silver nanoparticles to bacteria, yeast, algae, crustaceans and mammalian cells in vitro. *PLOS ONE*, *9*(7). Available from https://doi.org/10.1371/journal.pone.0102108.

Jou, P. C., Feldman, R. J., & Tomecki, K. J. (2012). UV protection and sunscreens: What to tell patients. *Cleveland Clinic Journal of Medicine*, *79*(6), 427–436. Available from https://doi.org/10.3949/ccjm.79a.11110.

Khan, M., Naqvi, A. H., & Ahmad, M. (2015). Comparative study of the cytotoxic and genotoxic potentials of zinc oxide and titanium dioxide nanoparticles. *Toxicology Reports*, *2*, 765–774. Available from https://doi.org/10.1016/j.toxrep.2015.02.004.

Kiss, B., Bíró, T., Czifra, G., Tóth, B. I., Kertész, Z., Szikszai, Z., ... Hunyadi, J. (2008). Investigation of micronized titanium dioxide penetration in human skin xenografts and its effect on cellular functions of human skin-derived cells. *Experimental Dermatology*, *17*(8), 659–667. Available from https://doi.org/10.1111/j.1600-0625.2007.00683.x.

Kulinowski, K. (2004). Nanotechnology: From "Wow" to "Yuck"? *Bulletin of Science, Technology and Society*, *24*(1), 13–20. Available from https://doi.org/10.1177/0270467604263112.

Kvam, E., & Tyrrell, R. M. (1997). Induction of oxidative DNA base damage in human skin cells by UV and near visible radiation. *Carcinogenesis*, *18*(12), 2379–2384. Available from https://doi.org/10.1093/carcin/18.12.2379.

Lademann, J., Weigmann, H. J., Rickmeyer, C., Barthelmes, H., Schaefer, H., Mueller, G., & Sterry, W. (1999). Penetration of titanium dioxide microparticles in a sunscreen formulation into the horny layer and the follicular orifice. *Skin Pharmacology and Applied Skin Physiology*, *12*(5), 247–256. Available from https://doi.org/10.1159/000066249.

Larese, F., Gianpietro, A., Venier, M., Maina, G., & Renzi, N. (2007). In vitro percutaneous absorption of metal compounds. *Toxicology Letters*, *170*(1), 49–56. Available from https://doi.org/10.1016/j.toxlet.2007.02.009.

Larese Filon, F., Crosera, M., Adami, G., Bovenzi, M., Rossi, F., & Maina, G. (2011). Human skin penetration of gold nanoparticles through intact and damaged skin. *Nanotoxicology*, *5* (4), 493–501. Available from https://doi.org/10.3109/17435390.2010.551428.

Larese Filon, F., Crosera, M., Timeus, E., Adami, G., Bovenzi, M., Ponti, J., & Maina, G. (2013). Human skin penetration of cobalt nanoparticles through intact and damaged skin. *Toxicology in Vitro*, *27*(1), 121–127. Available from https://doi.org/10.1016/j.tiv.2012.09.007.

Larese Filon, F., Mauro, M., Adami, G., Bovenzi, M., & Crosera, M. (2015). Nanoparticles skin absorption: New aspects for a safety profile evaluation. *Regulatory Toxicology and Pharmacology*, *72*(2), 310–322. Available from https://doi.org/10.1016/j.yrtph.2015.05.005.

Lee, C. C., Lin, Y. H., Hou, W. C., Li, M. H., & Chang, J. W. (2020). Exposure to ZnO/ TiO_2 nanoparticles affects health outcomes in cosmetics salesclerks. *International Journal of Environmental Research and Public Health*, *17*(17), 1–12. Available from https://doi.org/10.3390/ijerph17176088.

Linton, S. D., Aja, T., Armstrong, R. A., Bai, X., Chen, L. S., Chen, N., ... Zhang, C. Z. (2005). First-in-class pan caspase inhibitor developed for the treatment of liver disease. *Journal of Medicinal Chemistry*, *48*(22), 6779–6782. Available from https://doi.org/10.1021/jm050307e.

Maier, T., & Korting, H. C. (2005). Sunscreens—Which and what for? *Skin Pharmacology and Physiology*, *18*(6), 253–262. Available from https://doi.org/10.1159/000087606.

Manke, A., Wang, L., & Rojanasakul, Y. (2013). Mechanisms of nanoparticle-induced oxidative stress and toxicity. *BioMed Research International*, 2013. Available from https://doi.org/10.1155/2013/942916.

Marrot, L., Belaïdi, J. P., Lejeune, F., Meunier, J. R., Asselineau, D., & Bernerd, F. (2004). Photostability of sunscreen products influences the efficiency of protection with regard to UV-induced genotoxic or photoageing-related endpoints. *British Journal of Dermatology*, *151*(6), 1234–1244. Available from https://doi.org/10.1111/j.1365-2133.2004.06173.x.

Mauro, M., Crosera, M., Pelin, M., Florio, C., Bellomo, F., Adami, G., ... Filon, F. L. (2015). Cobalt oxide nanoparticles: Behavior towards intact and impaired human skin and keratinocytes toxicity. *International Journal of Environmental Research and Public Health*, *12*(7), 8263–8280. Available from https://doi.org/10.3390/ijerph120708263.

Mavon, A., Miquel, C., Lejeune, O., Payre, B., & Moretto, P. (2007). In vitro percutaneous absorption and in vivo stratum corneum distribution of an organic and mineral sunscreen. *Skin Pharmacology and Physiology*, *20*, 10–20.

Menzel, F., Reinert, T., Vogt, J., & Butz, T. (2004). Investigations of percutaneous uptake of ultrafine TiO_2 particles at the high energy ion nanoprobe LIPSION. *Nuclear Instruments and Methods in Physics Research, Section B: Beam Interactions with Materials and Atoms*, *219–220*(1–4), 82–86. Available from https://doi.org/10.1016/j.nimb.2004.01.032.

Midander, K., Wallinder, I. O., & Leygraf, C. (2007). In vitro studies of copper release from powder particles in synthetic biological media. *Environmental Pollution*, *145*(1), 51–59. Available from https://doi.org/10.1016/j.envpol.2006.03.041.

Morabito, K., Shapley, N. C., Steeley, K. G., & Tripathi, A. (2011). Review of sunscreen and the emergence of non-conventional absorbers and their applications in ultraviolet protection. *International Journal of Cosmetic Science*, *33*(5), 385–390. Available from https://doi.org/10.1111/j.1468-2494.2011.00654.x.

Nasir, A., Wang, S., & Friedman, A. (2011). The emerging role of nanotechnology in sunscreens: An update. *Expert Review of Dermatology*, *6*(5), 437–439. Available from https://doi.org/10.1586/edm.11.49.

Nasterlack, M., Zober, A., & Oberlinner, C. (2008). Considerations on occupational medical surveillance in employees handling nanoparticles. *International Archives of Occupational and Environmental Health*, *81*(6), 721–726. Available from https://doi.org/10.1007/s00420-007-0245-5.

Neale, R., Williams, G., & Green, A. (2002). Application patterns among participants randomized to daily sunscreen use in a skin cancer prevention trial. *Archives of Dermatology*, *138*(10), 1319–1325. Available from https://doi.org/10.1001/archderm.138.10.1319.

Ngoc, L. T. N., Tran, V. V., Moon, J. Y., Chae, M., Park, D., & Lee, Y. C. (2019). Recent trends of sunscreen cosmetic: An update review. *Cosmetics*, *6*(4), 64. Available from https://doi.org/10.3390/cosmetics6040064.

Niska, K., Zielinska, E., Radomski, M. W., & Inkielewicz-Stepniak, I. (2018). Metal nanoparticles in dermatology and cosmetology: Interactions with human skin cells. *Chemico-Biological Interactions*, *295*, 38–51. Available from https://doi.org/10.1016/j.cbi.2017.06.018.

Nosaka, Y., & Nosaka, A. Y. (2017). Generation and Detection of Reactive Oxygen Species in Photocatalysis. *Chemical Reviews*, *117*(17), 11302–11336. Available from https://doi.org/10.1021/acs.chemrev.7b00161.

Oberdörster, G., Oberdörster, E., & Oberdörster, J. (2005). Nanotoxicology: An emerging discipline evolving from studies of ultrafine particles. *Environmental Health Perspectives*, *113*(7), 823–839. Available from https://doi.org/10.1289/ehp.7339.

Olson, K. (2009). Cosmetics in Roman antiquity: Substance, remedy, poison. *Classical World*, *102*(3), 291–310. Available from https://doi.org/10.1353/clw.0.0098.

Papageorgiou, I., Brown, C., Schins, R., Singh, S., Newson, R., Davis, S., ... Case, C. P. (2007). The effect of nano- and micron-sized particles of cobalt-chromium alloy on human fibroblasts in vitro. *Biomaterials*, *28*(19), 2946–2958. Available from https://doi.org/10.1016/j.biomaterials.2007.02.034.

Pepić, I., Vujičić, M., Lovrić, J., & Filipović-Grčić, J. (2012). Nanočestice u dermatokozmetičkim pripravcima: Liposomi, mikroemulzije i polimerne micele. *Farmaceutski Glasnik*, *68*(12), 763–772.

Progress toward safe nanotechnology in the workplace—a report from the NIOSH Nanotechnology Research Center. (2007). DHHS (NIOSH) Publication No. https://www.cdc.gov/niosh/docs/2010-104/pdfs/2010-104.pdf?id = 10.26616/NIOSHPUB2010104

Raj, S., Jose, S., Sumod, U. S., & Sabitha, M. (2012). Nanotechnology in cosmetics: Opportunities and challenges. *Journal of Pharmacy and Bioallied Sciences*, *4*(3), 186–193. Available from https://doi.org/10.4103/0975-7406.99016.

Roco, M. C. (2011). The long view of nanotechnology development: The National Nanotechnology Initiative at 10 years. *Journal of Nanoparticle Research*, *13*(2), 427–445. Available from https://doi.org/10.1007/s11051-010-0192-z.

Sayes, C. M., Wahi, R., Kurian, P. A., Liu, Y., West, J. L., Ausman, K. D., ... Colvin, V. L. (2006). Correlating nanoscale titania structure with toxicity: A cytotoxicity and inflammatory response study with human dermal fibroblasts and human lung epithelial cells. *Toxicological Sciences*, *92*(1), 174–185. Available from https://doi.org/10.1093/toxsci/kfj197.

Schmid, K., & Riediker, M. (2008). Use of nanoparticles in swiss industry: A targeted survey. *Environmental Science and Technology*, *42*(7), 2253–2260. Available from https://doi.org/10.1021/es071818o.

Schulte, P., Geraci, C., Zumwalde, R., Hoover, M., & Kuempe, E. (2008). Occupational risk management of engineered nanoparticles. *Journal of Occupational and Environmental Hygiene*, *5*(4), 239–249. Available from https://doi.org/10.1080/15459620801907840.

Schulz, J., Hohenberg, H., Pflücker, F., Gartner, E., Will, T., PfeiVer, S., ... K. (2002). Wittern distribution of sunscreens on skin. *Advanced Drug Delivery Reviews*, 54.

Serpone, N., Salinaro, A., & Emeline, A. (2001). Deleterious effects of sunscreen titanium dioxide nanoparticles on DNA. Efforts to limit DNA damage by particle surface modification. *Proceedings of SPIE - The International Society for Optical Engineering*, *4258*, 86–98. Available from https://doi.org/10.1117/12.430765.

Sisubalan, N., Ramkumar, V. S., Pugazhendhi, A., Karthikeyan, C., Indira, K., Gopinath, K., ... Basha, M. H. G. (2018). ROS-mediated cytotoxic activity of ZnO and CeO_2 nanoparticles synthesized using the *Rubia cordifolia* L. leaf extract on MG-63 human osteosarcoma cell lines. *Environmental Science and Pollution Research*, *25*(11), 10482–10492. Available from https://doi.org/10.1007/s11356-017-0003-5.

Steckiewicz, K. P., & Inkielewicz-Stepniak, I. (2020). Modified nanoparticles as potential agents in bone diseases: Cancer and implant-related complications. *Nanomaterials*, *10* (4). Available from https://doi.org/10.3390/nano10040658.

Surai, P. F., Speake, B. K., & Sparks, N. H. C. (2001). Carotenoids in avian nutrition and embryonic development. 2. Antioxidant properties and discrimination in embryonic tissues. *Journal of Poultry Science*, *38*(2), 117–145. Available from https://doi.org/10.2141/jpsa.38.117.

Tan, M. H., Commens, C. A., Burnett, L., & Snitch, P. J. (1996). A pilot study on the percutaneous absorption of microfine titanium dioxide from sunscreens. *Australasian Journal of Dermatology*, *37*(4), 185–187. Available from https://doi.org/10.1111/j.1440-0960.1996.tb01050.x.

Tran, D. T., & Salmon, R. (2011). Potential photocarcinogenic effects of nanoparticle sunscreens. *Australasian Journal of Dermatology*, *52*(1), 1–6. Available from https://doi.org/10.1111/j.1440-0960.2010.00677.x.

Urbach, F. (2001). The historical aspects of sunscreens. *Journal of Photochemistry and Photobiology B: Biology*, *64*(2–3), 99–104. Available from https://doi.org/10.1016/S1011-1344(01)00202-0.

Wamer, W. G., Yin, J. J., & Wei, R. R. (1997). Oxidative damage to nucleic acids photosensitized by titanium dioxide. *Free Radical Biology and Medicine*, *23*(6), 851–858. Available from https://doi.org/10.1016/S0891-5849(97)00068-3.

Wang, D., Zhao, L., Guo, L. H., & Zhang, H. (2014). Online detection of reactive oxygen species in ultraviolet (UV)-irradiated nano-TiO_2 suspensions by continuous flow chemiluminescence. *Analytical Chemistry*, *86*(21), 10535–10539. Available from https://doi.org/10.1021/ac503213m.

Wang, D., Zhao, L., Ma, H., Zhang, H., & Guo, L. H. (2017). Quantitative analysis of reactive oxygen species photogenerated on metal oxide nanoparticles and their bacteria toxicity: The role of superoxide radicals. *Environmental Science and Technology*, *51*(17), 10137–10145. Available from https://doi.org/10.1021/acs.est.7b00473.

Williams, F. M. (2004). EDETOX. Evaluations and predictions of dermal absorption of toxic chemicals. *International Archives of Occupational and Environmental Health*, *77*(2), 150–151. Available from https://doi.org/10.1007/s00420-003-0484-z.

Wolf, R., Matz, H., Orion, E., & Lipozenčić, J. (2003). Sunscreens—The ultimate cosmetic. *Acta Dermatovenerologica Croatica*, *11*(3), 158–162.

Yin, J. J., Liu, J., Ehrenshaft, M., Roberts, J. E., Fu, P. P., Mason, R. P., & Zhao, B. (2012). Phototoxicity of nano titanium dioxides in HaCaT keratinocytes—Generation of reactive oxygen species and cell damage. *Toxicology and Applied Pharmacology*, *263*(1), 81–88. Available from https://doi.org/10.1016/j.taap.2012.06.001.

Yu, J. X., & Li, T. H. (2011). Distinct biological effects of different nanoparticles commonly used in cosmetics and medicine coatings. *Cell and Bioscience*, *1*(1). Available from https://doi.org/10.1186/2045-3701-1-19.

Yuan, J., Lipinski, M., & Degterev, A. (2003). Diversity in the mechanisms of neuronal cell death. *Neuron*, *40*(2), 401–413. Available from https://doi.org/10.1016/S0896-6273(03)00601-9.

Zs, K., S, Z., G, E., M, P., S-B, J.-E., K, B., ... K, Á. Z. (2005). Nuclear microprobe study of TiO_2-penetration in the epidermis of human skin xenografts. *Nuclear Instruments and Methods in Physics Research Section B: Beam Interactions with Materials and Atoms*, 280–285. Available from https://doi.org/10.1016/j.nimb.2005.01.071.

Tissue engineering

5

Yuhao Qiang
Florida Atlantic University, FL, United States

5.1 Metal oxides

Metal oxides are widely used in various fields of chemistry, physics, and materials science (Noguera, 1996). A large diversity of oxide compounds of metal elements allows them to form a large variety of nanostructures (Gleiter, 1995). Meanwhile, the physio-chemical properties of metal oxide show special relevance to their acute size, which makes them applicable in the industrial use, such as sensors, absorbents, ceramics, and catalysts, etc. The size effects of metal oxides could be summarized as in the influence of their chemical properties (Reddy, 2006), mechanical properties (Wang, Seal, Patil, Zha, & Xue, 2007), optical properties (Scott, Wirnsberger, & Stucky, 2001), transport properties, and so on (Ahmed, 2012).

5.2 Application of nanotechnology in tissue engineering

Tissue engineering is termed for the applications of bioengineering, chemical engineering, and nanotechnology to improve or restore impaired biological organs and tissues (Shi, Votruba, Farokhzad, & Langer, 2010). Tissue-engineered materials are of highly multifunctional properties to meet the requirements for implanting into the human body. Among those various tissue-engineered materials, metal oxide nanoparticles have been extensively developed and used thanks to their excellent and multifaceted properties, such as in the magnetic resonance imaging of tissue-engineered implants (Mertens et al., 2014), and permeability probing of tissues (Kim et al., 2014).

5.3 Application of metal oxides in tissue engineering

Metal oxides have been considerably employed in the field of tissue engineering. Nevertheless, the application is quite limited for the sake of its biotoxicity. For instance, the metal oxides were implemented as drug carriers for treating infection, while only limited availability of metal oxides, TiO_2, ZnO, CuO, ferric oxide

Metal Oxides for Biomedical and Biosensor Applications. DOI: https://doi.org/10.1016/B978-0-12-823033-6.00003-X

(Fe_2O_3), and ferrous oxide (Fe_3O_4) are relatively safe for human tissues (Chhabra et al., 2016).

References

Ahmed, A. S. (2012). *Structural, optical and electrical properties of some metal oxide nanomaterials*. Aligarh Muslim University.

Chhabra, H., Deshpande, R., Kanitkar, M., Jaiswal, A., Kale, V. P., & Bellare, J. R. (2016). A nano zinc oxide doped electrospun scaffold improves wound healing in a rodent model. *RSC Advances*, *6*(2), 1428–1439.

Gleiter, H. (1995). Nanostructured materials. *Acta Metallurgica Sinica*, *33*(2), 165–174.

Kim, Y., Lobatto, M. E., Kawahara, T., Chung, B. L., Mieszawska, A. J., Sanchez-Gaytan, B. L., ... Becraft, J. (2014). Probing nanoparticle translocation across the permeable endothelium in experimental atherosclerosis. *Proceedings of the National Academy of Sciences*, *111*(3), 1078–1083.

Mertens, M. E., Hermann, A., Bühren, A., Olde-Damink, L., Möckel, D., Gremse, F., ... Lammers, T. (2014). Iron oxide-labeled collagen scaffolds for non-invasive MR imaging in tissue engineering. *Advanced Functional Materials*, *24*(6), 754–762.

Noguera, C. (1996). *Physics and chemistry at oxide surfaces*. Cambridge University Press.

Reddy, B. (2006). Redox properties of oxides. In J. L. G. Fierro (Ed.), *Metal Oxides*. Boca Ratón: CRC.

Scott, B. J., Wirnsberger, G., & Stucky, G. D. (2001). Mesoporous and mesostructured materials for optical applications. *Chemistry of Materials*, *13*(10), 3140–3150.

Shi, J., Votruba, A. R., Farokhzad, O. C., & Langer, R. (2010). Nanotechnology in drug delivery and tissue engineering: From discovery to applications. *Nano Letters*, *10*(9), 3223–3230.

Wang, Z., Seal, S., Patil, S., Zha, C., & Xue, Q. (2007). Anomalous quasihydrostaticity and enhanced structural stability of 3 nm nanoceria. *The Journal of Physical Chemistry C*, *111*(32), 11756–11759.

Role of light-active metal oxide-based nanohybrids in biofilm annihilation devices

6

Suparna Dutta-Sinha[1] *and Alokmay Datta*[2]
[1]Living Systems Institute, University of Exeter, Exeter, United Kingdom, [2]CSIR-Central Glass and Ceramic Research Institute and University of Calcutta, Kolkata, India

6.1 What this chapter seeks to do

This chapter will try to review briefly the basic principle of action involved in the annihilation of a specific form of bacterial existence that has immense effect, negative more than positive, on our lives. These are the *bacterial biofilms* that form on surfaces under prolong exposure to organically laden water (Percival, Knottenbelt, & Cochrane, 2011). They are distinct from the *free-floating* or *planktonic* form not only in the vast difference in numbers (single or at best a few hundred vs billions) and disposition (freely floating in water vs attached to a surface) but also more fundamentally in the isolation versus the extensive exchange of genetic material that happens in biofilms (Cvitkovitch, 2004). It is also a striking fact that bacteria of different species can form a single biofilm with this genetic exchange (Yang et al., 2011). These differences make it apparent that the biofunctionalities of this form of bacterial existence will be very different from the planktonic form and indeed this has been established extensively. In particular, biofilms of pathogens pose a formidable challenge against medical intervention and are impervious to antibiotics to dosages potentially lethal to the patient involved (Sharma, Misba, & Khan, 2019). On the other hand, among other highly beneficial effects, microbial biofilms floating on water surfaces provide the essential food supply for a majority of aquatic living systems, either directly or as a part of the food chain (Weitere et al., 2018). Hence, the growth of this microbial life-form is an area of intense research. In particular, stopping the growth of pathogenic biofilms and enhancing that of the beneficial ones are in focus of such endeavor.

The most important aspect of a biofilm is obviously the role of the surface on which it grows. It has been well established that the crucial factor in this growth is the formation of the extracellular polymeric substance (EPS) on the surface (Flemming, Neu, & Wozniak, 2007). The EPS not only initiate the growth but also sustains it and perhaps protects the biofilm from adversities in the environment. In other words, the interactions between the surface and the EPS should decide whether the bacteria will develop into a biofilm or perish. This is an area where a large gap exists between the huge amount of experimental data that have been

Metal Oxides for Biomedical and Biosensor Applications. DOI: https://doi.org/10.1016/B978-0-12-823033-6.00004-1

collected over, especially, the last decade, and any proper understanding of the exact mechanism through which biofilm growth is affected by specific surfaces. We do not even have an idea about the exact composition of the EPS for specific bacterial strains under specific conditions of biofilm growth leave alone any notion about the particular components or molecular moieties that play the most important role in either the growth or the interaction with specific surfaces. The only general idea that we have about the EPS is that it consists of polysaccharides, proteins, and lipids (Costa, Raaijmakers, & Kuramae, 2018). Thus the major functional groups are expected to be OH, COOH, NH_2, and PO_4.

However, the electronic structures of such macromolecules is obviously complicated by orders of magnitude and understanding the interactions with even a restricted group of surfaces as commonly used metal oxide nanocrystals is a daunting task. Of course, the details of the electronic states participating in these interactions are finally to be arrived at by numerical procedures but these do not provide an overall understanding of the physics and chemistry of the process. Unless that is realized, even in its broad outlines, a cross-disciplinary approach to biofilm growth and stability cannot be developed.

This chapter essays to take the first steps toward such an understanding. Hence it will start from the basics of bonding with the bonding in the simplest molecules to build up the concept, will progress to the bonding in dimers and polymers, go over to transformation to bands in solids, the important modifications to the band structure in nanocrystals and nanocrystal–molecule hybrids, and will end with the possible extensions to biofilms on some common metal oxide nanocrystals and their photoinduced damage mechanisms (based on extant experimental data).

6.2 Formation of bonds: molecules

6.2.1 Why bonds form

The basis of permanent or chemical change is the formation of a *bond* between two atoms, that is, a pair of electrons shared between the two. The reason this happens is that through this sharing each of these electrons, and in turn each of these atoms, becomes a little delocalized, thereby lowering the momentum uncertainty in each atom and this goes to lower their ground-state energies. Once a bond is formed the original atomic ground states become excited states of the new system, requiring the balance energy to go back to them and it becomes less likely to find the system revert back to the atoms than to stay bonded. Thus the bonding *stabilizes* the system relative to the isolated atoms and this balance energy is the *bond enthalpy*.

6.2.2 Atomic and molecular orbitals

However, since atoms and electrons are quantum entities, the process through which the bond is formed is an interference of the ground-state electronic wave functions or *atomic orbitals* (*AOs*) of the two atoms. This is expressed mathematically as the linear superposition of the AOs of the two atoms A and B, that is, of wave functions ψ_A and ψ_B into a new wave function Ψ, the *molecular orbital* (*MO*).

$$\psi = c_A\psi_A + c_B\psi_B \quad \text{with} \quad |c_A|^2 + |c_B|^2 = 1 \tag{6.1}$$

where c_A and c_B are complex numbers whose square moduli give the respective probabilities of both electrons to be in ψ_A or ψ_B. Since the AOs add in the same phase, this equation expresses the situation stated above: both electrons occupy both atoms simultaneously. The fractional occupancy values $|c_A|^2$ and $|c_B|^2$ decide the nature of the bond. If $|c_A|^2 = |c_B|^2 = ½$, the bond is a perfect *covalent bond*, which can happen only between atoms of the same element giving rise to *homonuclear molecules* (Coulson, O'Leary, & Mallion, 1978).

For *heteronuclear molecules* two cases are possible. When either of $|c_A|^2$ or $|c_B|^2 = 1$, both electrons reside in either atom A or atom B, respectively, and that atom becomes negatively charged and turns into an *anion*. The other atom then turns into a positively charged *cation* and the atoms are held together by a Coulomb attraction which is termed an *ionic bond*. In the other case, neither is equal to 1 or ½ and of them the atom with a bigger fractional occupancy will have a fractional negative charge whereas the other will have a fractional positive charge. Since this will lead to a molecular dipole moment, the bond is called a *polar bond*.

Fig. 6.1 summarizes the three types of bonding in terms of the shifts in the energy levels of the bonding and antibonding MOs relative to those of the two participating AOs (http://nanowires.berkeley.edu/teaching/104a/201416.pdf). The quantity *electronegativity* of element A (denoted by $\chi(A)$) has been introduced as a measure of the "tendency" of an atom of A to acquire the pair of electrons involved in the bonding of this atom of A with an atom of element B. It has been defined quantitatively as

$$\left|\chi(A) - \chi(B)\right| = (\text{eV})^{\frac{1}{2}} \sqrt{E_d(AB) - \frac{E_d(AA) + E_d(BB)}{2}} \tag{6.2}$$

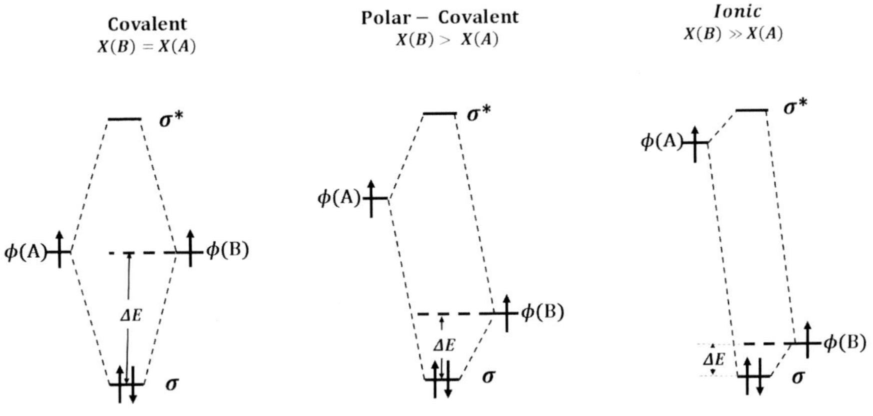

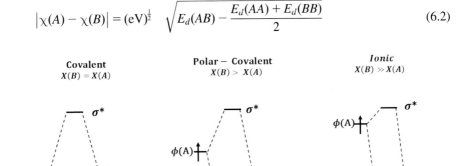

Figure 6.1 Three major types of bonding.

where $E_d(AB)$ stands for the dissociation energy for the AB bond and the prefactor makes χ dimensionless (Pauling, 1932). When A and B are identical, we have the perfect covalent bond. The drops in energy of the bonding σ MO as well as the rises in the energy of the antibonding σ^* MO from the AOs are equal and equal to the *covalent bond energy* (ΔE_{cov}).

$$E_{\varphi_B} = E_\sigma + \Delta E_{cov} = E_{\sigma*} - \Delta E_{cov} = E_{\varphi_A} \tag{6.3}$$

When A and B are not same, the atom of the more electronegative element, say B, will of course acquire a fractional negative charge in the bonding and the atom of A will have a fractional positive charge. Then $\chi(B) > \chi(A)$, $E_{\varphi B}$ moves closer to E_σ and $E_{\varphi A}$ moves closer to E_{σ^*}.

$$E_{\varphi_B} = E_\sigma + \Delta E_{cov} < E_{\varphi_A} = E_{\sigma*} - \Delta E_{cov} \tag{6.4}$$

We then have the polar bonding. This becomes the ionic bond when $\chi(B) >> \chi(A)$

$$E_{\varphi_B} = E_\sigma + \Delta E_{cov} \ll E_{\varphi_A} = E_{\sigma*} - \Delta E_{cov} \tag{6.5}$$

The physical significance of ΔE_{cov} in polar and ionic bonding is that it is the additional stabilization over and above the Coulomb forces provided by the delocalization of the electron pair between A and B.

6.2.3 Diatomic molecules

H_2: Let us start with the simplest covalent bond, that between two hydrogen atoms ("Hydrogen," 2005). The AOs are the $1s^1$ orbitals with one electron each and the MO, where the AOs are *in the same phase*, is given by a σ orbital

$$\sigma_{1s} = 1s^1 + 1s^1 \tag{6.6}$$

The s orbitals are spherically symmetric with *even* or *gerade* (g) wave functions and can be represented by white circles in two dimensions. The σ_{1s} orbital spreads along the axis connecting the centers of these circles. This axis is taken to be the z-axis by convention. MOs with this symmetry axis are called σ orbitals. As we have discussed earlier, this delocalization caused by the MO lowers the energy of the ground state of the bonded atoms below that of either atom, and the H_2 diatomic molecule is formed. This MO is thus called a *bonding* MO.

However, with two AOs there has to be another MO given by

$$\sigma_{1s}{}^* = 1s^1 - 1s^1 \tag{6.7}$$

Here the two s orbitals are again spherically symmetric but while one of them has an even wave function the other has an *odd* or *ungerade* (u) wave function (gray circle). This is a situation where both electrons occupy either one of the atoms or the other, that is, they become *more localized* than in the original atoms, which raises the ground-state energy of this MO *above* that of either of the atoms. This MO is thus less stable than either of the isolated atoms and is called an *antibonding* MO.

Since there are only two electrons, there is no need to put them in this antibonding MO and the H_2 molecule is more stable in its ground state than the two isolated atoms. The antibonding MO is an unstable excited state of this molecule.

Till now we have not discussed the spin of the electrons, which has two eigenvalues, namely $+\frac{1}{2}$ and $-\frac{1}{2}$, symbolized respectively as $\uparrow$ and $\downarrow$. Since electrons are fermions, the MO can have a $\uparrow\downarrow$ pair of electrons only, that is, from AOs having a $\uparrow$ and a $\downarrow$ only.

Taking all the above discussions into account we can now show the situation schematically in Fig. 6.2. This figure is the MO diagram of H_2.

He: What about two He atoms? The AOs are $1s^2$ for each atom with the orbital completely filled with two electrons having one $\uparrow$ and one $\downarrow$ spin. Here the four electrons have to occupy both the bonding σ_{1s} and antibonding $\sigma_{1s}{}^*$ MOs simultaneously. In other words, the highest occupied MO (HOMO) is antibonding. Though the bonding MO has less energy than the isolated atoms, yet that is offset by the high-energy antibonding MO and the total energy of the MOs becomes larger than the isolated atoms. This makes the He_2 "molecule" unstable relative to the He atoms, and as a consequence He remains in the form of atoms rather than forming molecules (http://www.nat.vu.nl/~wimu/EDUC/Werkcollege).

B_2: The first modification in the scheme comes with boron (B) atoms. Here there are five electrons in each atom and the AOs for each atom are $1s^2$, $2s^2$, and $2p^1$. The last AO is "dumbbell-shaped," can be along x-, y-, and z-axis and each of these $2p_x$, $2p_y$, and $2p_z$ AOs can accommodate two electrons, hence the single electron in B can be in any one of these three component orbitals. When these AOs combine to form MOs they can now have two kinds of symmetries—the $2p_z$ AOs forming the

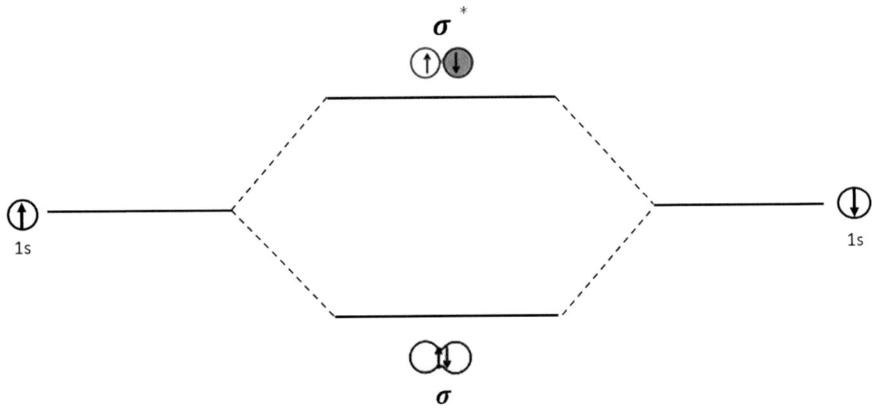

Figure 6.2 Orbitals of the hydrogen molecule. The simplest molecular orbital (MO) diagram.

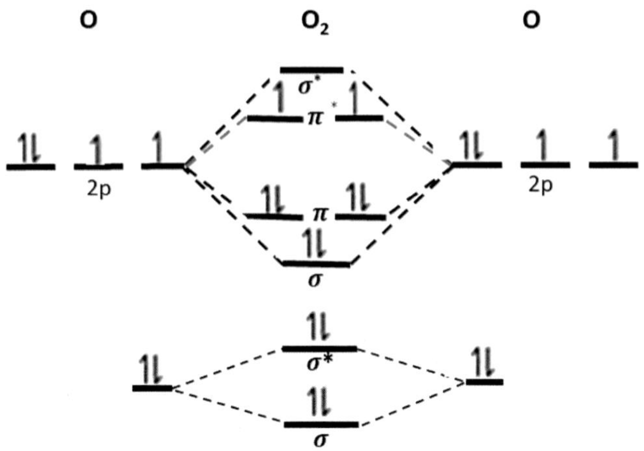

Figure 6.3 Molecular orbital diagrams of O_2 molecule.

σ_{2p} MO along the z-axis and the $2p_x$ and $2p_y$ AOs forming the two π_{2p} MOs along the x- and y-axis respectively with their nodes on z-axis and the dumbbells overlapping laterally. Here the HOMO consists the two-degenerateπ_{2p} MOs. Since these are bonding orbitals, the B_2 molecule is stable relative to the isolated B atoms. The lowest unoccupied MO (LUMO) is theσ_{2p} MO. The MO diagram, taking the spins into account, is shown in Fig. 6.3A (https://www.quora.com/What-is-the-number-of-molecular-bonding-orbitals-in-the-B2-atomic-number-5-molecule; https://en.wikipedia.org/wiki/Molecular_orbital_diagram).

O_2: With the O_2 molecule (Fig. 6.3B) the symmetries of the AOs involved are the same as in B_2 and the 12 valence electrons are distributed in σ_{2s}, σ^*_{2s}, σ_{2p}, π_{2p}, and π^*_{2p} MOs with the π^*_{2p} as the HOMO and σ^*_{2p} as the LUMO. In particular, the HOMO in both the diatomic molecules is in the spin triplet state, that is, with two electrons with parallel spins but in slightly different energy states to obey Pauli's Exclusion Principle. The magnetic interaction between them lowers the energy of the molecules.

6.2.4 Polyatomic molecules

BH_3: Boron hydride or borane is a planar molecule with the B atom at the center and the three B–H bonds spread out at 120 degrees to each other. The $2s$ AO of the B atom and a hybrid $(1/\sqrt{3}(1s + 1s + 1s))$ of the AOs of the three H atoms combine to form the bonding σ (a_1') MO and the antibonding σ^* $(a_1'^*)$ MO. The $2p_z$ AO of B forms an empty nonbonding MO. Finally, the $2p_x$ and $2p_y$ AOs of B and two other hybrids $(1/\sqrt{6}(2 \times 1s\text{-}1s\text{-}1s)$ and $1/\sqrt{2}(1s\text{-}1s))$ of the H AOs combine to form the two degenerate bonding σ (e') MOs and the two degenerate antibonding σ^* (e'^*) MOs. These are the HOMO and LUMO, respectively. The MO diagram is shown in Fig. 6.4 (http://www1.lasalle.edu/~prushan/IC-articles/Polyatomic%20Molecular%20Orbital%20Theory.pdf).

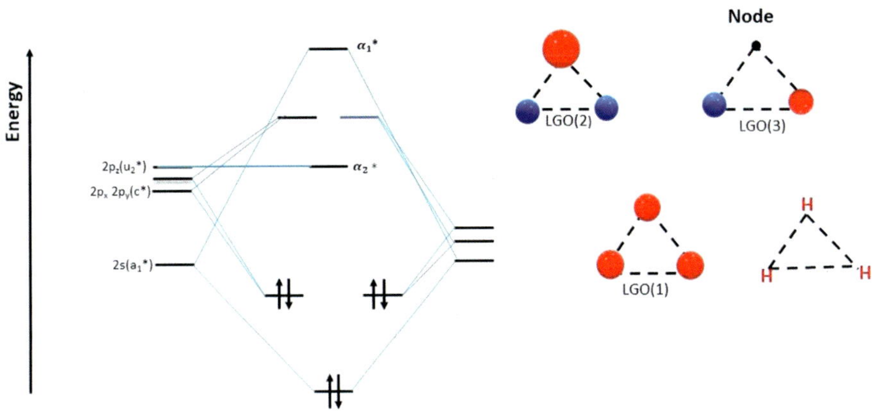

Figure 6.4 Molecular orbital diagram of borane (BH_3).
Source: http://www1.lasalle.edu/~prushan/IC-articles/Polyatomic%20Molecular%20Orbital%20Theory.pdf.

$\mathbf{B_2H_6}$: We end this section with the formation of MOs of the borane dimer. The most important point to take into account is the drastic change in molecular geometry due to the dimerization. The dimer is no longer planar. One H atom from each monomer is shared to act as a bridge between the monomers through the formation of two B–H–B links. This bridge is in a plane normal to the plane containing the two B atoms and the four B–H bonds of the monomers, and the axis through the bridging H atoms defines the z-axis. The x-axis passes through the two B atoms and bisects the two H–B–H angles. Hence there are two groups of MOs, those localized mainly on the bridge and those localized mainly on the other or terminal plane. The first group involves combinations of the hybrid AOs arising from two each of $2s$, $2p_x$, $2p_y$, and $2p_z$ orbitals of the two B atoms with hybrid AOs from the two $1s$ AOs of the bridging H atoms. Similarly, the second group involves combinations of hybrid AOs from the B atoms with hybrid AOs from the four $1s$ AOs of the H atoms on the terminal plane (https://www.staff.ncl.ac.uk/bruce.tattershall/teaching/maingrps/b2h6mo/b2h6.php). Finally, the dimer has six bonding, six antibonding, and two nonbonding MOs. The HOMO is bonding and the LUMO is nonbonding, as shown in Fig. 6.5. Comparison with Fig. 6.4 shows that dimerization has doubled all the bonding, nonbonding, and antibonding MOs. Hence an extrapolation to an N-mer would lead to an at least N-fold increase in all the MOs.

6.3 Bonds to bands: solids

6.3.1 Solids in the bond picture

Solids can be considered as "infinite polymers" following the discussions above. This is the so-called *Bond Picture* of solids (Hoffmann, 1989). According to the extrapolation mentioned in the last sentence of the previous section, in this picture there will be infinite numbers of bonding, nonbonding, and antibonding MOs.

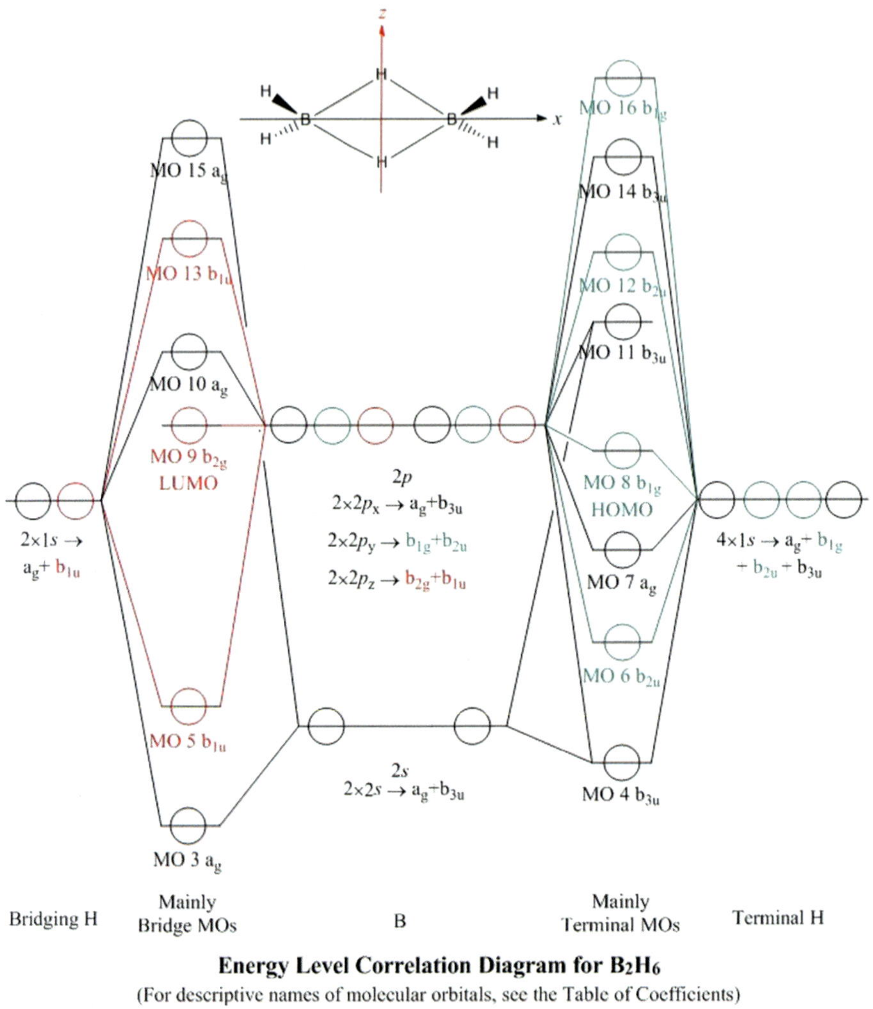

Figure 6.5 Energy level correlation diagram for B_2H_6.
Source: https://www.staff.ncl.ac.uk/bruce.tattershall/teaching/maingrps/b2h6mo/b2h6.php.

Obviously, the separation between the energy levels in each class of MOs will vanish in that limit, giving rise to *continua of energy levels*. There will be two such energy continua or *bands* in the ground state of the solid, one with the HOMO as the upper edge, comprising the bonding and nonbonding MOs, and the other with the LUMO as the lower edge, comprising the antibonding MOs.

This is the general picture of how bonds can give rise to energy bands in the limit of a solid viewed as a single crystal, that is, a system of perfect periodic arrangements of the bonded "repeating units" spreading to infinity in three dimensions. What the repeating units consist of depend on the type of bond that holds them together. For weak, short-range physical forces of bonding such as the van der

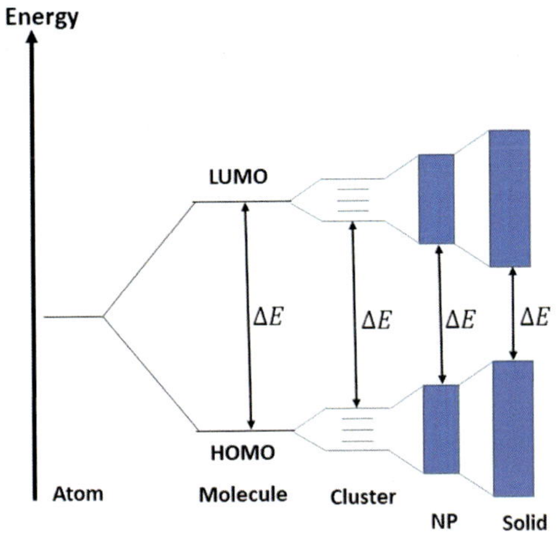

Figure 6.6 Bond to band representation in solids.
Source: Hoffmann, R. (1989). Solids and surfaces: A chemist's view of bonding in extended structures. *Wiley VCH.*

Waals force, the repeating unit is an organic molecule in most cases. For electrovalent, long-range Coulomb bonding they are ions, whereas for the strongest, covalent or polar bonds the units are molecular "fragments" that were the monomers before the bonding, as in the case of two borane molecules bonding to form the dimer.

In this bond-to-band picture, shown in Fig. 6.6, the band with the HOMO as the upper edge is said to be the *valence band* (VB) and the band with the LUMO as the lower edge is said to be the *conduction band* (CB). We shall look at this picture in more details for covalent and polar solids, of which the metal oxides fall in the latter group.

6.3.2 Covalent solid—Si

In Si the single 2*s* and three 2*p* AOs combine to form the four sp^3 hybrid AOs in order to match the tetrahedral symmetry of the five Si atoms bonding together to form the repeating unit of the crystal lattice. The hybrid AOs of these Si atoms then make up the bonding and the antibonding MOs resulting in the covalently bonded tetrahedral unit. Any such tetrahedron will have the corner atoms with bonds that are not satisfied. These *dangling bonds* give a hybrid AO between the bonding and antibonding MOs. The tetrahedral repeating units combine *ad infinitum* to build up the solid Si as the MOs transform to the VB and CB, respectively. In reality the crystal is truncated in surfaces where the last layers of Si tetrahedra end in dangling bonds. The corresponding AOs still remain isolated and therefore energetically in the same level as with the single tetrahedral unit. However, the CB has moved down closer to this level. If an electron is raised to this level it can move to the conducting state with relative ease, and it is free to undergo chemical bonding with other atoms or molecules. Hence this energy represents the *chemical potential* of the solid and is termed as *Fermi level* or *Fermi energy* (https://www.globalsino.com/EM/page2633.html). This is shown in Fig. 6.7.

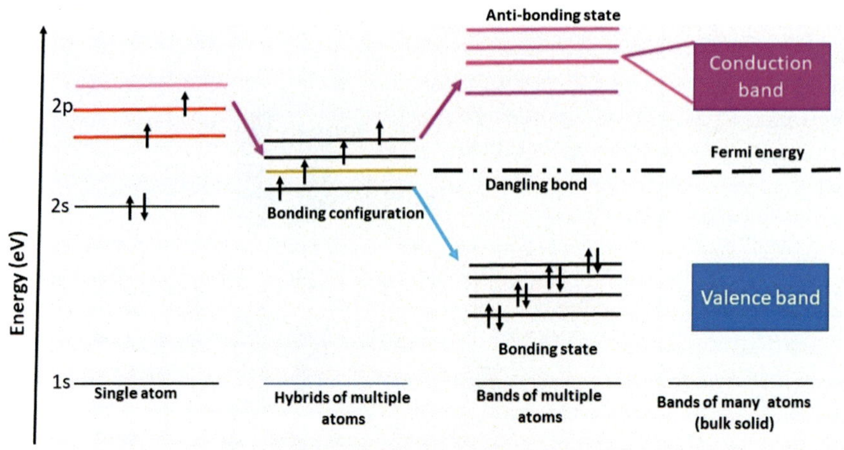

Figure 6.7 Representation of Fermi energy or Fermi level.
Source: https://www.globalsino.com/EM/page2633.html.

6.3.3 Polar solid—GaAs

In GaAs the lattice structure is the same as Si, only it can be thought of as *alternating* tetrahedra with the central atom as Ga or As and the corner atoms as As and Ga, correspondingly. Here the two 4*s* and one 4*p* AOs of Ga form three hybrid AOs whereas the two 4*s* and three 4*p* AOs of As form five hybrid AOs, that is, there are three valence-states for each Ga atom while there are five valence-states for each As atom. This makes GaAs a polar solid having a fractional positive charge on the Ga atoms and a fractional negative charge on the As atoms. Since the As AOs are lower in energy and there is a greater degree of hybridization in As, the hybrid AOs of As are lower in energy than those of Ga. These AOs now can form Ga–As bonding and antibonding MOs. The bonding MOs finally make up the VB and the antibonding MO makes up the CB of GaAs, with the dangling bonds of Ga forming the surface state close to the CB and the dangling bonds of As forming the surface state close to the VB. The Fermi level in this case is midway between VB and CB and does not coincide with any of the surface levels (Murdick, Zhou, Wadley, & Nguyen-Manh, 2005). This situation is shown in Fig. 6.8.

6.4 Nanocrystals: effects of surface and size

6.4.1 Effect of size

We have seen how MOs with discrete energies are converted to continuous energy bands as the repeating units tend to infinities in three dimensions. If this process is stopped after a finite number of steps, the resulting CB and VB will no longer be continuous but will consist of close-spaced but discrete levels. If this *atomic*

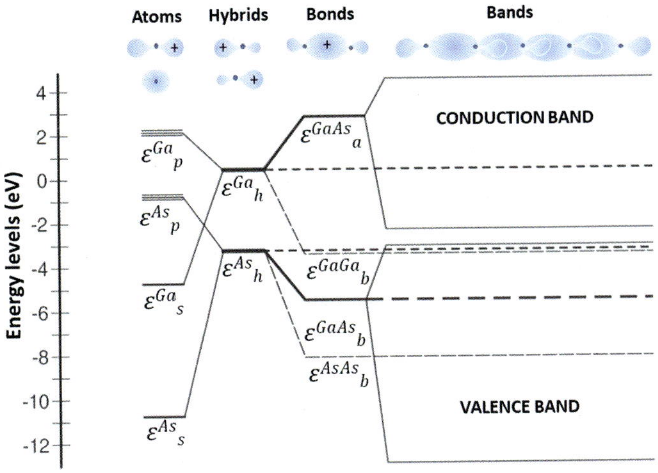

Figure 6.8 Representation of Fermi level midway between valence band and conduction band.
Source: Murdick, D. A., Zhou, X. W., Wadley, H. N. G., & Nguyen-Manh, D. (2005). *Journal of Physics: Condensed Matter: An Institute of Physics Journal*, 17, 6123.

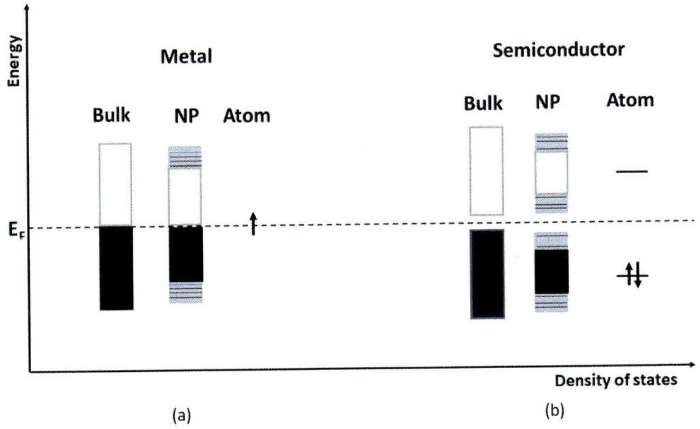

Figure 6.9 Decrease in number of discrete levels with decrease in nanocrystal size.
Source: Minelli, C. (2004). *Bottom-up approaches for organizing nanoparticles with polymers*, Thesis No. 3092, ÉÉcole Polytechnique Fédérale de Lausanne.

or *molecular cluster* is stabilized structurally, then a *nanocrystal* is achieved. The number of these levels will decrease and the gap between the successive levels will increase as the nanocrystal size is reduced. This is true for all solids (Fig. 6.9). However, there are additional differences between metals (no band gap) and other solids with band gap (Minelli, 2004).

6.4.2 Metal nanocrystals

In bulk metals the highest level at 0 K is the Fermi level. The free electrons oscillate and scatter light of all frequencies below a certain frequency proportional to the square root of the number density of the valence electrons and inversely proportional to the root of the effective mass of the electron, in the metal. This is the *plasma frequency* of the metal and it lies in the ultraviolet for all metals except Au and Cu. The nanocrystal still has the band structure of a metal as its highest level is the Fermi level. This situation changes below ~2 nm, when a discrete band appears above the Fermi level and the metal loses its metallic behavior (Fig. 6.9A).

6.4.3 Semiconductors and insulators

In semiconductors and insulators, which fall into the category of electrovalent, polar and covalently bonded solids, and have a band gap, no state in the CB or VB is found in the Fermi level and the latter is just an energy level midway between the bands (Fig. 6.9B). Here, when an electron is raised from the VB to the CB, a *hole* is created in the VB. The electron–hole pair is bound together by Coulomb force to form a "particle" called an *exciton*. To set the electron and the hole free for charge transport, the formation energy of the exciton, which is exactly equal to the band gap, has to be exceeded. In bulk materials, the electron and holes are generally located far apart, hence the Coulomb attraction and, in turn, the band gap, is small. As the nanocrystal becomes small in size, the electron–hole separation is reduced, and the Coulomb attraction is increased leading to the increase in the band gap (Fig. 6.10A). The band structures of some common metal oxide nanocrystals are shown in Fig. 6.10B (Ouyang, Huang, & Choi, 2019).

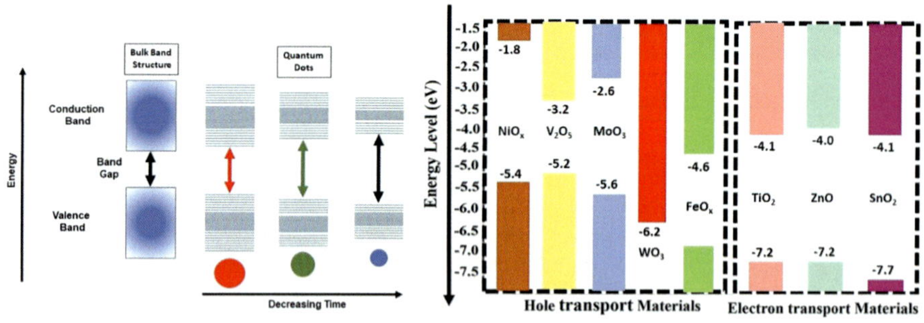

(a) (b)

Figure 6.10 (A) Increase in band gap with decrease in nanoparticle size. (B) Band structures of some common metal oxides.
Source: Boles, M. A., Ling, D., Hyeon, T., & Talapin, D. V. (2016). *Nature Materials*, 15, 141.

6.4.4 Surface states

The other major distinguishing feature of nanocrystals is the effect of the surface. Here, firstly, the surface-to-volume ratio is increased by orders of magnitude, causing a large number density of surface states or dangling bonds to occur, which gives rise to a large surface free energy. Thus the nanocrystal is much more reactive than the bulk. Again, to partially reduce this large free energy and satisfy the dangling bonds, the surface lattice structure undergoes a *reconstruction*, which changes the nature of the surface states and the position of not only these levels but also the top of the VB and bottom of the CB (Boles, Ling, Hyeon, & Talapin, 2016).

A model of the expected structural reconstructions at solid surfaces is achieved by representing the bulk and the surface as the smallest clusters of atoms that preserve the respective primitive unit cell and symmetry. In Fig. 6.11, the MO diagrams of such model clusters are presented. Fig. 6.11A shows the level structure of

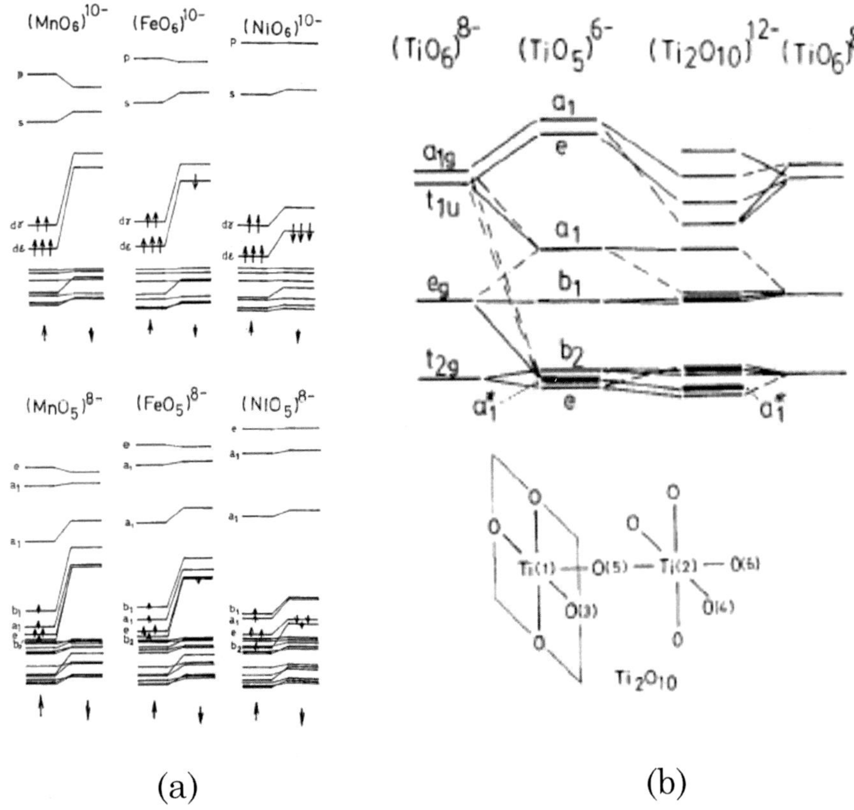

Figure 6.11 (A) MO diagrams of the level structure of the $(MO_6)^{10-}$ cluster that models the bulk and the level structure of the $(MO_5)^{8-}$ cluster that models the surface of the transition metal monoxides. (B) Energy level structures of the TiO_6 (bulk), TiO_5, and Ti_2O_{10} [(100) Ti surface] clusters of $SrTiO_3$. *MO,* Molecular orbital.
Source: Tsukada, M., & Shima, N. (1987). *Physics and Chemistry of Minerals*, 15, 35.

the $(MO_6)^{10-}$ cluster that models the bulk and the level structure of the $(MO_5)^{8-}$ cluster that models the surface of the transition metal monoxides (M = Mn, Fe, Ni). These models show that there is an effective field at the surface due to the reconstruction that splits the bulk threefold degenerate $d\varepsilon$ states into a doubly degenerate e state and a nondegenerate b_2 state, and the twofold degenerate $d\gamma$ states into the a_1 and b_1 states. They are also shifted relative to the top of the oxygen VB. This decreases the localized charge at the surface relative to the bulk and increases the covalent character of the metal–oxygen bond in these monoxides.

The energy level structures of the TiO_6 (bulk), TiO_5, and Ti_2O_{10} ((100) Ti surface) clusters of $SrTiO_3$ are shown in Fig. 6.11B. The distribution of level components on the top Ti (1) and the inner Ti (2) ion show that the LUMO consisting of the e and a_1^* states of the Ti_2O_{10} cluster are the intrinsic surface states. The e state has the character of d_{xz} or d_{yz} (where z is normal to the surface) and the a_1^* state of $4s + 4p_z + 3d_z^2$. These intrinsic surface states are shallow as the fraction of electrons transferred back to the top cation by the enhanced covalent character of the surface raises the surface state energy toward the CB bottom. Similarly, the surface state levels of the top oxygen are pulled down near the VB top (Tsukada & Shima, 1987). Nanocrystals of these oxides are very well modeled by these clusters of atoms representing the surfaces. Thus the major effect of surface reconstruction in these nanocrystals is an enhancement of the covalent character of the bonding.

6.4.5 Nanocrystal–molecule hybrids

In dielectric nanocrystals, the effect of size is to engineer the bandgap whereas the effects of enhanced surface-to-volume ratio are to introduce new localized states and to enhance the covalent nature of the surface bonds. All these attributes are utilized in creating hybrids of nanocrystals and organic molecules with band structures and electron transport properties that are both unique and are tunable through the nanocrystal size and shape as well as the organic molecule used. In the context of biofilm growth modification, electrons generated through photon absorption and higher-energy photons generated through frequency up shifting both affect the biofilm composition and chemistry quite drastically, as we shall see in a later section.

The photoemission of electrons in a hybrid of titania (TiO_2) nanocrystal and an azo-dye (Fig. 6.12A) is a good example (Prajongtata et al., 2017). The MO diagram and the band structure of anatase titania is shown in Fig. 6.12B, whereas the total band structure of the hybrid is shown in Fig. 6.12C. On irradiation of the surface with visible light, the dye molecule has a higher probability to be excited than the nanocrystal since its HOMO–LUMO gap is smaller than the bandgap (step 1 in Fig. 6.12C). Again, due to the proximity of the HOMO and the CB of titania, this excited electron is transferred through an ultrafast nonradiative process to the CB (step 2 in the Fig. 6.12C). This takes place through the bond between an O atom of one carboxyl group of the dye molecule to a fivefold coordinated titanium atom in TiO_2. The intensity of electrons in the CB becomes the output photocurrent density.

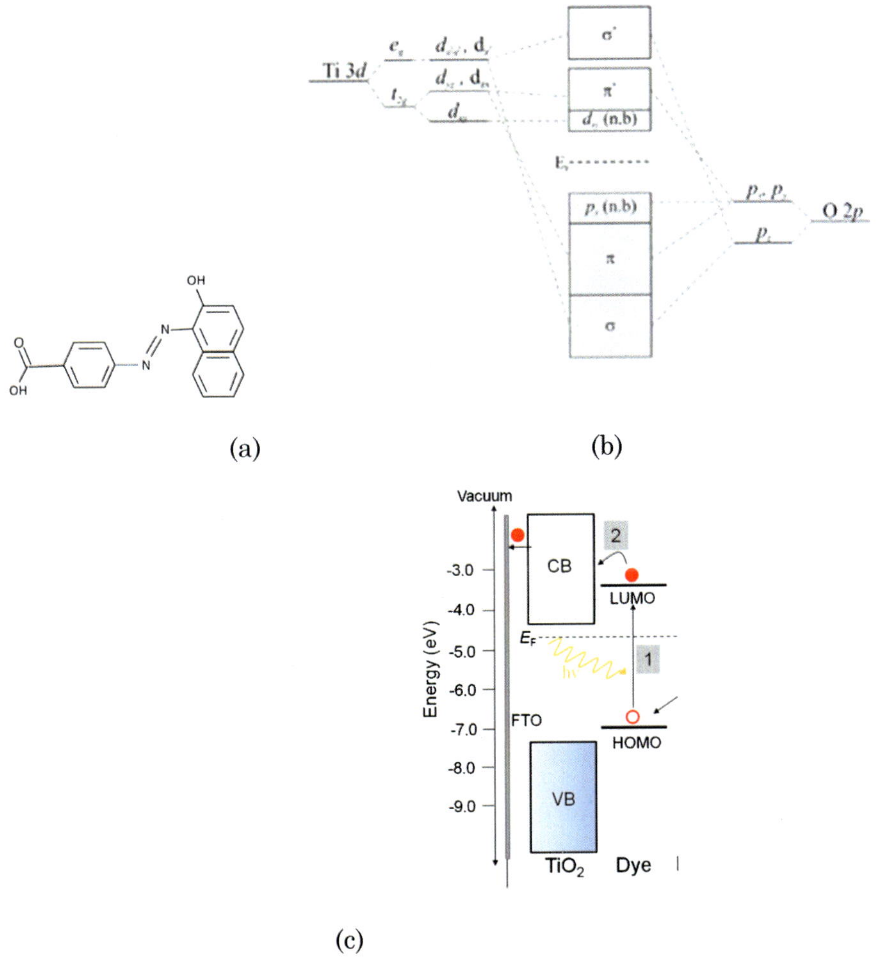

Figure 6.12 Photo-emission of electrons in a hybrid of titania (TiO_2) nanocrystal and an azo-dye.
Source: Prajongtat, P., Suramitr, S., Somkiat, Nokbin, S., Nakajima, K., Mitsuke, K., & Hannongbua, S. (2017). *Journal of Molecular Graphics and Modelling*, 76, 551.

6.5 Nanocrystal–molecule interactions—the microscopic picture

6.5.1 The basics

The physical basis of any interaction between two bodies lies, of course, in the exchange of energy or matter between them. From a microscopic or quantum mechanical point of view, this is through a transfer of a photon, virtual or real, or an electron from one body to the other. For either kind of transfer, the first body is a *donor* and the second is an *acceptor* and the transfer takes place from a quantum state of the donor to one of the acceptor. The states in question can be AO, HOMO, LUMO, VB, or CB, depending upon whether the bodies are atoms (or ions), molecules, or solids, in that order. In all cases the either the relevant upper state of the donor should be, close in energy to, but slightly above the relevant upper state of the acceptor or the same relation should exist between the lower states (Rodriguez, Chaturvedi, Kuhn, & Hrbek, 1998).

If we restrict ourselves to light-activated interactions, there are two major types of transfers that are found to occur (Ruiz-Hitzky, Ariga, & Lvov, 2018). In the first type, the two-level system of the donor is in resonance with that of the acceptor, that is, the energy difference between the upper and lower levels for the donor is the same (or nearly the same) as the energy difference between the two levels of the acceptor. When the exciting light, whose energy is equal to this energy gap, falls on the donor–acceptor pair, the energy of the photon is absorbed by the donor to raise the electron from lower to upper level. If the donor is a molecule or a solid (including a nanocrystal) then the transition is from a HOMO or VB to a LUMO or CB, respectively, creating an electron in the LUMO (CB) and a hole or deficiency in the HOMO (VB). Since the movements of the electron and the hole are correlated the pair constitutes an exciton. This exciton decays, emitting a virtual photon that creates an exciton in the acceptor. The exchange can happen through the coupling between the dipoles of the two excitons hence its probability decays with the spatial separation between the donor and the acceptor. The rate of energy transfer also depends on the relative dipole orientations of the donor and acceptor, the polarization of the medium given by its refractive index, the oscillator strength of the donor transition and, of course, the overlap between the relevant donor and acceptor levels. The main example of this is Förster resonance energy transfer (FRET) (Förster, 1948) and is shown in Fig. 6.13A.

The second type of interaction is of the more general kind since it does not require such resonance between the donor and the acceptor but only a higher degree of spatial proximity, since here an electron, rather than energy, is being transferred and it can get lost in the surroundings more easily. The result is a chemical reaction where the donor is oxidized and the acceptor is reduced (Fig. 6.13B) (Marcus, 1993). Such *electron transfer* (ET) can occur over long distances with multiple steps that might include multiple reduction–oxidation centers or via supramolecular exchange interactions such as π-stacking. Sometimes the first step maybe the excitation of the donor maybe through FRET and then the process is continued by electron transfer.

6.5.2 Resonance energy transfer in hybrids

Biologically relevant molecules hybridized with nanocrystals can either act directly as a resonant acceptor or donor or, when attached to an already hybrid system, as an additional two-state system facilitating energy transfer to the relevant biological junction. In most cases, the second role is employed through polypeptides or oligonucleotides attached to some organic dye, hybridized with the nanocrystal, through

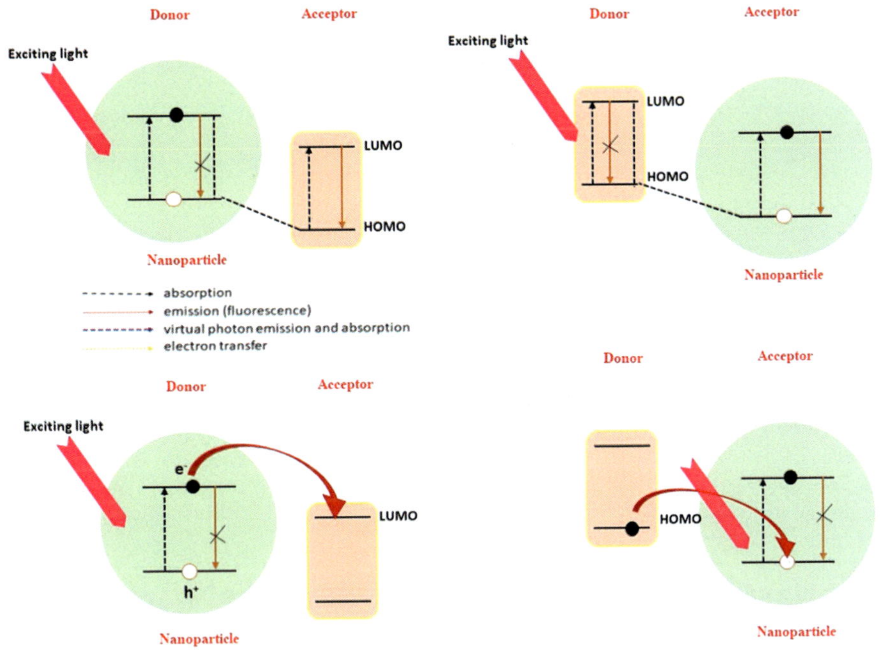

Figure 6.13 Representation of Förster resonance energy transfer.
Source: Marcus, R. A. (1993). *Reviews of Modern Physics*, 65, 599.

covalent bond or ligand. Due to the large surface-to-volume ratio of nanocrystals, a large number of surface states are available that can form hybrids with different types of molecules, both small and large, hence different biomolecules can play both roles on the same nanocrystal, making it multifunctional (Chen et al., 2012).

6.5.3 Electron transfer in hybrids

An example of ET in these hybrids is the enzymatic performance of glucose oxidase (GOx) in the formation of H_2O_2. Electron transfer from nanocrystal to attached flavin adenine dinucleotide cofactor (FAD) causes its reduction to $FADH_2$ which inhibits vicinal GOx. If the electron is transferred to the nanocrystal, it reoxidizes $FADH_2$ to FAD. Such a direct transfer may not always take place and intermediary molecules may be used for an ET cascade (Milton, 2013).

Again, since the abundance of surface states and the tunability of the bandgap in nanocrystals both FRET and ET can take place simultaneously if proper donors or acceptors are chosen. Even the same molecule can have two different two-level systems, one working as a FRET acceptor and the other as an ET acceptor, with the same nanocrystal (Sapsford, Berti, & Medintz, 2006).

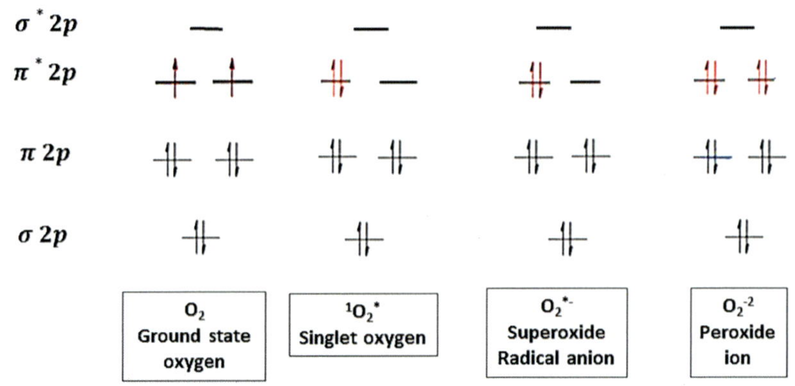

Figure 6.14 The MO structures of the various reactive oxygen species. *MO,* Molecular orbital.
Source: Krumova, K., & Cosa, G. (2016). Overview of reactive oxygen species. *Singlet oxygen: Applications in biosciences and nanosciences* (Vol. 1, p. 1). RSC Publications, Ch 1.

6.6 Nanocrystal–molecule interactions—antibacterial action

6.6.1 Production of reactive oxygen species

A crucial pathway of antibacterial action is through the formation of *reactive oxidation species* (ROS). This is a general name of all molecules or radicals that can undergo rapid and large-scale oxidation–reduction reactions (Auten & Davis, 2009). To have a better understanding of these species we have to look at the O_2 molecule with its HOMO composed of two electrons with same spin in separate AOs (Fig. 6.14). It can react with another molecule whose HOMO has likewise electrons, not only with unpaired spins, but also with spins opposite to its own. It can, however, have different reactive paths. One of them is to flip one of its spins by going to the energetically higher singlet state $^1O_2$. The other paths are to borrow one electron to become a superoxide radical O_2^-, two electrons to become hydrogen peroxide H_2O_2, or three to become a hydroxyl radical $\cdot OH$ (Krumova & Cosa, 2016). All these species are unstable and highly reactive oxidizing agents. Of these, H_2O_2 is the most stable (half-life $\sim$ 1 ms) and least reactive whereas the others have half-lives ranging in 2–4 μs.

Nanocrystals produce all four types of ROS. It has been found that the nanocrystals of metal oxides are much more effective in generating ROS than the nanocrystals of the respective metals and, in fact, the latter are more effective either when they oxidize or convert into ions (Shkodenko, Kassirov, & Koshel, 2020). Hence, in either case, the emergence of the bandgap is a necessary step in ROS generation. In fact, with nanocrystals the following reactions are most probably taking place (Pathakoti, Huang, Watts, He, & Hwang, 2014).

Under UV irradiation, electrons are promoted from VB to CB, creating energetic holes in the VB.

$$MO_{NP} + h\nu \rightarrow h^{+}_{VB} + e^{-}_{CB} \tag{6.8}$$

The holes react with OH^- or H_2O to produce $\cdot OH$,

$$h^{+}_{VB} + OH^{-} \rightarrow \cdot OH \tag{6.9}$$

$$h^{+}_{VB} + H_2O \rightarrow \cdot OH \tag{6.10}$$

The holes also react with O_2 to produce singlet O_2

$$h^{+}_{VB} + O_2 \rightarrow {}^{1}O_2 \tag{6.11}$$

While the electrons react with O_2 to produce the superoxide radical

$$e^{-}_{CB} + O_2 \rightarrow O_2^{-} \tag{6.12}$$

The main sites for these reactions seem to be the defects, the new surface states generated due to restructuring as seen above, and the oxygen vacancies in the nanocrystals (Gupta & Bahadur, 2018). In this process, the transfer of electrons and holes from nanocrystals to biomolecules through small molecules hybridized with the nanocrystal is supposed to play the key role. The major factors that will cause the transfer are matches in two quantities between the nanocrystal–molecule hybrid and the biomolecule involved. These are *chemical hardness* (η) and electronegativity (χ). In terms of the energies of the HOMO and LUMO, these are defined as

$$\chi = \frac{\varepsilon_{HOMO} + \varepsilon_{LUMO}}{2} \tag{6.13}$$

$$\eta = \frac{\varepsilon_{LUMO} - \varepsilon_{HOMO}}{2} \tag{6.14}$$

From the point of view of chemistry, a Lewis acid (base) with a particular value of hardness will react most strongly with a base (acid) with the same hardness. Since $\chi + \eta = \varepsilon_{LUMO}$, a higher LUMO of the donor will require it to be a less electronegative acid or base in the Lewis acid–base pair (Parsaee, Mohammadi, Ghahramaninezhad, & Hosseinzadeh, 2016). However, a clear correlation between the electronic structures and the reaction pathways are yet to be developed for either the nanocrystals themselves or, even more, for nanocrystal–molecule hybrids. For example, different metal oxide nanocrystals produce different ROS. CaO and MgO nanocrystals generate O_2^{-}, ZnO nanocrystals produce H_2O_2 and $\cdot OH$ but not O_2^{-}, and CuO nanocrystals can generate all four types of ROS (Szabó et al., 2008).

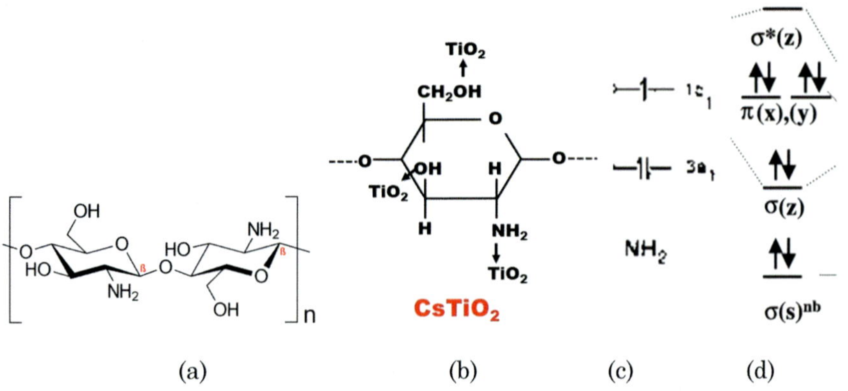

Figure 6.15 (A) Structure of chitosan, (B) chitosan hybridized with TiO_2 nanocrystals, (C) HOMO of chitosan, and (D) HOMO of chitosan hybridized with TiO_2 nanocrystals. *HOMO,* Highest occupied molecular orbital.
Source: Clemmer, D. E., & Armentrout, P. B. (1991). *The Journal of Physical Chemistry*, 95, 3084; Luther, G. W., III (2004). *Aquatic Geochemistry*, 10, 81.

The major complication lies in the dependence of the electronic structure on size, shape, and surface of the nanocrystal. While these aspects of nanocrystals comprise their most attractive properties, they also bring in computational difficulties, mainly involving calculation of band structures with finite but large number of states. Again, the chemistry of the process is posited to be dependent on the specific and practically irreproducible factors such as defects and vacancies —an idea that is inconsistent with a clear and generalized understanding of the mechanism.

As an example of a nanocrystal–molecule hybrid let us look at TiO_2–chitosan nanohybrids (Haldorai & Shim, 2014). Chitosan (Fig. 6.15A) is a polysaccharide with positively charged NH_2 and OH groups, which makes it interact with the negatively charged surfaces of the bacterial cell membrane, especially the double membrane of gram-negative bacteria. When hybridized with TiO_2 nanocrystals (Fig. 6.15B), the positive charge of the CS–TiO_2 nanohybrid is largely enhanced. This rigidifies the cell, blocking it from intake of nutrients and finally stops cell growth completely. The nanohybrid also generates ROS and causes the associated damages as mentioned.

The charge and energy transfer must be taking place between the TiO_2 and the NH_2 and OH but we do not have a clear idea about which path is more probable. Nor do we know whether the HOMOs of these groups, shown in Fig. 6.15C and D, respectively (Clemmer & Armentrout, 1991; Luther, 2004), combine with the VBs of the nanocrystals or with the surface states located on the O atoms. These latter are close to the VB according to Fig. 6.11 and the related discussion. If these are involved in the hybridization then the effects of the local structure on the nanocrystal surface will play important roles in the combination. Inspection of the symmetries of the HOMOs of the groups and these surface states suggest a preference for bonding of the former with the latter with a transfer of electrons to the HOMO leaving an energized hole to generate ROS through reactions (6.9)–(6.11). However, this suggested pathway needs to be experimentally verified.

6.6.2 Antibacterial action of reactive oxidation species

Under normal circumstances, the rates of production and removal of ROS from the bacterial cell are equal and there is no stress. If the production rate becomes large, however, the tendency of the system will be to restore balance through oxidation of these species to stable forms. This will produce a stress known as *oxidative stress* that can damage all parts of the cell since the energy released from the high rate of oxidation will be dissipated through these parts (Pizzino et al., 2017).

The ROS cause different levels of damage to the bacterial cells (Wang, Hu, & Shao, 2017). O_2^- and H_2O_2 cause less damage as they can be absorbed by antioxidants generated within the cell, such as superoxide enzymes and catalase but $\cdot OH$ and $^1O_2$ cause a huge oxidative stress. While O_2^- and $\cdot OH$, being negatively charged, remain confined on the anion-terminated cell membrane surface, H_2O_2 and $^1O_2$, being neutral, can penetrate the membrane and enter the cell. The oxidative stress changes permeability of the cell membrane, pushes the nanocrystals inside the cell, and eventually breaks it down (Muzammil et al., 2020). This causes a large degree of leakage of cellular proteins, DNA, especially double-stranded DNA, and sugars. The ROS also change the interaction between DNA and bacterial cells, which in turn cause more ROS to form, increase the production of proteins that cause further oxidation of the cell components, and at the same time they damage those proteins and deactivate enzymes that maintain cell integrity and normal functions (Tran et al., 2008).

The cells damaged by the ROS will be killed and removed by the lymphocytic T cells of bacteria, a process which is known as *cytotoxic response* (Burello & Worth, 2011). A large number of damaged bacterial cells due to the oxidative stress will cause a huge cytotoxic response, effectively removing the bacteria from the host. While this is the basis of nanomedicine, it also involves the risk of cytotoxic response from the host cells and has restricted the potential of this novel means of antibacterial action.

6.7 Nanocrystal-molecule hybrids against biofilms

6.7.1 How biofilms grow

Biofilms are formed in five stages as shown in Fig. 6.16 (Marić & Vraneš, 2007). Bacteria float freely in their planktonic form. These are also referred to as "swimmers," as they require aquatic media for transport. However, in the presence of suitable surfaces, they may undergo genetic transformation to a form that can attach to the surface or become "stickers." This is the first stage of growth, where the attachment is reversible. The stickers sense and attach to the surface using extracellular organelles and outer membrane proteins.

In the second stage of growth the attachment is irreversible. In this stage the crucial secretion of the EPS takes place facilitating adhesion between the cells and the

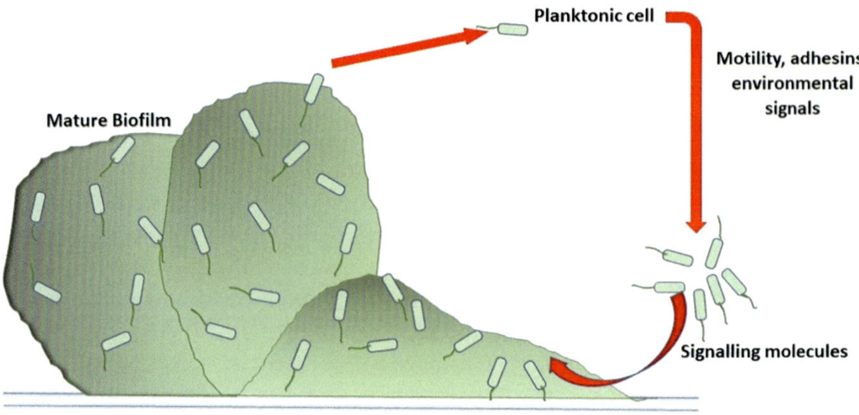

Figure 6.16 Stages of biofilm growth.
Source: Maric, S., & Vranés, J. (2007). *Periodicum Biologorum*, 109, 1.

surface. The third stage sees the replication of the cells and growth into colonies that are 10–100 μm in size, and are called *microcolonies*. The EPS varies in composition according to growth conditions and may have some dependence on bacterial species. At this stage, cells in the biofilm start intercommunication through the EPS since the close proximity within this matrix facilitates the exchange of signaling molecules. This is known as *quorum sensing* (QS) and it controls a variety of functions of the cells in the film such as pathogenesis, acquiring food, conjugation, mobility of the cells, and generation of secondary metabolite. Within the EPS there are different microenvironments with compositional and functional variations. In the fourth stage the bacterial community grows out of plane and becomes a biofilm with large-scale accumulation of cells through replication and of EPS through secretion. The EPS holds the cells together, works as a shield against all external stresses and prevents the biofilm from detachment from the surface. In the last stage, some individual cells detach from the biofilm to start new biofilms, a step crucial for propagation and self-renewal of the community.

6.7.2 Antibiotic resistance of biofilms

Biofilms protect the bacterial colonies inside them from all external stress such as variations in temperature and humidity, high-energy radiations, and toxic molecules, both natural and artificial (Pintucci, Corno, & Garotta, 2010). While a number of these protection mechanisms stem from the physical barrier imposed by the EPS hydrogel, the protective action against antibiotics is more involved and is our focus in this chapter. Biofilms show a remarkable resistance against antibiotics. In a majority of applications, biofilms require almost 1000 times higher concentrations of antibiotics to be destroyed than that required for planktonic bacteria (Chatterjee, Biswas, Datta, Dey, & Maiti, 2014; Donlan, 2001). This resistance to antibiotics by biofilms makes the infections caused by biofilms to acquire a chronic form (Spoering & Lewis, 2001). To understand this process, we first need to know a basic attribute of antibiotic action.

Antibiotics are molecules that maybe extracted from natural sources and modified or maybe totally synthetic. There are generally four mechanisms involved in their action against bacteria. They can block or control enzymes that synthesize the cell wall, they can impair nucleic acid metabolism and repair, they can stop protein synthesis, or they can disrupt membrane structure. Most of these mechanisms can work or at least work efficiently when the bacteria are multiplying (Dougherty & Pucci, 2011).

Initially it was thought that a biofilm has a structure that acts as a physical barrier to the diffusion of antibiotics inside the film (Taheri et al., 2014). Recent studies have shown that a gradient of nutrients and oxygen supply within the biofilm slows down and even stops the metabolism of bacterial cells at the periphery of the structure thus rendering the antibiotics essentially ineffective (Taylor et al., 2012). Coupled with this is the absorption of the antibiotic molecules in the EPS, where the polysaccharides, proteins, and the nucleic acids neutralize the biological activity of the antibiotics such that they are unable to reach the necessary bactericidal concentrations (Ramasamy, Lee, & Lee, 2016). The situation becomes more problematic when the biofilms consist of antibiotic-resistant bacteria (Muzammil et al., 2020).

6.7.3 Metal oxide nanoparticles against biofilms

The resistance mechanism of biofilms against antibiotics is completely ineffective against the high rate of ROS generation by photoactivated hybrids of metal oxide nanocrystals and suitable molecules that can transfer the energy and the free electrons and holes created in the nanocrystals to the films (Hu, 2017). As we have seen, though the exact mechanisms of transfer, of ROS generation, and the paths of damage caused by the ROS and the huge oxidative stress to the bacteria are unknown, the damage is nonspecific and global. Some of the most effective nanocrystals are of TiO_2, ZnO, CuO, and Fe_3O_4 (Sharma, Jandaik, Kumar, Chitkara, & Sandhu, 2016). The major functional groups of the possible EPS components, namely polysaccharides, proteins, lipids, and amino acids are OH, COOH, NH_2, and PO_4. The MOs of these groups terminating at the HOMOs are shown in Fig. 6.17A and B (Zhang & Sheng, 2014), C and D (https://socratic.org/questions/how-do-yo-write-the-orbital-diagram-for-phosphate), respectively. Inspection of the symmetries of these HOMOs may provide us clues to the specific states of the hybrids or, better, the ROS (Fig. 6.14A) that can combine with these. That will be the most important step in understanding the mechanism involved in the destruction of the EPS in particular and the biofilm in general.

There is another mechanism of antibiofilm action of these nanohybrids that has been demonstrated by combining β-cyclodextrin with SiO_2 nanocrystals (Papadimitriou et al., 2015). This molecule is made up of seven glucopyranose units linked so as to form a truncated cone with a hydrophilic outer surface and a slightly

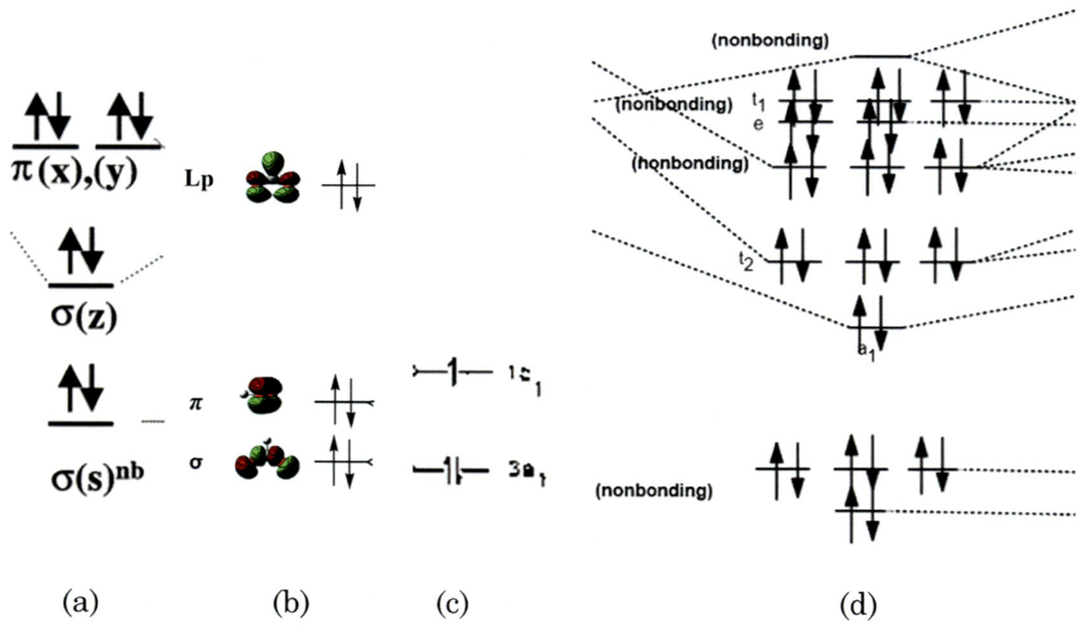

Figure 6.17 MOs terminating at the HOMOs for (A) OH, (B) COOH, (C) NH_2, and (D) PO_4. *HOMO,* Highest occupied molecular orbital; *MO,* molecular orbital.

hydrophobic inner surface. Gram-negative bacteria that are antibiotic-resistant to an extent in the planktonic form can largely enhance their resistance through QS in a biofilm, using *N*-acyl-L-homo serine lactone (HSL) molecules as signal. However, in the presence of the truncated cones, this signal molecule gets trapped in a complex with the cyclodextrin, thus ceding the molecular signal pathway to QS. The SiO_2 nanocrystals do not generate ROS and are highly biocompatible. They are used as carriers of the cyclodextrin traps at very high surface concentrations which synergistically enhance their performance.

6.8 Outlook

As we can perceive, photoinduced bactericidal action of metal oxide nanocrystal–molecule hybrids against biofilms is based on the general principles of their action against bacterial cells and, in fact, any cellular structure consisting of interconnected membranes of biopolymers and macromolecules. Therein lies the major stumbling block in the utilization of the immense medical potential of these materials since the same oxidative stress and cytotoxicity is bound to be generated in the host system as well as the pathogen (Qayyum & Khan, 2016). This is the main reason for our insistence in this chapter on the development of a cross-disciplinary approach based on the MO theory of interactions to build a comprehensive and detailed understanding of all the processes involved. We believe that such a task, though difficult, is achievable in the main and is now necessary to tap the potential of nanomedicines, not only against biofilms and multiple drug-resistant bacteria but also other pathogens including viruses. We would feel satisfied if this chapter can act as a motivator for that task.

Acknowledgment

Alokmay Datta would like to express thanks to the Department of Atomic Energy, India, for award of the Raja Ramanna Fellowship and to the Director, CSIR-Central Glass and Ceramic Research Institute for hosting.

References

Auten, R. L., & Davis, J. M. (2009). *Pediatric Research*, *66*, 121.
Boles, M. A., Ling, D., Hyeon, T., & Talapin, D. V. (2016). *Nature Materials.*, *15*, 141.
Burello, E., & Worth, A. P. (2011). *Nanotoxicology*, *5*, 228.
Chatterjee, S., Biswas, N., Datta, A., Dey, R., & Maiti, P. K. (2014). *Microscopy: The Journal of the Quekett Microscopical Club*, *63*, 269.
Chen, N.-T., Cheng, S.-H., Liu, C.-P., Souris, J. S., Chen-Tu Chen, C.-T., Mou, C, Y,, ... L.-W.. (2012). *International Journal of Molecular Sciences*, *13*, 16598.

Clemmer, D. E., & Armentrout, P. B. (1991). *The Journal of Physical Chemistry*, *95*, 3084.
Costa, O. Y. A., Raaijmakers, J. M., & Kuramae, E. E. (2018). *Frontiers in Microbiology*, *9*, 1636-1.
Coulson, C. A., O'Leary, B., & Mallion, R. B. (1978). *Hückel theory for organic chemists*. Academic Press.
Cvitkovitch, D. G. (2004). Genetic exchange in biofilms. In M. Ghannoum, & G. A. O'Toole (Eds.), *Microbial biofilms* (p. 192). ASM Science, Ch 11.
Donlan, R. M. (2001). *Emerging Infectious Diseases*, *7*, 277.
Dougherty, T. J., & Pucci, M. J. (2011). *Antibiotic discovery and development*. Springer.
Flemming, H.-C., Neu, T. R., & Wozniak, D. J. (2007). *Journal of Bacteriology*, *189*, 7945.
Förster, T. (1948). *Annals of Physics*, *437*, 55.
Gupta, J., & Bahadur, D. (2018). *ACS Omega*, *3*, 2956.
Haldorai, Y., & Shim, J.-J. (2014). *Polymer Composites*. Available from https://doi.org/10.1002/pc.
Hoffmann, R. (1989). *Solids and surfaces: A chemist's view of bonding in extended structures*. Wiley VCH.
HuL. (2017). The use of nanoparticles to prevent and eliminate bacterial biofilms. In *Antimicrobial research novel bioknowledge and educational programs* (pp. 344–350). Formatex: Badajoz, Spain..
'Hydrogen'. (2005). *Van Nostrand's Encyclopedia of chemistry* (p. 797) Wylie-Interscience.
Krumova, K., & Cosa, G. (2016). *Overview of reactive oxygen species*, . Singlet oxygen: Applications in biosciences and nanosciences (Vol. 1, p. 1). RSC Publications, Ch 1.
Luther, G. W., III (2004). *Aquatic Geochemistry*, *10*, 81.
Marcus, R. A. (1993). *Reviews of Modern Physics*, *65*, 599.
Marić, S., & Vraneš, J. (2007). *Periodicum Biologorum*, *109*, 1.
Milton, R. (2013). *Physical Chemistry Chemical Physics: PCCP*, *15*, 19371.
Minelli, C. (2004). *Bottom-up approaches for organizing nanoparticles with polymers*, Thesis No. 3092, École Polytechnique Fédérale de Lausanne.
Murdick, D. A., Zhou, X. W., Wadley, H. N. G., & Nguyen-Manh, D. (2005). *Journal of Physics. Condensed Matter: An Institute of Physics Journal*, *17*, 6123.
Muzammil, S., Khurshid, M., Nawaz, I., Siddique, M. H., Zubair, M., Nisar, M. A., ... Hayat, S. (2020). *Biofouling*, *36*, 492.
Ouyang, D., Huang, Z., & Choi, W. C. H. (2019). *Advanced Funcional Materials*, *29*, 1804660.
Papadimitriou, K. P., Zoumpopoulou, G., Foligne, B., Alexandraki, V., Kazou, M., Pot, B., & Tsakalidou, E. (2015). *Frontiers in Microbiology*, *6*, 189-1.
Parsaee, Z., Mohammadi, K., Ghahramaninezhad, M., & Hosseinzadeh, B. (2016). *New Journal of Chemistry*, *40*, 10569.
Pathakoti, K., Huang, M.-J., Watts, J. D., He, X., & Hwang, H.-M. (2014). *Journal of Photochemistry and Photobiology B: Biology*, *130*, 234.
Pauling, L. (1932). *Journal of the American Chemical Society*, *54*, 3570.
Percival, S. L., Knottenbelt, D. C., & Cochrane, C. A. (2011). Introduction to biofilms. In S. L. Percival, D. C. Knottenbelt, & C. A. Cochrane (Eds.), *Biofilms and veterinary medicine* (p. 41). Springer.
Pintucci, J. P., Corno, S., & Garotta, M. (2010). *European Review for Medical and Pharmacological Sciences*, *14*, 683.
Pizzino, G., Irrera, N., Cucinotta, M., Pallio, G., Mannino, F., Arcoraci, V., ... Bitto, A. (2017). *Oxidative Medicne and Cellular Longevity*, *2017*, 8416763-1.

Prajongtat, P., Suramitr, S., Somkiat, Nokbin, S., Nakajima, K., Mitsuke, K., & Hannongbua, S. (2017). *Journal of Molecular Graphics and Modelling*, *76*, 551.

Qayyum, S., & Khan, A. U. (2016). *MedChemComm.*, *7*, 1479.

Ramasamy, M., Lee, J.-H., & Lee, J. (2016). *Journal of Biomaterials Applications*, *31*, 366.

Rodriguez, J. A., Chaturvedi, S., Kuhn, M., & Hrbek, J. (1998). *The Journal of Physical Chemistry. B*, *102*, 5511.

Ruiz-Hitzky, E., Ariga, K., & Lvov, Y. (Eds.), (2018). *Bio-inorganic hybrid nanomaterials – Strategies, syntheses, characterization and application*. Wiley-VCH.

Sapsford, K. E., Berti, L., & Medintz, I. L. (2006). *Angewandte Chemie International Edition*, *45*, 4562.

Sharma, D., Misba, L., & Khan, A. U. (2019). *Antimicrobial Resistance & Infection Control*, *8*, 76.

Sharma, N., Jandaik, S., Kumar, S., Chitkara, M., & Sandhu, I. S. (2016). *Journal of Experimental Nanoscience*, *11*, 54.

Shkodenko, L., Kassirov, I., & Koshel, E. (2020). *Microorganisms*, *8*, 1545.

Spoering, A. L., & Lewis, K. (2001). *Journal of Bacteriology*, *183*, 6746.

Szabó, T., Németh, J., & Dékány, I. (2003). *Colloids and Surfaces. A, Physicochemical and Engineering Aspects*, *23*, 230, J. Sawai, et al. World Journal of Microbiology & Biotechnology 16, 187 (2000); S. Ikawa, K. Kitano, S. Hamaguchi, Polymer Reviews 7, 33 (2010); J. Sawai and T. Yoshikawa, Journal of Applied Microbiology 96, 803 (2004); N. Padmavathy and R. Vijayaraghavan, Science and Technology of Advanced Materials 9,035004 (2008).

Taheri, S., Baier, G., Majewski, P., Barton, M., Förch, R., Landfester, K., & Vasilev, K. (2014). *Nanotechnology*, *25*, 305102.

Taylor, E. N., Kummer, K. M., Durmus, N. G., Leuba, K., Tarquinio, K. M., & Webster, T. J. (2012). *Small (Weinheim an der Bergstrasse, Germany)*, *8*, 3016.

Tran, N., Mir, A., Mallik, D., Sinha, A., Nayar, S., & Webster, T. (2010). *International Journal of Nanomedicine*, *5*, 277, M. Noda et al., Scientific Reports 7, 8557-1 (2017); L.A. Rowe, N. Deghtyareva, and P.W. Doetsch, Free Radical Biology & Medicine 45, 1167 (2008).

Tsukada, M., & Shima, N. (1987). *Physics and Chemistry of Minerals*, *15*, 35.

Wang, L., Hu, C., & Shao, L. (2017). *International Journal of Nanomedicine*, *12*, 1227.

Weitere, M., Erken, M., Majdi, N., Arndt, H., Norf, H., Reinshagen, M., ... Wey, J. K (2018). *Ecological Monographs*, *1*.

Yang, L., Liu, Y., Wu, H., Høiby, N., Molin, S., & Song, Z.-J. (2011). *International Journal of Oral Science*, *3*, 74.

Zhang, M., & Sheng, L. (2014). *Physical Chemistry Chemical Physics: PCCP*, *16*, 196.

Section 2

Metal oxide-based biosensors

Introduction to metal oxide-based biosensing

7

Vinay Kishnani[1], *Kunal Mondal*[2] *and Ankur Gupta*[1]
[1]Department of Mechanical Engineering, Indian Institute of Technology, Jodhpur, India,
[2]Materials Science and Engineering Department, Idaho National Laboratory, Idaho Falls, ID, United States

7.1 Introduction

Over the last few decades, nanoparticles of metal oxides (NPMOs) have engrossed researchers due to its comprehensive technical applications and scientific capabilities. It has already been acknowledged that the chemical and physical properties of the materials at nanoscale changes drastically in comparison to the bulk materials due to the effect of quantum size, surface morphology, and macroscopic tunneling effect. Especially, NPMOs are at the center of attraction due to their electronic, magnetic, and optical properties. NPMOs are suitable for broad applications of gas sensors (Gupta et al., 2014), electrochemical energy storage, catalysis, light-emitting diode, enzymatic biosensors (Rahman, Ahammad, Jin, Ahn, & Lee, 2010; Shi et al., 2014), water filtration (Gupta, Mondal, Sharma, & Bhattacharya, 2015; Gupta, Saurav, & Bhattacharya, 2015), etc. due to its physical and chemical properties. NPMOs play a vital role in heterogeneous catalysis. To prepare and understand the catalytic materials, the sol-gel method is utilized due to its exceptional versatility. Its controlled composition, structure, and morphology texture nonhydrolytic sol-gel (NHSG) process is used for the catalytic materials (Debecker & Mutin, 2012). NPMOs is also utilized for electrodes for super capacitors, lithium-ion batteries, organic photovoltaic cells, and organic light-emitting diode (Jiang et al., 2012; Meyer et al., 2012). NPMOs show its good biocompatibility, chemical stability, high surface area, and the ability to transfer electrons at a higher speed. These characteristics of NPMOs make it perfect for immobilization matrices and transduction platforms. Recent research shows that the modification of electrodes through MOs can improve the sensitivity of the electrodes (Shaidarova & Budnikov, 2008; Shaidarova, Gedmina, Chelnokova, & Budnikov, 2011). In the emerging field of biosensing for the application of magnetic separation, thermal ablation therapy, high sensitivity, and early disease, diagnostic magnetic nanoparticles have been used very effectively (Haun, Yoon, Lee, & Weissleder, 2010; Xu & Wang, 2012). Among all the NPMOs, ZnO is preferred due to its sole fundamental materials properties for the biosensing applications (Arya et al., 2012; Gupta, Pandey, & Bhattacharya, 2013; Yakimova, 2012). The amalgamation of enzymes with NPMOs can amplify the efficient catalytic yield of the samples (Solanki, Kaushik, Agrawal,

Metal Oxides for Biomedical and Biosensor Applications. DOI: https://doi.org/10.1016/B978-0-12-823033-6.00005-3

& Malhotra, 2011), which can be readout through various strategies such as colorimetric, fluorescence, electrochemical, chemiluminescence, electrochemiluminescence, and so on. Fig. 7.1 represents the summary of various kinds of MOs and corresponding features investigated for biosensing application.

A sensor is a practical integrated miniaturized device that records a chemical, biological, or physical change and modifies that into a measurable signal. Schematic representation of a sensor and associated components have been shown in Fig. 7.2. The sensor is comprised of a recognition component (antibody, enzyme, a receptor protein, nucleic acid, whole cell or tissue section), which lets selective reaction to a specific analyte, and accordingly reduces interferences from other sample elements. The other critical area of a sensor is the transducer or the detector, which produces a signal that is further processed by a digital signal processor and exhibits on a digital display.

The ultrasensitive biosensors are fabricated with nanotechnology's help for the regulation of sensing at the molecular level. However, with the impressive success of detection with the use of various nanomaterials, there is still a requirement of continuous, noninvasive sensing for biological activities. The electroanalytical devices, coupled to biorecognition molecules fabricated on nanostructured MOs for high sensitivity, good selectivity, and low cost (Rahman et al., 2010). There are

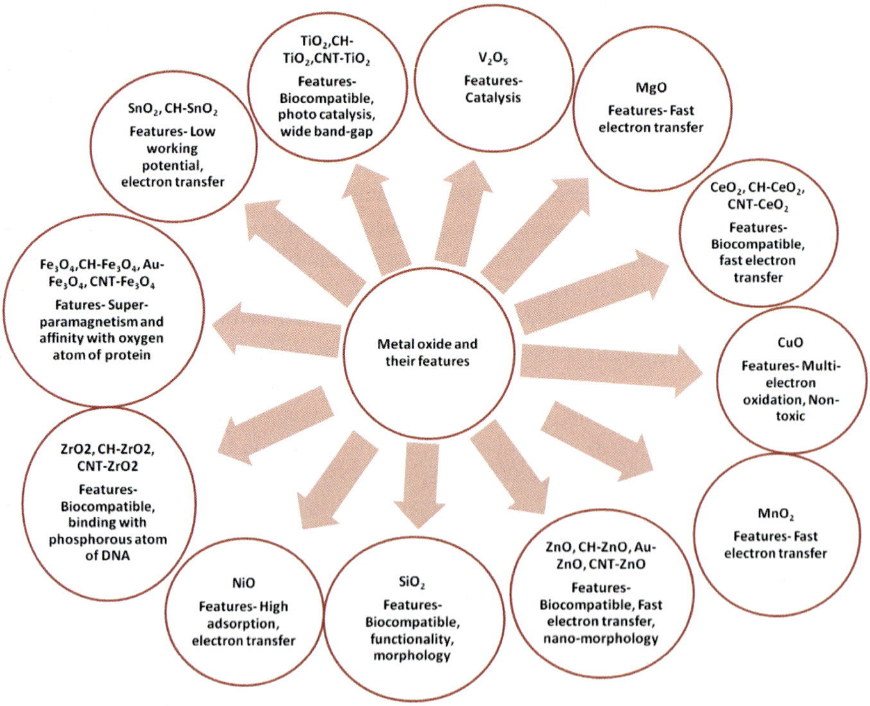

Figure 7.1 Representation of metal oxide and their features for biosensing application. Basics of metal oxide-based biosensing.

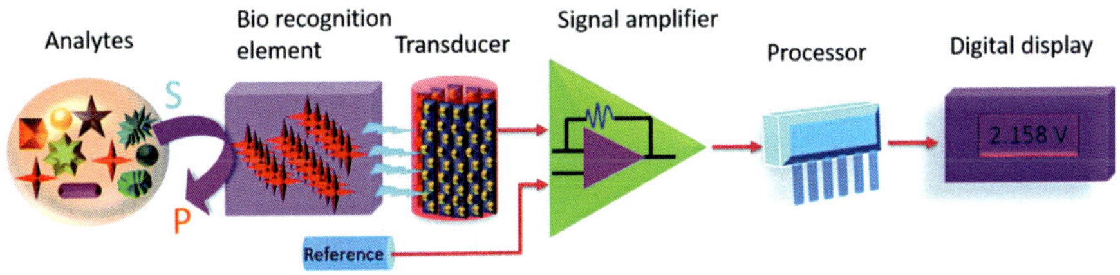

Figure 7.2 Representation of major components of a biosensor where the biocatalyst alters the substrate (S) to product (P) and then the transducer changes the biological response to an electrical signal. The transducer then sends the signal to a power amplifier followed by transistor processor, and ultimately the output is presented on a digital screen.
Source: Reproduced with permission from Mondal, K., & Sharma, A. (2016). Recent advances in electrospun metal-oxide nanofiber based interfaces for electrochemical biosensing. *RSC Advances*, *6*, 94595–94616. doi:10.1039/c6ra21477k. Copyright 2015, Royal Society of Chemistry.

various methods used to make NPMOs, which contain different morphologies. These NPMOs show various electrical and photochemical properties due to their high surface area, size, and stability. For the electroanalysis, NPMOs show three important functions, which include roughening of the conductive sensing interface, the increment of the catalytic properties relative to the metals and amplification in electrochemical detection of the metallic deposition as well as conductive properties of the NPMOs at nanodimension, which allow the contact of the electrode surface and protein's redox center electrically (Katz, Willner, & Wang, 2004). These NPMOs transfer fast electron between the analyte molecule and transducer due to its nanostructure and size, which is considered as electronic wires (Wei, Li, Zhang, Shi, & Jin, 2010). In general, biocompatible NPMOs are used to immobilization the biomolecules for the fabrication of enzyme sensors, DNA sensors, and immunosensors. In contrast, semiconductor nanoparticles are used as traces in electrochemical study and markers (Luo, Morrin, Killard, & Smyth, 2006). The immobilization of NPMOs as sensing elements on the working electrode can be achieved through various methods, namely, chemical covalent bonding, physical adsorption, electropolymerization with redox polymer, electrodeposition, etc. (Wang & Hu, 2009). Due to the wide band-gap of the NPMOs, it behaves as semiconductors or precisely insulator, and its ion transport kinetics is low (Li, Patrissi, Che, & Martin, 2000). Simultaneously, during the charging/discharging process, the electrode got pulverized through the expansion and contraction of the electrode (Li, Tan, & Wu, 2008). To overcome these drawbacks, hybridization is done in NPMOs with polymers, different nanoparticles, and carbonaceous materials, namely, silver and gold nanoparticles, quantum dots, graphene, carbon nanotubes, etc. (Gupta, Gangopadhyay, Gangopadhyay, & Bhattacharya, 2016; McNeil, 2005; Pan & Zhu, 2009; Wang, 2005a, 2005b). NPMOs are mostly used for biosensing and electrochemical sensing (George, Antony, & Mathew, 2018). There are various methods to obtain the NPMOs, namely, soft templating method for the grounding of the nanofibers, nanobundles, and nanorods (Gupta et al., 2014; Tenne, 2006), sol-gel methods for the derivation of rough nanostructures (Ansari, Solanki, & Malhotra, 2008; Choi, Kim, & Lee, 2005), sputtering of radiofrequency for the rough nanostructures (Singh et al., 2007), and controlled shape deposition through hydrothermal for nanoparticles (Zhao, Lei, Zhang, Wang, & Jiang, 2010).

7.2 Synthesis strategies of metal oxides for biosensors

Synthesis strategies of MOs play an important role in determining the metal oxides' toxicity and biological activity in a biosensor. All the possible synthesis strategies adopted for the MO fabrication have been listed out in Fig. 7.3. It can be prepared through solution precipitation (Chen & Chang, 2005), ball milling (Yadav & Srivastava, 2012), sonochemical (Yu, Zhang, & Lin, 2003), solvothermal,

Figure 7.3 Various synthesis methods available for metal oxide fabrication enzyme immobilization on metal oxides.

hydrothermal (Yan et al., 2012), spray pyrolysis (Feng et al., 2006), sol-gel method (He et al., 2012), thermal decomposition (Wang, Mori, Li, & Ikegami, 2002), thermal hydrolysis (Hirano, Fukuda, Iwata, Hotta, & Inagaki, 2000), biodirected methods of synthesis (Charbgoo, Bin Ahmad, & Darroudi, 2017), and electrospinning (Ali, Mondal, Singh, Dhar Malhotra, & Sharma, 2015; Ali et al., 2016, 2017; Mondal, Ali, Agrawal, Malhotra, & Sharma, 2014; Mondal et al., 2017). Gupta and Bhattacharya (2018) investigated the vertical and epitaxial growth rate of zinc oxide (ZnO) nanostructure along with the time function rate and side length through the aqueous chemical synthesis. Chauhan, Kant, Rai, Gupta, and Bhattacharya (2019) reported a novel method aqueous chemical method for the synthesis of ZnO nanoflower and graphene oxide (GO) composite for the uniform growth over a silicon substrate, for enhancing the photocatalytic decolorization of methyl blue (MB) dye and industrial wastewater under solar radiation. Kundu, Kumar, Sen, and Nilabh (2020) reported a novel ecofriendly and cost-effective biosynthesis method of SnO_2, through extracted water of kabuli chickpea seeds as a natural binder, and found that CO sensing ability of the biosynthesized SnO_2 is better than other conventional processes. Yakimova (2012) reviewed the synthesis and tailored surface of the ZnO nanoparticles and nanostructured film for the biosensor applications. Arya et al. (2012) reviewed the advancement and new strategies adopt for the synthesis of ZnO nanostructure used in biosensor technology.

The most critical task for the development of an enzymatic biosensor is the immobilization of enzymes with molecular detection skills on the biosensor. In biosensors, immobilization is completed through the three processes—covalent bonding, adsorption, cross-linking, and encapsulation. It must be stable at the surface, allow diffusion of products and substrate, and permit an outstanding electron transfer. The adsorption process is quite simple, but it has the drawback of percolation. This technique's lifetime is meager while comparing to the adsorption process cross-linking, and the covalent bonding technique shows a higher lifetime due to its property of strong bond formation in between the solid support and sensing molecule. Some factors cannot come within reach of covalent bonds, namely, ionic strength, pH, etc. But these processes are multifaceted and consume more time, and it involves harmful chemicals (Bhardwaj, 2014). Fig. 7.4 demonstrates the systematic immobilization representation and Fig. 7.5 illustrates the immobilization schemes explored for biosensor application. In general, there are two kinds of methods for immobilization, physical method and chemical method; encapsulation, and adsorption fit in the physical method whereas covalent bonding and cross-linking in the chemical method. For the development of enzymatic-based biosensors, the commonly used enzymes are polyphenol oxidase, lactate oxidase, horseradish, tyrosinase, peroxidase, cytochrome, urease, alcohol dehydrogenase, glucose oxidase (GOx), and glutamate dehydrogenase. Shakerian, Zhao, and Li (2020) reviewed the advancement in the application of immobilized oxidative enzymes for the remediation of harmful micropollutants. Liang, Wu, Xiong, Zong, and Lou (2020) summarized the metal–organic framework's perspectives for advancement, challenges, and future point of view as a supporting substrate for the immobilization of the enzyme. Ariaeenejad, Motamedi, and Hosseini Salekdeh (2021) reported a novel

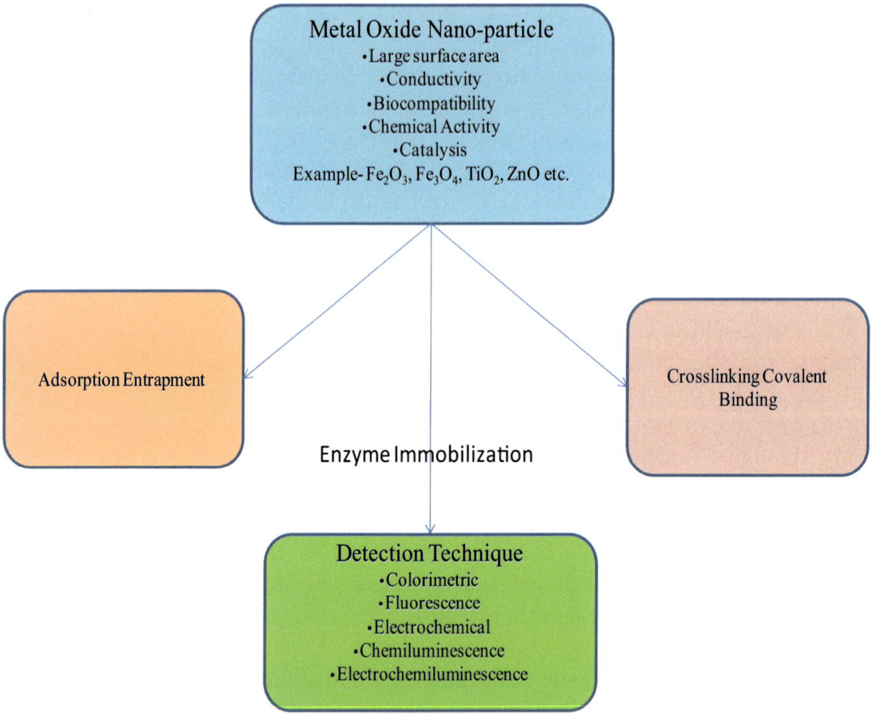

Figure 7.4 Systematic immobilization representation.

enzyme that covalently immobilized on the amine-functionalized magnetic GO and found it as a pioneering and competent bifunctionalized assembly for the elimination of dye from the water successfully. Soares et al. (2020) performed immobilization of papain enzyme over hybrid support, which contains ZnO/chitosan for the clinical application and found that enzyme had not lost its enzymatic activity and proteolytic activity. Qiu, Xiang, Liu, Huang, and Hu (2020) fabricated a fresh composite of organic-inorganic (SBA@CS), shows the high surface area, good biocompatibility, and rich functional groups, resulted that on comparison with immobilization property of mesoporous material SBA-15, it is relatively better for enzyme immobilization through metal–organic coordination. Kiran et al. (2020) fabricated a nanobiocatalyst for the dye decolorization through immobilization of the lignin peroxidase (LiP) on GO-functionalized $MnFe_2O_4$ superparamagnetic nanoparticle and resulted that its properties for thermostability, reusability, higher recovery rate, higher efficiency for the degradation of dye under the tremendous condition in industrial use. Gupta et al. (2014) fabricated a carpet-like structure of extremely dense nanobundles of ZnO with the support of self-assembled polymethylsilisesquoxane and resulted that it can be used as a highly dense platform for the immobilization of the antibody for immumosensors through its high aspect ratio and high surface roughness. Yang et al. (2020) demonstrated immobilization of

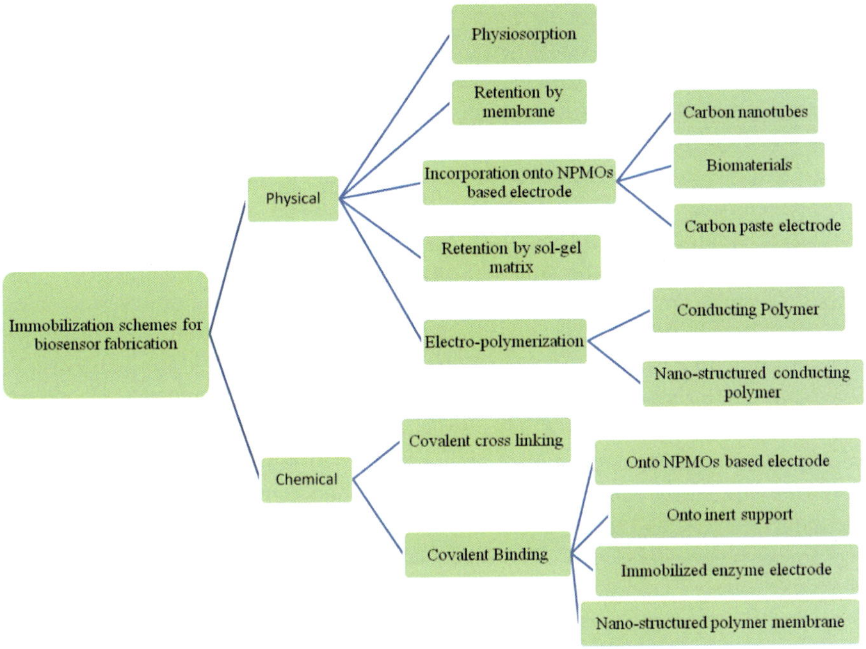

Figure 7.5 Immobilization schemes for biosensor application. *NPMOs*, Nanoparticles of metal oxide.

GOx over graphite-oxide/cobalthydroxide/chitosan (GCC) composite nanoparticle with the mediation through 1-ethyl-3(3-dimethylaminopropyl)-carbodiimide/*N*-hydroxysuccinimide (EDC:NHS) activation by using carbodiimide reaction and resulted in improvement in electron transfer through the catalytic reaction. Ameri et al. (2020) applied D-optimal design to optimize immobilization parameters of the lipase from *Bacillus atrophaeus* FSHM2(Bal) on maleic copolymer-coated amine-modified GO nanosheets.

7.3 Microsystems biosensing devices

The fabrication of microsystem biosensing devices cannot be achieved through conventional methods, but it can be achieved through advanced microfabrication techniques with precision manufacturing, namely, inkjet, laser-assisted patterning, photolithography, etc. These microsystems devices have broad uses in the medical application/ drug delivery system (Gupta & Pal, 2019). Nayak et al. (2013) reported a microchip platform for the real-time molecular identification by q-PCR process, utilized recognition through antibody with high sensitivity sensing and capturing through the mediated electric field for low detection limits. Bernasconi, Mangogna, and Magagnin (2018) fabricated a nonenzymatic electrochemical sensor by inkjet printing of CuO on platinum to detect glucose and resulted in a good sensitivity and adequate linear range. Jarošová et al. (2019) prepared carbon nanotube coated with polyacrylamide through inkjet printing to detect uric acid and pyocyanin in a wound. Zea et al. (2019) fabricated a Pt ink-based electrode through inkjet printing, anodic electrodeposited with iridium oxide for the pH sensing, and reported a highly sensitive, stable electrode. Kinnamon, Krishnan, Brosler, Sun, and Prasad (2018) developed a screen-printed biosensor on textiles by using a transduction film of GO over conductive silver electrode to detect influenza A virus in the environment, which reduces the economic and physical burden to the detection of disease. Kumar, Naveen Kumar, Aniley, Fernandez, and Bhansali (2019) fabricated an interdigitated electrode through the hydrothermal growth of ZnO nanorods to detect pH and found it highly sensitive for pH value of 4–10. Kim et al. (2016) demonstrated a wearable tattoo biosensing system for noninvasive alcohol monitoring in induced sweat, coupled with a flexible electronics board for transferring real-time data through Bluetooth communication. Chen et al. (2018) fabricated a sensor based on carbonized silk fabric coated with multiwalled carbon nanotube and decorated with Pt to detect glucose and found that this wearable, flexible electronic device shows a good linear range and sensitivity. Wang et al. (2019) fabricated a polyethylene terephthalate-based gold electrode by the technique of ultraviolet-mediated chemical plating for the sensing of glucose and found good selectivity, reproducibility, and long-term stability. Nag and Mukhopadhyay (2018) fabricated a laser-induced graphene electrode by using viable polymer films for taste sensing purpose.

7.4 Conclusions and future visions

The speedy materialization of NPMOs has led to the hybridization of nanoparticles and nanowires, nanodisks, nanobelts, nanorods, nanodisks, and nanotubes. It is widely used to fabricate enzymatic, electrochemical, and nonenzymatic biosensors due to its excellent nontoxicity characteristics, catalytic efficiency, large surface area, high adsorption efficiency, and electrical conductivity, reduction in potential, biocompatibility chemical, and mechanical stability. These characteristics of NPMOs support it for the fabrication of well-organized biosensing devices with high performance. The high stability in enzymatic biosensor can be achieved through the direct growth of nanostructure on the substrate, followed by the immobilization of particular biomolecules. In contrast, electrochemical biosensor achieves through the exploiting catalytic distinctiveness of different deliberate nanomaterials. NPMOs improve the sensitivity, stability, selectivity, and lower detection limit of the biosensors. In general, the tempting distinctiveness of the NPMOs biosensors endow with exceptional advantages for the fabrication of sensors in the field of environment, biomedical diagnostics, and food safety applications. For the accurate and rapid detection of the biomolecules, it is necessary to effectively select the NPMOs and the device's design. Therefore, contentious improvement is required in the engineering of protein/enzyme, immobilization strategies, and synthesis of the material to improve the functionality of biosensors. Further, an advanced manufacturing process is also required for the fabrication of the biosensors with at most precision with multifunctional properties detection designs. For real-time or near real-time sensing, futuristic research is going on in the advancement of functionalized bionanomaterials, NPMOs, to make the sensing devices cost-effective and readily available.

Acknowledgments

Dr. Kunal Mondal gratefully acknowledges the Energy & Environment S & T at the Idaho National Laboratory, United States for their support. Dr. Ankur Gupta acknowledges the Science and Engineering Research Board (SERB), Department of Science and Technology, India for their financial support under the grant number SRG/2020/001895.

References

Ali, M. A., Mondal, K., Singh, C., Dhar Malhotra, B., & Sharma, A. (2015). Anti-epidermal growth factor receptor conjugated mesoporous zinc oxide nanofibers for breast cancer diagnostics. *Nanoscale*, *7*(16), 7234–7245. Available from https://doi.org/10.1039/c5nr00194c.

Ali, M. A., et al. (2016). Microfluidic immuno-biochip for detection of breast cancer biomarkers using hierarchical composite of porous graphene and titanium dioxide nanofibers. *ACS Applied Materials & Interfaces*, *8*(32), 20570–20582. Available from https://doi.org/10.1021/acsami.6b05648.

Ali, M. A., et al. (2017). In situ integration of graphene foam-titanium nitride based bio scaffolds and microfluidic structures for soil nutrient sensors. *Lab on a Chip*, *17*(2), 274–285. Available from https://doi.org/10.1039/C6LC01266C.

Ameri, A., et al. (2020). Optimization of immobilization conditions of *Bacillus atrophaeus* FSHM2 lipase on maleic copolymer coated amine-modified graphene oxide nanosheets and its application for valeric acid esterification. *International Journal of Biological Macromolecules*, *162*, 1790–1806. Available from https://doi.org/10.1016/j.ijbiomac.2020.08.101.

Ansari, A. A., Solanki, P. R., & Malhotra, B. D. (2008). Sol-gel derived nanostructured cerium oxide film for glucose sensor. *Applied Physics Letters*, *92*(26), 263901. Available from https://doi.org/10.1063/1.2953686.

Ariaeenejad, S., Motamedi, E., & Hosseini Salekdeh, G. (2021). Application of the immobilized enzyme on magnetic graphene oxide nano-carrier as a versatile bi-functional tool for efficient removal of dye from water. *Bioresource Technology*, *319*, 124228. Available from https://doi.org/10.1016/j.biortech.2020.124228.

Arya, S. K., Saha, S., Ramirez-Vick, J. E., Gupta, V., Bhansali, S., & Singh, S. P. (2012). Recent advances in ZnO nanostructures and thin films for biosensor applications: Review. *Analytica Chimica Acta*, *737*, 1–21. Available from https://doi.org/10.1016/j.aca.2012.05.048.

Bernasconi, R., Mangogna, A., & Magagnin, L. (2018). Low cost inkjet fabrication of glucose electrochemical sensors based on copper oxide. *Journal of the Electrochemical Society*, *165*(8), B3176–B3183. Available from https://doi.org/10.1149/2.0241808jes.

Bhardwaj, T. (2014). A review on immobilization techniques of biosensors. *IJERT-International Journal of Engineering Research & Technology*. [Online]. Available from: http://www.ijert.org. Accessed November 04, 2020..

Charbgoo, F., Bin Ahmad, M., & Darroudi, M. (2017). Cerium oxide nanoparticles: Green synthesis and biological applications. *International Journal of Nanomedicine*, *12*, 1401–1413. Available from https://doi.org/10.2147/IJN.S124855.

Chauhan, P. S., Kant, R., Rai, A., Gupta, A., & Bhattacharya, S. (2019). Facile synthesis of ZnO/GO nanoflowers over Si substrate for improved photocatalytic decolorization of MB dye and industrial wastewater under solar irradiation. *Materials Science in Semiconductor Processing*, *89*, 6–17. Available from https://doi.org/10.1016/j.mssp.2018.08.022.

Chen, C., et al. (2018). An efficient flexible electrochemical glucose sensor based on carbon nanotubes/carbonized silk fabrics decorated with Pt microspheres. *Sensors & Actuators, B: Chemical*, *256*, 63–70. Available from https://doi.org/10.1016/j.snb.2017.10.067.

Chen, H. I., & Chang, H. Y. (2005). Synthesis of nanocrystalline cerium oxide particles by the precipitation method. *Ceramics International*, *31*(6), 795–802. Available from https://doi.org/10.1016/j.ceramint.2004.09.006.

Choi, H. N., Kim, M. A., & Lee, W. Y. (2005). Amperometric glucose biosensor based on sol-gel-derived metal oxide/Nafion composite films. *Analytica Chimica Acta*, *537*(1–2), 179–187. Available from https://doi.org/10.1016/j.aca.2005.01.010.

Debecker, D. P., & Mutin, P. H. (2012). Non-hydrolytic sol–gel routes to heterogeneous catalysts. *Chemical Society Reviews*, *41*(9), 3624–3650. Available from https://doi.org/10.1039/c2cs15330k.

Feng, X., et al. (2006). Converting ceria polyhedral nanoparticles into single-crystal nanospheres. *Science*, *312*(5779), 1504–1508. Available from https://doi.org/10.1126/science.1125767.

George, J. M., Antony, A., & Mathew, B. (2018). Metal oxide nanoparticles in electrochemical sensing and biosensing: A review. *Microchimica Acta*, *185*(7), 1–26. Available from https://doi.org/10.1007/s00604-018-2894-3.

Gupta, A., & Bhattacharya, S. (2018). On the growth mechanism of ZnO nano structure via aqueous chemical synthesis. *Applied Nanoscience*, *8*(3), 499–509. Available from https://doi.org/10.1007/s13204-018-0782-0.

Gupta, A., Gangopadhyay, S., Gangopadhyay, K., & Bhattacharya, S. (2016). Palladium-functionalized nanostructured platforms for enhanced hydrogen sensing. *Nanomaterials and Nanotechnology*, *6*, 40. Available from https://doi.org/10.5772/63987.

Gupta, A., Mondal, K., Sharma, A., & Bhattacharya, S. (2015). Superhydrophobic polymethylsilsesquioxane pinned one dimensional ZnO nanostructures for water remediation through photo-catalysis. *RSC Advances*, *5*(57), 45897–45907. Available from https://doi.org/10.1039/c5ra02938d.

Gupta, A., Nayak, M., Singh, D., & Bhattacharya, S. (2014). Antibody immobilization for ZnO nanowire based biosensor application. In *Materials research society symposium proceedings* (vol. 1675, pp. 33–39). doi: 10.1557/opl.2014.848.

Gupta, A., & Pal, P. (2019). *Micro-electro-mechanical system-based drug delivery devices. Bioelectronics and medical devices: From materials to devices - Fabrication, applications and reliability* (pp. 183–210). Elsevier.

Gupta, A., Pandey, S. S., & Bhattacharya, S. (2013). High aspect ZnO nanostructures based hydrogen sensing. *AIP Conference Proceedings*, *1536*(1), 291–292. Available from https://doi.org/10.1063/1.4810215.

Gupta, A., Pandey, S. S., Nayak, M., Maity, A., Majumder, S. B., & Bhattacharya, S. (2014). Hydrogen sensing based on nanoporous silica-embedded ultra dense ZnO nanobundles. *RSC Advances*, *4*(15), 7476–7482. Available from https://doi.org/10.1039/c3ra45316b.

Gupta, A., Saurav, J. R., & Bhattacharya, S. (2015). Solar light based degradation of organic pollutants using ZnO nanobrushes for water filtration. *RSC Advances*, *5*(87), 71472–71481. Available from https://doi.org/10.1039/c5ra10456d.

Gupta, A., Srivastava, A., Mathai, C. J., Gangopadhyay, K., Gangopadhyay, S., & Bhattacharya, S. (2014). Nano porous palladium sensor for sensitive and rapid detection of hydrogen. *Sensor Letters*, *12*(8), 1279–1285. Available from https://doi.org/10.1166/sl.2014.3307.

Haun, J. B., Yoon, T. J., Lee, H., & Weissleder, R. (2010). Magnetic nanoparticle biosensors. *Wiley Interdisciplinary Reviews: Nanomedicine and Nanobiotechnology*, *2*(3), 291–304. Available from https://doi.org/10.1002/wnan.84.

He, H. W., Wu, X. Q., Ren, W., Shi, P., Yao, X., & Song, Z. T. (2012). Synthesis of crystalline cerium dioxide hydrosol by a sol-gel method. *Ceramics International*, *38*(Suppl. 1), S501–S504. Available from https://doi.org/10.1016/j.ceramint.2011.05.063.

Hirano, M., Fukuda, Y., Iwata, H., Hotta, Y., & Inagaki, M. (2000). Preparation and spherical agglomeration of crystalline cerium (IV) oxide nanoparticles by thermal hydrolysis. *Journal of the American Ceramic Society*, *83*(5), 1287–1289. Available from https://doi.org/10.1111/j.1151-2916.2000.tb01371.x.

Jarošová, R., et al. (2019). Inkjet-printed carbon nanotube electrodes for measuring pyocyanin and uric acid in a wound fluid simulant and culture media. *Analytical Chemistry*, *91*(14), 8835–8844. Available from https://doi.org/10.1021/acs.analchem.8b05591.

Jiang, J., Li, Y., Liu, J., Huang, X., Yuan, C., & Lou, X. W. (2012). Recent advances in metal oxide-based electrode architecture design for electrochemical energy storage. *Advanced Materials*, *24*(38), 5166–5180. Available from https://doi.org/10.1002/adma.201202146.

Katz, E., Willner, I., & Wang, J. (2004). Electroanalytical and bioelectroanalytical systems based on metal and semiconductor nanoparticles. *Electroanalysis*, *16*(1–2), 19–44. Available from https://doi.org/10.1002/elan.200302930.

Kim, J., et al. (2016). Noninvasive alcohol monitoring using a wearable tattoo-based iontophoretic-biosensing system. *ACS Sensors*, *1*(8), 1011–1019. Available from https://doi.org/10.1021/acssensors.6b00356.

Kinnamon, D. S., Krishnan, S., Brosler, S., Sun, E., & Prasad, S. (2018). Screen printed graphene oxide textile biosensor for applications in inexpensive and wearable point-of-exposure detection of influenza for at-risk populations. *Journal of the Electrochemical Society*, *165*(8), B3084–B3090. Available from https://doi.org/10.1149/2.0131808jes.

Kiran, R. K., Rathour, R. K., Bhatia, R. K., Rana, D. S., Bhatt, A. K., & Thakur, N. (2020). Fabrication of thermostable and reusable nanobiocatalyst for dye decolourization by immobilization of lignin peroxidase on graphene oxide functionalized $MnFe_2O_4$ superparamagnetic nanoparticles. *Bioresource Technology*, *317*, 124020. Available from https://doi.org/10.1016/j.biortech.2020.124020.

Kumar, A., Naveen Kumar, S. K., Aniley, A. A., Fernandez, R. E., & Bhansali, S. (2019). Hydrothermal growth of zinc oxide (ZnO) nanorods (NRs) on screen printed IDEs for pH measurement application. *Journal of the Electrochemical Society*, *166*(9), B3264–B3270. Available from https://doi.org/10.1149/2.0431909jes.

Kundu, S., Kumar, A., Sen, S., & Nilabh, A. (2020). Bio-synthesis of SnO_2 and comparison its CO sensing performance with conventional processes. *Journal of Alloys and Compounds*, *818*, 152841. Available from https://doi.org/10.1016/j.jallcom.2019.152841.

Li, N., Patrissi, C. J., Che, G., & Martin, C. R. (2000). Rate capabilities of nanostructured $LiMn_2O_4$ electrodes in aqueous electrolyte. *Journal of the Electrochemical Society*, *147*(6), 2044. Available from https://doi.org/10.1149/1.1393483.

Li, Y., Tan, B., & Wu, Y. (2008). Mesoporous Co_3O_4 nanowire arrays for lithium ion batteries with high capacity and rate capability. *Nano Letters*, *8*(1), 265–270. Available from https://doi.org/10.1021/nl0725906.

Liang, S., Wu, X. L., Xiong, J., Zong, M. H., & Lou, W. Y. (2020). Metal-organic frameworks as novel matrices for efficient enzyme immobilization: An update review. *Coordination Chemistry Reviews*, *406*, 213149. Available from https://doi.org/10.1016/j.ccr.2019.213149.

Luo, X., Morrin, A., Killard, A. J., & Smyth, M. R. (2006). Application of nanoparticles in electrochemical sensors and biosensors. *Electroanalysis*, *18*(4), 319–326. Available from https://doi.org/10.1002/elan.200503415.

McNeil, S. E. (2005). Nanotechnology for the biologist. *Journal of Leukocyte Biology*, *78*(3), 585–594. Available from https://doi.org/10.1189/jlb.0205074.

Meyer, J., Hamwi, S., Kröger, M., Kowalsky, W., Riedl, T., & Kahn, A. (2012). Transition metal oxides for organic electronics: Energetics, device physics and applications. *Advanced Materials*, *24*(40), 5408–5427. Available from https://doi.org/10.1002/adma.201201630.

Mondal, K., Ali, M. A., Agrawal, V. V., Malhotra, B. D., & Sharma, A. (2014). Highly sensitive biofunctionalized mesoporous electrospun TiO_2 nanofiber based interface for biosensing. *ACS Applied Materials & Interfaces*, *6*(4), 2516–2527. Available from https://doi.org/10.1021/am404931f.

Mondal, K., Ali, M. A., Singh, C., Sumana, G., Malhotra, B. D., & Sharma, A. (2017). Highly sensitive porous carbon and metal/carbon conducting nanofiber based enzymatic biosensors for triglyceride detection. *Sensors & Actuators, B: Chemical*, *246*, 202–214. Available from https://doi.org/10.1016/j.snb.2017.02.050.

Mondal, K., & Sharma, A. (2016). Recent advances in electrospun metal-oxide nanofiber based interfaces for electrochemical biosensing. *RSC Advances*, *6*(97), 94595–94616. Available from https://doi.org/10.1039/c6ra21477k.

Nag, A., & Mukhopadhyay, S. C. (2018). Fabrication and implementation of printed sensors for taste sensing applications. *Sensors & Actuators, A: Physical*, *269*, 53–61. Available from https://doi.org/10.1016/j.sna.2017.11.023.

Nayak, M., et al. (2013). Integrated sorting, concentration and real time PCR based detection system for sensitive detection of microorganisms. *Scientific Reports*, *3*(1), 1–7. Available from https://doi.org/10.1038/srep03266.

Pan, C., & Zhu, J. (2009). The syntheses, properties and applications of Si, ZnO, metal, and heterojunction nanowires. *Journal of Materials Chemistry*, *19*(7), 869–884. Available from https://doi.org/10.1039/b816463k.

Qiu, X., Xiang, X., Liu, T., Huang, H., & Hu, Y. (2020). Fabrication of an organic–inorganic nanocomposite carrier for enzyme immobilization based on metal–organic coordination. *Process Biochemistry*, *95*, 47–54. Available from https://doi.org/10.1016/j.procbio.2020.05.007.

Rahman, M. M., Ahammad, A. J. S., Jin, J. H., Ahn, S. J., & Lee, J. J. (2010). A comprehensive review of glucose biosensors based on nanostructured metal-oxides. *Sensors*, *10*(5), 4855–4886. Available from https://doi.org/10.3390/s100504855.

Shaidarova, L. G., & Budnikov, G. K. (2008). Chemically modified electrodes based on noble metals, polymer films, or their composites in organic voltammetry. *Journal of Analytical Chemistry*, *63*(10), 922–942. Available from https://doi.org/10.1134/S106193480810002X.

Shaidarova, L. G., Gedmina, A. V., Chelnokova, I. A., & Budnikov, G. K. (2011). Selective determination of paracetamol and acetylsalicylic acid on electrode modified with a mixed-valent film of ruthenium oxide-ruthenium cyanide. *Russian Journal of Applied Chemistry*, *84*(4), 620–627. Available from https://doi.org/10.1134/S1070427211040112.

Shakerian, F., Zhao, J., & Li, S. P. (2020). Recent development in the application of immobilized oxidative enzymes for bioremediation of hazardous micropollutants – A review. *Chemosphere*, *239*, 124716. Available from https://doi.org/10.1016/j.chemosphere.2019.124716.

Shi, X., Gu, W., Li, B., Chen, N., Zhao, K., & Xian, Y. (2014). Enzymatic biosensors based on the use of metal oxide nanoparticles. *Microchimica Acta*, *181*(1–2), 1–22. Available from https://doi.org/10.1007/s00604-013-1069-5.

Singh, S. P., et al. (2007). Cholesterol biosensor based on rf sputtered zinc oxide nanoporous thin film. *Applied Physics Letters*, *91*(6), 063901. Available from https://doi.org/10.1063/1.2768302.

Soares, A. M. B. F., et al. (2020). Immobilization of papain enzyme on a hybrid support containing zinc oxide nanoparticles and chitosan for clinical applications. *Carbohydrate Polymers*, *243*, 116498. Available from https://doi.org/10.1016/j.carbpol.2020.116498.

Solanki, P. R., Kaushik, A., Agrawal, V. V., & Malhotra, B. D. (2011). Nanostructured metal oxide-based biosensors. *NPG Asia Materials*, *3*(1), 17–24. Available from https://doi.org/10.1038/asiamat.2010.137.

Tenne, R. (2006). Inorganic nanotubes and fullerene-like nanoparticles. *Nature Nanotechnology*, *1*(2), 103–111. Available from https://doi.org/10.1038/nnano.2006.62.

Wang, J. (2005a). Nanomaterial-based amplified transduction of biomolecular interactions. *Small (Weinheim an der Bergstrasse, Germany)*, *1*(11), 1036–1043. Available from https://doi.org/10.1002/smll.200500214.

Wang, J. (2005b). Nanomaterial-based electrochemical biosensors. *Analyst*, *130*(4), 421–426. Available from https://doi.org/10.1039/b414248a.

Wang, F., & Hu, S. (2009). Electrochemical sensors based on metal and semiconductor nanoparticles. *Microchimica Acta*, *165*(1–2), 1–22. Available from https://doi.org/10.1007/s00604-009-0136-4.

Wang, Y., Mori, T., Li, J. G., & Ikegami, T. (2002). Low-temperature synthesis of praseodymium-doped ceria nanopowders. *Journal of the American Ceramic Society*, *85* (12), 3105–3107. Available from https://doi.org/10.1111/j.1151-2916.2002.tb00591.x.

Wang, Y., Wang, X., Lu, W., Yuan, Q., Zheng, Y., & Yao, B. (2019). A thin film polyethylene terephthalate (PET) electrochemical sensor for detection of glucose in sweat. *Talanta*, *198*, 86–92. Available from https://doi.org/10.1016/j.talanta.2019.01.104.

Wei, Y., Li, Y., Zhang, N., Shi, G., & Jin, L. (2010). Ultrasound-radiated synthesis of PAMAM-Au nanocomposites and its application on glucose biosensor. *Ultrasonics Sonochemistry*, *17*(1), 17–20. Available from https://doi.org/10.1016/j.ultsonch.2009.06.017.

Xu, Y., & Wang, E. (2012). Electrochemical biosensors based on magnetic micro/nano particles. *Electrochimica Acta*, *84*, 62–73. Available from https://doi.org/10.1016/j.electacta.2012.03.147.

Yadav, T. P., & Srivastava, O. N. (2012). Synthesis of nanocrystalline cerium oxide by high energy ball milling. *Ceramics International*, *38*(7), 5783–5789. Available from https://doi.org/10.1016/j.ceramint.2012.04.025.

Yakimova, R. (2012). ZnO materials and surface tailoring for biosensing. *Frontiers in Bioscience: A Journal and Virtual Library*, *E4*(1), 254. Available from https://doi.org/10.2741/374.

Yan, Z., Wang, J., Zou, R., Liu, L., Zhang, Z., & Wang, X. (2012). Hydrothermal synthesis of CeO_2 nanoparticles on activated carbon with enhanced desulfurization activity. *Energy and Fuels*, *26*(9), 5879–5886. Available from https://doi.org/10.1021/ef301085w.

Yang, J. H., Kim, H. R., Lee, J. H., Jin, J. H., Lee, H. U., & Kim, S. W. (2020). Electrochemical properties of enzyme electrode covalently immobilized on a graphite oxide/cobalt hydroxide/chitosan composite mediator for biofuel cells. *International Journal of Hydrogen Energy*. Available from https://doi.org/10.1016/j.ijhydene.2020.03.084.

Yu, J. C., Zhang, L., & Lin, J. (2003). Direct sonochemical preparation of high-surface-area nanoporous ceria and ceria-zirconia solid solutions. *Journal of Colloid and Interface Science*, *260*(1), 240–243. Available from https://doi.org/10.1016/S0021-9797(02)00168-6.

Zea, M., Moya, A., Fritsch, M., Ramon, E., Villa, R., & Gabriel, G. (2019). Enhanced performance stability of iridium oxide-based pH sensors fabricated on rough inkjet-printed platinum. *ACS Applied Materials & Interfaces*, *11*(16), 15160–15169. Available from https://doi.org/10.1021/acsami.9b03085.

Zhao, Z., Lei, W., Zhang, X., Wang, B., & Jiang, H. (2010). ZnO-based amperometric enzyme biosensors. *Sensors*, *10*(2), 1216–1231. Available from https://doi.org/10.3390/s100201216, Multidisciplinary Digital Publishing Institute (MDPI).

Exploring the potential of metal oxides for biomedical applications

8

Jaba Mitra[1,2] *and Joyee Mitra*[3,4]
[1]Material Science & Engineering, University of Illinois, Urbana Champaign, IL, United States, [2]Thermo Fisher Scientific, South San Francisco, CA, United States, [3]Inorganic Materials & Catalysis Division, CSIR-CSMCRI, Bhavnagar, India, [4]Academy of Scientific and Innovative Research (AcSIR), Ghaziabad, India

8.1 Introduction

The burgeoning field of nanotechnology has enabled tremendous scientific and commercial developments in areas encompassing energy, environment, and medicine. In particular, biomedical applications of nanotechnology have catered to significant advances in diagnosis, therapy, drug delivery, bioimaging, and biosensing to name a few. Nanotechnology integrates interdisciplinary research into synthesis, characterization, and application of nanomaterials toward science and technology. At the heart of nanotechnology is the inherent physical and chemical properties of nanoparticles that are uniquely fashioned colloidal particles ranging in size from 10 nm to 1 μm (Kreuter, 2007). Biocompatibility is the first and foremost requirement of nanoparticles for biomedical applications. In addition, the engineered nanoparticles must be maneuverable in terms of composition, size, morphology, crystallinity, monodispersity, etc. (McNamara & Tofail, 2017). Typically, such particles have a preferential size of <200 nm. A wide variety of nanomaterials that have garnered immense interest for biomedical applications, including (1) metal nanoparticles of as gold, silver, copper, zinc, platinum, etc. (Mody, Siwale, Singh, & Mody, 2010), (2) metal oxide nanoparticles such as iron oxide, copper oxide, titanium dioxide, silver oxide, zinc oxide (ZnO), etc. (Nikolova & Chavali, 2020), (3) metal sulfide such as silver sulfide, iron sulfide nanoparticles (Fei et al., 2021; Yaqoob et al., 2020), (4) doped metal and metal oxide nanoparticles such as silver-titanium dioxide, silver-copper oxide nanoparticles, etc. (Stankic, Suman, Haque, & Vidic, 2016), and (5) metal organic frameworks templated on zinc, copper, manganese, etc. (Wang et al., 2018). Metal oxide nanoparticles have garnered immense interest, herein, because of their higher stability, lower toxicity, and optical properties. Furthermore, they can be tuned easily to the desired shape, size, and porosity, functionalized with various molecules and incorporated into hydrophilic or hydrophobic architectures (Patra et al., 2018). Size, shape, and surface properties govern interactions between a metal oxide nanoparticle and the in vivo environment, such that these properties underscore the suitability of nanoparticles in applications ranging from imaging of molecular markers, malignant cells, magnetic resonance

Metal Oxides for Biomedical and Biosensor Applications. DOI: https://doi.org/10.1016/B978-0-12-823033-6.00006-5

imaging (MRI), photodynamic therapy, and targeted delivery of drugs (Nikolova & Chavali, 2020). For example, efficacy of ZnO nanoparticles as drug delivery systems and bioimaging tools have been attributed to their morphology, crystallographic-orientation, and size-dependent electrical and thermal transport properties (Jiang, Pi, & Cai, 2018). Small-sized (10–20 nm) iron oxide nanoparticles (such as magnetite and maghemite) exhibit superparamagnetic properties and aid MRI, cell separation and detection, cancer theranostics, hyperthermia therapy, etc. (Arias et al., 2018). Table 8.1 summarizes the key biomedical applications of metal oxide nanoparticles.

Usage of metal oxide nanoparticles has greatly enriched modern medicine. In this review, we highlight the key physicochemical properties of nanoparticles that influence their performance as biomedical tools and later provide a summary of feasibility and applicability of some of the widely investigated metal oxide nanoparticles.

8.2 Physicochemical properties of metal oxide nanoparticles

As a nanoparticle enters the body, it interacts with cellular biomolecules and biofluids, prior to being transported into the inner cellular structures. Biological response to a nanoparticle is regulated by the latter's size, morphology, surface characteristics, etc. Below we summarize the key factors that influence the performance of nanoparticles for biomedical applications.

8.2.1 Chemical composition

The underlying chemical composition of a metal oxide nanoparticle is of paramount importance for understanding its suitability for specific biomedical applications. Essential trace metals such as iron, zinc, etc. exist in all body tissues including brain, muscle, bone and play crucial roles in nucleic acid and protein synthesis, hematopoiesis, and neurogenesis. Thus nanoparticles of such metal oxides are easily absorbed by the body. ZnO nanoparticles are widely studied for a plethora of biomedical applications including anticancer, antibacterial, antioxidant, antidiabetic, and antiinflammatory activities, drug delivery, and bioimaging (Jiang et al., 2018; Mirzaei & Darroudi, 2017). Iron oxide nanoparticles exhibit superparamagnetic phenomena and quantum tunneling of magnetization and have hence been found to be highly efficient for cellular therapy including labeling, targeting, and magnetic separation of cells, tissue repair, drug delivery, MRI, hyperthermia, magnetofection, etc. (Gupta & Gupta, 2005). Titanium dioxide nanoparticles are resistant to chemical degradation and swelling and are hence used for surface coating of implants. Moreover, inherent porosity of TiO_2 has been found to favor osseointegration and bone tissue growth (Ziental et al., 2020). Cerium oxide nanoparticles or nanoceria manifest multivalent cerium oxidation states in the form of surface defects. Thus

Table 8.1 Common biomedical applications of metal oxide nanoparticles.

Internal medicine: Therapeutically orchestrate molecular signaling pathways to regulate expression of growth factors, cell division, differentiation, and migration and apoptosis
• Iron oxide nanoparticles: hyperthermia, magnetically assisted drug delivery, cancer and antiviral therapy, gene and vaccine delivery systems, theranostics (Dadfar et al., 2020)
• Zinc oxide nanoparticles: cancer therapy, immune therapy, atherosclerosis (Jiang et al., 2018)
• Titanium dioxide nanoparticles: bone and tissue engineering, photodynamic therapy, tumor growth suppression, drug delivery (Ziental et al., 2020)
Diagnosis: Enable precise visualization and reliable quantification of disease viability and progression
• Fluorescent iron oxide and zinc oxide nanoparticles: targeted labeling of cancer cells, stem cells, cellular structures, and events (Jin, Yao, Ying, Lin, & Chen, 2020)
• Metal oxide quantum dots: in vivo ocular imaging, dynamic cellular imaging, single-cell analysis (Jin et al., 2020; Zhang, Yee, & Wang, 2008)
• Magnetic resonance imaging: iron oxide nanoparticles, alloy-based nanomaterials, for example, ferrites, core-shell nanoparticles (Abakumov et al., 2019)
Tissue regeneration: Functional restoration of tissues following disease or trauma
• Titanium dioxide nanoparticles: osseointegration, dentistry, wound healing (Ziental et al., 2020)
• Magnesium oxide nanoparticles: bone soft tissue engineering (Suryavanshi, Khanna, Sindhu, Bellare, & Srivastava, 2017), bone regeneration
• Iron oxide nanoparticles: neuronal tissue engineering, magnetomechanical actuation for bone tissue engineering, tissue engineering at vocal fold (Pöttler et al., 2019; Tomás, Gonçalves, Freitas, Domingues, & Gomes, 2019; Ziv-Polat, Margel, & Shahar, 2015)
• Cerium oxide nanoparticles: stem cell differentiation, angiogenesis, wound healing (Hosseini & Mozafari, 2020)
Biosensing: Signal transduction based on ligand–receptor binding
• Small-molecule electroanalysis: detection of glucose, dopamine, H_2O_2, etc. with iron oxide nanoparticles (Rahman, Ahammad, Jin, Ahn, & Lee, 2010; Solanki, Kaushik, Agrawal, & Malhotra, 2011)
• Biomolecule-based biosensors: immobilization of enzymes, DNA fragments, antigens, etc. as biorecognition motifs with ZnO nanorods, carbon-cerium oxide nanosystems, iron oxide nanospheres, titanium dioxide nanowires (Hahn, Ahmad, & Tripathy, 2012; Solanki et al., 2011)
Antimicrobial applications: Toxic effects engendered via interaction between microorganisms and nanoparticles
• Inhibition of bactericidal, fungal activities using copper oxide, iron oxide, ZnO, nickel oxide, silver oxide nanoparticles, etc. (Nikolova & Chavali, 2020)

nanoceria has the unique ability to switch between oxidation states and hence has garnered immense interest as biological antioxidants (Celardo, Pedersen, Traversa, & Ghibelli, 2011).

Although literature is replete with examples of a wide variety of metal nanoparticles, only TiO_2, ZnO, CuO, and iron oxide (ferrous and ferric oxides) are deemed safe for mammals (Najahi-Missaoui, Arnold, & Cummings, 2020). Cytotoxicity has been associated with chromosomal aberrations, enzyme degradation, changes in cell membrane, inadvertent activation of apoptotic caspase pathways, etc. In general, nanoparticle-induced toxicity has been associated with the ability of the nanoparticle to generate reactive oxygen species (ROS) and release metal ions into cells (Nikolova & Chavali, 2020). Similar sized metal oxide nanoparticles can induce differential toxic response in cells; for example, in human fetal lung fibroblasts were most susceptible to ZnO nanoparticles followed by TiO_2, Al_2O_3, and SiO_2 (Zhang, Yin, Tang, & Pu, 2011). Another study has shown that zinc and copper oxides have significantly higher toxicity toward pulmonary cells compared to zirconia, ceria, titania, and alumina nanoparticles irrespective of their sizes and surface areas (Ahamed, Akhtar, Alhadlaq, & Alrokayan, 2015; Lanone et al., 2009; Morimoto et al., 2016). Inherent toxicity of ZnO nanoparticles has been attributed to ROS generation which activates apoptotic signaling pathway. Thus ZnO nanoparticles are potential candidates for anticancer and antibacterial therapy (Jiang et al., 2018). Similarly, ultraviolet (UV) irradiation-induced generation of ROS from TiO_2 nanoparticles that can spread from nanoparticles to neighboring malignant tissues to promote cell death have been leveraged for photodynamic therapy against tumors and antibiotic-resistant bacteria (Ziental et al., 2020). Potential interactions between nanoparticles and a cell are depicted in Fig. 8.1.

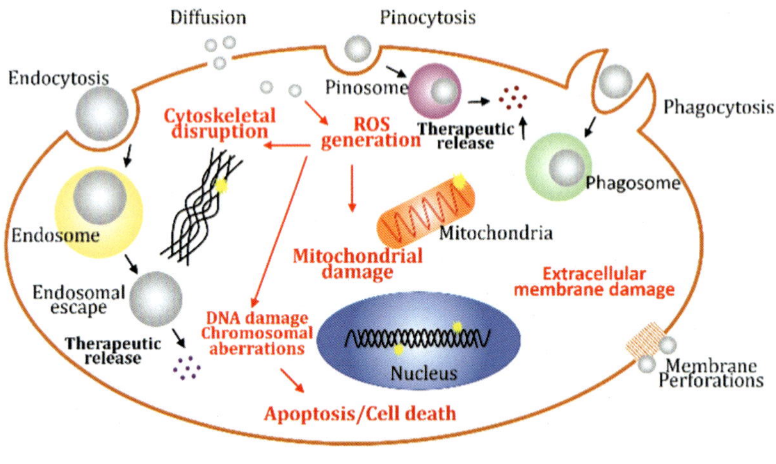

Figure 8.1 Interactions between nanoparticles and a cell: A nanoparticle is typically internalized by a cell via endocytosis, pinocytosis, or phagocytosis wherein after dissolution of the cellular vesicle, the nanoparticle releases its therapeutic components into the cell. Within the cell, nanoparticles may induce generation of reactive oxygen species (ROS) and subsequent damage and depolarization of mitochondria, chromosomal aberrations, segregations and DNA damage, structural changes and disruption of the microtubules of the cytoskeleton, culminating in apoptosis and cell death. In addition, nanoparticles may interact with the perforating cell membranes causing changes in its structure and integrity and prompt membrane dysfunctions and impaired cellular transport (Nikolova & Chavali, 2020).

8.2.2 Morphology and size

Metal oxide nanoparticles can be studied under zero-, one-, two-, and three-dimensional nanomaterials depending on the number of dimensions that are not constrained to the nanorange (Jeevanandam, Barhoum, Chan, Dufresne, & Danquah, 2018). Zero-dimensional materials such as nanospheres, nanoclusters, quantum dots, etc. manifest nanoconfinement in all dimensions. One-dimensional nanoparticles such as nanorods, nanotubes, nanowires, etc. have one dimension that is outside the nanorange. Two-dimensional nanomaterials include planar structures such as nanofilms, nanosheets, and nanoplatelets, whereas dendrimers, nanoflowers, bundles of nanowires and nanopillars borne out of controlled aggregation of nanomaterials in all three orthogonal directions constitute three-dimensional nanomaterials (Chavali & Nikolova, 2019; Jeevanandam et al., 2018) (Fig. 8.2). Differently shaped iron oxide nanoparticles including spheres, cubes, rods, octahedrons, prisms, rings, etc. designed for applications in magnetic theranostics have been shown to be biocompatible and easily up taken by cancer cells (Arias et al., 2018; Gupta, Naregalkar, Vaidya, & Gupta, 2007). Morphology of nanoparticles is of paramount

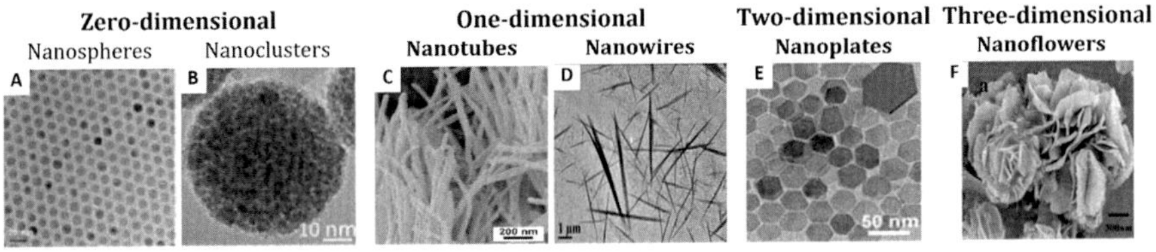

Figure 8.2 Morphological classification of nanomaterials. Electron microscopy images of iron oxide nanoparticles: (A) Nanospheres. (B) Nanoclusters. (C) Nanotubes. (D) Nanowires. (E) Nanoplates. (F) Nanoflowers.
Source: (A)Reused with permission from Park, J., An, K., Hwang, Y., Park, J. G., Noh, H. J., Kim J.-Y., et al. (2004). Ultra-large-scale syntheses of monodisperse nanocrystals. *Nature Materials*, *3*:891–895. © Springer Nature; (B) Reused with permission from Tadic, M., Kralj, S., Jagodic, M., Hanzel, D., & Makovec, D. (2014). Magnetic properties of novel superparamagnetic iron oxide nanoclusters and their peculiarity under annealing treatment. *Applied Surface Science*, *322*:255–264. © Elsevier; (C) Reused with permission from Kang, N., Park, J. H., Choi, J., Jin, J., Chun, J., Jung, I. G., et al. (2012). Nanoparticulate iron oxide tubes from microporous organic nanotubes as stable anode materials for lithium ion batteries. *Angewandte Chemie (International (Ed.) in English)*, *51*:6626–6630. © John Wiley & Sons; (D) Reused with permission from Lupan, O., Postica, V., Wolff, N., Polonskyi, O., Duppel, V., Kaidas, V., et al. (2017). Localized synthesis of iron oxide nanowires and fabrication of high performance nanosensors based on a single Fe2O3 nanowire. *Small (Weinheim an der Bergstrasse, Germany)*, *13*:1602868. © John Wiley & Sons; (E) Zhou, Z., Zhu, X., Wu, D., Huang, Q. C. D., Sun, C., & Xin, J., et al. (2015). Anisotropic shaped iron oxide nanostructures: Controlled synthesis and proton relaxation shortening effects. *Chemistry of Materials: A Publication of the American Chemical Society*, *27*, 3505–3515; (F) Sayed, F. N., Polshettiwar, V. (2015). Facile and sustainable synthesis of shaped iron oxide nanoparticles: Effect of iron precursor salts on the shapes of iron oxides. *Scientific Reports*, *5*, 9733.

importance as nanoparticles directly interact with cell membranes and biomolecules prior to internalization into cellular structure for specific functions. For example, larger contact area and higher aspect ratio of an iron oxide nanorod (Yue et al., 2011) and/or nanoflower (Andrade, Veloso, & Castanheira, 2020; Hugounenq et al., 2012) with respect to a nanosphere, facilitated cellular uptake and prolonged internalization. Fluorescence signal enhancing capability of ZnO nanorods have been harnessed for advanced bioassays for detecting nucleic acids and proteins (Hahm, 2014). Similarly, manganese oxide nanosheets have low cytotoxicity and superior histocompatibility and have been explored for bioimaging, biosensing, cancer therapy, and drug delivery (Chen et al., 2019).

In addition to nanoparticle morphology, its hydrodynamic size and hence the surface to volume ratio impacts distribution, cellular uptake, circulation time, and finally clearance from the organism. Furthermore, the nanoparticle sizes govern mechanisms of cellular uptake: smaller nanoparticles (~ 5 nm) can overcome cell barriers via translocation, whereas the larger ones undergo phagocytosis, pinocytosis, or other specific and nonspecific transport process. On similar lines, the former typically get eliminated via renal clearance and the latter potentially get trapped in the liver and spleen through macrophage phagocytosis (Elsabahy & Wooley, 2012). As common tumor vasculature has diameters <200 nm with pore sizes of 5 nm, and to avoid inadvertent size-dependent filtration by spleen and liver or kidney, nanoparticles with dimensions in the intermediate range of 20–100 nm are ideal for biomedical applications (Ankamwar, 2012; Toy, Hayden, Shoup, Baskaran, & Karathanasis, 2011).

8.2.3 Surface properties and crystallinity

Nanoparticle morphology determines surface area and hence surface energy, increase in which culminates in a significant reduction of thermodynamics stability and concomitantly nonuniform shape and size. Smaller the size of a nanoparticle, larger the fraction of atoms located on the surface, potentially eliciting a higher degree of interaction with cells, when compared to a larger particle. Particularly, the surface of metal oxide nanoparticles consists of oxygen atoms with lower coordination number, thereby increasing the reactive surface sites. Due to relative higher surface energy, ions are likely to break away from the corners and edges of a nanoparticle (Nikolova & Chavali, 2020). Moreover, crystallinity, phase, surface strain, size, and crystal defects determine the rate at which metal ions dissolve from the nanoparticle core (Avramescu, Rasmussen, Chénier, & Gardner, 2017). For example, the (0001) face of a ZnO nanoparticle had three times higher oxygen vacancy than the (1010) crystal surface (He, Cao, Fei, & Duan, 2019). Scanning tunneling microscopy (STM) has revealed that the surface of (0001) plane of ZnO is rough with triangular step edges, small terraces, holes and islands, whereas the (1010) plane has a well-defined terrace step structure that is associated with ZnO dimers, elongated and aligned along the (0001) or (1210) direction (Wöll, 2007). Depending on the underlying crystalline phase, ions released control bioavailability and cytotoxicity: rutile TiO_2 nanoparticles were found to induce chromosome segregation and membrane changes, whereas anatase TiO_2 was found to be non-toxic (Gurr, Wang, Chen, & Jan, 2005).

Surface defects on a nanoparticle are known to sequester holes, wherein chances of electron–hole recombination are minimized, culminating in photocatalytic and electrochemical activities (Nikolova & Chavali, 2020). Surface defects can engender single oxygen, hydroxyl, and superoxide radicals which are powerful ROS as well as •OOH peroxide radicals or hydrogen peroxide (Dayem et al., 2017). While the toxic effects of ROS have been well investigated, recent studies provide insight into the role of optimal ROS levels for mitogenic signaling and cell proliferation (Fu, Xia, Hwang, Ray, & Yu, 2014). ROS generated by ZnO nanoparticles have been implicated in decrease of mitochondrial potential, stress-mediated DNA damage, inhibition of DNA synthesis, and subsequent apoptosis (Dayem et al., 2017; Guo et al., 2013). Surface reactivity of nanoparticles has profound impact on how nanoparticles interact with the immune cells. For example, in TiO_2 photons migrate to the surface to react with oxygen, water or hydroxyl groups to form free radicals, whereas in CeO_2 nanoparticles photons are absorbed and hence isolated from the outside environment. Thus CeO_2 nanoparticles inhibit ROS production and have antioxidant, antiinflammatory potential, whereas TiO_2 nanoparticles are likely to induce an activated/proinflammatory state (Schanen et al., 2013).

In addition, surface charge on a metal oxide nanoparticle can inform on its interactions with cellular components and subsequent cellular uptake (Fröhlich, 2012). Positively charged nanoparticles have augmented capacity for interactions with plasma proteins and antibodies, with respect to negatively charged nanoparticles, and are hence more easily incorporated into cells (e.g., macrophages) (Alexis, Pridgen, Molnar, & Farokhzad, 2008). In contrast, negatively charged nanoparticles demonstrated higher tumor uptake, possibly through pinocytosis but lower levels of macrophage clearance (Xiao et al., 2011). A side-by-side comparison of positive and negatively charged nanoparticles by Kim et al. revealed that while positive nanoparticles are more proficient in delivering drugs into proliferating peripheral cells due to their enhanced uptake, negatively charged particles diffused quickly and were better adept in delivering drugs deeper into the tissue (Kim et al., 2010). Metal oxide nanoparticles with near-zero point of charge prevent nonspecific nanoparticle–biological interactions (Verma & Stellacci, 2010) but are susceptible to particle aggregation (Fröhlich, 2012).

8.2.4 Surface functionalization

Because of high surface energy, van der Waals forces, etc. nanoparticles in general have a strong tendency to agglomerate (Zhang, 2014). The degree of aggregation gleans biodistribution, biological, and biomedical activities of nanoparticles (Nikolova & Chavali, 2020). Due to the inverse ratio of aggregate size and surface area, efficiency of inadvertent agglomeration of metal oxide nanoparticles hinders interaction with cells. Moreover, aggregation of nanoparticles due to colloidal instability has been shown to impede clearance and eventually trigger toxicity (Pham et al., 2018). Hence is the need for surface functionalization of engineered nanoparticles with inorganic or organic moieties to ensure stable colloidal systems prior to applications in biomedicine. Surface modification facilitates homogeneous dispersion, improves surface activity, enhances physicochemical and mechanical properties and most importantly biocompatibility of nanoparticles. For example, silica coating of iron oxide nanoparticles has been found to be beneficial for enhancing stability, lowering aggregation, and reducing its cytotoxic effects (Abbas et al., 2013). Silver-coated iron oxide nanostructures have shown promise as multifunctional materials endowed with unique bactericidal characteristics, magneto-optical probes: bifunctional probes for MRI and two-photon fluorescence, near-infrared light-responsive drug delivery, in vivo computed tomography (CT) imaging and diagnosis based on surface-enhanced Raman Scattering (SERS) (Zhu et al., 2018). Iron oxide nanoparticles functionalized with a wide variety of organic macromolecules including both naturally occurring [e.g., chitosan (CS), dextran, gelatin] and synthetic polymers [e.g., poly(ethylene glycol) (PEG), poly(vinyl alcohol) (PVA), poly(lactide acid) (PLA)] have demonstrated improved stability, biocompatibility, blood circulation times, biodegradability, and low toxicity in cells (Wu, He, & Jiang, 2008). Simple physical immobilization of arginine-glycine-aspartic acid peptide (RGD) on to TiO_2 nanodot films have been harnessed for efficient light-induced cell detachment method for cell harvesting (Cheng et al., 2016). In addition, biocomposite of RGD-TiO_2 has shown promise as a therapeutic agent against integrin expressing tumor models (Dayan, Fleminger, & Ashur-Fabian, 2018).

Targeted functionalization of metal oxide nanoparticles promotes specificity and high payload at specific locations, multidrug conjugations with tunable release kinetics and multidrug resistance (Mout, Moyano, Rana, & Rotello, 2014). ZnO nanoparticles decorated with biological molecules comprising proteins, peptides, nucleic acids, folic acid, hyaluronan, etc. have shown increased targeting effects against cancer cells yet improved safety against nonmalignant normal cells (Jiang et al., 2018). For example, hyaluronan–ZnO nanocomposite engendered morphological changes and inhibited proliferation of adenocarcinoma cells (Namvar et al., 2016).

8.3 Commonly used metal oxides for biomedical applications

We have provided glimpses of some commonly used metal oxides for biomedical applications. The readers are requested to refer to some excellent reviews that are available covering the individual metal oxides (Arias et al., 2018; Verma & Kumar, 2019).

8.3.1 Oxides of iron (Fe_3O_4 and Fe_2O_3)

Magnetic nanoparticles have attractive properties that are beneficial for applications including but not limited to MRI as contrast agents, cell separation and detection, hyperthermia treatment, and drug delivery (Arias et al., 2018) (Fig. 8.3). These materials are biocompatible, environmentally benign and chemically stable; underlining their utility for clinical applications. Both magnetite (Fe_3O_4) and ɣ-maghemite (Fe_2O_3) have found utility in biomedical applications, especially at 10–20 nm sizes, when their superparamagnetic properties become important. Literature reports suggest that the size, shape, surface coatings on the nanoparticles are of immense importance while determining their efficacy for biomedical applications. Hence, synthetic strategies have to be designed keeping the end application in mind.

Several synthetic methodologies have been designed over the years for the synthesis of shape and size-controlled, soluble, and biocompatible iron oxide nanoparticles (Fig. 8.3). These can be broadly classified as chemical, physical, and biological methods (Wu et al., 2008). Coprecipitation method is the most common chemical route for nanomaterial synthesis, as this is an inexpensive method for the rapid scale-up of nanomaterial synthesis. However, this method suffers from drawbacks including a large size distribution owing to agglomeration of nanomaterials, and poor crystallinity. Microemulsion method overcomes the above-mentioned limitations by producing nanomaterials of low size distribution, but this method is not suitable for large-scale production. In addition, the removal of surfactants from the nanoparticles is cumbersome. Solvothermal methods, where the iron oxides are produced under high temperature and pressure conditions also provide a narrow size distribution and desired shape of the materials. However, prolonged reaction time limits their utility. Other methods such as sol-gel method, thermal decomposition of precursors, etc. have also been employed for the synthesis of iron oxide

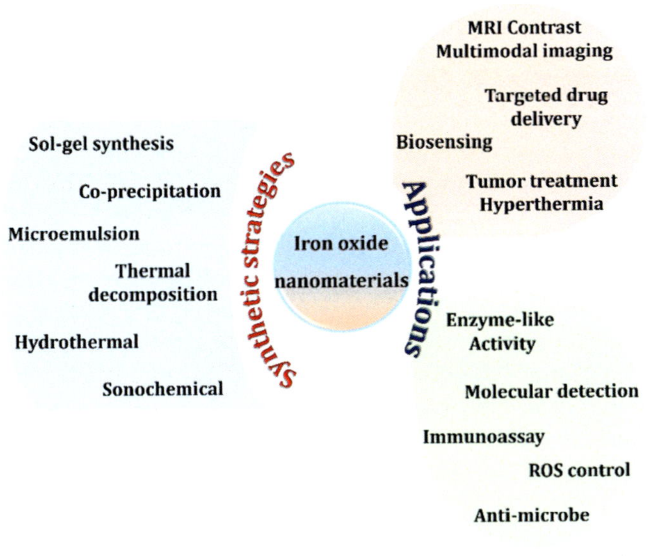

Figure 8.3 Synthetic strategies and applications of magnetic iron oxide nanomaterials. *ROS*, reactive oxygen species.

nanomaterials. Physical methods include pyrolysis of gaseous organometallic precursor and results in the desired control over size and morphology. Laser ablation synthesis has also been used in some cases, but here it has proved difficult in controlling the particle size of the resultant materials (Amendola & Meneghetti, 2013). Biosynthesis is also a cheap and ecofriendly method for the synthesis of iron oxide nanomaterials, where plant extracts or microbial materials with reducing abilities produce desired materials upon treatment with precursor solutions (Raghunath & Perumal, 2017).

Core-shell nanomaterials have proven beneficial in designing metal oxide species for biomedical applications. The metal oxide serves as the core, while surface coatings to improve stability and biocompatibility, surfactants, polymers, drug molecules for targeted delivery are present as shells (Prabhu, Mutalik, Rai, Udupa, & Rao, 2015). Polymers are commonly used surface coatings as these provide stability to the nanomaterials by resisting agglomeration of particles, preventing surface oxidation and enhancing their biocompatibility. Synthetic polymers such as PEG, naturally available CS and others have been employed for this purpose. PEG has the additional advantage of fooling the immune system of our body and increases the time the material remains in blood circulation (Prabhu et al., 2015). CS is intrinsically biocompatible, its positive charge makes it easier to breach the cell membrane and hence finds utility in vehicles for drug delivery. In addition, its low toxicity and water solubility, make CS the preferred coating material for metal oxide nanomaterials (Chen et al., 2019; Onnainty et al., 2016). However, partial protonation of CS under physiological pH limits its practical applications. Surfactants and stabilizers such as oleic acid, trisodium citrate, etc. can also stabilize nanomaterials by increasing steric repulsion between them. Similarly, inorganic materials including silica, carbon materials, and metal coating such as silver have been employed to increase the dispersity in solution, increasing stability, and to allow the conjugation of desired drugs to the iron oxide matrix (Ivashchenko et al., 2015). All these coatings have been reported to preserve the magnetic properties of the parent iron oxide species.

Magnetic nanoparticles have often been used as carriers for drug molecules owing to their ability to penetrate biological barriers. Desired drug molecules can either be adsorbed on to the surface of iron oxides, covalently attached to the surface, and sometimes adhered via electrostatic interactions. Various classes of drugs including immunosuppressants, anticancer medications, anticoagulants, antiinflammatory, and antibiotics have been delivered using magnetic iron oxides (Nigam & Bahadur, 2018; Nosrati, Sefidi, Sharafi, Danafar, Kheiri Manjili, 2018; Thomas et al., 2018). Iron oxides have also found use in photodynamic therapy by tagging with hypericine photosensitizer (Unterweger et al., 2015). Antibiotic resistant strains have prompted the use of antibiotic-carrier nanomaterials. These carriers often favor controlled release of the antibiotics compared to the conventional drugs, thereby increasing their efficacy as was reported in case of streptomycin-loaded iron oxide (Hussein-Al-Ali, El Zowalaty, Hussein, Ismail, & Webster, 2014). Various classes of drugs have been successfully loaded on Ag-coated iron oxides, which favored the linking of antibiotics to the oxide species via electrostatic interactions and in some cases hydrogen-bonding interactions (Ivashchenko et al., 2017). Reports suggest that the antimicrobial activities of the magnetic metal oxides are a cumulative effect of several factors including the production of ROS, DNA damage, membrane depolarization that impedes cell integrity, release of metal ions that affects homeostasis of the cell, to mention a few (Arias et al., 2018; Pan et al., 2010; Shah et al., 2017).

Toxicity of iron oxides has been subject to controversy both in vivo and in vitro studies (Xu et al., 2016). Modifications of the shape and size of nanoparticles under physiological were found to be directly related to their toxicity, as spherical particles were found to be less toxic compared to their rod-shaped counterparts (Feng et al., 2018). Literature reports state that iron oxides are often accumulated in the lysosomes, and can degrade into iron ions. These iron ions can react with peroxide and oxygen to generate ROS (Yarjanli, Ghaedi, Esmaeili, Rahgozar, & Zarrabi, 2017). Oxidative stress in the cells due to the production of ROS has been implicated in the cytotoxicity due to iron oxides. Concentration-dependent toxicity in vitro has been reported for magnetite in human lung alveolar epithelial cells (Dwivedi et al., 2014). Coating with polymers have been shown to significantly reduce their toxicity, by limiting their destructive influences on various cell organelles, as compared to uncoated particles, in most cases except for one report on the detrimental effect of D-mannose and poly-L-lysine coating (Pongrac et al., 2016).

8.3.2 Copper oxide (CuO)

Copper oxide is a low band gap oxide with high solution stability, and has promising applications as biosensors for clinically relevant analytes, antibiofouling materials, biocidal agents, and so on (Verma & Kumar, 2019). CuO has found applications in antitumor activity, and is approved by US Environmental Protection Agency (EPA) in biomedical devices to prevent bacterial infection (Meghana, Kabra, Chakraborty, & Padmavathy, 2015). This material also finds potential use in wound healing, targeted cancer therapy, etc. (Wu et al., 2018).

Different synthetic approaches including colloidal synthesis, solvothermal technique, template-assisted fabrication, etc. have been reported in order to tune the size and morphology of copper oxide simultaneously, resulting in various forms including nanowires, cubes, nanoflowers, nanofilms, etc. (Verma & Kumar, 2019). Electrochemical synthesis resulting in the anodic dissolution of copper in alkaline solutions has been widely reported owing to its simplicity, requirement of low temperature, commercial scalability, and control over desired morphology (Figueroa, Gana, Cooper, & Ji, 1993). Aside from the electrochemical route, chemical synthesis has been widely reported for metal oxide synthesis. Various inorganic salts of copper have been employed as precursors to obtain desired morphology of CuO. Choice of capping agents and concentration of the precursor also affects the final morphology of the oxide material (Siddiqui, Qureshi, & Haque, 2016). An increased precursor concentration results in an increment in the size of the final species. Microwave-assisted synthesis is gaining popularity in recent times owing to its efficiency and rapid reaction (Gajengi, Sasaki, & Bhanage, 2015). High temperature calcination step was necessary in few synthetic approaches, in order to improve the crystallinity and in case of sonochemical synthesis, the particle sizes of CuO. Environmental-friendly biological approaches for the synthesis of copper oxide have gained popularity, especially in case of biomedical end applications. Such green approach precludes the use of toxic chemicals, requirement of sophisticated instruments, and results in minimal waste generation (Upadhyay & Verma, 2015). In addition, this method eliminates the requirement of additional stabilizers, but the surface-adsorbed species prevents aggregation of particles.

Biocompatible nature of CuO paves the way for its use in the development of analytical devices having higher sensitivity, selectivity, and throughput compared to conventional methods. One of the most common applications of CuO is for the fabrication of sensors for physiologically relevant analytes including glucose, uric acid, dopamine, etc. (Verma & Kumar, 2019). This material finds utility in microbe-resistant devices, ointments and coatings owing to its intrinsic antimicrobial and antiinflammatory activities. Apoptosis of tumor cells can be induced by altering the intracellular levels of copper, and finds use in antitumor formulations. Thus CuO is a unique material that has applications encompassing both diagnosis and therapeutics.

CuO provides an efficient platform for glucose sensing as it electro-oxidizes glucose with a high rate of electron transfer. It has been reported that the sensitivity of the glucose sensor depends on the morphology of CuO, as CuO nanosheets reported an enhanced sensitivity as compared to nanowires (Zhong et al., 2016). Sensitivity toward glucose as well as the limit of detection has been improved by the researchers upon composite formation with carbon nanotubes/grapheme (Bao et al., 2014). Similarly, the detection of hydrogen peroxide, an important biomarker for oxidative stress in cells and tissues has been achieved with CuO, with an enhanced sensitivity as compared to the conventional "horseradish peroxidase"-based enzymatic sensor. Dopamine analysis in real samples was possible with CuO nanomaterials. To preclude the erroneous results due to coexisting compounds such as uric acid, ascorbic, etc., CuO composites with carbon nanomaterials were able to separate the peak potentials of dopamine and uric acid owing to a fast rate of electron transfer in the composite (Yang et al., 2013). Immunosensors, highly relevant for point-of-care diagnostics have also been designed incorporating CuO nanomaterials as nanocomposites with silver, ferrocene, etc. for the detection of tumor-specific biomarkers including prostate-specific antigen, carcinoembryonic antigen, and so on (Yang et al., 2018). Signal transduction accelerated by CuO resulted in the lowering of the detection limit to femtogram per millimeter.

CuO nanoparticles have been reported as therapeutic agents for the selective apoptosis of tumor cells and also to inhibit the growth and metastasis of melanoma. Some copper-containing drugs show selective antitumor activities as well. CuO nanomaterials have been found effective against various cancer cell lines including kidney, lung, liver, breast cancer, and so on (Siddiqui et al., 2013; Wang et al., 2013; Xue, Yu, Shan, & Li, 2018). CuO nanoparticles have been reported to target mitochondria specifically, to create mitochondrion-mediated apoptosis signaling pathway by triggering oxidative stress in the cell (Wang et al., 2013). Glutathione level was also found to reduce significantly in presence of CuO nanoparticles in the tumor cells. CuO and folic acid-conjugated CuO were found to improve the lifespan of tumor-affected mice. CuO provides a selective and targeted strategy toward the killing of cancerous cells (Verma & Kumar, 2019).

8.3.3 Zinc oxide (ZnO)

ZnO has been utilized in both industrial and biomedical applications owing to their optical, catalytic, and photochemical properties. It has been reported that ZnO can induce ROS-mediated DNA damage. Thus the unregulated use of ZnO has the potential to cause long-term health issues (Sruthi, Ashtami, & Mohanan, 2018). The size, morphology, and surface chemistry of ZnO nanomaterials can be tuned by modification of reaction parameters, which in turn can affect the toxicity of these materials. Physical methods such as physical vapor deposition, thermal evaporation, chemical methods including sol-gel method, precipitation, solvothermal synthesis, chemical vapor deposition, and green synthesis using plant extracts have gained popularity in the synthesis of ZnO nanoparticles (Sruthi et al., 2018). Green synthesis has gained importance in recent times owing to the expensive physical and chemical synthesis procedures, and also for the fact that the use of toxic chemicals can be avoided in this protocol. Compatibility with various pharmaceuticals and medicinal formulations also aids in the thrust for green synthetic protocols. Zinc is also an essential trace element, that is necessary for metabolic functions of the body and is recognized as an important food additive.

ZnO has garnered worldwide attention due to its antimicrobial ability. ZnO was reported to show a size-dependent bactericidal activities with both gram-positive and gram-negative strains, decrease in size enhanced its antibacterial activity (Yaqoob et al., 2020). It is also active against spores that are resistant to high temperature and pressure. Researchers suggest that the bactericidal activities of ZnO are due to the photochemical activity of ZnO and the generation of ROS. The positive charge on ZnO aids in its facile penetration of the negatively charged cell membrane by endocytosis. Release of Zn^{2+} ions within the cell wall disrupts cellular metabolism. One of the main applications of ZnO nanoparticles is for drug delivery (Mirzaei & Darroudi, 2017). In addition, sensing and bioimaging using ZnO have proven useful in monitoring patients. ZnO nanoparticles can induce the death of cancerous cells without any cytotoxicity on normal cells. However, extensive studies are required on the mechanism of cytotoxicity before the large-scale medicinal use of ZnO. Nanoparticle size is known to influence cell viability. However, dose-dependent cytotoxicity has been observed for ZnO on neural stem cells, without any influence of size. Some studies state that ZnO nanomaterials are effective against cancer cells. CS-encapsulated ZnO has found use as a drug delivery system for the delivery of doxorubicin to HeLa cells (Yuan, Hein, & Misra, 2010). This material has also found use as vehicles of gene delivery to tumor cells, ensuring a safe and efficient gene targeting to tissues (Nie et al., 2006). $ZnO-TiO_2$

composite was reported to show genotoxicity, without showing cytotoxicity (Osman, Baumgartner, Cemeli, Fletcher, & Anderson, 2010).

ZnO containing ointments have been effective in the treatment of sunburn, and other skin injuries caused by an exposure to UV sunrays. It has found utility as a broad-spectrum UVA, UVB reflector, and has been reported to be completely photostable, and thus features as a nonirritant, nonallergenic physical sunblock.

8.4 Conclusion and future perspectives

Metal oxide nanoparticles stand out for their wide spectrum of applications in nanomedicine, including regenerative medicine, wound healing, biosensing, tissue and immunotherapeutics, drug delivery, molecular imaging, biosensing, etc. Their cytotoxic potential has been effectively utilized in cancer therapy, antibacterial and antifungal applications, as highlighted in this review. The continued advances in engineering of novel metal oxide nanosystems have facilitated development multimodal approaches in therapy and diagnosis. In all, metal oxide nanoparticles hold immense promise as a new generation of biomaterials, with far reaching impact compared to their conventional bulk counterparts. However, the field of metal oxide-based nanomedicine needs to address challenges and emerging problems to enable large-scale clinical and industrial applications. A major drawback in this context, centers around ROS generation and concomitant cytotoxic effects. In addition, an extensive understanding of the mechanisms governing intra- and extracellular functioning of metal oxide nanoparticles is largely missing. This entails comprehensive investigations into biological effects including short- and long-term health risks based on the chemical composition, size, shape, synthesis method, etc. of nanoparticles. Molecular recognition mechanisms underlying signaling pathways associated with cell cycle, migration, differentiation, proliferation, etc. need to be mapped to fully realize the potential of metal oxide nanoparticles. This is of particular relevance for tumors, for treatments based on molecular profiling and selection of targeting agents. In turn, this will herald an era of personalized medicine for safe, more effective, customizable, and targeted therapeutic strategies and accelerate clinical trials and strengthen cancer diagnosis and therapy.

References

Abakumov, M. A., Ternovoi, S. K., Mazhuga, A. G., Chekhonin, V. P., Demikhov, E. I., Pistrak, A. G., et al. (2019). Contrast agents based on iron oxide nanoparticles for clinical magnetic resonance imaging. *Bulletin of Experimental Biology and Medicine*, *167*, 272–274.

Abbas, M., Rao, B. P., Islam, M. N., Naga, S. M., Takahashi, M., & Kim, C. (2013). Highly stable-silica encapsulating magnetite nanoparticles (Fe_3O_4/SiO_2) synthesized using single surfactantless-polyol process. *Ceramics International*, *40*, 1379–1385.

Ahamed, M., Akhtar, M. J., Alhadlaq, H. A., & Alrokayan, S. A. (2015). Assessment of the lung toxicity of copper oxide nanoparticles: Current status. *Nanomedicine (Lond), 10*, 2365–2377.

Alexis, F., Pridgen, E., Molnar, L. K., & Farokhzad, O. C. (2008). Factors affecting the clearance and biodistribution of polymeric nanoparticles. *Molecular Pharmaceutics, 5*, 501–515.

Amendola, V., & Meneghetti, M. (2013). What controls the composition and the structure of nanomaterials generated by laser ablation in liquid solution? *Physical Chemistry Chemical Physics: PCCP, 15*, 3027–3046.

Andrade, R. G. D., Veloso, S. R. S., & Castanheira, E. M. S. (2020). Shape anisotropic iron oxide-based magnetic nanoparticles: Synthesis and biomedical applications. *International Journal Molecular Sciences, 21*, 2455.

Ankamwar, B. (2012). *Size and shape effect on biomedical applications of nanomaterials. Biomedical Engineering-Technical Applications in Medicine*. London, United Kingdom: IntechOpen;.

Arias, L. S., Pessan, J. P., Vieira, A. P. M., Lima, T. M. T. D., Delbem, A. C. B., & Monteiro, D. R. (2018). Iron oxide nanoparticles for biomedical applications: A perspective on synthesis, drugs, antimicrobial activity, and toxicity. *Antibiotics (Basel), 7*, 46.

Avramescu, M. L., Rasmussen, P. E., Chénier, M., & Gardner, H. D. (2017). Influence of pH, particle size and crystal form on dissolution behaviour of engineered nanomaterials. *Environmental Science and Pollution Research International, 24*, 1553–1564.

Bao, J., Hou, C., Zhang, Y., Li, Q., Huo, D., Yang, M., et al. (2014). A non-enzymatic glucose sensor based on copper oxide nanowires-single wall carbon nanotubes. *Journal of the Electrochemical Society, 162*, B47–B51.

Celardo, I., Pedersen, J. Z., Traversa, E., & Ghibelli, L. (2011). Pharmacological potential of cerium oxide nanoparticles. *Nanoscale Horizons, 3*, 1411–1420.

Chavali, M. S., & Nikolova, M. P. (2019). Metal oxide nanoparticles and their applications in nanotechnology. *SN Applied Sciences, 1*, 607.

Chen, J., Meng, H., Tian, Y., Yang, R., Du, D., Li, Z., et al. (2019). Recent advances in functionalized MnO2 nanosheets for biosensing and biomedicine applications. *Nanoscale Horizons, 4*, 321–338.

Cheng, K., Wang, T., Yu, M., Wan, H., Lin, J., Weng, W., et al. (2016). Effects of RGD immobilization on light-induced cell sheet detachment from TiO_2 nanodots films. *Materials Science and Engineering C, Materials for Biological Applications, 63*, 240–246.

Dadfar, S. M., Roemhild, K., Drude, N. I., Stillfried, S., Knüchel, R., Kiessling, F., et al. (2020). Iron oxide nanoparticles: Diagnostic, therapeutic and theranostic applications. *Advanced Drug Delivery Reviews, 138*, 302–325.

Dayan, A., Fleminger, G., & Ashur-Fabian, O. (2018). RGD-modified dihydrolipoamide dehydrogenase conjugated to titanium dioxide nanoparticles-switchable integrin-targeted photodynamic treatment of melanoma cells. *RSC Advances, 8*, 9112–9119.

Dayem, A. A., Hossain, M. K., Lee, S. B., Kim, K., Saha, S. K., Yang, G. M., et al. (2017). The role of reactive oxygen species (ROS) in the biological activities of metallic nanoparticles. *International Journal of Molecular Sciences, 18*, 120.

Dwivedi, S., Siddiqui, M. A., Farshori, N. N., Ahamed, M., Musarrat, J., & Al-Khedhairy, A. A. (2014). Synthesis, characterization and toxicological evaluation of iron oxide nanoparticles in human lung alveolar epithelial cells. *Colloids and Surfaces B, Biointerfaces, 122*, 209–215.

Elsabahy, M., & Wooley, K. L. (2012). Design of polymeric nanoparticles for biomedical delivery applications. *Chemical Society Reviews*, *41*, 2545–2561.

Fei, W., Zhang, M., Fan, X., Ye, Y., Zhao, M., Zheng, C., et al. (2021). Engineering of bioactive metal sulfide nanomaterials for cancer therapy. *Journal of Nanobiotechnology*, *19*, 93.

Feng, Q., Liu, Y., Huang, J., Chen, K., Huang, J., & Xiao, K. (2018). Uptake, distribution, clearance, and toxicity of iron oxide nanoparticles with different sizes and coatings. *Scientific Reports*, *8*, 2082.

Figueroa, M. G., Gana, R. E., Cooper, W. C., & Ji, J. (1993). Electrochemical production of cuprous oxide using the anode-support system. *Journal of Applied Electrochemistry*, *23*, 308–315.

Fröhlich, E. (2012). The role of surface charge in cellular uptake and cytotoxicity of medical nanoparticles. *International Journal of Nanomedicine*, *7*, 5577–5591.

Fu, P. P., Xia, Q., Hwang, H. M., Ray, P. C., & Yu, H. (2014). Mechanisms of nanotoxicity: Generation of reactive oxygen species. *Journal of Food and Drug Analysis*, *22*, 64–75.

Gajengi, A. L., Sasaki, T., & Bhanage, B. M. (2015). NiO nanoparticles catalyzed three component coupling reaction of aldehyde, amine and terminal alkynes. *Catalysis Communications*, *72*, 174–179.

Guo, D., Bi, H., Liu, B., Wu, Q., Wang, D., & Cui, Y. (2013). Reactive oxygen species-induced cytotoxic effects of zinc oxide nanoparticles in rat retinal ganglion cells. *Toxicology In Vitro: An International Journal Published in Association with BIBRA*, *27*, 731–738.

Gupta, A. K., & Gupta, M. (2005). Synthesis and surface engineering of iron oxide nanoparticles for biomedical applications. *Biomaterials*, *26*, 3995–4021.

Gupta, A. K., Naregalkar, R. R., Vaidya, V. D., & Gupta, M. (2007). Recent advances on surface engineering of magnetic iron oxide nanoparticles and their biomedical applications. *Nanomedicine (London)*, *2*, 23–39.

Gurr, J. R., Wang, A. S. S., Chen, C. H., & Jan, K. Y. (2005). Ultrafine titanium dioxide particles in the absence of photoactivation can induce oxidative damage to human bronchial epithelial cells. *Toxicology*, *213*, 66–73.

Hahm, J. I. (2014). Zinc oxide nanomaterials for biomedical fluorescence detection. *Journal of Nanoscience and Nanotechnology*, *14*, 475–486.

Hahn, Y. B., Ahmad, R., & Tripathy, N. (2012). Chemical and biological sensors based on metal oxide nanostructures. *Chemical Communications (Cambridge, England)*, *48*, 10369–10385.

He, H., Cao, J., Fei, X., & Duan, N. (2019). High-temperature annealing of ZnO nanoparticles increases the dissolution magnitude and rate in water by altering O vacancy distribution. *Environment International*, *130*, 104930.

Hosseini, M., & Mozafari, M. (2020). Cerium oxide nanoparticles: Recent advances in tissue engineering. *Materials (Basel)*, *13*, 3072.

Hugounenq, P., Levy, M., Alloyeau, D., Lartigue, L., Dubois, E., Cabuil, V., et al. (2012). Iron oxide monocrystalline nanoflowers for highly efficient magnetic hyperthermia. *The Journal of Physical Chemistry C*, *16*, 15702–15712.

Hussein-Al-Ali, S. H., El Zowalaty, M. E., Hussein, M. Z., Ismail, M., & Webster, T. J. (2014). Synthesis, characterization, controlled release, and antibacterial studies of a novel streptomycin chitosan magnetic nanoantibiotic. *International Journal of Nanomedicine*, *9*, 549–557.

Ivashchenko, O., Coy, E., Peplinska, B., Jarek, M., Lewandowski, M., Zaleski, K., et al. (2017). Influence of silver content on rifampicin adsorptivity for magnetite/Ag/rifampicin nanoparticles. *Nanotechnology*, *28*, 055603.

Ivashchenko, O., Lewandowski, M., Peplinska, B., Jarek, M., Nowaczyk, G., Wiesner, M., et al. (2015). Synthesis and characterization of magnetite/silver/antibiotic nanocomposites for targeted antimicrobial therapy. *Materials Science & Engineering C, Materials for Biological Applications*, *55*, 343–359.

Jeevanandam, J., Barhoum, A., Chan, Y. S., Dufresne, A., & Danquah, M. K. (2018). Review on nanoparticles and nanostructured materials: History, sources, toxicity and regulations. *Beilstein Journal of Nanotechnology*, *9*, 1050–1074.

Jiang, J., Pi, J., & Cai, J. (2018). The advancing of zinc oxide nanoparticles for biomedical applications. *Bioinorganic Chemistry and Applications*, *2018*, 1062562.

Jin, K. T., Yao, J. Y., Ying, X. J., Lin, Y., & Chen, Y. F. (2020). Nanomedicine and early cancer diagnosis: Molecular imaging using fluorescence nanoparticles. *Current Topics in Medicinal Chemistry*, *20*, 2737–2761.

Kang, N., Park, J. H., Choi, J., Jin, J., Chun, J., Jung, I. G., et al. (2012). Nanoparticulate iron oxide tubes from microporous organic nanotubes as stable anode materials for lithium ion batteries. *Angewandte Chemie (International (Ed.) in English)*, *51*, 6626–6630.

Kim, B., Han, G., Toley, B. J., Kim, C. K., Rotello, V. M., & Forbes, N. S. (2010). Tuning payload delivery in tumour cylindroids using gold nanoparticles. *Nature Nanotechnology*, *5*, 465–472.

Kreuter, J. (2007). Nanoparticles—A historical perspective. *International Journal of Pharmaceutics*, *331*, 1–10.

Lanone, S., Rogerieux, F., Geys, J., Dupont, A., Maillot-Marechal, E., Boczkowski, J., et al. (2009). Comparative toxicity of 24 manufactured nanoparticles in human alveolar epithelial and macrophage cell lines. *Particle and Fibre Toxicology*, *6*, 14.

Lupan, O., Postica, V., Wolff, N., Polonskyi, O., Duppel, V., Kaidas, V., et al. (2017). Localized synthesis of iron oxide nanowires and fabrication of high performance nanosensors based on a single Fe_2O_3 nanowire. *Small (Weinheim an der Bergstrasse, Germany)*, *13*, 1602868.

McNamara, K., & Tofail, S. A. M. (2017). Nanoparticles in biomedical applications. *Advances in Physics: X*, *2*, 54–88.

Meghana, S., Kabra, P., Chakraborty, S., & Padmavathy, N. (2015). Understanding the pathway of antibacterial activity of copper oxide nanoparticles. *RSC Advances*, *5*, 12293–12299.

Mirzaei, H., & Darroudi, M. (2017). Zinc oxide nanoparticles: Biological synthesis and biomedical applications. *Ceramics International*, *43*, 907–914.

Mody, V. V., Siwale, R., Singh, A., & Mody, H. R. (2010). Introduction to metallic nanoparticles. *Journal of Pharmacy and Bioallied Sciences*, *2*, 282–289.

Morimoto, Y., Izumi, H., Yoshiura, Y., Tomonaga, T., Oyabu, T., Myojo, T., et al. (2016). Evaluation of pulmonary toxicity of zinc oxide nanoparticles following inhalation and intratracheal instillation. *International Journal of Molecular Sciences*, *17*, 1241.

Mout, R., Moyano, D. F., Rana, S., & Rotello, V. M. (2014). Surface functionalization of nanoparticles for nanomedicine. *Chemical Society Reviews*, *41*, 2539–2544.

Najahi-Missaoui, W., Arnold, R. D., & Cummings, B. S. (2020). Safe nanoparticles: Are we there yet? *International Journal of Molecular Sciences*, *22*, 385.

Namvar, F., Azizi, S., Rahman, H. S., Mohamad, R., Rasedee, A., Soltani, M., et al. (2016). Green synthesis, characterization, and anticancer activity of hyaluronan/zinc oxide nanocomposite. *OncoTargets and Therapy*, *9*, 4549–4559.

Nie, L., Gao, L., Feng, P., Zhang, J., Fu, X., Liu, Y., et al. (2006). Three-dimensional functionalized tetrapod-like ZnO nanostructures for plasmid DNA delivery. *Small (Weinheim an der Bergstrasse, Germany)*, *2*, 621–625.

Nigam, S., & Bahadur, D. (2018). Doxorubicin-loaded dendritic-Fe_3O_4 supramolecular nanoparticles for magnetic drug targeting and tumor regression in spheroid murine melanoma model. *Nanomedicine: Nanotechnology, Biology, and Medicine*, *14*, 759–768.

Nikolova, M. P., & Chavali, M. S. (2020). Metal oxide nanoparticles as biomedical materials. *Biomimetics (Basel)*, *5*, 27–73.

Nosrati, H., Sefidi, N., Sharafi, A., Danafar, H., & Kheiri Manjili, H. (2018). Bovine serum albumin (BSA) coated iron oxide magnetic nanoparticles as biocompatible carriers for curcumin-anticancer drug. *Bioorganic Chemistry*, *76*, 501–509.

Onnainty, R., Onida, B., Paez, P., Longhi, M., Barresi, A., & Granero, G. (2016). Targeted chitosan-based bionanocomposites for controlled oral mucosal delivery of chlorhexidine. *International Journal of Pharmaceutics*, *509*, 408–418.

Osman, I. F., Baumgartner, A., Cemeli, E., Fletcher, J. N., & Anderson, D. (2010). Genotoxicity and cytotoxicity of zinc oxide and titanium dioxide in HEp-2 cells. *Nanomedicine (London)*, *5*, 1193–1203.

Pan, X., Redding, J. E., Wiley, P. A., Wen, L., McConnell, J. S., & Zhang, B. (2010). Mutagenicity evaluation of metal oxide nanoparticles by the bacterial reverse mutation assay. *Chemosphere*, *79*, 113–116.

Park, J., An, K., Hwang, Y., Park, J.-G., Noh, H.-J., Kim, J.-Y., et al. (2004). Ultra-large-scale syntheses of monodisperse nanocrystals. *Nature Materials*, *3*, 891–895.

Patra, J. K., Das, G., Fraceto, L. F., Campos, E. V. R., Rodriguez-Torres, M. D. P., Acosta-Torres, L. S., et al. (2018). Nano based drug delivery systems: Recent developments and future prospects. *Journal of Nanobiotechnology*, *16*, 71.

Pham, B. T. T., Colvin, E. K., Pham, N. T. H., Kim, B. J., Fuller, E. S., Moon, E. A., et al. (2018). Biodistribution and clearance of stable superparamagnetic maghemite iron oxide nanoparticles in mice following intraperitoneal administration. *International Journal of Molecular Sciences*, *19*, 205.

Pongrac, I. M., Pavicic, I., Milic, M., Brkic Ahmed, L., Babic, M., Horak, D., et al. (2016). Oxidative stress response in neural stem cells exposed to different superparamagnetic iron oxide nanoparticles. *International Journal of Nanomedicine*, *11*, 1701–1715.

Pöttler, M., Fliedner, A., Bergmann, J., Bui, L. K., Mühlberger, M., Braun, C., et al. (2019). Magnetic tissue engineering of the vocal fold using superparamagnetic iron oxide nanoparticles. *Tissue Engineering Part A*, *25*, 470–1477.

Prabhu, S., Mutalik, S., Rai, S., Udupa, N., & Rao, B. S. S. (2015). PEGylation of superparamagnetic iron oxide nanoparticle for drug delivery applications with decreased toxicity: An in vivo study. *Journal of Nanoparticle Research*, 17.

Raghunath, A., & Perumal, E. (2017). Metal oxide nanoparticles as antimicrobial agents: A promise for the future. *International Journal of Antimicrobial Agents*, *49*, 137–152.

Rahman, M. M., Ahammad, A. J., Jin, J. H., Ahn, S. J., & Lee, J. J. (2010). A comprehensive review of glucose biosensors based on nanostructured metal-oxides. *Sensors (Basel)*, *10*, 4855–4886.

Sayed, F. N., & Polshettiwar, V. (2015). Facile and sustainable synthesis of shaped iron oxide nanoparticles: Effect of iron precursor salts on the shapes of iron oxides. *Scientific Reports*, *5*, 9733.

Schanen, B. C., Das, S., Reilly, C. M., Warren, W. L., Self, W. T., Seal, S., et al. (2013). Immunomodulation and T helper TH1/TH2 response polarization by CeO_2 and TiO_2 nanoparticles. *PLoS One*, *8*, e62816.

Shah, S. T., Wageeh, A. Y., Saad, O., Simarani, K., Chowdhury, Z., Abeer, A. A., et al. (2017). *Surface functionalization of iron oxide nanoparticles with gallic acid as potential antioxidant and antimicrobial agents, . Nanomaterials (Basel)* (7). .

Siddiqui, H., Qureshi, M. S., & Haque, F. Z. (2016). Effect of copper precursor salts: Facile and sustainable synthesis of controlled shaped copper oxide nanoparticles. *Optik*, *127*, 4726–4730.

Siddiqui, M. A., Alhadlaq, H. A., Ahmad, J., Al-Khedhairy, A. A., Musarrat, J., & Ahamed, M. (2013). Copper oxide nanoparticles induced mitochondria mediated apoptosis in human hepatocarcinoma cells. *PLoS One*, *8*, e69534.

Solanki, P. R., Kaushik, A., Agrawal, V. V., & Malhotra, B. D. (2011). Nanostructured metal oxide-based biosensors. *NPG Asia Materials*, *3*, 17–24.

Sruthi, S., Ashtami, J., & Mohanan, P. V. (2018). Biomedical application and hidden toxicity of zinc oxide nanoparticles. *Materials Today Chemistry (Weinheim an der Bergstrasse, Germany)*, *10*, 175–186.

Stankic, S., Suman, S., Haque, F., & Vidic, J. (2016). Pure and multi metal oxide nanoparticles: Synthesis, antibacterial and cytotoxic properties. *Journal of Nanobiotechnology*, *14*.

Suryavanshi, A., Khanna, K., Sindhu, K. R., Bellare, J., & Srivastava, R. (2017). Magnesium oxide nanoparticle-loaded polycaprolactone composite electrospun fiber scaffolds for bone-soft tissue engineering applications: In-vitro and in-vivo evaluation. *Biomedical Materials (Bristol, England)*, *2*, 055011.

Tadic, M., Kralj, S., Jagodic, M., Hanzel, D., & Makovec, D. (2014). Magnetic properties of novel superparamagnetic iron oxide nanoclusters and their peculiarity under annealing treatment. *Applied Surface Science*, *322*, 255–264.

Thomas, R. G., Unnithan, A. R., Moon, M. J., Surendran, S. P., Batgerel, T., Park, C. H., et al. (2018). Electromagnetic manipulation enabled calcium alginate Janus microsphere for targeted delivery of mesenchymal stem cells. *International Journal of Biological Macromolecules*, *110*, 465–471.

Tomás, A. R., Gonçalves, A. I., Freitas, E. P., Domingues, R. M. A., & Gomes, M. E. (2019). Magneto-mechanical actuation of magnetic responsive fibrous scaffolds boosts tenogenesis of human adipose stem cells. *Nanoscale Horizons*, *11*, 8255–18271.

Toy, R., Hayden, E., Shoup, C., Baskaran, H., & Karathanasis, E. (2011). The effects of particle size, density and shape on margination of nanoparticles in microcirculation. *Nanotechnology*, *22*, 115101.

Unterweger, H., Subatzus, D., Tietze, R., Janko, C., Poettler, M., Stiegelschmitt, A., et al. (2015). Hypericin-bearing magnetic iron oxide nanoparticles for selective drug delivery in photodynamic therapy. *International Journal of Nanomedicine*, *10*, 6985–6996.

Upadhyay, L. S. B., & Verma, N. (2015). Recent developments and applications in plant-extract mediated synthesis of silver nanoparticles. *Analytical Letters*, *48*, 2676–2692.

Verma, A., & Stellacci, F. (2010). Effect of surface properties on nanoparticle-cell interactions. *Small (Weinheim an der Bergstrasse, Germany)*, *6*, 12–21.

Verma, N., & Kumar, N. (2019). Synthesis and biomedical applications of copper oxide nanoparticles: An expanding horizon. *ACS Biomaterials Science & Engineering*, *5*, 1170–1188.

Wang, S., McGuirk, C. M., d'Aquino, A., Mason, J. A., & Mirkin, C. A. (2018). *Advanced Materials*, *30*, 1800202.

Wang, Y., Yang, F., Zhang, H. X., Zi, X. Y., Pan, X. H., Chen, F., et al. (2013). Cuprous oxide nanoparticles inhibit the growth and metastasis of melanoma by targeting mitochondria. *Cell Death & Disease*, *4*, e783.

Wöll, C. (2007). The chemistry and physics of zinc oxide surfaces. *Progress in Surface Science*, *82*, 55–120.

Wu, N., Zhang, C., Wang, C., Song, L., Yao, W., Gedanken, A., et al. (2018). Zinc-doped copper oxide nanocomposites reverse temozolomide resistance in glioblastoma by inhibiting AKT and ERK1/2. *Nanomedicine (Lond)*, *13*, 1303–1318.

Wu, W., He, Q., & Jiang, C. (2008). Magnetic iron oxide nanoparticles: Synthesis and surface functionalization strategies. *Nanoscale Research Letters*, *3*, 397–415.

Xiao, K., Li, Y., Luo, J., Lee, J. S., Xiao, W., Gonik, A. M., et al. (2011). The effect of surface charge on in vivo biodistribution of PEG-oligocholic acid based micellar nanoparticles. *Biomaterials*, *32*, 3435–3446.

Xu, H. L., Mao, K. L., Huang, Y. P., Yang, J. J., Xu, J., Chen, P. P., et al. (2016). Glioma-targeted superparamagnetic iron oxide nanoparticles as drug-carrying vehicles for theranostic effects. *Nanoscale*, *8*, 14222–14236.

Xue, Y., Yu, G., Shan, Z., & Li, Z. (2018). Phyto-mediated synthesized multifunctional Zn/CuO NPs hybrid nanoparticles for enhanced activity for kidney cancer therapy: A complete physical and biological analysis. *Journal of Photochemistry and Photobiology B, Biology*, *186*, 131–136.

Yang, S., Li, G., Yin, Y., Yang, R., Li, J., & Qu, L. (2013). Nano-sized copper oxide/multi-wall carbon nanotube/nafion modified electrode for sensitive detection of dopamine. *Journal of Electroanalytical Chemistry*, *703*, 45–51.

Yang, Y., Yan, Q., Liu, Q., Li, Y., Liu, H., Wang, P., et al. (2018). An ultrasensitive sandwich-type electrochemical immunosensor based on the signal amplification strategy of echinoidea-shaped Au@Ag-Cu_2O nanoparticles for prostate specific antigen detection. *Biosensors & Bioelectronics*, *99*, 450–457.

Yaqoob, A. A., Ahmad, H., Parveen, T., Ahmad, A., Oves, M., Ismail, I. M. I., et al. (2020). Recent advances in metal decorated nanomaterials and their various biological applications: A review. *Frontiers in Chemistry*, *8*, 341.

Yarjanli, Z., Ghaedi, K., Esmaeili, A., Rahgozar, S., & Zarrabi, A. (2017). Iron oxide nanoparticles may damage to the neural tissue through iron accumulation, oxidative stress, and protein aggregation. *BMC Neuroscience*, *18*, 51.

Yuan, Q., Hein, S., & Misra, R. D. (2010). New generation of chitosan-encapsulated ZnO quantum dots loaded with drug: Synthesis, characterization and in vitro drug delivery response. *Acta Biomaterialia*, *6*, 2732–2739.

Yue, Z. G., Wei, W., You, Z. X., Yang, Q. Z., Yue, H., Su, Z. G., et al. (2011). Iron oxide nanotubes for magnetically guided delivery and pH-activated release of insoluble anticancer drugs. *Advanced Functioncal Materials*, *21*, 3446–3453.

Zhang, H., Yee, D., & Wang, C. (2008). Quantum dots for cancer diagnosis and therapy: Biological and clinical perspectives. *Nanomedicine (London)*, *3*, 83–91.

Zhang, W. (2014). Nanoparticle aggregation: principles and modeling. *Advances in Experimental Medicine and Biology*, *11*, 19–43.

Zhang, X. Q., Yin, L. H., Tang, M., & Pu, Y. P. (2011). ZnO, $TiO_{(2)}$, $SiO_{(2)}$, and $Al_{(2)}O(3)$ nanoparticles-induced toxic effects on human fetal lung fibroblasts. *Biomedical and Environmental Sciences: BES*, *24*, 661–669.

Zhong, Y., Shi, T., Liu, Z., Cheng, S., Huang, Y., Tao, X., et al. (2016). Ultrasensitive non-enzymatic glucose sensors based on different copper oxide nanostructures by in-situ growth. *Sensors and Actuators B Chemical*, *236*, 326–333.

Zhou, Z., Zhu, X., Wu, D., Huang, Q. C. D., Sun, C., Xin, J., et al. (2015). Anisotropic shaped iron oxide nanostructures: Controlled synthesis and proton relaxation shortening effects. *Chemistry of Materials: A Publication of the American Chemical Society*, *27*, 3505–3515.

Zhu, N., Ji, H., Yu, P., Niu, J., Farooq, M. U., Akram, M. W., et al. (2018). Surface modification of magnetic iron oxide nanoparticle. *Nanomaterials (Basel)*, *8*, 810.

Ziental, D., Czarczynska-Goslinska, B., Mlynarczyk, D. T., Glowacka-Sobotta, A., Stanisz, B., Goslinski, T., et al. (2020). Titanium dioxide nanoparticles: Prospects and applications in medicine. *Nanomaterials (Basel)*, *10*, 387–407.

Ziv-Polat, O., Margel, S., & Shahar, A. (2015). Application of iron oxide anoparticles in neuronal tissue engineering. *Neural Regeneration Research*, *10*, 189–191.

Surface coating and functionalization of metal and metal oxide nanoparticles for biomedical applications

9

Raj Kumar[1,2], *Guruprasad Reddy Pulikanti*[1,3], *Konathala Ravi Shankar*[2,4], *Darsi Rambabu*[2,5], *Venkateswarulu Mangili*[2,6], *Lingeshwar Reddy Kumbam*[2,3], *Prateep Singh Sagara*[2], *Nagaraju Nakka*[2] *and Midathala Yogesh*[2]

[1]School of Basic Sciences and Advanced Material Research Center, Indian Institute of Technology Mandi, Mandi, India, [2]Department of Pharmaceutical Sciences, University of Michigan, Ann Arbor, MI, United States, [3]Department of Chemistry, Indian Institute of Science Education and Research Tirupati, Tirupati, India, [4]School of Nano Sciences, Central University of Gujrat, Gandhinagar, India, [5]Molecular Chemistry, Materials and Catalysis, Institute of Condensed Matter and Nanosciences, Universite Catholique de Louvain, Brussels, Belgium, [6]Department of Inorganic and Physical Chemistry, Indian Institute of Science, Bangalore, India

9.1 Introduction

Nanomaterials are materials at the nanoscale level with dimensions in the range from 1 to 100 nm. Nanomaterials follow the nanoscience and nanotechnology principles and their applications in a wide range of research areas are mostly interdisciplinary (Kumar, Dalvi, & Siril, 2020; Kumar, Kumar, & Gedanken, 2020; Kumar et al., 2020; Lingeshwar Reddy, Balaji, Kumar, & Krishnan, 2018). Nanomaterials show novel optical, electronic, magnetic, and many other properties with improved performance compared to their bulk counterparts (Kumar et al., 2020; Lingeshwar Reddy et al., 2018). Nanomaterials are broadly classified into two categories: organic and inorganic nanomaterials. Organic nanomaterials include carbon nanomaterials (carbon dots, carbon nanotubes, graphene family nanomaterials), polymeric nanomaterials, and small organic molecules or compounds such as pharmaceutics, drugs, stabilizers, dyes, and pigments. Inorganic nanomaterials are mainly composed of metals, nonmetals, and semiconducting materials (Kumar, 2016). The most widely studied and well-known inorganic nanomaterials include metals, metal oxides, alloys, bimetals, silica, iron oxides, and noble metal-based nanomaterials (Kumar, Mondal et al., 2020; Kumar, Aadil, Ranjan, & Kumar, 2020). Compared to organic nanomaterials, inorganic nanomaterials, especially metal and metal oxides, have gained significant interest in the last few decades. The

Metal Oxides for Biomedical and Biosensor Applications. DOI: https://doi.org/10.1016/B978-0-12-823033-6.00007-7

high surface area to volume ratio introduces several advantages over bulk materials, such as high catalytic performance, high loading, functionalization feasibility on the surface, drug delivery, biomedical applications, and flexible mechanical properties (Chawla, Kumar, & Siril, 2016; Dutt, Kumar, & Siril, 2015; Vats, Dutt, Kumar, & Siril, 2016).

Among the various inorganic nanomaterials, metal and metal oxides are more interesting and are widely used in a range of applications. This is because of the well-established facile synthesis strategies, feasibility to control the physicochemical properties, and moreover the interesting and tunable properties (Bhardwaj & Kaushik, 2017; Mishra, Murugan, Kotakoski, & Adam, 2017). For example, gold nanoparticles of different sizes enable different optical properties with differently shaped gold nanostructures such as gold nanorods, gold nanotubes, and gold nanoparticles with triangular shapes all showing different optical behaviors (Mejac, Bryan, Lee, & Tran, 2009). Similarly, magnetic nanoparticles at the nanoscale have shown different magnetic behaviors, such as iron oxide nanoparticles and aggregated magnetic nanoclusters that showed different magnetic behaviors (Tiwari, Kumar, Shefi, & Randhawa, 2020). Moreover, palladium nanoparticles, nanowires, nanocomposites with polyaniline, and graphene also showed different catalytic performances in different catalytic applications (Chawla et al., 2016; Dutt, Kumar et al., 2015). Hence, metals and metal oxides are more interesting, widely studied, commonly used, and easy to prepare and scale up for formulation, and also suitable for commercialization. There are several products based on metals and metal oxides that are already available in the market.

Although there are a number of metal and metal oxide nanomaterial applications only a few are of interest for biomedical applications such as drug delivery, tissue engineering, and bioimaging (Kumar, Mondal et al., 2020; Tiwari et al., 2020). The major challenges in metal and metal oxide nanomaterials are dispersibility, stability in physiological in vitro and in vivo environments, biocompatibility, biodegradability, and surface properties such as charge, nature, stabilizers, and functional group (Parveen, Misra, & Sahoo, 2012). Furthermore, their selectivity and specificity toward the targeted site requires disease location, sensitivity toward the local environment such as pH, ionic strength, temperature, pressure, and other stimuli including reactive oxygen species (ROS), which play a crucial role (Park, 2013). Moreover, metal and metal oxide nanomaterials are more promising if they have properties such as long-term blood circulation time, favorable interaction with blood constituents, and do not have unwanted or nonspecific interactions with other entities or candidates (Bhardwaj & Kaushik, 2017). Finally, efficient clearance is also important. To meet all the above-discussed requirements, researchers around the globe have increased their efforts and, so far, considerable progress has been achieved by bring about different functional properties into metals and metal oxides. To overcome the limitations and introduce the above-mentioned properties into metal and metal oxide nanoparticles several strategies have been developed with the functionalization of metal and metal oxide nanoparticle surfaces giving promising and simple results. In this chapter, we especially focus on the functionalization of metal and metal oxide nanomaterials or nanoparticles.

9.2 Metal nanoparticles

Metal nanoparticles are composed of metallic elements with a size range below 100 nm. Metals are abundant elements with a variety of properties. The properties of metal nanoparticles vary dependent on their composition, size, shape, and structure. Advanced synthesis strategies and processes enable the control of their optical, catalytic, electronic, and magnetic properties (Virkutyte & Varma, 2011). The most interesting properties of metal nanoparticles are superparamagnetic behavior (e.g., Fe_3O_4) and surface plasmon resonance (SPR) (e.g., Au nanoparticles) (Dutt, Siril, Sharma, & Periasamy, 2015; Mohapatra et al., 2008). Metal nanoparticles also show surface-enhanced Raman scattering (SERS) and hence can be used in diagnosis and imaging applications. Noble metal nanoparticles show SPR and SERS effects which can be tunable by changing the size, shape, and structure (Sharma, Sinha, Dutt, Chawla, & Siril, 2016). Metal nanoparticles with SPR behavior are used in biological imaging such as dark-field optical microscopy and optical coherence tomography (OCT). They also are used in photothermal cancer therapy. The most widely studied superparamagnetic metal nanoparticles include Co, Fe, Ni, and alloyed metal nanoparticles such as Fe-Pt, Co-Pt, and Co-Pt_3. These nanoparticles are interesting candidates and are widely used in imaging and therapy such as magnetic resonance imaging (MRI) and magnetic induction hyperthermia therapy (Tomitaka et al., 2019). Spherical gold nanoparticles show optical properties in the visible region, whereas gold nanorods, nanoshells, and nanocages show in the near infrared (NIR) region (Chen et al., 2007; Rengan, Kundu, Banerjee, & Srivastava, 2014; Vankayala, Lin, Kalluru, Chiang, & Hwang, 2014). Nanomaterials with optical properties in the NIR region are suitable for several biomedical and tissue engineering applications due to their several advantages such as deep tissue penetration efficacy (Zhao, Zhong, & Zhou, 2018).

9.3 Metal oxide nanoparticles

Compared to metals, metal oxides are more interesting and exhibit structural properties. Transition metals form different types of oxides, which are widely used in different technologies such as magnetic ferrites, ferroelectric oxides, superconductors, ionic conductors, phosphors, and photocatalysts. Various synthetic processes have been developed for the formulation of magnetic metal oxides, such as coprecipitation, thermal decomposition, reduction, micelle synthesis, and hydrothermal synthesis (Medhi, Marquez, & Lee, 2020). These synthesis strategies also are suitable for the formulation of other metal oxide nanoparticles such as cobalt oxide (CoO, Co_3O_4), nickel oxide (NiO), manganese oxide (MnO), metal ferrite ($NiFe_2O_4$, $MgFeO_4$, $MnFeO_4$, $CoFe_2O_4$), and doped magnetic aluminum oxide ($CoAl_2O_3$, $CuAl_2O_3$, and $NiAl_2O_3$) nanoparticles (Parham, Wicaksono, Bagherbaigi, Lee, & Nur, 2016). Magnetic metal oxides, silicon dioxide, titanium dioxide, Fe_2O_3, and Fe_3O_4 are the most widely studied magnetic metal oxide nanoparticles (Guo, Yao, Lin, & Nan, 2015). Metal oxides such as superparamagnetic iron oxide nanoparticles have been widely used in MRI signal enhancing applications, drug delivery, and tissue engineering (Haimov-Talmoud et al., 2019; Marcus et al., 2020; Tiwari, Verma, Singh, Nandi, & Randhawa, 2018; Tiwari et al., 2019).

9.4 Metal and metal oxide functionalization

Inorganic nanoparticle synthesis, including metal and metal oxide nanoparticles, can be broadly categorized into two types, hydrophobic environment and direct formulation in aqueous media, generating hydrophobic and hydrophilic nanoparticles, respectively. Most of the synthesis processes produce hydrophobic nanoparticles due to process conditions and the presence of surfactants (Paramasivam, Kayambu, Rabel, Sundramoorthy, & Sundaramurthy, 2017). This limits the solubility of nanoparticles in water or biological media such as phosphate buffer saline (PBS). Further functionalization of the surface of hydrophobic nanoparticles is challenging compared to hydrophilic nanoparticles. Hence, it limits their applications such as in targeted drug delivery and other biomedical and clinical studies. Hence, enhancing water solubility and functionalization are major challenges that need to be overcome to use nanoparticles in diverse applications. The functionalization of nanoparticles is a key step toward improving colloidal stability, water solubility, and introducing a flexible surface (Tiwari, Vig, Dennis, & Singh, 2011).

To use metal and metal oxide nanoparticles in biomedical applications, nanoparticles should have good stability, biocompatibility, and functionality. To achieve the desire properties, the surface of nanoparticles needs to be functionalized. There are a variety of functionalization techniques or approaches that have been developed such as coating, conjugation strategies, surface encapsulation, in situ synthesis, self-assembly, and creating core–shell nanoparticles (Erathodiyil & Ying, 2011). The most used functionalization materials include silica, synthetic polymers, biopolymers, dendrimers, and small molecules. Surface functionalization of gold nanoparticles, quantum dots, upconversion nanoparticles, polymeric nanoparticles, and magnetic nanoparticles have been well developed and standard protocols have been established and widely reported in the literature (Kumar, Kumar et al., 2020; Lingeshwar Reddy et al., 2018; Tiwari et al., 2018). In the following section, we discuss the functionalization of metal and metal oxide nanoparticles using different materials and various metal and metal oxide functionalizations and applications systematically.

9.4.1 Polymer coating

Nanoparticle surface modification can be done using hydrophilic ligands. It enhance water solubility but stability in different biological fluids is matter of concern (Lynge, Van Der Westen, Postma, & Städler, 2011). One of the best approaches to overcome this limitation is polymer coating. Polymer coating is the most simple and facile method, as it stabilizes most of inorganic nanoparticles including metal and metal oxide nanoparticles at physiological conditions. Further, it makes it feasible for further introduction of diverse functionalities (Chaudhary et al., 2019; Soliman, Pelaz, Parak, & Del Pino, 2015). The coating of polymer on metal and metal oxide nanoparticles is classified into two approaches: replacing the initial coating on the nanoparticle surface and directly coating the polymer on the surface of the nanoparticles. However, in many cases the replacement of ligand on the surface of nanoparticles means that there is no need to replace the functionalization with polymers. Moreover, maintaining ligands facilitates their solubility in water. Polymer coating makes the nanoparticle surface active toward biological interactions, and allows their use in biological applications (Chaudhary et al., 2019). Through coating of polymer on metal and metal oxide nanoparticles, the physicochemical properties such as hydrophobicity, surface charge, drug release kinetics, mechanism, and biological behavior will change. The most widely used polymers include poly(lactic acids), poly(ethylene glycol), poly(vinyl alcohol), poly(lactic-*co*-glycolic acid), poly(e-caprolactone), poly(vinylpyrrolidone), poly(methyl methacrylate), and poly(alkyl cyanoacrylates). At the same time, coating of the most suitable polymer selection is crucial based on the application. For example, to improve the blood compatibility of nanoparticles a sulfonic group containing a polymer coating is the best choice (Kim, Han, Park, & Kim, 2003). Similarly, cationic polymer such as a poly(ethyleneimine) (PEI) coating on the surface of nanoparticles is an interesting candidate for nonviral gene loading and delivery applications due to the strong interaction of PEI cationic with negative-charge siRNA (Tencomnao et al., 2011). The most widely used polymers as coating materials in drug and gene delivery are PEG, PEI, PAA, and amphiphilic coblock polymers. Coating of polymers also has several other advantages. For example, PEG coating on nanoparticles inhibits the formation of protein corona on the nanoparticle surface in biological fluids (Ban & Paul, 2016). This further improves biocompatibility and cellular uptake with a long circulation time. More interestingly, researchers have widely observed that PEG-coated quantum dots show the highest photoluminescence quantum yield in buffer (Daou, Li, Reiss, Josserand, & Texier, 2009). PEG also enables incorporation of functional materials such as targeting ligands, imaging reporters, and therapeutic agents (Pelaz et al., 2015). There are several reports that have shown enhancement of the ability to overcome the limitation to cross the blood–brain barrier (BBB) when nanoparticles were coated with PEG (D'souza & Shegokar, 2016). A PAA coating provides higher affinity toward the cellular membrane. Polymer poly(*N*-isopropylacrylamide) (PNIPAm) is a thermo- and pH-responsive polymer. The coating of the PNIPAm is interesting for stimuli-responsive drug delivery and cancer therapy (Zhan et al., 2015).

A polymer coating is more promising. However, most polymers are toxic in nature at higher concentrations and for long-term treatment. Natural polymers are comparatively less toxic. Biopolymers are further interesting candidate for coating, and are suitable for biomedical applications (Kumar et al., 2020). The most widely used biopolymers as coating materials are peptides, proteins, nucleic acids, dextran, chitosan, cellulose, heparin, and lignin. Biopolymer coating is the best way to formulate monodisperse metal and metal oxide nanoparticles. Proteins and nucleic acids also are used as templates for the synthesis of nanoparticles. It has been reported that templated proteins assisted synthesized gold nanoclusters to show excellent fluorescent behavior (Guo et al., 2020). The coating of biopolymers provides very low thickness and hence a change or improvement in particle size is very low compared with synthetic polymer-coated nanoparticles. Moreover, in vivo drug delivery, targeting a tumor environment, longer circulating time, facile bioconjugation, functionalization, greater cellular uptake, minimum nonspecific binding, specificity, selectivity, quick renal clearance, and negligible long-term cytotoxicity are the further advantages of biopolymer coating on inorganic nanomaterials including metal and metal oxide nanoparticles (Muddineti, Ghosh, & Biswas, 2015).

9.4.2 Silica coating

Among the organic materials polymers are the best choice, similarly among inorganic materials, to the best of our knowledge, silica is the most promising candidate as a coating material. More specially, silica is used as a shell material in most core–shell nanoparticles (Kumar et al., 2020). Silica provides further advantages such as ease of coating on a wide range on materials, several types of porosity enhancing the loading of any molecules, drugs, or genes, availability of a hydroxy function group on the surface of silica, which gives further flexibility to researchers for further functionalization or conjugation of targeting entities for targeted drug delivery and bioimaging (Kumar et al., 2020). Silica greatly enhance nanoparticle stability, biocompatibility, and surface functionality for biomedical applications. Silica is widely used as a coating material on gold nanoparticles (AuNPs), upconversion nanoparticles (UCNPs), quantum dots (QDs), and magnetic nanoparticles (MNPs) (Kumar et al., 2020; Lingeshwar Reddy et al., 2018). Silica coating can be done using classical techniques such as the well-known and widely studied Stobers' synthesis, through silane coupling agents, and the sodium silicate water glass process (Kobayashi et al., 2005). In recent years, this area has rapidly expanded, and several other silica coating strategies have been developed. Silica coating on metal and metal oxide nanoparticles can be done through three main approaches. (1) Silica coating in water-in-oil microemulsion. This has two types, presynthesized nanoparticle coating and coating in situ synthesis of nanoparticles. (2) Polymer- and surfactant-mediated silica coating. This is further categorized into three types, polymer aggregates, surfactant vesicles, and polymer- and surfactant-stabilized nanoparticles. (3) Assembly of silica colloids on nanoparticles. This is also into two types, layer-by-layer assembled composites and other physisorption strategies for self-assembly. Moreover, Guerrero-Martinez et al. comprehensively discussed and reviewed all silica coating strategies (Guerrero-Martínez, Pérez-Juste, & Liz-Marzán, 2010). Using a water-in-oil microemulsion approach a silica coating was achieved successfully on metal, alloys, and metal oxide core nanoparticles such as Ag, Au, Pd, Fe-Pt, Rh, CdS, CdTe, CdSe, PbSe, YF_3, Fe_2O_3, $CoFe_2O_4$, $MnFe_2O_4$, Au, Au, Fe_3O_4, and CdS: Mn/ZnS. The surfactants Igepal CO-520, CTAB, AOT, Triton X-100, SDS, Brij 52, and sodium decyl benzene sulfate have been used (Aubert et al., 2010). Silica coating through polymer- or surfactant-mediated strategies, with different types of polymers has been used, such as block copolymer (PEO-PPO-PEO, PDMA-PDPA, PS-PAA, Lys-Phe), homopolymers (Lys, PDMS, PAA), cationic surfactants (CTAB, DDAB, CTAOH), and flour surfactants (Zhang et al., 2019). More interestingly, a silica coating also enables control of several other parameters such as thickness of the coating, porosity, size of pores, further utilization of surface properties, and stimuli-responsive behavior which can be done by tuning the experimental conditions (Baeza, Colilla, & Vallet-Regí, 2015). A silica coating makes several nanomaterials, including metal and metal oxide nanoparticles, feasible for different applications which are not possible without a silica coating or core-alone nanoparticles (Liu & Han, 2009). Silica-coated nanoparticles are of greater interest in drug delivery, biosensing, electronics, and catalytic applications (Li, Barnes, Bosoy, Stoddart, & Zink, 2012). Overall, significant progress has been made in this area. However, further studies need to be carried out to overcome the existing limitations and explore their further uses.

Lipids are also used as surface coatings on metal and metal oxide nanoparticles. A lipid coating preserves the optical properties of nanoparticles, and also enhances their colloidal stability. The functionalized nanoparticle surface lipid bilayer is similar to the cell membrane structure and hence allows easier cellular internalization of nanoparticles.

9.4.3 Functionalization

Along with major coating strategies using polymer and silica as discussed above, there are several other types of coating or functionalization methods for metal and metal oxide nanoparticles that have been developed. However, they are less pronounced due to limitations in their successful use and applications. Bioconjugation is another approach for the functionalization of metal and metal oxide nanoparticles. The bioconjugation strategy is very common in biomedical applications in which nanoparticle surfaces are conjugated with biomolecules to induce specific interaction properties in a biological system (Sperling & Parak, 2010). Covalent binding of primary amines, carboxylic acid, and thiol are the most common types of bioconjugation. Bioconjugation is widely applied for AuNps, QDs, UCNPs, and MNPs (Sinha, Kim, Nie, & Shin, 2006). Nanoparticles functionalized with carboxyl group through bioconjugation increase specificity of the interaction toward the amine group with biomolecules such as avidin, peptides, antibodies, and nucleic acid (Fig. 9.1). Similarly, AuNPs have the best conjugation efficiency toward the thiol group or thiol group-containing molecules (Gao, Huang, Liu, Zan, & Ren, 2012). Bioconjugation plays a crucial role in drug delivery and cancer therapy. Anticancer drugs are unable to differentiate between normal cells and cancer cells, which is one of the reasons for their side effects and the requirement for a number of doses. Conjugation with targeting entities enables nanoparticles to reach and interact effectively with cancer cells at the site of activity (Farokhzad, Karp, & Langer, 2006; Lu, Shiah, Sakuma, Kopečková, & Kopeček, 2002). Antibodies, aptamer, and folate are the common targeting agents (Zhou & Rossi, 2017). The same strategies also are used in drug delivery, photodynamic therapy, and photothermal therapy (Dong et al., 2016). PEGylated AuNPs conjugated with cetuximab that showed good targetability toward epidermal growth factor receptor can be used as theragnostic nanoprobes (Kao et al., 2014). However, the bioconjugation system is limited due to specific matching between metals and metal oxides and the

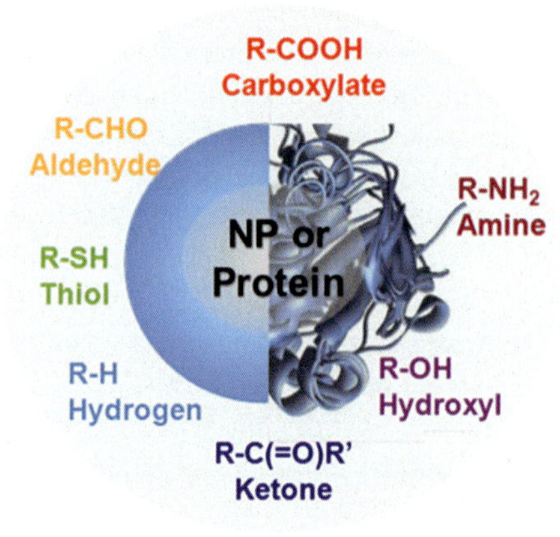

Figure 9.1 The widely used functional groups on nanoparticles or proteins which are useful for bioconjugation (Sapsford et al., 2013).
Source: Reproduced with permission from Sapsford, K. E., Algar, W. R., Berti, L., Gemmill, K. B., Casey, B. J., Oh, E. . . . Medintz, I. L. (2013). Functionalizing nanoparticles with biological molecules: Developing chemistries that facilitate nanotechnology. *Chemical Reviews*, *113*, 1904–2074. https://doi.org/10.1021/cr300143v. Copyright 2013, American Chemical Society.

respective entities such as thiol-based molecules that are best for Au nanoparticles which are not suitable for other nanoparticles. Therefore there is significant scope for researchers to explore and develop strategies to overcome these limitations.

In the following section, we briefly discuss the more promising and widely studied metal (Au) and metal oxide (iron oxide, aluminum oxide, zinc oxide) nanoparticles, and their functionalization and diverse applications.

9.5 Metal and metal oxide nanoparticles

9.5.1 Gold nanoparticles

The history of gold is well known; colloidal gold has been used in glass staining since ancient times. Due to its properties, such as optical, electronic, biocompatibility, feasibility to tune the size (3–200 nm) and shape, and chemistry of the gold surface with a functional group such as − SH, which makes the surface functionalization simple, gold nanoparticles have gained significant interest (Elahi, Kamali, & Baghersad, 2018). Scientific studies into gold nanoparticles began with Faraday's work in 1857 (Chang et al., 2018). There are several strategies that have been developed for the formulation of gold nanoparticles. However, many are solution phase synthesis processes. The best known method is the Turkevich method which was developed in 1951 (Kimling et al., 2006). In 1971, Frens made a significant improvement to the protocol (FRENS, 1973). Since then, gold nanoparticles have gained increasing interested and have become a promising material for new investigations and hence are used in diverse applications. This method has been most widely studied for Au nanoparticle preparation in the last 10 years. Among the reports on the synthesis of gold nanoparticles using a reducing agent, the presence of a stabilizer is well studied and widely reported, in which the stabilizer inhibits the aggregation of nanoparticles (Haume et al., 2016). By tuning the experimental conditions, the size and shape can be controlled easily. The synthesis of AuNPs is usually in organic solvents such as toluene, chloroform, and hexane. However, their stability and direct use in several applications are limited. Hence, further functionalization processes have been developed to overcome these limitations. However, further functionalization of nanoparticles depends on the chemistry of stabilizers on the surface (Haume et al., 2016). The most promising one is thiol molecules, due to the chemistry of the gold surface and thiol function group (Pensa et al., 2012). If stabilizers with a thiol group are the best choice of stabilizers to use in gold nanoparticle preparation, this would further allow functionalization of the nanoparticles. This results in monolayer-protected nanoparticles. Later, several other functionalization processes (Fig. 9.2) were developed, such as nucleic acid, proteins, and amphiphilic polymers (Han, Ghosh, & Rotello, 2007). Different ligands are commonly used for functionalization of gold nanoparticles for use in biomedical applications (Fig. 9.3). Biofunctionalization AuNPs are an excellent platform for biosensing and can be suitable for several other applications (Manickam et al., 2020; Pingarrón, Yáñez-Sedeño, & González-Cortés, 2008). Because of excellent

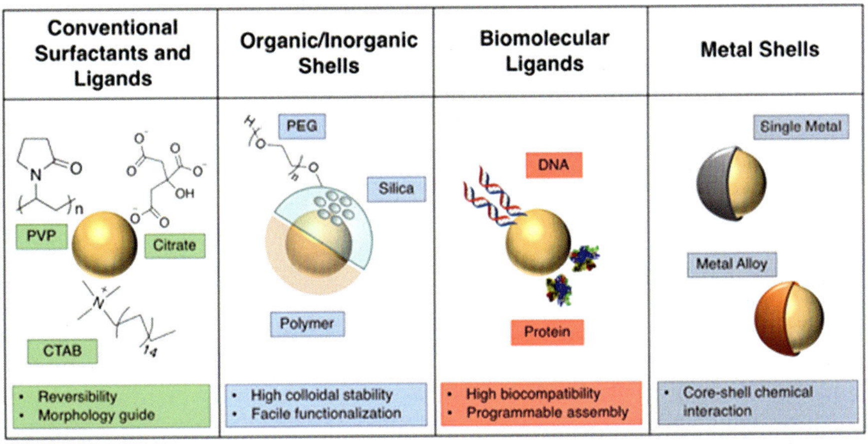

Figure 9.2 Different colloidal plasmonic gold nanoparticle preparations, with stabilization using different agents and their characteristics (Kang et al., 2019).
Source: Reproduced with permission from Kang, H., Buchman, J. T., Rodriguez, R. S., Ring, H. L., He, J., Bantz, K. C., & Haynes, C. L. (2019). Stabilization of silver and gold nanoparticles: Preservation and improvement of plasmonic functionalities. *Chemical Reviews*, *119*, 664–699. https://doi.org/10.1021/acs.chemrev.8b00341. Copyright 2018, American Chemical Society.

optical properties, AuNPs are used in sensing of various biomolecules and cells. AuNPs are widely used in targeting DNA, RNA, cell, metal ions, small organic compounds, and proteins (Fig. 9.4). For such applications, functionalization is a key step (Cai, 2008).

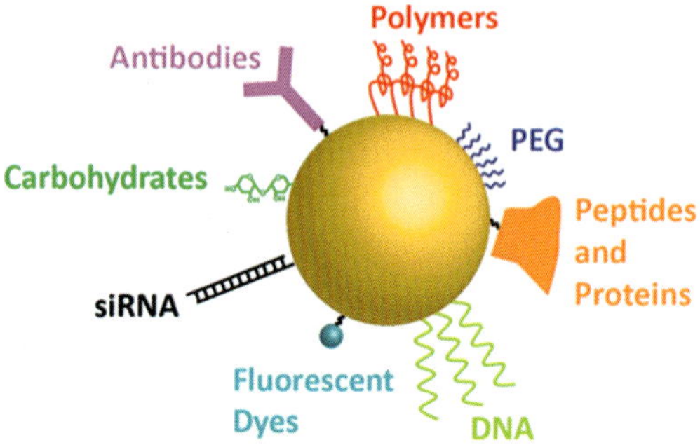

Figure 9.3 Different ligands such as antibodies, oligonucleotides, carbohydrates, proteins, polymers, and dyes commonly use gold nanoparticles for biomedical applications (Heuer-Jungemann et al., 2019).
Source: Reproduced with permission from Heuer-Jungemann, A., Feliu, N., Bakaimi, I., Hamaly, M., Alkilany, A., Chakraborty, I. ... Kanaras, A. G. (2019). The role of ligands in the chemical synthesis and applications of inorganic nanoparticles. *Chemical Reviews, 119*, 4819–4880. https://doi.org/10.1021/acs.chemrev.8b00733. Copyright 2019, American Chemical Society.

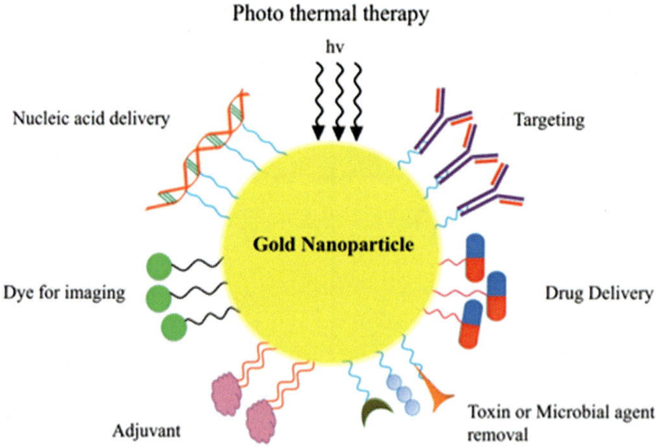

Figure 9.4 Diverse applications of gold nanoparticles (Bagheri et al., 2018).
Source: Reproduced with permission from Bagheri, S., Yasemi, M., Safaie-Qamsari, E., Rashidiani, J., Abkar, M., Hassani, M. ... Kooshki, H. (2018). Using gold nanoparticles in diagnosis and treatment of melanoma cancer. *Artificial Cells, Nanomedicine, and Biotechnology*, *46*, 462–471. https://doi.org/10.1080/21691401.2018.1430585. Copyright 2018, Informa, Taylor & Francis.

9.5.2 Iron oxides

Currently, nanomaterial development has moved from the discovery of new nanomaterials to the investigation and design of complex systems with combinations of known materials with new multifunctionalities. The development of multifunctional nanomaterials is speeding up for the development of novel applications for them to fit the desired needs. The conventional application of magnetic nanoparticles has also moved intensively into many different scientific applications (Fig. 9.5). In the last decade, it has been reported that magnetic nanoparticles' ability for acceleration of the proton relaxation of water molecules in tissues is a promising property. Magnetic nanoparticles are also promising materials for magnetic resonance imaging (MRI). Superparamagnetic iron oxide nanoparticles have been used as contrast agents in MRI for the last 20 years. The flexibility of tuning the particle size, shape, and feasibility of easily control of delivery using magnetic properties and introduction of functional properties with conjugation of biomolecules makes them excellent metal oxide nanoparticles compared with others. Further advantages of magnetic nanoparticles are the ability of cell wall penetration for delivery of drugs or

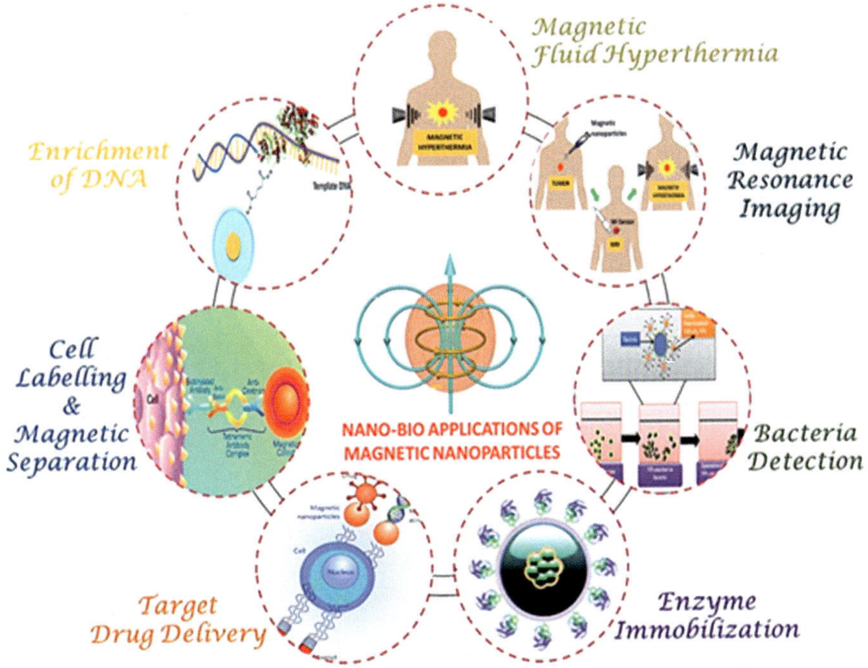

Figure 9.5 Diverse applications of magnetic nanoparticles (Bohara, Thorat, & Pawar, 2016). Source: Reproduced with permission from Bohara, R. A., Thorat, N. D., & Pawar, S. H. (2016). Role of functionalization: Strategies to explore potential nano-bio applications of magnetic nanoparticles. *RSC Advances*, *6*, 43989–44012. https://doi.org/10.1039/c6ra02129h. Copyright 2016, The Royal Society of Chemistry.

biomolecules and excellent biocompatibility, and so they are widely used in biomedical applications. Magnetic nanoparticles also have been approved by United States Food and Drug Administration (FDA).

Magnetic nanoparticles (MNPs) can be considered as a special class of metal oxide nanoparticle. MNPs are promising materials for cell labeling, bioimaging, hyperthermia, drug delivery, and neural tissue engineering. Among magnetic nanoparticles, iron oxide nanoparticles are most interesting due to their biocompatibility. Interestingly, iron oxide nanoparticles show superparamagnetic behavior at the nanoscale level. Briefly, superparamagnetic means that under an external magnetic field they are magnetized, and once the external magnetic field is removed, they are nonmagnetic. Noteworthy results on hyperthermia and clinical trials have stressed the effectiveness of MNPs. Magnetic nanoparticles can be prepared using a range of methods, and among them, chemical methods are most promising, such as coprecipitation and reverse micelles precipitation, which produced water-soluble magnetic nanoparticles, which are a good choice for further functionalization and biomedical applications. Different strategies for stabilization of magnetic nanoparticles are presented in Fig. 9.6. Moreover, different strategies have been developed for functionalization for post synthesis of MNPs such as amphiphilic polymers,

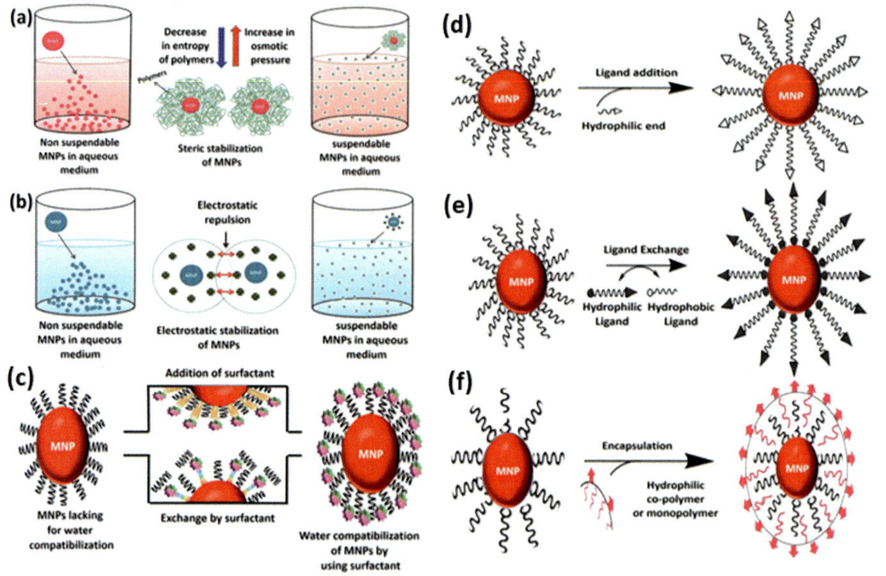

Figure 9.6 Magnetic nanoparticles: (A) steric stabilization, (B) electrostatic stabilization, (C) water compatibilization using surfactants; and mechanism of surface functionalization through (D) ligand addition, (E) ligand exchange, and (F) encapsulation.
Source: Reproduced with permission from Bohara, R. A., Thorat, N. D., & Pawar, S. H. (2016). Role of functionalization: Strategies to explore potential nano-bio applications of magnetic nanoparticles. RSC Advances, 6, 43989–44012. https://doi.org/10.1039/c6ra02129h. Copyright 2016, The Royal Society of Chemistry.

silanization, and modification or replacement of stabilizers. Silanization is one of the best options for functionalization, because MNPs coated with silica are extremely favorable for a wide variety of functionalization, conjugation, encapsulation purposes, and for introducing stimuli-responsive functional group and hence, can be employed for diverse applications. Fig. 9.7 presents a schematic view of the preparation of functional magnetic (DOX-MMSN/GQDs) nanoparticles with the incorporation optical properties through GQDs with the drug DOX loaded for delivery using an external magnetic field.

Small organic compounds, such as simple molecules and surfactants, used for the functionalization of iron oxide nanoparticles are classified into three categories, namely organic soluble, aqueous soluble, and amphiphilic. Molecules functionalized on the surface of iron oxide nanoparticles containing functional groups which have weaker and stronger attraction with solvents fall into the oil-soluble MNP and water-soluble MNP categories, respectively. Oil-soluble MNPs are highly stable and mono-dispersed. Fatty acids are an example for oil-soluble MNPs, and polyol, lycine, and ammonium salt are examples of water-soluble organic

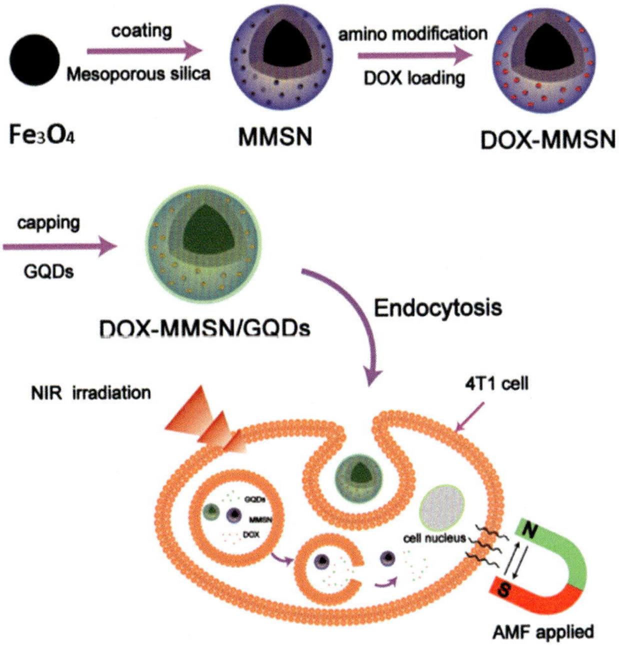

Figure 9.7 Illustration of DOX-MMSN/GQDs functional nanoparticle synthesis and drug delivery mechanism (Yao et al., 2017).
Source: Reproduced with permission from Yao, X., Niu, X., Ma, K., Huang, P., Grothe, J., Kaskel, S., & Zhu, Y. (2017). Graphene quantum dots-capped magnetic mesoporous silica nanoparticles as a multifunctional platform for controlled drug delivery, magnetic hyperthermia, and photothermal therapy. *Small (Weinheim an der Bergstrasse, Germany), 13*. https://doi.org/10.1002/smll.201602225. Copyright 2016, Wiley-VCH.

compounds. Molecules composed of both types of functional group coated on the surface of iron oxide nanoparticles fall in the amphiphilic MNPs category. Among them oil-soluble functionalization of iron oxide nanoparticles gives better stability over others with decreasing agglomeration. Hence, iron oxide nanoparticles in general are dispersed in oleic acid. Oleic acid has a *cis* double bond and tail of C18. The double bond forms a kink and the tail helps in the formation of a dense protective monolayer.

Shaoo et al. formulated iron oxide nanoparticles that were surface functionalized with various molecules, namely oleic acid, lauric acid, dodecyl phosphonate, hexadecyl phosphonate, and dihexadecyl phosphonate with the particle size ranging between 6 and 8 nm. Due to good biocompatibility and good binding efficiency, phosphate and phosphonate ligands are suitable for the encapsulation of iron oxide nanoparticles for biomedical use. Oil-soluble functionalized iron oxide nanoparticle formulation has been well established. However, the major challenge remains in the formulation of water-soluble iron oxide nanoparticles without agglomeration, that are biocompatible and biodegradable. One such approach is the use of amino acids, citric acid, cyclodextrin, and vitamin for functionalization. Another approach is formulated oil-soluble iron oxide nanoparticle transformation into the water-soluble type through a ligand exchange process. Sun et al. employed bipolar surfactants to transform oil-soluble iron oxide nanoparticles into water-soluble ones. Lattuada and Hatton also attempted a ligand exchange process using reactive hydroxyl moieties. APTS, APTES, and MPTES are the most commonly used small molecules to introduce amino and sulfhydryl groups, respectively. Polymers such as dextran, starch, gelatin, chitosan, PEG, PVA, PLA, PMMA, PAA, and alginate are the most commonly used polymers for the functionalization of iron oxide. Magnetic nanoparticles also have shown promising results in bioimaging. There are diverse types of magnetic fluorescent nanoparticles that have been developed which are suitable for bioimaging applications (Fig. 9.8). Compared to other metal oxides, iron oxide nanoparticle functionalization has been well studied and with standard protocols developed and several products commercialized and available in market.

9.5.3 Aluminum oxide

Aluminum oxide (Al_2O_3) nanoparticles are porous nanomaterials with a corundum-like structure in which each aluminum atom is surrounded by six oxygen atoms. They are easily synthesizable, cost-effective, and also possess extraordinary properties such as a vast porous surface area, mechanical strength, bio-inertness, and resistance toward chemicals that afford their utility in biomedical applications (Meder, Kaur, Treccani, & Rezwan, 2013; Rajan, Inbaraj, & Chen, 2015). Various synthetic methodologies have been deployed for Al_2O_3 nanoparticle synthesis including solid/gas/liquid phase-based methods ranging from mechanical ball-milling, solution reduction, decomposition, gas evaporation, and laser ablation. A review by Ghorbani et al. demonstrated in detail about various synthetic methods for Al_2O_3 nanoparticles (Ghorbani, 2014; Piriyawong, Thongpool, Asanithi, & Limsuwan, 2012). Considering the strategic importance of Al_2O_3 nanoparticles and their

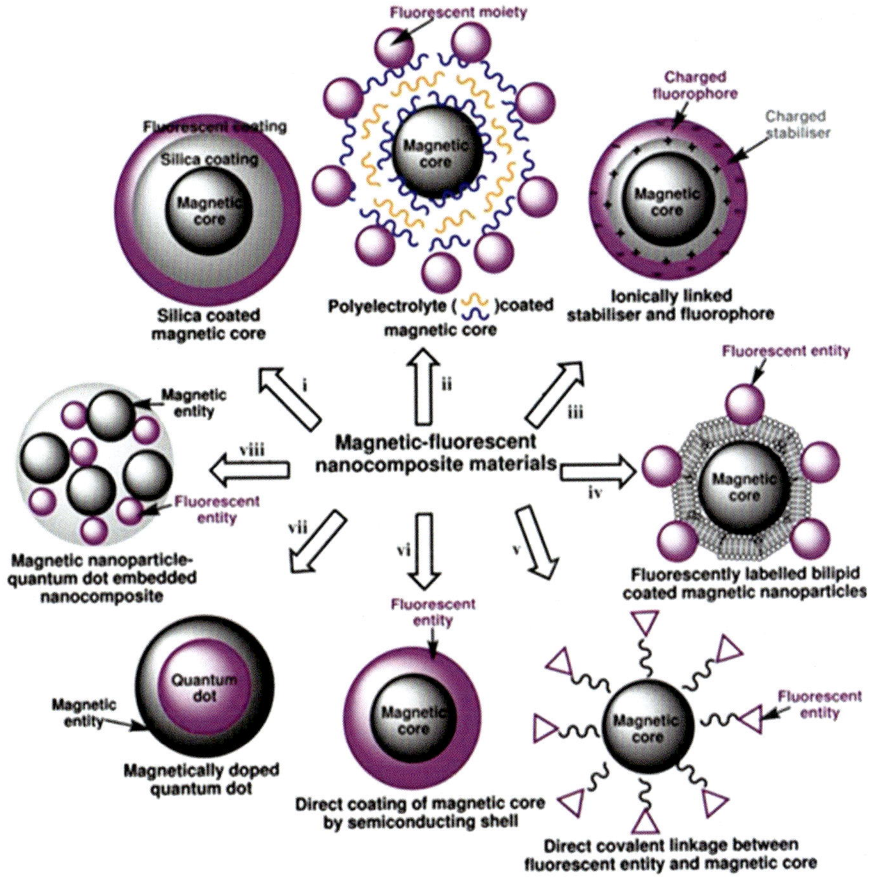

Figure 9.8 Prominent types of fluorescent magnetic nanoparticles (Corr, Rakovich, & Gun'Ko, 2008).
Source: Reproduced with permission from Corr, S. A., Rakovich, Y. P., & Gun'Ko, Y. K. (2008). Multifunctional magnetic-fluorescent nanocomposites for biomedical applications. *Nanoscale Research Letters*, *3*, 87–104. https://doi.org/10.1007/s11671-008-9122-8. Copyright 2018, Authors.

extraordinary properties, they have found diverse applications, particularly in biomedicine and biotechnology in which their utility was explored in drug delivery, biosensing, treatment of diseases, destruction of microbes, and biomolecular stabilization (Fig. 9.9) (Hassanpour et al., 2018). However, their applicability was hindered mostly due to their high specific surface energy, which essentially led to accumulation in the biological system. One of the possible ways to improve the biocompatibility of Al_2O_3 nanoparticles is by preventing their accumulation. Surface functionalization is one of the ways to address this issue related to nanoparticle accumulation. The most preferred biodegradable surface-modifying agents

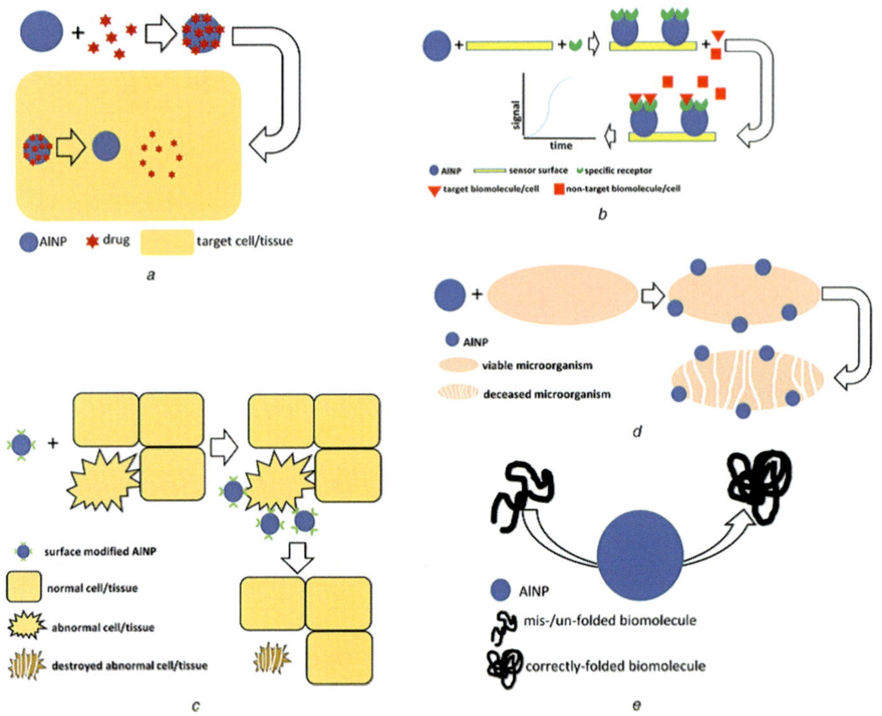

Figure 9.9 Biomedical applications of Al_2O_3 nanoparticles: (A) Al_2O_3 nanoparticles in drug delivery applications, (B) biosensing, (C) therapeutic application of Al_2O_3 nanoparticles, (D) antimicrobial effect of AlNPs, and (E) role of AlNPs in folding of biomolecules.
Source: Reproduced with permission from Hassanpour, P., Panahi, Y., Ebrahimi-Kalan, A., Akbarzadeh, A., Davaran, S., Nasibova, A. N. ... Kavetskyy, T. (2018). Biomedical applications of aluminium oxide nanoparticles. *Micro & Nano Letters, 13*, 1227–1231. https://doi.org/10.1049/mnl.2018.5070. Copyright 2018, The Institution of Engineering and Technology.

used for Al_2O_3 nanoparticles are citric acid, oleic acid, alkoxysilanes, and poly(ethylene glycol) (Mallakpour & Sadeghzadeh, 2017). These compounds consist of various functional groups such as amines, phosphine, and carboxylic acids which help them in surface functionalization on Al_2O_3 nanoparticles.

Some recent advancements in Al_2O_3 nanoparticles and their nanocomposite materials for various biomedical applications are as follows: Al_2O_3 nanoparticles improve the performance and biocompatibility of poorly water-soluble drugs. For example, Borbane et al. (Borbane, Pande, Vibhute, Kendre, & Dange, 2015) developed telmisartan (TEL)-loaded ordered mesoporous alumina (OMA) for enhanced solubility and dissolution rate of the blood pressure-lowering drug TEL. OMA was synthesized by an evaporation-induced self-assembly mechanism with a specific surface area and average pore diameter of 299 m^2/g and 26 nm, respectively, followed by loading of TEL drug ($\sim$45%) which led to a TEL:OME composite material. The resultant composite material showed an enhanced dissolution rate and drug-releasing properties as compared to the pure crystalline TEL. Similarly, Tarlani et al. (Tarlani, Isari, Khazraei, & Moghadam, 2017) developed a sol–gel of aluminum oxide–ibuprofen nanocomposite material, which has the potential to improve the bioavailability of the nonsteroidal antiinflammatory drug (NSAID) ibuprofen. In order to get this nanocomposite material, they used a sol–gel method for the fabrication of alumina by controlled hydrolysis of aluminum alkoxide followed by loading with the poorly water-soluble ibuprofen. The results reveal that due to the high surface area and high porosity of nano-aluminum oxide loaded in the nanocomposites, it afforded high loading, high solubility, and controlled release of ibuprofen at the specifically targeted sites in the physiological environment. Further, Al_2O_3 nanoparticles have had their utility explored in biosensing applications. For example, Ito et al. developed a modified localized surface plasmon resonance (LSPR) sensor with self-assembled anodic aluminum oxide for biosensing of bovine serum albumin (BSA) (Ito et al., 2017). The ordered aluminum oxide nanohole on an LSPR chip with diameter and depth, respectively, of 75 nm and 0.5 μm, was found to be the most sensitive sensing layer for BSA. In another study, the nanocarbon coatings were applied on the surface of aluminum metal oxide nanoparticles through a plasma-based fabrication technique (Aramesh, Tran, Ken Ostrikov, & Prawer, 2017). The resultant ultrathin carbon-layered alumina nanocrystals with new surface properties were found to be an effective carrier for DNA sensing in a competitive bioassay. They were also explored for their utility in simultaneous live cell imaging and intracellular DNA sensing. Owing to their large surface area and extraordinary properties Al_2O_3 nanoparticles have been used in cancer therapy. Chen and coworkers developed poly(γ-glutamic acid)-functionalized Al_2O_3 nanoparticles as cytotoxic agents to induce cancer cell death in the human prostate as well as for protein absorption ability using bovine serum albumin and lysozyme (Rajan et al., 2015). The mechanism for cancer cell death involves the following: the Al_2O_3 nanoparticles showed cytotoxicity towards PC-3 prostate cancer cells by inducing reactive oxygen species (ROS) which subsequently led to mitochondrial dysfunction.

9.5.4 Zinc oxide

Zinc oxide (ZnO) nanoparticles have received tremendous interest in biomedical applications because of their high surface area, low toxicity, biocompatibility, high chemical stability, and low price (Jiang, Pi, & Cai, 2018; Mishra, Mishra, Ekielski, Talegaonkar, & Vaidya, 2017). Especially, ZnO shows promising applications in anticancer and antibacterial fields as they trigger excess ROS production which subsequently induces cell apoptosis in the biological system (Jiang et al., 2018; Mishra, Mishra et al., 2017). It is well known that zinc ions maintain the structural integrity of insulin, taking into account that several ZnO-based nanoparticles were effectively developed for antidiabetic treatment (Asani, Umrani, & Paknikar, 2016; Bayrami, Parvinroo, Habibi-Yangjeh, & Rahim Pouran, 2018; Jiang et al., 2018; Mishra, Mishra et al., 2017; Umrani & Paknikar, 2014). In addition, their excellent luminescent properties make them useful as one of the important compounds for bioimaging applications

(Jiang et al., 2018; Mishra, Mishra et al., 2017; Zhang & Xiong, 2015). They were also used in personal care products such as cosmetics and sunscreens because of their excellent ultraviolet and visible light resistance properties (Jiang et al., 2018; Newman, Stotland, & Ellis, 2009). As a main and essential trace element, body tissues require Zn, including the brain, muscle, bone, and skin, and it also plays a major role in the metabolism of proteins and nucleic acid synthesis, hematopoiesis, and neurogenesis. ZnO nanoparticles with small particle size afforded zinc to be absorbed better in the body; this made ZnO nanoparticles more attractive for biomedical applications. The biological activity of ZnO nanoparticles depends on various factors including the size and dimensions of the nanoparticles, charge, surface chemistry, morphology, toxicity as well as particle reactivity in the solution (Jiang et al., 2018; Mishra, Mishra et al., 2017). Therefore it is essential to maintain nanomaterials with uniform size and morphology for improved biomedical applications. So far, various methods have been demonstrated for the synthesis of ZnO nanoparticles including the sol–gel process, solid-state pyrolytic method, chemical precipitation method, solution-free mechanochemical method, and biosynthesis method (Jiang et al., 2018). However, one of the common factors to diminish the bioactivity of ZnO nanoparticles is agglomeration in the biological media due to their high surface energy (Ghaffari, Sarrafzadeh, Fakhroueian, Shahriari, & Khorramizadeh, 2017; Simsikova, Antalík, Kaňuchová, & Škvarla, 2013). This issue can be addressed by surface coating or functionalization of capping agents on the surface of nanoparticles. Not only agglomeration, other characteristics including solubility, toxicity, and drug delivery properties also improved drastically when the nanomaterials are associated with bioactive probes or drugs. Various capping agents and polymeric compounds have come into existence to modify the surface functionalization of ZnO nanoparticles for improved biomedical applications. Some of the popularly used polymer materials as capping agents for this purpose are as follows: 3-aminopropyltriethoxysilane, monoethanolamine, mercaptoundecanoic acid, mercaptoacetic acid, and fatty acids (Fakhroueian et al., 2013; Kim, Kim, & Sung, 2012; Liao, Hu, Gu, & Xue, 2015; Saravanan, Jayamoorthy, & Ananda Kumar, 2015; Zhuang, Liu, & Liu, 2009). Some recent advancements of ZnO nanoparticles are as follows; Khorramizadeh et al. reported ZnO nanoparticles functionalized with 3-mercaptopropionic acid for water-soluble curcumin delivery applications (Ghaffari et al., 2017). They stabilized carboxyl-terminated ZnO nanoparticles with 3-mercaptopropionic acid followed by coupling with curcumin drug using 1,1′-carbonyldiimidazole (CDI) as a coupling agent that led to ZnO-MPA-curcumin nanoformulations. This nanoformulation showed improved solubility and anticancer properties toward MDA-MB-231 breast cancer cells compared with the curcumin drug alone. The calculated IC_{50} values for curcumin and ZnO-MPA-curcumin nanoformulations reported are 5 and 3.3 μg/mL, respectively. In another study by Salehzadeh and coworkers developed a novel nanoformulation called ZnO@Glu–TSC by the surface functionalization of ZnO nanoparticles with glutamic acid (Glu) followed by conjugation with thiosemicarbazid (TCS) for their antibacterial effect against ciprofloxacin-resistant Staphylococcus *aureus* (Nejabatdoust, Salehzadeh, Zamani, & Moradi-Shoeili, 2019). The developed nanoformulation was characterized by energy-dispersive X-ray spectroscopy, X-ray diffraction, FTIR, and TEM. According to their investigations, the synergistic interaction between the ZnO@Glu–TSC and ciprofloxacin (CIP) led to improved antimicrobial activity against drug-resistant *S. aureus*. Zhang et al. developed 3-(3,5-di-tert-butyl-4-hydroxyphenyl) propionic acid (DBHP)-functionalized ZnO nanoparticles (ZnO-DBHP) by the decomposition of organometallic precursor $Zn(DBHP)_2$ under alkaline conditions to improve the dispersibility and antioxidant performance of the lubricant di-iso-octylsebacate (DIOS) base oil (Huang, Zhou, Zhang, Zhang, & Zhang, 2019). According to their investigations based on a rotary oxygen bomb test, pressurized differential scanning calorimetry, and free-radical-scavenging method, the ZnO-DBHP acted as a good antioxidant and also showed improved thermal resistance as well as antiwear ability of DIOS as compared to the pure organic DBHP.

9.6 Conclusions

It is concluded that metal and metal oxide nanoparticles are significantly contributing to biomedical applications. The synthesis of metal and metal oxide is well established strategy and it is easy to control the size, shape, and other physicochemical properties. Metal and metal oxide nanoparticle coating on surfaces makes them more interesting, and polymer and silica are the major coating materials due to their physicochemical properties, simple coating strategies, and they provide opportunities for further functionalization. Polymer- and silica-coated metal and metal oxide nanoparticles can be an excellent system for drug delivery, tissue engineering, and bioimaging. Among the different functionalization methods, bioconjugation is more interesting and shows various advantages and promising results in biomedical applications including for controlled and targeted drug delivery. Functionalization with biomolecules is more interesting compared with the other candidates. Gold, iron oxide, aluminum oxide, and zinc oxides are widely studied metal and metal oxides in biomedical applications. Iron oxide nanoparticles also have shown amazing functional properties in MRI and other bioimaging techniques. It is concluded that several metals and metal oxide nanoparticles need to be explored further to develop strategies to formulate them with suitable properties to make them useful in biomedical applications.

Acknowledgments

The authors thank IIT Mandi and the respective supervisors, Raj Kumar (Dr Prem Felix Siril), Darsi Rambabu (Dr Abhimanew Dhir and Dr Pradeep Parameswaran), Mangili Venkateswarulu (Dr Subrata Ghosh and Dr Rik Rani Koner), Kumbam Lingeshwar Reddy (Dr Venkata Krishnan), Nagaraju Nakka (Dr Suman Kalyan Pal and Dr Subrata Ghosh), Yogesh Midathala (Dr Subrata Ghosh), Prateep Singh Sagara (Dr P.C. Ravi Kumar and Dr Prem Felix Siril), and Guruprasad Reddy (Dr Pradeep Parameshwaran), where the respective supervisors' names are in parentheses. All authors also thank the editor Dr Kunal Mondal for providing the opportunity for contribution of this chapter.

References

Aramesh, M., Tran, P. A., (Ken) Ostrikov, K., & Prawer, S. (2017). Conformal nanocarbon coating of alumina nanocrystals for biosensing and bioimaging. *Carbon*, *122*, 422–427. Available from https://doi.org/10.1016/j.carbon.2017.06.101.

Asani, S. C., Umrani, R. D., & Paknikar, K. M. (2016). In vitro studies on the pleotropic antidiabetic effects of zinc oxide nanoparticles. *Nanomedicine: Nanotechnology, Biology, and Medicine*, *11*, 1671–1687. Available from https://doi.org/10.2217/nnm-2016-0119.

Aubert, T., Grasset, F., Mornet, S., Duguet, E., Cador, O., Cordier, S., … Haneda, H. (2010). Functional silica nanoparticles synthesized by water-in-oil microemulsion processes. *Journal of Colloid and Interface Science*, *341*, 201–208. Available from https://doi.org/10.1016/j.jcis.2009.09.064.

Baeza, A., Colilla, M., & Vallet-Regí, M. (2015). Advances in mesoporous silica nanoparticles for targeted stimuli-responsive drug delivery. *Expert Opinion on Drug Delivery*, *12*, 319–337. Available from https://doi.org/10.1517/17425247.2014.953051.

Bagheri, S., Yasemi, M., Safaie-Qamsari, E., Rashidiani, J., Abkar, M., Hassani, M., … Kooshki, H. (2018). Using gold nanoparticles in diagnosis and treatment of melanoma cancer. *Artificial Cells, Nanomedicine, and Biotechnology*, *46*, 462–471. Available from https://doi.org/10.1080/21691401.2018.1430585.

Ban, D. K., & Paul, S. (2016). Protein corona over silver nanoparticles triggers conformational change of proteins and drop in bactericidal potential of nanoparticles: Polyethylene glycol capping as preventive strategy. *Colloids Surfaces B: Biointerfaces*, *146*, 577–584. Available from https://doi.org/10.1016/j.colsurfb.2016.06.050.

Bayrami, A., Parvinroo, S., Habibi-Yangjeh, A., & Rahim Pouran, S. (2018). Bio-extract-mediated ZnO nanoparticles: Microwave-assisted synthesis, characterization and antidiabetic activity evaluation. *Artificial Cells, Nanomedicine, and Biotechnology*, *46*, 730–739. Available from https://doi.org/10.1080/21691401.2017.1337025.

Bhardwaj, V., & Kaushik, A. (2017). Biomedical applications of nssanotechnology and nanomaterials. *Micromachines*, *8*, 298. Available from https://doi.org/10.3390/mi8100298.

Bohara, R. A., Thorat, N. D., & Pawar, S. H. (2016). Role of functionalization: Strategies to explore potential nano-bio applications of magnetic nanoparticles. *RSC Advances*, *6*, 43989–44012. Available from https://doi.org/10.1039/c6ra02129h.

Borbane, S., Pande, V., Vibhute, S., Kendre, P., & Dange, U. (2015). Design and fabrication of ordered mesoporous alumina scaffold for drug delivery of poorly water soluble drug. *Austin Therapeutics*, *2*, 1015. Available from https://austinpublishinggroup.com/therapeutics/fulltext/therapeutics-v2-id1015.php. (accessed September 13, 2020).

Cai, W. (2008). Applications of gold nanoparticles in cancer nanotechnology. *Nanotechnology, Science and Applications*, *1*, 17–32. Available from https://doi.org/10.2147/nsa.s3788.

Chang, M.-H., Pai, C.-L., Chen, Y.-C., Yu, H.-P., Hsu, C.-Y., & Lai, P.-S. (2018). Enhanced antitumor effects of epidermal growth factor receptor targetable cetuximab-conjugated polymeric micelles for photodynamic therapy. *Nanomaterials*, *8*, 121. Available from https://doi.org/10.3390/nano8020121.

Chaudhary, R. G., Bhusari, G. S., Tiple, A. D., Rai, A. R., Somkuvar, S. R., Potbhare, A. K., … Abdala, A. A. (2019). Metal/metal oxide nanoparticles: Toxicity, applications, and future prospects. *Current Pharmaceutical Design*, *25*, 4013–4029. Available from https://doi.org/10.2174/1381612825666191111091326.

Chawla, M., Kumar, R., & Siril, P. F. (2016). High catalytic activities of palladium nanowires synthesized using liquid crystal templating approach. *Journal of Molecular*

Catalysis A: Chemical, 423, 126–134. Available from https://doi.org/10.1016/j.molcata.2016.06.014.

Chen, J., Wang, D., Xi, J., Au, L., Siekkinen, A., Warsen, A., ... Li, X. (2007). Immuno gold nanocages with tailored optical properties for targeted photothermal destruction of cancer cells. *Nano Letters, 7*, 1318–1322. Available from https://doi.org/10.1021/nl070345g.

Corr, S. A., Rakovich, Y. P., & Gun'Ko, Y. K. (2008). Multifunctional magnetic-fluorescent nanocomposites for biomedical applications. *Nanoscale Research Letters, 3*, 87–104. Available from https://doi.org/10.1007/s11671-008-9122-8.

Daou, T. J., Li, L., Reiss, P., Josserand, V., & Texier, I. (2009). Effect of poly(ethylene glycol) length on the in vivo behavior of coated quantum dots. *Langmuir: The ACS Journal of Surfaces and Colloids, 25*, 3040–3044. Available from https://doi.org/10.1021/la8035083.

Dong, C., Liu, Z., Wang, S., Zheng, B., Guo, W., Yang, W., ... Chang, Jin (2016). A protein-polymer bioconjugate-coated upconversion nanosystem for simultaneous tumor cell imaging, photodynamic therapy, and chemotherapy. *ACS Applied Materials & Interfaces, 8*, 32688–32698. Available from https://doi.org/10.1021/acsami.6b11803.

D'souza, A. A., & Shegokar, R. (2016). Polyethylene glycol (PEG): A versatile polymer for pharmaceutical applications. *Expert Opinion on Drug Delivery, 13*, 1257–1275. Available from https://doi.org/10.1080/17425247.2016.1182485.

Dutt, S., Kumar, R., & Siril, P. F. (2015). Green synthesis of a palladium-polyaniline nanocomposite for green Suzuki-Miyaura coupling reactions. *RSC Advances, 5*, 33786–33791. Available from https://doi.org/10.1039/c5ra05007c.

Dutt, S., Siril, P. F., Sharma, V., & Periasamy, S. (2015). Goldcore-polyanilineshell composite nanowires as a substrate for surface enhanced Raman scattering and catalyst for dye reduction. *New Journal of Chemistry, 39*, 902–908. Available from https://doi.org/10.1039/c4nj01521e.

Elahi, N., Kamali, M., & Baghersad, M. H. (2018). Recent biomedical applications of gold nanoparticles: A review. *Talanta, 184*, 537–556. Available from https://doi.org/10.1016/j.talanta.2018.02.088.

Erathodiyil, N., & Ying, J. Y. (2011). Functionalization of inorganic nanoparticles for bioimaging applications. *Accounts of Chemical Research, 44*, 925–935. Available from https://doi.org/10.1021/ar2000327.

Fakhroueian, Z., Harsini, F. M., Chalabian, F., Katouzian, F., Shafickhani, A., & Esmaeilzadeh, P. (2013). Influence of modified ZnO quantum dots and nanostructures as new antibacterials. *Advances in Nanoparticles, 2*, 247–258. Available from https://doi.org/10.4236/anp.2013.23035.

Farokhzad, O. C., Karp, J. M., & Langer, R. (2006). Nanoparticle–aptamer bioconjugates for cancer targeting. *Expert Opinion on Drug Delivery, 3*, 311–324. Available from https://doi.org/10.1517/17425247.3.3.311.

FRENS, G. (1973). Controlled nucleation for the regulation of the particle size in monodisperse gold suspensions. *Nature Physical Science, 241*, 20–22. Available from https://doi.org/10.1038/physci241020a0.

Gao, J., Huang, X., Liu, H., Zan, F., & Ren, J. (2012). Colloidal stability of gold nanoparticles modified with thiol compounds: Bioconjugation and application in cancer cell imaging. *Langmuir: The ACS Journal of Surfaces and Colloids, 28*, 4464–4471. Available from https://doi.org/10.1021/la204289k.

Ghaffari, S. B., Sarrafzadeh, M. H., Fakhroueian, Z., Shahriari, S., & Khorramizadeh, M. R. (2017). Functionalization of ZnO nanoparticles by 3-mercaptopropionic acid for aqueous curcumin delivery: Synthesis, characterization, and anticancer assessment. *Materials Science and Engineering C, 79*, 465–472. Available from https://doi.org/10.1016/j.msec.2017.05.065.

Ghorbani, H. R. (2014). A review of methods for synthesis of Al nanoparticles. *Oriental Journal of Chemistry*, *30*, 1941–1949. Available from https://doi.org/10.13005/ojc/300456.

Guerrero-Martínez, A., Pérez-Juste, J., & Liz-Marzán, L. M. (2010). Recent progress on silica coating of nanoparticles and related nanomaterials. *Advanced Materials*, *22*, 1182–1195. Available from https://doi.org/10.1002/adma.200901263.

Guo, T., Yao, M. S., Lin, Y. H., & Nan, C. W. (2015). A comprehensive review on synthesis methods for transition-metal oxide nanostructures. *CrystEngComm*, *17*, 3551–3585. Available from https://doi.org/10.1039/c5ce00034c.

Guo, Y., Amunyela, H. T. N. N., Cheng, Y., Xie, Y., Yu, H., Yao, W., ... Qian, H. (2020). Natural protein-templated fluorescent gold nanoclusters: Syntheses and applications. *Food Chemistry*, *335*, 127657. Available from https://doi.org/10.1016/j.foodchem.2020.127657.

Haimov-Talmoud, E., Harel, Y., Schori, H., Motiei, M., Atkins, A., Popovtzer, R., ... Shefi, O. (2019). Magnetic targeting of mTHPC to improve the selectivity and efficiency of photodynamic therapy. *ACS Applied Materials & Interfaces*, *11*, 45368–45380. Available from https://doi.org/10.1021/acsami.9b14060.

Han, G., Ghosh, P., & Rotello, V. M. (2007). Functionalized gold nanoparticles for drug delivery. *Nanomedicine: Nanotechnology, Biology, and Medicine*, *2*, 113–123. Available from https://doi.org/10.2217/17435889.2.1.113.

Hassanpour, P., Panahi, Y., Ebrahimi-Kalan, A., Akbarzadeh, A., Davaran, S., Nasibova, A. N., ... Kavetskyy, T. (2018). Biomedical applications of aluminium oxide nanoparticles. *Micro & Nano Letters*, *13*, 1227–1231. Available from https://doi.org/10.1049/mnl.2018.5070.

Haume, K., Rosa, S., Grellet, S., Śmiałek, M. A., Butterworth, K. T., Solov'yov, A. V., ... Mason, N. J. (2016). Gold nanoparticles for cancer radiotherapy: A review. *Cancer Nanotechnolology*, *7*, 1–20. Available from https://doi.org/10.1186/s12645-016-0021-x.

Heuer-Jungemann, A., Feliu, N., Bakaimi, I., Hamaly, M., Alkilany, A., Chakraborty, I., ... Kanaras, A. G. (2019). The role of ligands in the chemical synthesis and applications of inorganic nanoparticles. *Chemical Reviews*, *119*, 4819–4880. Available from https://doi.org/10.1021/acs.chemrev.8b00733.

Huang, L., Zhou, C., Zhang, Y., Zhang, S., & Zhang, P. (2019). DBHP-functionalized ZnO nanoparticles with improved antioxidant properties as lubricant additives. *Langmuir: The ACS Journal of Surfaces and Colloids*, *35*, 4342–4352. Available from https://doi.org/10.1021/acs.langmuir.9b00093.

Ito, T., Matsuda, Y., Jinba, T., Asai, N., Shimizu, T., & Shingubara, S. (2017). Fabrication and characterization of nano porous lattice biosensor using anodic aluminum oxide substrate. *Japan Society of Applied Physics*, 06GG02. Available from https://doi.org/10.7567/JJAP.56.06GG02.

Jiang, J., Pi, J., & Cai, J. (2018). The advancing of zinc oxide nanoparticles for biomedical applications. *Bioinorganic Chemistry and Applications*, *2018*. Available from https://doi.org/10.1155/2018/1062562.

Kang, H., Buchman, J. T., Rodriguez, R. S., Ring, H. L., He, J., Bantz, K. C., & Haynes, C. L. (2019). Stabilization of silver and gold nanoparticles: Preservation and improvement of plasmonic functionalities. *Chemical Reviews*, *119*, 664–699. Available from https://doi.org/10.1021/acs.chemrev.8b00341.

Kao, H. W., Lin, Y. Y., Chen, C. C., Chi, K. H., Tien, D. C., Hsia, C. C., ... Wang, H. E. (2014). Biological characterization of cetuximab-conjugated gold nanoparticles in a tumor animal model. *Nanotechnology*, *25*, 295102. Available from https://doi.org/10.1088/0957-4484/25/29/295102.

Kim, K. E., Kim, T. G., & Sung, Y. M. (2012). Enzyme-conjugated ZnO nanocrystals for collisional quenching-based glucose sensing. *CrystEngComm*, *14*, 2859–2865. Available from https://doi.org/10.1039/c2ce06410c.

Kim, Y. H., Han, D. K., Park, K. D., & Kim, S. H. (2003). Enhanced blood compatibility of polymers grafted by sulfonated PEO via a negative cilia concept. *Biomaterials*, *24*, 2213–2223. Available from https://doi.org/10.1016/S0142-9612(03)00023-1.

Kimling, J., Maier, M., Okenve, B., Kotaidis, V., Ballot, H., & Plech, A. (2006). Turkevich method for gold nanoparticle synthesis revisited. *The Journal of Physical Chemistry B*, *110*, 15700–15707. Available from https://doi.org/10.1021/jp061667w.

Kobayashi, Y., Katakami, H., Mine, E., Nagao, D., Konno, M., & Liz-Marzán, L. M. (2005). Silica coating of silver nanoparticles using a modified Stöber method. *Journal of Colloid and Interface Science*, *283*, 392–396. Available from https://doi.org/10.1016/j.jcis.2004.08.184.

Kumar, R. (2016). A novel evaporation assisted solvent antisolvent interaction method for the nanocrystalization of organic compounds, IITMandi. http://odr.iitmandi.ac.in:8080/xmlui/handle/123456789/324 (accessed August 23, 2020).

Kumar, R., Aadil, K. R., Ranjan, S., & Kumar, V. B. (2020). Advances in nanotechnology and nanomaterials based strategies for neural tissue engineering. *Journal of Drug Delivery Science and Technology*, *57*, 101617. Available from https://doi.org/10.1016/j.jddst.2020.101617.

Kumar, R., Dalvi, S. V., & Siril, P. F. (2020). Nanoparticle-based drugs and formulations: Current status and emerging applications. *ACS Applied Nano Materials*, *3*, 4944–4961. Available from https://doi.org/10.1021/acsanm.0c00606.

Kumar, R., Gupta, A., Chawla, M., Aadil, K. R., Dutt, S., Kumar, V. B., & Chaudhary, A. (2020). *Advances in nanotechnology based strategies for synthesis of nanoparticles of lignin. Lignin* (pp. 203–229). Cham: Springer, 10.1007/978-3-030-40663-9_7.

Kumar, R., Kumar, V. B., & Gedanken, A. (2020). Sonochemical synthesis of carbon dots, mechanism, effect of parameters, and catalytic, energy, biomedical and tissue engineering applications. *Ultrasonics Sonochemistry*, *64*, 105009. Available from https://doi.org/10.1016/j.ultsonch.2020.105009.

Kumar, R., Mondal, K., Panda, P. K., Kaushik, A., Abolhassani, R., Ahuja, R., ... Mishra, Y. K. (2020). Core-shell nanostructures: Perspectives towards drug delivery applications. *Journal of Materials Chemistry B*. Available from https://doi.org/10.1039/D0TB01559H.

Li, Z., Barnes, J. C., Bosoy, A., Stoddart, J. F., & Zink, J. I. (2012). Mesoporous silica nanoparticles in biomedical applications. *Chemical Society Reviews*, *41*, 2590–2605. Available from https://doi.org/10.1039/c1cs15246g.

Liao, Y., Hu, Z., Gu, Q., & Xue, C. (2015). Amine-functionalized ZnO nanosheets for efficient CO_2 capture and photoreduction. *Molecules (Basel, Switzerland)*, *20*, 18847–18855. Available from https://doi.org/10.3390/molecules201018847.

Lingeshwar Reddy, K., Balaji, R., Kumar, A., & Krishnan, V. (2018). Lanthanide doped near infrared active upconversion nanophosphors: Fundamental concepts, synthesis strategies, and technological applications. *Small (Weinheim an der Bergstrasse, Germany)*, *14*, 1801304. Available from https://doi.org/10.1002/smll.201801304.

Liu, S., & Han, M.-Y. (2009). Silica-coated metal nanoparticles. *Chemistry: An Asian Journal*, *5*. Available from https://doi.org/10.1002/asia.200900228.

Lu, Z. R., Shiah, J. G., Sakuma, S., Kopečková, P., & Kopeček, J. (2002). Design of novel bioconjugates for targeted drug delivery. *Journal of Controlled Release: Official Journal of the Controlled Release Society*, 165–173. Available from https://doi.org/10.1016/S0168-3659(01)00495-3.

Lynge, M. E., Van Der Westen, R., Postma, A., & Städler, B. (2011). Polydopamine—A nature-inspired polymer coating for biomedical science. *Nanoscale*, *3*, 4916–4928. Available from https://doi.org/10.1039/c1nr10969c.

Mallakpour, S., & Sadeghzadeh, R. (2017). Surface functionalization of Al_2O_3 nanoparticles with biocompatible modifiers, preparation and characterization of poly(vinyl pyrrolidone)/modified Al_2O_3 nanocomposites. *Polymer-Plastics Technology and Engineering*, *56*, 1866–1873. Available from https://doi.org/10.1080/03602559.2017.1295311.

Manickam, P., Vashist, A., Madhu, S., Sadasivam, M., Sakthivel, A., Kaushik, A., & Nair, M. (2020). Gold nanocubes embedded biocompatible hybrid hydrogels for electrochemical detection of H_2O_2. *Bioelectrochemistry (Amsterdam, Netherlands)*, *131*, 107373. Available from https://doi.org/10.1016/j.bioelechem.2019.107373.

Marcus, M., Indech, G., Vardi, N., Levy, I., Smith, A., Margel, S., ... Sharoni, A. (2020). Magnetic organization of neural networks via micro-patterned devices. *Advanced Materials Interfaces*, 2000055. Available from https://doi.org/10.1002/admi.202000055.

Meder, F., Kaur, S., Treccani, L., & Rezwan, K. (2013). Controlling mixed-protein adsorption layers on colloidal alumina particles by tailoring carboxyl and hydroxyl surface group densities. *Langmuir: The ACS Journal of Surfaces and Colloids*, *29*, 12502–12510. Available from https://doi.org/10.1021/la402093j.

Medhi, R., Marquez, M. D., & Lee, T. R. (2020). Visible-light-active doped metal oxide nanoparticles: Review of their synthesis, properties, and applications. *ACS Applied Nano Materials*, *3*, 6156–6185. Available from https://doi.org/10.1021/acsanm.0c01035.

Mejac, I., Bryan, W. W., Lee, T. R., & Tran, C. D. (2009). Visualizing the size, shape, morphology, and localized surface plasmon resonance of individual gold nanoshells by near-infrared multispectral imaging microscopy. *Analytical Chemistry*, *81*, 6687–6694. Available from https://doi.org/10.1021/ac9007495.

Mishra, P. K., Mishra, H., Ekielski, A., Talegaonkar, S., & Vaidya, B. (2017). Zinc oxide nanoparticles: A promising nanomaterial for biomedical applications. *Drug Discovery Today*, *22*, 1825–1834. Available from https://doi.org/10.1016/j.drudis.2017.08.006.

Mishra, Y. K., Murugan, N. A., Kotakoski, J., & Adam, J. (2017). Progress in electronics and photonics with nanomaterials. *Vacuum*, *146*, 304–307. Available from https://doi.org/10.1016/j.vacuum.2017.09.035.

Mohapatra, S., Mishra, Y. K., Avasthi, D. K., Kabiraj, D., Ghatak, J., & Varma, S. (2008). Synthesis of gold-silicon core-shell nanoparticles with tunable localized surface plasmon resonance. *Applied Physics Letters*, *92*, 103105. Available from https://doi.org/10.1063/1.2894187.

Muddineti, O. S., Ghosh, B., & Biswas, S. (2015). Current trends in using polymer coated gold nanoparticles for cancer therapy. *International Journal of Pharmaceutics*, *484*, 252–267. Available from https://doi.org/10.1016/j.ijpharm.2015.02.038.

Nejabatdoust, A., Salehzadeh, A., Zamani, H., & Moradi-Shoeili, Z. (2019). Synthesis, characterization and functionalization of ZnO nanoparticles by glutamic acid (Glu) and conjugation of ZnO@Glu by thiosemicarbazide and its synergistic activity with ciprofloxacin against multi-drug resistant *Staphylococcus aureus*. *Journal of Cluster Science*, *30*, 329–336. Available from https://doi.org/10.1007/s10876-018-01487-3.

Newman, M. D., Stotland, M., & Ellis, J. I. (2009). The safety of nanosized particles in titanium dioxide- and zinc oxide-based sunscreens. *Journal of the American Academy of Dermatology*, *61*, 685–692. Available from https://doi.org/10.1016/j.jaad.2009.02.051.

Paramasivam, G., Kayambu, N., Rabel, A. M., Sundramoorthy, A. K., & Sundaramurthy, A. (2017). Anisotropic noble metal nanoparticles: Synthesis, surface functionalization and applications in biosensing, bioimaging, drug delivery and theranostics. *Acta Biomaterialia*, *49*, 45–65. Available from https://doi.org/10.1016/j.actbio.2016.11.066.

Parham, S., Wicaksono, D. H. B., Bagherbaigi, S., Lee, S. L., & Nur, H. (2016). Antimicrobial treatment of different metal oxide nanoparticles: A critical review. *Journal of the Chinese Chemical Society*, *63*, 385–393. Available from https://doi.org/10.1002/jccs.201500446.

Park, K. (2013). Facing the truth about nanotechnology in drug delivery. *ACS Nano*, *7*, 7442–7447. Available from https://doi.org/10.1021/nn404501g.

Parveen, S., Misra, R., & Sahoo, S. K. (2012). Nanoparticles: A boon to drug delivery, therapeutics, diagnostics and imaging. *Nanomedicine: Nanotechnology, Biology and Medicine*, *8*, 147–166. Available from https://doi.org/10.1016/j.nano.2011.05.016.

Pelaz, B., Del Pino, P., Maffre, P., Hartmann, R., Gallego, M., Rivera-Fernández, S., ... Parak, W. J. (2015). Surface functionalization of nanoparticles with polyethylene glycol: Effects on protein adsorption and cellular uptake. *ACS Nano*, *9*, 6996–7008. Available from https://doi.org/10.1021/acsnano.5b01326.

Pensa, E., Cortés, E., Corthey, G., Carro, P., Vericat, C., Fonticelli, M. H., ... Salvarezza, R. C. (2012). The chemistry of the sulfur-gold interface: In search of a unified model. *Accounts of Chemical Research*, *45*, 1183–1192. Available from https://doi.org/10.1021/ar200260p.

Pingarrón, J. M., Yáñez-Sedeño, P., & González-Cortés, A. (2008). Gold nanoparticle-based electrochemical biosensors. *Electrochimica Acta*, *53*, 5848–5866. Available from https://doi.org/10.1016/j.electacta.2008.03.005.

Piriyawong, V., Thongpool, V., Asanithi, P., & Limsuwan, P. (2012). Preparation and characterization of alumina nanoparticles in deionized water using laser ablation technique. *Journal of Nanomaterials*, *2012*. Available from https://doi.org/10.1155/2012/819403.

Rajan, Y. C., Inbaraj, B. S., & Chen, B. H. (2015). Synthesis and characterization of poly (γ-glutamic acid)-based alumina nanoparticles with their protein adsorption efficiency and cytotoxicity towards human prostate cancer cells. *RSC Advances*, *5*, 15126–15139. Available from https://doi.org/10.1039/c4ra10445e.

Rengan, A. K., Kundu, G., Banerjee, R., & Srivastava, R. (2014). Gold nanocages as effective photothermal transducers in killing highly tumorigenic cancer cells. *Particle & Particle Systems Characterization*, *31*, 398–405. Available from https://doi.org/10.1002/ppsc.201300173.

Sapsford, K. E., Algar, W. R., Berti, L., Gemmill, K. B., Casey, B. J., Oh, E., ... Medintz, I. L. (2013). Functionalizing nanoparticles with biological molecules: Developing chemistries that facilitate nanotechnology. *Chemical Reviews*, *113*, 1904–2074. Available from https://doi.org/10.1021/cr300143v.

Saravanan, P., Jayamoorthy, K., & Ananda Kumar, S. (2015). Switch-on fluorescence and photo-induced electron transfer of 3-aminopropyltriethoxysilane to ZnO: Dual applications in sensors and antibacterial activity. *Sensors & Actuators, B: Chemical*, *221*, 784–791. Available from https://doi.org/10.1016/j.snb.2015.05.069.

Sharma, V., Sinha, N., Dutt, S., Chawla, M., & Siril, P. F. (2016). Tuning the surface enhanced Raman scattering and catalytic activities of gold nanorods by controlled coating of platinum. *Journal of Colloid and Interface Science*, *463*, 180–187. Available from https://doi.org/10.1016/j.jcis.2015.10.036.

Simsikova, M., Antalík, M., Kaňuchová, M., & Škvarla, J. (2013). Anionic 11-mercaptoundecanoic acid capped ZnO nanoparticles. *Applied Surface Science*, *282*, 342–347. Available from https://doi.org/10.1016/j.apsusc.2013.05.130.

Sinha, R., Kim, G. J., Nie, S., & Shin, D. M. (2006). Nanotechnology in cancer therapeutics: Bioconjugated nanoparticles for drug delivery. *Molecular Cancer Therapeutics*, *5*, 1909–1917. Available from https://doi.org/10.1158/1535-7163.MCT-06-0141.

Soliman, M. G., Pelaz, B., Parak, W. J., & Del Pino, P. (2015). Phase transfer and polymer coating methods toward improving the stability of metallic nanoparticles for biological applications. *Chemistry of Materials: A Publication of the American Chemical Society*, *27*, 990–997. Available from https://doi.org/10.1021/cm5043167.

Sperling, R. A., & Parak, W. J. (2010). Surface modification, functionalization and bioconjugation of colloidal inorganic nanoparticles. *Philosophical Transactions of the Royal Society A-Mathematical Physical and Engineering Sciences*, *368*, 1333–1383. Available from https://doi.org/10.1098/rsta.2009.0273.

Tarlani, A., Isari, M., Khazraei, A., & Moghadam, M. E. (2017). New sol-gel derived aluminum oxide-ibuprofen nanocomposite as a controlled releasing medication. *Nanomedicine Research Journal*, *2*, 28–35. Available from https://doi.org/10.22034/NMRJ.2017.23256.

Tencomnao, T., Apijaraskul, A., Rakkhithawatthana, V., Chaleawlert-Umpon, S., Pimpa, N., Sajomsang, W., & Saengkrit, N. (2011). Gold/cationic polymer nano-scaffolds mediated transfection for non-viral gene delivery system. *Carbohydrate Polymers*, *84*, 216–222. Available from https://doi.org/10.1016/j.carbpol.2010.12.063.

Tiwari, A., Kumar, R., Shefi, O., & Randhawa, J. K. (2020). Fluorescent mantle carbon coated core–shell SPIONs for neuroengineering applications. *ACS Applied Bio Materials*, *3*, 4665–4673. Available from https://doi.org/10.1021/acsabm.0c00582.

Tiwari, A., Singh, A., Debnath, A., Kaul, A., Garg, N., Mathur, R., ... Randhawa, J. K. (2019). Multifunctional magneto-fluorescent nanocarriers for dual mode imaging and targeted drug delivery. *ACS Applied Nano Materials*, *2*, 3060–3072. Available from https://doi.org/10.1021/acsanm.9b00421.

Tiwari, A., Verma, N. C., Singh, A., Nandi, C. K., & Randhawa, J. K. (2018). Carbon coated core-shell multifunctional fluorescent SPIONs. *Nanoscale*, *10*, 10389–10394. Available from https://doi.org/10.1039/c8nr01941j.

Tiwari, P., Vig, K., Dennis, V., & Singh, S. (2011). Functionalized gold nanoparticles and their biomedical applications. *Nanomaterials*, *1*, 31–63. Available from https://doi.org/10.3390/nano1010031.

Tomitaka, A., Kaushik, A., Kevadiya, B. D., Mukadam, I., Gendelman, H. E., Khalili, K., ... Nair, M. (2019). Surface-engineered multimodal magnetic nanoparticles to manage CNS diseases. *Drug Discovery Today*, *24*, 873–882. Available from https://doi.org/10.1016/j.drudis.2019.01.006.

Umrani, R. D., & Paknikar, K. M. (2014). Zinc oxide nanoparticles show antidiabetic activity in streptozotocin-induced Type 1 and 2 diabetic rats. *Nanomedicine: Nanotechnology, Biology, and Medicine*, *9*, 89–104. Available from https://doi.org/10.2217/nnm.12.205.

Vankayala, R., Lin, C. C., Kalluru, P., Chiang, C. S., & Hwang, K. C. (2014). Gold nanoshells-mediated bimodal photodynamic and photothermal cancer treatment using ultra-low doses of near infra-red light. *Biomaterials*, *35*, 5527–5538. Available from https://doi.org/10.1016/j.biomaterials.2014.03.065.

Vats, T., Dutt, S., Kumar, R., & Siril, P. F. (2016). Facile synthesis of pristine graphene-palladium nanocomposites with extraordinary catalytic activities using swollen liquid crystals. *Scientific Reports*, *6*, 1–11. Available from https://doi.org/10.1038/srep33053.

Virkutyte, J., & Varma, R. S. (2011). Green synthesis of metal nanoparticles: Biodegradable polymers and enzymes in stabilization and surface functionalization. *Chemical Science*, *2*, 837–846. Available from https://doi.org/10.1039/c0sc00338g.

Yao, X., Niu, X., Ma, K., Huang, P., Grothe, J., Kaskel, S., & Zhu, Y. (2017). Graphene quantum dots-capped magnetic mesoporous silica nanoparticles as a multifunctional platform for controlled drug delivery, magnetic hyperthermia, and photothermal therapy.

Small (Weinheim an der Bergstrasse, Germany), 13. Available from https://doi.org/10.1002/smll.201602225.

Zhan, Y., Gonçalves, M., Yi, P., Capelo, D., Zhang, Y., Rodrigues, J., ... He, P. (2015). Thermo/redox/pH-triple sensitive poly(*N*-isopropylacrylamide-*co*-acrylic acid) nanogels for anticancer drug delivery. *Journal of Materials Chemistry B*, *3*, 4221–4230. Available from https://doi.org/10.1039/c5tb00468c.

Zhang, F., Shi, Z., Li, S., Ma, Z., Li, Y., Wang, L., ... Shan, C. (2019). Synergetic effect of the surfactant and silica coating on the enhanced emission and stability of perovskite quantum dots for anticounterfeiting. *ACS Applied Materials & Interfaces*, *11*, 28013–28022. Available from https://doi.org/10.1021/acsami.9b07518.

Zhang, Z.-Y., & Xiong, H.-M. (2015). Photoluminescent ZnO nanoparticles and their biological applications. *Materials (Basel)*, *8*, 3101–3127. Available from https://doi.org/10.3390/ma8063101.

Zhao, J., Zhong, D., & Zhou, S. (2018). NIR-I-to-NIR-II fluorescent nanomaterials for biomedical imaging and cancer therapy. *Journal of Materials Chemistry B*, *6*, 349–365. Available from https://doi.org/10.1039/c7tb02573d.

Zhou, J., & Rossi, J. (2017). Aptamers as targeted therapeutics: Current potential and challenges. *Nature Reviews Drug Discovery*, *16*, 181–202. Available from https://doi.org/10.1038/nrd.2016.199.

Zhuang, J., Liu, M., & Liu, H. (2009). MAA-modified and luminescence properties of ZnO quantum dots. *Science in China Series B: Chemistry*, *52*, 2125–2133. Available from https://doi.org/10.1007/s11426-009-0198-5.

Metal oxidesbased microfluidic biosensing

10

Agnivo Gosai[1] *and Md. Azahar Ali*[2]
[1]Corning Inc., Science & Technology, Painted Post, NY, United States, [2]Mechanical Engineering, Carnegie Mellon University, Pittsburgh, PA, United States

10.1 Introduction

In the last three decades, the microfluidic technology has revolutionized the manufacturing of biosensors and microelectronics (Kumar et al., 2013). This allows sample mixing, separation, transportation, and high-throughput chemical reactions in a single chip format. Owing to the low reagent volumes, low-power consumption, parallelization, and high-throughput screening, the microfluidic devices can be able to perform many biological processes simultaneously which make it suitable for biochemical analysis (Ali et al., 2018). In the diagnostics industry, it has a long history for the integration of microfluidic devices into biomedical sensors and other application including organ-on-chips (Bhatia & Ingber, 2014; Li, Ma, Cao, Pan, & Shi, 2020). For instance, a microfluidic-based lateral-flow immunoassay test (Oh, Joung, Kim, & Kim, 2013) can directly measure the presence of antibodies or protein biomarkers due to pathogenic or viral infections in human serums wherein a specific protein/antigen-immobilized paper trip will change the readout signal (Sajid, Kawde, & Daud, 2015; Wang, Ali, Chow, Dong, & Lu, 2018). The integration of microfluidic components such as microchannel, mixer, and valve with biosensing platforms led to improvement of device efficacy such as rapid detection, limit-of-detection (LOD), and high sensitivity which allows on-site screening of biomarkers and biomolecules at point-of-care (POC) level (Gervais, De Rooij, & Delamarche, 2011). Specifically, the introduction of microfluidic components into biosensors can help in situ immobilization of bioreceptors, control biofluids, initiating bioreactions, mixing labeling reagents, and manipulation of biological process (Malhotra & Ali, 2018). Further, microfluidic device has an important role in biosensor manufacturing as it significantly improves the final size and performance of the device.

Many methods of fabrication were exploited for the construction of microfluidic structures with two- and three-dimensional geometries. One of the popular techniques is soft lithography which was introduced by Whitesides et al. In this method, polydimethylsiloxane (PDMS) can be directly patterned from a silicon mold via standard photolithography. For 2D microfluidic devices, wax printing is commonly used via melting the wax onto a desire surface which allows the formation of hydrophobic barriers (Juang, Li, & Chen, 2017). In a recent advancement, instead

Metal Oxides for Biomedical and Biosensor Applications. DOI: https://doi.org/10.1016/B978-0-12-823033-6.00008-9

of using solid barrier, a rapid and direct printing method was utilized to microfluidic channels combined with biosensing recognition elements, and transducer elements for the analysis of an inflammatory response in human cells and chemotaxis in bacterial biofilms (Walsh et al., 2017). For 3D microfluidic devices, many additive manufacturing extrusion-based methods such as fused deposition modeling (FDM), stereolithography (SLA), two-photon polymerization (2PP), and selective laser sintering (SLS) combined with binders or photosensitive resins are utilized wherein the layer-by-layer printing of liquefied materials [acrylonitrile – butadiene – styrene (ABS) or PDMS] are performed to construct device in 3D space in minutes. The microstructures by 3D manufacturing can be sintered by thermally, UV curing or laser sintering. Recently, a proof-of-concept 3D-printed microfluidic biosensor was developed to monitor human tissue metabolite levels (glucose and lactate) by combining with FDA-approved clinical microdialysis probe (Gowers et al., 2015). Though these methods offer high resolution, they are not fully developed for the development of biosensors at POC level.

Nanostructures of metal oxides have received great attention to construct biosensors for healthcare monitoring (Malhotra & Ali, 2018; Malhotra, Srivastava, Ali, & Singh, 2014; Solanki, Kaushik, Agrawal, & Malhotra, 2011). Nanostructured metal oxides served as transduction elements by establishing nano–bio interfaces in the development of biosensors (Ansari, Alhoshan, Alsalhi, & Aldwayyan, 2010). These metal oxides provide higher surface area to immobilize enzymes, antibodies, antigens, nucleic acid and proteins, nontoxicity, high electrocatalytic properties, and biocompatibility (Rahman, Ahammad, Jin, Ahn, & Lee, 2010). These unique properties of metal oxides make a new generation of bioelectronic devices with superior performance such as improve sensitivity, detection limit, high electron transfer, and long-term stability. Many nanostructured metal oxides such as zinc oxide (ZnO) (Ali et al., 2009), cerium oxide (CeO_2) (Solanki, Ali, Kaushik, & Malhotra, 2013), titanium dioxide (TiO_2) (Azahar Ali et al., 2012; Mondal, Ali, Agrawal, Malhotra, & Sharma, 2014), titanium nitride (TiN) (Ali et al., 2017), nickel oxide (NiO) (Ali, Srivastava et al., 2013; Ali et al., 2014), zirconium oxide (ZrO_2) (Srivastava et al., 2013), and aluminum oxide (Al_2O_3) are not only served as immobilizing matrices but also provided direct electron transfer between electrode and electrolyte (Md, 2014). Specially, these materials are used to construct electrochemical biosensors which provided a number of benefits including device cost, miniaturized, rapid detection, and low-power requirements led to hold a great promise for a wide range of POC biomedical applications. Combining benefits of metal oxides and microfluidics, these metal oxides have recently increased interest for the development of microfluidic biosensors (Ali et al., 2016). However, it is critical to choose the specific metal oxide to integrate with microfluidic biosensor due to their different properties. Each material has their own characteristics in terms of surface charge, biocompatible, surface roughness, functional groups, porosity, and hygroscopic nature to form a biointerface. Due to the high isoelectric point of metal oxides such as ZnO, NiO, ZrO_2, and CeO_2, they can be easily bound with the enzymes of biomolecules having with low isoelectric point via electrostatic interaction. For instance, nanostructures of NiO were integrated into a microfluidic device to develop a high sensitive cholesterol biosensor (Ali, Solanki et al., 2013). In this sensor, the high isoelectric point (10.8) of nanostructured NiO allows direct conjugation of enzyme molecules specific to cholesterol oxidation during its detection resulting in a high Michaelis–Menten constant (k_m) of 0.67 mM and a sensitivity of 0.808 μA/mg/dL/cm^2. One-dimensional nanostructures of metal oxides such as nanorods, nanowires (NWs), and nanotubes can act as electronic wires which provide conduction channels for the transport of electrons from the active site enzyme's to the current collector (Ali et al., 2013; Solanki et al., 2013). Kim et al., demonstrated the bottom-up synthesis of patterned ZnO NWs in a microfluidic testing platform by controlled hydrothermal reaction and used for particle trapping and chemiresistive pH sensing (Kim, Li, & Park, 2011). Further, ZnO NWs are also shown excellent performance while integrated with a paper-based microfluidic glucose sensor (Li, Zhao, & Liu, 2015). Thus the integration of nanostructured metal oxides into the microfluidic biosensors provide high surface area for immobilization of biorecognition molecules, dimensions comparable to the Debye length, biocompatibility, nontoxicity, high electrocatalytic activity, and high electron transfer ability from electrolyte to current collector. These properties not only improve the detection ability of biomarkers with low LOD but also provide high sensitivity and stability of microfluidic device at POC level.

In this chapter, we focus to demonstrate the application of a variety of metal oxide nanostructures for detection of biomarkers within the microfluidic capability. We also cover the fundamentals of the microfluidic devices and nanostructured metal oxides with an emphasis on their electrochemical and optical properties that enhances the biomonitoring. Finally, we summarize the challenges and future prospccts of metal oxides for the next generation of biosensing devices.

10.2 Basics of microfluidics

A microfluidic device is the manipulation of fluids at the volume of milli (10^{-3}) to nano (10^{-9}) liter level via incorporating microchannels networks. Fluids can inject or evacuate through integrated microfluidic channels in a microdevice by several holes that linked to tubes and controlled by either syringe pump or capillary action. These fluidic networks allowed to mix, separate, and multiplex the many sensing elements by integrating different types of microfluidic components such as pumps, valves, filters, and mixers, and can be performed several tasks in a same miniature platform the so-called lab-on-a-chip. The microfluidic device offers many advantages such as (1) high-throughput, multiplexed, and parallelization, (2) rapid analyses due to the shorter reactions times, (3) portability for POC applications, (4) low reagent consumptions and low-cost, (5) whole biological process can be integrated and easy for the end-users, and (6) accurate measurement, low LOD and high sensitivity. At the microscale, the movement of fluidic behavior is easy to control and sample reaction resulting in minimizing the chemical waste. In addition, the sample analysis can be performed on spot rather being conducted to laboratory facility. To date, microfluidics applied for a variety of applications such as biosensors for many biomarkers monitoring, DNA sequencing, PCR amplification, immunoassays, cell shorting, protein analysis, and in vitro fertilization. For detection of biomarkers, the application of microfluidic device extended to multimodal analysis by implementing electrochemical, plasmonic, electrophoresis, fluorescence, and chromatography methods. In particular to electrochemical detection, the nanostructures of metal oxides are being integrated with microfluidic devices for rapid diagnostics for molecular markers.

The fluid flow through a microfluidic channel is defined by Reynolds number (the ratio of inertial forces to viscous forces) which is $R_e = LV_{ag}\rho/\mu$; wherein the L is the length of channel, V_{ag} is the average velocity, μ is the viscosity, and ρ is the fluid density. The low R_e in a fluid flow determines the fluidic pattern such as a low R_e ($\sim 1-100$) indicates a laminar flow. In a microchannel, the fluid can be controlled via different mechanisms including pressure-driven and electrokinetic flow. In pressure-driven, the fluid can be pump via syringe pump which results a parabolic velocity profile of the fluidic within a channel with no-slip boundary condition. In electrokinetic flow, the walls of a microfluidic channel are influenced by the electric field in which the ions in the double layer move toward the opposite polarity resulting in the movement of fluid near the walls. In an open channel the velocity of fluid is uniform, however, in a closed channel the velocity of the flow channel is 50% along the direction of centerline.

10.2.1 Microfabrication

In 1979 a miniaturized gas chromatographic analyzer with microelements was first realized by Terry on silicon substrate via chemical etching process (Terry, Jerman, & Angell, 1979). Later in 1980 the microfluidic components such as microvalves and micropumps were manufactured by silicon micromachining. During this period, the microfluidic devices were still fabricated on silicon or glass substrate. In the late 1990s soft lithography was exploited to fabricate microfluidic devices including microchannels, micropumps, microvalves, using polymeric molds to reduce the manufacturing costs (Xia & Whitesides, 1998).

The fabrication of microfluidic devices can be categorized by three techniques such as soft lithography (replica molding); injection molding, and hot embossing. In replica molding, a geometrically defined pattern can be transferred to a polymeric substrate using a photolithography technique. In this process, a photoresist (negative or positive such as SU-8) needs to pattern on a silicon or glass substrate upon exposing an UV light by introducing an optical mask (Xia & Whitesides, 1998). Then, this mold can be used as master wherein a PDMS polymeric material can copy an opposite pattern of master by thermal curing. This process is called "soft lithography." In addition to PDMS, biocompatible polymers such as agar or agarose can be patterned via soft lithography. The height and width of the pattern photoresist on the mold substrate determines the feature size of the microfluidic device. To fabricate the mold, the clean room facilities are required, especially for high-resolution feature sizes (5–200 μm). This method is commonly used to create microfluidic channel. Fig. 10.1 shows the schematic representation of the microfluidic channels using soft-lithography technique. The resolution down to sub-20 nM, a unique method of replica molding is used which is nanoimprinting lithography. The main advantages of replica molding are rapid prototyping, low cost, no optical diffraction limit, control surface chemistry, and board range of materials. This method is being widely used to POC diagnostic devices for biomonitoring (Bartlett, Markvicka, & Majidi, 2016; Whitesides, Ostuni, Takayama, Jiang, & Ingber, 2001). For example, 3D microfluidic paper analytical devices (μPADs) were created using soft-lithography technique for POC environmental monitoring and water analysis (Martinez, Phillips, & Whitesides, 2008).

In injection molding, a thermoplastic material is melted to form its liquid state within a compressed chamber (Lee et al., 2018). To create a mold cavity, the halves of the mold are compressed and filled the mold cavity by injecting a thermoplastic. Finally, the cast part is separated from the mold cavity by its cooling. Micromilling of metal molds can provide a resolution of ~25 μm, but electron beam lithography

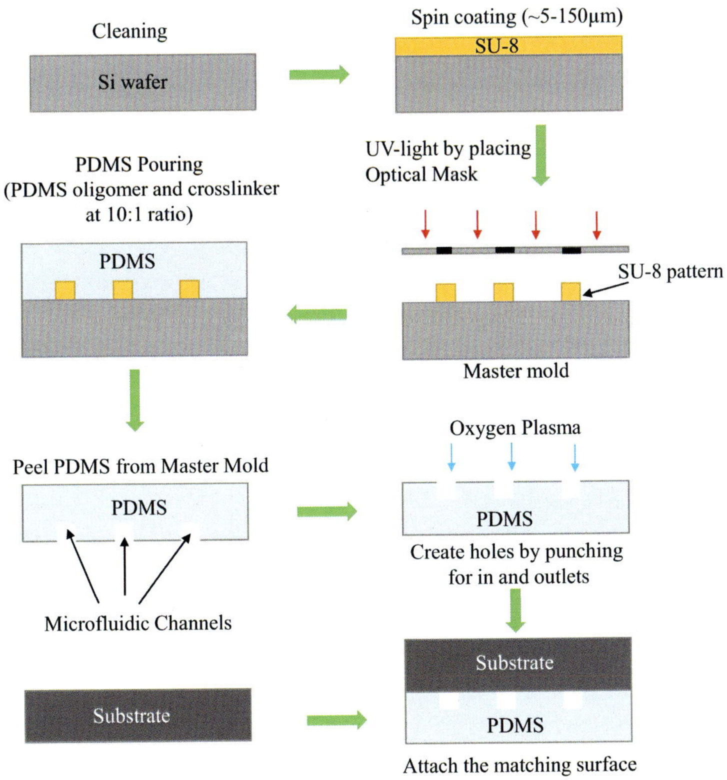

Figure 10.1 Schematic representation of soft lithography using photolithographic technique to create polydimethylsiloxane (PDMS) microfluidic channels.

can provide down to submicron resolution. Compared to replica molding, this method is relatively costly.

In hot embossing, a high temperature and elevated temperature are used to transfer patterns from master mold into polymers or thermoplastics by forming viscous liquids (Goral, Hsieh, Petzold, Faris, & Yuen, 2010). A variety type of thermoplastic materials such as cyclic olefin copolymer, polyethylene terephthalate, polycarbonate, and polymethylmethacrylate are used to create microfluidic structures. In this method of fabrication, first a thermoplastic film is positioned between two molds, mold chamber is then evacuated, compressed, and heated, resulting in a mold cast.

Though, soft lithography of PDMS pattering has significantly contributed to microfluidic community in terms of cost of manufacturing and ease process. However, this material does not translate well to industrial applications. In addition, hot embossing and injection molding are also limited with their feature sizes as they can produce only identical depth of the devices. Currently, additive manufacturing (i.e., 3D printing) is being used to fabricate polymeric microfluidic channels due to its high ability of layer-by-layer printing of polymeric materials such as PDMS and polymethyl methacrylate (PMMA) with reasonable resolution. This is an emerging technology of advanced manufacturing which uses the computer-aided design (CAD) program to transfer a pattern digitally from computer to substrate by a single click. This offers many advantages such as rapid prototypes, complex and new geometries, customizability, and materials combination with unique capabilities (Waheed et al. 2016). Many techniques such as SLA, inkjet, two photon polymerization (2PP) and extrusion printing (focusing on FDM) are being used to create polymeric materials-based microfluidic device based on thermal or UV-light sintering. Fig. 10.2 shows the 3D-printed microfluidic device with light-addressable potentiometric sensor (LAPS) chip to detect hydrogen ion concentration (Takenaga et al., 2015). In this device, a photopolymer (Asiga, PlasCLEAR) of polypropylene/acrylnitril-butadien-styrol was used as a base material to form microfluidic channels via a 3D-printer (Asiga, PicoPlus 27) layer-by-layer printing capacity followed by UV curing. The 3D printing thus not only allow to create microfluidic device but also allow to integrate biosensing elements such as metal oxide nanoparticles (NPs) printing to realize next generation of metal oxides-based microfluidic device at the industrial level.

10.3 Microfluidic metal oxide biosensors

10.3.1 Biosensor fundamentals

A biosensor can be described as an analytical device that is able to transform/convert biological interactions into measurable electronic signals (current/voltage) proportional to the target/analyte concentration. Biosensors have been demonstrated to detect a variety of substances such as proteins, nucleic acids, metals, metabolites, organic/inorganic pollutants, microbes/pathogens, viruses, toxins, etc. This sensing

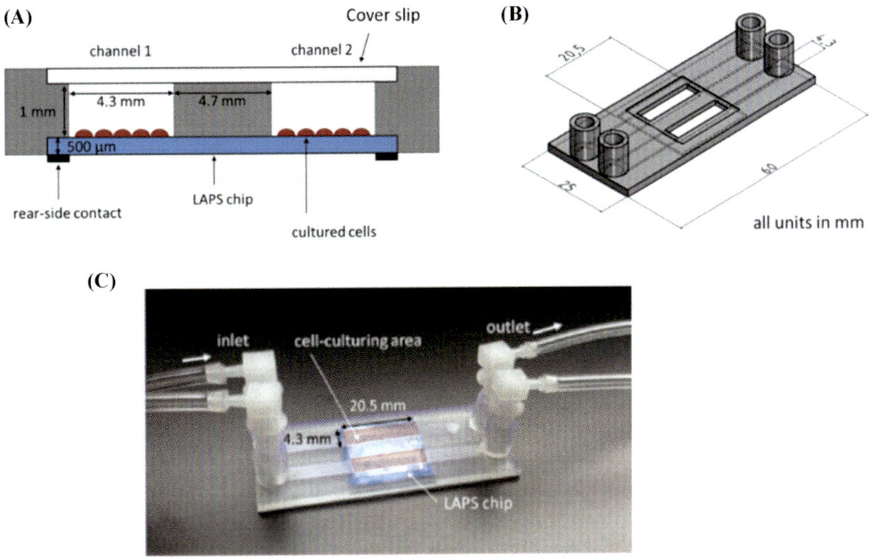

Figure 10.2 (A) Cross-sectional of a microfluidic device, (B) a computer-aided design of 3D layout for a microfluidic to be printed, and (C) the 3D-printed microfluidic device with two independent channels (Takenaga et al., 2015).

device is called a "bio"-sensor due to the fact that the sensing structure or architecture has a biological element, preferably the receptor at its heart. IUPAC defines the biosensor as follows: A biosensor is an integrated receptor-transducer device capable of providing selective quantitative or semiquantitative analytical information using a biological recognition element. A typical biosensor thus consists of two elements: biological sensing/recognition element and a transducer for the detection of analyte concentration (Grieshaber, MacKenzie, Vörös, & Reimhult, 2008). The outline of a typical biosensor is provided in Fig. 10.3.

As depicted in Fig. 10.3, the analyte is the target molecule (e.g., protein) and the biological recognition element (e.g., nucleic acid) is surface assembled or physiochemically attached onto the transducer. The transducer is made from a variety of different substrates, often involving nanostructured materials, depending on the type of biosensor. The interaction between the analyte and the recognition element generates a physiochemical signal which is transduced through the nano–bio interface of the transducer onto a detector (e.g., potentiostat system in case of electrochemical biosensor). The detector is coupled to a display (e.g., a desktop computer or laptop/smartphone, any digital reader) for reading of the result by the end user (e.g., healthcare professional/patient/operator).

The first use of the terminology "biosensor" may be credited to Leland C. Clarke, in 1962, when he demonstrated his pioneering work on a glucose oxidase-based electrochemical sensor for the detection of glucose (Grieshaber et al., 2008). In over 50 years since then, biosensors have seen rapid development and evolution

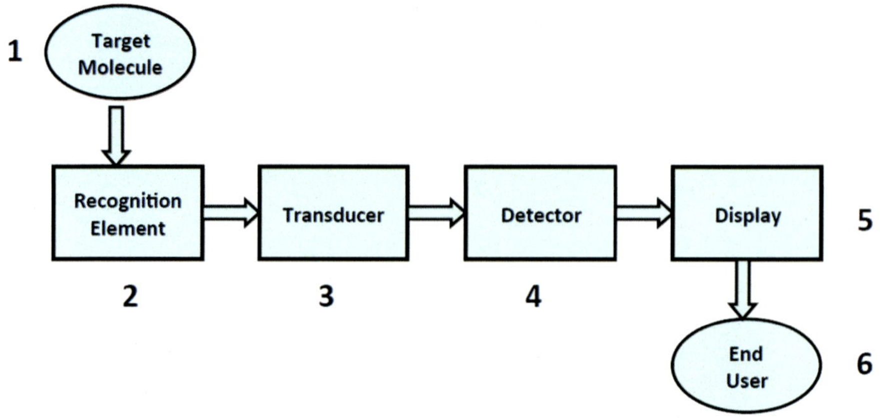

Figure 10.3 A typical schematic representation of a biosensor showing the individual components.
Source: Gosai, Agnivo, "Engineering of nano-bio interfaces towards the development of portable biosensors" (2018). Graduate Theses and Dissertations. 17194. https://lib.dr.iastate.edu/etd/17194 Copyright: Agnivo Gosai, 2018.

in diverse directions with the present thrust being in the field of portable diagnostics or POC devices. Still it may be said that in spite of the tremendous potential and the great expectations generated in the last 30 years, biosensors have found only limited commercial application. The commercial sector is mostly dominated by blood glucose measurement devices and the field of DNA-based biochips, which can detect viruses, bacteria, and toxins are yet to meet expectations. The lateral-flow immunoassay, for example, pregnancy test kits, is another technology that has seen commercial success and rapid adoption. It is mostly used for qualitative detection whereas the glucometer is able to quantify the amount of glucose. The impact of the COVID-19 pandemic on the world economy may provide the much needed thrust to commercialize the biosensor research and bring the devices out of academic/industrial research labs to the market.

The receptor or recognition element can be antibodies, nucleic acids such as DNA/RNA, "mini-enzymes," synzymes, aptamers, and molecularly imprinted polymers (MIPs), which are products of biological research and even whole cells, bacteria, and virus. The recognition element needs to be robust with a possibility of reuse or single use even after being stored for a long time. Some recognition elements can be stored in room temperature for a long time (e.g., aptamers, MIP). The recognition element is to be spatially integrated with the transducer which in turn makes a compact functional unit that allows usage of the biological component and miniaturization of the sensor body. Integration of the biochemical mechanisms and transducer signal processing has been realized in "intelligent biosensors" giving rise to electrochemical sensors and bio-FETs. Through the utilization of the recent advances in micro/nanoelectronics, molecular biology and nanotechnology, in the last decade, scientists and engineers can develop total microanalytical systems (μTAS), which are the lab-on-a-chip combination of biosensors with microfluidics and actuators.

Some of the important parameters/technical specifications for determining the performance of a biosensor are briefly mentioned:

Selectivity: This may be called as the most important specification for a biosensor. This is primarily determined by the recognition element/receptor which detects the analyte. A biosensor should be able to specifically pick out the analyte even in the presence of other substances in the sample. The transducer also plays a vital part in amplifying this selective response and thus increases the signal-to-noise ratio. Hence, for the same receptor/analyte combination, biosensors employing different transducers can have vastly different selectivity. In some literature, this is also called as specificity.

Sensitivity: It can be described as the second most important property of a biosensor and can be defined as the minimum amount of analyte that can be detected and is often designated by the LOD. A highly sensitive biosensor should be able to detect small/trace quantities of the analyte in real-world samples. The samples can be diluted or chemically processed to reduce the presence of interfering agents, but this also can reduce the concentration of the analyte itself. Hence, a highly sensitive biosensor should be able to generate adequate signal-to-noise ratio.

Reproducibility: This can be described as the ability of the biosensor to demonstrate identical/statistically significant response for identical experimental/test conditions. In this connection, we may also define the other two parameters: (1) accuracy: it deals with the deviation of the mean value, detected by the sensor, of the repeated tests on a particular sample from the true value of the analyte in the sample, (2) precision: it deals with the spread/standard deviation of the measured analyte value for the same sample under test.

Stability: A biosensor response should be stable over time, temperature, humidity, and other environmental factors. Stability of the response often depends on the structural/chemical stability of the recognition element and its strength of interaction/binding capabilities with the analyte. Other contributing factors can be the drift in the electronic/optical measurement systems. Stability is an important factor, for example, in continuous glucose monitoring devices.

Linearity: The response of a biosensor can quickly saturate with increasing analyte con centration and this may limit its usefulness where a greater demarcation of detected concentrations is required. Linearity is thus associated with the resolution of a biosensor response. Many sensor responses are limited by the reaction kinetics between the receptor and the target which prevents the proper discrimination at higher concentrations.

Traditional methods such as mass spectroscopy (Ms) and liquid chromatography (LC) are often used to detect toxins, pollutants, and contaminants in food samples, water, and work environment and mostly these techniques require considerable bulky instrumentation which can be costly and at times significant operator knowledge. Biosensors could be a viable alternative for cheaper and easier to use solution with the added advantage of portability. The most significant use of biosensors will still be to manage public health and stop the spread of infectious diseases in the event of epidemic/pandemic through early detection and improve patient outcomes for various diseases via timely intervention enabled by rapid diagnostics. All the biosensors described in the following sections intend to demonstrate cheaper solutions with fast turnaround times ($\sim$10 min to 1 h) in comparison to standard lab-based analysis.

10.3.2 Impact of microfluidics and metal oxides

Like the field of biosensors, microfluidics is also an interdisciplinary field and it depends on contributions from fluid mechanics, engineering, physics, chemistry, and nanotechnology. As biosensors deal with fluid-based samples and microfluidic devices have the advantage of miniaturization and automation, it's quite natural to have these two fields intersect. Microfluidics-based systems are often low cost and this is of particular benefit for POC diagnostics. It also offers the integration of miniaturized sensor chips with electrical readouts/connector pads which are essential to most of the sensors. Miniaturization of the sample handling and transducer system offers enhanced stability and reproducibility. This miniaturization also reduces the sensing area and simulation/experimental studies have demonstrated that this leads to greater sensitivity by reducing the time taken by the receptor to sense the target and by allowing favorable kinetics with lower sample volumes containing lesser concentration of the analyte. Smaller sensor area reduces the heterogeneity in receptor–target molecular interaction events and smaller interaction volume enhances transducer response, for example, for many electrochemical sensors the receptor–target binding should be within the Debye length for maximizing the signal value. In addition, microfluidic devices integrate several functionalities such as sample injection, pretreatment, and processing and are amenable to automation to control temperature, flow, and mixing. In recent years, paper-based microfluidic biosensors are being explored as a viable cheap POC device.

Metal oxides in particular the nanostructured metal oxides have over the years found increasingly useful applications in the field of biosensing (Solanki et al., 2011). Nanostructure oxides of metals such zinc, aluminum, silicon, titanium, magnesium, iron and tin have found many applications in biosensing devices. These materials serve as transduction materials and immobilization matrices in biosensors. These materials have useful morphological, catalytic, electronic, and optical properties and also demonstrate good biocompatibility. They are also suitable for physicochemical adsorption and are thus readily functionalized by biological receptors through easy to implement established chemical protocols. Taking advantage of bottom-up or top-down fabrication in nanotechnology as well as soft lithography, these metal oxide transducers can be readily integrated into microfluidic biosensing devices. In the following sections, detailed description of a selection of existing literature on the R&D of metal oxide-based biosensors is provided.

10.3.3 Al_2O_3-based microfluidic biosensors

Nanoporous anodized alumina (aluminum oxide) (NAAO) membranes have found application in many areas involving biosensors (Tan et al., 2011; Wang et al., 2009; Ye, Shi, Chan, Zhang, & Yang, 2014). Alumina membranes have several desirable properties which make them attractive for use in a variety of biosensing platforms. Some of the salient properties are nonconductivity, well-defined nanopores, small pore size, high pore density, and ease of functionalization. These features can be achieved through simple electrochemical techniques and are relatively inexpensive compared to lithographic methods to create micro/nano channels. NAAO membranes can be sputter coated with gold and thus become useful for applications involving surface-enhanced Raman spectroscopy. Alumina has been deposited on an electrode to form a nominal membrane structure (Rai, Deng, & Toh, 2012), or sputtered onto the surface and then anodized (Rai et al., 2012) to create the nanoporous structure. Commercially available NAAO membranes, particularly from Whatman, have been used in biosensor research (Deng & Toh, 2013; Gosai, Hau Yeah, Nilsen-Hamilton, & Shrotriya, 2019). Lipid bilayer-modified NAAO membranes have been shown to exhibit higher diffusivity than those formed on TiO_2 and SiO_2 substrates, and this is attributed to the presence of a thick hydration layer on Al_2O_3, This property is a key requirement to preserving the biological functionality of reconstituted membrane proteins (Venkatesan et al., 2011) and thus is a testament to the biocompatibility of NAAO membranes.

Due to the various advantages of NAAO, it has been used in microfluidics-enabled biosensing schemes and devices. One of the earliest mentions is found in the work of Cheow, Ting, Tan, and Toh (2008). They describe the use of NAAO membranes to separate proteins based on the principle of electrophoresis. Platinum (Pt) was sputter coated onto either side of the Whatman NAAO membrane with 100 nm average pore size, leaving a thin annular ring at the periphery as this will prevent short circuiting when voltage is applied across the membrane using the deposited Pt as the working electrode (WE) and counter electrode (CE) in an electrochemical cell. The thin platinum coating can achieve high electric field strength ($\sim$30 kV/m) with low applied potentials of $\pm$ 2 V that may not affect the structural integrity of the biomolecules. Pt sputtering was selected at a duration of 10 min with 20 mA resulting in 60 nm average pore size, optimizing for balance between porosity and electrical conductivity. The resulting average thickness of the Pt coating was 50 nm. The membrane was fixed to a permeation cell and a potentiostat was used in the two electrode modes to apply voltage across the membrane. Previous to the publication of this study, the authors have noted successful separation of two proteins of different sized using polymeric microfluidic devices, whereas complete resolution of similarly sized proteins remained difficult. Ku and Stroeve (2004) showed the differentiation of bovine serum albumin (BSA) and hemoglobin, which are similarly sized, by modifying the surface charge of porous polycarbonate track-etched membrane (Chun & Stroeve, 2002). In comparison, Cheow et al. (2008) show a simpler method with greater control through easily available electrochemical means to selectively differentiate BSA (66 kDa), chicken egg white lysozyme (Lys, 14.4 kDa) and horse heart myoglobin (Mb, 16.7 kDa). The net charges on the proteins are -15, $+20$ and 0 at pH $= 7$ for BSA, Lys and Mb, respectively. On introduction of the protein solution (single or in mixed form) from the WE side of the membrane, expectedly, the BSA anions moved more rapidly at $+1.5$ V and the transport rate of Lys cations were enhanced at -1.5 V. The neutral Mb was least influenced by a change in polarity.

Yang, Kim, Ahn, and Kim (2009) demonstrated a sensing system comprised of NAAO membrane grown on glass slide and incorporated into a PDMS microchannel structure to detect hepatitis B virus (HBV) surface antigen (HBsAg). Briefly,

NAAO membrane with pore opening diameter ~80 nm was grown using a two-step anodization of a 3 μm using oxalic and phosphoric acid. The PDMS microchannel was developed using standard soft lithography and PDMS replica molding scheme. The resulting mold was cut, fluidic ports were punched, and subsequently it was aligned to the NAAO substrate and bonded using oxygen plasma treatment. A thin film of silica was deposited on the wall of the nanochannels as well as on the outside of the porous NAAO substrate prior to attachment with the PDMS bold to improve bonding, following which HBV monoclonal antibody was immobilized onto the NAAO using glutaraldehyde chemistry. The microfluidic system is similar to a 96 well commercial ELISA conducted in a microwell plate and analyzed using the naked eye technique for a colorimetric assay. In comparison, the present sensor has a 15 min assay time against the 130 min taken by the commercial ELISA and has much smaller footprint thus making it a portable device.

Circulating tumor cells (CTCs) are mostly observed in patients with cancers having epithelial origin, for example, those of the bladder, pancreas, and melanoma. Detecting CTCs in the blood can diagnose initial stages of cancer and help in treatment. As, CTC count in 1 mL of blood is normally within few tens to thousands, even during metastasized state of cancer, a highly sensitive and rapid diagnostic platform can be a game changer in this field. Kumeria, Kurkuri, Diener, Parkinson, and Losic (2012), have described a reflectometric interference-based label free microchip biosensor to detect CTCs using a NAAO membrane. The epithelial cell adhesion molecule (EpCAM) which is overexpressed by human carcinoma cells was selected as a biomarker or target. The device is a POC optical diagnostic system and anti-EpCAM antibody which is specific to CTCs are immobilized on the NAAO membrane. Anodization of Al foil resulted in pores of 32 $\pm$ 3 nm diameter with depth of 4 $\pm$ 0.2 μm, satisfying the optical requirements for the detection procedure. Following this 8 nm thin film of Au was deposited on the NAAO membrane utilizing a metal vapor deposition process. Then biotinylted anti-EpCAM antibody was immobilized onto the membrane using streptavidin-mediated covalent attachment to activated carboxylated thiol SAM. The microfluidic device was fabricated in two reuseable halves, from a PMMA material and sealed using a bondless microfluidic device clamp for hybrid materials. The two halves allow the integration of two channels with simple mixers in order to evenly distribute fluid delivery to a cavity which incorporated the NAAO sensing platform. There is a gap of 100 μm between the substrate and borosilicate glass lid for passage of sample fluid, forcing all cells to pass in close proximity over the top of the functionalized AAO surface. The device clamp also facilitated integration of the spectrophotometer probe to a well-defined position above the substrate. Reflectometric interference spectroscopy (RIfS) is an optical method, used for detection in this case, which is based on the interference between incident and reflected light beams derived from the upper and lower surfaces of the NAAO membrane structure (Fig. 10.4). The interference produces a characteristic Fabry-Perot fringe pattern, which are dependent on the product of refractive index and film thickness. This gives rise to an effective optical thickness which is sensitive to minute changes in the local refractive index of the NAAO membrane as a result of binding of the biomarker with the antibody on the

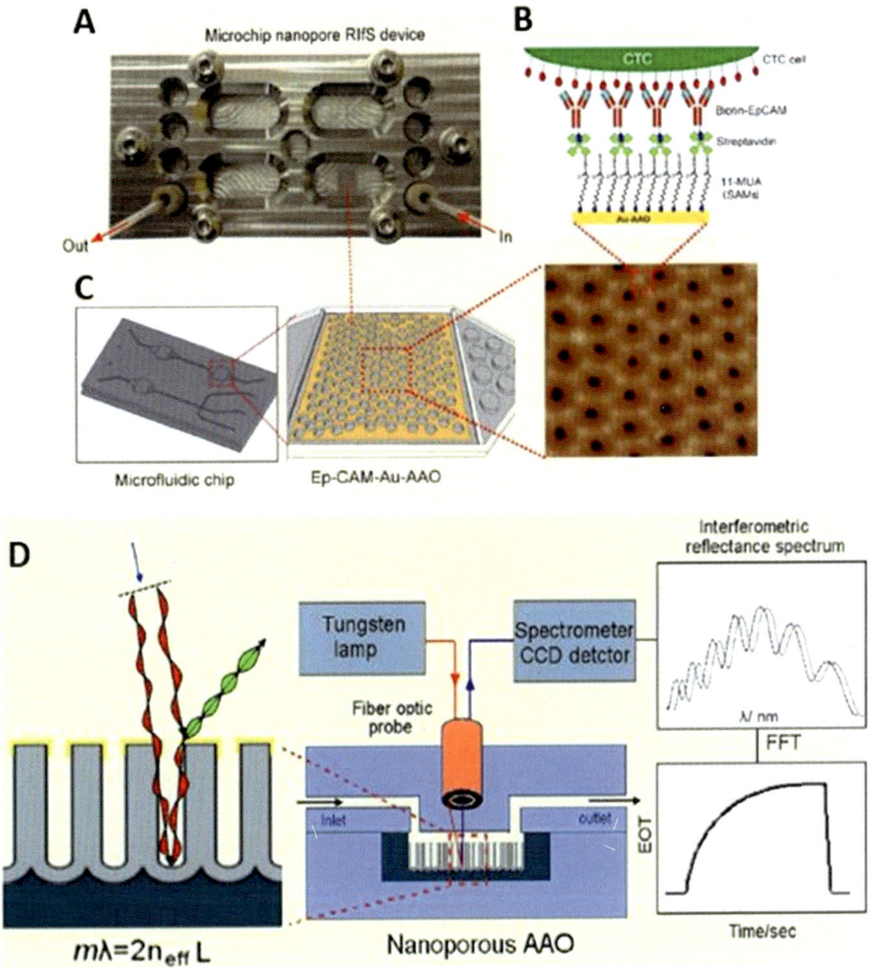

Figure 10.4 (A) Nanoporous anodized alumina (aluminum oxide) (NAAO)-based microfluidic biosensor for Circulating tumor cell (CTC) detection is shown. Inlet and outlet channels are shown. (B) Scheme of binding CTC cells by the biotinylated anti-EpCAM antibodies immobilized on the NAAO membrane through streptavidin-modified surface assembled monolayer. (C) Arrangement of the NAAO membrane inside the microfluidic channel is shown. (D) Schematic representation of the reflectometric interference spectroscopy (RIfS) detection system is shown: White light from a Tungsten lamp is used to produce Fabry-Perot fringes from the NAAO membrane which is detected by a spectrometer. The signal thus produced is analyzed using fast Fourier transform (FFT).
Source: Reprinted with permission from Kumeria, T., Kurkuri, M. D., Diener, K. R., Parkinson, L., & Losic, D. (2012). Label-free reflectometric interference microchip biosensor based on nanoporous alumina for detection of circulating tumour cells. *Biosensors and Bioelectronics*, 35(1), 167–173. Copyright by Elsevier, 2012.

surface. Whole blood or phosphate buffer saline spiked with low numbers of pancreatic cancer cells were detected using this biosensor and the authors report detection in the concentration range of 1000–100,000 cells/mL, with a detection limit of <1000 cells/mL, a response time of <5 min and sample volume of 50 μL. This rapid and low cost method avoids labor intensive fluorescence labeling and preenhancement of sample and thus may enable better diagnostics for metastatic cancer.

Tan et al. (2011) have described a PDMS microfluidic immunosensor containing specific antibody immobilized NAAO membrane aimed at rapid detection of foodborne pathogens such as *Escherichia coli* O157:H7 and *Staphylococcus aureus* with EIS. The antibodies specific to the bacteria were covalently immobilized on the membrane using (3-glycidoxypropyl)trimethoxysilane (GPMS) silane SAM. A combination of photolithography/soft lithography was utilized to create the microfluidic device. The membrane having the antibody was sandwiched between the top and bottom halves of the device and two platinum electrodes were inserted in the two halves, across the membrane, to measure the electrical response. When target bacteria are bound to the antibody on the membrane it reduces the electrolytic conductivity and thus changes the system impedance. EIS data were recorded for the frequency range of 1 Hz to 100 kHz and the maximum impedance amplitude change for the two food pathogens was determined to be occurring around 100 Hz. So, the impedance value at 100 Hz was used as the sensor response and detection could be completed within 2 h with a high sensitivity of 102 CFU/mL. Cross-bacteria experiments for *E. coli* O157:H7 and *S. aureus* were also carried out to check the specificity.

The use of live bacteria to detect potential toxins in water was demonstrated by Yagur-Kroll et al. (2015) using a microfluidic NAAO membrane device. They report a proof-of-concept study whereby recombinant bacteria were used as a live sensing element on a chip made of NAAO and integrated into a microfluidic flow cell. The porosity of the NAAO membrane enabled the retention of the microorganisms as wells as allowed diffusion of the chemical targets for biological reaction resulting in sensor response. Also, the biocompatibility and chemical inertness of the membrane allowed the use of bacteria as a sensing element. Dose-dependent demonstration of chemicals in water samples was possible by measuring the color and intensity of fluorescence. Fluorescent *E. coli* sensor strain was bioengineered to fluorescence in the presence of certain chemicals. The biosensor chip was placed between two glass substrates with the top having ports for holding the reporter elements and the bottom having microfluidic channels for sample entry and effluent exit. Selective detection was possible for the following model toxicants in water sample: nalidixic acid sodium salt (NA) which is a DNA damaging or genotoxic material, methyl viologen dichloridehydrate (MV, paraquat), cadmiumchloride ($CdCl_2$), and hydroquinone (HQ). For example, green fluorescence was observed for HQ. A cheap optical imaging system was utilized for the quantitative measurements, with results available in 90 min. The initial tests were done in a medium with neutral pH and effluent samples were treated with concentrated TGA medium with resultant pH equals to 7.0. The authors also show reusability of the chip for at least two times along with viable operation of lyophilized stored chips in cold/room temperature. Overall, the membrane-assisted sample enrichment in microfluidic systems is still in an infancy state, and therefore, extra efforts are needed to investigate new concepts that can be practically applied to design such miniaturized sensing devices.

10.3.4 Zinc oxide-based microfluidic biosensors

ZnO-based biosensors are being increasingly adopted in biosensor research over the last decade (Izyumskaya et al., 2017). ZnO demonstrates good electrical transport properties with wide band gap and high exciton-binding energy (Luqman et al., 2019; Moezzi, McDonagh, & Cortie, 2012). It is nontoxic, biocompatible, and can be manufactured/fabricated in a cost-effective way. The high isoelectric point of ZnO aids in the use of it as a transducer material in biosensors for easy and effective functionalization with DNA, antibodies and enzymes among a variety of recognition molecules. Also, the wide band gap of ZnO promotes high breakdown voltage which enables tolerance to high electric fields that makes it attractive in electrical sensors. As a result, ZnO-based biosensors have found application in sensing and detection of various chemicals such as glucose, cholesterol, lactic acid, uric acid, and cancer cells. Nanostructured ZnO may be characterized into four different configuration arrangements: (1) 0-dimensional including NPs and quantum dots (Luqman et al., 2019), (2) one-dimensional including NWs, nanotubes, and nanorods (Li et al., 2005), (3) nanoflakes, nanowalls, nanoforest, and (4) nanoflowers and agglomerated 2D ZnO (Luqman et al., 2019). In the following paragraphs some of the recent microfluidics-based ZnO biosensors are discussed.

Uranium is a very important element for nuclear applications (Bleise, Danesi, & Burkart, 2003) in the civil and military sectors and it also carries great environmental and human health risk due to its radioactivity (Bleise et al., 2003) and toxicity (Briner, 2010). Uranyl (UO_2^{2+}) ions are the most stable and prevalent form of uranium in aqueous environment and can quickly infect organs such as liver, kidney, and reproductive organs (Domingo, 2001). He, Zhou, Liu, and Wang (2020) have described the detection of UO_2^{2+} in tap/river water using microfluidic SERS sensor with femtomolar detection sensitivity and excellent selectivity. They report the use of synthetic double-stranded (ds) DNA structures also called as DNAzymes as bioreceptors in the sensor. DNAzymes have catalytic activities and their secondary structure can specifically recognize metal ions triggering the reversible DNA/RNA self-cleavage reactions (Liu et al., 2007; Xiang & Lu, 2011). The authors also inform that previously reported fluorescent, electrochemical or colorimetric detection techniques demonstrated moderate sensitivity and thus they chose surface-enhanced Raman spectroscopy (SERS) as the technique of biodetection in their work. The authors combined two separately designed microfluidic chips, one for reaction of the sample and the DNA receptor and the other chamber, consisting of three SERS substrates, was used for detection. The authors remark that most of the biosensors use ZnO nanrorods but they wanted to explore the usefulness of ZnO mesoporous nanosheet arrays (MSNs) as SERS substrate, as high surface area would help in loading of receptor—target complexes and the ordered configuration of MSNs would promote the continuity of SERS hotspots, thus improving signal-to-noise ratio. The MSNs were grown in the form of ordered nanoflower such as structures over Si columns, whereby AgNP-capped ZnO MSNs over two Si nanopillars showed maximum electromagnetic activity contributing to SERS hotspots. The sample and DNAZyme were mixed and introduced into the reaction chamber, following which the cleaved DNA, containing the Rhodamine label was hybridized with a complimentary strand, grafted onto the SERS substrate, in the detection chamber. Upon hybridization, the Rhodamine is brought closer to the surface which generates the SERS signal. The LOD in standard buffer was as low as 3.71×10^{-15} M. The described biosensor could detect UO_2^{2+} even when uranyl ion concentration was 0.005% among other 15 other toxic metal ions, which confirms the specificity. Five samples each of tap and river water, spiked with UO_2^{2+}, could detect uranyl concentration ranging from 10^{-12} to 10^{-8} M, with a relative standard deviation (RSD) of 3%−5%. This microfluidic biosensor was simple to operate, allowing high throughput and providing the capability of onsite detection. Reusability of the chip greatly reduced the assay costs and the authors determined that the biosensor could be used three times.

Gastroenteritis and diarrhea are the most common diseases caused by *Salmonella* (bacteria) infection in humans through contaminated food and water. As with most of the infectious diseases, RT-PCR and ELISA are the publicly available diagnostic techniques. In a recent publication, Huang et al. (2020) have described the use of ZnO-capped mesoporous silica nanoparticles (MSNs)-based microfluidic biosensor for the detection of *Salmonella* using a dual mode technique that relies on colorimetry as well as fluorescence, thus enhancing selectivity of the biosensor. Curcumin (CUR) is a natural hydrophobic, phenolic compound extracted from ginger family plants which is having poor water solubility and exhibits allochroic effect (color transformation) and fluorescence characteristics when it is dissolved in acidic or alkaline solution. Thus CUR could be used as a dual mode signal reporter. MSNs are promising candidate as nanocarriers due to their high loading capacity as result of large surface area, tunable pore size, biocompatibility, and physical/chemical stability. Acid decomposable ZnO NPs have been used in stimuli responsive-gated mesoporous sensing systems as their toxicity is very weak and they are easy to prepare, low cost and ultraresponsive. ZnO NPs are rapidly dissolved to Zn^{2+} at $pH < 5.5$. Huang et al. (2020), discuss the significance of mixing efficiency on the biological reactions that happen inside microfluidic circuits of biosensors and they identify Koch fractal structures (Chen & Zhang, 2018; Chen, Zhang, Wu, & Zheng, 2019) as one of the most efficient mixing architectures and use the same in their sensing platform. As shown in Fig. 10.5, the amino-modified MSNs (MSN-NH2 NPs) was initially incubated with the CUR to produce the MSN@CUR nanoparticles (MC NPs), following which it was capped by the amino group-modified ZnO NPs (ZnO-NH2 NPs) to obtain the MSN@CUR@ZnO nanoparticles (MCZ NPs). Afterward, this metal nanostructure complex was modified with the polyclonal antibodies (pAbs) that have been selected against *Salmonella Typhimurium* (*S. typhimurium*) using the small molecules, 1,2,4,5-tetrazine (Tz) and trans-cyclooctene (TCO), thus producing the pAbs-functionalized MCZ NPs (MCZP NPs) to aid in the specific

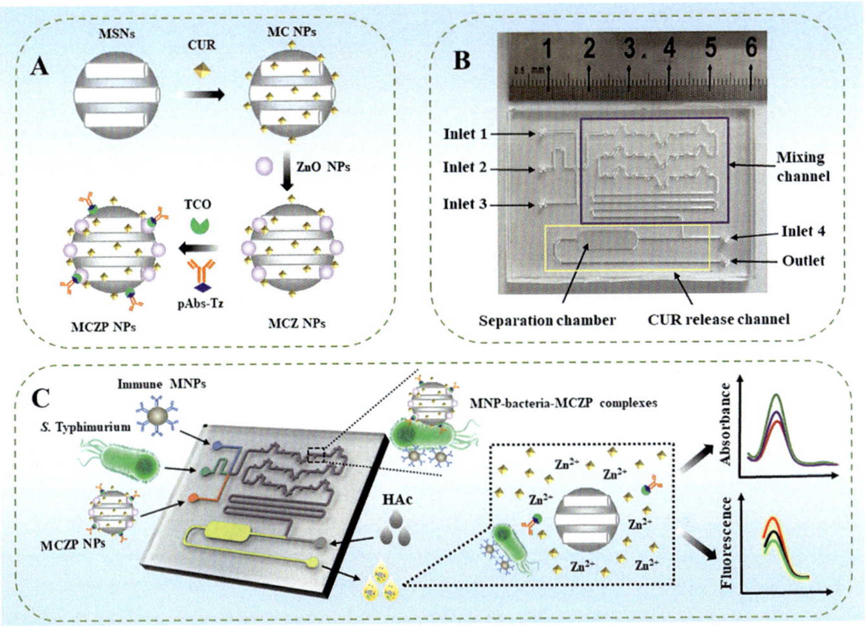

Figure 10.5 (A) Formation of antibody functionalization on ZnO nanoparticle (NP) – Curcumin-capped mesoporous silica nanoparticles (MSNs) is presented. (B) The microfluidic sensor consisting of inlet, outlet, mixing chamber with Koch fractal design and separation, release chamber is depicted. (C) Schematic representation of the microfluidic biosensor for rapid detection of Salmonella is shown (Huang et al., 2020).
Source: Huang, F., Guo, R., Xue, L., Cai, G., Wang, S., Li, Y., . . . Lin, J. (2020). An acid-responsive microfluidic Salmonella biosensor using curcumin as signal reporter and ZnO-capped mesoporous silica nanoparticles for signal amplification. *Sensors and Actuators B: Chemical*, *312*, 127958. Copyright Elsevier 2020.

recognition of *S. typhimurium*. During the detection steps, firstly, the magnetic nanoparticles (MNPs) modified with the monoclonal antibodies (mAbs) against *S. typhimurium*, the *S. typhimurium* cells and the MCZP NPs were injected into the inlet 1, inlet 2, and inlet 3 of the microfluidic chip, respectively, which were efficiently mixed by the Koch fractal micromixer microfluidic circuit. This resulted in the formation of the MNP-bacteria-MCZP sandwich complexes. The authors show that the mixing channel with three or more Koch fractal structures could efficiently mix the immune MNPs, the *S. typhimurium* cells and the MCZP NPs. The diffusion speed of the particle solution is slower than that of the water and thus the mixing channel with nine Koch fractal structures and a serpentine channel were used to mix the bacterial sample with the MNPs and the MCZP NPs. This prolonged the immune reaction time and thus ensured the complete forming of the sandwich complexes. Then sandwich complexes were captured in the separation chamber by using an external magnetic field. In the end, the acetic acid (HAc) solution was injected from the inlet to release the CUR signal reporter from the sandwich complexes. The HAc solution containing the released CUR was pushed out of the chip from the outlet port and collected in a centrifuge tube. Magnetic separation was utilized to remove the sandwich complexes and collect the released CUR. The absorbance and fluorescent intensity of the released CUR were determined using a microplate reader and an optical detector respectively for the determination of target bacteria. The authors determined that the absorbance and fluorescence values of the CUR increased as the pH of the HAc solution decreased. Higher concentrations of bacteria in the sample had more obvious color change and stronger fluorescence intensity. The LOD was calculated to be 63 CFU/mL and 40 CFU/mL for absorbance and fluorescence measurement, respectively. The authors could also successfully detect *Salmonella* in spiked chicken meat samples and negative controls such as *E. coli*, did not show any significant signal for the same concentration of bacteria.

Microfluidic paper-based analytical devices (μPADs) (Martinez et al., 2008; Martinez, Phillips, Whitesides, & Carrilho, 2010) are a promising candidate in the field of low-cost rapid POC diagnostic devices. They are being developed for use in resource limited settings by users who are not professionally skilled and thus could be a game changer for public health services in poor and developing countries (Martinez et al., 2010). Many μPADs adopt electrochemical detection technology (called as EμPAD) due to the high accuracy, sensitivity, and quantitative nature of electrochemical sensors (Dungchai, Chailapakul, & Henry, 2009; Nie, Deiss, Liu, Akbulut, & Whitesides, 2010). Metal NWs have shown great potential in different biosensors including electrochemical devices (Patolsky, Zheng, & Lieber, 2006). As discussed earlier, the most successful and commercially available electrochemical sensor is the electrochemical glucose sensor but still it is costly and not affordable by every diabetic even in the developed world and hence EμPAD can provide a viable alternative. Li, Zhao, and Liu (2015), developed a novel EμPAD with ZnO NWs directly grown on paper-based carbon ink WE, for high sensitivity glucose detection in human serum. The isoelectric point of ZnO NW is 9.5 which is higher than glucose oxidase (GO_x) enzyme thus enabling its functionalization on ZnO NW substrate for use as a receptor in the glucose sensor. To create the EμPAD, an array of circular paper channels (hydrophilic reaction zones) was formed on the paper substrate via wax patterning method. Electrodes were patterned on top of the reaction areas through stencil printing of conductive ink. The CE is formed by conductive ink on the layer with the reaction chamber. The same layer also incorporates a reference electrode (RE) made from Ag/AgCl ink on top of conductive ink. The WE is formed as a separate layer which is fabricated by covering a piece of paper with carbon ink and then growing ZnO NWs on the carbon ink by means of a low-cost hydrothermal process. GO_x is added to the WE through simple drop casting method and once the WE is dried; it is stacked along with the paper channel layer to form the complete 3-electrode EμPAD. A PDMS barrier is used to prepare thin films of hydrophobic barriers to confine the sample within the WE region. The authors confirmed that ZnO NWs can improve the performance of glucose sensing when compared to EμPADs which did not have ZnO NW on the WE. The improvement happens in two fronts: (1) the electrical potential required for the device to generate an electrical response is lowered and (2) the electrochemical current response is increased. This is due to the following reasons: (1) the electrical potential of the WE is lowered because of the fact that ZnO NWs have low redox potential which is close to that of glucose, (2) the high surface to volume ratio of the ZnO NW-modified WE enhances the electron transfer and generates a significant current response, and (3) ZnO NWs being n-type semiconductors, has excellent electron mobility and surface charge transport efficiency. Thus ZnO NW-based electrochemical sensors do not require charge transfer enhancing mediators such as $Fe(CN)_6{}^{3-/4-}$. The performance of the sensors were optimized for ZnO growth and WE area and the authors report glucose LOD of 94.7 μM in PBS and a linear range up to 15 mM in spiked human serum, which according to them are better than commercial glucometers and other published data on EμPADs.

Conventionally ELISA is utilized in early diagnosis of diseases and drug discovery wherein it is necessary to prepare 96 well samples, requiring a prolonged assay time, involving the use of expensive reagents and embedded complex procedures. However, the possibility of using nanomaterials may provide viable cost-effective alternatives and at the same time increase the fluorescence signal of ELISA-based assays. Many authors have reported the use of metal NWs, also called as "nano antennas" to show as much as 50-fold improvement in the amplification of fluorescence (Cheng & Xu, 2007; Dong et al., 2012; Rendl et al., 2011; Staiano et al., 2009). Sang, Chou, Pan, and Sheu (2016) have published a paper describing the use of various morphological ZnO structures such as sharp NWs, rod NWs, and hexahedral-puncheon nanostructures that were grown in a microfluidic channel on the same glass substrate and utilized to detect streptavidin and multiple immunoglobulin proteins. Their detection method relied on the measurement of fluorescence intensity and they showed that sharp ZnO NWs produced the best performance. ZnO NWs have quasi-1D geometry with large surface to volume ratio and hence surface-induced effects play an important role in its electrical properties. Consequently, it was shown that among the different nanostructures considered by Sang et al. (2016), ZnO NWs had the strongest fluorescence amplification with an 11-fold increment for the sensing of streptavidin. PDMS-based microfluidic channels were prepared using soft lithography on Si wafers wherein a microfluidic device had five channels, each of 100 μm width and height and having a 200 μm interchannel spacing. The microfluidic channels were attached to ZnO seed layer-coated glass substrates through compression sealing and consequently different morphologies could be chemically developed. As required, the ZnO nanostructures in the microfluidic channel could be functionalized using biotin followed by reaction with dye-conjugated streptavidin or mouse/rabbit IgG followed by reaction with dye-conjugated anti-mouse/anti-rabbit IgG. The IgG sensing could be done separately or combined and total reaction time in any scenario was about 30 min. The authors also performed fluid dynamics-based numerical simulations to ascertain the effect of ZnO morphology on the reaction rate between receptor and target to predict that sharp ZnO NWs provide the largest number of binding sites compared to rod and hexahedral-puncheon nanostructures for the same seed layer footprint. The authors could successfully detect streptavidin in the range of 417 fM to 417 pM. They also demonstrated that sharp ZnO NWs took about 10 min to achieve an intensity similar to a 30 min reaction time for the rod or hexahedral-puncheon nanostructures. For both of the anti-mouse/rabbit IgG the detection limit was determined to be 4.17 pM with a dynamic detection range of 5 orders of magnitude. It may be noted that target molecule in buffer sample was used for demonstration.

10.3.5 TiO_2-based microfluidic biosensors

The 1D TiO_2 metal oxide fiber or nanowire has certain advantages. The reduced size has a high aspect ratio and dimensions comparable to the Debye length of the solution system around TiO_2-based transducers in a biosensor promotes fast transfer of electrons resulting from molecular recognition events along the entire fiber. This is an important factor behind the choice of TiO_2 in sensors. Oxygen vacancy-induced point defects in TiO_2 are within the bridging oxygen rows of the $110-1 \times 1$ surface, resulting in the potential to be stabilized with adsorbates on the surface. This enables the robust functionalization of receptors. Also, TiO_2 nanofibers have excellent charge transfer resistance, biocompatibility, and long-term chemical stability.

Ali et al. (2016) have demonstrated a T-shaped porous graphene foam (GF) structure modified with carbon-doped $nTiO_2$ and anti-ErbB2 molecules for the detection of breast cancer. ErbB2 (also known as EGFR2 and HER2) is one of the four human epidermal growth factor receptors and is considered to be an important biomarker for noninvasive breast cancer diagnostics. However, the direct immobilization of biomolecules on the $nTiO_2$ surface via physical adsorption due to the low isoelectric points of $nTiO_2$ ($\sim$5.5, $-$ve) and anti-ErbB2 ($\sim$5.8, $-$ve) is a major limitation. The carbon available in the GF-$nTiO_2$ composite allows for covalent bonding with antibody molecules via the EDC$-$NHS coupling mechanism. The composite-based electrode integrated in the microfluidic device has a large surface area with 3D conductive pathways and provides improved detection of biomarker. Porous nature of this immunoelectrode provides necessary microscale passages for efficient handling of solutions during functionalization and detection inside a microfluidic channel.

Electrospinning technique was used to form $nTiO_2$, for which titanium isopropoxide [$Ti(OiPr)_4$] was the sol-gel precursor material. The electrospun polymeric nanofibers were heat treated at 350°C$-$400°C for 2.5 h in the presence of air. This produced residual carbon, which acts as the dopant and its amount was optimized using the calcination process. Soft lithography was used to fabricate a PDMS channel which was utilized to direct various solutions to the sensing area and further to the waste reservoir external to the device. E-beam evaporation and conventional photolithography was utilized to fabricate Au CE and Ag/AgCl RE. The 3D hierarchical porous GF structure was modified with the carbon-doped $nTiO_2$. The T-shaped WE was inserted into the PDMS channel in a way that the electrical contacts were accessible from outside. DPV and EIS were used to characterize the sensor and calibrate its response to ErbB2. The LOD from DPV and EIS were 1.0 fM and 0.1 pM, respectively in PBS solution and the sensor had negligible response to other receptors such as ErbB3 and ErbB4 which belong to the same tyrosine kinase family.

In another article published in 2012, Azahar Ali et al. (2012) have described using anatase-TiO_2 (ant-TiO_2) deposited onto ITO glass to create a microfluidic biosensor for detection of cholesterol. Ant-TiO_2 is a high purity single crystal with a considerable percentage of reactive 001 facets that may cause enhanced catalytic activity. The ant-TiO_2 when deposited onto a microelectrode surface may help in increased loading of the desired biomolecule due to its nanostructured morphology. Dip coating method was used to form the ant-TiO_2 microelectrode on the ITO glass substrate. Soft lithography enabled the formation of a microfluidic PDMS channel which is then tightly clamped to a glass substrate containing the ant-TiO_2 microelectrode. Bare ITO was used as the CE and Ag/AgCl wire was inserted from the outlet side to complete the 3-electrode electrochemical circuit. ChO_x enzyme is directly adsorbed onto the ant-TiO_2 WE facilitated by the electrostatic attraction. AFM images were analyzed to determine universal coverage. EIS was used to show that the ChO_x- ant-TiO_2-ITO microelectrode possessed improved electron transfer characteristics as depicted by decrease of more than 100% in charge transfer resistance (R_{CT}) compared to only the ant-TiO_2/ITO microelectrode. The monodispersive nature of the ant-TiO_2 enables electron mediation from active sites of the enzyme to the ITO layer and this enhances the electrochemical sensing ability. It may also be argued that the presence of strong electrostatic interactions and gibbosites on the ant-TiO_2 surface results in decreased tunneling distance between active sites of ChO_x and electrode. The biosensor was found to have an LOD of 0.14 mM in PBS and no significant response was observed for interfering agents such as glucose and ascorbic/uric acid.

Srivastava et al. (2013) demonstrated the usefulness of combining both ZrO_2 (zirconia) and TiO_2 into a composite nanostructure which transforms anatase phase of TiO_2 into rutile phase that improves the catalytic and electrochemical properties due to a modification in the electronic band structure and the interfacial state. Urease (Urs) and glutamate dehydrogenase (GLDH) were used as receptors and were coimmobilized on the nanostructured microelectrode. ZrO_2-TiO_2 binary oxide facilitates the direct electron transfer between active sites of the enzymes and the electrode and this nanocomposite was used to detect urea in an enzymatic microfluidic sensor. This binary oxide nanocomposite is formed through sol-gel chemical synthesis and is put onto the patterned ITO electrodes which are developed on a glass substrate using the dip-coating method. PDMS microchannel was fabricated using soft lithography and it was tightly clamped onto the glass substrate containing the binary oxide-modified ITO microelectrodes. The microelectrode exhibited higher oxidation peak current which could be attributed to the fact that ZrO_2-TiO_2 nanocomposite provides favorable microenvironment for enzyme to create electron mediation paths connecting the active sites. This increases the signal current by lowering the tunneling distance between active sites and transducer surface. Also, the microfluidic channel was designed to induce laminar flow which was optimized for retention time so that the biomolecular reaction between the target and the sensor surface/transducer can occur smoothly. The authors also observed that the efficiency of the biosensor was at 85% of full capacity after storing for 4 weeks at 4°C. The same sensor could also be used 12 times and the LOD of urea was determined to be 0.44 mM. The response of the sensor to interfering agents such as glucose, ascorbic acid, uric acid, and lactic acid at their usual physiological levels was found to be negligible.

A nanoporous microfluidic biochip based on chitosan grafted with dispersive nanostructured ant-TiO_2 NPs for the detection of cholesterol molecules has been demonstrated by Ali et al. (2014). Morphological and spectroscopic analysis revealed that chitosan (CH) could occupy about 57% of the volume of the nanoporous structure. The presence of defects (Ti^{3+}, Ti^{2+}) in the substrate may have a significant effect on the sensitivity of the device, by enhancing the electronic conductivity and catalytic activities. Chitosan (CH) is cost effective, biocompatible, biodegradable, and has an excellent film forming ability, whereby the porous structure of CH enhances the dispersion of TiO_2 NPs promoting sufficient electron conducting paths (useful in electrochemical sensors) and providing high surface area for the loading of biomolecular receptors which is a limitation if only TiO_2 NPs are used. The nanoporous ant-TiO_2−CH complex was deposited on the ITO microelectrode by dip-coating method and later functionalized using bienzyme complex of cholesterol esterase (ChEt) and cholesterol oxidase (ChOx). The microelectrode was designated as the WE and incorporated into a PDMS-glass microfluidic chip, having inlet and outlet, to form a 3-electrode electrochemical cell. EIS method was used to detect the presence of cholesterol in PBS within about 3 s of assay time. The LOD of 0.20 mg/dL was recorded for a linear range up to 5000 mg/dL along with a k_m of 1.3 mg/dL signifying increased catalytic activity.

Qiu et al. (2017) discussed the development of TiO_2 nanorod array which can be grown on a variety of substrates and utilized in a microfluidic biosensor for the detection of CTCs. CTCs are an important indicator of tumor metastasis during cancer and causes 90% of cancer-related deaths. CTCs are difficult to detect due to their low abundance in blood cells and in the past decade certain technologies such as immunomagnetic separation technology, flow cytometry, and microfluidics have been researched upon for the development of CTC detection. Microfluidics offers the advantage of low cost, portability, and low sample size accompanied by increased cell substrate contact but more needs to be investigated to increase the sensitivity and specificity of such devices. Previously Zhang et al. (2012) showed that electrospun TiO_2 nanofiber has good affinity for CTCs and TiO_2 nanofiber-coated substrate had improved efficiency over noncoated ones. In their paper, Qiu et al. (2017), describe the formation of vertical TiO_2 nanorod array on the surface of a microchannel with hexagonally patterned Si micropillars through a hydrothermal reaction. This resulted in the formation of a micro-nano 3D hierarchical structure within a microfluidic device. The microfluidic device was composed of a 40 mm × 20 mm Si substrate on which there was a microfluidic channel of 70 mm in length, 4 mm in width, and 115 μm in height as represented in Fig. 10.6. The TiO_2 nanorod array was formed on the Si micropillars through standard photolithographic and plasma etching process of the Si substrate, following which the microstructure was encapsulated in a PDMS cover having an inlet and outlet port. The device was functionalized by anti-EpCAM antibody through biotin-streptavidin chemistry. When sample containing CTCs is flown into the device, cells were found

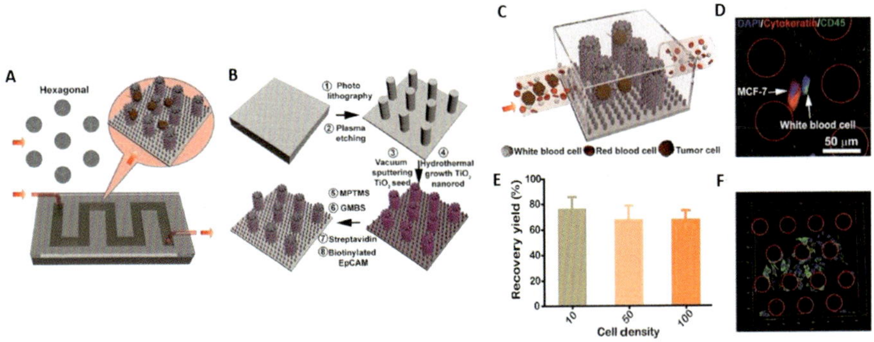

Figure 10.6 (A) The TiO_2 substrate is used as a means to capture and detect cell. A scheme of the microfluidic device for tumor cell capture is provided. (B) The scheme demonstrates the preparation of a TiO_2 nanorod array-modified microfluidic device. (C) The capture of target cells is shown schematically. (D) Immunostaining of MCF-7 cells and white blood cells for nuclei (DAPI, *blue*), for cytokeratin (*red*), and for CD45 (*green*). *Red circles* indicate Si micropillars. (E) Capture efficiency of the TiO_2 nanorod array-modified microfluidic device toward MCF-7 cells in a whole-blood sample ($n = 3$). (F) A fluorescence microscopy image of MCF-7 cells cultured for 1 week (*green*: cytoskeleton, *blue*: nuclei) captured by a TiO_2 nanorod array-modified microfluidic device from whole blood. *Red circles* indicate Si micropillars (Qiu et al., 2017).
Source: Qiu, J., Zhao, K., Li, L., Yu, X., Guo, W., Wang, S., ... Liu, H. (2017). A titanium dioxide nanorod array as a high-affinity nano-bio interface of a microfluidic device for efficient capture of circulating tumor cells. *Nano Research*, 10(3), 776–784. Copyright Tsinghua University Press and Springer-Verlag, Berlin, Heidelberg, 2016.

to adhere tightly onto the antibody modified sidewalls of the Si pillars. Confocal laser scanning microscopy (CLSM) was used to count the fluorescently labeled target cells. The 3D hierarchically structured microfluidic device demonstrated high efficiency of CTC capture to the tune of 76.7 $\pm$ 7.1% in an artificial whole blood sample.

10.3.6 Miscellaneous metal oxide biosensors

Alizadeh, Salimi, Sham, Bazylewski, and Fanchini (2020) have described an electrochemical 3-electrode PDMS microfluidic device with Au electrodes modified by CeO_2 nanosheets, exhibiting triple enzyme mimetic activity that could directly detect H_2O_2 secreted from living cells. CeO_2 NSs catalyzed the decomposition of H_2O_2 to produce OH radicals that oxidize the peroxide substrate represented by TMB, marked by solution color change from blue to yellow. Catalytic activity of the proposed nanozyme was demonstrated by monitoring the presence of OH radicals from this reaction via electron paramagnetic resonance (EPR) spectroscopy, facilitated by the use of DMPO as a spin trap for the solution. EPR results demonstrated that more DMPO-OH adducts were observed at lower pH which was also supported by increasing enzymatic activity at lower pH. CV results showed that with increasing H_2O_2 concentration, reduction current increased which is also corroborated by chronoamperometry. The sensor also works after 2 weeks and showed considerable improvement in sensitivity and detection limit over other H_2O_2 sensors.

Fungal mycotoxins such as Ochratoxin-A (OTA) are found in food products such as cereals, coffee beans, dried fruits, etc. Considering the possible harm to human health from consuming OTA contaminated food, it is imperative to detect it during the packaging and production step and that too in an efficient and scalable manner, possibly through on the spot tests which demands portability. Dhiman, Lakshmi, Roychoudhury, Jha, and Solanki (2019) demonstrated a metal oxide-based microfluidic biosensor to detect OTA in buffer samples using electrochemical method. The transducer is an ITO microelectrode modified by CeO_2/ceria nanoparticles (CeO_2 NPs) which provide the platform for the functionalization of OTA antibodies through EDC-NHS covalent crosslinking. The electrodes are also passivated using BSA to reduce nonspecific adsorption. CeO_2 NPs were synthesized through wet chemical methods and Raman spectroscopy/SEM/TEM was utilized to determine the fluorite cubic structure having 4–5 nm size. The authors note that the surface of CeO_2 NPs consist of the redox couple Ce^{3+}/Ce^{4+}, and Ce^{3+} concentration increases with smaller size of the NP. This promotes scavenging of free radicals and CeO_2 NPs show mimetic catalase properties. The ITO-modified WE is placed along with references/CEs inside the PDMS channels that had input/output ports for buffer injection using a syringe pump. Differential pulse voltammetry (DPV) was used to characterize the response of increasing concentration of OTA in buffer solution and the LOD was determined to be 350 pg/mL with a linear range up to 10 ng/mL. The total incubation time was 15 min and the response from other toxins such as Fumonisin and Aflatoxin were similar to that by the control buffer showing specificity for the sensor.

Field effect transistor (FET)-based POC device research has increased sharply in last decade with an emphasis on 1D structures such as NWs and nanotubes. Apart from SiO_2, other MOS devices based on ZnO, SnO_2, In_2O_3, and TiO_2 are emerging as possible alternatives for real-time and label-free detection of biomolecules. ZnO is the most popular choice but Jakob et al. (2017) demonstrated the use of SnO_2 in a microfluidic FET device in 2017. There are certain advantages in using SnO_2, one of them being the intrinsic n-type doping which avoids the need for intentional doping due to the presence of oxygen vacancies. In contrast SiO_2 NWs have surface segregation of dopants and the dielectric screening affects reliability. In addition, SnO_2 NWs exhibit high crystallinity with a band gap of 3.6 eV and the transparency in visible range allows the usage without the requirement of complex packaging. For biological applications involving a pH range of 3–10, the SnO_2 nanowire is chemically robust and inert facilitating its use in biosensors.

In a typical FET, a voltage is applied to the insulated metal top gate to regulate the current of the underlying semiconductor. For an n-type semiconductor, application of positive, negative gate voltage results in an attraction/repulsion of negative charge carriers within the drain-source channel facilitating an increase/decrease in current, respectively. When this concept is translated for use in a biosensor, the metal gate is replaced by an electrolytic solution/dielectric material. Typically, a microfluidic cell is normally used to establish a liquid gate. Jakob et al., used a PDMS channel for the liquid gate which could handle small volumes, in the range of a few hundreds of microliters. They utilized a flow through 3 M KCL Ag/AgCl RE to establish a well-defined reference potential to measure the drain current in a reliable and reproducible manner. It is customary to have Al_2O_3 as a dielectric material on top of the metal oxide FET so that a liquid gate can be created. The microfluidic circuit is established around this alumina layer. Normally Ti/Au pads are used to make the electrical contacts for the source and drain. A schematic of the microfluidic FET device is shown in Fig. 10.7. To detect a specific protein, the sensor surface has to be additionally functionalized with a receptor biomolecule. Depending on the protein isoelectric point as well as pH of the sample solution, proteins bear electrical charges (positive/neg). The receptor is typically placed close to the alumina layer so that the protein, when bound, is within the Debye length range. In this scenario, it can influence the source-drain current in the nanowire FET. The positive or negative protein electric field causes an attraction (accumulation) or repulsion (depletion) of negative charge carriers, respectively. Also, an FET needs to be set to its most sensitive operating regime to maximize the relative source-drain current change while the surface potential on the gate alumina layer changes due to the binding of the target biomarker with the receptor. To reduce the screening effect of salt in the sample solution, the Debye length is reduced by diluting the buffer to 0.01X PBS. In their paper, Jakob et al. performed ELISA tests to find out the optimized buffer concentration and they recommend washing microfluidic cell by 1X PBS followed by a second washing step of 0.01X PBS. They also passivated the gate-sensing surface using BSA. This demonstrated the detection of biotinylated tetracycline repressor bTetR using streptavidin functionalization of the gate surface.

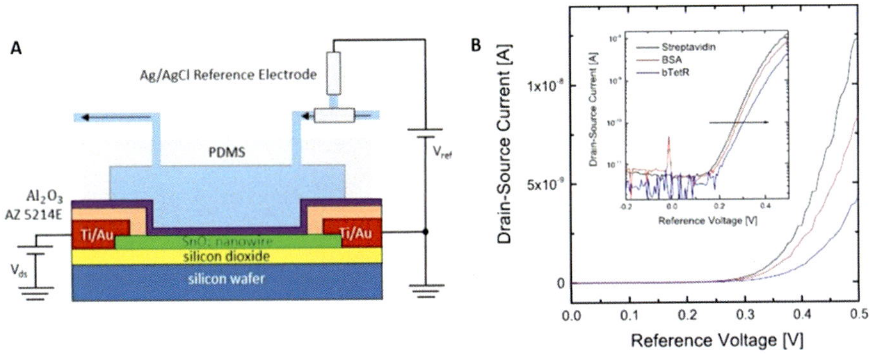

Figure 10.7 (A) Schematic side view diagram of a SnO_2 nanowire-FET with Al_2O_3 metallic top gate, used as a microfluidic biosensor is shown. (B) The shift in drain source current measured against the reference voltage is used as a signal for target binding with receptor. The bovine serum albumin (BSA) is used to passivate the streptavidin-coated device, once the streptavidin binds to the bTetR a large shift in current is observed compared to the one recorded for BSA.

10.4 Conclusion

In this chapter, we have detailed the application of a set of metal oxides with different nanostructures and geometries for microfluidic biosensing. The nanostructures of metal oxides serve as excellent biosensing transducer elements due to unique capabilities such as high surface area which can hold maximum biomolecules, high electron transfer kinetics between electrolyte and electrode, biocompatibility, high surface reactivity and feasibility of integration with microfluidic devices. We cover here some of the most applied metal oxides such as ZnO, TiO_2, Al_2O_3, etc. for biomarkers sensing toward many diseases. Overall, these metal oxides serve as excellent materials for healthcare monitoring and diagnostics. However, their applications still lie as 2D components for biosensor construction. In future, these metal oxides can be printed as 3D sensing elements into microfluidic devices due to the unique capabilities of additive manufacturing such as ability of producing complex new geometries and multimaterials combination in the same sensing platform leading to next generation of healthcare devices at industry level to solve many of the existing problems.

References

Ali, A., Ansari, A. A., Kaushik, A., Solanki, P. R., Barik, A., Pandey, M., & Malhotra, B. (2009). Nanostructured zinc oxide film for urea sensor. *Materials Letters*, *63*(28), 2473–2475.

Ali, M. A., Mondal, K., Jiao, Y., Oren, S., Xu, Z., Sharma, A., & Dong, L. (2016). Microfluidic immuno-biochip for detection of breast cancer biomarkers using hierarchical composite of porous graphene and titanium dioxide nanofibers. *ACS Applied Materials & Interfaces*, *8*(32), 20570–20582.

Ali, M. A., Mondal, K., Wang, Y., Jiang, H., Mahal, N. K., Castellano, M. J., … Dong, L. (2017). In situ integration of graphene foam–titanium nitride based bio-scaffolds and microfluidic structures for soil nutrient sensors. *Lab on a Chip*, *17*(2), 274–285.

Ali, M. A., Solanki, P. R., Patel, M. K., Dhayani, H., Agrawal, V. V., John, R., & Malhotra, B. D. (2013). A highly efficient microfluidic nano biochip based on nanostructured nickel oxide. *Nanoscale*, *5*(7), 2883–2891.

Ali, M. A., Srivastava, S., Mondal, K., Chavhan, P. M., Agrawal, V. V., John, R., … Malhotra, B. D. (2014). A surface functionalized nanoporous titania integrated microfluidic biochip. *Nanoscale*, *6*(22), 13958–13969.

Ali, M. A., Srivastava, S., Solanki, P. R., Reddy, V., Agrawal, V. V., Kim, C., … Malhotra, B. D. (2013). Highly efficient bienzyme functionalized nanocomposite-based microfluidics biosensor platform for biomedical application. *Scientific Reports*, *3*(1), 1–9.

Ali, M. A., Tabassum, S., Wang, Q., Wang, Y., Kumar, R., & Dong, L. (2018). Integrated dual-modality microfluidic sensor for biomarker detection using lithographic plasmonic crystal. *Lab on a Chip*, *18*(5), 803–817.

Alizadeh, N., Salimi, A., Sham, T.-K., Bazylewski, P., & Fanchini, G. (2020). Intrinsic enzyme-like activities of cerium oxide nanocomposite and its application for extracellular H_2O_2 detection using an electrochemical microfluidic device. *ACS Omega*, *5*(21), 11883–11894.

Ansari, A. A., Alhoshan, M., Alsalhi, M., & Aldwayyan, A. (2010). Nanostructured metal oxides based enzymatic electrochemical biosensors. *Biosensors*, 302.

Azahar Ali, M., Srivastava, S., Solanki, P. R., Varun Agrawal, V., John, R., & Malhotra, B. D. (2012). Nanostructured anatase-titanium dioxide based platform for application to microfluidics cholesterol biosensor. *Applied Physics Letters*, *101*(8), 084105.

Bartlett, M. D., Markvicka, E. J., & Majidi, C. (2016). Rapid fabrication of soft, multilayered electronics for wearable biomonitoring. *Advanced Functional Materials*, *26*(46), 8496–8504.

Bhatia, S. N., & Ingber, D. E. (2014). Microfluidic organs-on-chips. *Nature Biotechnology*, *32*(8), 760–772.

Bleise, A., Danesi, P. R., & Burkart, W. (2003). Properties, use and health effects of depleted uranium (DU): A general overview. *Journal of Environmental Radioactivity*, *64*(2), 93–112.

Briner, W. (2010). The toxicity of depleted uranium. *International Journal of Environmental Research and Public Health*, *7*(1), 303–313.

Chen, X., & Zhang, S. (2018). 3D micromixers based on Koch fractal principle. *Microsystem Technologies*, *24*(6), 2627–2636.

Chen, X., Zhang, S., Wu, Z., & Zheng, Y. (2019). A novel Koch fractal micromixer with rounding corners structure. *Microsystem Technologies*, *25*(7), 2751–2758.

Cheng, D., & Xu, Q.-H. (2007). Separation distance dependent fluorescence enhancement of fluorescein isothiocyanate by silver nanoparticles. *Chemical Communications*, *3*, 248–250.

Cheow, P.-S., Ting, E. Z. C., Tan, M. Q., & Toh, C.-S. (2008). Transport and separation of proteins across platinum-coated nanoporous alumina membranes. *Electrochimica Acta*, *53*(14), 4669–4673.

Chun, K.-Y., & Stroeve, P. (2002). Protein transport in nanoporous membranes modified with self-assembled monolayers of functionalized thiols. *Langmuir: The ACS Journal of Surfaces and Colloids*, *18*(12), 4653–4658.

Deng, J., & Toh, C.-S. (2013). Impedimetric DNA biosensor based on a nanoporous alumina membrane for the detection of the specific oligonucleotide sequence of dengue virus. *Sensors (Basel)*, *13*(6), 7774–7785.

Dhiman, T. K., Lakshmi, G., Roychoudhury, A., Jha, S. K., & Solanki, P. R. (2019). Ceria-nanoparticles-based microfluidic nanobiochip electrochemical sensor for the detection of ochratoxin-A. *ChemistrySelect*, *4*(17), 4867–4873.

Domingo, J. L. (2001). Reproductive and developmental toxicity of natural and depleted uranium: A review. *Reproductive Toxicology*, *15*(6), 603–609.

Dong, J., Qu, S., Zhang, Z., Liu, M., Liu, G., Yan, X., & Zheng, H. (2012). Surface enhanced fluorescence on three dimensional silver nanostructure substrate. *Journal of Applied Physics*, *111*(9), 093101.

Dungchai, W., Chailapakul, O., & Henry, C. S. (2009). Electrochemical detection for paper-based microfluidics. *Analytical Chemistry*, *81*(14), 5821–5826.

Gervais, L., De Rooij, N., & Delamarche, E. (2011). Microfluidic chips for point-of-care immunodiagnostics. *Advanced Materials*, *23*(24), H151–H176.

Goral, V. N., Hsieh, Y.-C., Petzold, O. N., Faris, R. A., & Yuen, P. K. (2010). Hot embossing of plastic microfluidic devices using poly (dimethylsiloxane) molds. *Journal of Micromechanics and Microengineering*, *21*(1), 017002.

Gosai, A., Hau Yeah, B. S., Nilsen-Hamilton, M., & Shrotriya, P. (2019). Label free thrombin detection in presence of high concentration of albumin using an aptamer-functionalized nanoporous membrane. *Biosensors and Bioelectronics*, *126*, 88–95.

Gowers, S. A., Curto, V. F., Seneci, C. A., Wang, C., Anastasova, S., Vadgama, P., ... Boutelle, M. G. (2015). 3D printed microfluidic device with integrated biosensors for online analysis of subcutaneous human microdialysate. *Analytical Chemistry*, *87*(15), 7763–7770.

Grieshaber, D., MacKenzie, R., Vörös, J., & Reimhult, E. (2008). Electrochemical biosensors—Sensor principles and architectures. *Sensors (Basel)*, *8*(3), 1400–1458.

He, X., Zhou, X., Liu, Y., & Wang, X. (2020). Ultrasensitive, recyclable and portable microfluidic surface-enhanced raman scattering (SERS) biosensor for uranyl ions detection. *Sensors and Actuators B: Chemical*, *311*, 127676.

Huang, F., Guo, R., Xue, L., Cai, G., Wang, S., Li, Y., ... Lin, J. (2020). An acid-responsive microfluidic Salmonella biosensor using curcumin as signal reporter and ZnO-capped mesoporous silica nanoparticles for signal amplification. *Sensors and Actuators B: Chemical*, *312*, 127958.

Izyumskaya, N., Tahira, A., Ibupoto, Z., Lewinski, N., Avrutin, V., Özgür, Ü., ... Morkoç, H. (2017). Review—Electrochemical biosensors based on ZnO nanostructures. *ECS Journal of Solid State Science and Technology*, *6*, Q84–Q100.

Jakob, M. H., Dong, B., Gutsch, S., Chatelle, C., Krishnaraja, A., Weber, W., & Zacharias, M. (2017). Label-free SnO_2 nanowire FET biosensor for protein detection. *Nanotechnology*, *28*(24), 245503.

Juang, Y.-J., Li, W.-S., & Chen, P.-S. (2017). Fabrication of microfluidic paper-based analytical devices by filtration-assisted screen printing. *Journal of the Taiwan Institute of Chemical Engineers*, *80*, 71–75.

Kim, J., Li, Z., & Park, I. (2011). Direct synthesis and integration of functional nanostructures in microfluidic devices. *Lab on a Chip*, *11*(11), 1946–1951.

Ku, J.-R., & Stroeve, P. (2004). Protein diffusion in charged nanotubes: "On – off" behavior of molecular transport. *Langmuir: The ACS Journal of Surfaces and Colloids*, *20*(5), 2030–2032.

Kumar, S., Kumar, S., Ali, M. A., Anand, P., Agrawal, V. V., John, R., ... Malhotra, B. D. (2013). Microfluidic-integrated biosensors: Prospects for point-of-care diagnostics. *Biotechnology Journal*, *8*(11), 1267–1279.

Kumeria, T., Kurkuri, M. D., Diener, K. R., Parkinson, L., & Losic, D. (2012). Label-free reflectometric interference microchip biosensor based on nanoporous alumina for detection of circulating tumour cells. *Biosensors and Bioelectronics*, *35*(1), 167–173.

Lee, U. N., Su, X., Guckenberger, D. J., Dostie, A. M., Zhang, T., Berthier, E., & Theberge, A. B. (2018). Fundamentals of rapid injection molding for microfluidic cell-based assays. *Lab on a Chip*, *18*(3), 496–504.

Li, Q., Kumar, V., Li, Y., Zhang, H., Marks, T. J., & Chang, R. P. H. (2005). Fabrication of ZnO nanorods and nanotubes in aqueous solutions. *Chemistry of Materials*, *17*(5), 1001–1006.

Li, S., Ma, Z., Cao, Z., Pan, L., & Shi, Y. (2020). Advanced wearable microfluidic sensors for healthcare monitoring. *Small (Weinheim an der Bergstrasse, Germany)*, *16*(9), 1903822.

Li, X., Zhao, C., & Liu, X. (2015). A paper-based microfluidic biosensor integrating zinc oxide nanowires for electrochemical glucose detection. *Microsystems & Nanoengineering*, *1*(1), 15014.

Li, X., Zhao, C., & Liu, X. (2015). A paper-based microfluidic biosensor integrating zinc oxide nanowires for electrochemical glucose detection. *Microsystems & Nanoengineering*, *1*(1), 1–7.

Liu, J., Brown, A. K., Meng, X., Cropek, D. M., Istok, J. D., Watson, D. B., & Lu, Y. (2007). A catalytic beacon sensor for uranium with parts-per-trillion sensitivity and millionfold selectivity. *Proceedings of the National Academy of Sciences*, *104*(7), 2056.

Luqman, M., Napi, M. L., Mohamed Sultan, S., Ismail, R., How, K., & Ahmad, M. K. (2019). Electrochemical-based biosensors on different zinc oxide nanostructures: A review. *Materials*, 12.

Malhotra, B. D., & Ali, M. A. (2018). Nanomaterials in biosensors: Fundamentals and applications. *Nanomaterials for Biosensors*, 1–74.

Malhotra, B. D., Srivastava, S., Ali, M. A., & Singh, C. (2014). Nanomaterial-based biosensors for food toxin detection. *Applied Biochemistry and Biotechnology*, *174*(3), 880–896.

Martinez, A. W., Phillips, S. T., & Whitesides, G. M. (2008). Three-dimensional microfluidic devices fabricated in layered paper and tape. *Proceedings of the National Academy of Sciences*, *105*(50), 19606–19611.

Martinez, A. W., Phillips, S. T., Whitesides, G. M., & Carrilho, E. (2010). Diagnostics for the developing world: Microfluidic paper-based analytical devices. *Analytical Chemistry*, *82*(1), 3–10.

Md, A. A. (2014). *Nanostructured metal oxide-based microfluidic biosensors for point-of-care diagnostics* (PhD thesis). IIT Hyderabad.

Moezzi, A., McDonagh, A. M., & Cortie, M. B. (2012). Zinc oxide particles: Synthesis, properties and applications. *Chemical Engineering Journal*, *185–186*, 1–22.

Mondal, K., Ali, M. A., Agrawal, V. V., Malhotra, B. D., & Sharma, A. (2014). Highly sensitive biofunctionalized mesoporous electrospun TiO_2 nanofiber based interface for biosensing. *ACS Applied Materials & Interfaces*, *6*(4), 2516–2527.

Nie, Z., Deiss, F., Liu, X., Akbulut, O., & Whitesides, G. M. (2010). Integration of paper-based microfluidic devices with commercial electrochemical readers. *Lab on a Chip*, *10* (22), 3163–3169.

Oh, Y. K., Joung, H.-A., Kim, S., & Kim, M.-G. (2013). Vertical flow immunoassay (VFA) biosensor for a rapid one-step immunoassay. *Lab on a Chip*, *13*(5), 768–772.

Patolsky, F., Zheng, G., & Lieber, C. M. (2006). Nanowire sensors for medicine and the life sciences. *Nanomedicine: Nanotechnology, Biology, and Medicine*, *1*(1), 51–65.

Qiu, J., Zhao, K., Li, L., Yu, X., Guo, W., Wang, S., ... Liu, H. (2017). A titanium dioxide nanorod array as a high-affinity nano-bio interface of a microfluidic device for efficient capture of circulating tumor cells. *Nano Research*, *10*(3), 776–784.

Rahman, M., Ahammad, A., Jin, J.-H., Ahn, S. J., & Lee, J.-J. (2010). A comprehensive review of glucose biosensors based on nanostructured metal-oxides. *Sensors (Basel)*, *10* (5), 4855–4886.

Rai, V., Deng, J., & Toh, C.-S. (2012). Electrochemical nanoporous alumina membrane-based label-free DNA biosensor for the detection of Legionella sp. *Talanta*, *98*, 112–117.

Rai, V., Hapuarachchi, H. C., Ng, L. C., Soh, S. H., Leo, Y. S., & Toh, C.-S. (2012). Ultrasensitive cDNA detection of dengue virus RNA using electrochemical nanoporous membrane-based biosensor. *PLoS One*, *7*(8), e42346.

Rendl, M., Bönisch, A., Mader, A., Schuh, K., Prucker, O., Brandstetter, T., & Rühe, J. (2011). Simple one-step process for immobilization of biomolecules on polymer substrates based on surface-attached polymer networks. *Langmuir: The ACS Journal of Surfaces and Colloids*, *27*(10), 6116–6123.

Sajid, M., Kawde, A.-N., & Daud, M. (2015). Designs, formats and applications of lateral flow assay: A literature review. *Journal of Saudi Chemical Society*, *19*(6), 689–705.

Sang, C.-H., Chou, S.-J., Pan, F. M., & Sheu, J.-T. (2016). Fluorescence enhancement and multiple protein detection in ZnO nanostructure microfluidic devices. *Biosensors and Bioelectronics*, *75*, 285–292.

Solanki, P. R., Ali, M. A., Agrawal, V. V., Srivastava, A., Kotnala, R., & Malhotra, B. (2013). Highly sensitive biofunctionalized nickel oxide nanowires for nanobiosensing applications. *RSC Advances*, *3*(36), 16060–16067.

Solanki, P. R., Ali, M., Kaushik, A., & Malhotra, B. (2013). Label-free capacitive immunosensor based on nanostructured cerium oxide. *Advanced Electrochemistry*, *1*(2), 92–97.

Solanki, P. R., Kaushik, A., Agrawal, V. V., & Malhotra, B. D. (2011). Nanostructured metal oxide-based biosensors. *NPG Asia Materials*, *3*(1), 17–24.

Srivastava, S., Ali, M. A., Solanki, P. R., Chavhan, P. M., Pandey, M. K., Mulchandani, A., ... Malhotra, B. D. (2013). Mediator-free microfluidics biosensor based on titania–zirconia nanocomposite for urea detection. *RSC Advances*, *3*(1), 228–235.

Staiano, M., Matveeva, E. G., Rossi, M., Crescenzo, R., Gryczynski, Z., Gryczynski, I., ... D'Auria, S. (2009). Nanostructured silver-based surfaces: New emergent methodologies for an easy detection of analytes. *ACS Applied Materials & Interfaces*, *1*(12), 2909–2916.

Takenaga, S., Schneider, B., Erbay, E., Biselli, M., Schnitzler, T., Schöning, M. J., & Wagner, T. (2015). Fabrication of biocompatible lab-on-chip devices for biomedical applications by means of a 3D-printing process. *Physica Status Solidi (a)*, *212*(6), 1347–1352.

Tan, F., Leung, P. H. M., Liu, Z.-b, Zhang, Y., Xiao, L., Ye, W., ... Yang, M. (2011). A PDMS microfluidic impedance immunosensor for *E. coli* O157:H7 and *Staphylococcus aureus* detection via antibody-immobilized nanoporous membrane. *Sensors and Actuators B: Chemical*, *159*(1), 328–335.

Terry, S. C., Jerman, J. H., & Angell, J. B. (1979). A gas chromatographic air analyzer fabricated on a silicon wafer. *IEEE Transactions on Electron Devices*, *26*(12), 1880–1886.

Venkatesan, B. M., Polans, J., Comer, J., Sridhar, S., Wendell, D., Aksimentiev, A., & Bashir, R. (2011). Lipid bilayer coated $Al(_2)O(_3)$ nanopore sensors: Towards a hybrid biological solid-state nanopore. *Biomedical Microdevices*, *13*(4), 671–682.

Waheed, S., Cabot, J. M., Macdonald, N. P., Lewis, T., Guijt, R. M., Paull, B., & Breadmore, M. C. (2016). 3D printed microfluidic devices: Enablers and barriers. *Lab on a Chip*, *16*(11), 1993–2013.

Walsh, E. J., Feuerborn, A., Wheeler, J. H., Tan, A. N., Durham, W. M., Foster, K. R., & Cook, P. R. (2017). Microfluidics with fluid walls. *Nature Communications*, *8*(1), 1–9.

Wang, L., Liu, Q., Hu, Z., Zhang, Y., Wu, C., Yang, M., & Wang, P. (2009). A novel electrochemical biosensor based on dynamic polymerase-extending hybridization for *E. coli* O157:H7 DNA detection. *Talanta*, *78*(3), 647–652.

Wang, Y., Ali, M. A., Chow, E. K., Dong, L., & Lu, M. (2018). An optofluidic metasurface for lateral flow-through detection of breast cancer biomarker. *Biosensors and Bioelectronics*, *107*, 224–229.

Whitesides, G. M., Ostuni, E., Takayama, S., Jiang, X., & Ingber, D. E. (2001). Soft lithography in biology and biochemistry. *Annual Review of Biomedical Engineering*, *3*(1), 335–373.

Xia, Y., & Whitesides, G. M. (1998). Soft lithography. *Annual Review of Materials Science*, *28*(1), 153–184.

Xiang, Y., & Lu, Y. (2011). Using personal glucose meters and functional DNA sensors to quantify a variety of analytical targets. *Nature Chemistry*, *3*(9), 697–703.

Yagur-Kroll, S., Schreuder, E., Ingham, C. J., Heideman, R., Rosen, R., & Belkin, S. (2015). A miniature porous aluminum oxide-based flow-cell for online water quality monitoring using bacterial sensor cells. *Biosensors and Bioelectronics*, *64*, 625–632.

Yang, K. S., Kim, H. J., Ahn, J. K., & Kim, D. H. (2009). Microfluidic chip with porous anodic alumina integrated with PDMS/glass substrate for immuno-diagnosis. *Current Applied Physics*, *9*(2, Supplement), e60–e65.

Ye, W. W., Shi, J. Y., Chan, C. Y., Zhang, Y., & Yang, M. (2014). A nanoporous membrane based impedance sensing platform for DNA sensing with gold nanoparticle amplification. *Sensors and Actuators B: Chemical*, *193*, 877–882.

Zhang, N., Deng, Y., Tai, Q., Cheng, B., Zhao, L., Shen, Q., ... Zhao, X.-Z. (2012). Electrospun TiO_2 nanofiber-based cell capture assay for detecting circulating tumor cells from colorectal and gastric cancer patients. *Advanced Materials*, *24*(20), 2756–2760.

Metal/metal oxides for electrochemical DNA biosensing

11

Ionela Cristina Nica[1], Miruna Silvia Stan[1] and Anca Dinischiotu[2]
[1]Research Institute of the University of Bucharest—ICUB, University of Buchares, Bucharest, Romania, [2]Department of Biochemistry and Molecular Biology, Faculty of Biology, University of Bucharest, Bucharest, Romania

11.1 Introduction in DNA biosensing

Biosensors are of great importance because they are capable to resolve a wide spectrum of analytical challenges in different areas such as food industry (Prasad, Ranjan, Lutfi, & Pandey, 2009), agriculture (Kundu, Krishnan, Kotnala, & Sumana, 2019), bioterrorism and homeland security (Nikoleli et al., 2016), environmental pollution control (Justino, Duarte, & Rocha-Santos, 2017), and medicine (health monitoring and clinical diagnostics) (Kim, Campbell, de Ávila, & Wang, 2019). Although the first biosensor was created in 1956 by Prof. Leland C. Clark (Clark, 1956), the foundation of this multidisciplinary "strange" field of research where biological recognition elements are associated with electrochemical sensors was built in 1962 when Clark and Lyons developed an amperometric enzyme electrode for glucose detection by adding a glucose oxidase enzyme transducer as "dialysis membrane enclosed sandwiches" to the Clark oxygen electrode (Clark & Lyons, 1962). Afterwards, Updike and Hicks designed a similar enzyme electrode for rapid and quantitative glucose measurement by covering the surface of the oxygen electrode with a polyacrylamide gel membrane containing glucose oxidase (Updike & Hicks, 1967), while Guilbault and Montalvo discovered the first potentiometric biosensor for urea detection using a glass electrode combined with immobilized urease (Guilbault & Montalvo, 1969). Finally, all this research contributed to the development of the first commercial biosensor for glucose detection produced by Yellow Spring Instruments (YSI) in 1975 (Yoo & Lee, 2010). Fig. 11.1 illustrates a brief overview of the most important steps in the history of biosensors.

Nowadays, after more than five decades of research, an electrochemical biosensor is defined as a self-contained integrated device that combines a biological recognition element (molecular receptor) capable of specific interaction with a target analyte, and a physicochemical transducer that converts this biochemical reaction into a measurable electrical signal (Rocchitta et al., 2016; Thévenot, Toth, Durst, & Wilson, 2001). Although the classification of biosensors is often done according to the signal transduction mechanisms they use, most electrochemical devices rely on the properties of the analyte of biological interest. Therefore there are two main classes of biosensors depending on the nature of the biorecognition element. The first class includes bioaffinity devices based on the highly selective binding of a

Metal Oxides for Biomedical and Biosensor Applications. DOI: https://doi.org/10.1016/B978-0-12-823033-6.00009-0

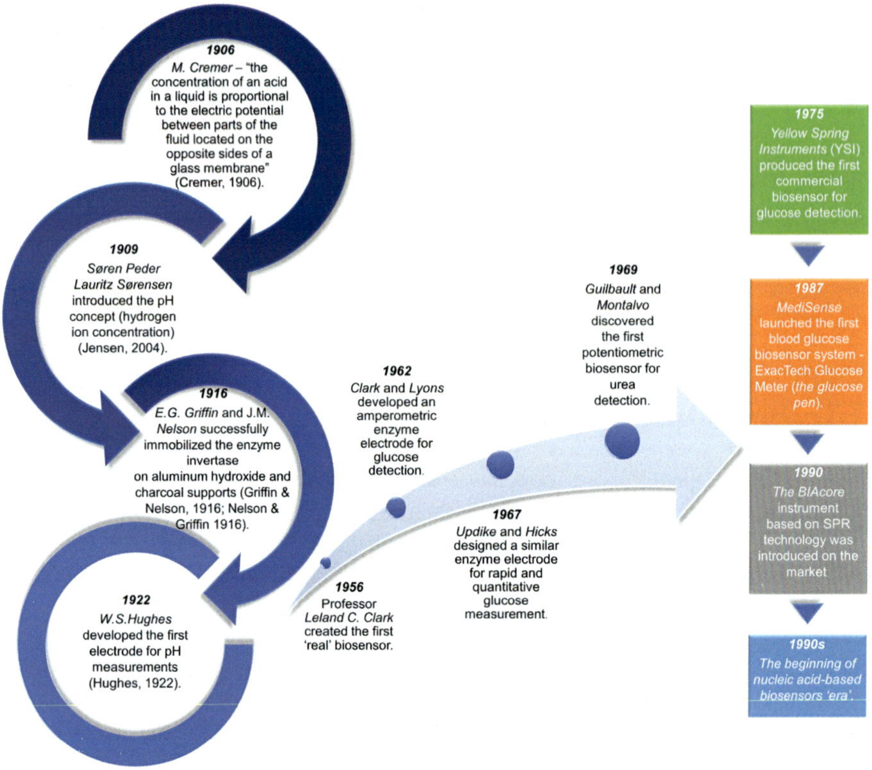

Figure 11.1 Important steps in the development of biosensors. (Cremer, 1906; Griffin & Nelson, 1916; Hughes, 1922; Jensen, 2004; Nelson & Griffin, 1916).

target analyte to a ligand trapped on the electrode surface (e.g., antibodies, oligonucleotides), while other sensors use immobilized enzymes as biocatalytic elements in order to recognize a target substrate (e.g., glucose oxidase) (Wang, 2000).

The database "Web of Science" has indexed over 56,000 reports on the topic of "biosensors" until 2020, more than 18,000 of them addressing deoxyribonucleic acid (DNA) biosensors. Nucleic acid-based biosensors began to gain interest in the early 1990s, but since then this concept has continued to evolve and expand because unlike antibodies or enzymes, DNA recognition layers can be readily synthesized and successfully regenerated for multiple use (Kavita, 2017).

A common DNA biosensor contains a DNA probe (biorecognition element), represented by a short single-stranded DNA segment (20–40 base pairs) that is able to hybridize with specific and complementary regions of the target DNA sequence, and an electrode as a signal transducer (Zhao, Xu, & Chen, 2014).

DNA biosensors are rapid and highly sensitive analysis tools developed to prevent and overcome the current limitations of DNA microarrays used for genotyping, gene expression measurements, and epigenetic markers detection (Hahn, Mergenthaler, Zimmermann, & Holzgreve, 2005). The most significant challenges for the new electrochemical DNA biosensing devices include: (1) automated detection; (2) integrated microelectronics into microchip-based DNA technologies through a process with high performance and scalability; (3) direct signal transduction, without image processing and statistical analysis required in the DNA microarray workflow (Cagnin et al., 2009).

11.2 Fundamental properties of nanostructured metal oxides used for DNA biosensing

The field of biosensors has experienced a rapid development in recent decades mainly due to the use of innovative biorecognition molecules and nanostructured materials as signal transducers and/or target recognition elements (Bhalla, Jolly, Formisano, & Estrela, 2016; Liu & Liu, 2019). With a high surface-to volume ratio and good biocompatibility, as well as enhanced electrocatalytic activity and superior electronic properties conferred by their nanoscale size, nanomaterials contributed to the design of novel ultrasensitive electrochemical DNA biosensors (Wang, 2005).

Metal oxides are one of the most interesting classes of materials with different properties including mechanical flexibility and oxidative tolerance (Platt et al., 2015), thermal conductivity and optical transparency (Said, Saidur, & Rahimb, 2014), excellent carrier mobility and electronic performance (Guo et al., 2019; Lin et al., 2015), fluorescence quenching (Liu & Liu, 2015), magnetic field (Díaz et al., 2017), heterogenous catalysis (Védrine, 2017). Due to these unique physicochemical characteristics, their low costs and high stability, nanostructured metal oxides (NMOs) are increasingly used as immobilization matrices for biosensing to the detriment of classical applications in ceramics, industrial catalysis, or optical coatings (Nunes et al., 2019; Prasanna, Balaji, Pandey, & Rana, 2019).

Some of the representative properties of NMOs that could influence their interaction with the biomolecules of interest, and consequently the efficiency of the biosensor, are described in more detail below.

11.3 Size and surface particularities

Besides the physical properties of metal oxides that make them the perfect candidates for biosensing devices, the size scale of biosensors may have a significant influence on the throughput and sensitivity of the biodetection process. It is well known that biosensors that contain nanomaterials (e.g., nanoparticles, nanotubes, nanosheets, nanowires, or nanorods) display a higher efficiency of electron transport and a more pronounced optical excitation compared to conventional devices with macroscopic sizes, even if they have the same chemical composition (Jeevanandam, Barhoum, Chan, Dufresne, & Danquah, 2018).

The high surface area of NMOs allows the loading of a larger number of biomolecules per unit mass of particles ensuring an increased detection sensitivity of DNA biosensors (Solanki, Kaushik, Agrawal, & Malhotra, 2011). Moreover, the use of nanosized materials creates the possibility to integrate more sensors into high-density arrays and lab-on-chip platforms, increasing the biodetection capacity for simple and faster in vivo analyses (Hahm, 2013). NMOs can also enable the manufacture of cost-effective, portable, and noninvasive miniaturized detection devices.

One of the most important criteria for the proper functioning of a DNA biosensor is the formation of an efficient interface between NMOs and biomolecules. This nano-biointerface depends on the nature and properties of the NMO used because a series of parameters such as surface area, porosity and roughness, functional groups and surface energy, physical states, and hygroscopicity could affect the performance of a DNA biosensor (Butterworth et al., 2019). This is the reason why the direct absorption of a biomolecule on the surface of a bulk material often leads to denaturation and loss of its biological activity, while the NMO's surface can create a biocompatible micro-environment in order to maintain its bioactivity. One possible explanation would be that the size scales of most important biological molecules are found in the nanometer range and fit well with those of nanomaterials (Malhotra, Das, & Solanki, 2012).

Therefore DNA biosensors involving NMOs represent an opportunity to obtain ultrasensitive devices with improved detection limits and extended shelf life, but without high costs (Malhotra et al., 2012).

11.4 Surface energy and electrical properties

As the redox site of a biomolecule is most often located deep inside its core, the electron transfer between this site and the electrode in the biosensor can be difficult to reach. Thus in order to remove these limitations and to achieve a functional nano-biointerface, the electrodes were successfully modified using NMOs (Katz, Willner, & Wang, 2004). Metal oxides possess two unique structural features that give them optical, magnetic, or electrical properties, namely oxygen vacancies and mixed cation valences (Liu, 2008). They also have wide bandgaps ranged between 1.22 and 6.48 eV, remarkable absorption in the ultraviolet region (at wavelengths up to $\sim 106\ cm^{-1}$), and high electron mobilities (up to $\sim 8540\ cm^2/V/s$) (Guo et al., 2019).

NMOs have different surface charges that are the result of protonation/deprotonation of surface hydroxides under different pH conditions. These hydroxide groups are formed in the process of hydration of metal oxides by water chemisorption (Blesa et al., 2000). Generally, lowering the pH value promotes the protonation of surface hydroxides, obtaining positively charged surfaces, while in basic solutions they are deprotonated to give them negative charges (Kosmulski, 2011). At physiological pH (~ 7), most NMOs possess a negative surface charge with few exceptions (ZnO, MgO, α-Fe_2O_3, NiO) (Tso et al., 2010). There is also a so-called point of zero charge (PZC) that represents the pH value (pH_{PZC}) for which the total charge on the particle surface is equal to zero (Bakatula, Richard, Neculita, & Zagury, 2018). This value is proportional to the nanocrystalline size, which means that the smaller particles have a lower pH_{PZC} (Finnegan, Zhang, Ridley, & Banfield, 2007).

Electrostatic interactions that inevitably occur between NMOs and ionized biomolecules play an essential role in biosensors (Limo et al., 2018). In addition, the metal ions on the surface of NMOs are able to coordinate with different ligands. In the early 1960s, Ralph Pearson introduced the Hard and Soft Acid and Bases principle which assumed that strong acids react faster and stronger with strong bases, while soft acids react faster and form stronger bonds with soft bases (Pearson, 1963, 1968a, 1968b). This concept widely used in chemistry can also be applied to metal oxides. Therefore it is assumed that hard metals are those that tend to form stable oxides. This category includes nickel and cobalt which form strong coordination interactions with ligands due to the low speed of the ligand exchange reactions. In contrast, soft metals such as copper, mercury, silver, and cadmium tend to form more stable compounds with sulfur and the high toxicity acquired in this way limits their applications in biosensing (Liu & Liu, 2019).

Electrical properties of NMOs are enhanced during their synthesis by introducing a trace element or "impurity" into the crystals in order to extend the wide range of applications that semiconductors can exhibit (Wisitsoraat, Tuantranont, Comini, Sberveglieri, & Wlodarski, 2009). For biomolecular sensing devices, NMOs are either *n*-doped with aluminum, indium, or gallium, or *p*-doped with nitrogen, phosphor, or arsenic (Hahm, 2013).

In the case of NMOs-based DNA biosensors, the electric charges of biomolecules can successfully function as gating components if they are connected to the surface of NMOs by covalent bindings. Among all metal oxides, ZnO, In_2O_3, and SnO_2 have the most remarkable surface sensitivity because even a small variation in charge carrier density can lead to changes in the electrical signal recorded through semiconductor oxide sensors (Zheng, Patolsky, Cui, Wang, & Lieber, 2005).

11.5 Stability and reactivity

Although NMOs are generally designed to have great stability, one of the most concerning toxicity paradigms is related to the release of toxic metal ions into biological suspensions or environments as a result of particle dissolution facilitated by several physicochemical and thermodynamic parameters (Xia et al., 2008). The dissolution rate can be accelerated under acidic conditions or reactions with ligands. In addition, CuO and ZnO are almost completely dissolved in the presence of peptides and amino acids in culture media (Odzak, Kistler, Behra, & Sigg, 2014). Also, biological thiol compounds with an important role in antioxidant defense such as reduced glutathione could dissolve MnO_2 to form free Mn^{2+} cations (Fan et al., 2015), while ethylenediaminetetraacetic acid (EDTA) can promote the dissolution of iron oxides by formation of Fe-ligand complexes (Ngwack & Sigg, 1997). Fortunately, a number of viable organic, inorganic, polymeric, or zwitterionic coatings have been designed in order to stabilize NMOs in aqueous and biological environments (Cartwright, Jackson, Morgan, Anderson, & Britt, 2020).

11.6 Biosensors based on DNA-functionalized nanostructured metal oxides

11.6.1 Configuration of DNA biosensors

According to IUPAC recommendations, an electrochemical DNA biosensor is an analytical device that integrates a single-stranded oligo- or polynucleotide (probe), representing the biochemical receptor, and an electrode as a physicochemical transducer (Labuda et al., 2010). When the biological recognition element pairs with the analyte of interest and the DNA hybrid is formed on the electrode surface, the event is translated into a measurable signal proportional to the concentration of complementary DNA target sequence (Thévenot et al., 2001).

11.6.2 Probe design

Since the recognition of the target oligonucleotide with high selectivity and sensitivity is essential for maximum efficiency of DNA hybridization reaction preventing nonspecific binding, probe design is by far the most important preanalytical step in the development of an electrochemical DNA biosensing device (Sohrabi, Valizadeh, Farkhani, & Akbarzadeh, 2016).

Single stranded DNA (ssDNA) molecules used as probes can be both natural and synthetic. Although the probes involved in DNA biosensors have different conformations (e.g., linear, hairpin) and chemical compositions; one of the essential criteria for their specificity is the number of bases they contain (Wang et al., 2016). In general, the optimal length of a probe is between 18 and 25 nucleotides; longer probes exceeding 30 bases cause a low hybridization efficiency due to the intramolecular hydrogen bonds that determine the formation of nonreactive hairpin structures (Lucarelli, Tombelli, Minunni, Marrazza, & Mascini, 2008).

The probe density is another important characteristic that controls the orientation of the recognition oligonucleotide and ensures the optimal performance of a DNA biosensing device, acting like a two-edged sword. A low value provides possible binding sites for the analyte, while an increased probe density promotes electrostatic repulsion of the target sequence and defective hybridization (Movilli, Rozzi, Ricciardi, Corradini, & Huskens, 2018).

Currently, the biggest challenge is to design a complete series of arrayed probes for the screening of closely related and unrelated pathogens, as well as their identification (Sohrabi et al., 2016).

11.6.3 Probe immobilization

After probe design and transducer selection, immobilization of the recognition oligonucleotide on the electrode surface is the next important step in the development of an electrochemical DNA biosensor. Depending on the nature of the electrode material and the anchoring process, probe immobilization strategies can be classified into three main categories: physical adsorption, covalent binding, and affinity binding (Zhao et al., 2014).

11.6.3.1 Physical adsorption

Physical adsorption is one of the simplest strategies for DNA immobilization, which does not involve any modification of the probe (Pividori, Merkoci, & Alegret, 2000). The recognition oligonucleotides are trapped on the electrode surface via different interactions (e.g., ionic, van der Waals, electrostatic) based on positive charges of the electrode surface obtained by using cationic polymer coatings and negatively charged DNA phosphate backbone (Paleček & Jelen, 2005; Qi et al., 2012; Solanki et al., 2011). This technique has the advantage of not using chemical reagents, being a simple and fast immobilization method that does not require laborious steps, but there are also several limitations, including random orientation of DNA probes on the electrode surface which can affect DNA hybridization yield, and weak attachment of the probe (Nimse, Song, Sonawane, Sayyed, & Kim, 2014). Because electrostatic forces are not as strong as covalent bonds, a number of parameters such as temperature, ionic strength, or pH value of the used buffers can alter these physical interactions leading to desorption of the DNA probes from the electrode surface (Rashid & Yusof, 2017).

11.6.3.2 Covalent binding

Covalent binding of DNA probe to NMOs-functionalized electrodes relies on the functional groups displayed on both elements. For this method, it is necessary to chemically modify the end of the recognition oligonucleotide by adding a thiol (—SH) or amino ($-NH_2$) group in order to form covalent bonds with the electrode surface also activated by inserting specific functional groups (Heise & Bier, 2006). The main advantage of this technique is related to the strong binding force between probe and immobilization matrix which ensures a high efficiency of DNA hybridization and allows easy removal of nonspecific bindings (Wang, Zhang, Lin, & Weng, 2011).

Chemisorption remains the most widely used among all methods of DNA probe covalent immobilization. The strong affinity between thiolated oligonucleotides and gold (Au) surfaces with the formation of Au—S covalent bonds, resulting in DNA monolayer assemblies that cover the electrode surface, is frequently used for the development of electrochemical DNA biosensors (Benvidi et al., 2015). In addition, Ma et al. revealed that DNA immobilization by chemisorption can be controlled by electric potential. This means that the application of a low positive potential accelerates the formation of a more compact layer of ssDNA on the electrode surface (Ma & Lennox, 2000).

Another covalent bond exploited for immobilizing probes in DNA biosensing involves the chemical interaction between amino-linked DNA and NMOs functionalized transducer surface showing oxidized groups such as —COOH (Zhang, Li, Jin, & Li, 2011). This technique is also known as carbodiimide binding. Also, based on the high affinity of phosphonic acids for TiO_2 surfaces, which leads to the condensation of phosphate groups with TiO_2 hydroxides, Tokudome et al. reported the immobilization of recognition DNA probe on TiO_2 surface through covalent interactions between phosphate backbone of nucleic acid and the $-OH$ groups of TiO_2 for electrochemical biosensing purposes (Tokudome et al., 2005).

11.6.3.3 Biochemical affinity binding

In the last decades, the streptavidin/avidin–biotin system has been widely used for enzyme-linked immunosorbent assay (ELISA), immunohistochemistry, molecular biology, and biosensing (Sakahara & Saga, 1999). When it is used for electrochemical biosensors, this strong affinity binding relies on the highly specific recognition between biotinylated probes and the streptavidin/avidin-modified electrode surface (Zhang et al., 2013). But unfortunately, this sensitive and efficient immobilization technique is rarely used in electrochemical DNA biosensing, being more suitable for DNA microarrays.

11.6.4 DNA hybridization

As described in 1953 by Watson and Crick, DNA biomolecules possess a double helix structure (Watson & Crick, 1953). Therefore the double strands should be firstly separated by thermal denaturation to allow the formation of hybrids with the ssDNA probe immobilized on the electrode surface (Mascini, Tombelli, & Palchetti, 2005).

In electrochemical DNA biosensing devices, DNA hybridization is the process in which the probe specifically recognizes and interacts with the target complementary DNA sequence from sample solution, forming a double-stranded DNA (dsDNA) hybrid (Zhao et al., 2014). The stability of the dsDNA duplex is maintained by hydrogen bonds established between complementary nucleic base pairs (adenosine/thymine and cytosine/guanine) (Sinden, 1994). Different electrochemical biosensors were developed for the detection of DNA hybridization using various signaling mechanisms in order to not only quantify the target DNA, but also for specific medical purposes such as highlighting DNA methylation and monitoring several cancer biomarkers (Rahman, Li, Lopa, Ahn, & Lee, 2015).

11.7 Metallic and semiconducting oxides used in DNA biosensing

The transducer is one of the most important elements of a DNA biosensor because at this level the immobilization of recognition biomolecules takes place and implicitly the interaction with the target analytes. Therefore the selection of appropriate materials and their quality play a special role for the analytical efficiency of a sensing device (Zhao et al., 2014). Given that genetic targets are found in very low concentrations in biological samples, an ultrasensitive DNA biosensor is needed to detect them. In the last decade, researchers have succeeded in functionalizing common transducers with various functional materials to obtain devices with a high sensitivity of nucleic acid detection (Xu, Huang, Ye, Ying, & Li, 2009). Among the most used metallic and semiconducting oxides in order to increase the performance of electrodes integrated in DNA biosensors are zirconium oxide (ZrO_2), iron oxide, titanium dioxide (TiO_2), and zinc oxide (ZnO).

11.7.1 Zirconium oxide

ZrO_2, also known as zirconia, is a chemically inert and thermostable white inorganic oxide, which has a high affinity for oxygen or phosphate-containing groups (Xu et al., 2009). Therefore these unique properties make it an ideal matrix for covalent immobilization of DNA strands in the manufacturing process of electrochemical DNA biosensors. Unfortunately, ZrO_2-based nanomaterials are susceptible to aggregation and cracking which considerably limits their applications in biosensing (Wang et al., 2015). This problem can be overcome by modifying the ZrO_2 surface with chitosan, a highly permeable biopolymer with great biocompatibility, excellent film-forming ability, and low costs (Malhotra et al., 2012).

Inspired by the specific characteristics of ZrO_2, Yang et al. developed a DNA biosensor using a composite containing ZrO_2, carbon nanotubes, and chitosan, which was able to accurately detect the complementary DNA by differential pulse voltammetry (Yang et al., 2007). Two years later, Das et al. designed a DNA biosensor for the detection of *Mycobacterium tuberculosis* by immobilizing a 21-mer ssDNA specific to this pathogen using a nano-ZrO_2 film electrodeposited on a gold-coated glass plate (Das, Sumana, Nagarajan, & Malhotra, 2010). In a similar way, Solanki et al. used a sol−gel derived nanostructured ZrO_2 to immobilize a 17-base ssDNA identified from the 16s rRNA coding region of *Escherichia coli*, thus obtaining a bioelectrode with high sensitivity and selectivity for complementary DNA hybridization (Solanki, Kaushik, Chavhan, Maheshwari, & Malhotra, 2009). Moreover, Zuo et al. created a DNA biosensing device using the high affinity of zirconia for oxygen atoms of the phosphate group in the structure of the DNA molecule. More specifically, they immobilized a ssDNA with a terminal 5′-phosphate group to a ZrO_2 surface in order to obtain a biosensor with a complementary DNA detection limit lower than 20 nM (Zuo et al., 2009).

11.7.2 Iron oxide

The three forms of iron oxides commonly found in nature are magnetite (Fe_3O_4), maghemite (γ-Fe_2O_3), and hematite (α-Fe_2O_3) (Sangaiya & Jayaprakash, 2018). Due to their good biocompatibility and superparamagnetic properties, iron oxides have a wide range of applications in many domains of life sciences, especially in agriculture, biomedicine, and environment studies (Ali et al., 2016). Besides these, iron oxides have also found applications in the development of electrochemical biosensors (Urbanova et al., 2014). Iron oxide-based nanomaterials were extensively used in order to modify bioelectrodes for the detection of several target analytes such as glucose, H_2O_2, heavy metals, nitrites and nitrates, dopamine, urea, or bisphenol A (George, Antony, & Mathew, 2018). Several achievements in the field of DNA biosensing have recently been added to these applications when Kaushik et al. managed to improve the sensitivity (0.1 fM) of a DNA biosensor designed for *M. tuberculosis* detection by immobilization of a 24-mer peptide nucleic acid (PNA) on 3-glycidoxypropyltrimethoxysilane/Fe_3O_4 nanocomposite (Kaushik, Solanki, Ansari, Malhotra, & Ahmad, 2009). Unlike DNA, PNAs contains a pseudo-peptide backbone, *N*-(2-aminoethyl) glycine. PNA–DNA complexes comply with Watson–Crick base-pairing rules but have higher thermal stability and sensitivity to mismatch (Smolina, Demidov, Soldatenkov, Chasovskikh, & Frank-Kamenetskii, 2005). Another iron oxide-based electrochemical DNA biosensor for the detection of the same pathogen was developed by Zaid et al. (2020) using a Fe_3O_4/3-mercaptopropionic acid/nanocellulose composite screen-printed carbon electrode.

11.7.3 Titanium dioxide

TiO_2 is a semiconductor material of the transition metal oxides family highly studied for its excellent photocatalytic properties (Gupta & Tripathi, 2011). Different forms of TiO_2 including nanoparticles, nanofibers, nanotubes, and nanoneedles began to be used in biosensing devices as "environment friendly electrode materials" (George et al., 2018). Thanks to their high surface areas, low toxicity, and great mechanical strength, nano-TiO_2 substrates are promising candidate materials for biomolecule immobilization matrices involved in sensors for biomedical applications. The electrochemical reactivity and behavior of TiO_2 depends equally on the surface and textural properties such as pore dimension and distribution and pore volume (Chen & Chatterjee, 2013).

A major concern about the use of TiO_2 in DNA biosensing is that it is not easily soluble, and it can be unstable on the electrode surface leading to low sensitivity and selectivity. For these reasons, TiO_2 nanostructures are often modified by loading with noble metals or enzymes with catalytic effects in order to increase the electrode performance (Luo et al., 2013).

For example, Nadzirah et al. developed a DNA biosensor based on TiO_2 nanoparticles that can detect different target concentrations in a sensitive, fast, and simple way. Briefly, they immobilized 30-mer ssDNA specific to *E. coli* O157:H7 on the surface of TiO_2 nanoparticles that were functionalized with (3-aminopropyl) triethoxysilane (Nadzirah, Azizah, Hashim, Gopinath, & Kashif, 2015; Nadzirah, Hashim, & Rusop, 2018). Moreover, another research led to the manufacture of a high-performance photoelectrochemical DNA biosensor using caffeic acid-modified TiO_2 photoelectrodes (Sakib, Pandey, Soleymani, & Zhitomirsky, 2020).

11.7.4 Zinc oxide

ZnO is a semiconductor metal oxide with low-costs synthesis, chemical stability, resistance against corrosion and oxidation, high electron mobility and isoelectric point, excellent film-forming ability, nontoxicity and catalytic efficiency (Chen,

Liu, Lin, Hsu, & Tsai, 2016). In addition to all these unique properties, ZnO-based nanomaterials are very suitable materials for highly sensitive and stable DNA sensor devices due to their good electrical conductivity (Tripathy & Kim, 2018). Wang et al. investigated the detection of sequence-specific target DNA using a gold electrode modified with ZnO nanowires and multiwalled carbon nanotubes (Wang, Li, & Zhang, 2010). Zhang et al. developed a novel DNA biosensor for acute promyelocytic leukemia detection by immobilizing a 18-mer oligonucleotides ssDNA sequence specific to promyelocytic leukemia/retinoic acid receptor alpha on a carbon ionic liquid electrode functionalized with ZnO nanoparticles (Zhang, Zheng, & Jiao, 2012). Even a DNA electrochemical sensor for hepatitis B virus detection and monitoring using a single-use graphite electrode modified with ZnO nanoparticles enriched with poly(vinylferrocenium) was designed (Yumak et al., 2011).

11.8 DNA biosensors applications

Since ssDNA oligonucleotides can bind with high specificity to complementary DNA sequences, and also to a broad range of target analytes from metal ions and small molecules to cells or microorganisms, they started to be used as novel affinity receptors in the development of biosensors for various analytical and biomedical applications. Two main classes of synthetic functional DNA are currently used for biosensing: aptamers and deoxyribose enzymes (DNAzymes) (Chambers, Arulanandam, Matta, Weis, & Valdes, 2008).

Aptamers are DNA or RNA ligands that are obtained from ssDNA libraries by systematic evolution of ligands by exponential enrichment (SELEX) in vitro (Ellington & Szostak, 1990; Tuerk & Gold, 1990). Due to the short length and self-annealing properties that allow them to fold into three-dimensional structures, it is believed that DNA aptamers would recognize their target first by conformation, and subsequently by sequence (Lim, Simpson, Kearns, & Kramer, 2005).

The use of aptamers in biosensing applications has several major advantages over antibodies that were carefully reviewed by Liu and his coworkers (Liu, Cao, & Lu, 2009). Unlike antibodies that cannot recognize small molecules such as metal ions, or nontoxic analytes with low immunogenicity, the number of possible targets of aptamers is virtually unlimited because they are synthesized in vitro. Another advantage provided by the synthesis mode is the fact that aptamers can be produced in large quantities, with lower costs, while the synthesis of antibodies is conditioned by the use of animals and cell cultures. In terms of size, aptamers are much smaller than antibodies, containing between 15 and 100 nucleotides, which allows them to be immobilized at a high density on the electrode surfaces. In addition, it is noteworthy that DNA aptamers-based biosensors have high thermal stability and are reusable; aptamers can be repeatedly denatured and renatured without losing their binding affinity, while antibodies are suitable for single-use only devices.

DNAzymes, also known as aptazymes, are aptamers with catalytic properties, which are not normally found in nature, but can be obtained in the laboratory using in vitro selection (Schlosser & Li, 2009). The first DNAzyme was created in order to cleave RNA strands (Breaker & Joyce, 1994), but since then a large number of analyte-dependent DNAzymes were incorporated as biorecognition elements in biosensors. In the clinic, they can be used alone or in combination with aptamers to diagnose and monitor several disorders such as diabetes and Alzheimer's disease, but they are also successfully used for many other biosensing applications. Aptazymes can also be denatured and renatured for many cycles without losing their catalytic properties (Morrison, Rothenbroker, & Li, 2018).

The detection of specific DNA sequences by electrochemical methods shows enormous potential for rapid diagnosis of almost any disease, drug screening, and forensic applications, but also allows the identification of metal ions, pathogenic microorganisms, and genetically modified organisms in environmental and food analysis (Kavita, 2017).

11.8.1 Biomedical applications

In recent years, there has been a growing demand for simple, sensitive, reliable, time efficient, user-friendly, and cheap analytical devices in modern healthcare systems. Biosensing technologies have responded to these needs through an interdisciplinary approach that combines biological principles with chemistry and nanotechnology (Abolhasan, Mehdizadeh, Rashidi, Aghebati-Maleki, & Yousefi, 2019; Gouvêa, 2011). DNA biosensors are considered the future of the medical diagnostics, being able to detect a wide range of genetic disorders and viruses, without requiring expensive equipment and highly qualified personnel (Kowalczyk, 2020).

Cancer is one of the most prevalent and deadly mankind diseases caused by different changes in gene sequences (e.g., specific mutation, translocation, amplification) (Metkar & Girigoswami, 2019). Therefore early diagnosis of cancer through sensitive methods is essential for patient survival and makes treatment easier and more efficient. DNA biosensing devices provide an optimal platform for DNA sequence analysis and detection of mutations and gene polymorphisms even before any symptoms of the disease appear.

The commonly known principle for detecting mutations is based on the fact that perfectly matched duplexes are more stable than those containing one or more noncomplementary base pairs. In this regard, Gao and Tansil reported the first electrochemical biosensor for ultrasensitive DNA hybridization detection (Gao & Tansil, 2005). This sensor was able to selectively detect genes in a DNA mixture and differentiate between perfectly matched and mismatched genes, but unfortunately the signal produced by the detection of a two-base mismatch could not be distinguished from the background noise. One year later, a new photoelectrochemical strategy for the detection of one or more base mismatches was developed using hairpin DNA hybridization on a TiO_2 electrode (Lu et al., 2006), while Okamoto and coworkers developed a new electrochemical single-nucleotide polymorphism typing method for gene diagnostic and pharmacogenomics (Okamoto, Kamei, & Saito, 2006).

In addition to the aforementioned applications, DNA biosensors were also successfully used for the detection of MXR7 liver cancer-related short gene and BRC1 gene which controls a person's susceptibility to developing breast cancer (Senel, Dervisevic, & Kokkokoglu, 2019; Wang, Tang, Chen, Cao, & Chen, 2019).

The biomedical applications of DNA biosensing devices we have described so far are based primarily on the DNA hybridization reaction, but there are three more categories of nucleic acid-related interactions that are responsible for their selective and specific use in clinical research.

11.8.2 Small molecules

Understanding the interaction between DNA and small molecules (e.g., drugs, chemicals) is usually of great importance in diagnosis and novel drug discovery. These small ligands can bind DNA via different noncovalent interactions such as intercalation, minor and major groove binding, or electrostatic interaction (Zhao et al., 2014).

Over the last years, increasingly more electrochemical DNA biosensors have been developed for the analysis and detection of different classes of medical drugs, some of these devices being improved to be able to detect target analytes directly from biological samples (Karadurmus, Kurbanoglu, Uslu, & Ozkan, 2016). For instance, vitamin B1 could be successfully analyzed from real serum, plasma, and urine samples in this way (Brahman, Dar, & Pitre, 2013).

Anticancer drugs are a special group because one of their mechanisms of action involves direct interaction with DNA. Commonly used chemotherapeutic agents are DNA intercalators that have the ability to insert between two stacked base pairs in dsDNA in order to stop DNA replication in rapidly dividing tumor cells (Liu & Sadler, 2011). Many DNA intercalators were intensively studied through electrochemical methods, including daunomycin, epirubicin, doxorubicin, actinomycin D, and thalidomide (Erdem, Eksin, & Kesici, 2017). Since there is a resemblance between biological and electrochemical processes, and consequently the oxidative mechanisms that take place in the human body are similar to those on the surface of the electrodes, DNA-anticancer drug interactions are further analyzed in order to elucidate the affinity of binding, the mechanism and nature of the interaction, and to design novel DNA-targeted drugs for genetic diseases.

Antibiotics are among the most widely used medicines in healthcare facilities. Because they are intensely prescribed and not applied properly, they have spread everywhere and reached the environment and food, leading to a global phenomenon of antibiotic resistance. In this context, electrochemical DNA biosensors display a huge potential for a fast, versatile, and sensitive detection of antibacterial traces in food products, water, and soil pollution by antibiotics and even antibiotic resistance (Reder-Christ & Bendas, 2011).

Although there are several variants of recognition principles that can be used in the design of biosensors, only a few of them were selected for antibacterial applications. The interaction between antibiotics and aptamers is frequently detected by cyclic or square wave voltammetry. Kim et al. reported the specific detection of oxytetracycline by immobilization of a thiolated ssDNA aptamer on a gold electrode surface via covalent interactions (Kim, Niazi, & Gu, 2009). Further, Zhang et al. developed a biosensing device to quantify the unknown concentration of tetracycline in milk using a tetracycline binding ssDNA-aptamer immobilized on a glassy carbon electrode by EDC/NHS covalent crosslinking and a cyclic voltammogram of defined tetracycline concentration (Zhang, et al., 2010). However, the most innovative approach in this field is the "bioluminescent whole-cell biosensor" for tetracycline detection introduced by Virolainen et al. For this study they used *E. coli* cells containing a luciferase operon placed under control of the tetracycline responsive elements (Virolainen, Pikkemaat, Elferink, & Karp, 2008).

11.8.3 Proteins

DNA aptamers can bind to their protein targets due to their specific 3D structures and their ability to recognize a specific epitope on the target molecule (Strehlitz, Nikolaus, & Stoltenburg, 2008). The first isolated aptamer was for thrombin detection, an important serine protease involved in the blood coagulation cascade (Bock, Griffin, Latham, Vermaas, & Toole, 1992). Recently, aptamers designed to recognize viral proteins attracted a lot of interest since they started to be used as rapid virus detection tools and for antiviral therapy. While DNA aptamers targeting the envelop protein of all four serotypes of dengue virus are able to treat the infection (Chen, Hsiao, Lee, Wu, & Cheng, 2015), in the field of biosensing NS1 protein-DNA aptamers were successfully used for medical diagnosis of Zika virus infection (Lee & Zeng, 2017). Also, different strains of influenza virus (e.g., H1N1, H3N2, H5N1, H9N2) were detected by using aptamers for the hemagglutinin proteins (Zou, Wu, Gu, Shen, & Mao, 2019). In the last two decades, plenty of protein binding DNA aptamers were obtained for the treatment of other human viruses including human immunodeficiency virus, human cytomegalovirus, hepatitis viruses, herpes simplex virus, and Ebola virus (Davydova et al., 2016). Even in the current context of the global pandemic of coronavirus disease (COVID-19), Chen et al. reported an innovative strategy for the early stage diagnosis of SARS-CoV-2 infection using DNA aptamers against nucleocapsid (N) protein of the virus (Chen et al., 2020).

In addition to deadly viruses, human health is threatened every day by a series of toxins with a protein structure. Therefore electrochemical DNA aptasensors can be a promising solution for the rapid detection of protein-based toxins such as botulinum neurotoxin produced by *Clostridium botulinum* bacterial strain or ricin toxins produced by the castor oil plant (*Ricinus communis*) (Fetter et al., 2015).

11.9 Eukaryotic cells and microorganisms

Various cancer-involved proteins such as vascular endothelial growth factor, platelet-derived growth factor, receptor tyrosine-protein kinase erbB-3, nuclear factor-kB, or prostate-specific membrane antigen, as well as certain bacterial components (e.g., proteins, lipopolysaccharides) were intensively used for DNA aptamers selection (Ireson & Kelland, 2006). Lately, due to the fact that preserving the native conformation of transmembrane proteins or cell surface receptors is essential for their functions, another approach has gained interest, namely the use of living whole eukaryotic cells to designate DNA aptamers with high specificity and selectivity known as cell-SELEX (Fang & Tan, 2010).

Cell-SELEX technique allows the identification of aptamers that specifically bind a certain cell type without knowing precisely the nature and structure of membrane proteins. Shangguan et al. used a liquid tumor model represented by acute lymphocytic leukemia cells (CCRF-CEM cell line) as target in order to develop a method of cell-based selection of DNA aptamers (Shangguan et al., 2006). A negative selection step was also required to remove DNA sequences that have affinity for common molecules normally found on the surface of leukemic cells. First, CCRF-CEM cells were incubated with the ssDNA library, then, the unbound oligonucleotides were removed by washing, while the sequences that have bound to the cell surface were eluted by heating. Next, these sequences are allowed to interact with B cells (negative control), and after this step only the oligonucleotides that remained unattached were collected, amplified by PCR and used in the next selection cycle. Normally, it takes about 20 rounds to select DNA aptamers with high selectivity for the cell of interest. These ssDNA molecules are subsequently isolated, sequenced, and analyzed because, as in the case of DNA probes obtained against proteins, the DNA aptamers selected by cell-SELEX can also be modified in order to increase their affinity and to obtain the binding motives that ensure their cell recognition function. A typical cell-based selection process is schematically illustrated in Fig. 11.2.

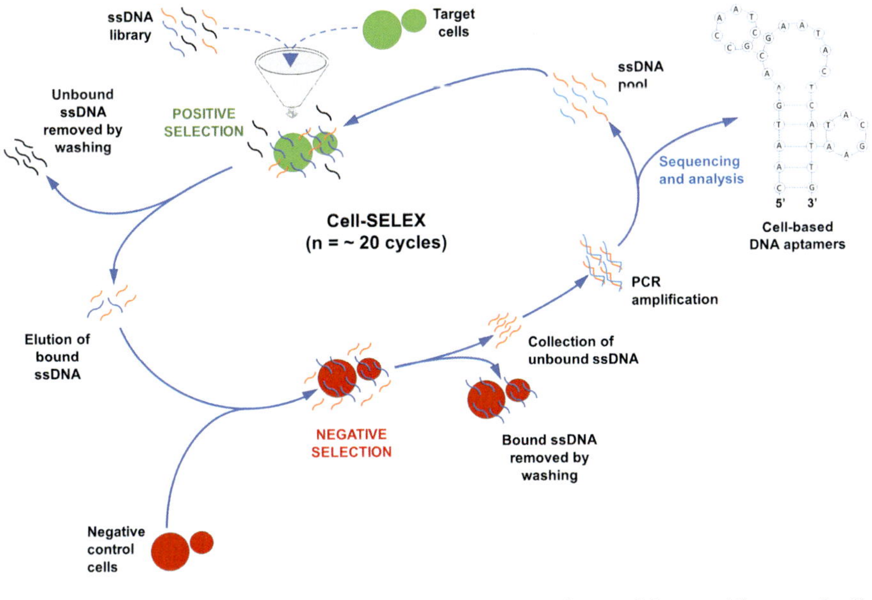

Figure 11.2 Cell-based selection of DNA aptamers capable of recognizing specific type of cells.

Given that each tumor cell is characterized by a specific molecular profile, the main goal of this technology is to obtain a set of aptamers for the diagnosis, monitoring, and treatment of a particular type of cancer. So far cancer cell-specific aptamers panels were selected against liver cancer, lymphocytic leukemia, myeloid leukemia, small and nonsmall cell lung cancer, prostate cancer, and breast cancer metastasis (Li, Zhou, & Fang, 2018; Tang et al., 2007; Wang et al., 2014).

This category of cell-specific probes also includes probes capable of recognizing virus-infected cells. For instance, Tang et al. obtained DNA aptamers against A549 human lung cancer cells infected with the vaccinia virus, showing that these aptamers are also capable to recognize other cell lines infected with the same virus, but they have no affinity to human normal cells (Tang, Parekh, Turner, Moyer, & Tan, 2009).

These DNA aptamers-based biosensors could be useful for a very fast and reliable detection of pathogens in biological samples without the need for any other processing steps if the probes are selected using whole bacterial cells. During the last decades, several DNA aptamers were selected against different dangerous pathogens that threaten human life and can be transmitted through contaminated surfaces, food or water, as well as inadequate hygiene. These include both Gram-positive (*Bacillus anthracis, Bacillus thuringiensis, Listeria monocytogenes, Streptococcus pyogenes, Staphylococcus aureus, Lactobacillus acidophilus*) and Gram-negative (*Salmonella enterica* serotype *Typhimurium* and *S. enterica* serotype *Enteritidis, E. coli, Campylobacter jejuni, Vibrio parahaemolyticus, Shigella dysenteriae, Francisella tularensis, Pseudomonas aeruginosa*) bacteria with the mention that all these aptamers can successfully be used as diagnostic tools and infection treatment agents (Davydova et al., 2016).

11.10 DNA-based metal sensing

The two strands of DNA and the double helical structure are held together by hydrogen bonds, base complementarity, $\pi-\pi$ stacking, and balance between hydrophobic and hydrophilic groups (Du et al., 2018). Both natural and synthetic nucleotides can form stable pairs in which the hydrogen bond is replaced by a coordination interaction mediated by different metal ions (Sigel & Sigel, 2010). This means that each nucleotide is considered to have a high selectivity for a particular metal ion. For example, cytosine specifically chelates Ag^+ (Ono et al., 2008), and thymine has a high affinity for Hg^{2+} (Miyake et al., 2006). But, DNA has much more valuable properties that make it a good candidate for metal detection, including the selective binding of metals due to 3D pockets that form during tertiary DNA folding. In addition, DNA oligonucleotides do not lose their metal binding affinity even after several denaturation–renaturation cycles.

Considering the low costs of DNA synthesis and its highly programmable nature (Jones, Seeman, & Mirkin, 2015), the use of oligonucleotides as metal ligands has been extensively studied. Several DNA-based biosensors were developed for the detection of alkali metal ions (Na^+, K^+, Cs^+), alkaline earth metal ions (Mg^{2+}, Ca^{2+}, Sr^{2+}, Ba^{2+}), lanthanide and actinide ions (Ln^{3+}, Gd^{3+}, $UO_2{}^{2+}$), posttransition metal ions (Pb^{2+}, Tl^+,

Al^{3+}), transition metals (Cu^{2+}, Zn^{2+}, Hg^{2+}, Cd^{2+}, Co^{2+}, Mn^{2+}, Ni^{2+}, Fe^{2+}, Cr^{3+}), and noble metals (Ag^{+}, Au^{3+}, Pt^{2+}) (Zhou, Saran, & Liu, 2017).

DNA-based metal sensing can be achieved by direct metal binding using aptamers or by metal-assisted DNA catalysis involving DNAzymes. Since it is almost impossible to immobilize the aptamers on the surface of the electrodes without covering their coordination sites, Nutiu et al. proposed a novel strategy for small molecules selection that requires a switch from DNA−DNA duplex to DNA-metal ion complex (Nutiu & Li, 2005). According to this method, if a DNA sequence binds Zn^{2+} ions, this interaction destabilizes the duplex causing the release of that sequence, which was then collected and amplified by PCR. Also, fluorophore-labeled DNA aptamers and short oligonucleotide-based quenchers were used for Zn^{2+} direct sensing (Nutiu & Li, 2003).

While direct detection of metal ions is still remaining a continuous challenge, DNAzymes offers an alternative and indirect method for metal sensing with high specificity. Unlike their natural equivalents (ribozymes), synthetic DNAzymes need metal ions for their activity. Among all the DNAzymes variants that can be obtained in the laboratory, those intended for RNA cleavage are the most suitable for metal sensing devices. The first selected RNA-cleavage DNAzyme aforementioned proved to be extremely selective for Pb^{2+} (Lan, Furuya, & Lu, 2010). Zhang et al. reported that DNAzymes have also an important advantage over aptamers, namely catalytic turnovers, which allows a metal to participate in multiple cleavage reactions (Zhang, Wang, Xing, Xiang, & Lu, 2010).

11.11 Conclusions and future perspectives

Metal oxides represent versatile materials for a fast and proper integration into the DNA biosensor technology made by the engineers. Possessing multiple unique physicochemical properties that include mechanical flexibility, oxidative tolerance, thermal conductivity, optical transparency, electronic performance, fluorescence quenching, magnetic field, and heterogenous catalysis, the DNA biosensors based on NMOs gained an amazing promising future in the detriment of classical applications in ceramics, industrial catalysis, or optical coatings. In addition to these specific characteristics, their low costs and high stability have brought MOs on the DNA biosensors market in a highly competitive way. Recent advancements indicate that DNA-based metal sensing using aptamers could be improved in order to remove the existing limitations and to obtain a successful nanobiointerface for a fast and reliable detection of pathogens in biological samples without any other processing steps. By achieving this goal, NMOs will become extremely useful in this pandemic COVID-19 era when there is a great demand of valuable coronavirus tests.

References

Abolhasan, R., Mehdizadeh, A., Rashidi, M. R., Aghebati-Maleki, L., & Yousefi, M. (2019). Application of hairpin DNA-based biosensors with various signal amplification strategies in clinical diagnosis. *Biosensors & Bioelectronics*, *129*, 164−174.

Ali, A., Zafar, H., Zia, M., Haq, I., Phull, A. R., Ali, J. S., & Hussain, A. (2016). Synthesis, characterization, applications, and challenges of iron oxide nanoparticles. *Nanotechnology, Science and Applications*, *9*, 49–67.

Bakatula, E. N., Richard, D., Neculita, C. M., & Zagury, G. J. (2018). Determination of point of zero charge of natural organic materials. *Environmental Science and Pollution Research*, *25*, 7823–7833.

Benvidi, A., Firouzabadi, A. D., Tezerjani, M. D., Moshtaghiun, S., Mazloum-Ardakani, M., & Ansarin, A. (2015). A highly sensitive and selective electrochemical DNA biosensor to diagnose breast cancer. *Journal of Electroanalytical Chemistry*, *750*, 57–64.

Bhalla, N., Jolly, P., Formisano, N., & Estrela, P. (2016). Introduction to biosensors. *Essays in Biochemistry*, *60*, 1–8.

Blesa, M. A., Weisz, A. D., Morando, P. J., Salfity, J. A., Magaz, G. E., & Regazzoni, A. E. (2000). The interaction of metal oxide surfaces with complexing agents dissolved in water. *Coordination Chemistry Reviews*, *196*, 31–63.

Bock, L. C., Griffin, L. C., Latham, J. A., Vermaas, E. H., & Toole, J. J. (1992). Selection of single-stranded DNA molecules that bind and inhibit human thrombin. *Nature*, *355*, 564–566.

Brahman, P. K., Dar, R. A., & Pitre, K. S. (2013). DNA-functionalized electrochemical biosensor for detection of vitamin B1 using electrochemically treated multiwalled carbon nanotube paste electrode by voltammetric methods. *Sensors and Actuators, B: Chemical*, *177*, 807–812.

Breaker, R. R., & Joyce, G. F. (1994). A DNA enzyme that cleaves RNA. *Chemistry & Biology*, *1*, 223–229.

Butterworth, A., Blues, E., Williamson, P., Cardona, M., Gray, L., & Corrigan, D. K. (2019). SAM composition and electrode roughness affect performance of a DNA biosensor for antibiotic resistance. *Biosensors*, *9*, 22.

Cagnin, S., Caraballo, M., Guiducci, C., Martini, P., Ross, M., Santaana, M., … … Lanfranchi, G. (2009). *Overview of electrochemical DNA biosensors: New approaches to detect the expression of life*, . *Sensors (Basel)* (9, pp. 3122–3148). .

Cartwright, A., Jackson, K., Morgan, C., Anderson, A., & Britt, D. W. (2020). A review of metal and metal-oxide nanoparticle coating technologies to inhibit agglomeration and increase bioactivity for agricultural applications. *Agronomy*, *10*, 1018.

Chambers, J. P., Arulanandam, B. P., Matta, L. L., Weis, A., & Valdes, J. J. (2008). Biosensor recognition elements. *Current Issues in Molecular Biology*, *10*, 1–12.

Chen, A., & Chatterjee, S. (2013). Nanomaterials based electrochemical sensors for biomedical applications. *Chemical Society Reviews*, *42*, 5425.

Chen, C. Y., Liu, Y. R., Lin, S. S., Hsu, L. J., & Tsai, S. L. (2016). Role of annealing temperature on the formation of aligned zinc oxide nanorod arrays for efficient photocatalysts and photodetectors. *Science of Advanced Matererials*, *8*, 2197–2203.

Chen, H. L., Hsiao, W. H., Lee, H. C., Wu, S. C., & Cheng, J. W. (2015). Selection and characterization of DNA aptamers targeting all four serotypes of dengue viruses. *PLoS One*, *10*, e0131240.

Chen, Z., Wu, Q., Chen, J., Ni, X., & Dai, J. (2020). A DNA aptamer based method for detection of SARS-CoV-2 nucleocapsid protein. *Virologica Sinica*, *35*, 351–354.

Clark, L. C. (1956). Monitor and control of blood and tissue oxygenation. *Transactions - American Society for Artificial Internal Organs*, *2*, 41–48.

Clark, L. C., & Lyons, C. (1962). Electrode systems for continuous monitoring cardiovascular surgery. *Annals of the New York Academy of Sciences*, *102*, 29–45.

Cremer, M. (1906). Über die ursache der elektromotorischen eigenschaften der gewebe, zugleich ein beitrag zur lehre von den polyphasischen elektrolytketten. *Zeitschrift fur Biologie*, *47*, 562–608.

Das, M., Sumana, G., Nagarajan, R., & Malhotra, B. D. (2010). Zirconia based nucleic acid sensor for *Mycobacterium tuberculosis* detection. *Applied Physics Letters*, *96*, 133703.

Davydova, A., Vorobjeva, M., Pyshnyi, D., Altman, S., Vlassov, V., & Venyaminova, A. (2016). Aptamers against pathogenic microorganisms. *Critical Reviews in Microbiology*, *42*, 847–865.

Du, Z. H., Li, X. Y., Tian, J. J., Zhang, Y. Z., Tian, H. T., & Xu, W. T. (2018). Progress on detection of metals ions by functional nucleic acids biosensor. *Chinese Journal of Analytical Chemistry*, *46*, 995–1004.

Díaz, C., Valenzuela, M. L., Laguna-Bercero, M. A., Orera, A., Bobadilla, D., Abarca, S., & Peña, O. (2017). Synthesis and magnetic properties of nanostructured metallic Co, Mn and Ni oxide materials obtained from solid-state metal–macromolecular complex precursors. *RSC Advances*, *7*, 27729.

Ellington, A. D., & Szostak, J. K. (1990). In vitro selection of RNA molecules that bind specific ligands. *Nature*, *346*, 818–822.

Erdem, A., Eksin, E., & Kesici, E. (2017). *Biosensors and nanotechnology: Applications in health care diagnostics: Biosensors for detection of anticancer drug–DNA interactions*. Wiley Online Library.

Fan, H., Zhao, Z., Yan, G., Zhang, X., Yang, C., Meng, H., ... Tan, W. (2015). A smart DNAzyme–MnO_2 nanosystem for efficient gene silencing. *Angewandte Chemie International Edition*, *54*, 4801–4805.

Fang, X., & Tan, W. (2010). Aptamers generated from Cell-SELEX for molecular medicine: A chemical biology approach. *Accounts of Chemical Research*, *43*, 48–57.

Fetter, L., Richards, J., Daniel, J., Roon, L., Rowland, T. J., & Bonham, A. J. (2015). Electrochemical aptamer scaffold biosensors for detection of botulism and ricin toxins. *Chemical Communications*, *51*, 15137–15140.

Finnegan, M. P., Zhang, H. Z., Ridley, M. K., & Banfield, J. F. (2007). Phase stability and transformation in titania nanoparticles in aqueous solutions dominated by surface energy. *Journal of Physical Chemistry C*, *111*, 1962–1968.

Gao, Z. Q., & Tansil, N. C. (2005). An ultrasensitive photoelectrochemical nucleic acid biosensor. *Nucleic Acids Research*, *33*, e123.

George, J. M., Antony, A., & Mathew, B. (2018). Metal oxide nanoparticles in electrochemical sensing and biosensing: A review. *Microchimica Acta*, *185*, 358.

Gouvêa, C. M. C. P. (2011). *Biosensors for health, environment and biosecurity*. IntechOpen.

Guilbault, G. G., & Montalvo, J. (1969). A urea specific enzyme electrode. *JACS*, *91*, 2164–2169.

Guo, Y., Ma, L., Mao, K., Ju, M., Bai, Y., Zhao, J., & Zeng, X. C. (2019). Eighteen functional monolayer metal oxides: wide bandgap semiconductors with superior oxidation resistance and ultrahigh carrier mobility. *Nanoscale Horizons*, *4*, 592–600.

Gupta, S. M., & Tripathi, M. (2011). A review of TiO_2 nanoparticles. *Chinese Science Bulletin*, *56*, 1639–1657.

Griffin, E. G., & Nelson, J. M. (1916). The influence of certain substances on the activity of invertase. *Journal of the American Chemical Society*, *38*, 722–730.

Hahm, J. I. (2013). Biomedical detection via macro- and nano-sensors fabricated with metallic and semiconducting oxides. *Journal of Biomedical Nanotechnology*, *9*, 1–25.

Hahn, S., Mergenthaler, S., Zimmermann, B., & Holzgreve, W. (2005). Nucleic acid based biosensors: The desires of the user. *Bioelectrochemistry (Amsterdam, Netherlands)*, *67*, 151–154.

Heise, C., & Bier, F. F. (2006). Immobilization of DNA on microarrays. *Topics in Current Chemistry*, *261*, 1–25.

Hughes, W. S. (1922). The potential difference between glass and electrolytes in contact with the glass. *Journal of the American Chemical Society*, *44*, 2860–2867.

Ireson, C. R., & Kelland, L. R. (2006). Discovery and development of anticancer aptamers. *Molecular Cancer Therapeutics*, *5*, 2957–2962.

Jeevanandam, J., Barhoum, A., Chan, Y. S., Dufresne, A., & Danquah, M. K. (2018). Review on nanoparticles and nanostructured materials: History, sources, toxicity and regulations. *Beilstein Journal of Nanotechnology*, *9*, 1050–1074.

Jensen, W. B. (2004). The symbol for pH. *Journal of Chemical Education*, *81*, 21.

Jones, M. R., Seeman, N. C., & Mirkin, C. A. (2015). Programmable materials and the nature of the DNA bond. *Science (New York, N.Y.)*, *347*, 1260901.

Justino, C. I. L., Duarte, A. C., & Rocha-Santos, T. A. P. (2017). Recent progress in biosensors for environmental monitoring: A review. *Sensors (Basel)*, *17*, 2918.

Karadurmus, L., Kurbanoglu, S., Uslu, B., & Ozkan, S. A. (2016). Electrochemical DNA biosensors in drug analysis. *Current Pharmaceutical Analysis*, *12*, 1–14.

Katz, E., Willner, I., & Wang, J. (2004). Electroanalytical and bioelectroanalytical systems based on metal and semiconductor nanoparticles. *Electroanalysis*, *16*, 19.

Kaushik, A., Solanki, P. R., Ansari, A. A., Malhotra, B. D., & Ahmad, S. (2009). Iron oxide-chitosan hybrid nanobiocomposite based nucleic acid sensor for pyrethroid detection. *Biochemical Engineering Journal*, *46*, 132–140.

Kavita, V. (2017). DNA biosensors—A review. *Journal of Bioengineering and Biomedical Sciences*, *7*, 222.

Kim, J., Campbell, A. S., de Ávila, B. E., & Wang, J. (2019). Wearable biosensors for healthcare monitoring. *Nature Biotechnology*, *37*, 389–406.

Kim, Y. S., Niazi, J. H., & Gu, M. B. (2009). Specific detection of oxytetracycline using DNA aptamer-immobilized interdigitated array electrode chip. *Analytica Chimica Acta*, *634*, 250–254.

Kosmulski, M. (2011). The pH-dependent surface charging and points of zero charge: V. Update. *Journal of Colloid and Interface Science*, *353*, 1–15.

Kowalczyk, A. (2020). Trends and perspectives in DNA biosensors as diagnostic devices. *Current Opinion in Electrochemistry*, *23*, 36–41.

Kundu, M., Krishnan, P., Kotnala, R. K., & Sumana, G. (2019). Recent developments in biosensors to combat agricultural challenges and their future prospects. *Trends in Food Science and Technology*, *88*, 157–178.

Labuda, J., Oliveira Brett, A. M., Evtugyn, G., Fojta, M., Mascini, M., Ozsoz, M., … … Wang, J. (2010). Electrochemical nucleic acid-based biosensors: Concepts, terms, and methodology (IUPAC Technical Report). *Pure and Applied Chemistry. Chimie Pure et Appliquee*, *82*, 1161–1187.

Lan, T., Furuya, K., & Lu, Y. (2010). A highly selective lead sensor based on a classic lead DNAzyme. *Chemical Communications*, *46*, 3896–3898.

Lee, K. H., & Zeng, H. (2017). Aptamer-based ELISA assay for highly specific and sensitive detection of Zika NS1 protein. *Analytical Chemistry*, *89*, 12743–12748.

Li, W. M., Zhou, L. L., & Fang, J. (2018). Selection of metastatic breast cancer cell-specific aptamers for the capture of CTCs with a metastatic phenotype by Cell-SELEX. *Molecular Therapy – Nucleic Acids*, *12*, 707–717.

Lim, D. V., Simpson, J. M., Kearns, E. A., & Kramer, M. F. (2005). Current and developing technologies for monitoring agents of bioterrorism and biowarfare. *Clinical Microbiology Reviews*, *18*, 583–607.

Limo, M. J., Sola-Rabada, A., Boix, E., Thota, V., Westcott, Z. C., Puddu, V., & Perry, C. C. (2018). Interactions between metal oxides and biomolecules: From fundamental understanding to applications. *Chemical Reviews*, *118*, 11118–11193.

Lin, Y. H., Faber, H., Labram, J. G., Stratakis, E., Sygellou, L., Kymakis, E., ... Anthopoulos, T. D. (2015). High electron mobility thin-film transistors based on solution-processed semiconducting metal oxide heterojunctions and quasi-superlattices. *Advancement of Science*, *2*, 1500058.

Liu, A. (2008). Towards development of chemosensors and biosensors with metal-oxide-based nanowires or nanotubes. *Biosensors & Bioelectronics*, *24*, 167–177.

Liu, B., & Liu, J. (2015). Comprehensive screen of metal oxide nanoparticles for DNA adsorption, fluorescence quenching, and anion discrimination. *ACS Applied Materials & Interfaces*, *7*, 24833–24838.

Liu, B., & Liu, J. (2019). Sensors and biosensors based on metal oxide nanomaterials. *Trends in Analytical Chemistry*, *121*, 115690.

Liu, H. K., & Sadler, P. (2011). Metal complexes as DNA intercalators. *Journal of Accounts of Chemical Research*, *44*, 349–359.

Liu, J., Cao, Z., & Lu, Y. (2009). Functional nucleic acid sensors. *Chemical Reviews*, *109*, 1948–1998.

Lu, W., Wang, G., Jin, Y., Yao, X., Hu, J. Q., & Li, J. H. (2006). Label-free photoelectrochemical strategy for hairpin DNA hybridization detection on titanium dioxide electrode. *Applied Physics Letters*, *89*, 263902.

Lucarelli, F., Tombelli, S., Minunni, M., Marrazza, G., & Mascini, M. (2008). Electrochemical and piezoelectric DNA biosensors for hybridisation detection. *Analytica Chimica Acta*, *609*, 139–159.

Luo, Z., Ma, X., Yang, D., Yuwen, L., Zhu, X., Weng, L., & Wang, L. (2013). Synthesis of highly dispersed titanium dioxide nanoclusters on reduced graphene oxide for increased glucose sensing. *Carbon*, *7*, 470–476.

Ma, F. Y., & Lennox, R. B. (2000). Potential-assisted deposition of alkanethiols on Au: Controlled preparation of single- and mixed-component SAMs. *Langmuir: The ACS Journal of Surfaces and Colloids*, *16*, 6188.

Malhotra, B. D., Das, M., & Solanki, P. R. (2012). Opportunities in nano structured metal oxides based biosensors. *Journal of Physics: Conference Series*, *358*, 012007.

Mascini, M., Tombelli, S., & Palchetti, I. (2005). New trends in nucleic acid based biosensor. *Bioelectrochemistry (Amsterdam, Netherlands)*, *67*, 131–133.

Metkar, S. K., & Girigoswami, K. (2019). Diagnostic biosensors in medicine – A review. *Biocatalysis and Agricultural Biotechnology*, *17*, 271–283.

Miyake, Y., Togashi, H., Tashiro, M., Yamaguchi, H., Oda, S., Kudo, M., ... Ono, A. (2006). MercuryII-mediated formation of thymine-HgII thymine base pairs in DNA duplexes. *Journal of the American Chemical Society*, *128*, 2172–2173.

Morrison, D., Rothenbroker, M., & Li, Y. (2018). DNAzymes: Selected for applications. *Small Methods*, *2*, 1700319.

Movilli, J., Rozzi, A., Ricciardi, R., Corradini, R., & Huskens, J. (2018). Control of probe density at DNA biosensor surfaces using poly(L-lysine) with appended reactive groups. *Bioconjugate Chemistry*, *29*, 4110–4118.

Nadzirah, S., Azizah, N., Hashim, U., Gopinath, S. C. B., & Kashif, M. (2015). Titanium dioxide nanoparticle-based interdigitated electrodes: A novel current to voltage DNA biosensor recognizes *E. coli* O157:H7. *PLoS One*, *10*, e0139766.

Nadzirah, S., Hashim, U., & Rusop, M. (2018). Development of DNA biosensor based on TiO_2 nanoparticles. *AIP Conference Proceedings*, *1963*, 020062.

Nelson, J. M., & Griffin, E. G. (1916). Adsorption of invertase. *Journal of the American Chemical Society*, *38*, 1109–1115.

Ngwack, B., & Sigg, L. (1997). Dissolution of Fe(III) (hydr) oxides by metal-EDTA complexes. *Geochimica et Cosmochimica Acta*, *61*, 951–963.

Nikoleli, G. P., Karapetis, S., Bratakou, S., Nikolelis, D. P., Tzamtzis, N., Psychoyios, V. N., & Psaroudakis, N. (2016). *Biosensors for security and bioterrorism applications: Biosensors for security and bioterrorism: Definitions, history, types of agents, new trends and applications*. Berlin, Heidelberg: Springer.

Nimse, S. B., Song, K., Sonawane, M. D., Sayyed, D. R., & Kim, T. (2014). Immobilization techniques for microarray: Challenges and applications. *Sensors*, *14*, 22208–22229.

Nunes, D., Pimentel, A., Santos, L., Barquinha, P., Pereira, L., Fortunato, E., & Martins, R. (2019). *Metal oxides nanostructures synthesis, properties and applications: Structural, optical, and electronic properties of metal oxide nanostructures*. Elsevier Science.

Nutiu, R., & Li, Y. (2003). Structure-switching signaling aptamers. *Journal of the American Chemical Society*, *125*, 4771–4778.

Nutiu, R., & Li, Y. (2005). In vitro selection of structure-switching signaling aptamers. *Angewandte Chemie International Edition*, *44*, 1061–1065.

Odzak, N., Kistler, D., Behra, R., & Sigg, L. (2014). Dissolution of metal and metal oxide nanoparticles in aqueous media. *Environmental Pollution (Barking, Essex: 1987)*, *191*, 132–138.

Okamoto, A., Kamei, T., & Saito, I. (2006). DNA hole transport on an electrode: Application to effective photoelectrochemical SNP typing. *Journal of the American Chemical Society*, *128*, 658–662.

Ono, A., Cao, S., Togashi, H., Tashiro, M., Fujimoto, T., Machinami, T., … … Tanaka, Y. (2008). Specific interactions between silver(I) ions and cytosine-cytosine pairs in DNA duplexes. *Chemical Communications*, *2008*, 4825–4827.

Paleček, E., & Jelen, F. (2005). *Electrochemistry of nucleic acids and proteins. Towards electrochemical sensors for genomics and proteomics*. Amsterdam: Elsevier.

Pearson, R. G. (1968a). Hard and soft acids and bases, HSAB, part 1: Fundamental principles. *Journal of Chemical Education*, *45*, 581–586.

Pearson, R. G. (1968b). Hard and soft acids and bases, HSAB, part II: Underlying theories. *Journal of Chemical Education*, *45*, 643–648.

Pearson, R. G. (1963). Hard and soft acids and bases. *Journal of the American Chemical Society*, *85*, 3533–3539.

Pividori, M., Merkoci, A., & Alegret, S. (2000). Electrochemical genosensor design: Immobilization of oligonucleotides onto transducer surfaces and detection methods. *Biosensors & Bioelectronics*, *15*, 291–303.

Platt, P., Polatidis, E., Frankel, P., Klaus, M., Gass, M., Howells, R., & Preuss, M. (2015). A study into stress relaxation in oxides formed on zirconium alloys. *Journal of Nuclear Materials*, *456*, 415–425.

Prasad, K., Ranjan, R. K., Lutfi, Z., & Pandey, H. (2009). Biosensors: Applications and overview in industrial automation. *International Journal on Applied Bioengineering*, *3*, 66–70.

Prasanna, S. R. V. S., Balaji, K., Pandey, S., & Rana, S. (2019). *Nanomaterials and polymer nanocomposites raw materials to applications: Metal oxide based nanomaterials and their polymer nanocomposites*. Elsevier Science.

Qi, X., Gao, H., Zhang, Y., Wang, X., Chen, Y., & Sun, W. (2012). Electrochemical DNA biosensor with chitosan-Co_3O_4 nanorod-graphene composite for the sensitive detection of *Staphylococcus aureus* nuc gene sequence. *Bioelectrochemistry (Amsterdam, Netherlands)*, *88*, 42–47.

Rahman, M. M., Li, X. B., Lopa, N. S., Ahn, S. J., & Lee, J. J. (2015). Electrochemical DNA hybridization sensors based on conducting polymers. *Sensors*, *15*, 3801–3829.

Rashid, J. I. A., & Yusof, N. A. (2017). The strategies of DNA immobilization and hybridization detection mechanism in the construction of electrochemical DNA sensor: A review. *Sensing and Bio-Sensing Research*, *16*, 19–31.

Reder-Christ, K., & Bendas, G. (2011). Biosensor applications in the field of antibiotic research – A review of recent developments. *Sensors*, *11*, 9450–9466.

Rocchitta, G., Spanu, A., Babudieri, S., Latte, G., Madeddu, G., Galleri, G., ... Serra, P. A. (2016). *Enzyme biosensors for biomedical applications: Strategies for safeguarding analytical performances in biological fluids*, . *Sensors* (16, p. 780). .

Said, Z., Saidur, R., & Rahimb, N. A. (2014). Optical properties of metal oxides based nanofluids. *International Communications in Heat and Mass Transfer*, *59*, 46–54.

Sakahara, H., & Saga, T. (1999). Avidin–biotin system for delivery of diagnostic agents. *Advanced Drug Delivery Reviews*, *37*, 89–101.

Sakib, S., Pandey, R., Soleymani, L., & Zhitomirsky, I. (2020). Surface modification of TiO_2 for photoelectrochemical DNA biosensors. *Medical Devices & Sensors*, *3*, e10066.

Sangaiya, P., & Jayaprakash, R. (2018). A review on iron oxide nanoparticles and their biomedical applications. *Journal of Superconductivity and Novel Magnetism*, *31*, 3397–3413.

Schlosser, K., & Li, Y. (2009). Biologically inspired synthetic enzymes made from DNA. *Chemistry & Biology*, *16*, 311–322.

Senel, M., Dervisevic, M., & Kokkokoglu, F. (2019). Electrochemical DNA biosensors for label-free breast cancer gene marker detection. *Analytical and Bioanalytical Chemistry*, *411*, 2925–2935.

Shangguan, D., Li, Y., Tang, Z., Cao, Z., Chen, H., Mallikaratchy, P., ... Tan, W. (2006). Aptamers evolved from live cells as effective molecular probes for cancer study. *Proceedings of the National Academy of Sciences of the United States of America*, *103*, 11838–11843.

Sigel, R. K., & Sigel, H. A. (2010). Stability concept for metal ion coordination to single-stranded nucleic acids and affinities of individual sites. *Accounts of Chemical Research*, *43*, 974–984.

Sinden, R. R. (1994). *DNA structure and function*. Amsterdam: Elsevier.

Smolina, I. V., Demidov, V. V., Soldatenkov, V. A., Chasovskikh, S. G., & Frank-Kamenetskii, M. D. (2005). End invasion of peptide nucleic acids (PNAs) with mixed-base composition into linear DNA duplexes. *Nucleic Acids Research*, *33*, e146.

Sohrabi, N., Valizadeh, A., Farkhani, S. M., & Akbarzadeh, A. (2016). Basics of DNA biosensors and cancer diagnosis. *Artificial Cells, Nanomedicine and Biotechnology*, *44*, 654–663.

Solanki, P. R., Kaushik, A., Agrawal, V. V., & Malhotra, B. D. (2011). Nanostructured metal oxide-based biosensors. *NPG Asia Materials*, *3*, 17–24.

Solanki, P. R., Kaushik, A., Chavhan, P. M., Maheshwari, S. N., & Malhotra, B. D. (2009). Nanostructured zirconium oxide based genosensor for *Escherichia coli* detection. *Electrochemistry Communications*, *11*, 2272–2277.

Strehlitz, B., Nikolaus, N., & Stoltenburg, R. (2008). Protein detection with aptamer biosensors. *Sensors*, *8*, 4296–4307.

Tang, Z., Parekh, P., Turner, P., Moyer, R. W., & Tan, W. (2009). Generating aptamers for recognition of virus-infected cells. *Clinical Chemistry*, *55*, 813–822.

Tang, Z., Shangguan, D., Wang, K., Shi, H., Sefah, K., Mallikratchy, P., ... Tan, W. (2007). Selection of aptamers for molecular recognition and characterization of cancer cells. *Analytical Chemistry*, *79*, 4900–4907.

Thévenot, D. R., Toth, K., Durst, R. A., & Wilson, G. S. (2001). Electrochemical biosensors: Recommended definitions and classification. *Biosensors & Bioelectronics*, *16*, 121–131.

Tokudome, H., Yamada, Y., Sonezaki, S., Ishikawa, H., Bekki, M., Kanehira, K., & Miyauchi, M. (2005). Photoelectrochemical deoxyribonucleic acid sensing on a nanostructured TiO_2 electrode. *Applied Physics Letters*, *87*, 213901.

Tripathy, N., & Kim, D. K. (2018). Metal oxide modified ZnO nanomaterials for biosensor applications. *Nano Convergence*, *5*, 272018.

Tso, C. P., Zhung, C. M., Shih, Y. H., Tseng, Y. M., Wu, S. C., & Doong, R. A. (2010). Stability of metal oxide nanoparticles in aqueous solutions. *Water Science and Technology: A Journal of the International Association on Water Pollution Research*, *61*, 127–133.

Tuerk, C., & Gold, L. (1990). Systematic evolution of ligands by exponential enrichment: RNA ligands to bacteriophage T4 DNA polymerase. *Science (New York, N.Y.)*, *249*, 505–510.

Updike, S. J., & Hicks, G. P. (1967). The enzyme electrode. *Nature*, *214*, 986–988.

Urbanova, V., Magro, M., Gedanken, A., Baratella, D., Vianello, F., & Zboril, R. (2014). Nanocrystalline iron oxides, composites and related materials as a platform for electrochemical, magnetic, and chemical biosensors. *Chemistry of Materials: A Publication of the American Chemical Society*, *26*, 6653–6673.

Védrine, J. C. (2017). Heterogeneous catalysis on metal oxides. *Catalysts*, *7*, 341.

Virolainen, N. E., Pikkemaat, M. G., Elferink, J. W. A., & Karp, M. T. (2008). Rapid detection of tetracyclines and their 4-epimer derivatives from poultry meat with bioluminescent biosensor bacteria. *Journal of Agricultural and Food Chemistry*, *56*, 11065–11070.

Wang, J. (2000). From DNA biosensors to gene chips. *Nucleic Acids Research*, *28*, 3011–3016.

Wang, J. (2005). Nanomaterial-based electrochemical biosensors. *Analyst*, *130*, 421–426.

Wang, J., Li, S., & Zhang, Y. (2010). A sensitive DNA biosensor fabricated from gold nanoparticles, carbon nanotubes, and zinc oxide nanowires on a glassy carbon electrode. *Electrochimica Acta*, *55*, 4436–4440.

Wang, M., Tang, Y., Chen, Y., Cao, Y., & Chen, G. (2019). Catalytic hairpin assembly-programmed formation of clickable nucleic acids for electrochemical detection of liver cancer related short gene. *Analytica Chimica Acta*, *1045*, 77–84.

Wang, Q., Gao, F., Ni, J., Liao, X., Zhang, X., & Lin, Z. (2016). Facile construction of a highly sensitive DNA biosensor by in-situ assembly of electro-active tags on hairpin-structured probe fragment. *Scientific Reports*, *6*, 22441.

Wang, Q., Zhang, B., Lin, X., & Weng, W. (2011). Hybridization biosensor based on the covalent immobilization of probe DNA on chitosan–mutiwalled carbon nanotubes nanocomposite by using glutaraldehyde as an arm linker. *Sensors and Actuators, B: Chemical*, *156*, 599–605.

Wang, Y., Jin, J., Yuan, C., Zhang, F., Ma, L., Qin, D., ... Lu, X. (2015). A novel electrochemical sensor based on zirconia/ordered macroporous polyaniline for ultrasensitive detection of pesticides. *Analyst*, *140*, 560–566.

Wang, Y., Luo, Y., Bing, T., Chen, Z., Lu, M., Zhang, N., ... Gao, X. (2014). DNA aptamer evolved by Cell-SELEX for recognition of prostate cancer. *PLoS One*, *9*, e100243.

Watson, J. D., & Crick, F. H. C. (1953). Molecular structure of nucleic acids: A structure for deoxyribose nucleic acid. *Nature*, *171*, 737–738.

Wisitsoraat, A., Tuantranont, A., Comini, E., Sberveglieri, G., & Wlodarski, W. (2009). Characterization of n-type and p-type semiconductor gas sensors based on NiO_x doped TiO_2 thin films. *Thin Solid Films*, *517*, 2775–2780.

Xia, T., Kovochich, M., Liong, M., Mädler, L., Gilbert, B., Shi, H., ... Nel, A. E. (2008). Comparison of the mechanism of toxicity of zinc oxide and cerium oxide nanoparticles based on dissolution and oxidative stress properties. *ACS Nano*, *2*, 2121–2134.

Xu, K., Huang, J., Ye, Z., Ying, Y., & Li, Y. (2009). Recent development of nano-materials used in DNA biosensors. *Sensors*, *9*, 5534–5557.

Yang, Y., Wang, Z., Yang, M., Li, J., Zheng, F., Shen, G., & Yu, R. (2007). Electrical detection of deoxyribonucleic acid hybridization based on carbon-nanotubes/nano zirconium dioxide/chitosan-modified electrodes. *Analytica Chimica Acta*, *584*, 268–274.

Yoo, E. H., & Lee, S. Y. (2010). Glucose biosensors: An overview of use in clinical practice. *Sensors*, *10*, 4558–4576.

Yumak, T., Kuralay, F., Muti, M., Sinag, A., Erdem, A., & Abaci, S. (2011). Preparation and characterization of zinc oxide nanoparticles and their sensor applications for electrochemical monitoring of nucleic acid hybridization. *Colloids and Surfaces B: Biointerfaces*, *86*, 397–403.

Zaid, M. H. M., Che-Engku-Chik, C. E. N., Yusof, N. A., Abdullah, J., Othman, S. S., Issa, R., ... Wasoh, H. (2020). DNA electrochemical biosensor based on iron oxide/nanocellulose crystalline composite modified screen-printed carbon electrode for detection of *Mycobacterium tuberculosis*. *Molecules (Basel, Switzerland)*, *25*, 3373.

Zhang, H. Q., Li, F., Dever, B., Wang, C., Li, X. F., & Le, X. C. (2013). Assembling DNA through affinity binding to achieve ultrasensitive protein detection. *Angewandte Chemie International Edition*, *52*, 10698.

Zhang, J., Zhang, B., Wu, Y., Jia, S., Fan, T., Zhang, Z., & Zhang, C. (2010). Fast determination of the tetracyclines in milk samples by the aptamer biosensor. *Analyst*, *135*, 2706–2710.

Zhang, W., Zheng, X., & Jiao, K. (2012). Label-free and enhanced DNA sensing platform for PML/RARA fusion gene detection based on nano-ZnO functionalized carbon ionic liquid electrode. *Sensons and Actuators, B*, *162*, 396–399.

Zhang, X. R., Li, S. G., Jin, X., & Li, X. M. (2011). Aptamer based photoelectrochemical cytosensor with layer-by-layer assembly of CdSe semiconductor nanoparticles as photoelectrochemically active species. *Biosensors & Bioelectronics*, *26*, 3674–3678.

Zhang, X. B., Wang, Z., Xing, H., Xiang, Y., & Lu, Y. (2010). Catalytic and molecular beacons for amplified detection of metal ions and organic molecules with high sensitivity. *Analytical Chemistry*, *82*, 5005–5011.

Zhao, W. W., Xu, J. J., & Chen, H. Y. (2014). Photoelectrochemical DNA biosensors. *Chemical Reviews*, *114*, 7421–7441.

Zheng, G., Patolsky, F., Cui, Y., Wang, W. U., & Lieber, C. M. (2005). Multiplexed electrical detection of cancer markers with nanowire sensor arrays. *Nature Biotechnology*, *23*, 1294.

Zhou, W., Saran, R., & Liu, J. (2017). Metal sensing by DNA. *Chemical Reviews*, *117*, 8272–8325.

Zou, X., Wu, J., Gu, J., Shen, L., & Mao, L. (2019). Application of aptamers in virus detection and antiviral therapy. *Frontiers in Microbiology*, *10*, 1462.

Zuo, S. H., Zhang, L. F., Yuan, H. H., Lan, M. B., Lawrance, G. A., & Wei, G. (2009). Electrochemical detection of DNA hybridization by using a zirconia modified renewable carbon paste electrode. *Bioelectrochemistry (Amsterdam, Netherlands)*, *74*, 223–226.

Metal oxides and their composites as flow-through biosensors for biomonitoring 12

Rudra Kumar, Gaurav Chauhan and Sergio O. Martinez-Chapa
School of Engineering and Sciences, Tecnologico de Monterrey, Monterrey, Mexico

12.1 Introduction

Rapid industrialization and increasing pollution have created serious health issues for humans and other species. The extensive amount of carcinogenic toxic substances that have been produced globally and released in the environment causes various types of manmade diseases such as cancer, Alzheimer's, heart and liver diseases, etc. (Chen & Liao, 2006; Manisalidis, Stavropoulou, Stavropoulos, & Bezirtzoglou, 2020). Therefore the early detection of such chronic diseases and the monitoring of toxic materials including carcinogen and mutagen are required.

Recently, biosensors technology has been gaining enormous attention in the healthcare to diagnose and detect various kinds of diseases (Ronkainen, Halsall, & Heineman, 2010; Thévenot, Toth, Durst, & Wilson, 2001). Biosensors are devices which have biological recognizing elements that can detect chemicals and biomolecules and are also equipped with transducers which can transform the concentrations of these analytes into measurable signals (Kimmel, LeBlanc, Meschievitz, & Cliffel, 2012; Wongkaew, Simsek, Griesche, & Baeumner, 2019). The amplitude of the measurable signal is directly proportional to the concentration of specific chemical/biological species. There are a lot of applications of biosensors in our daily life such as monitoring of glucose, urea, cholesterol, lipoproteins; in blood serum, saliva, or sweat (Maduraiveeran, 2020). The electrodes of biosensors are prepared of various kinds of nanomaterials and in the form of micro/nanoelectrodes or probes. The advantage of nanomaterials is their unique chemical and physical properties such as high surface area, and high surface to volume ratio with respect to the bulk. The nanosensors or nanoprobes enable the rapid and precise monitoring of biological elements.

Microfluidics is an emerging area of science and technology that deals with the flow of liquid through micrometer-sized channels for a variety of applications such as drug delivery, biomedical devices, portable electronics, lab-on-a-chip (LOC), micro heat exchangers for microelectronics cooling, micro total analysis systems (μTAS), and so on (Mark, Haeberle, Roth, von Stetten, & Zengerle, 2010). Microfluidic devices work with small fluid volume and short reaction time. They can be used in the parallel operation of separation and detection. Recently, biosensors and microfluidic technology

Metal Oxides for Biomedical and Biosensor Applications. DOI: https://doi.org/10.1016/B978-0-12-823033-6.00010-7

have been combined to create microfluidic biosensors that have a huge advantage over the conventional biosensors (Choi, Goryll, Sin, Wong, & Chae, 2011; Zaytseva, Goral, Montagna, & Baeumner, 2005). The microfluidic biosensors are the advanced and miniaturized version of traditional sensing devices, integrating various functions in single devices (Loo, Ho, Turner, & Mak, 2019). Microfluidic biosensors require lower sample volumes and shorter times of analysis. They have also low cost and enhanced sensitivity. Further, the small size of microfluidic biosensors can enhance the reaction rate, real-time analysis, perform the full analysis together with the continuous injecting, mixing, preconcentration, treatment, as needed in point-of-care (POC) devices. Thus the integration of microfluidic with biosensing can provide a robust analytical tool that has the benefits of detecting and monitoring the disease at home.

In this chapter, we will discuss the various kinds of metal oxide and their nanocomposites incorporated with biomarkers in microfluidic channels to detect/sense/monitor the biomolecules by various methods. We first study the basic concepts of microfluidics, various types of microfluidic devices, microfluidic devices in biomedical applications, and biosensors. After that, we will demonstrate the various methods of detection/monitoring of the biomolecules. Finally, we will study the recent progress and future perspective in the biosensor's technology and POC devices.

12.2 Microfluidic devices

Microfluidics is the science which can precisely manipulate and control the fluids in micro- to nanoliter range through the micrometer-sized channel. The microfluidic devices have numerous applications including LOC, biomedical device, drug delivery, μTAS, microelectronics cooling, and so on. The main advantage of microfluidic devices is the reduction of the sample volume which in turn reduces the cost of expensive analytes.

The governing continuity equation of fluid flow in micro- or nanochannels for the Newtonian, isotropic, and incompressible fluid is given by Panton (2013)

$$\frac{D\rho}{Dt} + \rho \mathrm{div}\, \nabla = 0$$

where D/Dt is the total derivative, ρ is the density of the fluid, and ∇ is the del operator.

The total derivative in the Cartesian coordinate is given as

$$\frac{D}{Dt} = \frac{\partial}{\partial t} + u\frac{\partial}{\partial x} + v\frac{\partial}{\partial y} + w\frac{\partial}{\partial z} = \frac{\partial}{\partial t} + (V.\nabla)$$

where $V = (u,v,w)$ is the velocity vector.

The Navier– Stokes equation derived from Newton's second law of motion from the fluid flow is given by Byron Bird, Stewart, and Lightfoot (2012)

$$\rho\frac{Du}{Dt} = \rho\left(\frac{\partial u}{\partial t} + u.\nabla u\right) = -\nabla p + \mu\nabla^2 u + \frac{1}{3}\mu\nabla(\nabla .u) + \rho g$$

where μ is the dynamic viscosity of the fluid, g is the acceleration due to gravity, and p is the pressure.

The flowing of fluids in various geometries is characterized by a nondimensional number known as Reynolds number. The Reynolds number is defined as the ratio of inertia force to viscous force within a fluid (Byron Bird et al., 2012)

$$\text{Reynolds number} = \text{Inertia force}/\text{viscous force}$$

$$\text{Re} = \frac{\rho v d}{\mu}$$

where Re, ρ, v, d, and μ are the Reynolds number, density, average fluid velocity, characteristics length scale, and dynamic viscosity of the fluid.

12.2.1 Scaling effects

Changes in the physical phenomenon occur as the devices are miniaturized. The change in length scale causes the fundamental change in the physics of fluids leading to the new fluidic behavior, for example, domination of surface properties such as surface forces (viscous and surface tension). The fluid properties such as surface tension are dominated over the gravitational forces, mass transport is subjected to viscous dissipation, and molecular diffusion is more dominant than the bulk fluid movements. Moreover, the electric fields are more prominent than the pressure gradients in microfluidic devices. Inertia forces and its effects are comparatively negligible which causes nonlinearity. In addition, the fluid properties have varied including surface to volume ratio, diffusion time, and distance. The scaling factor is introduced to characterize the size of the fluidic component or the change in the three dimensions of the device. The scaling factor of the miniaturization is less than unity ($S < 1$). The scaling is the ratio of surface forces to the volume forces (weight, inertial force, buoyancy force) (Tian & Finehout, 2009)

$$\frac{\text{Surface force}}{\text{Volume force}} = \frac{A}{V} = \frac{L^2}{L^3} = \frac{1}{L} = L^{-1}$$

where L, V, and A are the characteristic length, volume, and surface area, respectively.

12.2.2 Integration of nanomaterial with microfluidics

The integration of nanostructured materials in the microfluidic devices has received increasing attention due to enhanced immobilization of biomolecules which leads to boost the signals and can detect a very tiny amount of substance present in the analytes. Metal nanoparticles such as gold, silver, transition metals have been widely applied in the fabrication of electrodes for various types of flow-through biosensors (Bae et al., 2019; SadAbadi, Badilescu, Packirisamy, & Wüthrich, 2013). The small size of nanomaterials offers the intensification of signal transduction and thus improves the biosensor's performance. The nanoparticles have a high surface to volume ratio, high surface area, and very good catalytic properties toward the biomolecules. Further, the integration of nanoparticles on the electrode surface improves the electron transfer which enhances the microfluidic biosensor's proficiency. Also, metal/metal oxide nanostructures can be easily integrated into microfluidic and nanofluidic channels of POC devices.

12.2.3 Integration of microfluidics with biosensor technology

The integration of microfluidic devices with biosensor technology has a great benefit to humans. The amalgamation of both technologies offers the miniaturization of traditional laboratory-based diagnosis. This miniaturization proposes fast and accurate detection, good selectivity and excellent sensitivity, less energy consumption and waste production, small sample size, and low cost. In addition, the miniaturized microfluidic flow-through biosensors can be operated as single devices that perform the full analysis such as continuous sampling, sample separation, mixing, preconcentration, and detection (Nikoleli et al., 2018). Further, the flow-through microfluidic biosensors have the advantage of high throughput, enhanced diagnostic performance, real-time detection, fast and accurate diagnosis, and portability (Marzban, Dargahi, & Packirisamy, 2019). Therefore the integration of biosensor technology with microfluidics devices offers a lot of advantages which can revolutionize the diagnosis technology.

12.3 Methods for fabrication of microfluidic channels

The following methods are being used to fabricate the microfluidic channels for the flow-through biosensing applications.

12.3.1 Polymer laminates techniques

Microfluidic devices can be fabricated from polymer laminates that are cut individually and then stacked to form the microfluidic circuits (Fiorini & Chiu, 2005). The thin sheets are bonded or laminated together by adhesive. This technique offers the low cost, of scalable microfluidic components (Aeinehvand et al., 2017). In addition, a wide choice of medical-grade polymeric materials has been utilized such as polyethylene ester terephthalate, polyimides, polymethylmethacrylate, fluoropolymers, polypropylene, and silicone sheets. Thermally sensitive materials are also used in this technique because of the room temperature operating condition. Also, the polymer laminate fabrication process allows the incorporation of various types of functionalities including recirculating pumps, filtration components, pneumatic valves, etc. with a simple interface.

12.3.2 Lithographic techniques

Lithography methods are an attractive approach to fabricate microfluidic channels of various shapes and sizes. Numerous types of lithography methods are available such as photolithography, electron beam lithography (EBL), helium ion beam lithography (HIBL), extreme ultraviolet lithography (EUVL), deep ultraviolet lithography (DUVL), nanoimprint lithography (NIL), and immersion lithography to fabricate microfluidic devices (Li, Morton, Veres, & Cui, 2011). Photolithography is performed on photosensitive materials or photoresist via a partially transparent photomask exposed through the ultraviolet light source ranging from 248 to 435 nm. The image of the photomask is projected on the resist to fabricate the desired pattern after the development of photoresist in the developer. The feature size generated in this method is in the micron range. With photolithography, sub-250 nm features are very difficult to fabricate. The photolithography technique is quite simple, fast, and can fabricate bulk microfluidic channels, wells of desired shape and sizes. To fabricate nanometer-sized features, EBL, EUVL, and NIL are employed (Ha, Hong, Shin, & Kim, 2016).

In EBL, a focused high electron beam is used to expose the electron active resist, increasing the crosslinking or solubility of the resist. After the development process, nano range features are formed. As compared to the photolithography, a mask-based lithography technique, EBL is a mask-less technique and can fabricate a defect-free pattern. Moreover, it does not require any optical elements, therefore it overcomes the diffraction limitations in photolithography and can pattern sub-10 nm feature size. However, the throughput in EBL is very low as compared to photolithography (Kumar, Chauhan, Moinuddin, Sharma, & Gonsalves, 2020). EBL is an ideal lithography technique to fabricate nanofluidic structures such as nanochannels for biosensors (Fouad, Yavuz, & Cui, 2010). EUVL is another optical lithography technique to fabricate nanosized features using 13.5 nm extreme ultraviolet light sources. The processing condition in EUVL is similar to photolithography. EUVL can fabricate various types of micro- and nanofluidic channels with high throughput similar to photolithography. A microfluidic device containing nanopillars fabricated by EUVL has been used for cell study. Unfortunately, the use of EUVL has not been widely spread in other applications because of its high cost (Gale et al., 2018).

12.3.3 Three-dimensional printing nanofabrication

Microfluidic device fabrication using three-dimensional (3D) printing technology is a novel technique where the material is added layer by layer on top of the previous layer to fabricate desired shape and size (Raoufi et al., 2020). A 3D digital model can be designed in the computer and the model is interpreted and formed by the printer. Various 3D printing technologies have been used such as fused deposition modeling (FDM), stereolithography (SL), multijet modeling, and two-photon polymerization.

12.4 Microfluidic and nanofluidic biosensor platforms

A microfluidic platform provides various types of unit operations such as fluid transport, metering, valving, mixing, separation, reagent storage and release, incubation, etc. (Fig. 12.1). These platforms employ miniaturization, integration, and parallelization of chemical or biochemical processes. The microfluidic devices are more beneficial over conventional methods of macroscopic liquid handling systems such as petri dishes, culture bottles, and microtiter plates for performing analytical and diagnostic assays (Carnero, 2006). The advantages of microfluidic platforms are higher sensitivity, lower cost, portability, wearability, shorter time to result, small size, etc. Microfluidic and nanofluidic platforms have varieties of applications such as cell cultures (Coluccio et al., 2019), LOC devices, POC testing, centrifugal microfluidics, and so on (Mark et al., 2010).

The fabrication of microfluidic platforms requires precise control in design and accuracy. Generally, the most common materials used in the fabrication of microfluidic biosensor platforms are silicon and glass (Ng, Gitlin, Stroock, & Whitesides, 2002). The biocompatibility and wettability of the materials are important parameters to design the

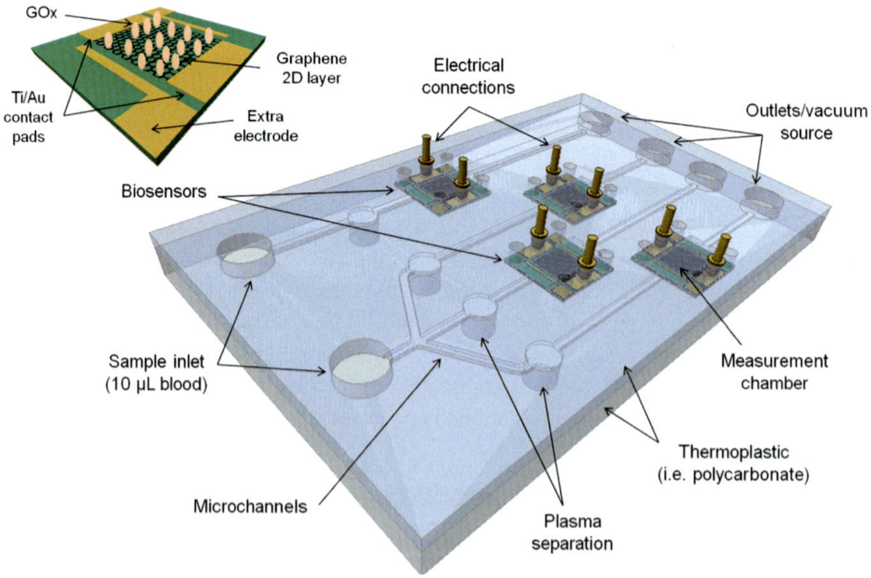

Figure 12.1 Example of a plastic-based design for a microfluidic platform for multiple, standardized biosensor chips (Viswanathan et al., 2015).

microfluidic biosensors. Apart from conventional glass and silicon substrates, polymers are widespread used in sensing platforms owing to low cost; flexibility; good mechanical, chemical, and physical properties; and biocompatibility. Polymethylmethacrylate (PMMA), and polydimethylsiloxane (PDMS) are common polymers to fabricate the microfluidic devices in recent times (Sia & Whitesides, 2003). The figure shows a CAD design of a microfluidic device, in which four silicon-based biosensors are integrated. Electrical connections are allowed through the top plastic layer providing the necessary sealing of the sensor (Viswanathan et al., 2015)

12.4.1 The electric double layer

The electric double layer (EDL) plays an important role in microfluidic-based biosensors (Prakash, Pinti, & Bhushan, 2012). Here we start with the origin of net surface charge on surfaces in contact with an aqueous solution. The surface charge is obtained by the dissociation of protons from surface-bound carboxylic acid groups or dissociation of silane groups to form negative charges (anions). The negatively charged surface attracts the opposite polar ions (cations) which are placed along themselves to form a fixed layer called the Stern layer. A mobile diffuse layer is also formed with the majority of cations. The combination of the Stern and diffuse layers is called the EDL (Fig. 12.2). A slip plane is located between the Stern layer and the diffuse layer. The potential at this plane is defined as zeta potential. The bulk solution which is screened by the EDL is electrically neutral, and the screening length is known as Debye length. The distance is tracked from the wall to the bulk solution, where the electrolyte solution is electrically neutral. The range of Debye length is about 0.1–100 nm. A smaller Debye length means a higher concentration of ions. The Debye length can be easily tuned by changing the ion concentration or by varying the surface charge density. The Debye length (λ_D) is defined as Steffen, Silva, Evangelista, and Cardozo-Filho (2018)

$$\lambda_D = \sqrt{\frac{\varepsilon_e RT}{F^2 z_i^2 C_i}}$$

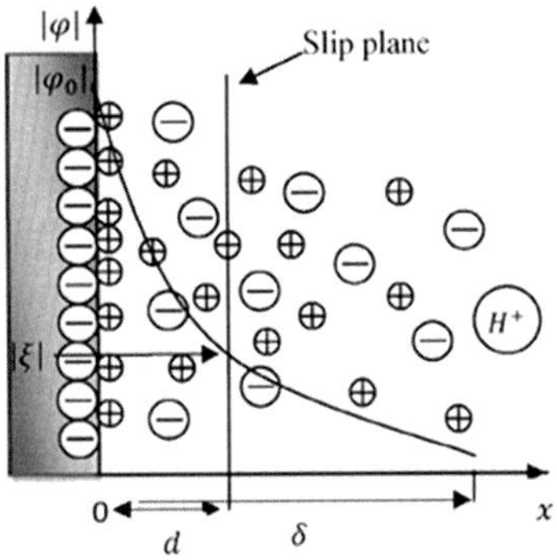

Figure 12.2 Electrical double layer (Suleimanov & Abbasov, 2017).

where ε_e, R, T, F, Z_i, and C_i are the electrical permittivity of the medium, gas constant, temperature, Faraday constant, the valence of the ith species, and the concentration of the ith species, respectively.

The EDL plays a significant role in microfluidic and nanofluidic devices. The transport of fluid is affected by the relative size of the double layer and microfluidic channel diameter. The product of inverse Debye length k and the channel radius a is the main parameter to consider for the fluid transport behavior. Debye length is inversely proportional to the square root of the bulk ionic concentration and its value lies between 0.1 and 100 nm, the larger Debye length the smaller ion concentration. Moreover, the electric potential decreases by magnitude of $1/e$ for every Debye length and the distance is measured from the wall toward the bulk solution. When the channel size is decreased, the EDL spreads across the channel and overlaps. In this situation, the electric potential in the bulk solution is not neutral. On the other hand, when the EDL is collapsed near the channel wall or the microfluidic channel diameter is large, and bulk is neutral. In this case, the diffuse layer which is charged is moved along the channel under the applied electric potential. This charged diffuse layer also drags the bulk liquid due to the viscous effect, which is responsible of the electroosmotic flow.

12.4.2 Electrokinetics in microfluidics and nanofluidics

The understanding of electrokinetics is an important aspect of microfluidic-based electrochemical biosensors.

When the external electric field applies across the channel, resulting in a net force exerted on the ions of EDL. This net force causes the motion of the diffuse layer along with the dragging of bulk fluid, resulting in the flow of fluid called electroosmotic flow (EOF). The flow velocity in EOF is determined by solving the Navier–Stokes equation. The distribution of ions in EDL derives the net charge density (ρ_e) from Poisson's equation (Mathews & Walker, 1970)

$$-\varepsilon\nabla^2\phi = \rho_e$$

where ε and $\nabla^2\phi$ are the relative permittivity and potential distribution of EDL. When the channel walls are too far and the bulk fluid is nearly neutral, the net charge distribution outside the shear plane is given by Boltzmann distribution for the symmetric electrolyte.

$$\rho_e = -2zec_0\sinh\left(\frac{ze\phi}{k_bT}\right)$$

where z, c_0, e, k_b, and T are valence, concentration of the bulk electrolyte, electron charge, Boltzmann constant, and temperature of fluid, respectively. This equation is not valid when the bulk solution is not electrically neutral that means the EDLs are overlapped. Combining the equations derive the governing equation of potential distribution as

$$\nabla^2 \phi = \frac{-2zec_0}{\varepsilon} \sinh\left(\frac{ze\phi}{k_b T}\right)$$

Debye–Huckel approximation is used to linearize the equation and valid when I $ze\phi$ I $< k_bT$, or wall potential is nearly 26 mV. In the microfluidic channel, the equations described above with all assumptions are valid. When the EDLs are thin (i.e., $\lambda_D << h$), the Helmholtz–Smoluchowski approximation gives the electroosmotic velocity expressed as (Craven, Rees, & Zimmerman, 2008; Mondal, Misra, & De, 2014)

$$u = \frac{\varepsilon \zeta}{\mu} E = \mu_{EO} E$$

where u, ζ, μ, and μ_{EO} are the electroosmotic velocity, zeta potential, dynamic viscosity, and electroosmotic mobility, respectively. This is a plug flow distribution in the channel.

12.4.3 Analyte transport regimes

Diffusion is the rate-limiting step of target molecule detection in the microfluidic biosensors. The diffusion of the target molecule across the width of the channel without bulk flow is given by Fick's law. The equation for microchannel or large nanochannel in one-dimensional (1D) form is expressed as Backhurst, Harker, Richardson, and Coulson (1999)

$$J_i = -D_i \frac{\partial C_i}{\partial x}$$

where J_i, D_i, C_i, and x is the molar flux, diffusion coefficient, concentration of the jth species, and x-direction along the length of the channel, respectively. When the channel dimension is decreased, the diffusion of molecules becomes hindered. In this case, the convection of target molecules is also considered. The Nernst–Plank equation considers both the diffusion and concentration gradient. The 1D form of the equation is

$$J_i = -D_i \frac{\partial C_i}{\partial x} + C_i u - \frac{z_i F}{RT} D_i C_i \frac{\partial \phi}{\partial x}$$

where u, z_i, ϕ, F, R, and T are the bulk fluid velocity, valence of the ith species, the electrical potential (sum of the applied potential plus the local potential from the EDL), Faraday constant, gas constant, and absolute temperature, respectively (Kuo & Chiu, 2011).

The depletion zone is created around the sensing area with time when the target molecules are collected by the biosensors, which impede the effective analyte transport. A concentration gradient is formed along the length of the channel which is shown in the first term of the Nernst–Plank equation. When the flow rate is increased, there is no sufficient time for the target molecules to diffuse across the width of the channel to the detector before being transferred downstream past the detector. In this situation, fluid mechanics should allow enough time for the target molecules to bind the detector. If the flow rate is too slow, the depletion zone grows and very few target molecules are close and bind to the detector, decreasing the detection and resolution of detection. If the flow rate is increased with decreasing the depletion zone, the sensitivity of the detector will be reduced.

12.4.4 Slip flow considerations

The biomolecules can stick the micro/nanochannel surfaces and resist the flow when the adhesion between the biofluids and microfluidic channels is high. Therefore low bio-adhesion microfluidic surfaces are required to assist the flow. Generally, the polymers have high triboelectric surface potential and can vary with types of polymer used, which may affect the flow. To reduce the triboelectric effect, a conducting surface layer is deposited on top of the polymer. Also, superhydrophobic surfaces could be used to reduce the fluid drag and exhibit the slip flow. At the solid–liquid interface, the superhydrophobic surface interacts with the aqueous fluid solution during wetting, resulting in the formation of spherical cap-shaped bubbles with 100 nm diameters. These nanobubbles are charged species with the formation of EDL at the gas–liquid interface, revealing the electrostatic stabilization. The surface heterogeneity and roughness, bias at the surface, and charged species are major parameters to control the size and tendency of the nanobubbles. Fig. 12.3 shows the formation of nanobubbles at water repellent hydrophobic surfaces, liquid adhesion, and fluid flow characteristics. The fluid slip is influenced by the presence of nanobubbles. Also, the presence of the electric field can affect the contact angle, size of the nanobubbles, and fluid drag behavior. When the contact angle is almost negligible, that is, hydrophilic surfaces, the water slip is nearly zero. When the contact angle is greater than 90 degrees or more than that, that is, hydrophobic surfaces, the slip length is increasing with the increase in contact angle (Duan & Muzychka, 2007).

12.4.5 CD microfluidics technology

Centrifugally driven microfluidics is another category of microfluidic technology which deals with manipulation of fluids in the microdomain. This technology is different as it uses minimalistic instrumentation to control the fluid with the contactless nature of propulsion to perform complete assays in a single CD. This technology, called lab-on-a-disc/CD, is applied as a tool for POC diagnostics. Based on a plastic substrate, this CD has embedded fluid microchannels and reservoirs designed according to a particular assay or diagnostic application. The movement of fluid is controlled by placing the CD on a spinning motor platform to manage the whole range of fluidic functions. Various such fluidic functions implemented on the CD platform include valving, decanting, calibration, mixing, metering, sample splitting, and separation (Aeinehvand et al., 2018, 2019).

CD microfluidics is based on three pseudo forces: Centrifugal force ($F\omega$), Euler force (FE), and Coriolis forces (FC) (Fig. 12.4). Assuming a body of mass m at position r, rotating at angular speed ω, the centrifugal force is given by:

$$\vec{F}_{\omega} = -m\vec{\omega} \times (\vec{\omega} \times \vec{r})$$

Centrifugal force always points away from the rotation axis, acting as a gravitational force in this reference frame. This is the main force of centrifugal microfluidics, which can be carefully manipulated to perform different operations, such as capillary burst valves (van den Berg, 2000), droplet generation (Liu, Sun, Yang, & Xu, 2016; Roy et al., 2015), and fractionation (Kim, Hwang, Gorkin, Madou, & Cho, 2013; Morijiri, Sunahiro, Senaha, Yamada, & Seki, 2011).

Coriolis force is the combined effect of fluid velocity and system rotation. It is described by:

$$\vec{F}_c = -2m\vec{\omega} \times \frac{d\vec{r}}{dt}$$

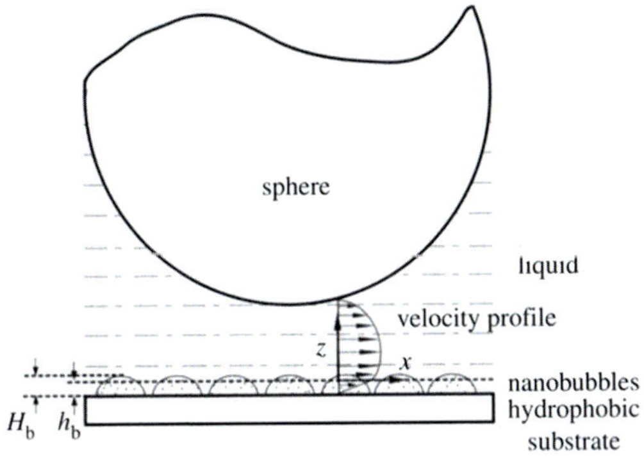

Figure 12.3 A schematic of a sphere–flat system with nanobubbles covering the flat surface. The presence of the nanobubbles on the flat surface changes the velocity profile between the sphere and the plane surface, which results in an increase of slip length (Wang, Bhushan, & Maali, 2009).

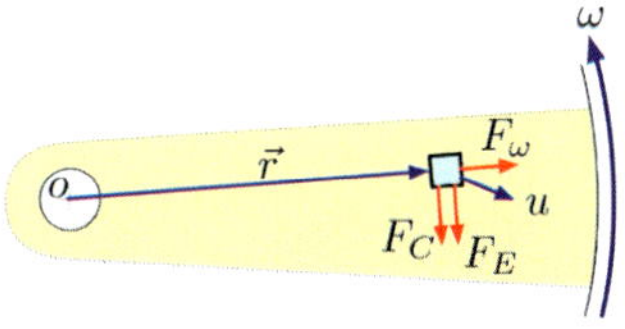

Figure 12.4 Pseudo forces involved in centrifugal microfluidic platforms (Smith et al., 2016).

Coriolis force always points in the azimuthal direction. For the common case of a fluid that moves away from the center of rotation, F_C points in the opposite direction of ω. This force is usually manipulated to perform fluid routing at junctions (Kazemzadeh, Ganesan, Ibrahim, Kulinsky, & Madou, 2015). It can also be applied for cell separation and is very useful when performing mixing (Ducrée, Haeberle, Brenner, Glatzel, & Zengerle, 2006; Kuo & Li, 2014).

Euler force appears when there is angular acceleration. It is described by:

$$\vec{F_E} = -m\frac{\mathrm{d}\omega}{\mathrm{d}t} \times r$$

which points in the opposite direction of rotation and its magnitude is proportional to the radius. This force has been used as the main actuation mechanism for siphon valves, and to induce flow reversal and vortical flow for mixing (Deng et al., 2014).

Centrifugal microfluidic technology allows the automation of a sequence of unit operations which usually require several laboratory platforms and technicians. This technology can replace bulky and expensive machinery, manpower, and evade the human error. Role of this automated technology as POC diagnostic tool can overcome various such challenges as mentioned in Fig. 12.5. By miniaturizing laboratory functions, the required sample and reagent volumes are smaller, and the cost of a fluid analysis is reduced. Many laboratory functions have already been miniaturized on this centrifugal microfluidic platform including cell separation, cell lysis, purification, mixing, protein amplification, and detection. This technology is now combined with analytical measurement techniques, such as optical imaging, absorbance, and fluorescence spectroscopy, and mass spectrometry, to make the centrifugal platform a powerful solution for medical and clinical diagnostics and high throughput screening (HTS) in drug discovery. It has already shown great promise as a POC device, as complex protocols such as enzyme-linked immunosorbent assay and reverse transcriptase polymerase chain reaction have already been integrated for detection of pathogens, showing total automated control from sample preparation to detection (Kim, Abi-Samra et al., 2013; Stumpf et al., 2016).

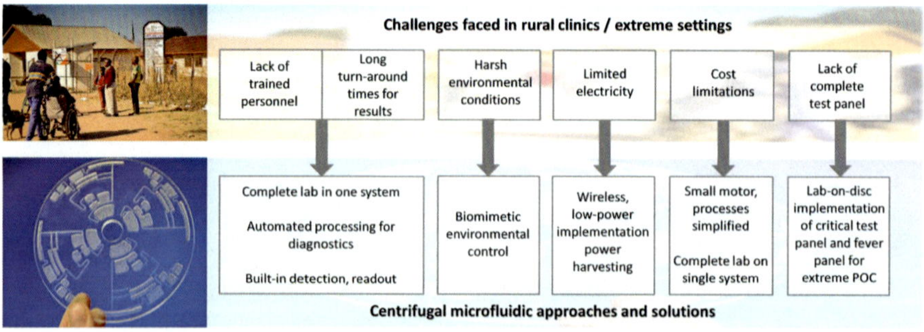

Figure 12.5 Centrifugal microfluidic approaches to tackle challenges faced in rural areas/extreme settings (Smith et al., 2016).

Advantages of these centrifugal microfluidic platforms can be summarized as a simple, versatile, automatable, and compatible platform with high throughput, which can be easily prototyped and customized at very low cost.

12.5 Biosensors: how they work? What are the benefits of biosensors in modern life?

Biosensors are devices that can sense/detect/monitor the wide range of biological or chemical materials of interest through the biological recognition elements, covert the biological response through the transducer and measure the signal in user-friendly methods such as electrochemical, optical, or other methods. The intensity of the measured signal is directly proportional to the concentration of biological/chemical species. Fig. 12.6 shows the biological targets, recognition molecules, microfluidic platforms, and signal output elements. In general, the biosensors comprise a biological recognizing element and a signal output element.

Various types of target analytes have been used such as small molecules, nucleic acid, proteins, bacteria, cell, etc. The interaction of analytes with the biological recognition element can be classified in many ways. Several types of interactions between the target analytes and the recognition molecules are possible such as

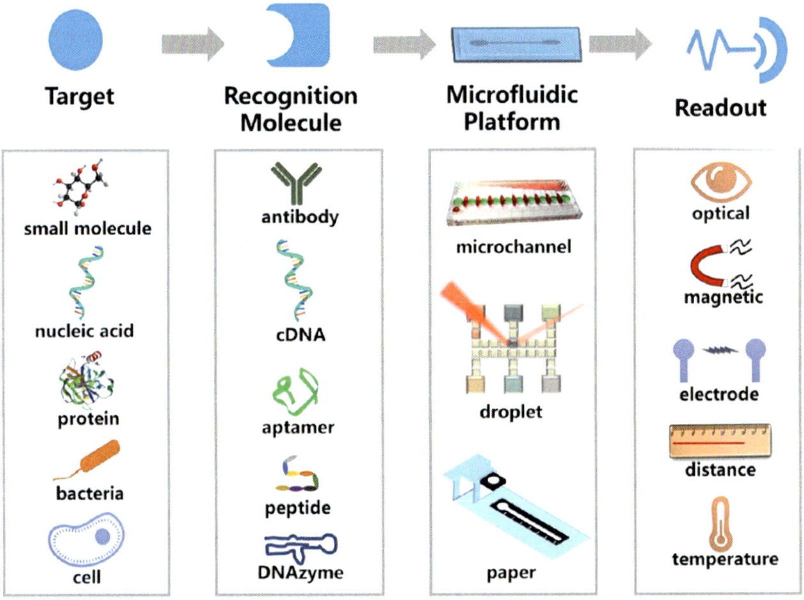

Figure 12.6 Schematic of different parts of a microfluidics-based biosensor including biological recognition element, microfluidic device platform, and signal readout (Song et al., 2019).

antibody–antigen, nucleic acid–oligonucleotides, aptamers, peptides and corresponding target, enzyme–substrate, ligand–receptor, and host–guest interactions.

Thus the numerous interaction methods of biosensing allow for the design and development of accurate and target-specific molecules to sense with high precision. Apart from these, various types of signal transduction and output methods are available such as electrochemical, optical, surface plasmon resonance (SPR), magnetic, and mass-sensitive transducer. Biosensors have been widely applied in health monitoring, environmental assessments, medical diagnosis, food safety surveillance, and other fields owing to low cost, fast response, ease of use, and reliability.

12.5.1 Detection or monitoring methods

In this section, we will discuss various methods to detect the biomolecules in flow-through biosensors. The transducer elements can be either electrochemical, optical, colorimetric, mass sensitive, or thermal to detect the biosignals.

12.5.1.1 Electrochemical methods

The electrochemical method for the detection of biomolecules is an attractive and cost-effective technique in the microfluidic system. The high signal-to-noise ratio leading to high sensitivity, the low limit of detection, and the high ratio of faradic to charging current are the major advantages of the electrochemical methods (Goral, Zaytseva, & Baeumner, 2006; Rackus, Shamsi, & Wheeler, 2015). Moreover, the electrochemical methods can detect the signal without affecting the system. An electrochemical biosensor consists of a reference electrode, a working electrode, and a counter electrode. The working principle involves the reaction between target analytes and recognition molecules located on the electrode surface to catalyze the reaction and produce electrons. This reaction on the sensor surface causes the electron transfer through the double layer which either produces current or can change the double-layer potential to generate a voltage. The analyte concentration plays an important role to measure the current or voltage. At a fixed potential, the change in analyte concentration is directly related to the electron flow rate or current. Electrochemical sensors are classified into various categories such as potentiometric, amperometric, impedimetric, and capacitive biosensors.

Amperometric and potentiometric detection: Amperometric electrochemical sensors measure the change in the electrical current at constant applied potential when the chemical reaction takes place between electroactive materials and the bioanalytes on the surface of working electrodes (La Rosa, Pariente, Hernández, & Lorenzo, 1995; Sadeghi, 2013). The change in current is directly proportional to the concentration of bioanalytes in the solution. The working electrode is made from the conducting carbon, noble metals, screen-printed layer coated by bioreceptor elements, or conducting glass such as ITO. Potentiometric biosensors are used at constant current to measure the change in voltage. Ion-selective field-effect transistors (ISFETs), bio-FETs, and carbon-based FETs such as graphene FETs are used as potentiometric biosensors.

Impedimetric and capacitive biosensors: An impedimetric biosensor measures the electrical impedance between the electrode and the target biomolecule at DC bias (Yang & Guiseppi-Elie, 2008). The electrical properties such as dielectric constant or electrical resistance of the electrode surface change during the interaction of biomolecules which in turn modify the impedance. The change in the impedance as a function of frequency is typically plotted by Nyquist and Bode plots. In the electrochemical cell, the flow of electrons is influenced by electrode kinetics, diffusion, molecular interaction, redox reaction, etc. at the electrode surface. Capacitive biosensors are based on the interaction between the target biomolecules with the electrode surface which causes the change in dielectric properties and the thickness of the EDL at the electrode–electrolyte interface (Ertürk & Mattiasson, 2017). Antigens, antibodies, DNA fragments, heavy metal ions, and proteins are detected by this method with low limit of detection.

12.5.1.2 Optical methods

Optical-based biosensors have numerous advantages including high selectivity, remote sensing, specificity, label-free detection, multiparameter detection, and real-time measurements (Chen & Wang, 2020). During in vivo analysis, the noninvasive sensing technique is another advantage in the optical method. Optical biosensors involve the physical or chemical change in the same optical properties such as absorption, reflection, refraction, amplitude, transmission, phase, polarization, or frequency during the biorecognition process. Various types of optical biosensors have been developed such as luminescence-, fluorescence-, absorbance-, and surface plasmon-based biosensors.

12.6 Metal oxides and their composites in microfluidic biosensing

Nanomaterials have been broadly used in various types of biosensors. There are various shapes and orientations of nanostructured materials such as nanofibers, nanobelts, nanorods, nanospheres, nanocages, nanotubes, and other nanosized structures of metal oxides which are enormously used as transducer elements for biosensors. Nanostructured materials have been functionalized with various functional groups to serve in the catalysis of the biochemical reactions. Therefore these materials have been widely explored in many applications including flow-through biosensors, tissue engineering, drug delivery, electrochemical detection, biomedical devices, health care, and sustainability, in vivo and in vitro thermal and cellular imaging, magnetic resonance imaging, etc. There are various types of metal oxides such as ZnO, TiO_2, MgO, CeO_2, NiO, Fe_3O_4, SiO_2, CuO, V_2O_5, Al_2O_3, ZrO_2, Co_3O_4, Mn_3O_4, SnO_2, etc. which have been applied in biosensor applications due to better immobilization of biomolecules on the surface (Zhu, Yang, Li, Du, & Lin, 2015). In addition, the better electron transfer between the electrodes and the analytes makes metal oxides important candidates for biosensors. Metal oxides have also been employed in flow-through devices such as microfluidic biosensing due to the high surface area and porosity, biocompatibility, possibility to be functionalized with various types of functional groups, surface charge, hygroscopic nature, and apposite for the immobilization of numerous types of biomolecules, etc. The immobilization or binding of biomolecules with metal oxides occurs through two processes, for example, physical adsorption via van der Waals force or chemical interaction via covalent bonding. Van der Waals forces are weak and depend on the electrostatic surface charge and the morphology of metal oxides where the biomolecules adhere. Chemical interaction between biomolecules and metal oxide occurs via covalent linkage. The surface functional groups present in metal oxides have covalently bound the biomolecules. The surface functional groups have been introduced in metal oxides by various chemical reactions and are very important for the immobilization of biomolecules and the construction of bioelectrodes. Also, the rate of electron and charge transport is mainly depending on the functionalization of metal oxide–biomolecules which lead to better signal transduction, sensitivity, and stability. Furthermore, the surface functionalization of metal oxides can prevent the agglomeration on the transducer surface. Next, we will discuss various types of metal oxide in different shapes and their use in microfluidic biosensors.

12.6.1 Types of metal oxides used in flow-through devices for biosensing and other biomonitoring devices

12.6.1.1 Nickel oxide

Nickel and its oxide forms are widely studied in numerous applications including electrochemical detection, biosensors, energy storage devices, optoelectronic devices, and so on. Various morphologies of nickel-modified electrodes have been synthesized such as nanorods, nanoparticles, hollow nanospheres, nanobelts, etc. to immobilize the biomolecules which improve electron transfer and the intensity of electrochemical signals (Cao et al., 2011; Zhao et al., 2016). In addition, nanostructured nickel oxide has a bandgap of 3.6 eV, high isoelectric point (10.8), antiferromagnetic in nature, and very good electrochemical properties. 1D nickel oxide nanostructure shows excellent electrode material due to highly stable in air and immobilization of enzyme or biomolecules without degradation. Further 1D NiO nanostructure has high charge storage on the surface, selective binding of biomolecules, and comparable Debye length to the nanostructures. Recently, Ali et al. demonstrated the fabrication of microfluidic biochip based on nanostructured nickel oxide nanorods which can measure total cholesterol in human blood via the electrochemical method (Ali, Solanki et al., 2013). Fig. 12.7 shows the microfluidic channel fabricated from PDMS with the inlet and outlet reservoirs, a nickel oxide nanorod electrode as a working electrode, Ag/AgCl wire as a reference electrode, and indium tin oxide (ITO) as a counter electrode, respectively. The working electrode was immobilized with cholesterol esterase (ChEt) and cholesterol oxidase (ChOx). The device was operated in a laminar flow regime with a flow rate of

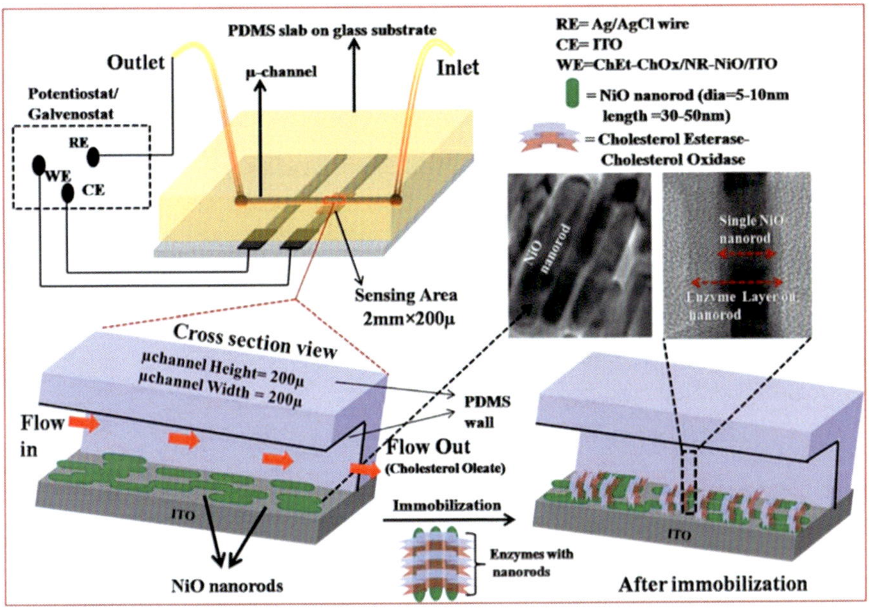

Figure 12.7 NiO nanorods in the microfluidic chip used for the detection of total cholesterol (Ali, Solanki et al., 2013).

10 mL/min and a low Reynolds number of 0.82 through the microchannel. The fabricated microfluidic sensor exhibits the sensitivity of 0.12 mA/mM/cm^2, linear in the range of 1.5–10.3 mM, and low value of Michaelis–Menten constant (km = 0.16 mM). Moreover, there are no significant changes in the peak current during the incorporation of interference substances along with the analyte.

12.6.1.2 *Mixed metal oxides*

Apart from single metal oxides, mixed metal oxides are also used in microfluidic biosensors due to multiple oxidation states and sites, higher electronic conductivity, and fast heterogeneous electron transfer (Salehabadi & Enhessari, 2019). These materials have been recently used in electrochemical biosensing applications owing to enhanced electrocatalytic activity. The following mixed metal oxides are used in flow-through biosensors.

Hollow nickel vanadate ($Ni_3V_2O_8$) nanospheres: Nickel vanadate has tunable oxidation states (V^{5+}, V^{4+}, V^{3+}), higher conductivity than nickel oxide, and extremely enhanced electrochemical properties (Kumar, Rai, & Sharma, 2016). Nickel vanadate has an orthorhombic crystal structure with the Adam symmetry group. The hierarchical structure of metal oxides particularly hollow architecture has demonstrated enhanced electrochemical properties due to their homogeneous porosity, higher stability, and resistance to aggregation. The hollow structure and mesoporous topography of nickel vanadate can assist to bind the antibody molecules. In addition, the hollow architecture can load a large number of antibody molecules inside and outside areas of the spheres due to the availability of more active sites, thereby enhancing the electrochemical biorecognition event (Singh et al., 2019). However, the high surface area and large active sites present in nickel vanadate hollow structure, the weak conjugation of biomolecules is the major concern for the development of efficient flow-through biosensors. To overcome this problem, a nanocomposite of metal oxides with biocompatible polymers is prepared. A composite of mesoporous nickel vanadate hollow nanosphere with chitosan (Ch-$Ni_3V_2O_8$) is synthesized and integrated with microfluidic flow-through device for the early biomonitoring of cardiac troponin I. Fig. 12.8 shows the proposed mechanism of the Ch-$Ni_3V_2O_8$ nanocomposite preparation. This nanocomposite is functionalized with cardiac antibodies I (cAb) which can amplify the electrochemical biomonitoring response signals.

12.6.2 Metal oxides and its composites in biosensing

There are various types of metal oxides and their composites have been utilized as an electrode material for biosensing of numerous types of biomolecules.

Titanium oxide (TiO_2)-based biosensors: TiO_2 is a family of transition metal oxides which present in three types of phase structures known as anatase, rutile, and brookite. Anatase and rutile are tetragonal crystal structures and more stable than the brookite phase which is an orthorhombic crystal structure. Rutile is the more stable phase as compared to anatase. The as-prepared TiO_2 is mostly amorphous and the annealing at a high temperature can play an important role to convert the phase structure. The amorphous phase is converted into the anatase phase at an annealing temperature of 300–500°C whereas the rutile phase is formed at 600–700°C annealing temperature. TiO_2 crystal possesses an oxygen vacancy,

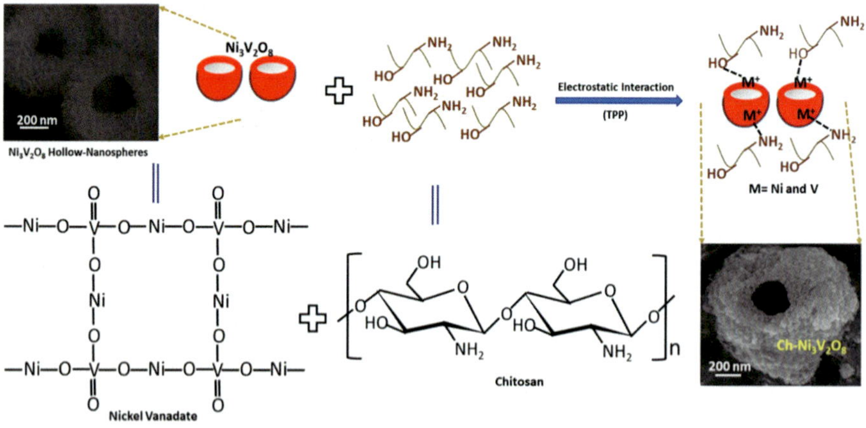

Figure 12.8 Proposed mechanism for the modification of $Ni_3V_2O_8$ with chitosan (Singh et al., 2019).

indicating more plus charge from Ti as compared to minus charge from oxygen. Thus the crystal is electron-rich and an n-type semiconductor. The conductivity of TiO_2 can be tuned by doping the metal elements. Also, n-type TiO_2 can be transformed into p-type TiO_2 by controlling the doping amount and heating temperature. TiO_2 is widely used in biosensing due to biocompatible, environmentally friendly, and a potential interface for the immobilization of various types of biomolecules (Bai & Zhou, 2014). The oxygen-containing anatase TiO_2 surface shows heterogeneous electron transfer, leading to fast sensing of biomolecules in microfluidic biosensors. Also, the TiO_2 phases, compositions, and crystal structures play an important role to sense the biological analytes. In addition, titanium maintains the biocatalytic activity through the formation of coordination bond with carboxylic groups of enzymes and amine (Evans & Hanson, 1976; Kurokawa, Sano, Ohta, & Nakagawa, 1993). Moreover, TiO_2 harvested the electrons produced by the reaction between analyte and biomolecules as the electron-accepting character of TiO_2, and this electron flows through the external circuits to detect the reaction. In electrochemical biosensing, anatase TiO_2 can stimulate the active sites of enzyme, leading to enhance electrochemical sensing performance. The stimulation of the active site of the enzyme is due to the presence of gibbosites on anatase TiO_2 surface, leading to a decrease in the tunneling distance between the active site of the enzyme and the electrode. A porous nanocrystalline TiO_2 film provides a large number of active sites leading to an abundance of loading capacity of enzyme molecules and retains the biological activity of enzymes. Azahar Ali et al. (2012) fabricated microfluidic cholesterol biosensor based on nanocrystalline anatase TiO_2 on ITO glass. Cholesterol enzyme is immobilized on anatase TiO_2/ITO microfluidic electrodes, showing thc sensitivity of 94.65 μA/mM/cm (Chen & Liao, 2006), due to the high surface to volume ratio of nanocrystalline TiO_2 and small geometry of the microfluidic system.

TiO_2–chitosan composite: Chitosan is a linear amino polysaccharide composed of D-glucosamine and *N*-acetyl-D-glucosamine units. It is derived from chitin after deacetylation and also easily extracted from the shrimp and crabs. Chitosan is a cost-effective, biocompatible biomaterial that has outstanding film-forming properties. The nanoporous structure and electroactive properties of chitosan lead to the high surface area, the better binding ability of nanoparticles, enhancing the adsorption of target biomolecules, and provide better electron-conducting path. Moreover, the availabilities of various oxygen- and nitrogen-based functional groups such as $-OH$, $-NHCOCH_3$, and $-NH_2$ in chitosan leads to the better immobilization of biomolecules (Suginta, Khunkaewla, & Schulte, 2013). Porous TiO_2 shows good binding property with chitosan to make TiO_2–chitosan nanocomposite which offers better enzyme functionalization, enhances electrochemical performance, and biosensing efficiency. However, mesoporous anatase TiO_2 alone does not attach enough biomolecules on its surface due to weak intermolecular interaction with the biomolecules. Therefore the amalgamation of TiO_2 into chitosan in the microfluidic devices improves the sensitivity and electrochemical response (Azahar Ali et al., 2012).

Manganese oxide-reduced graphene oxide (Mn_3O_4-RGO) nanocomposite: Mn_3O_4 is a tetragonal arrangement (I41/amd space group) with the cell parameters of $\alpha = \beta = \gamma = 90°$, $a = b = 5.762$ Å, and $c = 9.442$ Å (Garcês Gonçalves, De Abreu, & Duarte, 2018; Jarosch, 1987). Mn_3O_4 is a normal spinel structure that presents the formula $A^{2+}B_2^{3+}O_4$, where the tetrahedral sites are occupied by Mn^{2+} and the octahedral sites by Mn^{3+} cations. Mn_3O_4 attracted interest due to its excellent electrochemical properties, catalytic properties, good biocompatibility, chemical stability, high carrier mobility, and fascinating transducer material. Moreover, the nanosized domain can prevent the restacking of RGO and reduce the agglomeration of RGO nanosheets. In addition, the combination of Mn_3O_4 with RGO nanosheets enhances the bio interface for better conveyor transport and increased mobility for improved biomolecule sensing. Further, the porous structure of Mn_3O_4-RGO nanocomposite provides a greater number of antibody molecules to attach on the surface, enabling to enhance the limit of detection, sensitivity, and stability. Such remarkable properties of Mn_3O_4-RGO nanocomposite is used as a functional material in microfluidic biosensing POC devices (Singh et al., 2017).

CNT-NiO nanocomposite: Carbon is widely used electrode materials owing to its remarkable properties such as good electrical conductivity, nonreactive, easy preparation, reproducibility, uniform distribution of catalyst, and robustness in aqueous solution. 1D CNTs possess exceptional electrical, chemical, mechanical, and optical properties for the utilization in microfluidic biosensors. In addition, CNTs have good tensile strength, high aspect ratio, wide potential window, and high carrier mobility that leads to quantum electron transport in the microfluidic chip (Ghasemi et al., 2017). The fast electron transfer occurs when the CNTs are used as electrode materials in microfluidic biosensors. Further, CNTs provide a conducting pathway for the transportation of charge carriers, resulting in high sensitivity. The presence of CNTs in biosensors enhances the effect because of $\pi - \pi$ interactions between CNT and the immobilized biomolecules. This immobilization of biomolecules hinders the CNT agglomeration due to the van der Waal effect, causing poor film-forming properties. The noncovalent modification in CNTs maintains electrical conductivity, contributing to the low detection limit and fast response. The covalent functionalization of CNTs and the high surface area offers a better platform for biofunctionalization. Moreover, the covalent bonding of biomolecules such as proteins, nucleic acids, and enzymes may provide improved stability and reproducibility. Another way to prevent CNT agglomeration is by making CNT–nanostructured metal oxide nanocomposites. When the CNT is combined with conducting polymer or metal oxides to form CNT hybrid materials, a synergistic effect, such as enzymeless biosensing with greater durability and is cost-effective, is observed. Nickel oxide (NiO) is a poor electrical conductivity electrode material which can be used as noncovalent immobilization of enzyme to detect DNA, glucose, and cholesterol. To improve the biosensing performance, nanostructured nickel oxide is incorporated with multiwalled CNTs to make nanocomposite material that can exhibit high sensitivity and good selectivity toward cholesterol detection (Ali, Srivastava et al., 2013).

Iron oxide–graphene oxide composite: Iron oxide (Fe_3O_4) is a cubic inverse spinel structure that exists in the both octahedral and tetrahedral sites (Movlaee, Ganjali, Norouzi & Neri, 2017; Sun et al., 2014). Fe^{2+} and Fe^{3+} ions are surrounded by six oxygen atoms in the octahedral site whereas Fe^{3+} ions are surrounded by four oxygen atoms in the tetrahedral site. Iron oxide nanoparticles possess peroxidase-like catalytic properties that can be used in hydrogen peroxide reduction. Moreover, this peroxidase-like activity could be further enhanced by the incorporation of other materials such as palladium, graphene, platinum, and gold. Recently, Sharafeldin et al. reported the iron oxide nanoparticles decorated on graphene oxide (GO) sheets for the amperometric detection of cancer biomarker protein using a microfluidic biosensing technique. The incorporation of poly(diallyl dimethylammonium chloride) PDDA-functionalized Fe_3O_4 NPs on to the GO surface not only reduced the nanoparticle agglomeration, and graphene sheet restacking but also improved the wettability and dispersion of composite materials. In addition, the electrostatic interaction between positively charged PDDA-functionalized Fe_3O_4 and negatively charged GO sheets formed core-shell Fe_3O_4@GO particles and also offered precise control over the number of iron oxide NPs on graphene sheets. Ab_2-Fe_3O_4@GO composite was prepared using the reaction between antibodies (Ab_2) with Fe_3O_4@GO. The antibodies-decorated Fe_3O_4@GO composite was used as a dual function to isolate the biomarker protein from the sample under magnetic control and then inherent peroxidase-like catalytic activity was applied to electrochemically detect with ultra-high sensitivity. Also, the intrinsic peroxidase-like property of Ab_2-Fe_3O_4@GO particles can be used as an alternative for HRP-Ab_2-magnetic beads (MB) (1 μm diameter). The antibodies loaded Fe_3O_4@GO composite was designed to offline capture the prostate-specific antigen (PSA) and prostate-specific membrane antigen (PSMA). The sensitivity of Fe_3O_4@GO composite can be varied from 0.036 nA/log (pg/mL) to 10.5 nA/log (pg/mL) for PSA while the sensitivity of PSMA from 0.061 nA/log (pg/mL) to 25.9 nA/log (pg/mL), with the limit of detection was 15 fg/mL and 4.8 fg/mL for PSA and PSMA, respectively (Sharafeldin et al., 2017).

MoS_2-$CuFe_2O_4$ nanocomposite-based microfluidic biosensors: Molybdenum sulfide is a two-dimensional (2D) transition metal dichalcogenide material having a direct bandgap of 1.8 eV, and the monolayer thickness of ~0.65 nm. MoS_2 nanosheets are sandwiched of molybdenum atoms between two sulfur atoms which have good electrical, chemical, and physical properties for electrochemical biosensing (Chhowalla et al., 2013). A small bandgap and ultrathin monolayer MoS_2 offers fast heterogeneous electron-transfer capabilities, and allow highly sensitive electrochemical sensing. Moreover, endowing a high surface area, MoS_2 nanosheets can be used as a base/matrix material to load various types of organic and inorganic molecules for making a nanocomposite and nanocarriers. Fig. 12.9A shows the synthesis process of copper ferrite nanoparticle ($CuFe_2O_4$ NP)-loaded MOS_2 nanosheets composite electrode using a chemical reaction between MoS_2 and (3-mercaptopropyl) trimethoxysilane (MPTS) and silane chemistry between $CuFe_2O_4$ NPs and MPTS. The synthesized nanocomposite possesses outstanding electrochemical properties, excellent electrocatalytic properties due to the copper-based NPs, great biocompatibility, and high surface to volume ratio. Moreover, the synthesized nanocomposite retains the fundamental properties of individual components as well as introduces novel properties of the nanocomposite (Gawande et al., 2016). The MoS_2-$CuFe_2O_4$ nanocomposite was electrodeposited and electropolymerized with amine-functionalized 2-aminobenzylamine (ABA) homopolymer on the screen-printed electrode surface, simultaneously (Chand, Ramalingam, &

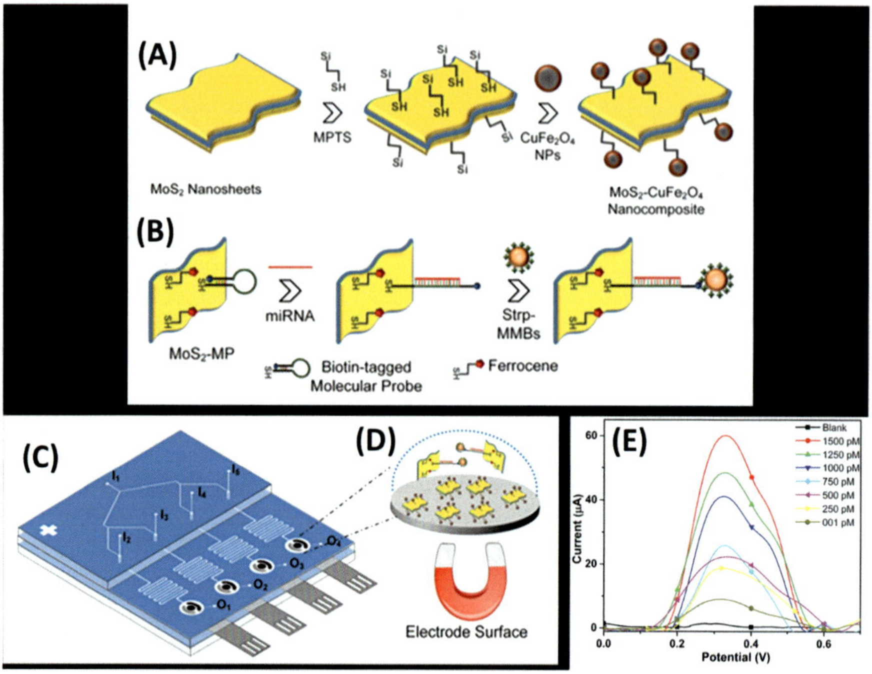

Figure 12.9 (A) Synthesis of $MoS_2-CuFe_2O_4$ nanocomposite. (B) MoS_2-MP nanocarriers for miRNA detection principle. (C) PDMS-based microfluidic platform showing the inlets (I1-I5) and outlets (O1–O4), and integrated electrodes. (D) Enlarged view of miRNA detection on the $MoS_2-CuFe_2O_4$ nanocomposite electrode. (E) Square-wave voltammetry curves of various concentrations of miRNA in PBS buffer solution (Chand et al., 2018).

Neethirajan, 2018). MoS_2 nanosheets-based nanocarrier was functionalized with biotin-tagged molecular probe (MoS_2-MP). Fig. 12.9B shows the label-free miRNAs detection on MoS_2-MP and the miRNA interaction in a single step.

Fig. 12.9C shows the design of a 3D microfluidic platform using lithography on PDMS to fabricate microchannels. The top layer of PDMS consists of five inlet zones, the middle layer has mixing zone, incubation zone, sensing zone, and outlet zone, each of four in numbers, while the bottom layer contains four screen-printed electrodes. The microfluidic platform was integrated the $MoS_2-CuFe_2O_4$ nanocomposite screen-printed electrode with the sample-processing zone. miRNA (20 μL) was inserted through I1 into the sensing zone of the microfluidic platform and the mixture of MoS_2-MP and streptavidin-coated magnetic microbeads (Strp-MMBs) were injected through I2–I5. The miRNA and MoS_2-MP mixture are mixed in the mixing zone and passed through the incubation zone where the molecular probes immobilized on the MoS_2-MP captured the miRNA. The magnets trapped all the MoS_2-MP and Strp-MMB complex and leave the unreacted MoS_2-MP through the outlet. (Fig. 12.9D). The microfluidic electrochemical miRNA biosensors can detect Paratuberculosis (pTb) very rapidly with high selectivity and sensitivity (Chand et al., 2018). The high selectivity of miRNA is due to the biotin-tagged thiolated molecular probe and functionalized MoS_2 as a nanocarrier. Furthermore, the MoS_2 serves as a matrix for the maximum loading of the molecular probe leads to an increase in the electrochemical signals. The fabricated microfluidic biosensor was detected the miRNA concentration of 1 pM to 1.5 nM with the limit of detection was 0.48 pM using square-wave voltammetry (Fig. 12.9E) (Medina-Sánchez, Miserere, & Merkoçi, 2012).

12.6.3 Metal oxides incorporated in microfluidic biosensors for point-of-care-devices

Microfluidic biosensors have many advantages that make them suitable solutions for POC devices: less reagent consumption, less biorecognition time, portability, small size, design versatility, multiplexing, and cost-effective (Zarei, 2017). Fully integrated POC devices do not require sophisticated instrumentation or expert technicians, which allows their use in resource-limited areas. POC devices for various types of testing including blood glucose monitoring, pregnancy, cardiac biomarker, urine analysis, etc. have been realized (Mohammadniaei, Nguyen, Van Tieu, & Lee, 2019). Various types of 2D materials such as graphene, transition metal dichalcogenide, Mxene, and metal oxides have been incorporated in POC devices to enhance the signal. Optical and electrochemical sensing methods are generally used for POC screening devices. The integration of various technologies, such as microfluidics, smartphone readouts, or paper strip, with the existing sensing methods, leads to developing portable and user-friendly POC sensing devices. Recently, Ko et al. (2019) developed an electrochemical microfluidic biosensor POC device for the monitoring of the H_2O_2 level in artificial urine. A bimetallic nanoparticle-decorated GO (Au@PtNP/GO) nanostructure shows catalytic activity toward 3,3′,5,5′-tetramethylbenzidine (TMB) substrate in the presence of H_2O_2, with the limit of detection of 1.62 μM. The inexpensive polymeric thin film-based device understands the electrochemical POC cancer screening for small molecule detection.

12.7 New advancement in microfluidic biosensors

12.7.1 Wearable microfluidic biosensing

The advancement in microfluidics, fabrication techniques, material design, and digital electronics paves the way toward the developing of wearable biosensors. These devices measure noninvasively and in real time the biochemical markers in biofluids such as saliva, tears, sweat, and interstitial fluid (Bandodkar, Jeerapan, & Wang, 2016; Kim, Campbell, de Ávila, & Wang, 2019). Apart from biosensing, wearable biosensors can measure the physical parameters such as motion, temperature, potential or strain (Choi, Kang, Han, Kim, & Rogers, 2017; Koh et al., 2016). The wearable microfluidic devices consist of an adhesion layer for sweat collecting, a microfluidic channel equipped with reagents for colorimetric analysis, and a magnetic loop antenna as an electronics communication for transmitting the signals to the smartphone. In a reported literature, thin and soft microfluidic devices are designed to harvest the sweat from the pores of skins from different locations and route this sweat to different channels for multiparametric sensing. The microchannel layer was fabricated which consists of four circular reservoirs and an orbicular serpentine channel surrounded them. An independent guiding channel in the skin-compatible adhesive layer is connected with four reservoirs and channels. The reservoirs are filled with a chromogenic reagent for glucose, lactate, pH, and chloride concentration detection. If the backpressure is created to block the fluid, an outlet channel on top layer of the device is linked with all the reservoirs and channels (Koh et al., 2016). Similar wearable microfluidic biosensors are also used to detect in situ measurements of zinc and sodium in sweat (Sekine et al., 2018).

12.7.2 Microfluidic paper-based device

Microfluidic paper-based device (μPAD) is an ideal platform for the POC testing due to low cost, portability, and easy to use (Dungchai, Chailapakul, & Henry, 2009; Sekine et al., 2018). First μPAD-based biosensor is reported by Whiteside's group. A microfluidic channel is fabricated on μPAD to control the flow direction of liquid which can detect multiple targets in a single biological sample. An electrochemical method is used on the paper-based device to detect uric acid, lactate, and glucose by oxidase-based enzymatic reactions (Chen et al., 2012). A paper-based microfluidic POC device is fabricated for drug metabolism monitoring.

12.7.3 Continuous microfluidic-based biosensing

Continuous microfluidic-based biosensing consists of lithographically fabricated microchannels, syringe pump, inlet, and outlet chamber as shown in Fig. 12.10A. The motion of continuous flow in the microchannel is the main operating principle of such a microfluidic system. The advantages of a continuous microfluidic system include ease of fabrication and operation, operation of large sample volume in a continuous flow, and compatibility with the current screening and sensing mechanisms.

12.7.4 Discrete microfluidic-based biosensing

Discrete or droplet-based microfluidic biosensing methods have improved throughput, fast and efficient mixing, creation of isolated reaction sites, better screening, reduced sample volumes, and single-cell analysis capabilities (Fig. 12.10B) (Brouzes et al., 2009). This microfluidic technique uses a two-compartment where the aqueous droplet of 1 pL to 10 nL volume is surrounded by immiscible oil. The operating principle of the discrete microfluidic device is the motion of droplets in microchannels using streams of immiscible fluids. The chemical and physical mixing of droplets is fully eliminated which reduces the risk of cross-contamination. Moreover, this technique can be suitable for stem cells from patients due to the analysis of a small number of cells.

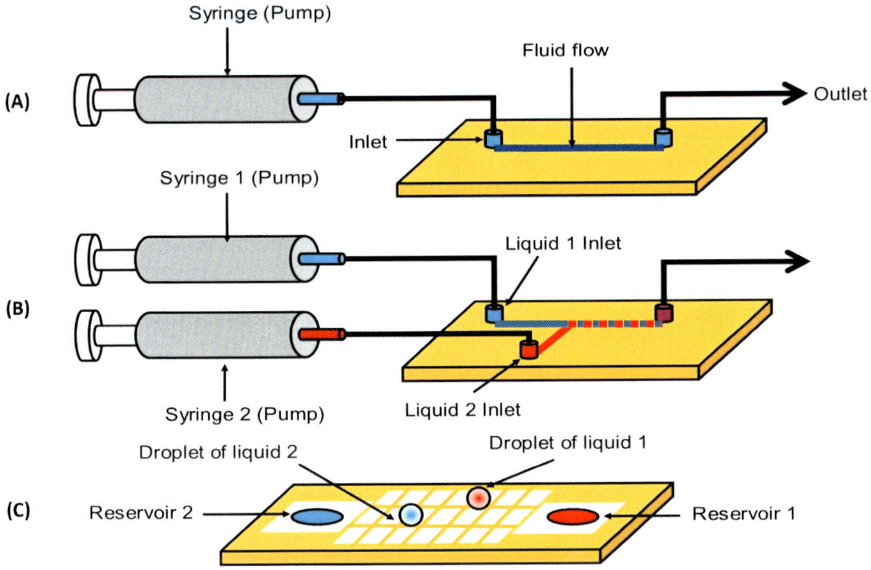

Figure 12.10 Schematic presentation of (A) continuous; (B) drop-base, and (C) digital microfluidic systems (Nikoleli et al., 2018).

12.7.5 Digital microfluidic biosensing

This type of microfluidic device is based on the principle of the electrowetting-on-dielectric (EWOD) technique and uses an array of electrostatically actuated electrodes. In the electrowetting (EW) method, the contact angle of the liquid droplet is tuned by applying the external voltage. The applied voltage reduces the surface tension at the liquid–solid interface, decreases the contact angle of the droplet, and wets the substrate surface. The EWOD is the EW method applied on the dielectric surface to modify the wettability of inert hydrophobic liquid droplets using an external voltage (Kim et al., 2018). The advantages of the digital microfluidic devices are low sample volume, better localization, scalability, portability, and capability to process multiple samples on a single device. However, the complicated fabrication process, evaporation, and bioadsorption are the major challenges in this system (Fig. 12.10C) (Nikoleli et al., 2018).

12.8 Future perspective

From the last decade, microfluidic biosensor technology has progressed very fast, producing complex microfluidic sensors with various fabricating techniques, and also developing handheld easy to use robust devices. However, it is still a long way to go before we can fulfill real-time detection with greater accuracy. We are still using conventional microfluidic devices which have external bulky detectors. To enhance the sensing performance, fully integrated and compact microfluidic systems are still required. The concurrent design including materials, sensing methods, fabrication techniques, and microfluidics is expected to produce more reliable and user-friendly devices, showing ultra-high sensitivity and very low limit of detection. Nanomaterials and nanocomposites incorporating bio/chemically functionalization could be integrated into the microfluidic channel to better recognize the target biomolecules. Apart from the hardware, software development and connectivity are also essential for the more sophisticated microfluidic biosensors. A deeper theoretical understanding of microfluidics is still required. In the case of wearable microfluidic biosensors, the continuous long-term real-time monitoring is still a challenge.

References

Aeinehvand, M. M., Magaña, P., Aeinehvand, M. S., Aguilar, O., Madou, M. J., & Martinez-Chapa, S. O. (2017). *RSC Advances*, *7*, 55400–55407.

Aeinehvand, M. M., Martins Fernandes, R. F., Jiménez Moreno, M. F., Lara Díaz, V. J., Madou, M., & Martinez-Chapa, S. O. (2018). *Sensors & Actuators, B: Chemical*, *276*, 429–436.

Aeinehvand, M. M., Weber, L., Jiménez, M., Palermo, A., Bauer, M., Loeffler, F. F., ... Martínez-Chapa, S. O. (2019). *Lab on a Chip*, *19*, 1090–1100.

Ali, M. A., Solanki, P. R., Patel, M. K., Dhayani, H., Agrawal, V. V., John, R., & Malhotra, B. D. (2013). *Nanoscale*, *5*, 2883–2891.

Ali, M. A., Srivastava, S., Solanki, P. R., Reddy, V., Agrawal, V. V., Kim, C., ... Malhotra, B. D. (2013). *Scientific Reports*, *3*, 2661.

Azahar Ali, M., Srivastava, S., Solanki, P. R., Varun Agrawal, V., John, R., & Malhotra, B. D. (2012). *Applied Physics Letters*, *101*, 84105.

Backhurst, J. R., Harker, J. H., Richardson, J. F., & Coulson, J. M. (1999). *Chemical engineering, Vol. 1: Fluid flow, heat transfer and mass transfer* (6th ed.).

Bae, C. W., Toi, P. T., Kim, B. Y., Il Lee, W., Lee, H. B., Hanif, A., ... Lee, N.-E. (2019). *ACS Applied Materials & Interfaces*, *11*, 14567–14575.

Bai, J., & Zhou, B. (2014). *Chemical Reviews*, *114*, 10131–10176.

Bandodkar, A. J., Jeerapan, I., & Wang, J. (2016). *ACS Sensors*, *1*, 464–482.

Brouzes, E., Medkova, M., Savenelli, N., Marran, D., Twardowski, M., Hutchison, J. B., ... Samuels, M. L. (2009). *Proceedings of National Academy of Sciences of the United States of America*, *106*, 14195–14200.

Byron Bird, R., Stewart, W. E., & Lightfoot, E. N. (2012). *Transport phenomena* (2nd ed.)..

Cao, F., Guo, S., Ma, H., Shan, D., Yang, S., & Gong, J. (2011). *Biosensors & Bioelectronics*, *26*, 2756–2760.

Carnero, A. (2006). *Clinical & Translational Oncology: Official Publication of the Federation of Spanish Oncology Societies and of the National Cancer Institute of Mexico*, *8*, 482–490.

Chand, R., Ramalingam, S., & Neethirajan, S. (2018). *Nanoscale*, *10*, 8217–8225.

Chen, X., Chen, J., Wang, F., Xiang, X., Luo, M., Ji, X., & He, Z. (2012). *Biosensors & Bioelectronics*, *35*, 363–368.

Chen, S.-C., & Liao, C.-M. (2006). *The Science of the Total Environment*, *366*, 112–123.

Chen, C., & Wang, J. (2020). *Analyst*, *145*, 1605–1628.

Chhowalla, M., Shin, H. S., Eda, G., Li, L.-J., Loh, K. P., & Zhang, H. (2013). *Nature Chemistry*, *5*, 263–275.
Choi, S., Goryll, M., Sin, L. Y. M., Wong, P. K., & Chae, J. (2011). *Microfluid Nanofluidics*, *10*, 231–247.
Choi, J., Kang, D., Han, S., Kim, S. B., & Rogers, J. A. (2017). *Advanced Healthcare Materials*, *6*, 1601355.
Coluccio, M. L., Perozziello, G., Malara, N., Parrotta, E., Zhang, P., Gentile, F., ... Di Fabrizio, E. (2019). *Microelectronic Engineering*, *208*, 14–28.
Craven, T. J., Rees, J. M., & Zimmerman, W. B. (2008). *Physics of Fluids*, *20*, 43603.
Deng, Y., Fan, J., Zhou, S., Zhou, T., Wu, J., Li, Y., ... Wu, Y. (2014). *Biomicrofluidics*, *8*, 24101.
Duan, Z., & Muzychka, Y. S. (2007). *Microfluid Nanofluidics*, *3*, 473–484.
Ducrée, J., Haeberle, S., Brenner, T., Glatzel, T., & Zengerle, R. (2006). *Microfluid Nanofluidics*, *2*, 97–105.
Dungchai, W., Chailapakul, O., & Henry, C. S. (2009). *Analytical Chemistry*, *81*, 5821–5826.
Ertürk, G., & Mattiasson, B. (2017). *Sensors (Basel)*, *17*, 390. doi:10.3390/s17020390.
Evans, R., & Hanson, J. R. (1976). *Journal of the Chemical Society, Perkin Transactions*, *1*, 326–329.
Fiorini, G. S., & Chiu, D. T. (2005). *Biotechniques*, *38*, 429–446.
Fouad, M., Yavuz, M., & Cui, B. (2010). *Journal of the Vacuum Science & Technology B*, *28*, C6I11–C6I13.
Gale, B. K., Jafek, A. R., Lambert, C. J., Goenner, B. L., Moghimifam, H., Nze, U. C., & Kumar Kamarapu, S. (2018). *Inventions*, *3*, 60.
Garcês Gonçalves, P. R., De Abreu, H. A., & Duarte, H. A. (2018). *Journal of Physical Chemistry C*, *122*, 20841–20849.
Gawande, M. B., Goswami, A., Felpin, F.-X., Asefa, T., Huang, X., Silva, R., ... Varma, R. S. (2016). *Chemical Reviews*, *116*, 3722–3811.
Ghasemi, A., Amiri, H., Zare, H., Masroor, M., Hasanzadeh, A., Beyzavi, A., ... Hamblin, M. R. (2017). *Microfluid Nanofluidics*, *21*, 151.
Goral, V. N., Zaytseva, N. V., & Baeumner, A. J. (2006). *Lab on a Chip*, *6*, 414–421.
Ha, D., Hong, J., Shin, H., & Kim, T. (2016). *Lab on a Chip*, *16*, 4296–4312.
Jarosch, D. (1987). *Mineralogy and Petrology*, *37*, 15–23.
Kazemzadeh, A., Ganesan, P., Ibrahim, F., Kulinsky, L., & Madou, M. J. (2015). *RSC Advances*, *5*, 8669–8679.
Kim, T.-H., Abi-Samra, K., Sunkara, V., Park, D.-K., Amasia, M., Kim, N., ... Cho, Y.-K. (2013). *Lab on a Chip*, *13*, 3747–3754.
Kim, J., Campbell, A. S., de Ávila, B. E.-F., & Wang, J. (2019). *Nature Biotechnology*, *37*, 389–406.
Kim, T.-H., Hwang, H., Gorkin, R., Madou, M., & Cho, Y.-K. (2013). *Sensors & Actuators, B: Chemical*, *178*, 648–655.
Kim, J.-H., Lee, J.-H., Kim, J.-Y., Mirzaei, A., Wu, P., Kim, H. W., & Kim, S. S. (2018). *Journal of Materials Chemistry C*, *6*, 6808–6815.
Kimmel, D. W., LeBlanc, G., Meschievitz, M. E., & Cliffel, D. E. (2012). *Analytical Chemistry*, *84*, 685–707.
Ko, E., Tran, V.-K., Son, S. E., Hur, W., Choi, H., & Seong, G. H. (2019). *Sensors & Actuators, B: Chemical*, *294*, 166–176.
Koh, A., Kang, D., Xue, Y., Lee, S., Pielak, R. M., Kim, J., ... Rogers, J. A. (2016). *Science Translational Medicine*, *8*, 366ra165 LP-366ra165.
Kumar, R., Chauhan, M., Moinuddin, M. G., Sharma, S. K., & Gonsalves, K. E. (2020). *ACS Applied Materials & Interfaces*, *12*, 19616–19624.

Kumar, R., Rai, P., & Sharma, A. (2016). *Journal of Materials Chemistry A*, *4*, 9822–9831.
Kuo, J. S., & Chiu, D. T. (2011). *Annual Review of Analytical Chemistry*, *4*, 275–296.
Kuo, J.-N., & Li, B.-S. (2014). *Biomedical Microdevices*, *16*, 549–558.
Kurokawa, Y., Sano, T., Ohta, H., & Nakagawa, Y. (1993). *Biotechnology and Bioengineering*, *42*, 394–397.
La Rosa, C., Pariente, F., Hernández, L., & Lorenzo, E. (1995). *Analytica Chimica Acta*, *308*, 129–136.
Li, K., Morton, K., Veres, T., & Cui, B., (Eds.) (2011). M. B. T. In M. Moo-Young (Ed.), *Comprehensive biotechnology* (2nd ed., pp. 125–139). Burlington: Academic Press.
Liu, M., Sun, X.-T., Yang, C.-G., & Xu, Z.-R. (2016). *Journal of Colloid and Interface Science*, *466*, 20–27.
Loo, J. F. C., Ho, A. H. P., Turner, A. P. F., & Mak, W. C. (2019). *Trends in Biotechnology*, *37*, 1104–1120.
Maduraiveeran, G. (2020). *Analytical Methods*, *12*, 1688–1701.
Manisalidis, I., Stavropoulou, E., Stavropoulos, A., & Bezirtzoglou, E. (2020). *Front Public Health*, *8*, 14.
Mark, D., Haeberle, S., Roth, G., von Stetten, F., & Zengerle, R. (2010). *Chemical Society Reviews*, *39*, 1153–1182.
Marzban, M., Dargahi, J., & Packirisamy, M. (2019). *Electrophoresis*, *40*, 388–400.
Mathews, J., & Walker, R. L. (1970). *Mathematical methods of physics*. New York: W.A. Benjamin.
Medina-Sánchez, M., Miserere, S., & Merkoçi, A. (2012). *Lab on a Chip*, *12*, 1932–1943.
Mohammadniaei, M., Nguyen, H. V., Van Tieu, M., & Lee, M.-H. (2019). *Micromachines*, *10*, 662.
Mondal, M., Misra, R. P., & De, S. (2014). *International Journal of Thermal Sciences*, *86*, 48–59.
Morijiri, T., Sunahiro, S., Senaha, M., Yamada, M., & Seki, M. (2011). *Microfluid Nanofluidics*, *11*, 105–110.
Movlaee, K., Ganjali, M. R., Norouzi, P., & Neri, G. (2017). *Nanomaterials (Basel, Switzerland)*, *7*, 406. doi:10.3390/nano7120406.
Ng, J. M. K., Gitlin, I., Stroock, A. D., & Whitesides, G. M. (2002). *Electrophoresis*, *23*, 3461–3473.
Nikoleli G. -P., Siontorou C. G., Nikolelis D. P., Bratakou S., Karapetis S. & Tzamtzis N., (2018). In *Advanced nanomaterials*, (eds.) D.P. Nikolelis, G.-P. B. T.-N., B. Nikoleli, Elsevier, pp. 375–394.
Panton, R. L. (2013). *Incompressible Flow*, 111–126.
Prakash, S., Pinti, M., & Bhushan, B. (2012). *Philosophical Transactions of the Royal Society a-Mathematical Physical and Engineering Sciences*, *370*, 2269–2303.
Rackus, D. G., Shamsi, M. H., & Wheeler, A. R. (2015). *Chemical Society Reviews*, *44*, 5320–5340.
Raoufi, M. A., Razavi Bazaz, S., Niazmand, H., Rouhi, O., Asadnia, M., Razmjou, A., & Ebrahimi Warkiani, M. (2020). *Soft Matter*, *16*, 2448–2459.
Ronkainen, N. J., Halsall, H. B., & Heineman, W. R. (2010). *Chemical Society Reviews*, *39*, 1747–1763.
Roy, E., Stewart, G., Mounier, M., Malic, L., Peytavi, R., Clime, L., ... Veres, T. (2015). *Lab on a Chip*, *15*, 406–416.
SadAbadi, H., Badilescu, S., Packirisamy, M., & Wüthrich, R. (2013). *Biosensors & Bioelectronics*, *44*, 77–84.
Sadeghi, S. J. (Ed.) (2013). In G. C. K. Roberts (Ed.), *Encyclopedia of biophysics* (pp. 61–67). Springer Berlin Heidelberg, Berlin, Heidelberg.

Salehabadi, A., & Enhessari, M. (Eds.). (2019). V. Grumezescu and A. M. B. T.-M. for B. E. Grumezescu (pp. 357–396). Elsevier.

Sekine, Y., Kim, S. B., Zhang, Y., Bandodkar, A. J., Xu, S., Choi, J., ... Rogers, J. A. (2018). *Lab on a Chip*, *18*, 2178–2186.

Sharafeldin, M., Bishop, G. W., Bhakta, S., El-Sawy, A., Suib, S. L., & Rusling, J. F. (2017). *Biosensors & Bioelectronics*, *91*, 359–366.

Sia, S. K., & Whitesides, G. M. (2003). *Electrophoresis*, *24*, 3563–3576.

Singh, N., Ali, M. A., Rai, P., Sharma, A., Malhotra, B. D., & John, R. (2017). *ACS Applied Materials & Interfaces*, *9*, 33576–33588.

Singh, N., Rai, P., Ali, M. A., Kumar, R., Sharma, A., Malhotra, B. D., & John, R. (2019). *Journal of Materials Chemistry B*, *7*, 3826–3839.

Smith, S., Mager, D., Perebikovsky, A., Shamloo, E., Kinahan, D., Mishra, R., ... Korvink, J. G. (2016). *Micromachines*, *7*, 22.

Song, Y., Lin, B., Tian, T., Xu, X., Wang, W., Ruan, Q., ... Yang, C. (2019). *Analytical Chemistry*, *91*, 388–404.

Steffen, V., Silva, E. A., Evangelista, L. R., & Cardozo-Filho, L. (2018). *Surfaces and Interfaces*, *10*, 144–148.

Stumpf, F., Schwemmer, F., Hutzenlaub, T., Baumann, D., Strohmeier, O., Dingemanns, G., ... Mark, D. (2016). *Lab on a Chip*, *16*, 199–207.

Suginta, W., Khunkaewla, P., & Schulte, A. (2013). *Chemical Reviews*, *113*, 5458–5479.

Suleimanov, B. A., & Abbasov, H. F. (2017). *Journal of Dispersion Science Technology*, *38*, 1103–1109.

Sun, S.-N., Wei, C., Zhu, Z.-Z., Hou, Y.-L., Venkatraman, S. S., & Xu, Z.-C. (2014). *Chinese Physics B*, *23*, 37503.

Thévenot, D. R., Toth, K., Durst, R. A., & Wilson, G. S. (2001). *Biosensors & Bioelectronics*, *16*, 121–131.

Tian, W. -C., & Finehout, E. (2009). Springer United States, Boston, MA, 2009, pp. 1–34.

van den Berg, A., Olthuis, W., & Bergveld, P. (2000) *Proceedings of the μTAS 2000 symposium, held in Enschede, the Netherlands, 14–18 May 2000.*

Viswanathan, S., Narayanan, T. N., Aran, K., Fink, K. D., Paredes, J., Ajayan, P. M., ... Renugopalakrishanan, V. (2015). *Materials Today*, *18*, 513–522.

Wang, Y., Bhushan, B., & Maali, A. (2009). *Journal of Vacuum Science & Technology. A, Vacuum, Surfaces, and Films: an Official Journal of the American Vacuum Society*, *27*, 754–760.

Wongkaew, N., Simsek, M., Griesche, C., & Baeumner, A. J. (2019). *Chemical Reviews*, *119*, 120–194.

Yang, L., & Guiseppi-Elie, A. (Eds). (2008). Impedimetric biosensors for nano- and microfluidics. In D. Li (Ed), Encyclopedia of microfluidics and nanofluidics (pp. 811–823). Boston, MA: Springer US.

Zarei, M. (2017). *TrAC Trends in Analytical Chemistry*, *91*, 26–41.

Zaytseva, N. V., Goral, V. N., Montagna, R. A., & Baeumner, A. J. (2005). *Lab on a Chip*, *5*, 805–811.

Zhao, B., Wang, T., Jiang, L., Zhang, K., Yuen, M. M. F., Xu, J.-B., ... Wong, C.-P. (2016). *Electrochimica Acta*, *192*, 205–215.

Zhu, C., Yang, G., Li, H., Du, D., & Lin, Y. (2015). *Analytical Chemistry*, *87*, 230–249.

Nanomaterials of metal and metal oxides for optical biosensing application

13

Sunil Dutt[1], Abhishek Kumar Gupta[2,3], Keshaw Ram Aadil[4], Naveen Bunekar[5], Vivek K. Mishra[6], Raj Kumar[7], Abhishek Gupta[8], Abhishek Chaudhary[9], Ashwani Kumar[10], Mohit Chawla[11] and Kishan Gugulothu[12]

[1]Department of Chemistry, Government Post Graduate College Una, Una, India, [2]Organic Semiconductor Centre, EaStCHEM School of Chemistry, University of St. Andrews, Fife, United Kingdom, [3]Organic Semiconductor Centre, SUPA, School of Physics and Astronomy, University of St. Andrews, Fife, United Kingdom, [4]Center for Basic Sciences, Pt. Ravishankar Shukla University, Raipur, India, [5]Department of Chemistry, Center for Nanotechnology, Chung Yuan Christian University, Chung Li, Taiwan (ROC), [6]Independent Researcher, Groningen, The Netherlands, [7]Department of Pharmaceutical Sciences, University of Michigan, Ann Arbor, MI, United States, [8]Department of Desalination and Water Treatment, Zuckerberg Institute for Water Research, The Jacob Blaustein Institutes for Desert Research, Ben-Gurion University of the Negev, Beer Sheva, Israel, [9]Department of Biotechnology and Bioinformatics, Jaypee University of Information Technology, Solan, India, [10]Department of Chemistry, Government College Anni, Kullu, India, [11]Gyan Sankul, Gyan Sankul Complex, Kullu, India, [12]Department of chemistry, Osmania University, Hyderabad, India

13.1 Introduction

Nanomaterial relates to nanotechnology and the term was first introduced by Norio Taniguchi in 1974 (Guisbiers, Mejía-Rosales, & Deepak, 2012). Later, K. Eric Drexler popularized and invented the concept of nanotechnology and molecular nanotechnology (Morrow, Bawa, & Wei, 2007). Nanomaterials were defined by the *International Organization for Standardization and European Commission*, they can be natural (glucose, DNA, hemoglobin, enzymes and antibodies, etc.), incidental (dust storm, cosmic dust, volcanic eruptions, etc.) manufactured (combustion of fuel, oil, and coal in airplane engines, welding, ore refining, chemical manufacturing, and smelting, etc.), and having at least 50% or more particles in a range of 1.0–100 nm, present in an unbound, aggregate, or an agglomerated state (Bleeker et al., 2013). Nanoparticles are intrinsically small flecks, bridging bulk material (75,000 nm human hair) with atomic (0.00001 nm hydrogen atom) or molecular

Metal Oxides for Biomedical and Biosensor Applications. DOI: https://doi.org/10.1016/B978-0-12-823033-6.00011-9

structures, and present in various shapes such as cylindrical, spherical, conical, tubular, hollow core, spiral, flat, and can be organic, inorganic, and carbon-based in origin (Jeevanandam, Barhoum, Chan, Dufresne, & Danquah, 2018). Dendrimers, micelles, liposomes, and ferritins are biodegradable organic and nontoxic nanomaterials (Kumar, 2019a, 2019b, 2020; Kumar, Dalvi, & Siril, 2020; Kumar, Soni, & Felix Siril, 2019; Tiwari, Kumar, Shefi, & Kaur Randhawa, 2020), whereas carbon-based nanoparticles are exclusively composed of carbon and include fullerenes, graphene, carbon nanotubes (CNTs), carbon nanofibers, and carbon black (Vats, Dutt, Kumar, & Siril, 2016). Inorganic nanoparticles of metals and transition metals [aluminum (Al), cadmium (Cd), cobalt (Co), copper (Cu), gold (Au), iron (Fe), lead (Pb), silver (Ag), and zinc (Zn)] have extensively been studied and efficiently been synthesized utilizing either destructive or constructive methods for various applications due to their pore size, surface charge, high surface area to volume ratio and surface charge density (Erathodiyil & Ying, 2011; Kumar, Kumar, Gedanken, & Shefi, 2018; Kumar, Kumar, Friedman et al., 2019; Kumar, Kumar, & Gedanken, 2020; Kumar, Kumar, Marcus, Gedanken, & Shefi, 2019; Sanchez, Belleville, Popall, & Nicole, 2011). To change the physical or chemical properties of metal nanoparticles (MNPs) for the desired application, metal oxide nanoparticles were developed (Vayssieres, Hagfeldt, & Lindquist, 2000).

MNPs are transparent to light and are capable to change optical functions dependent on factors such as shape, size, surface area, doping, and interaction with the surroundings (Chawla, Kumar, & Siril, 2016; Dutt, Siril, Sharma, & Periasamy, 2015; Sharma, Chawla, & Randhawa, 2016; Sharma, Sinha, Dutt, Chawla, & Siril, 2016). A significant color change has been observed in Surface Plasmon Resonance (SPR) of Ag and Au nanoparticles on changing the shape, size, and rate of condensation (Dutt et al., 2015; Dutt, Siril, & Remita, 2017; Siril, Lehoux, Ramos, Beaunier, & Remita, 2012). Among these, specifically electronic and geometric properties make MNPs peculiar to be used in imaging sensors, displays, solar cells, photocatalysis, biomedicine, and an optical detector (Lingeshwar Reddy, Balaji, Kumar, & Krishnan, 2018; Tiwari et al., 2020). Two technologies have been used to probe the electronic properties of MNPs such as nonlocal technique photoelectron spectroscopy and a local approach scanning tunneling spectroscopy.

Nanomaterial equipped analytical devices are being used to derive the events happening at the nanoscale level in biomaterials (enzymes, antigens, blood, etc.) are called nanobiosensors (Parolo & Merkoçi, 2013). Conceptually a biosensor has three components namely bioreceptor (material which analyzed), transducer (modulate signals), and a detector (reproduction of responses). Where the transducer is a main segment of the device involves in the conversion of specific biochemical reaction energy into identifiable signal as a responses from interactions of a bio analyte with receptors (Erathodiyil & Ying, 2011; Parolo & Merkoçi, 2013; Sanchez et al., 2011; Vayssieres et al., 2000). In recent years, a huge number of biosensors have been produced and named (e.g., enzyme biosensor, antigen biosensor) based on analyte detected as per convention (Chawla, Kumari, & Siril, 2018; Dutt et al., 2015; Sharma, Chawla et al., 2016). Moreover, they were further classified based on a kind of nanomaterial used, the major classes being nanoparticle, nanotube, and nanowire-based nanobiosensors.

Here in this chapter, we discuss the metal-based and metal oxide-based optical biosensing techniques especially based on fluorescence, SPR, and surface-enhanced Raman spectroscopy (SERS). We have briefly discussed the scientific concept and working principle of each technique, use of metal, and metal oxide-based nanoparticles and nanocomposites, and their application in optical biosensing of various entities along with recent advancement and literature.

13.2 Optical biosensing strategies

Biosensors are of various types such as electrochemical, mass-based, biological element based, and optical (Crosley & Yip, 2018). This section mainly focuses on optical biosensors. Optical biosensors are again can be classified as fluorescence, SPR, surface enhanced Raman scattering (SERS), Fourier transform infrared, fluorescence, and others (Gupta, Sarkar, Katranidis, & Bhattacharya, 2019).

13.2.1 Optical biosensors based on fluorescence technique

The field of optical biosensors have been an accelerating research field over the past few decades (Kirsch, Siltanen, Zhou, Revzin, & Simonian, 2013; McDonagh, Burke, & MacCraith, 2008). The interaction of surfaces of substrates and biomolecules can be studied by optical properties and corresponding measurements such as fluorescence (Fig. 13.1) (Long, Zhu, & Shi, 2013), Raman scattering, photoluminescence, transmittance, absorbance, and reflectance (Li, Cushing, & Wu, 2015; Zhou

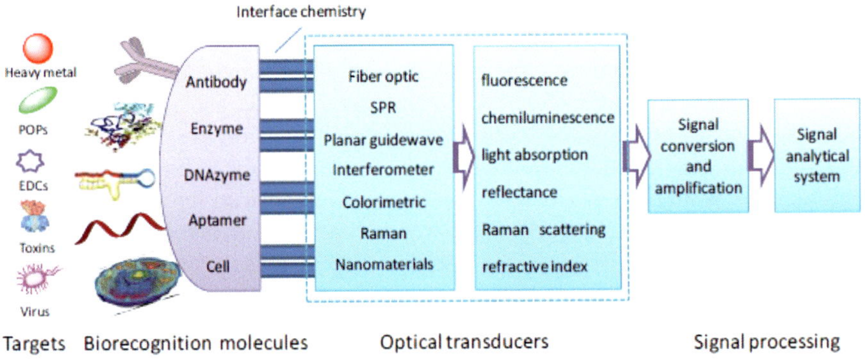

Figure 13.1 Schematic view of concept, working principle, mechanism, and design of optical biosensing technique for detection of various analyte.
Source: Reproduced with permission from Long, F., Zhu, A., & Shi, H. (2013). Recent advances in optical biosensors for environmental monitoring and early warning. *Sensors*, *13*, 13928–13948. doi:10.3390/s131013928. Open access, MDPI.

et al., 2016). In particular, optical methods of detection using fluorescence has potential to construct simple and robust device due to their great versatility, sensitivity, nondestructive measurement, multianalyte detection and solvent sensitivity which offer high signal-to-noise ratios (Jeong, Kook, Lee, & Koh, 2018; Walsh, 2011). In this section, we deal with fluorescence-based optical biosensors for metal/metal oxide and their interaction with biomaterials. In the following subsection, we have schematically discussed the fluorescence-based biosensor, utility of metal and metal oxide in fluorescence biosensors, and its biomedical applications.

13.2.1.1 Fluorescence-based optical biosensor

The fluorescence sensor became popular as optical sensor in the 1980s for the first chemically synthesized fluorescence probe based on quinoline derivative which showed fluorescence quenching for specific analyte such as chloride (Staiano, Bazzicalupo, Rossi, & D'Auria, 2005; Verkman, Sellers, Chao, Leung, & Ketcham, 1989). Fluorescence sensing technique offers numerous advantages over other optical sensing techniques because of their improved selectivity, sensitivity, and low detection limit (He, Feng, Kang, & Li, 2017; Jeong et al., 2018; Lakowicz, 1999). Generally, other optical techniques also worked on similar phenomena such as response to light behavior through interaction. However, fluorescence is a photoluminescence process where a substance emits light with a different wavelength from that of the absorbed wavelength. Therefore fluorescent materials can effectively change some intrinsic properties which makes fluorescent materials as excellent sensing probe. Typically, optical biosensors consist of three components as presented in Fig. 13.1. In fluorescence-based biosensor, fluorescent probes act as effective transducers that transfer respective changes into optical signals which can be analyzed by detector (Viter & Iatsunskyi, 2019; Walsh, 2011). The performance of fluorescence-based biosensor depends on two different components, that is, sensing components and transducing components. Labeling and nonlabeling assays can be used for sensing component modification. There is requirement of one kind of fluorophore and/or chromophore for labeled sensing, and commonly utilize fluorescence resonance energy transfer (FRET) to detect various analytes. Normally, a FRET-based assay contains a variety of donors and acceptors fluorophore and utilize overlaps between donor emission and acceptor excitation to detect analytes. By contrast, there is no need of any chromophores or fluorophores for label-free assays for the detection process. Recently, label-free aptamer sensors involving DNA intercalators, aptamer-binding dyes, and metal nanomaterials have much attraction (Feng, Dai, & Wang, 2014).

Furthermore, the nanomaterials are often used as alternatives for dye-labeled assay due to its very small size variations and unique optical and electronic properties which provide several ways to overcome limitations (low chemical stability, photobleaching, photo blinking, short lifetime, low color purity, and broad emission bands) associated with fluorescent dyes (Solanki, Kaushik, Agrawal, & Malhotra, 2011). In this regard, many improvements were made to construct fluorescence-based sensor by developing fluorescent assay based on nanomaterials such as quantum dot,

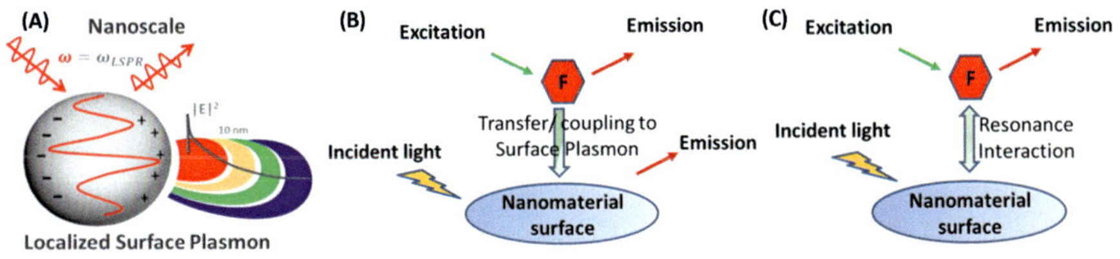

Figure 13.2 (A) Plasmon-coupling effect; (B) radiative decay engineering (RDE). *Source*: Reproduced with permission from Aslan, K., Leonenko, Z., Lakowicz, J. R., & Geddes, C. D. (2005). Annealed silver-island films for applications in metal-enhanced fluorescence: Interpretation in terms of radiating plasmons. *Journal of Fluorescence*, *15*, 643–654. doi:10.1007/s10895–005–2970-z; Jeong, Y., Kook, Y. M., Lee, K., & Koh, W. G. (2018). Metal enhanced fluorescence (MEF) for biosensors: General approaches and a review of recent developments. *Biosensors & Bioelectronics*, *111*, 102–116. doi:10.1016/j.bios.2018.04.007. Copyright 2005, 2018 Springer Science + Business Media Inc and the authors published by Elsevier, respectively.

upconversion nanoparticles, carbon dots, and 2D nanomaterials (Wongkaew, Simsek, Griesche, & Baeumner, 2019). In comparison to conventional fluorophores, functionalised fluorescent assay with nanomaterials shows improved optical response and the sensing mechanism and have great potentials in terms of brightness, photostability, and tunable emission spectrum (Fig. 13.2) (Aslan, Leonenko, Lakowicz, & Geddes, 2005). Depending on intrinsic nature of materials, some fluorescent nanomaterials showed high sensitivity toward target materials (Zheng et al., 2015; Zhu, Du, & Lin, 2017). Among various nanomaterials, metal and metal oxide significantly improved the biosensing ability by using fluorescence technique. On interactions of metal with the fluorophore increase the rate of excitation and emission by opening additional electron configuration of fluorophores and can improve the sensitivity of fluorescence detection as well as reduce the photodegradation of fluorophore in biosensor system. This phenomenon known as metal-enhanced fluorescence is originated through three mechanisms: (1) localized surface plasmon resonance (LSPR) near to metallic surface; (2) plasmon-coupling effect mediated by nonradiative interaction between the metal and fluorophores; and (3) radiative decay engineering (Fig. 13.2) (Jeong et al., 2018). Therefore fluorescent-based optical biosensors are very efficient tools to detect biomaterials and are useful in biomedical industry. Tremendous progress have been achieved in last decade in terms of advancement, use of various nanotechnologies, applications in different fields, and improvement in sensitivity and selectivity (Wongkaew et al., 2019). In the next section, we discuss the significance of metal and metal oxide in fluorescence biosensor.

13.2.1.2 Metal and metal oxides in fluorescence-based optical biosensor and its applications

In past three decades, metal-based and metal oxide-based (M/MO) fluorescent optical biosensors have significantly improved human life in many aspects such as environmental monitoring, healthcare, and food safety by determining the heavy metals and biologically important compounds (Aragay, Pons, & Merkoçi, 2011; Feng et al., 2014; Liang et al., 2017). Generally, trace amount of nonessential heavy metals, such as Cd, mercury (Hg), arsenic (As), and Pb are highly toxic and carcinogenic, while excessive amount of essential metals, such as nickel (Ni), Fe, Cu, and Zn can be dangerous to human health as well (Aragay et al., 2011). Therefore detecting and quantifying these metal ions at trace level is necessary for the environment and human health. Very often all metal ions have specific properties such as unique binding sites, metal-dependent activity, complex formation, and diverse catalytic mechanisms toward biological materials. Therefore these metals act as analyte that specially binds to bioreceptor or specific fluorophore and transfer signal to detector through signal transducer mechanisms such as fluorescent sensor. In this regard, DNAzymes/DNA molecule-based sensors became a trendy fluorescence biosensor due to its high metal ions selectivity and have been used to detect a wide range of metal ions such as Mg^{2+}, Zn^{2+}, Hg^{2+}, Cu^{2+}, Pb^{2+}, and UO^{2+} (Zhou et al., 2016). However, certain DNA bases can selectively bind with Hg^{2+} and Ag^{+} and form strong metal-base duplexes, that is, thymine$-Hg^{2+}-$thymine ($T-Hg^{2+}-T$) (Tang, Zhao, He, & Yin, 2010) or cytosine$-Ag^{+}-$cytosine ($C-Ag^{+}-C$) (Wang et al., 2014) complexes, while the Pb^{2+} can stabilize G-quadruplex (Li, Liu, Lin, & Chang, 2011). Therefore much effort has been made to develop nanomaterials-attached DNA molecules/DNAzymes as fluorescent biosensors for these metal ions. Based on the above technique, Ye and Yin developed a gold nanoparticle (GNP) enhanced fluorescence polarization-based assay $T-Hg^{2+}-T$ mismatch biosensor to detect heavy metal ion (Hg^{2+}) with detection limit 0.2 ppb (Ye & Yin, 2008). Similarly, Zeng group utilized the $T-Hg^{2+}-T$ complex and reported the functionalized DNA molecules with GNPs and quantum dots (QDs) as efficient "turn-on" fluorescent biosensor for Hg^{2+} ion (Huang, Niu, Wang, Lv, & Zeng, 2013). The sensing strategy of these nano-structured fluorescent biosensors depends on the FRET and time-gated fluorescence resonance energy transfer (TGFRET), whereas nanomaterials such as GNPs, QD, graphene oxide (GO), and CNTs were used as labeled quencher, fluorophore, and linked with DNAzymes for heavy metal detection. Inspired by this strategy, many researchers designed and reported fluorescent biosensor based on nanomaterial-functionalized DNA molecules for the detection of Hg^{2+}, Ag^{+}, Pb^{2+}, and UO^{2+} with very low detection limit (Aragay et al., 2011). For example, DNAzyme labeled with single-wall carbon nanotube (SWCNT) was used for Pb^{2+} ion detection and became first intracellular metal ion fluorescent sensors based on DNAzyme immobilized with GNPs (Liggett, 2014).

Furthermore, Ag, Au, and Cu nanoclusters-based DNA molecules are also recognized as fluorescent biosensor for heavy metal (Zhao, Wang, Mi, Jiang, & Wang, 2019). Due to excellent optical and electrical properties, Ag, Au, Cu, and carbon-based materials are frequently used for sensor. However, Ag, Au, Ag, and many more metals are quite expensive, and their surface properties limit their performance. To overcome these problems, metal oxides are the best alternatives due to their excellent physicochemical properties (Medhi, Marquez, & Lee, 2020). For example, Fe_3O_4 is a biocompatible magnetic material suitable for separation and drug delivery, TiO_2 is a photocatalyst, and CeO_2 is used as oxidative catalyst (Ganesh et al., 2012; Medhi et al., 2020). To enhance the sensing ability, metal oxides need to be exploited from macroscopic to nanoscopic surface and form metal oxide nanomaterials (MONs) because at nanoscopic level the surface-to-volume ratio is high which helps to manipulate the fluorescence signals (enhancement, quenching). In 2007 magnetic Fe_3O_4 nanoparticle was reported as a first catalyst to mimic peroxidase activity and it oxidized several chromogenic and fluorogenic substrates such as 3,30,5,50-tetramethylbenzidine (TMB) and 2,20-azino-bis(3-ethylbenzothiazoline-6-sulfonic acid) (ABTS) in the presence of H_2O_2 (Gao et al., 2007). Inspired by this work, other groups also reported many MONs, such as V_2O_5, Co_3O_4, Fe_2O_3, and CuO as catalysts to mimic peroxidase (Aragay et al., 2011; Medhi et al., 2020).

In addition, the surface of MONs can be hydrolyzed by the chemisorption of water and produce hydroxide. Through this process the surface of MONs can be easily modified and decorated with desired charge (positive or negative) by varying pH. For example, at physiological pH and basic buffer, most oxides are negatively charged while in acidic buffer the surface hydroxide becomes positively charged. Therefore surface modification plays a significant role in sensing ability because it not only affects the electrostatic interaction between nanomaterials and charged biomolecules but also alters the adsorption behavior, which affect the intrinsic properties of sensor and can enhance reactivity, stability, and catalytic activity (enzyme-like catalytic activity) of the system. Thus there are several types of interactions possible between the MO and biomacromolecules. By using these strategies, not only anions such as arsenate but also some biologically important molecules such as dopamine and catechol can strongly bind on the surface of any MONs and can serve as a bridge to anchor functional biomolecules (e.g., DNA) on the surface of oxides (Shultz, Ulises Reveles, Khanna, & Carpenter, 2007). For example, Pratsinis et al. reported an Eu^{3+}-doped nanoceria (CeO_2:Eu^{3+}) as a turn-off fluorescence biosensor for the detection of H_2O_2 in buffer solution and bacterial cell culture (Pratsinis et al., 2017). Moreover, CeO_2:Eu^{3+}materials not only possess as catalase-like activity but also act as luminescent materials and the mechanism depends on the adsorption of H_2O_2 on nanoceria, resulting in the decrease of CeO_2:Eu^{3+} luminescence. Also, TiO_2 nanomaterials are reported to be novel photoluminescence biosensor for retroviral leucosis detection based on adsorption of antibodies on the surface of TiO_2 nanoparticles (Viter et al., 2012). Study of adsorption of DNA oligonucleotides by a large number of MONs have been performed, including TiO_2, In_2O_3, ZnO, CeO_2, Fe_3O_4, NiO, CoO(OH), MnO_2, HfO_2, Al_2O_3, and Mg/Fe double hydroxide (Aragay et al., 2011). MONs has selectivity toward range of molecules and hence they are potential candidates for sensing. In the following section, finally we discuss the utilization of M/MO in the fluorescence-based optical biosensor in medical sciences.

13.2.1.3 Fluorescence optical biosensors applications

Biosensors play a significant role in human life because of their capability to resolve many analytical challenges in several fields (Medhi et al., 2020).

Biosensors are used to detect biotoxins, biological pathogens, and neurotoxin compounds, and used to develop medical biosensors. The biosensors accelerated their progress since the development of glucose biosensing technology. As we discussed in previous sections, metal and metal oxide have great significance to develop fluorescent optical biosensors. Zhang et al. utilized Au nanoparticles for the development of bio-barcode DNS assay. This technique showed promising results in the detection of salmonella bacteria with the 1 ng/mL limit of detection (Fig. 13.3) (Zhang, Carr, & Alocilja, 2009). Boujday et al. (2008) used antibody-conjugated gold to detect *Staphylococcus aureus* using fluorescence transduction technique. Donaldson, Kramer, and Lim (2004) proposed using antibodies against the Vaccinia virus labeled with fluorescent tracers for the detection of smallpox virus in throat culture swabs. Asati, Kaittanis, Santra, and Perez (2011) used ploy(acrylicacid) (PAA) modified nanoceria (PAA-CeO_2) for the oxidation of Ampliflu to form stable fluorescent product resorufin at pH 7.0 which was further used to detect lung cancer cell line A-549 and breast cancer cell MCF-7 (Fig. 13.4). There are several companies developed and commercialized the biosensors based on fluorescence technology such as PerkinElmer and Horiba. Finally, we can understand that metal/ metal oxide in fluorescence-based optical biosensors are playing a very important

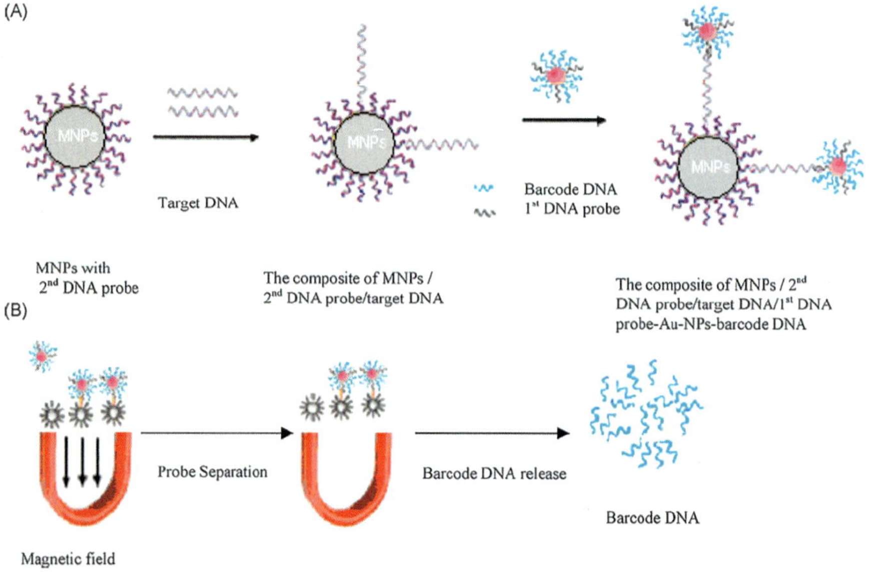

Figure 13.3 Schematic of the bio-barcode assay: (A) formation of the DNA probe complex of magnetic nanoparticles (MNPs) including the target DNA and the DNA probe-Au-NPs-barcode DNA and (B) the separation of the complex of the barcode DNA and release. *Source*: Reproduced with permission from Zhang, D., Carr, D. J., & Alocilja, E. C. (2009). Fluorescent bio-barcode DNA assay for the detection of *Salmonella enterica* serovar Enteritidis, *Biosensors & Bioelectronics*, *24*, 1377–1381. doi:10.1016/j.bios.2008.07.081. Copyright, 2008 Elsevier.

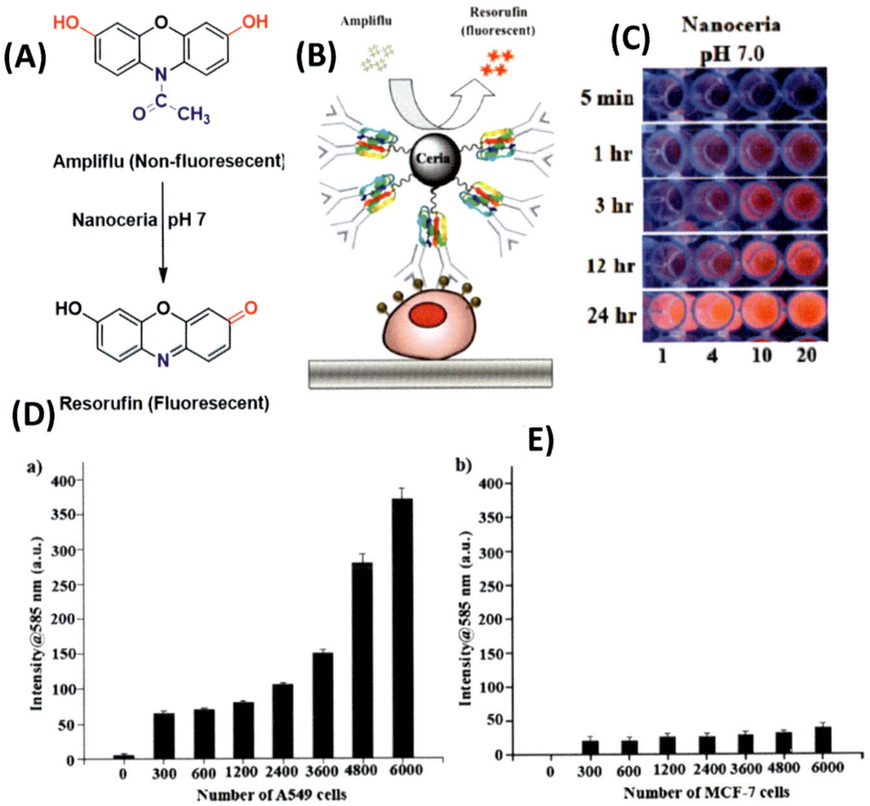

Figure 13.4 (A) Nanoceria mediated oxidation of ampliflu to resorufin at pH 7. (B) The ability of nanoceria-involved enzyme-linked immunosorbent assay (ELISA) can oxidize ampliflu into stable fluorescent products in pH range 6–8. (C) Images showing the oxidation of ampliflu at pH 7 by using nanoceria with respect of time and varying concentration of nanoceria. (D) Lung carcinoma cells (A-549). (E) MCF-7 breast carcinoma cells.
Source: Reproduced with permission from Asati, A., Kaittanis, C., Santra, S., & Perez, J. M. (2011). PH-tunable oxidase-like activity of cerium oxide nanoparticles achieving sensitive fluorigenic detection of cancer biomarkers at neutral pH. *Analytical Chemistry*, *83*, 2547–2553. doi:10.1021/ac102826k. Copyright 2011, American Chemical Society.

role to detect biological molecules and having vast applications in biomedical industries.

13.2.2 Optical biosensors based on surface plasma resonance

In this section, we have discussed the SPR-based optical biosensors. There are various materials used for SPR biosensors, but our major objective of this chapter is on metal-based and metal oxide-based optical biosensors. In the following subsection, we have schematically discussed the optical SPR technique, metal-based and metal

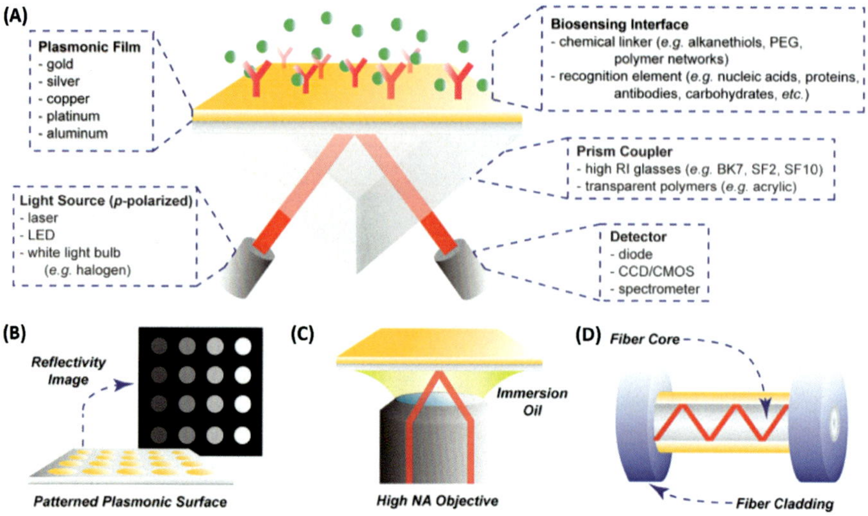

Figure 13.5 General instrumnetal surface plasma resonance (SPR) system. (A) Spectroscopic SPR setup in the Kretschmann configuration composed of several components. (B) In SPR imaging, interactions with a patterned plasmonic array are visualized and monitored for changes in reflected light intensity, indicating molecular interactions. (C) SPR microscopy utilized a high numerical aperture objective to guide incident light toward the gold film at a high angle of incidence. (D) Single-mode fiber SPR setup in which an area of optical fiber cladding is removed and a gold film is deposited.
Source: Reproduced with permision from Hinman, S. S., McKeating, K. S., & Cheng, Q. (2018). Surface plasmon resonance: Material and interface design for universal accessibility. *Analytical Chemistry*, *90*, 19–39. doi:10.1021/acs.analchem.7b04251. Copyright 2017, American Chemical Society.

oxide-based SPR optical biosensors, and its applications. Fig. 13.5 presents the common instrumental SPR system (Hinman, McKeating, & Cheng, 2018).

13.2.2.1 Surface plasma resonance sensing technology

In the past few decades, various optical biosensor methods have been developed and have found practical applications in various chemical fields such as analytical chemistry and biochemistry (Prakash, Chakrabarty, Singh, & Shahi, 2013). Among the various optical sensing methodologies such as SPR (Singh, 2016), quartz crystal microbalance (Mannelli, Minunni, Tombelli, & Mascini, 2003), and ellipsometry (Wang et al., 2019), SPR is a promising technique for optical biosensors (Zhou et al., 2019). In 1902 R.W. Wood observed SPR phenomenon and in 1968 it was explained (Li, 2019; Singh, Kaler, & Sharma, 2019). Then later Otto, Kretschmann, and Raether proposed the optical excitation of SPR (Heckmann et al., 2018). After almost 20 years, in 1983 using SPR technique, gas detection and biosensing were reported successfully (Omar & Fen, 2018). Since then, SPR has been widely employed as a sensing technique for detection of chemical and

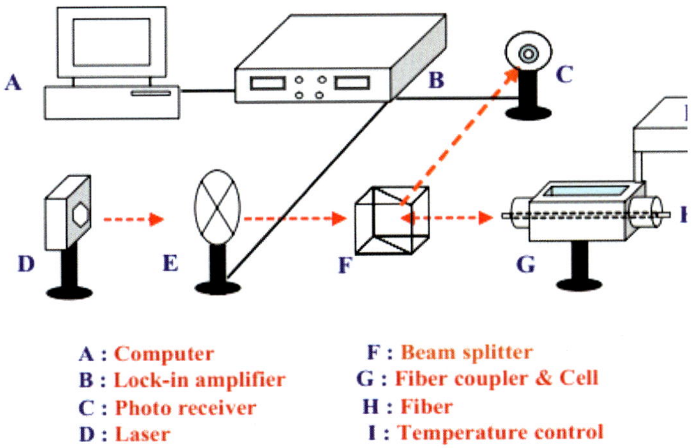

Figure 13.6 The systemic view of localized surface plasmon resonance (LSPR) biosensing system.
Source: Reproduced with permision from Lin, T.-J., & Chung, M.-F. (2008). Using monoclonal antibody to determine lead ions with a localized surface plasmon resonance fiber-optic biosensor. *Sensors*, *8*, 582–593. doi:10.3390/s8010582. Open access 2008 MDPI.

biological entities. Excitation of surface plasmon by using light is called SPR. Plasmonic is a phenomenon which is based on interaction of free electrons and electromagnetic fields in various materials including metal and metal oxides. SPR is an interface of metals and dielectric materials. SPR is again classified into two types: propagating SPR (PSPR) and localized SPR (LSPR). Fig. 13.6 presents the schematic view of LSPR-based biosensing system (Lin & Chung, 2008). Surface plasmon excitation occurs on the surface of the material/nanoparticle and is moved to thin film called as PSPR, on the other hand, nonpropagating SPR excited on the surface of the MNPs is called as LSPR (Nishikawa et al., 2006; Shin et al., 2007). Both PSPR and LSPR also improve SERS and fluorescence-based optical biosensors detection through change in refractive index (RI). The simple SPR is composed of source of light, optical system, coupling agent, imaging component, and detector (Fig. 13.7). The advantages of SPR optical biosensors are dependent on very high sensitivity of SPR excitation at metal–dielectric interface to tune the RI of the dielectric. The medium senses the RI change which results in the change in propagation constant of the SPR. This affects the condition for resonance, and based on conditions, SPR sensors can be classified as sensors with angular, intensity, wavelength, and/or phase modulation. Surface plasmon excitation by light waves is called optical sensors. Detection and sensing of biomolecules using optical sensors called optical biosensors. Hence, using light waves, and SPR chemistry detecting or sensing special biomolecules is called SPR-based optical biosensors. Further advantage of SPR optical biosensors is that it provides biosensing without requiring fluorescent, and radioactive moreover, it is interesting due to label-free technique for monitoring of biomolecular

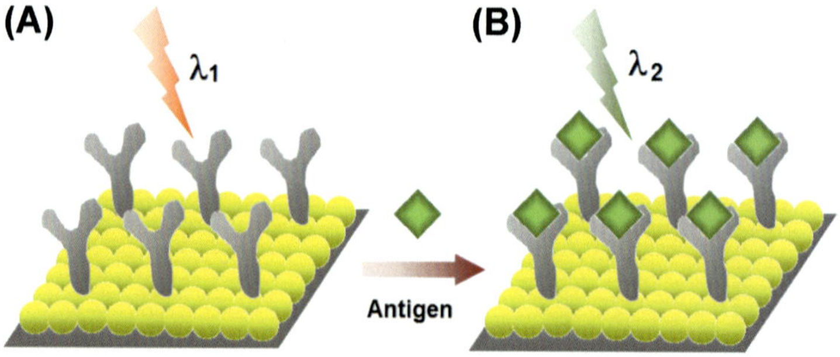

Figure 13.7 Schematic presentation of label-free optical detection of targeted biomolecules using localized surface plasmon resonance (LSPR) of metal nanoparticles. (A) LSPR absorbance of the metal nanoparticles (λ1) changes as the layer of antibodies is formed on the surface. (B) The binding of antigens with antibodies increases the layers on the surface and absorbance intensity of the metal nanoparticles increases (λ2), leading to sensitive detection of biorecognition events.
Source: Reproduced with permision from Vestergaard, M., Kerman, K., & Tamiya, E. (2007). An overview of label-free electrochemical protein sensors. *Sensors*, *7*, 3442–3458. doi:10.3390/s7123442. Open access 2007 MDPI.

interactions (Helmerhorst, Chandler, Nussio, & Mamotte, 2012; Lin et al., 2002). Fig. 13.6 presents the label-free optical detection of biomolecules using LSPR of MNPs (Vestergaard, Kerman, & Tamiya, 2007). SPR-based optical biosensing is extensively studied and commercially available. SPR is limited to inorganic materials especially noble metals. Properties of materials also affects SPR, and nanomaterials showed enhanced SPR with improved SPR selectivity and sensitivity. Among the different nanomaterials, metal and metal oxides are special classes of materials with unique properties for SPR applications. In the following section, we discuss the metal-based and metal oxide-based optical biosensors.

13.2.2.2 Metal and metal oxides in surface plasmon resonance based optical biosensor

Meanwhile, in recent years, inorganic nanomaterials, especially metal-based and metal oxide-based nanomaterials have been attracting more interest because of their unique properties without much effort, which provide an effective surface for better adsorption of the proteins, and properties such as optical, mechanical, magnetic, and chemical; and metal nanostructure allows strong coupling between the electron resonances with the light (Liao, Nehl, & Hafner, 2006; Si, Huang, Wang, & Ma, 2013; Wang, 2005). Simultaneously metal particles and biomolecule immobilization provide high biological activity which results in enhanced sensing characteristics (Haes, Zou, Schatz, & Van Duyne, 2003). With this regard the metal and metal oxides nanoparticles has been used for biosensing applications such as gold (Cao, Sun, & Grattan, 2014), silver (Gao et al., 2012), iron oxide (Zhang et al., 2014), copper oxide (Chan, Zhao, Hicks, Schatz, & Van Duyne, 2007), zinc oxide (Chiu, Tu, & Huang, 2013), and titanium oxide (Dominik et al., 2017; Solanki et al., 2011). Due to unique properties of materials at nanoscale, nanomaterials are promising candidates for sensing applications including SPR-based optical biosensors. Metal and metal oxides are major nanomaterials/nanoparticles that have been used in the process of detection of chemical and biological moieties using electrochemical sensing (Rhodes et al., 2006; Tamada, Nakamura, Ito, Li, & Baba, 2007). Recently, SPR techniques have been developed and used for heavy metal ion detection in water. Through using nanomaterials, the sensitivity of SPR detection significantly enhanced. This is due to enhanced surface area and high RI. Semiconductors, metals, alloys, and materials having large negative dielectric constant are suitable for SPR applications. However, silver and gold are the most widely used metal in SPR and SPR-based optical biosensors (Cao et al., 2014; Gao et al., 2012). Among the metal oxide nanoparticles, ZnO and TiO_2 are the metal oxides extensively employed in SPR-based optical biosensing (Chiu et al., 2013; Dominik et al., 2017). Compare to other nanomaterials, in metal and metal oxide nanoparticles electrons move 10–100 nm at ambient conditions, hence shrink corresponding direction and different effects can be observed. Due to size at nanoscale level, such properties will be enhanced further. The use of gold nanoparticles has been extensively studied and well established. Gold showed promising results in medicine, biotechnology, drug delivery, and sensing applications (Daniel & Astruc, 2003). Gold is also a promising candidate for SPR-based optical biosensors due to plasmonic and optical properties. It can also be used simultaneously for the detection and therapy of various diseases. Gold nanoparticles based SPR showed enhanced properties such as stability, sensitivity, and selectivity (Mustafa et al., 2010). More interestingly, the size and shape of gold nanoparticle affect its optical and electrical properties as well. By tuning the shape of gold nanostructures, the selectivity of SPR optical biosensors can be changed. The sensing of metal ion using gold nanoparticle is well established and studied. Furthermore, due to excellent chemistry between gold nanoparticle surface with thiol, they can also be used for detection of biomolecules composed of thiol functionality. Thiol-containing biomolecules such as glucose, glutathione, carcinoembryonic antigen, prolactin hormone, and biomolecules containing mercapto group can be detected using gold nanoparticles. Hence, due to the combination of two different properties, noble metals are promising candidates for the development of biosensors. Due to large number of reports on gold nanoparticles based SPR biosensing studies, several researchers reviewed the literature thoroughly. Cao, Ye, and Liu (2011) reviewed the signal amplification using gold nanoparticles. Cao et al. (2014) recently reviewed the gold nanorod-based LSPR biosensors. Li, Schluesener, and Xu (2010) and Zeng et al. (2011) reviewed gold-based biosensors and functionalized gold nanoparticles based biosensing applications, respectively. Other noble metals are also employed in SPR biosensing; however, their use is limited due to their physicochemical properties being not as good as gold. Similarly, in metal oxide category, zinc oxide and titanium oxide are the extensively used metal oxide nanoparticles investigated in SPR optical biosensors.

13.2.2.3 Surface plasmon resonance optical biosensors applications

Sensors are gaining more and more interest in recent years. Most advanced technology and self-performance devices, electronics, and robotics mainly depend on the sensor and their performance. Among different sensors, optical biosensors are gaining more interest in the field of drug delivery, nanomedicine, tissue engineering, and biomedical applications (Austin, Mackey, Dreaden, & El-Sayed, 2014). Fig. 13.8 presents various types of plasmonically engineered nanoprobes applied for biosensing and biomedical applications (Kumar, Kim, & Nam, 2016). It has been proven biosensors to monitor and provide information about biological processes with very promising results especially in studying biomolecular interactions. SPR has been proven to be one of the powerful technologies to detect the change in dielectric constants, affinity, and kinetic parameters during the binding of biomolecules in many types of bonds, including protein–DNA (Pollet et al., 2009), protein–protein (Karlsson & Fält, 1997), enzymes–substrate or inhibitor (Xu et al., 2005), antibodies–drug (Yu, Blankert, Viré, & Kauffmann, 2005), and virus–protein (Park et al., 2012). Most of the available commercial SPR instrument technologies are expanded to biomedical, environmental, and industrial areas. In the last 20 years, various researchers employed SPR-based optical sensing technology for different biosensing applications. Nowadays, SPR imaging process is one of the interesting methods applied for immune-sensing and measurement of thickness of thin films in the range of 2–20 Å. de Juan-Franco, Rodríguez-Frade, Mellado, and Lechuga (2013) employed SPR and detected successfully hGH isoforms 22K and 20K hGH in serum with a detection limit of 0.9 ng/mL using isoform-specific

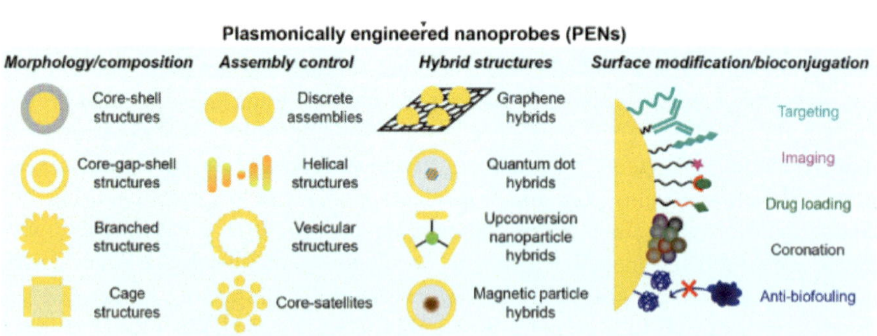

Figure 13.8 Various plasmonically engineered nanoprobes composed of different particles morphology, particle composition, assembled structure, hybrid structure, surface modification, and bioconjugation for biosensing and biomedical applications (Kumar et al., 2016).
Source: Reproduced with permission from Kumar, A., Kim, S., & Nam, J. M. (2016). Plasmonically engineered nanoprobes for biomedical applications, *Journal of the American Chemical Society*, 138, 14509–14525. doi:10.1021/jacs.6b09451. Copyright 2016 American Chemical Society.

monoclonal antibodies. Law, Yong, Baev, and Prasad (2011) successfully detected antigen (tumor necrosis factor-alpha, TNF-alpha) at very low concentration, that is, femtomolar range using nanoparticles (bio-conjugated gold nanorods), immunoassay method with the help of SPR system. The performance is nearly 40-fold higher compared to traditional SPR biosensing technique. Larsson, Alegret, Kăll, and Sutherland (2007) further try to study the effect of unique nanostructure of gold such as gold nanorings of 75–150 nm diameters. Interestingly, nanoring's sensitivity is high and it successfully demonstrates the real-time protein binding without any label (Fig. 13.9). Kumar et al. thoroughly investigated the sensitivity of SPR biosensors with Kretschmann configuration using alternative layers. They studied the use of prism, zinc oxide, silver, gold, graphene, and biomolecule ss-DNA. Interestingly, they observed that the key role of the intermediate layers of zinc

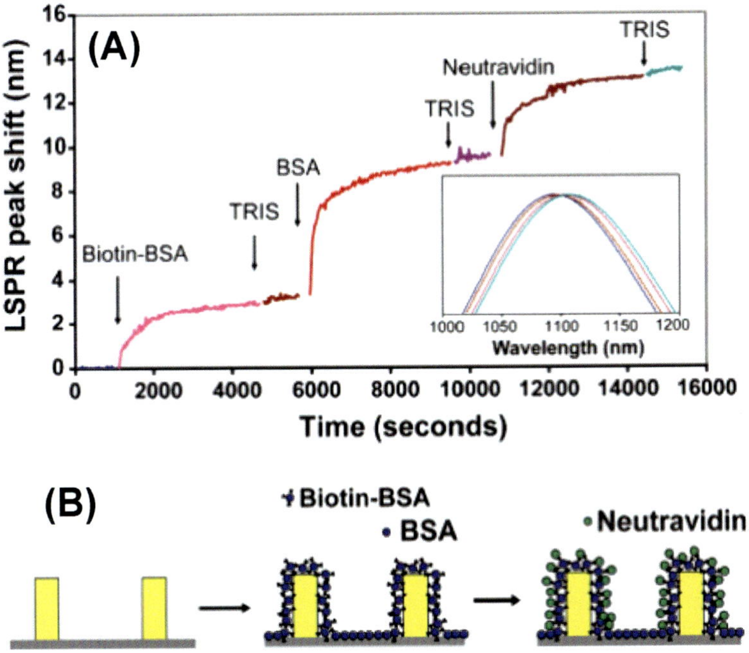

Figure 13.9 (A) Localized surface plasmon resonance (LSPR) peak shift as a function of time of biotin-BSA adsoption and BSA (Bovin Serum Albumin) onto 75 nm gold nanorings on a soda-lime glass substrates. Peak shifting is observed due to binding of neutravidin to biotinylated BSA. Its binding on substrate is blocked by BSA. The inset shows LSPR peak before and after binding of protein. LSPR in buffer was 1096 nm. (B) simple illustration of biotin-BSA, BSA, and neutravidin biding onto Au nanorings immobilized on a glass substrate.

Source: Reproduced with permision from Larsson, E. M., Alegret, J., K□ll, M., & Sutherland, D. S. (2007). Sensing characteristics of NIR localized surface plasmon resonances in gold nanorings for application as ultrasensitive biosensors. *Nano Letters*, *7*, 1256–1263. doi:10.1021/nl0701612. Copyright 2007 American Chemical Society.

oxide is to achieve high sensitivity. With higher number of graphene layers (up to nine) SPR showed enhanced sensitivity (Kumar, Kushwaha, Srivastava, Mishra, & Srivastava, 2018). Gao et al. used Ag@Au nanoplates as enhancer for high-performance SPR biosensing. This technique can be feasible for the development of a wide range of biosensors for many biomolecule detections and interesting in biological entities labeling and imaging (Gao et al., 2012). Zhang et al. reported the development of wavelength modulation SPR biosensors using aldehyde-functionalized gold nanorod-coated magnetic iron oxide nanoparticles. It showed promising results in the detection of mouse IgG (0.15−40 μg/mL) for immunoassay with improved sensitivity (Zhang et al., 2014). Detection of highly sensitive biomolecules, such as peptide, hormone, and insulin, demands ultrasensitive technique due to very low molecular weight (5800 Da). Still, sensitivity of SPR technique needs to be improved which can make it feasible to work with molecules having low molecular weight such as mycotoxins, vitamins, drugs, and pathogens such as bacteria and viruses at very low concentrations (Puiu & Bala, 2016). SPR technique gained significant attention in gas sensing, food industries, biology, and medical diagnostics. The SPR-based system is commercially available. SPR has been used in biosensing extensively. SPR is an optical technique widely used in the detection of biomolecules. Several companies have developed and commercialized the biosensors based on SPR technology such as Autolab, Biocore (GE Healthcare), Bio-Rad, Graffinity Pharmaceuticals, and IBIS (Narsaiah, Jha, Bhardwaj, Sharma, & Kumar, 2012). However, further advanced biosensors may be at different stages of development and commercialization.

13.2.3 Optical biosensor based on Surface-enhanced Raman spectroscopy

13.2.3.1 Surface-enhanced Raman spectroscopy

SERS is one of the most effective spectroscopy techniques used in the field of sensor due to its exceptional sensitivity and easy performance. Generally, for detection of molecules or analyte with very low concentration SERS technique is promising over others. In SERS through excitation of localized surface plasmons which produce the electromagnetic filed used for the detection of targeted molecules (Gu, Trujillo, Olson, & Camden, 2018; Sharma, Balaji, Kumari, & Krishnan, 2018). Since the last two decades, SERS has been extensively studied as a plausible solution for the detection of different compounds at the sample origin point. Owing to huge progress and diverse ranging of analytical potential, SERS is nowadays applied in different fields such as biological sample analysis, diagnosis, imaging, food safety and surface science (Gu et al., 2018).

SERS can efficiently detect various types of molecules or chemicals by molecular vibrations. The molecular vibrations are unique for each molecule and compound; thus SERS detection is considered as fingerprinting recognition for the detection of molecules (Yan, Li, & Su, 2018).

13.2.3.2 Mechanism of surface-enhanced Raman scattering

It has been widely reported that SERS phenomena depend mainly on electromagnetic enhancement. It involves amplification of signal through LSPR. The second mechanism describes signal enhancement. In general, optical sensor contains recognition molecules that can react particularly with desired target compound, which is connected to transducer part that is employed for signaling the reaction or binding event. In order to improve the sensitivity and analytical performance of optical sensor, different recognition elements such as antibody, aptamer, enzyme, molecular imprinted polymers, metal/metal oxide nanoparticles and host−guest recognizer can be used (Yan et al., 2018).

13.2.3.3 Metal/Metal Oxide Surface-enhaced Raman spectroscopy optical sensors

Metals such as Ag, Au, and Cu have conventionally been used as the substrate material in SERS sensor (Karthick Kannan, Shankar, Blackman, & Chung, 2019). In the process to enhance the SERS signal initial requirement is the presence of highly tenanted hot spots on a metallic surface. Therefore much research has been focused toward the development of high-density hot spot materials. To achieve this objective, several materials such as metal oxide, composite nanoparticles, core-shell nanoparticles, nanometal-based hybrid materials, and single-element semiconductors have been studied for the detection of different analytes (Chen et al., 2016; Karthick Kannan et al., 2019). Recently, SERS signals could be improved by applying hybrids known as nanocomposites of noble metals and metal oxides. This mechanism enhanced the electromagnetic signal arising from hybrid systems (Lin et al., 2002). The widely used metal oxides include CuO, Fe_2O_3, SiO_2, TiO_2, ZnO, and GO, which have been used for the enhancement of SERS signals. M/MO-based SERS systems, at the interface of substrate surface (M/MO) and target molecule the charge transfer takes place (Lin et al., 2002). The schematic diagram of optical sensor working is displayed in Fig. 13.10.

Among the different M/MO, TiO_2 is one of the most commonly used metal oxides. Recently, Li et al. (2020) utilized metal oxide-based nanomaterials SiO_2@TiO_2@Ag and designed the sensors for detection of pollutant in water such as pyrethroids. The main advantages of this sensor are self-cleaning capacity and feasibility to recycle. To overcome the sensitivity limitation, molecularly imprinted technique (MIT) was utilized (Li, Huang, & Lu, 2020).

In another report, Zhao et al. (2019) fabricated a SERS sensor with improved sensitivity (10^{-9} M) toward the detection of rhodamine 6G (R6G) molecules based on TiO_2/AuNWA (nanowire arrays) as a Raman-sensitive substrate. Moreover, it showed good recyclability. This study opens a new window of development of SERS sensors with a combination of noble metals and semiconductors (Zhao et al., 2019).

Similarly, several researchers developed various SERS sensors using ZnO, Fe_3O_4, and MnO_2. These MOs provide more flexibility to control the various parameters to achieve desired physicochemical properties and performance

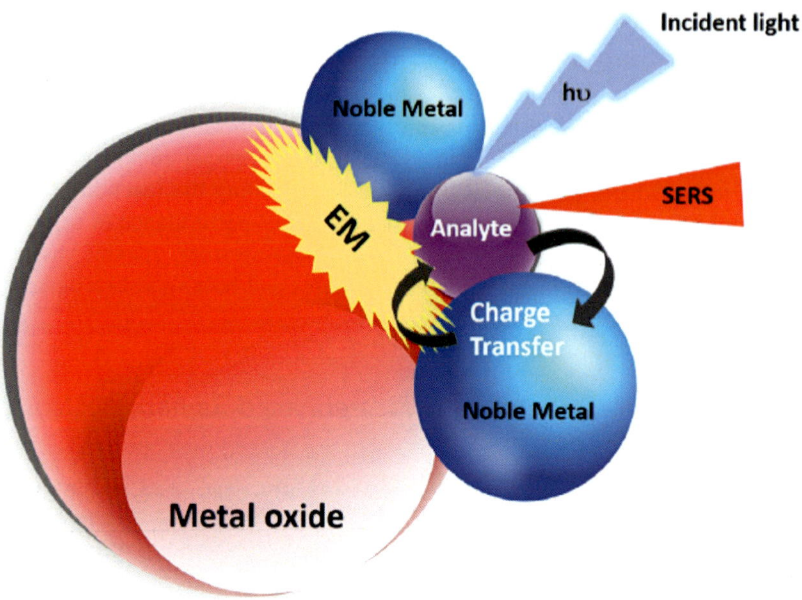

Figure 13.10 Mechanism of the noble metal/metal oxide-based (M/MO-based) nanocomposites for surface enhanced Raman spectroscopy enhancement.
Source: Reproduced with permission from Sharma, V., Balaji, R., Kumari, N., & Krishnan, V. (2018). SERS application of noble metal-metal oxide hybrid nanoparticles. In *Noble metal-metal oxide hybrid nanoparticles: Fundamentals and applications* (pp. 457–486). Elsevier. doi:10.1016/B978–0–12–814134-2.00021–8. Copyright, 2019 Elsevier.

(Abdulrahman, Kołątaj, Lenczewski, Krajczewski, & Kudelski, 2016; Wang, Shi, She, & Mu, 2012).

13.2.4 Applications

Optical sensor is a facile, rapid, and low-cost technique for the detection of a wide range of compounds such as drugs, toxin, pesticide, chemicals, explosive, environmental motoring, and heavy metals. Currently, SERS and M/MO SERS-based optical sensors are successfully used in various fields. The schematic diagram of the major application of SERS-based sensor is displayed in Fig. 13.11.

Among the above-mentioned fields, biomedical is one of the most important fields where SERS and M/MO-based sensors are successfully applied. Fig. 13.12 shows the development of flexible SERS platforms for different applications.

Further, SERS-based sensors are widely used in the field of food and agriculture industries for the detection of various pesticides and antibiotics. SERS has many advantages such as low production cost, high speed and sensitivity, and portability, which permit this technique to be highly used for the detection of pesticides and antibiotics. Carrasco et al. (2016) fabricated a nanocomposite-based multibranched

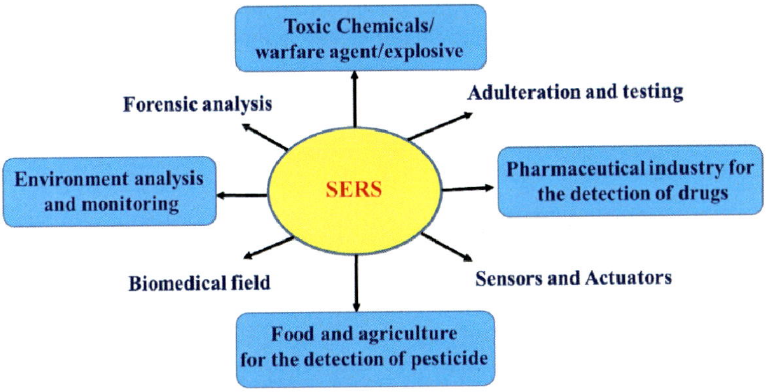

Figure 13.11 Schematic diagram showing the major application of surface enhanced Raman spectroscopy-based sensor.

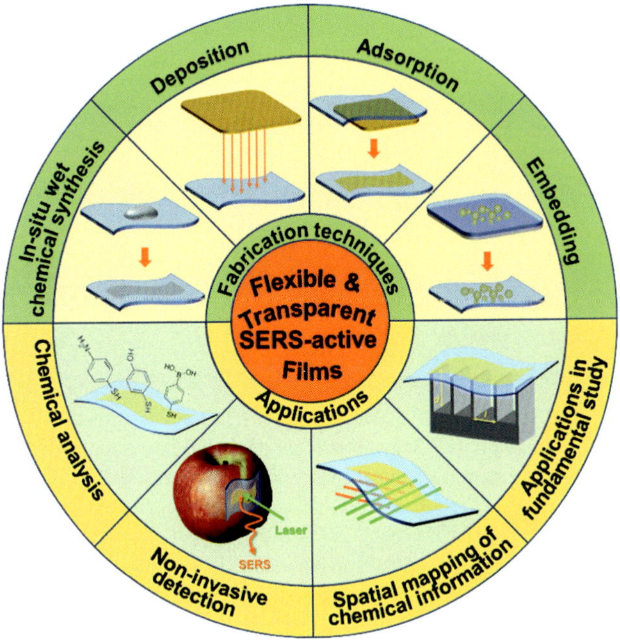

Figure 13.12 Recent developments of flexible and transparent SERS substrates.
Source: Reproduced with permission from Li, H., Wang, Y., Li, Y., Zhang, J., Qiao, Y., Wang, Q., & Che, G. (2020). Fabrication of pollutant-resistance SERS imprinted sensors based on SiO2@TiO2@Ag composites for selective detection of pyrethroids in water. *The Journal of Physics and Chemistry of Solids*, *138*, 109254. doi:10.1016/j.jpcs.2019.109254. Copyright 2012 The Royal Chemical Society.

($bAu@mSiO_2@MIP$), Au nanoparticle as the core, mesoporous silica as the shell, and MIP as the recognition component for the detection of enrofloxacin (ENR). They reported that use of nanocomposites significantly enhanced the Raman scattering of ENR upon binding to the MIP. Moreover, many SERS-based sensors have been reported for the detection of pesticides and antibiotics (Yan et al., 2018). The overview of the detection mechanism is summarized in Fig. 13.13 (Kumar, Goel, & Singh, 2017; Xu, Zhou, Takei, & Hong, 2019).

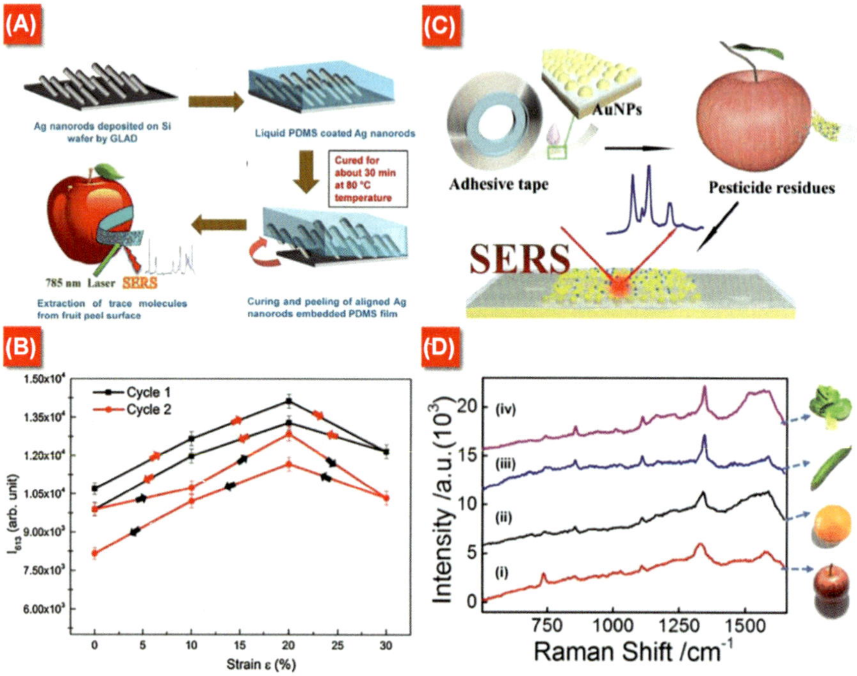

Figure 13.13 (A) Schematic illustration of the fabrication of SERS substrate composed of AgNRs. (B) The SERS response of AgNR arrays on flexible Polydimethylsiloxane substrate as a function of tensile strain. (C) Schematic view of the fabrication of SERS tape and extraction of targets from fruit peel surface for SERS analysis. (D) Raman spectra of parathion-methyl extracted from the surfaces of apple, orange, cucumbers, and green vegetables using SERS tape.

Source: Reproduced with permission from Kumar, S., Goel, P., & Singh, J. P. (2017). Flexible and robust SERS active substrates for conformal rapid detection of pesticide residues from fruits. *Sensors & Actuators, B: Chemical, 241*, 577–583. doi:10.1016/j.snb.2016.10.106. Copyright 2016 Elsevier; Chen, J., Huang, Y., Kannan, P., Zhang, L., Lin, Z., Zhang, J., . . . Guo, L. (2016). Flexible and adhesive surface enhance Raman scattering active tape for rapid detection of pesticide residues in fruits and vegetables. *Analytical Chemistry*, *88*, 2149–2155. doi:10.1021/acs.analchem.5b03735. Copyright 2016 American chemical Society; Xu, K., Zhou, R., Takei, K., & Hong, M. (2019). Toward flexible surface-enhanced Raman scattering (SERS) sensors for point-of-care diagnostics. *Advancement of Science*, *6*. doi:10.1002/advs.201900925. Copyright 2019 Wiley-VCH.

Researchers developed SERS adhesive tape coated with AuNP/Ag NS (nanostructure) based sensor for the pesticide detection in apples (Chen et al., 2016; Li et al., 2010). M/MO-based SERS optical sensors were also used for the detection of heavy metal ions. In this context, Du, Cui, and Jing (2014) developed the rapid in situ SERS sensor comprising Fe_3O_4@Ag nanoparticles for the detection of arsenite and arsenate. The detection mechanism mainly depends on the SERS peak position and vibration of the As–O bond (Du et al., 2014). Likewise, Yang et al. utilized Au@Ag nanoparticles. However, the nanoparticles tend to get aggregation or dispersion. To overcome this issue, they have used different substrates such as PDDA, aptamer, and arsenite. Then utilizing enhancement of Raman signal of 4-MPY they detected the arsenite (Yang, Chen, Liu, Liu, & Jiang, 2015).

SERS substrates are well explored for the detection of different dyes using Raman spectroscopy. Composite nanowires with gold core (Au_{core}) and polyaniline shell ($PANI_{shell}$) (as shown in Fig. 13.14) as SERS substrate was prepared using swollen liquid crystals as a structure-directing soft template and used as substrate for the detection of methylene blue (MB) dyes (Dutt, Kumar, & Siril, 2015).

The Au_{core}–$PANI_{shell}$ nanocomposite formation was confirmed by XPS (X-ray Photoelectron Spectroscopy) study as shown in Fig. 13.15A. SERS performance was evaluated by using MB. Raman spectra of MB on Au_{core}–$PANI_{shell}$ nanocomposite is shown in Fig. 13.15B. Fig. 13.15B presents the Raman spectrum of MB pure and MB on bulk-PANI. It concludes the SERS property of prepared substrates. Raman peak for MB was observed at $1620\ cm^{-1}$ corresponding to ring stretch (ν CC). The peak at $1396\ cm^{-1}$ signifies the symmetric CN stretching (ν-sym CN). Peaks at $1164\ cm^{-1}$, $1037\ cm^{-1}$ (β CH), $861\ cm^{-1}$, $668\ cm^{-1}$ (γ CH), $449\ cm^{-1}$ (δ CNC) of MB were also observed to be enhanced (Dutt et al., 2015).

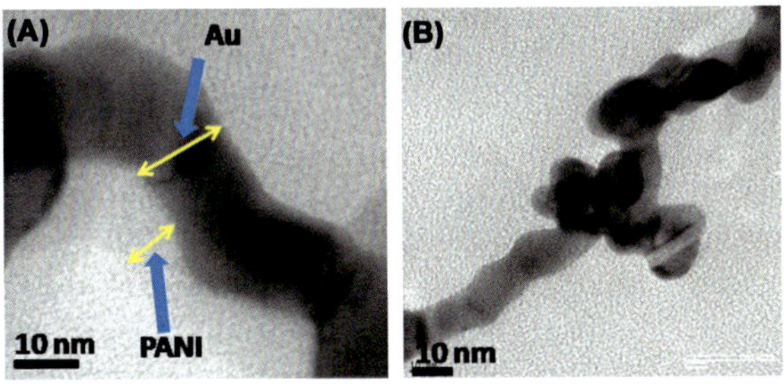

Figure 13.14 High Resolution Tranmission Electron Microscopy. images of Au_{core}–$PANI_{shell}$ nanocomposite.
Source: Reproduced with permission from Dutt, S., Kumar, R., & Siril, P. F. (2015). Green synthesis of a palladium–polyaniline nanocomposite for green Suzuki–Miyaura coupling reactions. *RSC Advances*, *5*, 33786–33791. doi:10.1039/C5RA05007C. Copyright 2015 The royal Society of Chemistry and the Centre National de la Recherche Sientifique.

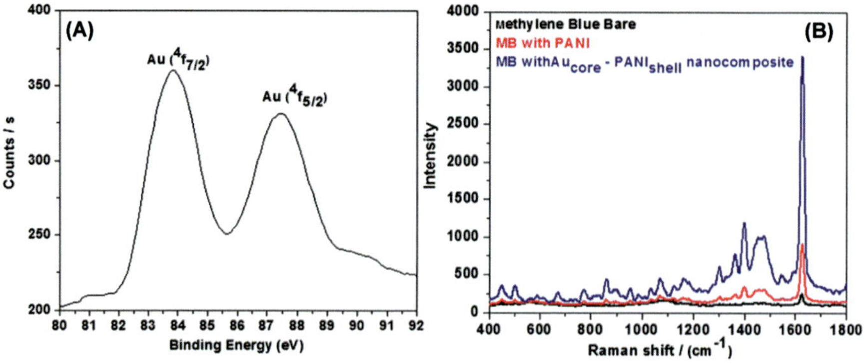

Figure 13.15 (A) X-ray Photoelectron spectroscopy spectra of Au. (B) Surface enhanced Raman Spectroscopy activity of the Au_{core}−$PANI_{shell}$ nanocomposite.
Source: Reproduced with permission from Dutt, S., Siril, P. F., Sharma, V., & Periasamy, S. (2015). Gold core−polyaniline shell composite nanowires as a substrate for surface enhanced Raman scattering and catalyst for dye reduction. *New Journal of Chemistry*, *39*, 902−908. doi:10.1039/C4NJ01521E. Copyright 2015 The royal Society of Chemistry and the Centre National de la Recherche Sientifique.

13.3 Conclusion

In conclusion, this chapter presented various optical biosensing techniques. We have discussed the concept of all the three optical biosensing techniques, namely, fluorescence, SPR, and SERS. The unique properties of metal and metal oxide nanoparticles allow them to be applied as optical sensors that provide better sensitivity, accuracy, and detection limit. We also discussed the mechanism of each technique and their applications in especially optical biosensors compare to other nanomaterials metal and metal oxides are has several advantages due to optical and other physicochemical properties and compatibility to their use in several application rage form heavy metal ions, water treatment, and biomedical. Compared to other techniques, fluorescence-based M/MO optical biosensing technique is widely studied, easy, and simple. However, it is limited by detection limitation. SPR technique is suitable for the detection of a range of biomolecules, and their performance mainly depends on optical properties and hence their sizes and shapes. Due to great progress in optical sensors, mainly SERS-based sensors are currently used in various fields such as diagnosis, imaging, toxin detection, and biological sample analysis. We also discussed the latest progress on the development of M/MO nanomaterial-based optical, fluorescence, SPR, and SERS-based sensors. The optical sensors are flexible with less expensive substrates and low maintenance techniques. It is easy to operate, portable, and can be easily integrated with spectroscopes for diagnostic applications. M/MO-based optical sensors have the potential to enter the global market and can be a next-generation wearable sensor in the future.

Acknowledgment

All authors are very thankful for kind support and help from their Department of Chemistry, Govt. Post Graduate College Una, Una, HP, India and Department of Higher Education, Govt. of HP (Shimla), India. RK thanks to BINA (Nanotechnology Institute for Nanotechnologies and Advanced Materials), Bar Ilan University, and Council of Higher Education, Israel for PBC Outstanding Postdoctoral Fellowship. KRA is grateful to the Science and Engineering Research Board (SERB), New Delhi, India for funding the NPDF project (PDF/2016/001156/LS).

References

Abdulrahman, H. B., Kołątaj, K., Lenczewski, P., Krajczewski, J., & Kudelski, A. (2016). MnO_2-protected silver nanoparticles: New electromagnetic nanoresonators for Raman analysis of surfaces in basis environment. *Applied Surface Science*, *388*, 704–709. Available from https://doi.org/10.1016/j.apsusc.2016.01.262.

Aragay, G., Pons, J., & Merkoçi, A. (2011). Recent trends in macro-, micro-, and nanomaterial-based tools and strategies for heavy-metal detection. *Chemical Reviews*, *111*, 3433–3458. Available from https://doi.org/10.1021/cr100383r.

Asati, A., Kaittanis, C., Santra, S., & Perez, J. M. (2011). PH-tunable oxidase-like activity of cerium oxide nanoparticles achieving sensitive fluorigenic detection of cancer biomarkers at neutral pH. *Analytical Chemistry*, *83*, 2547–2553. Available from https://doi.org/10.1021/ac102826k.

Aslan, K., Leonenko, Z., Lakowicz, J. R., & Geddes, C. D. (2005). Annealed silver-island films for applications in metal-enhanced fluorescence: Interpretation in terms of radiating plasmons. *Journal of Fluorescence*, *15*, 643–654. Available from https://doi.org/10.1007/s10895-005-2970-z.

Austin, L. A., Mackey, M. A., Dreaden, E. C., & El-Sayed, M. A. (2014). The optical, photothermal, and facile surface chemical properties of gold and silver nanoparticles in biodiagnostics, therapy, and drug delivery. *Archives of Toxicology*, *88*, 1391–1417. Available from https://doi.org/10.1007/s00204-014-1245-3.

Bleeker, E. A. J., de Jong, W. H., Geertsma, R. E., Groenewold, M., Heugens, E. H. W., Koers-Jacquemijns, M., ... Oomen, A. G. (2013). Considerations on the EU definition of a nanomaterial: Science to support policy making. *Regulatory Toxicology and Pharmacology: RTP*, *65*, 119–125. Available from https://doi.org/10.1016/j.yrtph.2012.11.007.

Boujday, S., Briandet, R., Salmain, M., Herry, J. M., Marnet, P. G., Gautier, M., & Pradier, C. M. (2008). Detection of pathogenic *Staphylococcus aureus* bacteria by gold based immunosensors. *Microchimica Acta*, *163*, 203–209. Available from https://doi.org/10.1007/s00604-008-0024-3.

Cao, J., Sun, T., & Grattan, K. T. V. (2014). Gold nanorod-based localized surface plasmon resonance biosensors: A review. *Sensors & Actuators, B: Chemical*, *195*, 332–351. Available from https://doi.org/10.1016/J.SNB.2014.01.056.

Cao, X., Ye, Y., & Liu, S. (2011). Gold nanoparticle-based signal amplification for biosensing. *Analytical Biochemistry*, *417*, 1–16. Available from https://doi.org/10.1016/J.AB.2011.05.027.

Carrasco, S., Benito-Peña, E., Navarro-Villoslada, F., Langer, J., Sanz-Ortiz, M. N., Reguera, J., ... Moreno-Bondi, M. C. (2016). Multibranched gold-mesoporous silica nanoparticles coated with a molecularly imprinted polymer for label-free antibiotic surface-enhanced Raman scattering analysis. *Chemistry of Materials: A Publication of the American Chemical Society*, *28*, 7947–7954. Available from https://doi.org/10.1021/acs.chemmater.6b03613.

Daniel, M.-C., & Astruc, D. (2003). Gold nanoparticles: Assembly, supramolecular chemistry, quantum-size-related properties, and applications toward biology, catalysis, and nanotechnology. doi:10.1021/CR030698 + .

Chan, G. H., Zhao, J., Hicks, E. M., Schatz, G. C., & Van Duyne, R. P. (2007). Plasmonic properties of copper nanoparticles fabricated by nanosphere lithography. doi:10.1021/NL070648A.

Chawla, M., Kumar, R., & Siril, P. F. (2016). High catalytic activities of palladium nanowires synthesized using liquid crystal templating approach. *Journal of Molecular Catalysis A: Chemical*, *423*, 126–134. Available from https://doi.org/10.1016/J.MOLCATA.2016.06.014.

Chawla, M., Kumari, A., & Siril, P. F. (2018). Exceptional catalytic activities and sensing performance of palladium decorated anisotropic gold nanoparticles. *ChemistrySelect*, *3*, 9071–9083. Available from https://doi.org/10.1002/slct.201801426.

Chen, J., Huang, Y., Kannan, P., Zhang, L., Lin, Z., Zhang, J., ... Guo, L. (2016). Flexible and adhesive surface enhance Raman scattering active tape for rapid detection of pesticide residues in fruits and vegetables. *Analytical Chemistry*, *88*, 2149–2155. Available from https://doi.org/10.1021/acs.analchem.5b03735.

Chiu, N.-F., Tu, Y.-C., & Huang, T.-Y. (2013). Enhanced sensitivity of anti-symmetrically structured surface plasmon Resonance sensors with zinc oxide intermediate layers. *Sensors*, *14*, 170–187. Available from https://doi.org/10.3390/s140100170.

Crosley, M. S., & Yip, W. T. (2018). Kinetically doped silica sol–gel optical biosensors: Expanding potential through dip-coating. *ACS Omega*, *3*, 7971–7978. Available from https://doi.org/10.1021/acsomega.8b00897.

de Juan-Franco, E., Rodríguez-Frade, J. M., Mellado, M., & Lechuga, L. M. (2013). Implementation of a SPR immunosensor for the simultaneous detection of the 22K and 20K hGH isoforms in human serum samples. *Talanta*, *114*, 268–275. Available from https://doi.org/10.1016/J.TALANTA.2013.04.042.

Dominik, M., Leśniewski, A., Janczuk, M., Niedziółka-Jönsson, J., Hołdyński, M., Wachnicki, Ł., ... Śmietana, M. (2017). Titanium oxide thin films obtained with physical and chemical vapour deposition methods for optical biosensing purposes. *Biosensors & Bioelectronics*, *93*, 102–109. Available from https://doi.org/10.1016/J.BIOS.2016.09.079.

Donaldson, K. A., Kramer, M. F., & Lim, D. V. (2004). A rapid detection method for Vaccinia virus, the surrogate for smallpox virus. *Biosensors & Bioelectronics*, *20*, 322–327. Available from https://doi.org/10.1016/j.bios.2004.01.029.

Du, J., Cui, J., & Jing, C. (2014). Rapid in situ identification of arsenic species using a portable Fe3O4@Ag SERS sensor. *Chemical Communications*, *50*, 347–349. Available from https://doi.org/10.1039/c3cc46920d.

Dutt, S., Kumar, R., & Siril, P. F. (2015). Green synthesis of a palladium–polyaniline nanocomposite for green Suzuki–Miyaura coupling reactions. *RSC Advances*, *5*, 33786–33791. Available from https://doi.org/10.1039/C5RA05007C.

Dutt, S., Siril, P. F., & Remita, S. (2017). Swollen liquid crystals (SLCs): A versatile template for the synthesis of nano structured materials. *RSC Advances*, *7*, 5733–5750. Available from https://doi.org/10.1039/C6RA26390A.

Dutt, S., Siril, P. F., Sharma, V., & Periasamy, S. (2015). Gold $_{core}$−polyaniline $_{shell}$ composite nanowires as a substrate for surface enhanced Raman scattering and catalyst for dye reduction. *New Journal of Chemistry*, *39*, 902−908. Available from https://doi.org/10.1039/C4NJ01521E.

Erathodiyil, N., & Ying, J. Y. (2011). Functionalization of inorganic nanoparticles for bioimaging applications. *Accounts of Chemical Research*, *44*, 925−935. Available from https://doi.org/10.1021/ar2000327.

Feng, C., Dai, S., & Wang, L. (2014). Optical aptasensors for quantitative detection of small biomolecules: A review. *Biosensors & Bioelectronics*, *59*, 64−74. Available from https://doi.org/10.1016/j.bios.2014.03.014.

Ganesh, I., Gupta, A. K., Kumar, P. P., Chandra Sekhar, P. S., Radha, K., Padmanabham, G., & Sundararajan, G. (2012). Preparation and characterization of Co-doped TiO_2 materials for solar light induced current and photocatalytic applications. *Materials Chemistry and Physics*, *135*, 220−234. Available from https://doi.org/10.1016/j.matchemphys.2012.04.062.

Gao, C., Lu, Z., Liu, Y., Zhang, Q., Chi, M., Cheng, Q., & Yin, Y. (2012). Highly stable silver nanoplates for surface plasmon resonance biosensing. *Angewandte Chemie International Edition*, *51*, 5629−5633. Available from https://doi.org/10.1002/anie.201108971.

Gao, L., Zhuang, J., Nie, L., Zhang, J., Zhang, Y., Gu, N., ... Yan, X. (2007). Intrinsic peroxidase-like activity of ferromagnetic nanoparticles. *Nature Nanotechnology*, *2*, 577−583. Available from https://doi.org/10.1038/nnano.2007.260.

Gu, X., Trujillo, M. J., Olson, J. E., & Camden, J. P. (2018). SERS sensors: Recent developments and a generalized classification scheme based on the signal origin. *Annual Review of Analytical Chemistry*, *11*, 147−169. Available from https://doi.org/10.1146/annurev-anchem-061417-125724.

Guisbiers, G., Mejía-Rosales, S., & Deepak, F. L. (2012). Nanomaterial properties: Size and shape dependencies. *Journal of Nanomaterials*, 180976. Available from https://www.hindawi.com/journals/jnm/2012/180976/, Accessed 01.08.20.

Gupta, S., Sarkar, S., Katranidis, A., & Bhattacharya, J. (2019). Development of a cell-free optical biosensor for detection of a broad range of mercury contaminants in water: A plasmid DNA-based approach. *ACS Omega*, *4*, 9480−9487. Available from https://doi.org/10.1021/acsomega.9b00205.

Haes, A. J., Zou, S., Schatz, G. C., & Van Duyne, R. P. (2003). A nanoscale optical biosensor: The long range distance dependence of the localized surface plasmon resonance of noble metal nanoparticles. doi:10.1021/JP0361327.

He, Y. W., Feng, Y., Kang, L. W., & Li, X. L. (2017). A turn-on fluorescent sensor for Hg^{2+} based on graphene oxide. *Journal of Chemistry*, *2017*. Available from https://doi.org/10.1155/2017/9431605.

Heckmann, J., Pufahl, K., Franz, P., Grosse, N. B., Li, X., & Woggon, U. (2018). Plasmon-enhanced nonlinear yield in the Otto and Kretschmann configurations. *Physical Review B*, *98*, 115415. Available from https://doi.org/10.1103/PhysRevB.98.115415.

Helmerhorst, E., Chandler, D. J., Nussio, M., & Mamotte, C. D. (2012). Real-time and label-free bio-sensing of molecular interactions by surface plasmon resonance: A laboratory medicine perspective. *The Clinical Biochemist. Reviews / Australian Association of Clinical Biochemists*, *33*, 161−173. Available from http://www.ncbi.nlm.nih.gov/pubmed/23267248, Accessed 14.09.19.

Hinman, S. S., McKeating, K. S., & Cheng, Q. (2018). Surface plasmon resonance: Material and interface design for universal accessibility. *Analytical Chemistry*, *90*, 19−39. Available from https://doi.org/10.1021/acs.analchem.7b04251.

Huang, D., Niu, C., Wang, X., Lv, X., & Zeng, G. (2013). "Turn-on" fluorescent sensor for Hg^{2+} based on single-stranded DNA functionalized Mn:CdS/ZnS quantum dots and gold nanoparticles by time-gated mode. *Analytical Chemistry*, *85*, 1164–1170. Available from https://doi.org/10.1021/ac303084d.

Jeevanandam, J., Barhoum, A., Chan, Y. S., Dufresne, A., & Danquah, M. K. (2018). Review on nanoparticles and nanostructured materials: History, sources, toxicity and regulations. *Beilstein Journal of Nanotechnology*, *9*, 1050–1074. Available from https://doi.org/10.3762/bjnano.9.98.

Jeong, Y., Kook, Y. M., Lee, K., & Koh, W. G. (2018). Metal enhanced fluorescence (MEF) for biosensors: General approaches and a review of recent developments. *Biosensors & Bioelectronics*, *111*, 102–116. Available from https://doi.org/10.1016/j.bios.2018.04.007.

Karlsson, R., & Fält, A. (1997). Experimental design for kinetic analysis of protein-protein interactions with surface plasmon resonance biosensors. *Journal of Immunological Methods*, *200*, 121–133. Available from https://doi.org/10.1016/S0022-1759(96)00195-0.

Karthick Kannan, P., Shankar, P., Blackman, C., & Chung, C. H. (2019). Recent advances in 2D inorganic nanomaterials for SERS sensing. *Advanced Materials*, *1803432*, 1–27. Available from https://doi.org/10.1002/adma.201803432.

Kirsch, J., Siltanen, C., Zhou, Q., Revzin, A., & Simonian, A. (2013). Biosensor technology: Recent advances in threat agent detection and medicine. *Chemical Society Reviews*, *42*, 8733–8768. Available from https://doi.org/10.1039/c3cs60141b.

Kumar, A., Kim, S., & Nam, J. M. (2016). Plasmonically engineered nanoprobes for biomedical applications. *Journal of the American Chemical Society*, *138*, 14509–14525. Available from https://doi.org/10.1021/jacs.6b09451.

Kumar, R. (2019a). Lipid-based nanoparticles for drug-delivery systems. *Nanocarriers for Drug Delivery*, 249–284. Available from https://doi.org/10.1016/B978-0-12-814033-8.00008-4.

Kumar, R. (2019b). Nanotechnology based approaches to enhance aqueous solubility and bioavailability of griseofulvin: A literature survey. *Journal of Drug Delivery Science and Technology*, *53*, 101221. Available from https://doi.org/10.1016/j.jddst.2019.101221.

Kumar, R. (2020). Solubility and bioavailability of fenofibrate nanoformulations. *ChemistrySelect*, *5*, 1478–1490. Available from https://doi.org/10.1002/slct.201903647.

Kumar, R., Dalvi, S. V., & Siril, P. F. (2020). Nanoparticle-based drugs and formulations: Current status and emerging applications. *ACS Applied Nano Materials*, *3*, 4944–4961. Available from https://doi.org/10.1021/acsanm.0c00606, acsanm.0c00606.

Kumar, R., Kumar, V. B., & Gedanken, A. (2020). Sonochemical synthesis of carbon dots, mechanism, effect of parameters, and catalytic, energy, biomedical and tissue engineering applications. *Ultrasonics Sonochemistry*, *64*, 105009. Available from https://doi.org/10.1016/j.ultsonch.2020.105009.

Kumar, R., Kumar, V. B., Marcus, M., Gedanken, A., & Shefi, O. (2019). Element (B, N, P) doped carbon dots interaction with neural cells: promising results and future prospective. In W. J. Parak, & M. Osiński (Eds.), *Colloid. Nanoparticles Biomed. Appl. XIV* (p. 22). SPIE. Available from http://doi.org/10.1117/12.2509610.

Kumar, R., Kushwaha, A. S., Srivastava, M., Mishra, H., & Srivastava, S. K. (2018). Enhancement in sensitivity of graphene-based zinc oxide assisted bimetallic surface plasmon resonance (SPR) biosensor. *Applied Physics A*, *124*, 235. Available from https://doi.org/10.1007/s00339-018-1606-5.

Kumar, R., Soni, P., & Felix Siril, P. (2019). Engineering the morphology and particle size of high energetic compounds using drop-by-drop and drop-to-drop solvent–antisolvent interaction methods. *ACS Omega*, *4*, 5424–5433. Available from https://doi.org/10.1021/acsomega.8b03214.

Kumar, S., Goel, P., & Singh, J. P. (2017). Flexible and robust SERS active substrates for conformal rapid detection of pesticide residues from fruits. *Sensors & Actuators, B: Chemical*, *241*, 577–583. Available from https://doi.org/10.1016/j.snb.2016.10.106.

Kumar, V. B., Kumar, R., Friedman, O., Golan, Y., Gedanken, A., & Shefi, O. (2019). One-pot hydrothermal synthesis of elements (B, N, P)-doped fluorescent carbon dots for cell labelling, differentiation and outgrowth of neuronal cells. *ChemistrySelect*, *4*, 4222–4232. Available from https://doi.org/10.1002/slct.201900581.

Kumar, V. B., Kumar, R., Gedanken, A., & Shefi, O. (2018). Fluorescent metal-doped carbon dots for neuronal manipulations. *Ultrasonics Sonochemistry*. Available from https://doi.org/10.1016/J.ULTSONCH.2018.11.017.

Lakowicz, J. R. (1999). *Fluorescence sensing. Princ. Fluoresc. Spectrosc* (pp. 531–572). Boston, MA: Springer United States. Available from http://doi.org/10.1007/978-1-4757-3061-6_19.

Larsson, E. M., Alegret, J., Kăll, M., & Sutherland, D. S. (2007). Sensing characteristics of NIR localized surface plasmon resonances in gold nanorings for application as ultrasensitive biosensors. *Nano Letters*, *7*, 1256–1263. Available from https://doi.org/10.1021/nl0701612.

Law, W.-C., Yong, K.-T., Baev, A., & Prasad, P. N. (2011). Sensitivity improved surface plasmon resonance biosensor for cancer biomarker detection based on plasmonic enhancement. *ACS Nano*, *5*, 4858–4864. Available from https://doi.org/10.1021/nn2009485.

Li, C. L., Liu, K. T., Lin, Y. W., & Chang, H. T. (2011). Fluorescence detection of lead(II) ions through their induced catalytic activity of DNAzymes. *Analytical Chemistry*, *83*, 225–230. Available from https://doi.org/10.1021/ac1028787.

Li, H., Wang, Y., Li, Y., Zhang, J., Qiao, Y., Wang, Q., & Che, G. (2020). Fabrication of pollutant-resistance SERS imprinted sensors based on SiO2@TiO2@Ag composites for selective detection of pyrethroids in water. *The Journal of Physics and Chemistry of Solids*, *138*, 109254. Available from https://doi.org/10.1016/j.jpcs.2019.109254.

Li, J. F., Huang, Y. F., Ding, Y., Yang, Z. L., Li, S. B., Zhou, X. S., ... Tian, Z. Q. (2010). Shell-isolated nanoparticle-enhanced Raman spectroscopy. *Nature*, *464*, 392–395. Available from https://doi.org/10.1038/nature08907.

Li, M., Cushing, S. K., & Wu, N. (2015). Plasmon-enhanced optical sensors: A review. *Analyst*, *140*, 386–406. Available from https://doi.org/10.1039/c4an01079e.

Li, X. (2019). *Nucleic acid amplification strategies in surface plasmon resonance technologies. Nucleic Acid Amplif. Strateg. Biosensing, Bioimaging Biomed* (pp. 111–128). Singapore: Springer Singapore. Available from http://doi.org/10.1007/978-981-13-7044-1_6.

Li, Y., Schluesener, H. J., & Xu, S. (2010). Gold nanoparticle-based biosensors. *Gold Bulletin*, *43*, 29–41. Available from https://doi.org/10.1007/BF03214964.

Li, Z., Huang, X., & Lu, G. (2020). Recent developments of flexible and transparent SERS substrates. *Journal of Materials Chemistry C*, *8*, 3956–3969. Available from https://doi.org/10.1039/d0tc00002g.

Liang, L., Lan, F., Ge, S., Yu, J., Ren, N., & Yan, M. (2017). Metal-enhanced ratiometric fluorescence/naked eye bimodal biosensor for lead ions analysis with bifunctional nanocomposite probes. *Analytical Chemistry*, *89*, 3597–3605. Available from https://doi.org/10.1021/acs.analchem.6b04978.

Liao, H., Nehl, C. L., & Hafner, J. H. (2006). Biomedical applications of plasmon resonant metal nanoparticles. *Nanomedicine: Nanotechnology, Biology, and Medicine*, *1*, 201–208. Available from https://doi.org/10.2217/17435889.1.2.201.

Liggett. (2014). NIH public access, 基因的改变*Bone.*, *23*, 1–7. Available from https://doi.org/10.1038/jid.2014.371.

Lin, B., Qiu, J., Gerstenmeier, J., Li, P., Pien, H., Pepper, J., & Cunningham, B. (2002). A label-free optical technique for detecting small molecule interactions. *Biosensors & Bioelectronics*, *17*, 827–834. Available from https://doi.org/10.1016/S0956-5663(02)00077-5.

Lin, T.-J., & Chung, M.-F. (2008). Using monoclonal antibody to determine lead ions with a localized surface plasmon resonance fiber-optic biosensor. *Sensors*, *8*, 582–593. Available from https://doi.org/10.3390/s8010582.

Lingeshwar Reddy, K., Balaji, R., Kumar, A., & Krishnan, V. (2018). Lanthanide doped near infrared active upconversion nanophosphors: Fundamental concepts, synthesis strategies, and technological applications. *Small (Weinheim an der Bergstrasse, Germany)*, *14*, 1801304. Available from https://doi.org/10.1002/smll.201801304.

Long, F., Zhu, A., & Shi, H. (2013). Recent advances in optical biosensors for environmental monitoring and early warning. *Sensors*, *13*, 13928–13948. Available from https://doi.org/10.3390/s131013928.

Mannelli, I., Minunni, M., Tombelli, S., & Mascini, M. (2003). Quartz crystal microbalance (QCM) affinity biosensor for genetically modified organisms (GMOs) detection. *Biosensors & Bioelectronics*, *18*, 129–140. Available from https://doi.org/10.1016/S0956-5663(02)00166-5.

McDonagh, C., Burke, C. S., & MacCraith, B. D. (2008). Optical chemical sensors. *Chemical Reviews*, *108*, 400–422. Available from https://doi.org/10.1021/cr068102g.

Medhi, R., Marquez, M. D., & Lee, T. R. (2020). Visible-light-active doped metal oxide nanoparticles: Review of their synthesis, properties, and applications. *ACS Applied Nano Materials*, *3*, 6156–6185. Available from https://doi.org/10.1021/acsanm.0c01035.

Morrow, K. J., Bawa, R., & Wei, C. (2007). Recent advances in basic and clinical nanomedicine. *The Medical Clinics of North America*, *91*, 805–843. Available from https://doi.org/10.1016/j.mcna.2007.05.009.

Mustafa, D. E., Yang, T., Xuan, Z., Chen, S., Tu, H., & Zhang, A. (2010). Surface plasmon coupling effect of gold nanoparticles with different shape and size on conventional surface plasmon resonance signal. *Plasmonics*, *5*, 221–231. Available from https://doi.org/10.1007/s11468-010-9141-z.

Narsaiah, K., Jha, S. N., Bhardwaj, R., Sharma, R., & Kumar, R. (2012). Optical biosensors for food quality and safety assurance—A review. *Journal of Food Science and Technology*, *49*, 383–406. Available from https://doi.org/10.1007/s13197-011-0437-6.

Nishikawa, T., Yamashita, H., Nakamura, M., Hasui, R., Matsushita, T., & Aoyama, S. (2006). Development of new localized surface plasmon resonance sensor with nanoimprinting technique. In *2006 First IEEE international conference on nano/micro engineered and molecular systems* (pp. 262–265). IEEE. doi:10.1109/NEMS.2006.334718.

Omar, N. A. S., & Fen, Y. W. (2018). Recent development of SPR spectroscopy as potential method for diagnosis of dengue virus E-protein. *Remote Sensing*, *38*, 106–116. Available from https://doi.org/10.1108/SR-07-2017-0130.

Park, T. J., Lee, S. J., Kim, D.-K., Heo, N. S., Park, J. Y., & Lee, S. Y. (2012). Development of label-free optical diagnosis for sensitive detection of influenza virus with genetically engineered fusion protein. *Talanta*, *89*, 246–252. Available from https://doi.org/10.1016/J.TALANTA.2011.12.021.

Parolo, C., & Merkoçi, A. (2013). Paper-based nanobiosensors for diagnostics. *Chemical Society Reviews*, *42*, 450–457. Available from https://doi.org/10.1039/c2cs35255a.

Pollet, J., Delport, F., Janssen, K. P. F., Jans, K., Maes, G., Pfeiffer, H., ... Lammertyn, J. (2009). Fiber optic SPR biosensing of DNA hybridization and DNA–protein interactions. *Biosensors & Bioelectronics*, *25*, 864–869. Available from https://doi.org/10.1016/J.BIOS.2009.08.045.

Prakash, S., Chakrabarty, T., Singh, A. K., & Shahi, V. K. (2013). Polymer thin films embedded with metal nanoparticles for electrochemical biosensors applications. *Biosensors & Bioelectronics*, *41*, 43–53. Available from https://doi.org/10.1016/J.BIOS.2012.09.031.

Pratsinis, A., Kelesidis, G. A., Zuercher, S., Krumeich, F., Bolisetty, S., Mezzenga, R., ... Sotiriou, G. A. (2017). Enzyme-mimetic antioxidant luminescent nanoparticles for highly sensitive hydrogen peroxide biosensing. *ACS Nano*, *11*, 12210–12218. Available from https://doi.org/10.1021/acsnano.7b05518.

Puiu, M., & Bala, C. (2016). SPR and SPR imaging: Recent trends in developing nanodevices for detection and real-time monitoring of biomolecular events. *Sensors*, *16*, 870. Available from https://doi.org/10.3390/s16060870.

Rhodes, C., Franzen, S., Maria, J.-P., Losego, M., Leonard, D. N., Laughlin, B., ... Weibel, S. (2006). Surface plasmon resonance in conducting metal oxides. *Journal of Applied Physics*, *100*, 054905. Available from https://doi.org/10.1063/1.2222070.

Sanchez, C., Belleville, P., Popall, M., & Nicole, L. (2011). Applications of advanced hybrid organic–inorganic nanomaterials: From laboratory to market. *Chemical Society Reviews*, *40*, 696–753. Available from https://doi.org/10.1039/c0cs00136h.

Sharma, V., Balaji, R., Kumari, N., & Krishnan, V. (2018). *SERS application of noble metal-metal oxide hybrid nanoparticles. Noble metal-metal oxide hybrid nanoparticles: Fundamentals and applications* (pp. 457–486). Elsevier. Available from http://doi.org/10.1016/B978-0-12-814134-2.00021-8.

Sharma, V., Chawla, M., & Randhawa, J. K. (2016). Enhanced sensitivity of nanostructured copper oxide for non-enzymatic glucose biosensing. *Journal of the Electrochemical Society*, *163*, B594–B600. Available from https://doi.org/10.1149/2.0301613jes.

Sharma, V., Sinha, N., Dutt, S., Chawla, M., & Siril, P. F. (2016). Tuning the surface enhanced Raman scattering and catalytic activities of gold nanorods by controlled coating of platinum. *Journal of Colloid and Interface Science*, *463*, 180–187. Available from https://doi.org/10.1016/J.JCIS.2015.10.036.

Shin, Y.-B., Lee, J.-M., Park, M.-R., Kim, M.-G., Chung, B. H., Pyo, H.-B., & Maeng, S. (2007). Analysis of recombinant protein expression using localized surface plasmon resonance (LSPR). *Biosensors & Bioelectronics*, *22*, 2301–2307. Available from https://doi.org/10.1016/J.BIOS.2006.12.028.

Shultz, M. D., Ulises Reveles, J., Khanna, S. N., & Carpenter, E. E. (2007). Reactive nature of dopamine as a surface functionalization agent in iron oxide nanoparticles. *Journal of the American Chemical Society*, *129*, 2482–2487. Available from https://doi.org/10.1021/ja0651963.

Si, P., Huang, Y., Wang, T., & Ma, J. (2013). Nanomaterials for electrochemical non-enzymatic glucose biosensors. *RSC Advances*, *3*, 3487. Available from https://doi.org/10.1039/c2ra22360k.

Singh, P. (2016). SPR biosensors: Historical perspectives and current challenges. *Sensors & Actuators, B: Chemical*, *229*, 110–130. Available from https://doi.org/10.1016/J.SNB.2016.01.118.

Singh, S., Kaler, R. S., & Sharma, S. (2019). Resonance effect of bimetallic diffraction grating on the sensing characteristics of surface plasmon resonance sensor with COMSOL multiphysics. *Journal of Nanoelectronics and Optoelectronics*, *14*, 669–674. Available from https://doi.org/10.1166/jno.2019.2523.

Siril, P. F., Lehoux, A., Ramos, L., Beaunier, P., & Remita, H. (2012). Facile synthesis of palladium nanowires by a soft templating method. *New Journal of Chemistry*, *36*, 2135. Available from https://doi.org/10.1039/c2nj40342k.

Solanki, P. R., Kaushik, A., Agrawal, V. V., & Malhotra, B. D. (2011). Nanostructured metal oxide-based biosensors. *NPG Asia Materials*, *3*, 17–24. Available from https://doi.org/10.1038/asiamat.2010.137.

Staiano, M., Bazzicalupo, P., Rossi, M., & D'Auria, S. (2005). Glucose biosensors as models for the development of advanced protein-based biosensors. *Molecular Biosystems*, *1*, 354–362. Available from https://doi.org/10.1039/b513385h.

Tamada, K., Nakamura, F., Ito, M., Li, X., & Baba, A. (2007). SPR-based DNA detection with metal nanoparticles. *Plasmonics*, *2*, 185–191. Available from https://doi.org/10.1007/s11468-007-9035-x.

Tang, C. X., Zhao, Y., He, X. W., & Yin, X. B. (2010). A "turn-on" electrochemiluminescent biosensor for detecting Hg^{2+} at femtomole level based on the intercalation of Ru(phen) 32 + into ds-DNA. *Chemical Communications*, *46*, 9022–9024. Available from https://doi.org/10.1039/c0cc03495a.

Tiwari, A., Kumar, R., Shefi, O., & Kaur Randhawa, J. (2020). Fluorescent mantle carbon coated core-shell SPIONs for neuroengineering applications. *ACS Applied Bio Materials*, *3*. Available from https://doi.org/10.1021/acsabm.0c00582.

Vats, T., Dutt, S., Kumar, R., & Siril, P. F. (2016). Facile synthesis of pristine graphene-palladium nanocomposites with extraordinary catalytic activities using swollen liquid crystals. *Scientific Reports*, *6*, 33053. Available from https://doi.org/10.1038/srep33053.

Vayssieres, L., Hagfeldt, A., & Lindquist, S. E. (2000). *Purpose-built metal oxide nanomaterials. The emergence of a new generation of smart materials. Pure Applied Chemistry* (pp. 47–52). Walter de Gruyter GmbH. Available from http://doi.org/10.1351/pac200072010047.

Verkman, A. S., Sellers, M. C., Chao, A. C., Leung, T., & Ketcham, R. (1989). Synthesis and characterization of improved chloride-sensitive fluorescent indicators for biological applications. *Analytical Biochemistry*, *178*, 355–361. Available from https://doi.org/10.1016/0003-2697(89)90652-0.

Vestergaard, M., Kerman, K., & Tamiya, E. (2007). An overview of label-free electrochemical protein sensors. *Sensors*, *7*, 3442–3458. Available from https://doi.org/10.3390/s7123442.

Viter, R., & Iatsunskyi, I. (2019). *Metal oxide nanostructures in sensing*. Elsevier Inc. Available from http://doi.org/10.1016/B978-0-12-814505-0.00002-3.

Viter, R., Smyntyna, V., Starodub, N., Tereshchenko, A., Kusevitch, A., Doychoa, I., ... Spigulis, J. (2012). Novel immune TiO_2 photoluminescence biosensors for leucosis detection. *Procedia Engineering*, *47*, 338–341. Available from https://doi.org/10.1016/j.proeng.2012.09.152.

Walsh, P. F. (2011). , Chapter 8*Intell. Intell. Anal.*, *875*, 235–254. Available from https://doi.org/10.1007/978-1-61779-806-1.

Wang, J. (2005). Nanomaterial-based electrochemical biosensors. *Analyst*, *130*, 421–426. Available from https://doi.org/10.1039/b414248a.

Wang, M., Leung, K. H., Lin, S., Chan, D. S. H., Leung, C. H., & Ma, D. L. (2014). A G-quadruplex-based, label-free, switch-on luminescent detection assay for Ag^{+} ions based on the exonuclease III-mediated digestion of C-Ag + -C DNA. *Journal of Materials Chemistry B*, *20*, 6467–6471. Available from https://doi.org/10.1039/c4tb01140f.

Wang, X., Shi, W., She, G., & Mu, L. (2012). Surface-enhanced Raman scattering (SERS) on transition metal and semiconductor nanostructures. *Physical Chemistry Chemical Physics*, *14*, 5891–5901. Available from https://doi.org/10.1039/c2cp40080d.

Wang, Z., Xianyu, Y., Liu, W., Li, Y., Cai, Z., Fu, X., ... Chen, Y. (2019). Nanoparticles-enabled surface-enhanced imaging ellipsometry for amplified biosensing. *Analytical Chemistry*, *91*, 6769–6774. Available from https://doi.org/10.1021/acs.analchem.9b00846, acs.analchem.9b00846.

Wongkaew, N., Simsek, M., Griesche, C., & Baeumner, A. J. (2019). Functional nanomaterials and nanostructures enhancing electrochemical biosensors and lab-on-a-chip performances: Recent progress, applications, and future perspective. *Chemical Reviews*, *119*, 120–194. Available from https://doi.org/10.1021/acs.chemrev.8b00172.

Xu, F., Zhen, G., Yu, F., Kuennemann, E., Textor, M., & Knoll, W. (2005). Combined affinity and catalytic biosensor: In situ enzymatic activity monitoring of surface-bound enzymes. doi:10.1021/JA050818Q.

Xu, K., Zhou, R., Takei, K., & Hong, M. (2019). Toward flexible surface-enhanced Raman scattering (SERS) sensors for point-of-care diagnostics. *Advancement of Science*, *6*. Available from https://doi.org/10.1002/advs.201900925.

Yan, X., Li, H., & Su, X. (2018). Review of optical sensors for pesticides. *Trends in Analytical Chemistry*, *103*, 1–20. Available from https://doi.org/10.1016/j.trac.2018.03.004.

Yang, B., Chen, X., Liu, R., Liu, B., & Jiang, C. (2015). Target induced aggregation of modified Au@Ag nanoparticles for surface enhanced Raman scattering and its ultrasensitive detection of arsenic(III) in aqueous solution. *RSC Advances*, *5*, 77755–77759. Available from https://doi.org/10.1039/c5ra15954g.

Ye, B. C., & Yin, B. C. (2008). Highly sensitive detection of mercury(II) ions by fluorescence polarization enhanced by gold nanoparticles. *Angewandte Chemie International Edition*, *47*, 8386–8389. Available from https://doi.org/10.1002/anie.200803069.

Yu, D., Blankert, B., Viré, J., & Kauffmann, J. (2005). Biosensors in drug discovery and drug analysis. *Analytical Letters*, *38*, 1687–1701. Available from https://doi.org/10.1080/00032710500205659.

Zeng, S., Yong, K.-T., Roy, I., Dinh, X. Q., Yu, X., & Luan, F. (2011). A review on functionalized gold nanoparticles for biosensing applications. *Plasmonics*, *6*, 491–506. Available from https://doi.org/10.1007/s11468-011-9228-1.

Zhang, D., Carr, D. J., & Alocilja, E. C. (2009). Fluorescent bio-barcode DNA assay for the detection of Salmonella enterica serovar Enteritidis. *Biosensors & Bioelectronics*, *24*, 1377–1381. Available from https://doi.org/10.1016/j.bios.2008.07.081.

Zhang, H., Sun, Y., Gao, S., Zhang, H., Zhang, J., Bai, Y., & Song, D. (2014). Studies of gold nanorod-iron oxide nanohybrids for immunoassay based on SPR biosensor. *Talanta*, *125*, 29–35. Available from https://doi.org/10.1016/J.TALANTA.2014.02.036.

Zhao, X., Wang, W., Liang, Y., Fu, J., Zhu, M., Shi, H., ... Tao, C. (2019). Visible-light-driven charge transfer to significantly improve surface-enhanced Raman scattering (SERS) activity of self-cleaning TiO_2/Au nanowire arrays as highly sensitive and recyclable SERS sensor. *Sensors & Actuators, B: Chemical.*, *279*, 313–319. Available from https://doi.org/10.1016/j.snb.2018.10.010.

Zhao, Y., Wang, X., Mi, J., Jiang, Y., & Wang, C. (2019). Metal nanoclusters-based ratiometric fluorescent probes from design to sensing applications. *Particle & Particle Systems Characterization*, *36*, 1900298. Available from https://doi.org/10.1002/ppsc.201900298.

Zheng, M., Ruan, S., Liu, S., Sun, T., Qu, D., Zhao, H., ... Sun, Z. (2015). Self-targeting fluorescent carbon dots for diagnosis of brain cancer cells. *ACS Nano*, *9*, 11455–11461. Available from https://doi.org/10.1021/acsnano.5b05575.

Zhou, J., Qi, Q., Wang, C., Qian, Y., Liu, G., Wang, Y., & Fu, L. (2019). Surface plasmon resonance (SPR) biosensors for food allergen detection in food matrices. *Biosensors & Bioelectronics*, *142*, 111449. Available from https://doi.org/10.1016/J.BIOS.2019.111449.

Zhou, Y., Tang, L., Zeng, G., Zhang, C., Zhang, Y., & Xie, X. (2016). Current progress in biosensors for heavy metal ions based on DNAzymes/DNA molecules functionalized nanostructures: A review. *Sensors & Actuators, B: Chemical*, *223*, 280–294. Available from https://doi.org/10.1016/j.snb.2015.09.090.

Zhu, C., Du, D., & Lin, Y. (2017). Graphene-like 2D nanomaterial-based biointerfaces for biosensing applications. *Biosensors & Bioelectronics*, *89*, 43–55. Available from https://doi.org/10.1016/j.bios.2016.06.045.

Metal oxides for detection of cardiac biomarkers

14

Deepika Sandil[1] *and Nitin Puri*[2]
[1]Bhagwan Parshuram Institute of Technology, Delhi, India, [2]Delhi Technological University, Delhi, India

14.1 Introduction

According to the World Health Organization (WHO), cardiovascular disease (CVD) is a leading cause of mortality worldwide. CVD is defined as a class of diseases related to heart and blood vessel disorders. The diseases namely coronary heart disease, stroke, congenital, and inflammatory heart disease occur due to cardiovascular reasons. In particular, acute coronary syndrome (ACS) and heart failure (HF) contribute largely to CVD deaths (Corti, Fuster, & Badimon, 2003). The ACS is a disorder that results due to the formation of atherosclerotic plaques, tissue inflammation, rupturing of plaque, or myocardial ischemia (Wu, 2015). In the last few decades, the daily routine of people from every section has switched toward an unbalanced nutrition diet, mental stress, and less physical activities. Such a change has highly elevated the incidence of CVD and particularly the heart attack occurrence.

The heart attack in biomedical science is also defined as acute myocardial infarction (AMI). The word "infarction" signifies the death of tissue due to the lack of blood flow in some specific area resulting in a deficiency of oxygen. While "Myo" refers to the muscle and "cardinal" refers to the heart tissue, thus AMI is defined as the death of tissues in the heart muscles due to sudden blockage of blood flow process called necrosis. Consequently, timely and efficient diagnosis of the AMI disease is a major challenge to facilitate immediate treatment to a cardiac patient at an early stage. At present, the clinical diagnostics for AMI are based on conventional techniques such as cardiac stress test, coronary angiography, electrocardiogram, and transthoracic echocardiogram (Szunerits, Mishyn, Grabowska, & Boukherroub, 2019). All these techniques are time-consuming, costly, need expertise, and are unable to present desired information for detection of the CVD. However, the recent developments provide an alternative in early diagnosis as *biomarkers monitoring*. Biomarkers are molecular indicators in the blood serum that can be correlated to the occurrence of cardiac diseases. In biomedicine, a biomarker is defined as a qualitative and quantitative biological characteristic that evaluates the pathological condition and can give indications on disease activity. Cardiac biomarkers are protein biomolecules that are released into the blood circulation with the onset of cardiac injury/disease. Cardiac disease is related to the contusion of myocardial muscle, disruption of a heart valve, or rupture of a cardiac chamber. Further, these

Metal Oxides for Biomedical and Biosensor Applications. DOI: https://doi.org/10.1016/B978-0-12-823033-6.00012-0

diseases are asymptomatic and are confused with other diseases including respiratory disease.

The analysis of cardiac biomarkers in the blood serum is crucial for the identification of the CVD. For the generalization, measurement of the concentration of these CVD biomarkers in the bloodstream with the onset of disease can enable the early diagnosis of the disease. Further for the clinical applications, the cardiac biomarkers for CVDs risk must present high specificity and sensitivity, wide time window for the detection, and get quickly release into the bloodstream (Tiwari et al., 2012).

14.2 Biomarkers for diagnosis

Conventionally, the number of biomarkers associated with ACS and HF plays a significant role in routine clinical laboratories. In principle, an ideal biomarker for cardiac detection needs to be specifically present at a high concentration within cardiac muscles, along with high sensitivity and specificity. The numerous cardiac biomarkers that have found significance in cardiac diagnostics have been discussed in the following section.

14.2.1 Myoglobin

Myoglobin is a cardiac biomarker released in blood with the onset of the myocardial infarction (MI) or skeletal muscle injury. It is a cytoplasmic serum protein having a low molecular weight which is released more rapidly than other cardiac biomarkers (Bhayana & Henderson, 1995). Within 1 h after the occurrence of the MI, the concentration of myoglobin starts increasing and achieves the peak value in roughly 4–8 h. A clinical laboratory reported on the examination of one of the patients, who was admitted with severe chest pain, had elevated myoglobin level within 2 h (Montague & Kircher, 1995). Thus this biomarker is the first choice for the early diagnosis of MI. Clinically, myoglobin assays are easily accessible, but they do suffer from the lack of cardiospecificity due to the presence of noncardiac-related causes. Some of the noncardiac causes include cardiac bypass surgery, skeletal muscle, renal failure, and neuromuscular disorders, all of which are found to elevate myoglobin level in serum (Möckel et al., 2001). Further, with the variation in reference ranges (50–120 μg/mL) at which myoglobin becomes symptomatic, sensitivity tests also vary resulting in poor diagnosis (Möckel et al., 2001). Though myoglobin is not considered as an ideal cardiac biomarker for the diagnosis of MI, it is helpful when used in combination with other cardiac biomarkers for the diagnosis of MI.

14.2.2 Creatine kinase

Before the discovery of troponin assays, evaluation of creatine kinase (CK) activity was preferred for the diagnosis of AMI up to 20 years. CK also known as creatine phosphokinase is an enzyme that facilitates the reversible transformation of creatine and adenosine triphosphate to creatine phosphate and adenosine diphosphate (McLeish & Kenyon, 2005). This procedure is likely to be found in muscle tissues as well as in the brain of humans. With the damage of cell membranes due to insufficient flow of oxygen, CK gets released in the blood circulation (Singh, Martinezclark, Pascual, Shaw, & O'Neil, 2010). On account of this, an elevated level of CK in the serum becomes a sensitive biomarker for the detection of MI. However, the predominance of CK in many tissues apart from the myocardium reflects the poor specificity of CK.

CK is a dimeric enzyme and is composed of two subunits namely CK-M (muscle type) and CK-B (brain type), which result in the formation of three isoenzymes: CK-MM, CK-MB, and CK-BB. Among these, CK-MM is mostly found in all tissues, while CK-BB can be traced in the brain, gastrointestinal, and kidney (Aydin, Ugur, Aydin, Sahin, & Yardim, 2019; Schlattner, Tokarska-Schlattner, & Wallimann, 2006). However, the appearance of CK-MB traces in the heart, skeletal muscle, tongue, or uterus considered it as a more sensitive marker for myocardial damage than the whole CK molecule due to its significant appearance in the myocardium and lower basal level (Ingwall et al., 1985). The normal reference range for serum CK-MB is 0–5 ng/mL. With the onset of MI, the CK-MB level gets elevated within 4–6 h and reaches its peak in 14–28 h (Singh, Martinezclark, Pascual, Shaw, & O'Neil, 2010). However, CK-MB gets dissipated from serum at a faster rate than CK. Thus due to the narrow window of detection, CK-MB might not be considered an ideal biomarker for cardiac detection. However, an elevated value of CK and CK-MB can be used in determining infarct size and considered as important predictor of prognosis of disease.

14.2.3 Troponin

Currently, the most prevalent cardiac biomolecule used in the diagnosis of AMI is troponin I (cTnI) and troponin T (cTnT) (Park, Gaze, Collinson, & Marber, 2017). Cardiac troponin (cTn) is a regulatory complex protein molecule associated with the contractile mechanism of the cardiac muscle by regulating the calcium-dependent interaction of actin and myosin (Aydin, Ugur, Aydin, Sahin, & Yardim, 2019). A cTn complex consists of three subunit proteins, namely troponin C (cTnC), cTnT, and cTnI located between actin filaments of muscle tissue.

cTnC is a calcium ions binding protein (18 kD) which regulates the activation of actin filaments and has no cardiac specificity as the same isoform found in smooth muscle (Bodor, 2016). However, cTnI and cTnT exhibit different genes for skeletal and cardiac forms and have higher sensitivity and selectivity for the diagnosis of cardiac injuries (Korff, Katus, & Giannitsis, 2006). The release of troponin into blood circulation is considered as an indication for the early events in the heart such as damage of cardiac myocytes, necrosis, or tissue degeneration. The cardiac isoform, cTnI has an additional amino acid residue on its N-terminal that does not exist in skeletal muscles, and thus it can only be traced in the myocardium while cTnT is traced both in the myocardium and a very small amount in skeletal muscle.

The quantification of cTnI and cTnT concentration is excellent in terms of specificity and sensitivity toward the detection of cardiac muscle damage and hence considered as a standard cardiac biomarker for the diagnosis of AMI.

14.2.4 Brain-type natriuretic peptides

Natriuretic peptides are another class of cardiac biomarkers used in the diagnosis of HF disease. Both brain-type natriuretic peptides (BNP) and N-terminal proBNP (NT-proBNP) are important biomarkers for HF. BNP is a hormone synthesized by heart ventricles and get a release by the heart in response to pressure variation inside the heart. The cut-off level of BNP and that of NT-proBNP is 100 and 125 pg/mL, respectively. Moreover, the fast release kinetics of BNP and well-defined cut-off make it a prominent biomarker for the diagnosis and assessment of HF (Horii et al., 2013).

14.2.5 C-reactive protein

C-reactive protein (CRP) is one of the cardiac biomarkers useful in the prognosis of ACS disease. The elevated level of CRP predicts the occurrence of death of cardiac tissues, AMI, or HF (Berton et al., 2003). Compared to cTn, CRP is a less sensitive and specific biomarker for MI but is widely used as an inflammatory marker in clinical laboratories.

14.2.5.1 Methods of cardiac detection

The recognition and quantification of cardiac biomarkers have been a crucial concern for clinical diagnostics and researchers. The conventional detection techniques used in clinical laboratories include colorimetric sensors, radioimmunoassay, fluoroimmunoassay, and enzyme-linked immunosorbent assay (ELISA) (Cummins, Auckland, & Cummins, 1987; Heeschen, Goldmann, Langenbrink, Matschuck, & Hamm, 1999; Song, Han, Kim, Yang, & Yoon, 2011). All these assays and tests are capable of showing good sensitivity and selectivity; however, these techniques are susceptible; require tedious methods of sample preparation are time-consuming, need experts to analyze data, and are not appropriate for early detection of cardiac disease and its severity. Biosensors, a bioanalytical device are considered as an ideal platform for cardiac detection which has shown an unparalleled advantage in diagnostic and clinical applications.

14.2.6 Biosensor

A biosensor is defined as an analytical system comprising of a biologically active layer (also known as sensing layer) in contact with an appropriate transduction unit capable of analyzing biointeractions and detecting the concentrations of analytes. The main components of a biosensor device shown in Fig. 14.1 include (1) a biorecognition element (e.g., nucleic acid, enzymes, cells, antibody, or other proteins);

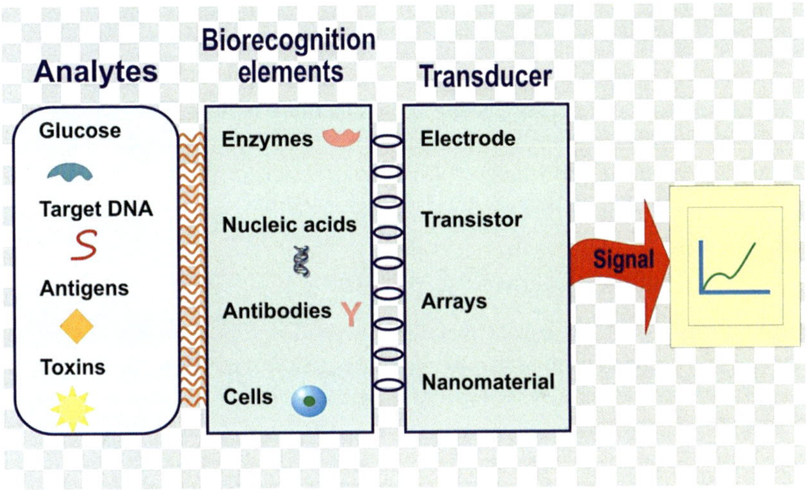

Figure 14.1 Schematic of biosensor.

(2) a transducer (sensing element), and (3) a signal processor. The working principle of a biosensor involves the recognition of a specific target analyte by a biorecognition element which produces the physicochemical response converted into a measurable signal via a transducer unit.

The crucial characteristics of a biosensor are its selectivity and sensitivity which contribute toward an efficient biosensor. Selectivity is the ability of a biosensor in the selective analysis of target analyte in presence of nonselective biomolecules. Selectivity depends solely on the type of biorecognition elements used in designing a biosensor, whereas sensitivity is defined as the ratio of variation in output response to a unit change in input value and it depends on both the biorecognition element and the type of transducer. Subsequently, the choice of the biorecognition element plays a vital role in determining the sensitivity and selectivity of a biosensor.

The biosensors can be classified based on the type of biorecognition element immobilized onto transducer surface and means of transduction mechanism. Based on the choice of the biorecognition element, biosensors are classified as catalytic biosensors and affinity-based biosensors. Catalytic biosensors utilize enzymes, microorganisms, tissues, or cells as the biorecognition element which recognizes the specific analyte by catalyzing the bioreactions significantly. Glucose biosensor, lactate, and xanthine biosensors are a few of the catalytic biosensor-based devices. Affinity-based biosensor works on the principle of selective and strong binding mainly a covalent binding of biorecognition molecules such as DNA, antibodies, receptors, or nucleic acid with a target analyte resulting in a measurable electrical signal (Slaughter, 2006). Based on measurable output signals using variant transduction techniques, biosensors are classified as electrical-based biosensors, optical-based biosensors, and piezoelectric-based biosensors. In particular, many studies have been made toward the development of cardiac detection systems using field-effect transistor (FET), electrical, electrochemical, surface plasmon resonance method, and fluorescence-based detection method (Qureshi, Gurbuz, & Niazi, 2012). The underlying approach using these methods is their ease alignment with the functional entities especially when metal oxide nanostructures are incorporated for the enhancement of performance of biosensor in terms of sensitivity, selectivity, simplicity in processing, low cost, and fast response time.

14.2.6.1 Metal oxides-based biosensor

Among the variety of biosensors that have been developed so far, metal oxides-based biosensors are the one that has shown remarkable and practical significance in the field of both chemical as well as biosensors. Metal oxide-based nanostructures exhibit many distinctive properties such as high surface to volume ratio, chemical stability, variant nanomorphology, enhanced catalytic kinetics, biocompatibility, and nontoxicity (Şerban & Enesca, 2020). Such materials have also evinced strong adsorption capability and have reinforced electron transfer kinetics, providing a suitable microenvironment for the biomolecule immobilization which results in improved biosensor's characteristics namely sensitivity, selectivity, and stability. The variant morphologies of metal oxide nanomaterials have been obtained utilizing different techniques namely hydrothermal method, sol–gel process, soft template techniques, and others. Metal oxide-based nanomaterials such as zinc oxide, titanium oxide, tungsten oxide, magnesium and zirconium oxides, and many others as an immobilizing matrix have shown remarkable development in the designing of a biosensor. These matrices enable high loading of biomolecules and retain the bioactivity of molecules by facilitating a large number of active sites and providing a microenvironment.

The unique features of metal oxide nanomaterials present the insightful analysis for a combination of biorecognition elements with transducing units, thus developing novel biosensing devices (Rahman, Saleh Ahammad, Jin, Ahn, & Lee, 2010). Furthermore, with the incorporation of other semiconducting materials (oxides, carbon nanomaterials, and quantum dots), modified metal oxide nanomaterials enhanced the biosensor performance (Hu, Lu, Chen, & Zhang, 2013). Thus these nanomaterials as single or hybrid architectures with engineered functionalities and characteristics can be efficiently used to enhance the interfacial binding properties in the designing of a biosensor. The variant parameters such as effective surface area, roughness, surface charge physical state, additional functional groups, and the hygroscopic nature of the chosen metal oxide nanomaterial play a principal role in the formation of an efficient biointerface. The so-formed biointerface helps the biomolecules to retain their biological activity along with high stability.

The numerous metal oxide nanomaterials such as TiO_2, ZnO, SnO_2, In_2O_3, WO_3, ZrO_2, and others are being employed as an immobilizing matrix for the development of an efficient biosensor for cardiac detection. The fusion of metal oxides-based matrix with different transduction mechanism enables the enhancement in characteristics of diagnostic biosensors for cardiac biomarkers detection.

The details of biosensors developed with the use of variant metal oxide nanomaterials based on transduction mechanisms are mentioned in the next section.

14.2.6.2 Platforms for cardiac detection

The different detection methods of cardiac biomarkers are based on the type of transduction method. In this section, we present the most studied metal oxide-based nanomaterials for cardiac detection using different types of electrical methods.

14.2.6.3 Electrical-based detection platform

Cardiac detection using electrical methods presents a platform consisting of simple instrumentation and low power operation ensuring a facile, cost-effective, and sensitive analysis of cardiac biomarkers. The different electrical means of operation such as electrochemical methods and FET are considered ideal for the label-free detection and real-time analysis of biomolecules (Fig. 14.2). The electrochemical approach uses the electrochemical interaction occurring at the transducer surface where in oxidation and reduction process takes place. Usually, the analyses of biomarkers using electrochemical methods are performed using different modes such as voltammetric, amperometric, and potentiometric methods. The FET approach consists of a semiconducting channel and a source and drain region. The electrical charges transport across the channel and are optimized with the application of different gate voltages.

14.2.6.4 Electrochemical-based detection

Electrochemical-based sensors are well known for application in the field of health as well as clinical laboratory investigations due to its easier usability, low cost,

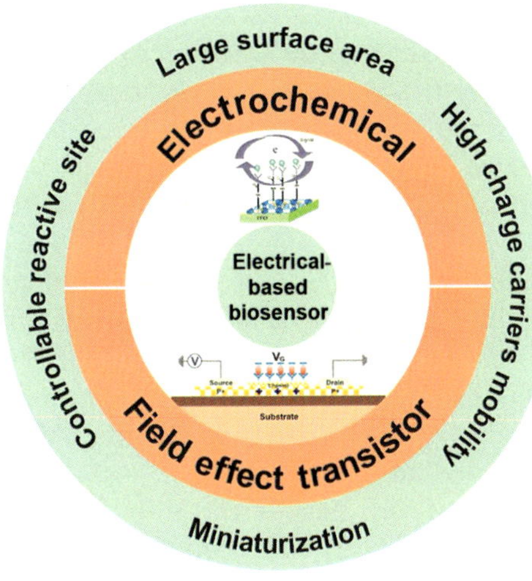

Figure 14.2 Representation of electrical-based biosensor illustrating its features.

simple instrumentation, and most importantly being a surface phenomenon. There are various means of electrochemical measurements: first is voltammetric, in which current is measured in response to the applied potential (e.g., cyclic voltammetry, differential pulse voltammetry, or chronoamperometry); second is impedimetric in which impedance is measured in the response to small amplitude perturbation [e.g., electrochemical impedance spectroscopy (EIS)], and third is conductometric which measures electrical conductivity in the response to change in the chemical reaction. The electrochemical detection technique enables enhanced sensitivity and low detection limits as the performance is directly linked to its electrical or electrochemical response occurring at the interface of fabricated bioelectrode and electrolyte in the electrochemical cell (Ronkainen, Halsall, & Heineman, 2010). Principally, the sensitivity of a biosensor depends on the sensing platform which ideally exhibits fast electron transfer kinetics and provides a micro bioenvironment to retain the bioactivity of the molecules. Metal oxide nanomaterials as immobilizing matrices in designing biosensor have shown promising results in terms of sensitivity, stability, reproducibility, and other parameters due to its distinctive features such as large surface to volume ratio, chemical active sites, biocompatible, ease functionalization, and strong adsorption capability. Zinc oxide (ZnO), tungsten oxide (WO_3), titanium oxide (TiO_2), manganese oxide (Mn_3O_4), and other oxides have shown remarkable performance in the diagnosis of cardiac biomarkers as discussed.

Researcher Munje and his coworkers had designed a nonfaradaic electrochemical biosensor for the diagnosis of cardiac protein biomarker (cTnT) employing nanostructured ZnO thin film as a transducing matrix due to its tunable electrical characteristics (Munje et al., 2015). The fabricated matrix surface was functionalized using two types of linker one as dithiobis succinimidyl propionate (DSP) and another as 3-aminopropyl triethoxy silane (APTES) which demonstrated linear dynamic range for cTnT concentration (10–300 pg/mL) in human serum. The validation of the functionalization of linkers was also studied using fluorescence studies which provide the visual validation of the DSP and APTES binding to the ZnO surface. The utilization of the EIS technique for electrical quantification analysis of the fabricated biosensor showed enhanced performance for the DSP linker with limit of detection (LOD) as low as 10 fg/mL.

In another study, the sputter-deposited ZnO film was utilized for the development of an ultra-sensitive electrical biosensor for the detection of cardiac biomarker cTnT (Jacobs, Muthukumar, Selvam, Craven, & Prasad, 2014). The resultant variation in the electrical signal indicates the amount of cTnT adsorbed onto the surface of ZnO and allowing a highly sensitive surface for the detection of cTnT. As the sputtering process results in oxygen deficiency in deposited ZnO film enhanced electrical biosensing characteristics were observed along with a lower LOD (10 fg/mL) and wide dynamic detection range.

Recently, researchers used ZnO nanorods as a diagnostic sensing platform for the detection of cardiac biomarker cTnT using an electrochemical transduction mechanism (Shanmugam, Muthukumar, Selvam, & Prasad, 2016). Hydrothermally grown ZnO nanorods were functionalized with cross-linker DSP for the immobilization of cTnT antibodies. In this work, the authors studied the response of fabricated nanoelectrode robustness based on the varied density of ZnO nanorods. Later using EIS measurements, modulation in charge transfer resistance (R_{CT}), and effective charge density with respect to density of ZnO nanorods was observed based on the presence of zinc interstitials sites and oxygen vacancies onto sensing surface. The selectivity and sensitivity of the fabricated ZnO biosensor as studied using the EIS and Mott–Schottky method results in an LOD as low as 0.1 pg/mL.

Also, the authors further developed a rapid and flexible nanostructured ZnO array-based electrochemical biosensor platform for the simultaneous detection of cardiac proteins cTnI and cTnT (Shanmugam, Muthukumar, Chaudhry, Anguiano, & Prasad, 2017). Optimization of hydrothermal grown conditions of the ZnO nanostructure onto the working electrode of the polyimide substrate verified the radial diffusion of the electrolyte. The evaluation of the sensitivity and selectivity of the cardiac isoforms (cTnI and cTnT) studied simultaneously using EIS and Mott–Schottky results in the LOD as 1 pg/mL. These favorable results of the fabricated nanostructured ZnO sensing platforms prompt authors to study the sensor's analytical performance for the detection of different cardiac biomarkers simultaneously (Shanmugam, Muthukumar, Tanak, & Prasad, 2018). The authors have reported linear signal response of the fabricated biosensor for the detection of three multiplexed cardiac biomarkers: cTnT, cTnI, and BNP in the dynamic range (1 pg/mL–100 ng/mL) along with an LOD as 1 pg/mL.

Similarly, various nanomorphologies of tungsten trioxide (WO_3) such as nanoparticles, nanosheets, nanorods, and nanocomposite have been studied as an immobilizing matrix for electrochemical detection of cardiac biomarker cTnI. The modified structures of WO_3 are nontoxic, easily synthesized, environmentally friendly, highly electrocatalytic, and chemically stable. Tungsten trioxide nanoparticles (WO_3) functionalized with APTES demonstrate as a potential matrix for the immobilization of cTnI antibodies and detection of analyte due to its fast electron transfer kinetics, large surface to volume ratio, and strong adsorption. The electrochemical response studies of the fabricated immunosensor as a function of antigen cTnI exhibit sensitivity 26.56 Ω ng/mL/cm^2 in the range of 1–250 ng/mL using the EIS technique (Sandil et al., 2017). The interaction between the cTnI antibody and its antigen results in the formation of an electrically insulated complex resulting in an increase in R_{CT} with the increase in the concentration of cTnI antigen.

In another study, Sandil and coauthors used the tungsten trioxide nanorods (WO_3 NRs) as the immobilizing matrix toward the fabrication of a sensitive and label-free immunosensing platform for cTnI detection. EDC–NHS linker was used for the successful covalent immobilization of the cTnI antibody. The as-fabricated immunosensor presents high sensitivity [6.81 kΩ mL/(ng cm^2)] and LOD as 0.01 ng/mL along with good reproducibility and selectivity (Sandil, Sharma, & Puri, 2019).

To enhance the performance of the WO_3-based sensor, incorporation of a semiconducting element such as graphene can contribute toward improved biosensing characteristics. In this reference, authors have demonstrated the fabrication of a sensitive electrochemical immunosensor for cTnI detection using tungsten trioxide-reduced graphene oxide nanocomposite (WO_3–RGO) (Sandil, Srivastava, Malhotra, Sharma,

& Puri, 2018). The immunosensor platform comprises of in situ hydrothermally grown WO_3−RGO nanocomposite as an immobilizing matrix functionalized with APTES linker. The application of APTES molecule onto the WO_3−RGO matrix activates amino functional groups (−NH_2) that covalently binds with cTnI antibodies and was validated with FT-IR and contact angle studies. The mutually reinforcing characteristic of WO_3−RGO nanocomposite results in enhanced heterogeneous electron transfer rate constant, K_o as 2.4×10^{-4} cm/s of the fabricated biosensor influencing its efficiency. The superior performance of the WO_3−RGO-based sensor platform as demonstrated by its sensitivity as 58.24 μA/cm^2 per decade in a wide range 0.01−250 ng/mL indicates the strong covalent coupling of antibodies to matrix leading to high stability; second, the presence of oxygen moieties in WO_3 which enhanced loading capacity of antibodies and lastly WO_3−RGO nanocomposite that enhanced electron transfer kinetics.

Titanium oxide (TiO_2) is one of the important metal oxide materials for biosensor applications owing to its strong oxygen capability, good chemical activity, and abundance availability. Gao and his coworkers demonstrated a sensitive and portable bioassay using a microfluidic paper-based photoelectrochemical (PEC) analyzer for the detection of cTnI (Gao et al., 2019). In this work, the authors employed TiO_2 material which is capable of producing a good photocurrent signal. The photocurrent intensity of the fabricated sensor decreases with the increase of cTnI antigen in the range of 1.2 fg/mL−20 ng/mL and demonstrated the detection limit as low as 0.47 fg/mL. The remarkable performance of the fabricated paper-based single-crystalline TiO_2 arrays for the detection of cTnI is attributed to its short diffusion length for charge carriers and favorable charge separation pathways.

Among other metal oxides, manganese oxide (Mn_3O_4) is also a promising immobilization matrix for the development of cardiac biosensor. The salient features such as high charge carrier mobility, good chemical stability, and excellent electrochemical and catalytic properties of Mn_3O_4 enable its applications in electrochemical biosensing. In this context, Singh and his coworkers demonstrated the fabrication of a highly sensitive microfluidic biochip using a microporous manganese oxide−graphene oxide nanocomposite for the detection of cardiac biomarker cTnI (Singh et al., 2017). The authors employed the photolithography technique for the fabrication of a microfluidic channel which offers high sensitivity and fast response in sensing applications. Also, the synthesis of a nanostructured Mn_3O_4 onto RGO sheets as a matrix provides a large surface area that imparts high loading of cTnI antibodies and enhances the electrochemical process at the sensor surface. The fabricated microfluidic platform for cardiac detection presents high sensitivity as log [87.58] kΩ/(ng/mL)/cm^2 in a wide detection range 0.008−20 ng/mL along with high stability and reproducibility.

Further, Fan et al. studied hydrothermally synthesized flower-like morphology of SnO_2 toward the construction of PEC immunosensor based on SnO_2/NCQDs/BiOI composite for cTnI detection (Fan et al., 2020). The surface modification with nitrogen-doped carbon quantum dots and then with bismuth oxyiodide enhanced the photochemical activity of SnO_2 as demonstrated by photocurrent measurement and EIS study. The fabricated immunosensor demonstrates high sensitivity with a wide linear detection range (0.001–100 ng/mL) and a low detection limit as 0.3 pg/mL which can be attributed to porous structured SnO_2 microflowers providing stability and large surface area for the immobilization of biomolecules.

Researchers have also reported the work on iridium oxide (IrOx) nanomaterial as a cardiac detection platform due to its high charge transfer ratio ability which enables detection of small change in surface charge. In this regard, Venkatraman and his coworkers reported the design of the IrOx nanowires-based platform toward the detection of two cardiac protein molecules CRP and myeloperoxidase (MPO) (Venkatraman et al., 2009). The fabricated device incorporated the electrochemical cell technology worked on the formation and perturbation of the electrical double layer. The facile and rapid charge transfer across IrOx nanowires generates distinct capacitance variations associated with the binding process and presented LOD as low as 1 ng/mL for CRP and 500 pg/mL for MPO in a linear detection range: 10 ng/mL–100 μg/mL and 1 ng/mL–1 μg/mL, respectively.

In an attempt toward the development of novel functional nanomaterials as a transducing element for the superior performance of a biosensor, researchers have explored and studied the ternary oxide semiconductor material such as zinc tin oxide ($ZnSnO_3$). Supraja and his coworkers demonstrated the electrochemical detection of cardiac biomarker cTnT utilizing hydrothermally synthesized $ZnSnO_3$ nanomaterial as an immobilized matrix (Supraja, Sudarshan, Tripathy, Agrawal, & Singh, 2019). The authors have employed EIS study to analyze the response study of the target antigen in the wide detection range 1 fg/mL–1 μg/mL for two different substrate materials including glassy carbon electrode (GCE) and polyethylene terephthalate (PET) coated onto ITO. The fabricated platforms present the sensitivity as 35.25 and 8.813 kΩ (μg/mL)/cm^2 for GCE and ITO/PET-based bioelectrodes along with LOD as 0.187 and 0.571 fg/mL respectively.

14.2.6.5 FET-based detection platform

An FET-based sensing platform can be accomplished using metal oxides-based matrix. The electrostatic charges carried by biomolecules and generation of electrical potential change on bioreactions are the reasons for using FET approach in biosensing. The following approach promises parallel sensing, economical power consumption, miniaturization, and compact electronic read-out systems.

The working principle of an FET-based biosensor is based on the measurement of ions set out at the gate electrode of the FET. In this approach, diagnosis is based on the generation of surface potential change onto the transducer surface with the onset of biomolecules interactions at the gate electrode followed by variation in the channel's current located in between drain and source (Arshad, Fathil, & Hashim, 2017). The flow of current, increasing, or decreasing across the transducer's channel depends on the type of semiconductor used (n-type or p-type). The FET-based biosensors are further classified based on different gate approaches including ion-sensitive FET, Si nanowires-based biosensor, and organic FET. All these surface designs have demonstrated an enhancement in the performance of the biosensor in terms of sensitivity and selectivity.

In recent times, Fathil et al. (2017) incorporated a thin film of ZnO nanoparticles as a matrix at the microsized channel on silicon-on-insulator wafer for the detection of cTnI biomarker using FET-based mechanism. A wet chemical method, sol−gel synthesis approach was used to produce a uniform ZnO nanoparticles thin film via spin coating technique. To enhance the sensitivity of the fabricated sensor, the author used a substrate-gate approach in which controlled conduction current enables more current across the channel (ZnO nanoparticles matrix) between the p-type source and drain region. The fabricated biosensor has been reported having high sensitivity and selectivity and low LOD (3.24 pg/mL) which attribute toward the high loading capacity and large surface area of the immobilization matrix (ZnO NPs) enabling a suitable microenvironment for the cTnI antibodies biomolecules to retain their activity.

Arshad, Adzhri, Fathil, Gopinath, & N M, (2018) used TiO_2 thin film as a transducing agent for the development of the FET-based biosensor for the diagnosis of elevated cTn levels. He explored the label-free detection of cTn by employing the FET-coupled back gate concept which amplifies the signal using ambipolar conduction. The deposited TiO_2 thin film was observed to have less roughness with an average grain size of 200 nm. The proposed FET-based biosensor depicted the sensitivity as 459.2 nA/(g/mL) and LOD as 1 fg/mL for cTnI marker.

Tin oxide (SnO_2) is one of the metal oxide materials exhibiting a high isoelectric point (4.5−5) which makes it a suitable platform for the immobilization of biomolecules. The strong adsorption capability, good biocompatibility, and high electrocatalytic potential make SnO_2 a promising candidate for the high performance of the biosensor. A nanostructured SnO_2-based FET biosensor has been utilized for cTnI detection using an electrical technique where antibodies of cTnI were selectively assembled on the SnO_2 nanobelts channel of FET (Cheng et al., 2011).

Similarly, indium oxide (In_2O_3) can be used as a sensing material due to its high mobility rate [160 $cm^2/(V\ S)$], large bandgap (3.5−3.7 eV), and chemical stability. In_2O_3 nanoribbons have been successfully applied to highly sensitive and scalable FET-based biosensors using a shadow mask technique for the detection of different three cardiac biomarkers, namely, cTnI, CK-MB, and BNP (Aroonyadet et al., 2015). The fabricated device was well studied for pH sensing experiment, reusability, and real-time sensing. Liu et al. have fabricated a facile and rapid technique to detect cardiac biomarkers by employing In_2O_3 nanoribbons-based biosensor.

14.3 Future prospects and conclusion

The development of a rapid and reliable platform for the detection of cardiac disease AMI has stimulated biomedicine research due to the limitations in the performance of the device in terms of time, sensitivity, selectivity, and cost. To meet the challenge of developing an early stage diagnostic device along with superior performance, the design of biosensors has got even more complex and time-consuming. However, the advancement in the emergence of a novel wide range of metal oxides-based nanomaterials has open new doors in biosensing toward a rapid and facile diagnosis of AMI. They have reformed the designs of a biosensor in terms of detection techniques. Herein, we addressed electrical-based biosensors mainly electrochemical and FETs using metal oxide nanomaterials demonstrating the outstanding potential for cardiac detection. Although, there are no commercial kits available for cardiac detection using nano-based biosensor soon emphasis on monitoring of cardiac biomarkers using the point-of-care kit will be demonstrated. The coupling of variant biorecognition elements with novel metal oxide nanomaterials and matrices surface along with variant transducing units advances the biosensor's design and will revolutionize the clinical procedures for cardiac biomarkers detection.

References

Aroonyadet, N., Wang, X., Song, Y., Chen, H., Chote, R. J., Thompson, M. E., ... Zhou, C. (2015). Highly scalable, uniform, and sensitive biosensors based on top-down indium oxide nanoribbons and electronic enzyme-linked immunosorbent assay. *Nano Letters, 15*(3), 1943–1951.

Arshad, M. K. M., Fathil, M. F. M., & Hashim, U. (2017). FET-biosensor for cardiac troponin biomarker. In *EPJ web of conferences*. EDP Sciences.

Arshad, M. K., Adzhri, R., Fathil, M. F., Gopinath, S. C., & N M, N. M. (2018). Field-effect transistor-integration with TiO_2 nanoparticles for sensing of cardiac troponin I biomarker. *Journal of Nanoscience and Nanotechnology, 18*(8), 5283–5291.

Aydin, S., Ugur, K., Aydin, S., Sahin, I., & Yardim, M. (2019). Biomarkers in acute myocardial infarction: Current perspectives. *Vascular Health and Risk Management, 15*, 1.

Berton, G., Cordiano, R., Palmieri, R., Pianca, S., Pagliara, V., & Palatini, P. (2003). C-reactive protein in acute myocardial infarction: Association with heart failure. *American Heart Journal, 145*(6), 1094–1101.

Bhayana, V., & Henderson, A. R. (1995). Biochemical markers of myocardial damage. *Clinical Biochemistry, 28*(1), 1–29.

Bodor, G. S. (2016). Biochemical markers of myocardial damage. *Electronic Journal of the International Federation of Clinical Chemistry and Laboratory Medicine, 27*(2), 95.

Cheng, Y., Chen, K. S., Meyer, N. L., Yuan, J., Hirst, L. S., Chase, P. B., & Xiong, P. (2011). Functionalized SnO2 nanobelt field-effect transistor sensors for label-free detection of cardiac troponin. *Biosensors and Bioelectronics, 26*(11), 4538–4544.

Corti, R., Fuster, V., & Badimon, J. J. (2003). Pathogenetic concepts of acute coronary syndromes. *Journal of the American College of Cardiology, 41*(4 Suppl), S7–S14.

Cummins, B., Auckland, M. L., & Cummins, P. (1987). Cardiac-specific troponin-l radioimmunoassay in the diagnosis of acute myocardial infarction. *American Heart Journal, 113*(6), 1333–1344.

Fan, D., Liu, X., Shao, X., Zhang, Y., Zhang, N., Wang, X., ... Ju, H. (2020). A cardiac troponin I photoelectrochemical immunosensor: Nitrogen-doped carbon quantum dots–bismuth oxyiodide–flower-like SnO_2. *Microchimica Acta, 187*, 1–10.

Fathil, M., Arshad, M. M., Ruslinda, A. R., Gopinath, S. C., Nuzaihan, M., Adzhri, R., ... Lam, H. Y. (2017). Substrate-gate coupling in ZnO-FET biosensor for cardiac troponin I detection. *Sensors and Actuators B: Chemical, 242*, 1142–1154.

Gao, C., Xue, J., Zhang, L., Zhao, P., Cui, K., Ge, S., & Yu, J. (2019). Paper based modification-free photoelectrochemical sensing platform with single-crystalline aloe like TiO_2 as electron transporting material for cTnI detection. *Biosensors and Bioelectronics*, *131*, 17–23.

Heeschen, C., Goldmann, B. U., Langenbrink, L., Matschuck, G., & Hamm, C. W. (1999). Evaluation of a rapid whole blood ELISA for quantification of troponin I in patients with acute chest pain. *Clinical Chemistry*, *45*(10), 1789–1796.

Horii, M., Matsumoto, T., Uemura, S., Sugawara, Y., Takitsume, A., Ueda, T., & Nakagawa, H. (2013). Prognostic value of B-type natriuretic peptide and its amino-terminal proBNP fragment for cardiovascular events with stratification by renal function. *Journal of Cardiology*, *61*(6), 410–416.

Hu, C., Lu, T., Chen, F., & Zhang, R. (2013). A brief review of graphene–metal oxide composites synthesis and applications in photocatalysis. *Journal of the Chinese Advanced Materials Society*, *1*(1), 21–39.

Ingwall, J. S., Kramer, M. F., Fifer, M. A., Lorell, B. H., Shemin, R., Grossman, W., & Allen, P. D. (1985). The creatine kinase system in normal and diseased human myocardium. *New England Journal of Medicine*, *313*(17), 1050–1054.

Jacobs, M., Muthukumar, S., Selvam, A. P., Craven, J. E., & Prasad, S. (2014). Ultrasensitive electrical immunoassay biosensors using nanotextured zinc oxide thin films on printed circuit board platforms. *Biosensors and Bioelectronics*, *55*, 7–13.

Korff, S., Katus, H. A., & Giannitsis, E. (2006). Differential diagnosis of elevated troponins. *Heart (British Cardiac Society)*, *92*(7), 987–993.

McLeish, M. J., & Kenyon, G. L. (2005). Relating structure to mechanism in creatine kinase. *Critical Reviews in Biochemistry and Molecular Biology*, *40*(1), 1–20.

Möckel, M., Gerhardt, W., Heller, Jr G., Klefisch, F., Danne, O., Maske, J., ... Stork, T. (2001). Validation of NACB and IFCC guidelines for the use of cardiac markers for early diagnosis and risk assessment in patients with acute coronary syndromes. *Clinica chimica acta*, *303*(1–2), 167–179.

Montague, C., & Kircher, T. (1995). Myoglobin in the early evaluation of acute chest pain. *American Journal of Clinical Pathology*, *104*(4), 472–476.

Munje, R. D., Jacobs, M., Muthukumar, S., Quadri, B., Shanmugam, N. R., & Prasad, S. (2015). A novel approach for electrical tuning of nano-textured zinc oxide surfaces for ultra-sensitive troponin-T detection. *Analytical Methods*, *7*(24), 10136–10144.

Park, K. C., Gaze, D. C., Collinson, P. O., & Marber, M. S. (2017). Cardiac troponins: From myocardial infarction to chronic disease. *Cardiovascular Research*, *113*(14), 1708–1718.

Qureshi, A., Gurbuz, Y., & Niazi, J. H. (2012). Biosensors for cardiac biomarkers detection: A review. *Sensors and Actuators B: Chemical*, *171*, 62–76.

Rahman, M., Saleh Ahammad, A. J., Jin, J.-H., Ahn, S. J., & Lee, J.-J. (2010). A comprehensive review of glucose biosensors based on nanostructured metal-oxides. *Sensors*, *10*, 4855–4886.

Ronkainen, N. J., Halsall, H. B., & Heineman, W. R. (2010). Electrochemical biosensors. *Chemical Society Reviews*, *39*(5), 1747–1763.

Sandil, D., Kumar, S., Arora, K., Srivastava, S., Malhotra, B. D., Sharma, S. C., & Puri, N. K. (2017). Biofunctionalized nanostructured tungsten trioxide based sensor for cardiac biomarker detection. *Materials Letters*, *186*, 202–205.

Sandil, D., Sharma, S. C., & Puri, N. K. (2019). Protein-functionalized WO3 nanorods–based impedimetric platform for sensitive and label-free detection of a cardiac biomarker. *Journal of Materials Research*, *34*(8), 1331–1340.

Sandil, D., Srivastava, S., Malhotra, B. D., Sharma, S. C., & Puri, N. K. (2018). Biofunctionalized tungsten trioxide-reduced graphene oxide nanocomposites for sensitive electrochemical immunosensing of cardiac biomarker. *Journal of Alloys and Compounds*, *763*, 102–110.

Schlattner, U., Tokarska-Schlattner, M., & Wallimann, T. (2006). Mitochondrial creatine kinase in human health and disease. *Biochimica et Biophysica Acta (BBA)-Molecular Basis of Disease*, *1762*(2), 164–180.

Shanmugam, N. R., Muthukumar, S., Chaudhry, S., Anguiano, J., & Prasad, S. (2017). Ultrasensitive nanostructure sensor arrays on flexible substrates for multiplexed and simultaneous electrochemical detection of a panel of cardiac biomarkers. *Biosensors and Bioelectronics*, *89*, 764–772.

Shanmugam, N. R., Muthukumar, S., Selvam, A. P., & Prasad, S. (2016). Electrochemical nanostructured ZnO biosensor for ultrasensitive detection of cardiac troponin-T. *Nanomedicine: Nanotechnology, Biology, and Medicine*, *11*(11), 1345–1358.

Shanmugam, N. R., Muthukumar, S., Tanak, A. S., & Prasad, S. (2018). Multiplexed electrochemical detection of three cardiac biomarkers cTnI, cTnT and BNP using nanostructured ZnO-sensing platform. *Future Cardiology*, *14*(2), 131–141.

Singh, N., Ali, M. A., Rai, P., Sharma, A., Malhotra, B. D., & John, R. (2017). Microporous nanocomposite enabled microfluidic biochip for cardiac biomarker detection. *ACS applied materials & interfaces*, *9*(39), 33576–33588.

Singh, V., Martinezclark, P., Pascual, M., Shaw, E. S., & O'Neil, W. W. (2010). Cardiac biomarkers–the old and the new: A review. *Coronary Artery Disease*, *21*(4), 244–256.

Slaughter, G. (2006). Current advances in biosensor design and fabrication. *Encyclopedia of Analytical Chemistry: Applications, Theory and Instrumentation*, 1–25.

Song, S. Y., Han, Y. D., Kim, K., Yang, S. S., & Yoon, H. C. (2011). A fluoro-microbead guiding chip for simple and quantifiable immunoassay of cardiac troponin I (cTnI). *Biosensors and Bioelectronics*, *26*(9), 3818–3824.

Supraja, P., Sudarshan, V., Tripathy, S., Agrawal, A., & Singh, S. G. (2019). Label free electrochemical detection of cardiac biomarker troponin T using $ZnSnO_3$ perovskite nanomaterials. *Analytical Methods (San Diego, CA)*, *11*(6), 744–751.

Szunerits, S., Mishyn, V., Grabowska, I., & Boukherroub, R. (2019). Electrochemical cardiovascular platforms: Current state of the art and beyond. *Biosensors and Bioelectronics*, *131*, 287–298.

Tiwari, R. P., Jain, A., Khan, Z., Kohli, V., Bharmal, R. N., Kartikeyan, S., & Bisen, P. S. (2012). Cardiac troponins I and T: Molecular markers for early diagnosis, prognosis, and accurate triaging of patients with acute myocardial infarction. *Molecular Diagnosis & Therapy*, *16*(6), 371–381.

Venkatraman, V. L., Reddy, R. K., Zhang, F., Evans, D., Ulrich, B., & Prasad, S. (2009). Iridium oxide nanomonitors: Clinical diagnostic devices for health monitoring systems. *Biosensors and Bioelectronics*, *24*(10), 3078–3083.

Wu, A. H. (2015). Analytical validation of novel cardiac biomarkers used in clinical trials. *American Heart Journal*, *169*(5), 674–683.

Şerban, I., & Enesca, A. (2020). Metal oxides-based semiconductors for biosensors applications. *Frontiers in Chemistry*, *8*, 354.

Section 3

Specific metal oxides and their biomedical applications

Regulating cell function through micro- and nanostructured transition metal oxides

15

Miguel Manso Silvan
Departamento de Física Aplicada and Instituto de Ciencia de Materiales Nicolás Cabrera, Universidad Autónoma de Madrid, Madrid, Spain

15.1 Introduction

15.1.1 Regulating cell adhesion

The interaction of cells with surfaces is a complex process in which different molecular structures have to find an equilibrium in a complex matrix of force fields (Missirlis & Spiliotis 2002; Yamamoto, Mishima, Maruyama, & Sumita, 2000). A passive surface tension contribution from the cell membrane and the material interface is dominant at the first interaction stages (Bacakova, Filova, Parizek, Ruml, & Svorcik, 2011; Sales, Holle, & Kemkemer, 2017). In a physiological environment, the original natue of the material surface is however not preserved, and adsorbed proteins, sugars, etc., mediate the interaction (Rabe, Verdes, & Seeger, 2011; Wilson, Clegg, Leavesley, & Pearcy, 2005). This affects drastically the surface properties of the material in a few seconds. The final conformation of the adsorbed biomolecules (mainly the hydrophilic/hydrophobic nature of the medium exposed factors) determines the dynamics and kinetics of the ulterior cell-surface interaction (Dou, Zhang, Feng, & Jiang, 2015; Yang et al., 2012). Although most transition metal oxide (TMO) surfaces are mildly hydrophilic, the protein adsorption rates are high and consequently, dense biomolecular layers are formed on their surfaces (Fukuzaki, Urano, & Nagata, 1996). The surface tension nature of the material changes notably and, upon interaction with the hydrophobic phospholipid bilayer membrane, initiates the passive stage of cell adhesion. A restricted combination of interactions can be described at the first stage of cell−surface contact (few minutes in general): one shall take into account hydrophilic (surface)/hydrophobic (cell membrane) interactions leading to repulsive phenomena (against cell spreading) and hydrophobic/hydrophobic attractive interactions (favoring cell spreading) (Karakecili & Gumusderelioglu, 2002).

However, cell-directed mechanisms take control of the process in a time scale varying from a few minutes to a couple of hours. On hydrophobic surfaces one would expect to observe well-spread cells, but interaction with cell transmembrane proteins (mainly integrins) is highly inhibited (Altankov, Groth, Krasteva,

Metal Oxides for Biomedical and Biosensor Applications. DOI: https://doi.org/10.1016/B978-0-12-823033-6.00013-2

Albrecht, & Paul, 1997). Consequently, the membrane spreading is not followed by an intense and rapid development of adhesions (Bacakova et al., 2011). These may take place with time, eventually favored by a catalytic activity of the cell, which modifies the surface chemistry facilitating focal adhesion formation (Ma, Mao, & Gao, 2007). Inversely, though the cell membrane extension on mild hydrophilic surfaces is not drastically favored, the action of integrins toward the formation of focal adhesions is much faster (Altankov et al., 1997; Ko et al., 2013). Also the strength of the adhesion is much stronger and well-extended filopodia are much more often observed in mild hydrophilic surfaces (Ma et al., 2007). At this stage, the effect of the mechanical properties of the substrate emerges as extremely relevant. The formation of focal adhesions implies intense local tensions, which must be externally compensated by the materials toughness (Wong, Leach, & Brown, 2004). The study of the tensile and compressive stress distributions observed as a result of the interaction of a cell with its environment has been categorized within tensegrity principles (Ingber et al., 2003).

This description implies that cells respond to surface features with a hierarchical dependence (see Fig. 15.1). In fact, the exploratory system of the cell includes lamellipodia and filopodia, correlated with preadhesion phenomena at the micro- and nanoscale, respectively (Sales et al., 2017). It derives then that, the control of nano- and microscale features on surfaces is a resource to control cell topography and function (Dou et al., 2015; Kim et al., 2010). Cells exhibit in fact "contact guidance" (a conditioning of adhesion due to topographic motives) in a material, design, and cell-specific manner (Abiko et al., 1993). The engineered surface features present contrasts of surface-free energy due to topographic or chemical modification and these lead to different cell adhesion modes and consequently, to

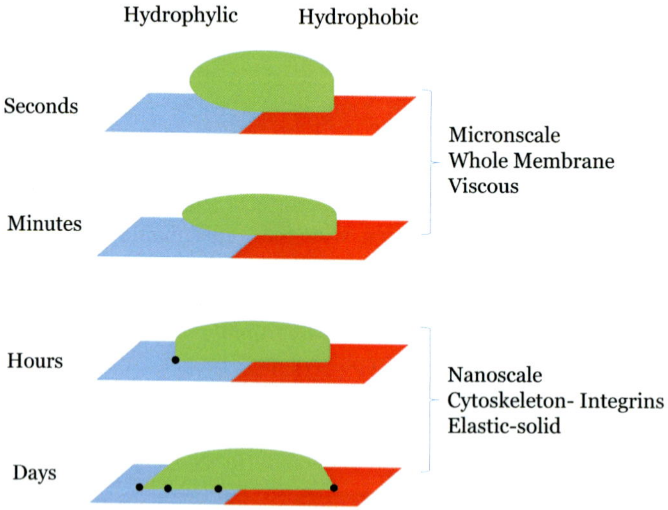

Figure 15.1 Chronogram of the generic interaction of cells with hydrophilic–hydrophobic surfaces and their relationship to the cell-material micro/nanostructure.

changes in gene expression (Lee, Kang, & Lee, 2012). The effects are observable in the specific polarization, migration, and differentiation responses, which are the objective of many biomedical studies (Wilson et al., 2005).

The biological implications and biomedical applications of surface contrasts exposed to cell cultures are very wide and out of the scope of this chapter. However, it is of high importance to try to categorize the applications according to simple principles, even if cell lines and tissues may respond differently to the same material. We underline three potential families of applications in the present chapter (see Fig. 15.2), which highlight that, in the presence of surface structures, cells exhibit an exacerbated fluid behavior (Ventre, Valle, Bianchi, Biscarini, & Netti, 2012). In fact, they exhibit cohesion forces, flow, and can compress to a certain limit. Consequently, the term "Cell Fluidics" has been used to describe the families of systems designed to control cell behavior through constricted, guided, or halted migration (Ventre et al., 2012). The particularities of the "molecular units" in these fluids may make it convenient to rather speak about "celluidics."

The first group of devices (Fig. 15.2A) stems from the use of cell cultures from a specific lineage, which are transformed during the culture process by the induced migratory stage or morphogenic transformation of the cells (Ynsa et al., 2014). The most common cell line for these devices are human mesenchymal cells (MSCs), a pluripotential cell line, isolated principally from the bone marrow, that can get differentiated into bone, cartilage, and other functional tissues (Subramani & Birch 2006; Sundelacruz & Kaplan 2009). These transformations are also relevant for neural cells, which are promoted to neurogenic stages (Su, Liao, Wu, Wang, & Shih, 2013). This kind of reactors can be defined as migration reactors since the active dynamic state of the cell is accompanied by a signaling and secreting activity that drastically influences the own and neighboring cellular environment (Salasznyk, Klees, Hughlock, & Plopper, 2004; Ynsa et al., 2014). The devices that allow the reverse operation, that is, the reprogramming of specific functional cells toward a multipotent stage, can be also considered within this categorization of devices (Tseng, Kunze, Kittur, & Di Carlo, 2014).

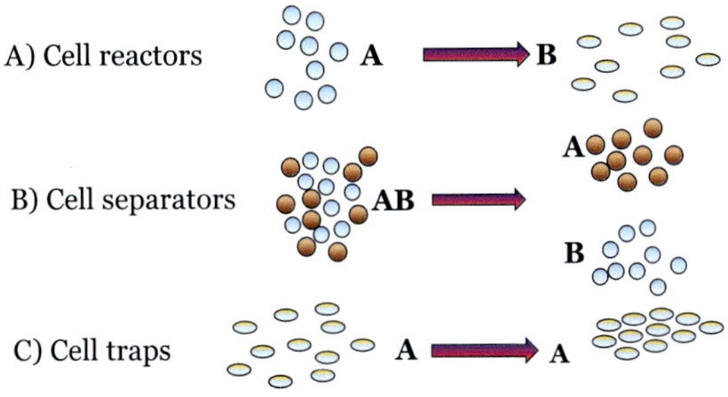

Figure 15.2 Migration reactors, cell separators, and structured tissue traps. (A) cell function transformation in a reactor. (B) Separation of heterogeneous cultures in vitro. (C) Cell structuring into tissue-like forms through traps.

The second group of devices considers coculture environments (in vitro systems in which two different cell lines are analyzed together) and the presence of surface features with engineered cell affinity (Hui & Bhatia, 2007a) (see Fig. 15.2B). These devices allow to selectively condition the two cell lines so that they respond differently to the substrate and organize, communicate, or separate accordingly. These kinds of substrates are ideal candidates for the study of the influence of the cellular microenvironment of secreting cells, which are determinants of the secreting ability (Hui & Bhatia, 2007b). In addition, tumor—healthy cell interactions with implication in tumor progressing processes (metastasis) (Ibrahim et al., 2012; Phillips & Birnby 2004) and tumor therapies (Langhans, 2018; Liu, Netzel-Arnett, Birkedal-Hansen, & Leppla, 2000) can be also studied.

The third group of devices comprises the surfaces engineered to locally structure cells by promoting fixation in specific sites that bear adhesion promotion features (see Fig. 15.2C). This family of devices is the one that favors from an easier expansion to a 3D environment (Wong, Perez-Castillejos, Love, & Whitesides, 2008). Some devices aim at few (single) cells immobilization (Liu & Sun, 2009; Maruyama, Arai, Fukuda, & Katsuragi, 2005), for instance to facilitate physioelectrochemical analysis (Ribeiro et al., 2015) or studies of mechanical properties (Thomas, Lhoest, Castner, McFarland, & Healy, 1999). Alternatively, for many cell systems these devices can assemble cells in a tissue-specific structure [tissue or extracellular matrix (ECM) mimicry] (Klymov, Prodanov, Lamers, Jansen, & Walboomers, 2013; Martinez-Calderon et al., 2020) with alternative properties to those obtained by flat culture dishes (Luo, Weiss, Liu, & Tian, 2018).

15.1.2 Surface micro- and nanostructuring techniques

The surface structuring of TMOs has been explored for many applications prior to their expansion to the biomedical field (Huey & Bonnell 2000; Xiang, Masuda, & Koumoto, 2004). It is thus worth mentioning that the biomedical applications of surface structures of TMOs benefit from decades of developments in deposition and lithography. It is not the scope of this chapter to provide an exhaustive description of all the available surface modification techniques. We highlight that, since the response of cells to surfaces is hierarchical, the most recent state of the art looks at the precise description of structural details at both the micro and the nanoscale (Jaeggi, Kern, Michler, Zehnder, & Siegenthaler, 2005). From a historical point of view, it is widely accepted that nanoscale features have always been present on the surfaces of microstructured biomaterials, but researchers have only been aware of few nanometer surface structuring since the application of scanning electron (Daculsi, Rohanizadeh, Weiss, & Bouler, 2000) and probe (Ektessabi & Kimura 1995) microscopes. Control of microscale features has been primarily been possible through techniques developed during the microelectronics revolution. In fact, TMOs are compatible with both, wet and gas phase deposition and etching processes, which can be sequentially applied after a lithographic process.

In this way, one can define on the one hand the processes used to determine the TMO chemistry of the surface features and, on the other hand, the surface micro- and nanostructure of the system. Some of the processes are related and transformation of the chemistry is inherently linked to the modification of the nanostructure, as in the case of anodization of titanium to produce nanoporous titania (Hernandez-Velez, 2006).

Starting by wet processes, anodic etching is a traditional method to produce TMOs from thin elemental foils (Vega et al., 2007). The environmental conditions, such as low pH of the electrolyte and temperature are controlled in parallel to the electrochemical parameters to produce thin nanoporous oxide films on the original metal foil (Punzon-Quijorna et al., 2012). Other wet modification techniques exploit high temperature or pressure to reshape an alloy structure, as induced by hydrothermal processes, in which additional chemical parameters play a decisive role in determining the final shape (Mohan et al., 2012), size (Guo et al., 2017), or multiwall characteristics (Brammer et al., 2011) of the generated surface features. Alternatively, TMO films can be deposited by sol casting methods (dip, spin, drop, aerosol, etc.) on generic substrates by the sol–gel process (Kaciulis et al., 1998; Manso, Langlet, & Martinez-Duart, 2001). The production of conformal optical-quality titania films by this method is highly reliable by controlling sol parameters (precursor concentration, water content, pH) and serves as a model process for sol–gel deposition of TMOs (Langlet et al., 2001). By forming nonconfluent structures through evaporation-induced self-assembly of pure Ti (Bass et al., 2010) or hybrid Ti-silane (Manso-Silvan et al., 2007) sols, researchers have found ways to create biofunctional TiO_x structures. Annealing of the films is necessary if polycrystalline films are the objective of the study (Exarhos & Aloi 1990).

Gas phase deposition processes are also compatible with any kind of substrate and offer specific microstructural characteristics related to the nucleation processes and growth of clusters that can give an added value (Poncin-Epaillard & Legeay 2003). Physical processes, such as pulsed-laser deposition (Tanaka et al., 1996), electron-thermal evaporation (Seike & Nagai 1991), and magnetron sputtering (Ngaruiya, Kappertz, Mohamed, & Wuttig, 2004) are available to deposit TMO films from metallic or oxide targets that react with oxygen species in vacuum conditions, leading in most cases to conformal polycrystalline films on planar substrates. In spite of the apparently artificial conditions, the sputtering–clustering processes are emerging and have been oriented in this technological field of cluster assembling to produce TMO nanotopographies mimicking the ECM (Schulte, Podesta, Lenardi, Tedeschi, & Milani, 2017) or sustaining relevant musculoskeletal differentiation of stem cells (Belicchi et al., 2010). Chemical vapor deposition and atomic layer deposition (ALD) work also in controlled vacuum conditions although substantial thermal or plasma energy is required to decompose the flowing metal–organic molecules that most often lead to amorphous TMO films (Ivanova, Gesheva, Popkirov, Ganchev, & Tzvetkova, 2005). ALD outstands as the method allowing the best conformal TMO coverage on nonplanar substrates, even with submicroscale topographic features (Blanquart, Niinisto, Ritala, & Leskela, 2014; Willinger et al., 2009). Nevertheless, the most direct and straightforward method to produce TMO surfaces is the thermal oxidation in controlled conditions at atmospheric pressure of a metallic foil. The process can be as simple as the thermal processing [atmospheric annealing (Dinan, Gallego-Perez, Lee, Hansford, & Akbar, 2013) in reducing conditions (Yoo, Dregia, & Akbar, 2006), hydrothermal process (Jiang et al., 2011)] of a foil or the local application of thermal energy, which can be easily achieved by laser ablation (Beigi, Safaie, Nasr-Esfahani, & Kiani, 2019; Huerta-Murillo et al., 2019; Tavangar, Tan, & Venkatakrishnan, 2011). This brings as back to processes that intrinsically modify both surface chemistry and structure since lasers concentrate locally high amounts of energy, which destabilize materials chemistry and structure. Relevantly, the optical and environmental control of laser-induced processes leads to wide experimental conditions that allow speaking about specific "laser surface engineering" opening the path to the formation of micro- and nanoscale hierarchical features (see for instance laser-induced periodic surface structures) (Martinez-Calderon et al., 2020). Finally, in this latter group of intrinsic modification of surface chemistry and topography, we mention additionally processes developed by controlling the thermal history of materials processed by plasma spraying. Playing with different compounds and structuring effects has allowed defining paths for a better bioactivity and biocompatibility of Ti alloy implants (Zhao et al., 2013).

From the point of view of specific techniques for the topographic modification of TMOs, one must remind again that most techniques modify intrinsically both, chemistry and topography. Every TMO surface modification study should follow-up how composition and topography at different scales relate, registering thus a set of system parameters with well-defined controllable structures. That is certainly the case for electrochemical processing of TMO surfaces (Hernandez-Velez, 2006; Punzon-Quijorna et al., 2012), which suffer deep modifications at the nanoscale, but also for laser-processed materials, which are transformed at the micro- and nanoscale simultaneously (Huerta-Murillo et al., 2019; Martinez-Calderon et al., 2020). In any case, the traditional way to microprocess the surfaces of TMOs was by wet or gas phase chemical etching (Ramadan, Manso-Silvan, & Martin-Palma, 2020) after photolithographic protocols from the microelectronics industry (Clark, Connolly, Curtis, Dow, & Wilkinson, 1987; Ding et al., 2013). Another traditional but always in progress process is mechanical structuring, such as in attrition treatments, which have been adapted down to the nanograin scale for biomedical applications (Azadmanjiri et al., 2016).

From the point of view of low dimensional structures, TMOs have been extensively formed in the form of nanowires (Tan, Pingguan-Murphy, Ahmad, & Akbar, 2013), nanopillars (Moxey et al., 2015), and other kinds of supported structures (Lao, Wen, & Ren, 2002). Some methods, such as electron (Donthu et al., 2005; Shimada, Matsuu, Miyazawa, & Kuwabara, 2002) or ion beam (Juodkazis et al., 2011; Pan & Xu 2020) lithographies, are however not appropriate for the applications focused in this chapter since they are not easily scalable to the centimeter scale and would adapt only for very punctual "single-cell" applications. At this point, a very flexible technique for the production of supported nanowires and nanofibers is electrospinning (Lim, Yu, Woo, & Lee, 2008).

It is convenient to underline that, irrespective of the micro- or nanoscale, lithographic techniques especially adapted for biological applications (soft lithographies) have expanded to the patterning of TMO substrates. The main exponent of these techniques, namely, microcontact printing, has been used to transfer different molecular structures through stamps to Ti. Even more, nanoimprint lithography has been used to produce sub-100 nm structures on TiO_2 and the shape of the transferred features was observed to influence different behavior on endothelial and muscle cells (Muhammad et al., 2014). Even lower scale ordered features could be obtained by using self-assembled block copolymers as templates for a subsequent anodization (Sjostrom, McNamara, Yang, Dalby, & Su, 2012).

From this starting point, TMOs, as many other materials, have benefitted from hybrid processes that have allowed controlled lithographic/structuring processes compatible with hierarchical scalability. In these approaches, bottom-up processes are exploited for the growth of nanoscale structures whereas more traditional mask or photolithography is used to create the microscale design. This approach has evolved enormously in the search for multifunctional structures (Choi, Sofranko, & Dionysiou, 2006) and has expanded to the field of TMO interaction with cells (Gallach Perez et al., 2015). We highlight the processes in which high-energy ion beams are used to condition the growth of mesopores or nanospikes in microscaled areas of a substrate (Munoz Noval et al., 2012; Sanz, Jaafar, Hernandez-Velez, Asenjo, & Vazquez, 2010) or low-energy ion beams or plasmas (Wang, Zang, Chua, & Fonstad, 2006) are used through micromasks to etch nanostructures previously grown on the desired substrates. This hybrid processes benefit from a wide range of parameters to independently design the shape of the features pertaining to each metric scale. However, laser-processing techniques emerge as more effective processes for the simultaneous processing of the micro- and nanoscale features in very different optical configurations such as direct laser writing, laser-induced periodic surface structuring or laser interference micropatterning of nanostructured materials (Martinez-Calderon et al., 2020; Pelaez et al., 2013; Xiang et al., 2016) (Table 15.1).

15.2 Cellular response to transition metal oxide micro- and nanostructures

15.2.1 Biocompatibility of transition metal oxide surface structures

In order to bioengineer a device based on patterned TMO surfaces, it is fundamental to make a good choice of the starting materials. The physicochemical properties that one could exploit (optical, electrical, mechanical, etc.) may be attractive for an application, but the subsequent aspect to inquire is knowing whether the material of choice is biocompatible with the cell line desired for the application. In general, the response is cell specific, so one should test the cell line on the desired oxide even if positive or negative results were obtained in published research with other cell lines. General trends can be extracted with caution, so that protective oxides generally lead to good biocompatibility. If the oxide is not protective and generates particles, the response is however not always a bad biocompatibility. However, attention must be paid to the release of ions from the surface, which may lead to toxicity

Table 15.1 Processes for the production of micro-, nano-, and hierarchical transition metal oxide (TMO) structures for cellular and tissue biomedical devices.

Scale	Process	TMO	Structures	References
MICRO	Photolithography	TiO_2, Nb_2O_5, V_2O_5	Steps, grooves, stripes, pillars	Clark et al. (1987), Ding et al. (2013), and Scotchford et al. (2003)
	Sol−gel (direct resist photolithography)	TiO_2 (hybrid)	Stripes	Ramadan et al. (2020)
	Sol−gel (phase separation)	TiO_2 (hybrids)	Stripes/Islands	Manso-Silvan et al. (2007)
	Attrition	TiO_2	Spikes	Azadmanjiri et al. (2016)
	Microcontact printing	TiO_2/PEG	Stripes, dots	Csucs, Michel, Lussi, Textor, and Danuser (2003)
	Photocatalysis assisted lithography	TiO_2	Spots	Sekine et al. (2015) and Yamamoto et al. (2014)
	Embossing	TiO_2 composite	Grooves	Jiang et al. (2012)
NANO	Anodization	TiO_2, ZrO_2, Ta_2O_5	Tubes	Frandsen et al. (2011), Punzon-Quijorna et al. (2012), and Wang et al. (2012)
	Anodized masks	Ta_2O_5	Dots	Dhawan et al. (2015)
	Heterogeneous hydrothermal nucleation	TiO_2	Wires	Brammer et al. (2011), Guo et al. (2017), and Mohan et al. (2012)
	Sol−gel (evaporation-induced self-assembly)	TiO_2	Dots	Bass et al. (2010)
	Sputtering−clustering	TiO_2, ZrO_2	Dots	Belicchi et al. (2010) and Schulte et al. (2017)
	Electrospinning	TiO_2, TiO_2 composites, Fe_3O_4 composites	Fibers	Kim et al. (2010) Lim et al. (2008), and Wei et al. (2011)
	Laser ablation	TiO_2	Fibers, wires	Beigi et al. (2019), Tavangar et al. (2011), and Vijayakumar, Venkatakrishnan, and Tan (2015)
	Thermal annealing	TiO_2	Roughness	Dinan et al. (2013)
	Imprint lithography	TiO_2	Dots, gratings	Muhammad et al. (2014) and Sjostrom et al. (2012)

Hierarchical	Laser-induced periodic surface structures	TiO_x	Rippled stripes/gratings	Martinez-Calderon et al. (2020)
	Direct laser writing/ablation	TiO_2/composite	Nanospiked corrals	Xiang et al. (2016)
	Laser interference	CoCr (O)	Wavy stripes	Schieber et al. (2017)
	Photolithography/hydrothermal growth	ZnO	Flower arrays	Park et al. (2010)
	Mask lithography/ion beam implantation/wet etching	TiO_2, TiO_2/TiAlV	Nanospiked stripes, nanospiked gratings	Gallach Perez et al. (2015), Lopez, Ynsa, de Pablo, Lim, and Manso Silvan (2020), and Sanz et al. (2010)
	Mask lithography/anodization	TiO_2	Nanotube perforated grooves	Zhou et al. (2020)
	Colloidal lithography/cluster deposition	TiO_2	Nanostructured triangle arrays	Singh et al. (2011)
	Photolithography/sol−gel	TiO_2	Mesoporous microwells	Park et al. (2013)
	Magnetic nanoparticle sedimentation/stretch relaxation corrugation	Fe_3O_4	Nanorough-magnetic/grooves	Lin, Chang, Yung, Huang, and Chen (2020)

PEG, Polyethyleneglycol.

even for metals with well-defined metabolization paths such as Zn and Fe. A generic limit for the interference of cell activity by TMO ions can be established in the 10–50 μM range. In this sense, high concentrations of Zn and Fe ions are cytotoxic. We will not refer to levels or doses of ion or particle administration to tissues in this generic description but considering them for each TMO selected for a biomedical application is a highly recommended caution.

A brief summary of the biocompatibility of TMOs follows, as it can help the reader identifying strengths or limitations of a particular TMO for biomedical structuring of surfaces. Starting by 3d orbital metals, scandia (Sc_2O_3) has been identified as a promising biomaterial with good inertness and biocompatibility face to osteoblast-like cells (Herath, Di Silvio, & Evans, 2005). In addition, its protective nature was evidenced in applications for the development of Mg biodegradable alloys, where it contributed to an overall biocompatibility of the alloy as confirmed with osteoblastic cells (Brar, Ball, Berglund, Allen, & Manuel, 2013). The protective nature of titania is well known and has sustained the superior performance of Ti alloys in different orthopedic applications (Brunette, Tengvall, Textor, & Thomsen, 2001; Massaro et al., 2002). The performance of Ti oxides and their additional physicochemical properties are so attractive (Moreno, Carballo, Jurado, & Chinarro, 2009) that we will devote a significant space to describe the advances achieved with the derived nano–micro structures. Although vanadia has been described as a poor biocompatible material (Chang, Zhang, & Huang, 2018) and contributes to lowering the biocompatibility of titania (Jarrell et al., 2007) an interplay has been reached to exploit its properties as an antitumor material (Li et al., 2019). The impact of chromium oxide is known to depend strongly on its oxidation state. As a solid, Cr provides anticorrosion properties to stainless steel and other alloys, but care should be taken in the physiological conditions of the place of implantation to avoid formation of chromate species in view of their strong genotoxicity. Diffusion barriers and anticorrosion treatments are often applied to minimize the risk of exposure to oxidized chromium ions (Lutz, Diaz, Garcia, Blawert, & Mandl, 2011). Mn is rarely used as a pure oxide in biomedical applications but appears often in alloys (Fe, mainly), where Mn species have been identified as products of implant degradation (Huang, Nauman, & Stanciu, 2019; Kao, Ding, Chen, & Huang, 2002). In fact, the natural tendency of MnO to produce particles has been exploited recently in multifunctional biomedical applications, with no relevant cytotoxic effects (Pradhan, Giri, Banerjee, Bellare, & Bahadur, 2007). The same applies to Fe materials, whose oxides (if not protected) form multiscale particles, which can have detrimental effects if released by an implant (Basle, Bertrand, Guyetant, Chappard, & Lesourd, 1996). Nevertheless, imaging, diagnostic, and therapeutic properties are outstanding when iron oxide particles are appropriately designed though nanoscaling (Perigo et al., 2015), shaping (Ayachi, Mechakra, Manso Silvan, Boudjaadar, & Achour, 2015), and surface modification (Zverev, Pyatakov, Shtil, & Tishin, 2018). Cobalt oxides are generally released from implant alloys (Co–Cr). In spite of the poor predictions on biocompatibility for the derived compounds (Howie & Vernonroberts 1988), they have been used intensively in biomedical applications. The use of Ni as a biomedical material has also been controversial in spite of the attractive shape memory properties of its alloys with Ti. Coating has been a traditional route to avoid corrosion and oxidation of Ni (Wong, Cheng, & Man, 2007), which leads to particle formation and release with potential genotoxic effects, especially for NiO type particles (Akerlund, Islam, McCarrick, Alfaro-Moreno, & Karlsson, 2019). The cytotoxic effects of Cu materials have been related previously to its oxides but have been implemented in material designs with bactericidal properties (Rtimi, Sanjines, Bensimon, Pulgarin, & Kiwi, 2014) and have been even described as a proangiogenic factor (Li et al., 2016). The strong ionic character of Zn, favors its presence as a divalent ion in physiological environments, which limits lifetime and stability of devices. This has prevented a deep implementation of ZnO in biomedicine, farther than descriptions as antibacterial agent in nanocomposites (Espitia et al., 2013), where its release is exploited as working principle. Its high impact as optoelectronic material has derived in research of ZnO compounds incorporating Ti as covalent stabilizer (Ramadan et al., 2019).

For 4d orbital transition metals, the discussion can be drastically reduced to Y, Zr, Nb, Mo, Pd, and Ag in view of the acute toxicity of Cd and the low impact of Tc, Ru, and Rh oxides in biomedical applications. Yttria has a long trajectory in the field of biomaterials as a stabilizer of Zirconia ceramics. Its low cytotoxicity has been evidenced in other matrix materials (Radu, Chiriac, Popescu, Simon, & Simon, 2013) or independently, considering even its potential accumulation in vital organs (Sonmez et al., 2015). Zirconia stands as one of the most reliable ceramic materials due to both outstanding biocompatibility, tailorable porosity, and mechanical properties (Afzal, 2014). This has sustained its integration in biomedical devices, mainly in composite bioceramics (Silva, Lameiras, & Lobato, 2002) and as protective coating material (Nagarajan & Rajendran 2009). Even if not extensively used, the anticorrosion/protective properties of Nb_2O_5 have been exploited for the improvement of Ti (Mazur et al., 2015) or stainless steel (PremKumar, Duraipandy, Kiran, & Rajendran, 2018) prosthetic alloys, with Cu or Zn doping, respectively, to induce antibacterial effects. Even if Mo is present as a secondary element in biomedical alloys, the presence of MoO_x oxides increases upon physiological aging of these alloys (Hanawa, Hiromoto, & Asami, 2001). Although its conversion to oxide is evident in physiological environment, even metallic Mo nanoparticles induce strong cytotoxic activity (Siddiqui et al., 2015) with an acute damage of MoO_x nanoparticles to cells with respect to that induced by Fe_3O_4 or MnO_2 counterparts (Hussain, Hess, Gearhart, Geiss, & Schlager, 2005). These properties have been however exploited in specific designs, including targeting agents to propose tumor-specific therapies (Hu et al., 2020). In the case of palladium oxide (PdO) the biocompatibility is rarely mentioned in related reports due to its low reduction potential, which is exploited for the biosustained production of Pd metal nanoparticles (Rana, Yadav, & Jagadevan, 2020). Biomedical applications relate thus to the metallic state of Pd, often in dental alloys (Cai, Vermilyea, & Brantley, 1999) or in the form of nanoalloys with Au or Ag for the exploitation of catalytic properties in sensors (Yang, Deng, Lei, Ju, & Gunasekaran, 2011) or tumor imaging and therapy (Guo et al., 2018). With regard to silver, its high oxidation rate in physiological conditions and the subsequent release of ions have been allied with its mild toxic effects to propose antibacterial coatings on polymer (Stobie, Duffy, Hinder, McHale, & McCormack, 2009) and metallic–alloy (Marsich et al., 2013) prosthetic materials.

As for the case of 4d metals, the presence of 5d TMOs in biomedical applications is mostly restricted to Hf and Ta. In fact most of the biomedical applications of the elements in this period emerge from their metallic state (Pt, Au), exploited mostly in bioelectrode applications (Lee & Schmidt 2010). Briefly, Hf oxides present extremely similar protective properties as those of Ti and Zr oxides and derived particles have been described as biocompatible (Field et al., 2011). However, the high processing costs of raw Hf containing materials are a drawback for its biomedical development (Mohammadi, Esposito, Cucu, Ericson, & Thomsen, 2001). Ta is a common minor element in last generation Ti-based prosthetic alloys (Yamaguchi, Takadama, Matsushita, Nakamura, & Kokubo, 2011). The anticorrosion properties and biocompatibility of Ta_2O_5 have been exploited in antibacterial protective layers on TiAlV by doping the films with Cu (Ding et al., 2020). In the form of hollow particles, Ta_2O_5 has been loaded with anticancer drugs and infrared fluorescent dyes to develop theranostic systems (Jin et al., 2017).

From the above listing, one can resume the transition metal elements that can be used for the fabrication of supported structures. It shall be reminded that the biocompatibility is a material cell line-specific issue and that long-term exposure of cells and tissues to materials must be studied ad hoc. Fig. 15.3 shows an extraction from the periodic table of the elements of the most relevant transition metals that can be used for the production of oxide biomedical micro- and nanostructures. The extraction is based on the interest of their properties, their processability, their stability or biodegradability, and their generic biocompatibility. Color codes have been used to categorize relevant properties from the biomedical point of view, such as

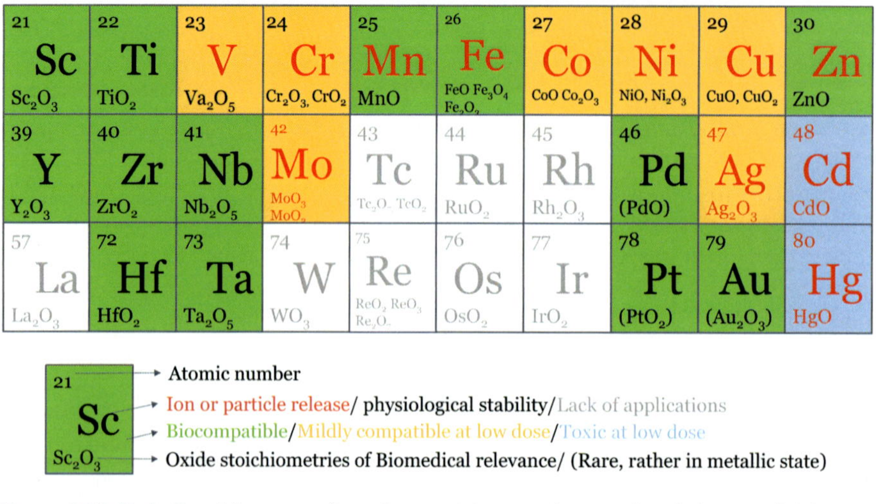

Figure 15.3 Periodic table extract from the transition metals resuming their potential for biomedical applications derived from surface structuring.

biodegradability and biocompatibility. The extraction does not pretend to exclude, but is rather based on the current state of the art, so that the table is also relevant for the identification of future research niches.

15.2.2 Control of cell adhesion on micropatterned transition metal oxides

The research of the effects of microtopographic features had a blooming period in the early 1980s on different families of materials, extending in the 1990s to pure metals and alloys (Damji, Weston, & Brunette, 1996). The degree of surface oxidation of these metal and alloy surfaces has not been described in detail, but in view of the alkaline conditions and temperature of cell culture media, the cell interactions can be considered to happen with a few nanometers thick metal oxide surface (precise thickness depends on the transition metal itself and the nanoscale surface finishing) (see Fig. 15.4A). In any case, this study on microtopographic features was extended to metal oxide surfaces (Scotchford et al., 2003) produced by electron beam evaporation, photolithography, and oxygen-plasma annealing. In this work, the authors were able to make analysis of the adhesion of osteoblasts on complementary surface contrasts of Ti, Nb, and V oxide dots and stripes exhibiting negligible nanotopography (<2 nm), as illustrated in Fig. 15.4B. The study included cellular and biomolecular analyses leading to in situ observation of comparative adhesion patterns to the different metal oxides, all frequently used in prosthetic alloys. Cells presented a more favorable adhesion on the areas of these TMOs when compared in microtopographic contrasts with aluminum oxide. The positive surface charge of this latter surface at the pH of the cell culture medium was pointed as the determinant factor, which influenced reduced adsorption of intermediate proteins,

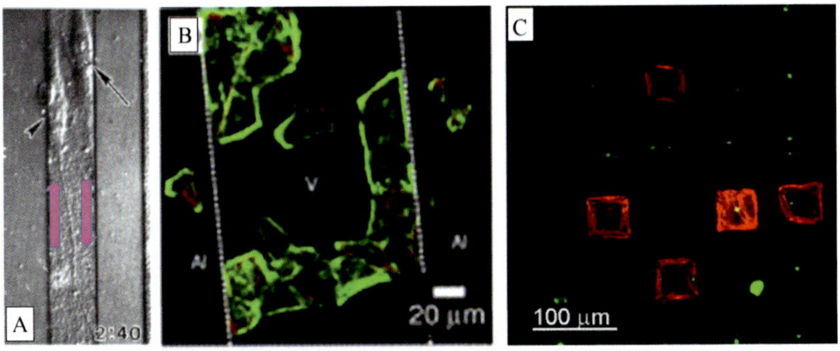

Figure 15.4 Examples of images from cell interactions with microstructured transition metal oxide surfaces. (A) Optical image of two fibroblasts crossing in a microchannel grooved on Ti with native oxide. Dimensions are 40 μm width and 3 μm depth (Damji et al., 1996). (B) Confocal microscopy images of osteoblasts cultured on V/Al micropatterned surfaces with actin (*green*) and tubulin (*red staining*) (Scotchford et al., 2003). (C) Fluorescence images of fibroblasts on fibrinogen traps with polylysine-polyethyleneglycol antifouling contrasts produced by microcontact printing on TiO_2 (Csucs et al., 2003).

especially human serum albumin. In addition, Csucs et al. created molecular micropatterns on TiO_2 films deposited by sputtering and treated by oxygen-plasma cleaning (Csucs et al., 2003). Microcontact printing was used to create cell antifouling areas by binding polylysine-polyethyleneglycol (PLL-PEG) conjugate, which allowed concentrating fibroblast and keratocyte spreading and motility on the complementary TiO_2 areas (Fig. 15.4C). The photocatalytic properties of TiO_2 have been used to control cell adhesion by exploiting intermediate self-assembled organosilane features that control initial cell adhesion and can be selectively transformed (Yamamoto et al., 2014). In fact, pheochromocytes colonized through neurite extension the neighboring transformed surface features created in situ (during cell culture) by selective UV irradiation. Even though with reduced surface selectivity with respect to the previous methodology, the same photocatalytic approach was used to trigger a transfer of laminin, a binding promoter protein, which allowed controlled binding of primary neurons (Sekine et al., 2015).

Microgrooved anatase patterns were obtained by a soft lithographic embossing of TiO_2 gels and subsequent annealing at 400°C (Jiang et al., 2012). The one-dimensional microgrooves were successful in the induction of a cell orientation irrespective of their periodicity. Such phenomenon could explain the reduced proliferation and differentiation rates observed in osteoblast cells.

15.2.3 Nanoscale surface features and derived biomedical applications

In the present section we will refer to the preparation and cell organization applications of TMOs with well-defined nanostructured features. This includes different kinds of nanotubes, nanowires, etc. but excludes every kind of continuous film or laminar material, even if the microstructural features of this latter (such as crystallites) present typical nanoscale sizes. This restriction opens a clear differentiation aspect in the adhesion of cells since well-defined nanostructural features have a high impact in the development of filopodia and formation of focal adhesions.

The implications of nanotopography on cell behavior have been observed to extend to molecular secretions. In particular, the production of nitric oxide, a relevant messenger in the cardiovascular system, has been observed to depend on induced nanotopography. Tantalum oxide nanodot surfaces have been observed to increase the secretion of NO in cardiomyocytes with a dot size-dependent profile. In the first 24 h, the 200 nm nanodots were observed to drastically influence NO secretion to a higher rate than any of the smaller nanodots (Dhawan et al., 2015). This observation can be extrapolated to cellular expression. For instance, nanotopographic substrata have been identified as ideal scaffolds to identify the heterogeneity of MSC populations. This was observed by following Wnt signaling-associated markers as an indication of differentiation into osteogenic cells (Khan et al., 2019). To emphasize the impact of nanotopography on cellular expression, the enhanced gene expression of the Runx2 differentiation pathway was confirmed by mRNA detection (Komasa, Taguchi, Nishida, Tanaka, & Kawazoe, 2012).

The effects of different TiO_2 surface nanotopographies on the differentiation of MSCs were studied by using different wet etchants on Ti surfaces (Bajpai, Rukini, Jung, Song, & Kim, 2017). The expression of osteocalcin and osteoponting was simultaneously activated on micron–submicron crater-like TiO_2 surfaces (issuing from an etching step in HCl) with respect to nanowire and nanoporous TiO_2. By tailoring nanostructured TMO surface properties, it has been demonstrated that alternative sources of cells for tissue engineering can be modulated. For instance, stem cells from adipose tissue have been demonstrated to become osteodifferentiated (Malec et al., 2016), which provides in many occasions an autologous source for tissue repair after appropriate expansion and scaffolding. Following a nonelectrolytic modification, significant progress has been achieved in the confirmation of biocompatibility with simultaneous bactericidal character for dental, orthopedic, laryngeal (Ferraris et al., 2019), or vascular (Mohan et al., 2012) implants (see illustrative examples in Fig. 15.5A–C). Remarkably, impressive wound healing results have been obtained using MSCs on hierarchical rutile patterns. The results show a remarkably low cell adhesion, even after 72 h incubation, on nanospiked areas, which promotes cell migration and wound closing through narrow flat rutile areas (see Fig. 15.6A) (Gallach Perez et al., 2015).

In spite of the great attraction of anodically obtained TiO_2 nanotubes, the results with respect to their cell-inductive properties are controversial with reports in favor (Demetrescu, Pirvu, & Mitran, 2010; Oh & Jin, 2006; Popat, Leoni, Grimes, & Desai, 2007) (see Fig. 15.6B) and reports against (Punzon-Quijorna et al., 2012; Zhao et al., 2011), which have promoted further research aiming at a functionalization of the nanoporous surface (Mutreja, Kumar, Boyd, & Meenan, 2013) (see Fig. 15.6C). In fact, the smaller nanopores inhibit focal adhesions and preosteoblasts tend to form stable bonding by contacting the walls of bigger pores (Yu, Jiang, Zhang, & Xu, 2010). This behavior at cellular level is however compatible with increasing protein adhesion after nanostructuring (Zhao et al., 2010). This increased protein adhesion is also responsible for an increased rate of blood clotting kinetics (Smith, Yoriya, Grissom, Grimes, & Popat, 2010). In fact,

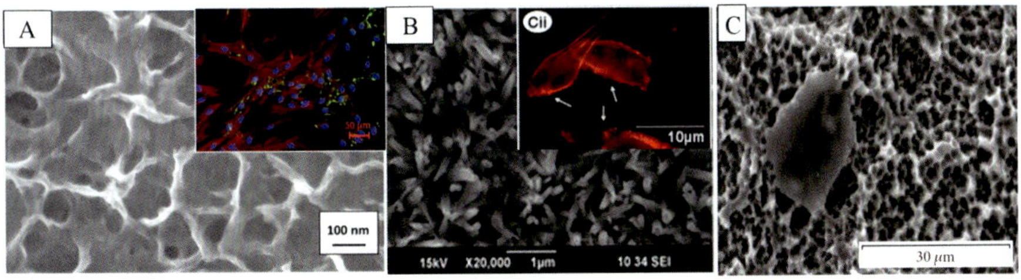

Figure 15.5 Examples of cellular interaction with nanoscale titania features obtained by electroless methods. (A) Scanning electron microscopy (SEM) image of the sponge-like titania structure obtained after sequential HF, H_2O_2 etching of TiAlV alloys. *Inset*, fluorescence images denoting collagen I formation by progenitor osteoblasts (Ferraris et al., 2019). (B) SEM images of TiO_2 nanorods and florescence image of proliferating endothelial cells after 24 h culture with *arrows* indicating extending filopodia (Mohan et al., 2012). (C) Mid stage (30 min etching) adaptation of mesenchymal stem cells to titania surfaces produced by sandblasting and nitric acid etching (Khan et al., 2019).

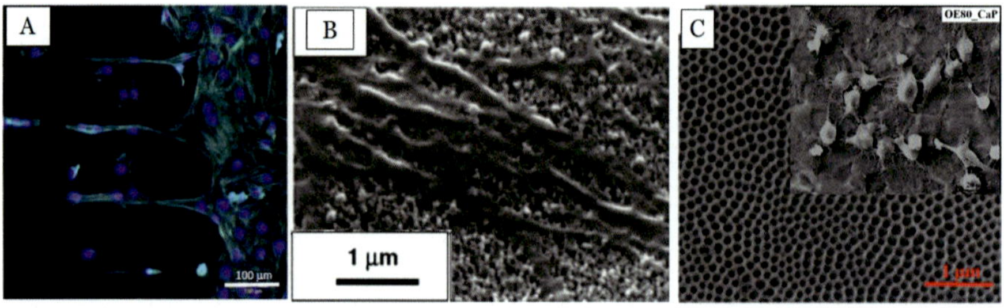

Figure 15.6 (A) Fluorescence images of mesenchymal stem cells on HF-etched TiO_2 surfaces (rutile) after high-energy ion beam irradiation, which creates a hierarchical pattern of stripes on a nanospiked background (unreleased image from Gallach Perez et al., 2015). Examples of cell interactions with anodic TiO_2 nanotubes. (B) Strong interaction of filopodia from osteoblast with anodic titania nanotubes (Oh & Jin 2006). (C) Scanning electron microscopy images of osteosarcoma cells (*inset*) on CaP-sputtered films decorating titania nanotube surfaces prepared by anodization at 80 V (image shows surface prior to surface functionalization) (Mutreja et al., 2013).

in many occasions the attractive physical properties of the TiO_2 nanotubes are made compatible with their cell or tissue applications by surface modification through surface polymerization (Punzon-Quijorna et al., 2012; Vasilev et al., 2010) or intermediate protein [such as for vinculin (Das, Bose, & Bandyopadhyay, 2009)] or peptide (Balasundaram, Yao, & Webster, 2008) adhesive layers.

A similar approach was used to create arrays of TiO_2 nanotubes with the aim of designing surfaces for an appropriate endothelialization of cardiovascular implants. The so-formed biomimetic substratum was efficient in favoring MSC differentiation into smooth muscle cells (Luo et al., 2020). In this identical direction, stent modification with TiO_2 nanotubes has been proposed, being an oxygen-plasma posttreatment critical for a positive response of human coronary artery endothelial cells (Junkar et al., 2020). Previous plasma treatments of similar TiO_2 nanotubes had confirmed an enhanced behavior for osteogenic applications by using mixed O_2-N_2 discharges (Mahmood et al., 2011). In fact, the attractiveness of the TiO_2 nanotube structures has evolved considerably reaching in vivo testing, at least for bone implant applications (Salou, Hoornaert, Louarn, & Layrolle, 2015). The attractiveness of the TiO_2 nanotube arrays for biomedical applications may have an origin in the low immune response induced with respect to the surface of implantable grade Ti. Smith et al. studied the such response after culture of a whole blood lysate and confirmed a lower level of activation of monocytes, macrophages, and neutrophils (Smith, Capellato, Kelley, Gonzalez-Juarrero, & Popat, 2013). Another attractive point from the point of view of their clinical management is their intrinsic antibacterial effect (Peng et al., 2013).

Further control of the biocompatibility of anodized TiO_2 was achieved by modifying the cell–Titania interface with hydroxyapatite (HAP) (Bai et al., 2011; Kar, Raja, & Misra, 2006; Narayanan, Lee, Kwon, & Kim, 2011), HAP-$CaTiO_3$ and chitosan–HAP–graphene interface compounds. The culture of mouse preosteoblasts on the HAP-$CaTiO_3$ surfaces induced an increase of the osteoconductivity (Lin et al., 2019). In the latter case, the biocompatibility of the functionalized nanotube layer increased with no compromise of the physical integrity of the layer (Karimi, Kharaziha, & Raeissi, 2019). In another example of functionalized TiO_2 nanotube arrays, Bilek et al. immobilized Se nanoparticles to induce selective adhesion of cancerous versus noncancerous cell lines. They demonstrated that, not only Se-decorated TiO_2 nanotubes exert an effective antibacterial effect, but also they selectively favored the adhesion of healthy NIH/3T_3 fibroblasts and prevented the growth of cancerous osteoblast such as MG-63 cells (Bilek, Fohlerova, & Hubalek, 2019). Similarly, the functionalization of the TiO_2 nanotubes with F and P led to a decrease in the proliferation of five different bacterial lines while maintaining an increased differentiation signaling in MCT3T3-E1 osteoblastic cells (Aguilera-Correa et al., 2019). This latter report agrees with other Ca (Lim, Chai, & Loh, 2017), Ca-P (Roguska et al., 2016), F (Lozano et al., 2015), F-HAP (Huang et al., 2018), or alkalescent MgO (Zhang et al., 2017) functionalization of TiO_2 nanotubes leading to increased osteoinductivity as compared with planar Ti. Other relevant functionalizations of TiO_2 nanotube arrays include also bioactive Sr-based surface modification (Zhang, Huang, Hang, Zhang, & Tang, 2018) or antibacterial Ag (Lan et al., 2014) and ZnO formulations (Elizabeth, Baranwal, Krishnan, Menon, & Nair, 2014; Liu et al., 2015), which showed increased cytocompatibility with nasal epithelial cells and allowed decreasing the inflammatory response in macrophage cultures (Yao et al., 2018), respectively. Additional strategies to biofunctional TiO_2 nanotubes stem from biofunctionalization protocols providing sustained drug release.

Another attractive element of nanotube approach is the possibility to load drugs within the cavities, which can reinforce the biocompatibility in different aspects. This approach was initially validated by loading penicillin (Yao & Webster 2009), bivalirudin (as a thrombin inhibitor to prevent thrombosis and restenosis) (Yang et al., 2014), and fluorescent drug Rhodamine B (Rahman et al., 2016). In order to reinforce the antibacterial behavior of the surface, Wang et al. filled tubes of optimized sizes with ibuprofen, occluding with poly(lactic-co-glycolic acid) to control the release. Such antibacterial behavior could increase the rates of successful implantation of titanium implants (Wang et al., 2017). In the projection of results to real 3D implant structures, the nanotube porous approach has been extrapolated to functionalize volumetric designs with an impressive increase of mechanical properties and osteogenic behavior (Jin, Yao, Wang, Qiao, & Volinsky, 2016).

In opposition to the cavity structures generated by anodization, high aspect ratio titania structures have been grown by electrospinning and applied in cell culture applications (Lim et al., 2008). The method is especially attractive for the production of composites, for instance with HAP (Kim et al., 2010), polycaprolactone (Gupta et al., 2012), silk fibroin (Jao, Yang, Lin, & Hsu, 2012), or poly(vinyl pyrrolidone) (Wang et al., 2012). In this latter report, the processing parameters were varied to study the influence of the surface microroughness of the mats on the osteogenic response of TiO_2 fibers, finding a significance in osteocalcin expression in patterned fibers with 10% of the polymer. Apart from the bone tissue applications (Kiran, Balu, & Kumar, 2012), this kind of mats have been found attractive for wound healing(Jao et al., 2012) and myoblasts networks (Amna et al., 2013). One of the most recent and attractive applications emerges after functionalization of

TiO_2 nanofibers with specific aptamers allowing the capture of circulating tumor cells (Liu et al., 2020). The compositing nanofiber approach through electrospinning has been also exploited with other TMOs, such as for compositing Fe_3O_4 with polyvinyl alcohol with potential interest for bone tissue engineering (Wei et al., 2011).

By using ultrafast laser processing, Vijayakumar et al. controlled different phases of TiO_x in nanostructured surfaces. The structuring process was successfully oriented to modulate the response of HeLa cells (cancer cells) face to NIH3T3 fibroblasts (Vijayakumar et al., 2015).

In spite of the dominant attention paid to TiO_2 compounds, the bibliography extends to other TMOs with very attractive results. By using the same anodization approach to produce quaternary TMOs, Qadri et al. studied the most appropriate aspect ratio for activation of attachment and growth of osteoblast-like cells on Ti35Zr28Nb alloys (Qadir, Lin, Biesiekierski, Li, & Wen, 2020) (as illustrated in Fig. 15.7A) following previous validation results on Ti6Al7Nb (Li, Zhao, Tang, Li, & Chi, 2016), Ti-25Nb-25Zr (Huang, Wu, Sun, Huang, & Lee, 2013), or even in the pentanary Ti-5Zr-3Sn-5Mo-15Nb (Mei, Zhao, Wang, Ma, & Zhang, 2012). Previously, nanotube films with different pore size had been prepared in the Ti-Zr system (Sista et al., 2013) by using electrochemical parameters (Grigorescu et al., 2014; Hao, Li, Hao, Zhao, & Ai, 2013) or composition (Minagar, Li, Berndt, & Wen, 2015) as modulation of the structure and determining the most adapted biofunctional properties by culturing osteoblasts. In fact, the confirmed biocompatibility of ZrO_2 nanotubes (Frandsen et al., 2011) inspired their role as shell onto vertically standing ZnO nanowires. In this way, authors prepared surfaces with attractive Zn delivery properties functional for antibiosis and osteogenesis (Cheng et al., 2015) as illustrated in Fig. 15.7B. There are additional successful examples of processing of nanotubular structures on Ta. In fact, Ta_2O_5 nanotubes were prepared and analyzed in terms of their osteoinduction potential, confirming a superior response with respect to Ta foil (Wang et al., 2012). Other authors have followed the electrochemical anodization route to create rods, rather than tubes, using HfO_2 as biomaterial. The modification affected also to the surface chemistry, with presence of

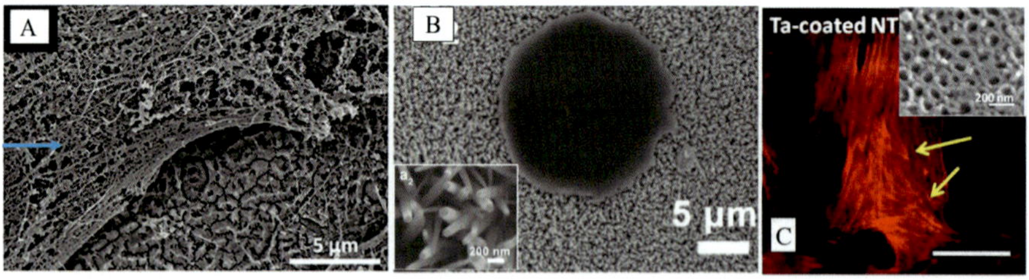

Figure 15.7 Examples of interactions of cells with nanoscale-structured transition metal oxides. (A) SEM images of osteoblastic cell membrane on the corrugated surfaces of Ti35Zr28Nb after anodic etching at 40 V during 2 h (Qadir et al., 2020). (B) Spheroidal osteoblastic cells on the surfaces of ZnO nanorods (*inset*) as observed by SEM (Cheng et al., 2015). (C) Surfaces of anodic titania nanotubes decorated with Ta coating (SEM image, *inset*) and fluorescence image of the actin cytoskeleton of osteoblast cells showing a characteristic tensegrity pattern (Frandsen et al., 2014).

suboxide and hydroxide compounds, and led to an improvement of antibacterial effect in compatibility with better attachment and proliferation of MG-63 osteoblast-like cells (Fohlerova & Mozalev 2019). In spite of the relatively low biocompatibility of MoO_x materials, a recent study showed that the formation of mixed Ti-Mo oxides and its nanostructuring could highly improve the cell attachment (Hsu, Hsu, Wu, & Ho, 2019). The Ti-Ta system has also been approached by deposition of a thin metallic Ta thin film on TiO_2 nantubes. With this process, an increase in bone formation could be achieved as compared to bare TiO_2 nanotubes (Frandsen, Brammer, Noh, Johnston, & Jin, 2014) (see attractive actin cytoskeleton structures in Fig. 15.7C).

15.2.4 Hierarchical micro- and nanostructured transition metal oxides, toward biomedical devices

Laser interference micropatterning of CoCr alloys enhances oxidation and changes in the wettability that reduce platelet adhesion with no significant reduction of endothelial cell functionality (Schieber et al., 2017). Analog alloys have been also textured with microscale circles, squares, and triangles by laser surface structuring, although the internal nanostructuring of the irradiated area was not studied in detail (Qin, Zeng, Wang, Zhang, & Dong, 2014) (see illustrative image on Fig. 15.8A). It was demonstrated that osteoblast proliferation and differentiation is influenced by the different topographies as suggested by gene expression. The study suggests that particular designs are attractive for texturing microenvironments on implant alloys. The effect of surface oxidation induced by laser texturing was also evidenced on micro–nanostructures produced by laser surface texturing (Tiainen et al., 2019). Relevantly, these structures were observed to be stable when tested by friction with

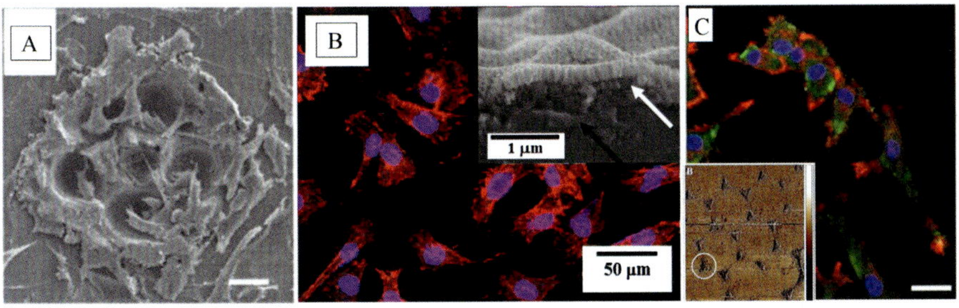

Figure 15.8 Examples of cell interaction with hierarchical structures prepared with transition metal oxides. (A) SEM image of osteoblasts response to laser-drilled holes on CoCrMo alloy (Qin et al., 2014). (B) SEM image of a TiO_2 nanocoating on a microprocessed wavy surface (*inset*) and fluorescence microscopy image of the response of osteoblastic cells after 3 h of culture (*red actin, blue nuclei*) (Han et al., 2014). (C) Phase contrast atomic force microscopy image of TiO_2 nanostructures prepared by colloidal lithography (*inset*) and fluorescence image of neural cells and filopodia–lamelipodia formation (*actin red, vinculin green, nuclei blue*) (Singh et al., 2011).

bovine bone, which supports the reliability of the surface treatment after implantation.

The electrochemical formation of TiO_2 nanotubes was combined with photolithography to create hierarchical micropatterns on Ti implants (Zhou et al., 2020). The so-formed structures could help designing the body of a dental implant with an osteogenic ECM-like environment favoring osteoblasts progress, but deterring growth of fibroblasts, epithelial cells, and *Porphyromonas gingivalis* bacteria. In a parallel approach, a wavy surface at the microscale was decorated with nanopores by two sequent processes; electrolytic microarc oxidation was applied prior to a deposition and anodic oxidation of a Ti film. These surfaces presented increased protein affinity and biocompatibility with respect to only micro or only nano-TiO_2 (Han, Kim, & Koh, 2014), as illustrated in Fig. 15.8B.

Another alternative technique to create hierarchical scale patterns was developed by Singh, Lenardi, Gailite, Gianfelice, and Milani (2009). A first step of microscale patterning was performed on a PMMA/bovine serum albumin (BSA) bilayer, and a selective adhesion of TiO_2 nanoclusters was then induced on PMMA. The nanotopographic TiO_2 behaved as a fouling surface for primary neurons, whereas BSA efficiently bloqued their spreading for periods over 72 h. The same group modified the fabrication protocol of the hierarchical structures by substituting the photolithographic step by a colloidal lithography process, thus obtaining hexagonal distributions of nanostructured TiO_2 triangles (Singh et al., 2011). By this approach, they were able to stimulate neuritogenesis and proposed their structures as candidates for drug screening arrays (see Fig. 15.8C).

A relevant study from the point of view of the inclusion of bioceramic counterparts on TiO_2 surfaces was completed by Huang et al. (Huang, Lai, Lin, Sun, & Lin, 2010). They created by electrochemical methods a series of nanostructured porous-like calcium phosphate motives on the flat TiO_2 surface. Relevantly, osteosarcoma cells exhibited a high selectivity toward the porous calcium phosphate surface.

The sol–gel process has been used to induce hierarchical structures by producing first mesoporous structures induced by the presence of amphiphilic copolymers in the sol (Park et al., 2013). The condensation and annealing at 450°C induced a spongiform structure with pores below 50 nm and undetermined crystalline structure. The higher scale micropatterning was succeeded by creating negative photoresists containing antifouling PEG, which concentrated cell adhesion on mesoporous $TiO_{2\text{counterparts.}}$

The effects of high aspect ratio of microstructures on platelet adhesion were studied on model TiO_2 surfaces (Ding et al., 2013). The authors prepared first submicrometer (lateral size) stripe and pillar motives on Si by reactive ion etching. These sculptured structures were then coated with Ti by magnetron sputtering and finally oxidized by annealing in air at 700°C. The effects on platelet adhesion were evident with enhanced adhesion on stripes and pillars with respect to flat TiO_2. The effects spanned to platelet activation, which could be correlated with two different adhesion modes: bridging on top of submicron features (increased platelet activation) and full contact of the cell with deep TiO_2 surface (inhibited platelet activation).

Additional micropatterns (in practice, macropatterns) containing nanotopographic contrasts were prepared by alternating series of compact and porous TiO_2 (rutile) (Mou et al., 2016). The patterns did not lead to adhesion contrasts of neural stem cells, but differences in the differentiation patterns were observed: differentiation to astrocytes was preferential on porous TiO_2 whereas differentiation to neurons was dominant on flat TiO_2.

Making use of an original magnetic–topographic patterning approach, Lin et al. managed to localize nerve growth factors conjugated to iron oxide nanoparticles at the bottom of strain induced one-dimensional corrugations on silk-gelatin (Lin et al., 2020). These patterns allowed inducing an alignment of neuronal cultures with selective directionality of extended neurites, which could be exploited for the generation of nerve conduits.

Using lithography and hydrothermal growth, Park et al. decorated ZnO surfaces with nanoflowers in well-defined square arrangement. The surfaces demonstrated a more effective osteoblast growth with respect to flat ZnO as indicated by the analysis of three different signals (Park et al., 2010).

Self-assembled hierarchical TiO_2 structures have been observed on Ti6Al4V alloys, where the selective etching of surrounding beta grains was exploited to generate microscale steps on the surfaces. In addition, a contrasting texture could be observed between the nanoscale topography at the alpha and beta grains. The so-formed surface induced the growth of osteoblasts but inhibited the progress of fibroblasts, thus proving the specific properties of cellular microenvironments (Variola et al., 2008). A similar acid etching process but with a prior ion beam irradiation step induced contrasting adhesion properties of neuronal cells (Lopez et al., 2020). Hierarchical structuring processes and surface control of TMOs offer a diversity of attractive combinations of properties that are expected to accutely contribute in the coming decade to new cell-based devices and tissue engineering supports.

References

Abiko, Y., Brunette, D. M., Jensen, J. A., Nanci, A., Todescan, R., & Heath, J. P. (1993). Immunohistochemical investigation of tracks left by the migration of fibroblasts on titanium surfaces. *Cells and Materials*, *3*, 161–170.

Afzal, A. (2014). Implantable zirconia bioceramics for bone repair and replacement: A chronological review. *Materials Express*, *4*, 1–12.

Aguilera-Correa, J. J., Mediero, A., Conesa-Buendia, F. M., Conde, A., Arenas, M. A., de-Damborenea, J. J., ... Esteban, J. (2019). Microbiological and cellular evaluation of a fluorine-phosphorus-doped titanium alloy, a novel antibacterial and osteostimulatory biomaterial with potential applications in orthopedic surgery. *Applied and Environmental Microbiology*, *85*, 17.

Akerlund, E., Islam, M. S., McCarrick, S., Alfaro-Moreno, E., & Karlsson, H. L. (2019). Inflammation and (secondary) genotoxicity of Ni and NiO nanoparticles. *Nanotoxicology*, *13*, 1060–1072.

Altankov, G., Groth, T., Krasteva, N., Albrecht, W., & Paul, D. (1997). Morphological evidence for a different fibronectin receptor organization and function during fibroblast adhesion on hydrophilic and hydrophobic glass substrata. *Journal of Biomaterials Science-Polymer Edition*, *8*, 721–740.

Amna, T., Hassan, M. S., Shin, W. S., Ba, H. V., Lee, H. K., Khil, M. S., ... Hwang, I. H. (2013). TiO_2 nanorods via one-step electrospinning technique: A novel nanomatrix for mouse myoblasts adhesion and propagation. *Colloids and Surfaces B—Biointerfaces*, *101*, 424–429.

Ayachi, A. A., Mechakra, H., Manso Silvan, M., Boudjaadar, S., & Achour, S. (2015). Monodisperse alpha-Fe_2O_3 nanoplatelets: Synthesis and characterization. *Ceramics International*, *41*, 2228–2233.

Azadmanjiri, J., Wang, P. Y., Pingle, H., Kingshott, P., Wang, J., Srivastava, V. K., ... Kapoor, A. (2016). Enhanced attachment of human mesenchymal stem cells on nanograined titania surfaces. *RSC Advances*, *6*, 55825–55833.

Bacakova, L., Filova, E., Parizek, M., Ruml, T., & Svorcik, V. (2011). Modulation of cell adhesion, proliferation and differentiation on materials designed for body implants. *Biotechnology Advances*, *29*, 739–767.

Bai, Y., Park, I. S., Lee, S. J., Bae, T. S., Duncan, W., Swain, M., ... Lee, M. H. (2011). One-step approach for hydroxyapatite-incorporated TiO_2 coating on titanium via a combined technique of micro-arc oxidation and electrophoretic deposition. *Applied Surface Science*, *257*, 7010–7018.

Bajpai, I., Rukini, A., Jung, K. J., Song, I. H., & Kim, S. (2017). Surface morphological influence on the in vitro bioactivity and response of mesenchymal stem cells. *Materials Technology*, *32*, 535–542.

Balasundaram, G., Yao, C., & Webster, T. J. (2008). TiO_2 nanotubes functionalized with regions of bone morphogenetic protein-2 increases osteoblast adhesion. *Journal of Biomedical Materials Research. Part A*, *84A*, 447–453.

Basle, M. F., Bertrand, G., Guyetant, S., Chappard, D., & Lesourd, M. (1996). Migration of metal and polyethylene particles from articular prostheses may generate lymphadenopathy with histiocytosis. *Journal of Biomedical Materials Research*, *30*, 157–164.

Bass, J. D., Belamie, E., Grosso, D., Boissiere, C., Coradin, T., & Sanchez, C. (2010). Nanostructuration of titania films prepared by self-assembly to affect cell adhesion. *Journal of Biomedical Materials Research. Part A*, *93A*, 96–106.

Beigi, M.-H., Safaie, N., Nasr-Esfahani, M.-H., & Kiani, A. (2019). 3D titania nanofiber-like webs induced by plasma ionization: A new direction for bioreactivity and osteoinductivity enhancement of biomaterials. *Scientific Reports*, *9*.

Belicchi, M., Erratico, S., Razini, P., Meregalli, M., Cattaneo, A., Jacchetti, E., ... Torrente, Y. (2010). Ex vivo expansion of human circulating myogenic progenitors on cluster-assembled nanostructured TiO_2. *Biomaterials*, *31*, 5385–5396.

Bilek, O., Fohlerova, Z., & Hubalek, J. (2019). Enhanced antibacterial and anticancer properties of Se-NPs decorated TiO_2 nanotube film. *PLOS ONE*, *14*.

Blanquart, T., Niinisto, J., Ritala, M., & Leskela, M. (2014). Atomic layer deposition of groups 4 and 5 transition metal oxide thin films: Focus on heteroleptic precursors. *Chemical Vapor Deposition*, *20*, 189–208.

Brammer, K. S., Kim, H., Noh, K., Loya, M., Frandsen, C. J., Chen, L. H., ... Jin, S. (2011). Highly bioactive 8 nm hydrothermal TiO_2 nanotubes elicit enhanced bone cell response. *Advanced Engineering Materials*, *13*, B88–B94.

Brar, H. S., Ball, J. P., Berglund, I. S., Allen, J. B., & Manuel, M. V. (2013). A study of a biodegradable Mg-3Sc-3Y alloy and the effect of self-passivation on the in vitro degradation. *Acta Biomaterialia*, *9*, 5331–5340.

Brunette, D. M., Tengvall, P., Textor, M., & Thomsen, P. (2001). *Titanium in medicine*. Berlin, Heidelberg: Springer Verlag.

Cai, Z., Vermilyea, S. G., & Brantley, W. A. (1999). In vitro corrosion resistance of high-palladium dental casting alloys. *Dental Materials*, *15*, 202–210.

Chang, Y. Y., Zhang, J. H., & Huang, H. L. (2018). Effects of laser texture oxidation and high-temperature annealing of TiV alloy thin films on mechanical and antibacterial properties and cytotoxicity. *Materials*, *11*.

Cheng, H. Y., Mao, L., Xu, X., Zeng, Y., Lan, D. N., Hu, H., ... Zhu, Z. H. (2015). The bifunctional regulation of interconnected Zn-incorporated ZrO_2 nanoarrays in antibiosis and osteogenesis. *Biomaterials Science*, *3*, 665–680.

Choi, H., Sofranko, A. C., & Dionysiou, D. D. (2006). Nanocrystalline TiO_2 photocatalytic membranes with a hierarchical mesoporous multilayer structure: Synthesis, characterization, and multifunction. *Advanced Functional Materials*, *16*, 1067–1074.

Clark, P., Connolly, P., Curtis, A. S. G., Dow, J. A. T., & Wilkinson, C. D. W. (1987). Topographical control of cell behavior. 1. Simple step cues. *Development (Cambridge, England)*, *99*, 439–448.

Csucs, G., Michel, R., Lussi, J. W., Textor, M., & Danuser, G. (2003). Microcontact printing of novel co-polymers in combination with proteins for cell-biological applications. *Biomaterials*, *24*, 1713–1720.

Daculsi, G., Rohanizadeh, R., Weiss, P., & Bouler, J. M. (2000). Crystal polymer interaction with new injectable bone substitute; SEM and Hr TEM study. *Journal of Biomedical Materials Research*, *50*, 1–7.

Damji, A., Weston, L., & Brunette, D. M. (1996). Directed confrontations between fibroblasts and epithelial cells on micromachined grooved substrata. *Experimental Cell Research*, *228*, 114–124.

Das, K., Bose, S., & Bandyopadhyay, A. (2009). TiO_2 nanotubes on Ti: Influence of nanoscale morphology on bone cell-materials interaction. *Journal of Biomedical Materials Research. Part A*, *90A*, 225–237.

Demetrescu, I., Pirvu, C., & Mitran, V. (2010). Effect of nano-topographical features of Ti/TiO_2 electrode surface on cell response and electrochemical stability in artificial saliva. *Bioelectrochemistry (Amsterdam, Netherlands)*, *79*, 122–129.

Dhawan, U., Lee, C. H., Huang, C. C., Chu, Y. H., Huang, G. S., Lin, Y. R., & Chen, W. L. (2015). Topological control of nitric oxide secretion by tantalum oxide nanodot arrays. *Journal of Nanobiotechnology*, *13*.

Dinan, B., Gallego-Perez, D., Lee, H., Hansford, D., & Akbar, S. A. (2013). Thermally grown TiO_2 nanowires to improve cell growth and proliferation on titanium based materials. *Ceramics International*, *39*, 5949–5954.

Ding, Y. H., Leng, Y., Huang, N., Yang, P., Lu, X., Ge, X., ... Guo, X. (2013). Effects of microtopographic patterns on platelet adhesion and activation on titanium oxide surfaces. *Journal of Biomedical Materials Research. Part A*, *101*, 622–632.

Ding, Z. L., Wang, Y., Zhou, Q., Ding, Z. Y., Liu, J., He, Q. G., & Zhang, H. B. (2020). Microstructure, wettability, corrosion resistance and antibacterial property of Cu-MTa_2O_5 multilayer composite coatings with different Cu incorporation contents. *Biomolecules*, *10*.

Donthu, S., Pan, Z. X., Myers, B., Shekhawat, G., Wu, N. G., & Dravid, V. (2005). Facile scheme for fabricating solid-state nanostructures using e-beam lithography and solution precursors. *Nano Letters*, *5*, 1710–1715.

Dou, X. Q., Zhang, D., Feng, C. L., & Jiang, L. (2015). Bioinspired hierarchical surface structures with tunable wettability for regulating bacteria adhesion. *ACS Nano*, *9*, 10664–10672.

Ektessabi, A. M., & Kimura, H. (1995). Characterization of the surface of bio-ceramic thin films. *Thin Solid Films*, *270*, 335–340.

Elizabeth, E., Baranwal, G., Krishnan, A. G., Menon, D., & Nair, M. (2014). ZnO nanoparticle incorporated nanostructured metallic titanium for increased mesenchymal stem cell response and antibacterial activity. *Nanotechnology*, *25*, 12.

Espitia, P. J. P., Soares, N. D. F., Teofilo, R. F., Coimbra, J. S. D., Vitor, D. M., Batista, R. A., ... Medeiros, E. A. A. (2013). Physical-mechanical and antimicrobial properties of nanocomposite films with pediocin and ZnO nanoparticles. *Carbohydrate Polymers*, *94*, 199–208.

Exarhos, G. J., & Aloi, M. J. (1990). Crystallite growth-kinetics in isothermally annealed sol-gel films. *Thin Solid Films*, *193*, 42–50.

Ferraris, S., Cochis, A., Cazzola, M., Tortello, M., Scalia, A., Spriano, S., & Rimondini, L. (2019). Cytocompatible and anti-bacterial adhesion nanotextured titanium oxide layer on titanium surfaces for dental and orthopedic implants. *Frontiers in Bioengineering and Biotechnology*, *7*, 12.

Field, J. A., Luna-Velasco, A., Boitano, S. A., Shadman, F., Ratner, B. D., Barnes, C., & Sierra-Alvarez, R. (2011). Cytotoxicity and physicochemical properties of hafnium oxide nanoparticles. *Chemosphere*, *84*, 1401–1407.

Fohlerova, Z., & Mozalev, A. (2019). Anodic formation and biomedical properties of hafnium-oxide nanofilms. *Journal of Materials Chemistry B*, *7*, 2300–2310.

Frandsen, C. J., Brammer, K. S., Noh, K., Connelly, L. S., Oh, S., Chen, L. H., & Jin, S. J. (2011). Zirconium oxide nanotube surface prompts increased osteoblast functionality and mineralization. *Materials Science & Engineering C-Materials for Biological Applications*, *31*, 1716–1722.

Frandsen, C. J., Brammer, K. S., Noh, K., Johnston, G., & Jin, S. (2014). Tantalum coating on TiO_2 nanotubes induces superior rate of matrix mineralization and osteofunctionality in human osteoblasts. *Materials Science & Engineering C—Materials for Biological Applications*, *37*, 332–341.

Fukuzaki, S., Urano, H., & Nagata, K. (1996). Adsorption of bovine serum albumin onto metal oxide surfaces. *Journal of Fermentation and Bioengineering*, *81*, 163–167.

Gallach Perez, D., Punzon Quijorna, E., Sanz, R., Torres-Costa, V., Garcia Ruiz, J. P., & Manso Silvan, M. (2015). Nanotopography enhanced mobility determines mesenchymal stem cell distribution on micropatterned semiconductors bearing nanorough areas. *Colloids and Surfaces B-Biointerfaces*, *126*, 146–153.

Grigorescu, S., Pruna, V., Titorencu, I., Jinga, V. V., Mazare, A., Schmuki, P., & Demetrescu, I. (2014). The two step nanotube formation on TiZr as scaffolds for cell growth. *Bioelectrochemistry (Amsterdam, Netherlands)*, *98*, 39–45.

Guo, Z. D., Chen, M., Peng, C. Y., Mo, S. G., Shi, C. R., Fu, G. F., ... Zhang, X. Z. (2018). pH-sensitive radiolabeled and superfluorinated ultra-small palladium nanosheet as a high-performance multimodal platform for tumor theranostics. *Biomaterials*, *179*, 134–143.

Guo, Z. J., Jiang, N., Chen, C., Zhu, S. S., Zhang, L., & Li, Y. B. (2017). Surface bioactivation through the nanostructured layer on titanium modified by facile HPT treatment. *Scientific Reports*, *7*.

Gupta, K. K., Kundan, A., Mishra, P. K., Srivastava, P., Mohanty, S., Singh, N. K., ... Maiti, P. (2012). Polycaprolactone composites with TiO_2 for potential nanobiomaterials: Tunable properties using different phases. *Physical Chemistry Chemical Physics*, *14*, 12844–12853.

Han, C. M., Kim, H. E., & Koh, Y. H. (2014). Creation of hierarchical micro/nano-porous TiO_2 surface layer onto Ti implants for improved biocompatibility. *Surface & Coatings Technology*, *251*, 226–231.

Hanawa, T., Hiromoto, S., & Asami, K. (2001). Characterization of the surface oxide film of a Co-Cr-Mo alloy after being located in quasi-biological environments using XPS. *Applied Surface Science*, *183*, 68–75.

Hao, Y. Q., Li, S. J., Hao, Y. L., Zhao, Y. K., & Ai, H. J. (2013). Effect of nanotube diameters on bioactivity of a multifunctional titanium alloy. *Applied Surface Science*, *268*, 44–51.

Herath, H., Di Silvio, L., & Evans, J. R. G. (2005). Scandia—A potential biomaterial? *Journal of Materials Science-Materials in Medicine*, *16*, 1061–1065.

Hernandez-Velez, M. (2006). Nanowires and 1D arrays fabrication: An overview. *Thin Solid Films*, *495*, 51–63.

Howie, D. W., & Vernonroberts, B. (1988). Synovial macrophage response to aluminum-oxide ceramic and cobalt chrome alloy wear particles in rats. *Biomaterials*, *9*, 442–448.

Hsu, H. C., Hsu, S. K., Wu, S. C., & Ho, W. F. (2019). Formation of nanotubular structure on low-modulus Ti-7.5Mo alloy surface and its bioactivity evaluation. *Thin Solid Films*, *669*, 329–337.

Hu, X., Li, F. Y., Xia, F., Guo, X., Wang, N., Liang, L. L., ... Ling, D. S. (2020). Biodegradation-mediated enzymatic activity-tunable molybdenum oxide nanourchins for tumor-specific cascade catalytic therapy. *Journal of the American Chemical Society*, *142*, 1636–1644.

Huang, H. H., Wu, C. P., Sun, Y. S., Huang, H. M., & Lee, T. H. (2013). Enhanced corrosion resistance and biocompatibility of beta-type Ti-25Nb-25Zr alloy by electrochemical anodization. *Thin Solid Films*, *549*, 87–92.

Huang, S. M., Nauman, E. A., & Stanciu, L. A. (2019). Investigation of porosity on mechanical properties, degradation and in-vitro cytotoxicity limit of Fe30Mn using space holder technique. *Materials Science & Engineering C-Materials for Biological Applications*, *99*, 1048–1057.

Huang, Y., Song, G. Q., Chang, X. T., Wang, Z. H., Zhang, X. J., Han, S. G., ... Yang, D. D. (2018). Nanostructured Ag^{+}-substituted fluorhydroxyapatite-TiO_2 coatings for enhanced bactericidal effects and osteoinductivity of Ti for biomedical applications. *International Journal of Nanomedicine*, *13*, 2665–2684.

Huang, Y.-X., Lai, Y.-K., Lin, L.-X., Sun, L., & Lin, C.-J. (2010). Electrochemical construction and biological performance of micropatterned CaP films. *Acta Physico-Chimica Sinica*, *26*, 2057–2060.

Huerta-Murillo, D., Garcia-Giron, A., Romano, J. M., Cardoso, J. T., Cordovilla, F., Walker, M., ... Ocana, J. L. (2019). Wettability modification of laser-fabricated hierarchical surface structures in Ti-6Al-4V titanium alloy. *Applied Surface Science*, *463*, 838–846.

Huey, B. D., & Bonnell, D. A. (2000). Spatially localized dynamic properties of individual interfaces in semiconducting oxides. *Applied Physics Letters*, *76*, 1012–1014.

Hui, E. E., & Bhatia, S. N. (2007a). Microscale control of cell contact and spacing via three-component surface patterning. *Langmuir: The ACS Journal of Surfaces and Colloids*, *23*, 4103–4107.

Hui, E. E., & Bhatia, S. N. (2007b). Micromechanical control of cell-cell interactions. *Proceedings of the National Academy of Sciences of the United States of America*, *104*, 5722–5726.

Hussain, S. M., Hess, K. L., Gearhart, J. M., Geiss, K. T., & Schlager, J. J. (2005). In vitro toxicity of nanoparticles in BRL 3A rat liver cells. *Toxicology in Vitro*, *19*, 975–983.

Ibrahim, J., Nguyen, A. H., Rehman, A., Ochi, A., Jamal, M., Graffeo, C. S., ... Miller, G. (2012). Dendritic cell populations with different concentrations of lipid regulate tolerance and immunity in mouse and human liver. *Gastroenterology*, *143*, 1061–1072.

Ingber, D., & Tensegrity, I. (2003). Cell structure and hierarchical systems biology. *Journal of Cell Science*, *116*, 1157–1173.

Ivanova, T., Gesheva, K. A., Popkirov, G., Ganchev, M., & Tzvetkova, E. (2005). Electrochromic properties of atmospheric CVD MoO_3 and MoO_3-WO_3 films and their application in electrochromic devices. *Materials Science and Engineering B—Solid State Materials for Advanced Technology*, *119*, 232–239.

Jaeggi, C., Kern, P., Michler, J., Zehnder, T., & Siegenthaler, H. (2005). Anodic thin films on titanium used as masks for surface micropatterning of biomedical devices. *Surface & Coatings Technology*, *200*, 1913–1919.

Jao, W. C., Yang, M. C., Lin, C. H., & Hsu, C. C. (2012). Fabrication and characterization of electrospun silk fibroin/TiO_2 nanofibrous mats for wound dressings. *Polymers for Advanced Technologies*, *23*, 1066–1076.

Jarrell, J. D., Eun, T. H., Samale, M., Briant, C., Sheldon, B. W., & Morgan, J. R. (2007). Metal oxide coated cell culture arrays for rapid biological screening. *Journal of Biomedical Materials Research. Part A*, *83A*, 853–860.

Jiang, J., Zhu, J. H., Ding, R. M., Li, Y. Y., Wu, F., Liu, J. P., & Huang, X. T. (2011). Co-Fe layered double hydroxide nanowall array grown from an alloy substrate and its calcined product as a composite anode for lithium-ion batteries. *Journal of Materials Chemistry*, *21*, 15969–15974.

Jiang, L., Lu, X., Leng, Y., Qu, S., Feng, B., Weng, J., & Watari, F. (2012). Osteoblast behavior on TiO_2 microgrooves prepared by soft-lithography and sol-gel methods. *Materials Science & Engineering C—Materials for Biological Applications*, *32*, 742–748.

Jin, M., Yao, S. L., Wang, L. N., Qiao, Y., & Volinsky, A. A. (2016). Enhanced bond strength and bioactivity of interconnected 3D TiO_2 nanoporous layer on titanium implants. *Surface & Coatings Technology*, *304*, 459–467.

Jin, Y. S., Ma, X. B., Zhang, S., Meng, H., Xu, M., Yang, X., . . . Tian, J. (2017). A tantalum oxide-based core/shell nanoparticle for triple-modality image-guided chemo-thermal synergetic therapy of esophageal carcinoma. *Cancer Letters*, *397*, 61–71.

Junkar, I., Kulkarni, M., Bencina, M., Kovac, J., Mrak-Poljsak, K., Lakota, K., . . . Iglic, A. (2020). Titanium dioxide nanotube arrays for cardiovascular stent applications. *ACS Omega*, *5*, 7280–7289.

Juodkazis, S., Rosa, L., Bauerdick, S., Peto, L., El-Ganainy, R., & John, S. (2011). Sculpturing of photonic crystals by ion beam lithography: Towards complete photonic bandgap at visible wavelengths. *Optics Express*, *19*, 5802–5810.

Kaciulis, S., Mattogno, G., Napoli, A., Bemporad, E., Ferrari, F., Montenero, A., & Gnappi, G. (1998). Surface analysis of biocompatible coatings on titanium. *Journal of Electron Spectroscopy and Related Phenomena*, *95*, 61–69.

Kao, C. T., Ding, S. J., Chen, Y. C., & Huang, T. H. (2002). The anticorrosion ability of titanium nitride (TiN) plating on an orthodontic metal bracket and its biocompatibility. *Journal of Biomedical Materials Research*, *63*, 786–792.

Kar, A., Raja, K. S., & Misra, M. (2006). Electrodeposition of hydroxyapatite onto nanotubular TiO_2 for implant applications. *Surface & Coatings Technology*, *201*, 3723–3731.

Karakecili, A. G., & Gumusderelioglu, M. (2002). Comparison of bacterial and tissue cell initial adhesion on hydrophilic/hydrophobic biomaterials. *Journal of Biomaterials Science-Polymer Edition*, *13*, 185–196.

Karimi, N., Kharaziha, M., & Raeissi, K. (2019). Electrophoretic deposition of chitosan reinforced graphene oxide-hydroxyapatite on the anodized titanium to improve biological and electrochemical characteristics. *Materials Science & Engineering C-Materials for Biological Applications*, *98*, 140–152.

Khan, M. R., Mordan, N., Parkar, M., Salih, V., Donos, N., & Brett, P. M. (2019). Atypical mesenchymal stromal cell responses to topographic modifications of titanium biomaterials indicate cytoskeletal- and genetic plasticity-based heterogeneity of cells. *Stem Cells International, 2019*.

Kim, D. H., Lipke, E. A., Kim, P., Cheong, R., Thompson, S., Delannoy, M., ... Levchenko, A. (2010). Nanoscale cues regulate the structure and function of macroscopic cardiac tissue constructs. *Proceedings of the National Academy of Sciences of the United States of America*, *107*, 565–570.

Kim, H. M., Chae, W. P., Chang, K. W., Chun, S., Kim, S., Jeong, Y., & Kang, I. K. (2010). Composite nanofiber mats consisting of hydroxyapatite and titania for biomedical applications. *Journal of Biomedical Materials Research Part B—Applied Biomaterials*, *94B*, 380–387.

Kiran, A. S. K., Balu, R., & Kumar, T. S. S. (2012). Vero cell viability and human osteoblast cell response to electrospun phase controlled titania nanofibers. *Journal of Biomaterials and Tissue Engineering*, *2*, 292–298.

Klymov, A., Prodanov, L., Lamers, E., Jansen, J. A., & Walboomers, X. F. (2013). Understanding the role of nano-topography on the surface of a bone-implant. *Biomaterials Science*, *1*, 135–151.

Ko, T. J., Kim, E., Nagashima, S., Oh, K. H., Lee, K. R., Kim, S., & Moon, M. W. (2013). Adhesion behavior of mouse liver cancer cells on nanostructured superhydrophobic and superhydrophilic surfaces. *Soft Matter*, *9*, 8705–8711.

Komasa, S., Taguchi, Y., Nishida, H., Tanaka, M., & Kawazoe, T. (2012). Bioactivity of nanostructure on titanium surface modified by chemical processing at room temperature. *Journal of Prosthodontic Research*, *56*, 170–177.

Lan, M. Y., Lee, S. L., Huang, H. H., Chen, P. F., Liu, C. P., & Lee, S. W. (2014). Diameter selective behavior of human nasal epithelial cell on Ag-coated TiO_2 nanotubes. *Ceramics International*, *40*, 4745–4751.

Langhans, S. A. (2018). Three-dimensional in vitro cell culture models in drug discovery and drug repositioning. *Frontiers in Pharmacology*, *9*.

Langlet, M., Burgos, M., Coutier, C., Jimenez, C., Morant, C., & Manso, M. (2001). Low temperature preparation of high refractive index and mechanically resistant sol-gel TiO_2 films for multilayer antireflective coating applications. *Journal of Sol-Gel Science and Technology*, *22*, 139–150.

Lao, J. Y., Wen, J. G., & Ren, Z. F. (2002). Hierarchical ZnO nanostructures. *Nano Letters*, *2*, 1287–1291.

Lee, J. Y., & Schmidt, C. E. (2010). Pyrrole-hyaluronic acid conjugates for decreasing cell binding to metals and conducting polymers. *Acta Biomaterialia*, *6*, 4396–4404.

Lee, M. H., Kang, J. H., & Lee, S. W. (2012). The significance of differential expression of genes and proteins in human primary cells caused by microgrooved biomaterial substrata. *Biomaterials*, *33*, 3216–3234.

Li, G. Z., Zhao, Q. M., Tang, H. P., Li, G., & Chi, Y. D. (2016). Fabrication, characterization and biocompatibility of TiO_2 nanotubes via anodization of Ti6Al7Nb. *Composite Interfaces*, *23*, 223–230.

Li, J. H., Jiang, M., Zhou, H. J., Jin, P., Cheung, K. M. C., Chu, P. K., & Yeung, K. W. K. (2019). Vanadium dioxide nanocoating induces tumor cell death through mitochondrial electron transport chain interruption. *Global Challenges*, *3*.

Li, J. Y., Zhai, D., Lv, F., Yu, Q. Q., Ma, H. S., Yin, J. B., ... Wu, C. T. (2016). Preparation of copper-containing bioactive glass/eggshell membrane nanocomposites for improving angiogenesis, antibacterial activity and wound healing. *Acta Biomaterialia*, *36*, 254–266.

Lim, J. I., Yu, B., Woo, K. M., & Lee, Y. K. (2008). Immobilization of TiO_2 nanofibers on titanium plates for implant applications. *Applied Surface Science*, *255*, 2456–2460.

Lim, S. S., Chai, C. Y., & Loh, H. S. (2017). In vitro evaluation of osteoblast adhesion, proliferation and differentiation on chitosan-TiO2 nanotubes scaffolds with Ca2 + ions. *Materials Science & Engineering C-Materials for Biological Applications*, *76*, 144–152.

Lin, C. C., Chang, J. J., Yung, M. C., Huang, W. C., & Chen, S. Y. (2020). Spontaneously micropatterned silk/gelatin scaffolds with topographical, biological, and electrical stimuli for neuronal regulation. *ACS Biomaterials Science & Engineering*, *6*, 1144–1153.

Lin, Q. X., Huang, D., Du, J. J., Wei, Y., Hu, Y. C., Lian, X. J., ... Zhang, Y. S. (2019). Nano-hydroxyapatite crystal formation based on calcified TiO_2 nanotube arrays. *Applied Surface Science*, *478*, 237–246.

Liu, H., Sun, N., Ding, P., Chen, C. C., Wu, Z., Zhu, W. P., ... Pei, R. J. (2020). Fabrication of aptamer modified TiO_2 nanofibers for specific capture of circulating tumor cells. *Colloids and Surfaces B-Biointerfaces*, *191*.

Liu, S. H., Netzel-Arnett, S., Birkedal-Hansen, H., & Leppla, S. H. (2000). Tumor cell-selective cytotoxicity of matrix metalloproteinase-activated anthrax toxin. *Cancer Research*, *60*, 6061–6067.

Liu, W. W., Su, P. L., Gonzales, A., Chen, S., Wang, N., Wang, J. S., ... Webster, T. J. (2015). Optimizing stem cell functions and antibacterial properties of TiO_2 nanotubes incorporated with ZnO nanoparticles: Experiments and modeling. *International Journal of Nanomedicine*, *10*, 1997–2019.

Liu, X. Y., & Sun, Y. (2009). Microfabricated glass devices for rapid single cell immobilization in mouse zygote microinjection. *Biomedical Microdevices*, *11*, 1169–1174.

Lopez, R., Ynsa, M. D., de Pablo, P. J., Lim, F., & Manso Silvan, M. (2020). Engineering nanostructured cell micropatterns on Ti6Al4V by selective ion-beam inhibition of pitting. *Corrosion Science*, *167*, 108528.

Lozano, D., Hernandez-Lopez, J. M., Esbrit, P., Arenas, M. A., Gomez-Barrena, E., de Damborenea, J., ... Conde, A. (2015). Influence of the nanostructure of F-doped TiO_2 films on osteoblast growth and function. *Journal of Biomedical Materials Research. Part A*, *103*, 1985–1990.

Luo, X., Yang, P., Zhao, A. S., Jiang, L., Zou, D., Han, C. Z., ... Yin, B. L. (2020). The self-organized differentiation from MSCs into SMCs with manipulated micro/Nano two-scale arrays on TiO_2 surfaces for biomimetic construction of vascular endothelial substratum. *Materials Science & Engineering C-Materials for Biological Applications*, *116*, 12.

Luo, Z. Q., Weiss, D. E., Liu, Q. Y., & Tian, B. Z. (2018). Biomimetic approaches toward smart bio-hybrid systems. *Nano Research*, *11*, 3009–3030.

Lutz, J., Diaz, C., Garcia, J. A., Blawert, C., & Mandl, S. (2011). Corrosion behaviour of medical CoCr alloy after nitrogen plasma immersion ion implantation. *Surface & Coatings Technology*, *205*, 3043–3049.

Ma, Z. W., Mao, Z. W., & Gao, C. Y. (2007). Surface modification and property analysis of biomedical polymers used for tissue engineering. *Colloids and Surfaces B—Biointerfaces*, *60*, 137–157.

Mahmood, M., Fejleh, P., Karmakar, A., Fejleh, A., Xu, Y., Kannarpady, G., ... Biris, A. S. (2011). Enhanced bone cells growth and proliferation on TiO_2 nanotubular substrates treated by RF plasma discharge. *Advanced Engineering Materials*, *13*, B95–B101.

Malec, K., Goralska, J., Hubalewska-Mazgaj, M., Glowacz, P., Jarosz, M., Brzewski, P., ... Wybranska, I. (2016). Effects of nanoporous anodic titanium oxide on human adipose derived stem cells. *International Journal of Nanomedicine*, *11*, 5349–5360.

Manso, M., Langlet, M., & Martinez-Duart, J. (2001). Mechanical properties of biocompatibe titania layers produced in an aerosol-gel reactor. *Revista De Metalurgia*, *37*, 260–263.

Manso-Silvan, M., Valsesia, A., Hasiwa, M., Rodriguez-Navas, C., Gilliland, D., Ceccone, G., ... Rossi, F. (2007). Micro-spot, UV and wetting patterning pathways for applications of biofunctional aminosilane-titanate coatings. *Biomedical Microdevices*, *9*, 287–294.

Marsich, E., Travan, A., Donati, I., Turco, G., Kulkova, J., Moritz, N., ... Paoletti, S. (2013). Biological responses of silver-coated thermosets: An in vitro and in vivo study. *Acta Biomaterialia*, *9*, 5088–5099.

Martinez-Calderon, M., Martin-Palma, R. J., Rodriguez, A., Gomez-Aranzadi, M., Garcia-Ruiz, J. P., Olaizola, S. M., & Manso-Silvan, M. (2020). Biomimetic hierarchical micro/nano texturing of TiAlV alloys by femtosecond laser processing for the control of cell adhesion and migration. *Physical Review Materials*, *4*.

Maruyama, H., Arai, F., Fukuda, T., & Katsuragi, T. (2005). Immobilization of individual cells by local photo-polymerization on a chip. *Analyst*, *130*, 304–310.

Massaro, C., Rotolo, P., De Riccardis, F., Milella, E., Napoli, A., Wieland, M., ... Brunette, D. M. (2002). Comparative investigation of the surface properties of commercial titanium dental implants. Part I: Chemical composition. *Journal of Materials Science-Materials in Medicine*, *13*, 535–548.

Mazur, M., Kalisz, M., Wojcieszak, D., Grobelny, M., Mazur, P., Kaczmarek, D., & Domaradzki, J. (2015). Determination of structural, mechanical and corrosion properties of Nb_2O_5 and (NbyCu1-y)O-x thin films deposited on Ti6Al4V alloy substrates for dental implant applications. *Materials Science & Engineering C—Materials for Biological Applications*, *47*, 211–221.

Mei, S. L., Zhao, L. Z., Wang, W., Ma, Q. L., & Zhang, Y. M. (2012). Biomimetic titanium alloy with sparsely distributed nanotubes could enhance osteoblast functions. *Advanced Engineering Materials*, *14*, B166–B174.

Minagar, S., Li, Y. C., Berndt, C. C., & Wen, C. (2015). The influence of titania-zirconia-zirconium titanate nanotube characteristics on osteoblast cell adhesion. *Acta Biomaterialia*, *12*, 281–289.

Missirlis, Y. F., & Spiliotis, A. D. (2002). Assessment of techniques used in calculating cell-material interactions. *Biomolecular Engineering*, *19*, 287–294.

Mohammadi, S., Esposito, M., Cucu, M., Ericson, L. E., & Thomsen, P. (2001). Tissue response to hafnium. *Journal of Materials Science—Materials in Medicine*, *12*, 603–611.

Mohan, C. C., Sreerekha, P. R., Divyarani, V. V., Nair, S., Chennazhi, K., & Menon, D. (2012). Influence of titania nanotopography on human vascular cell functionality and its proliferation in vitro. *Journal of Materials Chemistry*, *22*, 1326–1340.

Moreno, B., Carballo, M., Jurado, J. R., & Chinarro, E. (2009). Applications of TiO_2 in cellular therapies and tissue engineering: A review. *Boletin De La Sociedad Espanola De Ceramica Y Vidrio*, *48*, 321–328.

Mou, X. N., Wang, S., Guo, W. B., Ji, S. Z., Qiu, J. C., Li, D. S., ... Liu, H. (2016). Localized committed differentiation of neural stem cells based on the topographical regulation effects of TiO_2 nanostructured ceramics. *Nanoscale*, *8*, 13186–13191.

Moxey, M., Johnson, A., El-Zubir, O., Cartron, M., Dinachali, S. S., Hunter, C. N., ... Leggett, G. J. (2015). Fabrication of self-cleaning, reusable titania templates for nanometer and micrometer scale protein patterning. *ACS Nano*, *9*, 6262–6270.

Muhammad, R., Lim, S. H., Goh, S. H., Law, J. B. K., Saifullah, M. S. M., Ho, G. W., & Yim, E. K. F. (2014). Sub-100 nm patterning of TiO_2 film for the regulation of endothelial and smooth muscle cell functions. *Biomaterials Science*, *2*, 1740–1749.

Munoz Noval, A., Sanchez Vaquero, V., Punzon Quijorna, E., Torres Costa, V., Gallach Perez, D., Gonzalez Mendez, L., ... Manso Silvan, M. (2012). Aging of porous silicon in physiological conditions: Cell adhesion modes on scaled 1D micropatterns. *Journal of Biomedical Materials Research. Part A*, *100A*, 1615–1622.

Mutreja, I., Kumar, D., Boyd, A. R., & Meenan, B. J. (2013). Titania nanotube porosity controls dissolution rate of sputter deposited calcium phosphate (CaP) thin film coatings. *RSC Advances*, *3*, 11263–11273.

Nagarajan, S., & Rajendran, N. (2009). Sol-gel derived porous zirconium dioxide coated on 316L SS for orthopedic applications. *Journal of Sol-Gel Science and Technology*, *52*, 188–196.

Narayanan, R., Lee, H. J., Kwon, T. Y., & Kim, K. H. (2011). Anodic TiO_2 nanotubes from stirred baths: Hydroxyapatite growth & osteoblast responses. *Materials Chemistry and Physics*, *125*, 510–517.

Ngaruiya, J. M., Kappertz, O., Mohamed, S. H., & Wuttig, M. (2004). Structure formation upon reactive direct current magnetron sputtering of transition metal oxide films. *Applied Physics Letters*, *85*, 748–750.

Oh, S., & Jin, S. (2006). Titanium oxide nanotubes with controlled morphology for enhanced bone growth. *Materials Science & Engineering C-Biomimetic and Supramolecular Systems*, *26*, 1301–1306.

Pan, Y. S., & Xu, K. (2020). Recent progress in nano-electronic devices based on EBL and IBL. *Current Nanoscience*, *16*, 157–169.

Park, J. K., Kim, Y. J., Yeom, J., Jeon, J. H., Yi, G. C., Je, J. H., & Hahn, S. K. (2010). The topographic effect of zinc oxide nanoflowers on osteoblast growth and osseointegration. *Advanced Materials*, *22*, 4857- + .

Park, S., Ahn, S. H., Lee, H. J., Chung, U. S., Kim, J. H., & Koh, W.-G. (2013). Mesoporous TiO(2) as a nanostructured substrate for cell culture and cell patterning. *RSC Advances*, *3*, 23673–23680.

Pelaez, R. J., Afonso, C. N., Vega, F., Recio-Sanchez, G., Torres-Costa, V., Manso-Silvan, M., ... Martin-Palma, R. J. (2013). Laser fabrication of porous silicon-based platforms for cell culturing. *Journal of Biomedical Materials Research Part B—Applied Biomaterials*, *101*, 1463–1468.

Peng, Z. X., Ni, J. H., Zheng, K., Shen, Y. D., Wang, X. Q., He, G., ... Tang, T. T. (2013). Dual effects and mechanism of TiO_2 nanotube arrays in reducing bacterial colonization and enhancing C3H10T1/2 cell adhesion. *International Journal of Nanomedicine*, *8*, 3093–3105.

Perigo, E. A., Hemery, G., Sandre, O., Ortega, D., Garaio, E., Plazaola, F., & Teran, F. J. (2015). Fundamentals and advances in magnetic hyperthermia. *Applied Physics Reviews*, *2*.

Phillips, P. G., & Birnby, L. M. (2004). Nitric oxide modulates caveolin-1 and matrix metalloproteinase-9 expression and distribution at the endothelial cell/tumor cell interface. *American Journal of Physiology-Lung Cellular and Molecular Physiology*, *286*, L1055–L1065.

Poncin-Epaillard, F., & Legeay, G. (2003). Surface engineering of biomaterials with plasma techniques. *Journal of Biomaterials Science-Polymer Edition*, *14*, 1005–1028.

Popat, K. C., Leoni, L., Grimes, C. A., & Desai, T. A. (2007). Influence of engineered titania nanotubular surfaces on bone cells. *Biomaterials*, *28*, 3188–3197.

Pradhan, P., Giri, J., Banerjee, R., Bellare, J., & Bahadur, D. (2007). Preparation and characterization of manganese ferrite-based magnetic liposomes for hyperthermia treatment of cancer. *Journal of Magnetism and Magnetic Materials*, *311*, 208–215.

PremKumar, K. P., Duraipandy, N., Kiran, M. S., & Rajendran, N. (2018). Antibacterial effects, biocompatibility and electrochemical behavior of zinc incorporated niobium oxide coating on 316L SS for biomedical applications. *Applied Surface Science*, *427*, 1166–1181.

Punzon-Quijorna, E., Sanchez Vaquero, V., Rodriguez-Lopez, S., de la Prida, V. M., Climent Font, A., Garcia Ruiz, J. P., ... Manso Silvan, M. (2012). Polymerized nanoporous titania surfaces: Modification of cell adhesion by acrylic acid functionalization. *Composite Interfaces*, *19*, 251–258.

Qadir, M., Lin, J. X., Biesiekierski, A., Li, Y. C., & Wen, C. E. (2020). Effect of anodized TiO_2-Nb_2O_5-ZrO_2 nanotubes with different nanoscale dimensions on the biocompatibility of a Ti35Zr28Nb alloy. *ACS Applied Materials & Interfaces*, *12*, 6776–6787.

Qin, L. G., Zeng, Q. F., Wang, W. X., Zhang, Y. L., & Dong, G. N. (2014). Response of MC3T3-E1 osteoblast cells to the microenvironment produced on Co-Cr-Mo alloy using laser surface texturing. *Journal of Materials Science*, *49*, 2662–2671.

Rabe, M., Verdes, D., & Seeger, S. (2011). Understanding protein adsorption phenomena at solid surfaces. *Advances in Colloid and Interface Science*, *162*, 87–106.

Radu, T., Chiriac, M. T., Popescu, O., Simon, V., & Simon, S. (2013). In vitro evaluation of the effects of yttria-alumina-silica microspheres on human keratinocyte cells. *Journal of Biomedical Materials Research. Part A*, *101*, 472–477.

Rahman, S., Gulati, K., Kogawa, M., Atkins, G. J., Pivonka, P., Findlay, D. M., & Losic, D. (2016). Drug diffusion, integration, and stability of nanoengineered drug-releasing implants in bone ex-vivo. *Journal of Biomedical Materials Research. Part A*, *104*, 714–725.

Ramadan, R., Romera, D., Carrascon, R. D., Cantero, M., Aguilera-Correa, J. J., Ruiz, J. P. G., ... Silvan, M. M. (2019). Sol-gel-deposited Ti-doped ZnO: Toward cell fouling transparent conductive oxides. *ACS Omega*, *4*, 11354–11363.

Ramadan, R., Manso-Silvan, M., & Martin-Palma, R. J. (2020). Hybrid porous silicon/silver nanostructures for the development of enhanced photovoltaic devices. *Journal of Materials Science*, *55*, 5458–5470.

Rana, A., Yadav, K., & Jagadevan, S. (2020). A comprehensive review on green synthesis of nature-inspired metal nanoparticles: Mechanism, application and toxicity. *Journal of Cleaner Production*, *272*.

Ribeiro, A. J. S., Ang, Y. S., Fu, J. D., Rivas, R. N., Mohamed, T. M. A., Higgs, G. C., ... Pruitt, B. L. (2015). Contractility of single cardiomyocytes differentiated from pluripotent stem cells depends on physiological shape and substrate stiffness. *Proceedings of the National Academy of Sciences of the United States of America*, *112*, 12705–12710.

Roguska, A., Pisarek, M., Belcarz, A., Marcon, L., Holdynski, M., Andrzejczuk, M., & Janik-Czachor, M. (2016). Improvement of the bio-functional properties of TiO_2 nanotubes. *Applied Surface Science*, *388*, 775–785.

Rtimi, S., Sanjines, R., Bensimon, M., Pulgarin, C., & Kiwi, J. (2014). Accelerated *Escherichia coli* inactivation in the dark on uniform copper flexible surfaces. *Biointerphases*, *9*.

Salasznyk, R. M., Klees, R. F., Hughlock, M. K., & Plopper, G. E. (2004). ERK signaling pathways regulate the osteogenic differentiation of human mesenchymal stem cells on collagen I and vitronectin. *Cell Communication and Adhesion*, *11*, 137–153.

Sales, A., Holle, A. W., & Kemkemer, R. (2017). Initial contact guidance during cell spreading is contractility-independent. *Soft Matter*, *13*, 5158–5167.

Salou, L., Hoornaert, A., Louarn, G., & Layrolle, P. (2015). Enhanced osseointegration of titanium implants with nanostructured surfaces: An experimental study in rabbits. *Acta Biomaterialia*, *11*, 494–502.

Sanz, R., Jaafar, M., Hernandez-Velez, M., Asenjo, A., & Vazquez, M. (2010). Patterning of rutile TiO_2 surface by ion beam lithography through full-solid masks. *Nanotechnology*, *21*, 235301.

Schieber, R., Lasserre, F., Hans, M., Fernandez-Yague, M., Diaz-Ricart, M., Escolar, G., ... Pegueroles, M. (2017). Direct laser interference patterning of CoCr alloy surfaces to control endothelial cell and platelet response for cardiovascular applications. *Advanced Healthcare Materials*, *6*.

Schulte, C., Podesta, A., Lenardi, C., Tedeschi, G., & Milani, P. (2017). Quantitative control of protein and cell interaction with nanostructured surfaces by cluster assembling. *Accounts of Chemical Research*, *50*, 231–239.

Scotchford, C. A., Ball, M., Winkelmann, M., Voros, J., Csucs, C., Brunette, D. M., ... Textor, M. (2003). Chemically patterned, metal-oxide-based surfaces produced by photolithographic techniques for studying protein- and cell-interactions. II: Protein adsorption and early cell interactions. *Biomaterials*, *24*, 1147–1158.

Seike, T., & Nagai, J. (1991). Electrochromism of 3d-transition metal-oxides. *Solar Energy Materials*, *22*, 107–117.

Sekine, K., Yamamoto, H., Kono, S., Ikeda, T., Kuroda, A., & Tanii, T. (2015). Surface modification of cell scaffold in aqueous solution using TiO_2 photocatalysis and linker protein L2 for patterning primary neurons. *E-Journal of Surface Science and Nanotechnology*, *13*, 213–218.

Shimada, S., Matsuu, M., Miyazawa, K., & Kuwabara, M. (2002). Fabrication of two-dimensional gel photonic crystals by sol-gel method using high concentration of alkoxide solution. *Journal of the Ceramic Society of Japan*, *110*, 391–394.

Siddiqui, M. A., Saquib, Q., Ahamed, M., Farshori, N. N., Ahmad, J., Wahab, R., ... Pant, A. B. (2015). Molybdenum nanoparticles-induced cytotoxicity, oxidative stress, G2/M arrest, and DNA damage in mouse skin fibroblast cells (L929). *Colloids and Surfaces B —Biointerfaces*, *125*, 73–81.

Silva, V. V., Lameiras, F. S., & Lobato, Z. I. P. (2002). Biological reactivity of zirconia-hydroxyapatite composites. *Journal of Biomedical Materials Research*, *63*, 583–590.

Singh, A. V., Lenardi, C., Gailite, L., Gianfelice, A., & Milani, P. (2009). A simple lift-off-based patterning method for micro- and nanostructuring of functional substrates for cell culture. *Journal of Micromechanics and Microengineering*, *19*.

Singh, A. V., Gailite, L., Vyas, V., Lenardi, C., Forti, S., Matteoli, M., & Milani, P. (2011). Rapid prototyping of nano- and micro-patterned substrates for the control of cell neuritogenesis by topographic and chemical cues. *Materials Science & Engineering C—Materials for Biological Applications*, *31*, 892–899.

Sista, S., Nouri, A., Li, Y. C., Wen, C. E., Hodgson, P. D., & Pande, G. (2013). Cell biological responses of osteoblasts on anodized nanotubular surface of a titanium-zirconium alloy. *Journal of Biomedical Materials Research. Part A*, *101*, 3416–3430.

Sjostrom, T., McNamara, L. E., Yang, L., Dalby, M. J., & Su, B. (2012). Novel anodization technique using a block copolymer template for nanopatterning of titanium implant surfaces. *ACS Applied Materials & Interfaces*, *4*, 6354–6361.

Smith, B. S., Capellato, P., Kelley, S., Gonzalez-Juarrero, M., & Popat, K. C. (2013). Reduced in vitro immune response on titania nanotube arrays compared to titanium surface. *Biomaterials Science*, *1*, 322–332.

Smith, B. S., Yoriya, S., Grissom, L., Grimes, C. A., & Popat, K. C. (2010). Hemocompatibility of titania nanotube arrays. *Journal of Biomedical Materials Research. Part A*, *95A*, 350–360.

Sonmez, E., Turkez, H., Aydin, E., Ozgeris, F. B., Oztetik, E., Kerli, S., ... Di, A. (2015). Stefano, Hepatic effects of yttrium oxide nanoflowers: In vitro risk evaluation. *Toxicological and Environmental Chemistry*, *97*, 599–608.

Stobie, N., Duffy, B., Hinder, S. J., McHale, P., & McCormack, D. E. (2009). Silver doped perfluoropolyether-urethane coatings: Antibacterial activity and surface analysis. *Colloids and Surfaces B-Biointerfaces*, *72*, 62–67.
Su, W. T., Liao, Y. F., Wu, T. W., Wang, B. J., & Shih, Y. Y. (2013). Microgrooved patterns enhanced PC12 cell growth, orientation, neurite elongation, and neuritogenesis. *Journal of Biomedical Materials Research. Part A*, *101*, 185–194.
Subramani, K., & Birch, M. A. (2006). Fabrication of poly(ethylene glycol) hydrogel micropatterns with osteoinductive growth factors and evaluation of the effects on osteoblast activity and function. *Biomedical Materials*, *1*, 144–154.
Sundelacruz, S., & Kaplan, D. L. (2009). Stem cell- and scaffold-based tissue engineering approaches to osteochondral regenerative medicine. *Seminars in Cell & Developmental Biology*, *20*, 646–655.
Tan, A. W., Pingguan-Murphy, B., Ahmad, R., & Akbar, S. A. (2013). Advances in fabrication of TiO_2 nanofiber/nanowire arrays toward the cellular response in biomedical implantations: A review. *Journal of Materials Science*, *48*, 8337–8353.
Tanaka, M., Mukai, M., Fujimori, Y., Kondoh, M., Tasaka, Y., Baba, H., & Usami, S. (1996). Transition metal oxide films prepared by pulsed laser deposition for atomic beam detection. *Thin Solid Films*, *281*, 453–456.
Tavangar, A., Tan, B., & Venkatakrishnan, K. (2011). Synthesis of bio-functionalized three-dimensional titania nanofibrous structures using femtosecond laser ablation. *Acta Biomaterialia*, *7*, 2726–2732.
Thomas, C. H., Lhoest, J. B., Castner, D. G., McFarland, C. D., & Healy, K. E. (1999). Surfaces designed to control the projected area and shape of individual cells. *Journal of Biomechanical Engineering-Transactions of the ASME*, *121*, 40–48.
Tiainen, L., Abreu, P., Buciumeanu, M., Silva, F., Gasik, M., Guerrero, R. S., & Carvalho, O. (2019). Novel laser surface texturing for improved primary stability of titanium implants. *Journal of the Mechanical Behavior of Biomedical Materials*, *98*, 26–39.
Tseng, P., Kunze, A., Kittur, H., & Di Carlo, D. (2014). Research highlights: Microtechnologies for engineering the cellular environment. *Lab on a Chip*, *14*, 1226–1229.
Variola, F., Yi, J. H., Richert, L., Wuest, J. D., Rosei, F., & Nanci, A. (2008). Tailoring the surface properties of Ti6Al4V by controlled chemical oxidation. *Biomaterials*, *29*, 1285–1298.
Vasilev, K., Poh, Z., Kant, K., Chan, J., Michelmore, A., & Losic, D. (2010). Tailoring the surface functionalities of titania nanotube arrays. *Biomaterials*, *31*, 532–540.
Vega, V., Prida, V. M., Hernandez-Velez, M., Manova, E., Aranda, P., Ruiz-Hitzky, E., & Vazquez, M. (2007). Influence of anodic conditions on self-ordered growth of highly aligned titanium oxide nanopores. *Nanoscale Research Letters*, *2*, 355–363.
Ventre, M., Valle, F., Bianchi, M., Biscarini, F., & Netti, P. A. (2012). Cell fluidics: Producing cellular streams on micropatterned synthetic surfaces. *Langmuir: The ACS Journal of Surfaces and Colloids*, *28*, 714–721.
Vijayakumar, C. C., Venkatakrishnan, K., & Tan, B. (2015). Harmonizing HeLa cell cytoskeleton behavior by multi-Ti oxide phased nanostructure synthesized through ultrashort pulsed laser. *Scientific Reports*, *5*.
Wang, N., Li, H. Y., Wang, J. S., Chen, S., Ma, Y. P., & Zhang, Z. T. (2012). Study on the anticorrosion, biocompatibility, and osteoinductivity of tantalum decorated with tantalum oxide nanotube array films. *ACS Applied Materials & Interfaces*, *4*, 4516–4523.
Wang, T. T., Weng, Z. Y., Liu, X. M., Yeung, K. W. K., Pan, H. B., & Wu, S. L. (2017). Controlled release and biocompatibility of polymer/titania nanotube array system on titanium implants. *Bioactive Materials*, *2*, 44–50.

Wang, X. K., Gittens, R. A., Song, R., Tannenbaum, R., Olivares-Navarrete, R., Schwartz, Z., ... Boyan, B. D. (2012). Effects of structural properties of electrospun TiO2 nanofiber meshes on their osteogenic potential. *Acta Biomaterialia*, *8*, 878–885.

Wang, Y. D., Zang, K. Y., Chua, S. J., & Fonstad, C. G. (2006). Catalyst-free growth of uniform ZnO nanowire arrays on prepatterned substrate. *Applied Physics Letters*, *89*.

Wei, Y., Zhang, X. H., Song, Y., Han, B., Hu, X. Y., Wang, X. Z., ... Deng, X. L. (2011). Magnetic biodegradable Fe_3O_4/CS/PVA nanofibrous membranes for bone regeneration. *Biomedical Materials*, *6*.

Willinger, M. G., Neri, G., Bonavita, A., Micali, G., Rauwel, E., Herntrich, T., & Pinna, N. (2009). The controlled deposition of metal oxides onto carbon nanotubes by atomic layer deposition: Examples and a case study on the application of V_2O_4 coated nanotubes in gas sensing. *Physical Chemistry Chemical Physics*, *11*, 3615–3622.

Wilson, C. J., Clegg, R. E., Leavesley, D. I., & Pearcy, M. J. (2005). Mediation of biomaterial-cell interactions by adsorbed proteins: A review. *Tissue Engineering*, *11*, 1–18.

Wong, A. P., Perez-Castillejos, R., Love, J. C., & Whitesides, G. M. (2008). Partitioning microfluidic channels with hydrogel to construct tunable 3-D cellular microenvironments. *Biomaterials*, *29*, 1853–1861.

Wong, J. Y., Leach, J. B., & Brown, X. Q. (2004). Balance of chemistry, topography, and mechanics at the cell-biomaterial interface: Issues and challenges for assessing the role of substrate mechanics on cell response. *Surface Science*, *570*, 119–133.

Wong, M. H., Cheng, F. T., & Man, H. C. (2007). Characteristics, apatite-forming ability and corrosion resistance of NiTi surface modified by AC anodization. *Applied Surface Science*, *253*, 7527–7534.

Xiang, J. H., Masuda, Y., & Koumoto, K. (2004). Fabrication of super-site-selective TiO2 micropattern on a flexible polymer substrate using a barrier-effect self-assembly process. *Advanced Materials*, *16*, 1461.

Xiang, Y. C., Martinez-Martinez, R. M., Torres-Costa, V., Agullo-Rueda, F., Garcia-Ruiz, J. P., & Manso-Silvan, M. (2016). Direct laser writing of nanorough cell microbarriers on anatase/Si and graphite/Si. *Materials Science & Engineering C—Materials for Biological Applications*, *66*, 8–15.

Yamaguchi, S., Takadama, H., Matsushita, T., Nakamura, T., & Kokubo, T. (2011). Preparation of bioactive Ti-15Zr-4Nb-4Ta alloy from HCl and heat treatments after an NaOH treatment. *Journal of Biomedical Materials Research. Part A*, *97A*, 135–144.

Yamamoto, A., Mishima, S., Maruyama, N., & Sumita, M. (2000). Quantitative evaluation of cell attachment to glass, polystyrene, and fibronectin- or collagen-coated polystyrene by measurement of cell adhesive shear force and cell detachment energy. *Journal of Biomedical Materials Research*, *50*, 114–124.

Yamamoto, H., Demura, T., Morita, M., Kono, S., Sekine, K., Shinada, T., ... Tanii, T. (2014). In situ modification of cell-culture scaffolds by photocatalytic decomposition of organosilane monolayers. *Biofabrication*, *6*.

Yang, J., Deng, S. Y., Lei, J. P., Ju, H. X., & Gunasekaran, S. (2011). Electrochemical synthesis of reduced graphene sheet-AuPd alloy nanoparticle composites for enzymatic biosensing. *Biosensors & Bioelectronics*, *29*, 159–166.

Yang, J. C., Zhao, C., Hsieh, I. F., Subramanian, S., Liu, L. Y., Cheng, G., ... Zheng, J. (2012). Strong resistance of poly (ethylene glycol) based L-tyrosine polyurethanes to protein adsorption and cell adhesion. *Polymer International*, *61*, 616–621.

Yang, Z. L., Zhong, S., Yang, Y., Maitz, M. F., Li, X. Y., Tu, Q. F., ... Huang, N. (2014). Polydopamine-mediated long-term elution of the direct thrombin inhibitor bivalirudin from TiO_2 nanotubes for improved vascular biocompatibility. *Journal of Materials Chemistry B*, *2*, 6767–6778.

Yao, C., & Webster, T. J. (2009). Prolonged antibiotic delivery from anodized nanotubular titanium using a co-precipitation drug loading method. *Journal of Biomedical Materials Research Part B—Applied Biomaterials*, *91B*, 587–595.

Yao, S. L., Feng, X. J., Lu, J. J., Zheng, Y. D., Wang, X. M., Volinsky, A. A., & Wang, L. N. (2018). Antibacterial activity and inflammation inhibition of ZnO nanoparticles embedded TiO_2 nanotubes. *Nanotechnology*, *29*, 10.

Ynsa, M. D., Dang, Z. Y., Manso-Silvan, M., Song, J., Azimi, S., Wu, J. F., … Garcia-Ruiz, J. P. (2014). Reprogramming hMSCs morphology with silicon/porous silicon geometric micro-patterns. *Biomedical Microdevices*, *16*, 229–236.

Yoo, S., Dregia, S. A., & Akbar, S. A. (2006). Kinetic mechanism of TiO_2 nanocarving via reaction with hydrogen gas. *Journal of Materials Research*, *21*, 1822–1829.

Yu, W. Q., Jiang, X. Q., Zhang, F. Q., & Xu, L. (2010). The effect of anatase TiO_2 nanotube layers on MC3T3-E1 preosteoblast adhesion, proliferation, and differentiation. *Journal of Biomedical Materials Research. Part A*, *94A*, 1012–1022.

Zhang, M., Huang, X. B., Hang, R. Q., Zhang, X. Y., & Tang, B. (2018). Effect of a biomimetic titania mesoporous coating doped with Sr on the osteogenic activity. *Materials Science & Engineering C—Materials for Biological Applications*, *91*, 153–162.

Zhang, Y. X., Dong, C. F., Yang, S. F., Wu, J. S., Xiao, K., Huang, Y. H., & Li, X. G. (2017). Alkalescent nanotube films on a titanium-based implant: A novel approach to enhance biocompatibility. *Materials Science & Engineering C—Materials for Biological Applications*, *72*, 464–471.

Zhao, L. Z., Hu, L. S., Huo, K. F., Zhang, Y. M., Wu, Z. F., & Chu, P. K. (2010). Mechanism of cell repellence on quasi-aligned nanowire arrays on Ti alloy. *Biomaterials*, *31*, 8341–8349.

Zhao, L. Z., Mei, S. L., Wang, W., Chu, P. K., Zhang, Y. M., & Wu, Z. F. (2011). Suppressed primary osteoblast functions on nanoporous titania surface. *Journal of Biomedical Materials Research. Part A*, *96A*, 100–107.

Zhao, X. B., Wang, G. C., Zheng, H., Lu, Z. F., Zhong, X., Cheng, X. B., & Zreiqat, H. (2013). Delicate refinement of surface nanotopography by adjusting TiO_2 coating chemical composition for enhanced interfacial biocompatibility. *ACS Applied Materials & Interfaces*, *5*, 8203–8209.

Zhou, P., Long, S. Q., Mao, F. F., Huang, H. X., Li, H. J., He, F., … Wei, S. C. (2020). Controlling cell viability and bacterial attachment through fabricating extracellular matrix-like micro/nanostructured surface on titanium implant. *Biomedical Materials*, *15*, 12.

Zverev, V. I., Pyatakov, A. P., Shtil, A. A., & Tishin, A. M. (2018). Novel applications of magnetic materials and technologies for medicine. *Journal of Magnetism and Magnetic Materials*, *459*, 182–186.

Biomedical application of ZnO nanoscale materials

16

Anshul Yadav[1], *Kunal Mondal*[2] *and Ankur Gupta*[3]
[1]Membrane Science and Separation Technology, CSIR-Central Salt & Marine Chemicals Research Institute, Bhavnagar, India, [2]Idaho National University, Pocatello, ID, United States, [3]Department of Mechanical Engineering, Indian Institute of Technology, Jodhpur, India

16.1 Introduction

Advances in nanotechnology play a noteworthy role in developing nanomaterials (NMs) with unique functional properties capable of solving diagnostic and therapeutic deficiencies. NMs have created substantial opportunities in the drug industry and have developed many innovations which led to a colossal extension in the biomedical industries (Banerjee, 2017; Vishnukumar, Vivekanandhan, Misra, & Mohanty, 2018). Due to their distinctive performance in electronics, optics and photonics, nanostructured zinc oxide (ZnO) materials have received extensive attention (Gupta et al., 2020). The benefits of NMs are primarily because of their nanoscale and significant regions, where the target ligands, therapeutic monomers that can be functionalized (Chandra, Barick, & Bahadur, 2011; Ramos, Cruz, Tovani, & Ciancaglini, 2017; Saji, Choe, & Yeung, 2010). Due to its excellent biocompatibility, economical and low toxicity, ZnO NMs have been in the last two decades, one of the most common metal oxide NMs in biological applications. ZnO NMs are extensively utilized in numerous industrial areas such as UV light-emitting devices (Jood et al., 2011; Rajalakshmi et al., 2012), gas sensors (Gupta, Nayak, Singh, & Bhattacharya, 2014; Gupta, Parida, & Pal, 2019; Yakimova, 2012), photocatalysts (Bhatia & Verma, 2017; Khalafi, Buazar, & Ghanemi, 2019; Ong, Ng, & Mohammad, 2018), pharmaceutical (Sridar, Ramanane, & Rajasimman, 2018), energy storage (Wang, Liu, Song, & Wang, 2007; Yang et al., 2011), nanooptical devices (Zhou, Xu, & Wang, 2006), biomedical systems (Jiang, Pi, & Cai, 2018), and cosmetic industries (Lu, Huang, Chen, Chiueh, & Shih, 2015). Other industries, including concrete manufacturing, the photocatalysis, telecommunications, telecommunications, and other factories are also able to employ ZnO, apart from those listed above. ZnO has exciting properties such as high direct bandwidth of 3.3 eV and high activation energy of 60 meV among all these types of metal oxides attract much attention (Dong, Han, Qian, & Chen, 2012).

ZnO NMs have demonstrated promising biomedical potential, in particular in the anticancer fields and antibacterial fields, involving their potent capacity to cause the growth, discharge, and induction of excessive reactive oxygen species (ROS),

Metal Oxides for Biomedical and Biosensor Applications. DOI: https://doi.org/10.1016/B978-0-12-823033-6.00014-4

cell apoptosis [zinc (Zn) ions] (Gupta, Pandey, et al., 2014). Zn is also well known for preserving insulin's intrinsic integrity. ZnO NMs for antidiabetic care has, therefore, also been adequately established. Also, ZnO NMs show superb light properties, good mechanical strength, biocompatibility, etc. which make them a significant bioimagery nominee (Dong et al., 2012; Gupta, Singh, et al., 2015). It ties together the synthesis and advantages of ZnO NMs in the area of biomedicine that will lead to their potential scientific advancements and their emphasis on biomedical fields.

Work in the medicinal use of NMs in recent years has seen a spike due to its remarkable potential to cure diseases. While there is the extensiveness of assessments on the advancements of NMs from manufacturing to use, there are relatively fewer assessments, especially on ZnO have been done in the past.

16.2 ZnO metal oxide

16.2.1 Physical and chemical properties of ZnO

Oxide and hybrid nanostructures depend heavily on their physicochemical properties, such as permeability, stability, morphology (size, shape, features, and surfaces), and biocompatibility in the therapeutic applications (Gupta, Patel, Kant, & Bhattacharya, 2016). The forms, compositions, and directions of the compounds forming the oxide and hybrid nanostructures define these physicochemical properties. ZnO is known as Group II–VI semiconductor because Zn and O are numbered in Groups II and VI, respectively in the periodic table. ZnO has distinct mechanical, chemical, semiconducting, electrical, and piezoelectrical properties.

16.2.2 Crystal structures of ZnO

ZnO crystallizes in two main types, that is, hexagonal wurtzite and cubic unit. At ambient conditions, the wurtzite structure is most common and stable. ZnO crystallized in the wurtzite (B4 type) structure at atmospheric pressure and temperature (Özgür et al., 2005). The ZnO of wurtzite has a lattice parameter configuration (space group C6mc) $a = 0.3296$ nm and $c = 0.520$ nm. Several alternating planes composed of tetrahedrally coordinated O^2 and Zn^{2+} ions, alternatively stacked along a c-axis, can simply describe the structure of ZnO (as depicted in Fig. 16.1). The tetrahedral alignment in ZnO contributes to noncentralized symmetry and thus piezoelectricity and pyroelectricity. The polar surfaces are another function of ZnO. The basal plane is the most popular polar surface (Kong & Wang, 2004). Different loading ions produce Zn– (0001) positive and O– $(000\bar{1})$ negative surfaces that cause a normal dipole moment and spontaneous c-axis polarization as well as surface energy discrepancies. The polar surfaces have general faces or display large surface reconstructions in order to preserve a consistent shape, but ZnO ± (0001) were exceptions: they are atomically smooth, solid, and unconstructed. The two other facets most commonly observed in Zno polar are the nonpolar surfaces $<2\bar{1}1\bar{0}>$ and $<011\bar{0}>$, have less energies than $<0001>$ facets (Kong & Wang,

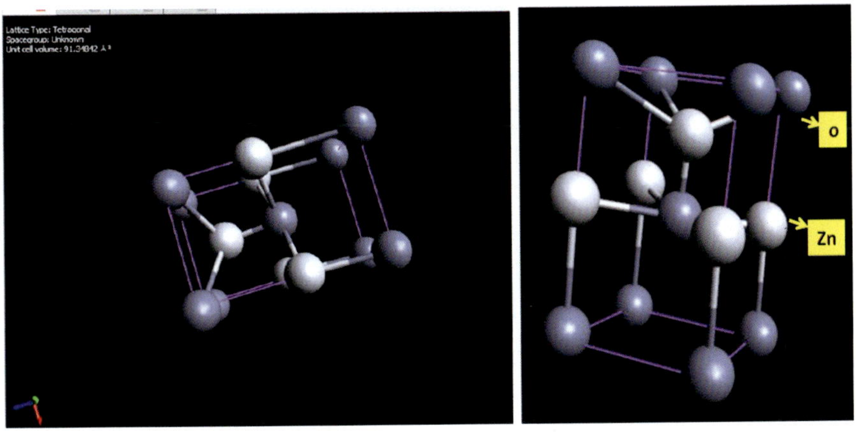

Figure 16.1 The wurtzite structure model of ZnO. The tetrahedral coordination of Zn−O is shown.

2004). The facet of higher surface energy is typically small under thermodynamic equilibrium conditions, whereas the lower energy faces are wide.

16.3 Nanostructures and the growth processes

ZnO is a common substance with two distinct properties: semiconducting and piezoelectric. The growth behavior of ZnO nanostructures can be changed by manipulating growth kinetics. Various ZnO nanostructures have been obtained here by monitoring synthesized parameters such as deposition and pressure and the flux of carrier gas. Nanocombs, nanorods (NRs), nanohelixes, nanobrushes, nanobelts, nanosprings, and nanocages ZnO have been synthesized under different growth conditions using a solid-vapor phase thermal sublimation process (Gupta, Saurav, & Bhattacharya, 2015). These unique nanostructures clearly demonstrate that ZnO is probably the best among all the materials in terms of its structures and properties. New applications in optoelectronics, cameras, transducers, and biomedical science may be found in the nanostructures. Zn with spherical or faceted high-surface metal particles are NMs, nanodots, or nanopowder (Fig. 16.2).

A range of ZnO nanostructures has been successfully synthesized with different morphologies for growth, such as NRs, nanoporous films, nanowires, and nanobowls by the author. Glimpses of the structures are shown in Fig. 16.3. One-dimensional (1D) ZnO NRs, as contrasted with other ZnO nanostructures, have a high surface−volume ratio, crystallinity and load power. Nanobelts are geometrically formed nanowires with well-defined side surfaces. ZnO nanobelts are typically produced without introducing a catalyst by sublimation of ZnO powder (Pan, 2001). The nanobelt cross-section is triangular, with standard ranges of thickness and width from ~5 to 10 (Wang, 2004a). ZnO nanowires 1D nanostructured/

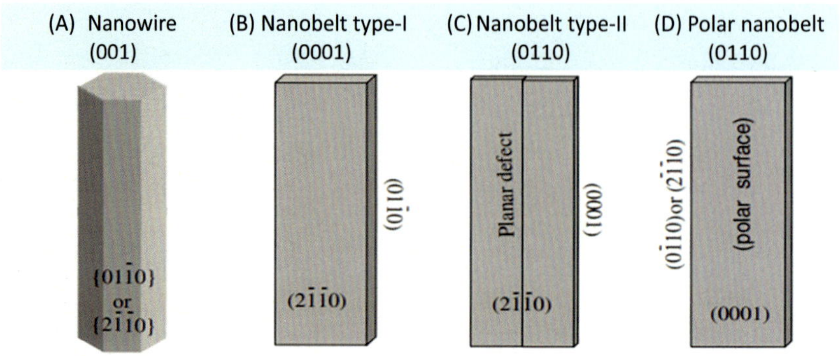

Figure 16.2 Typical growth morphologies of one-dimensional ZnO nanostructures and the corresponding facets.

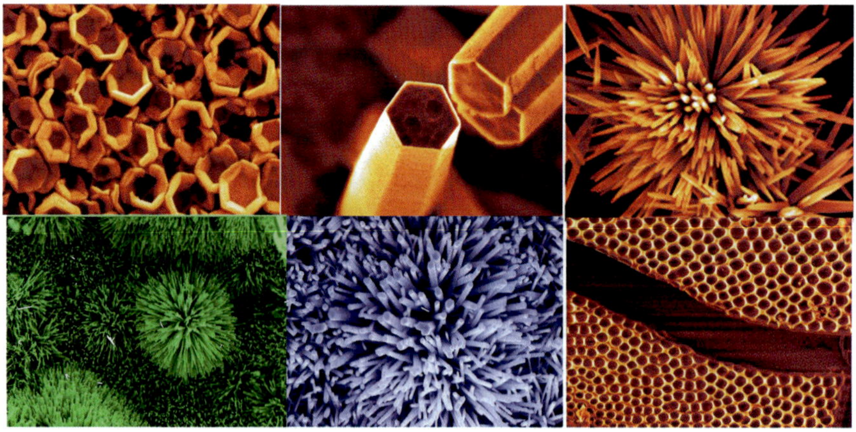

Figure 16.3 A collection of nanostructures of ZnO fabricated by Ankur Gupta (the Author of this article).

microstructured have been successfully synthesized and used for effective antimicrobial activities. Nanowires have a fairly large surface area in order to generate oxygenated reactive materials (Gupta, Srivastava, et al., 2014). The proper and regulated homogeneous nucleation is achieved in a high aspect ratio for ZnO nanowires (Chen et al., 2014; Kumar, Umar, Kumar, & Nalwa, 2017).

16.4 Synthesis techniques for ZnO nanomaterials

The design of effective and environmentally sustainable NMs engineering methods is a core factor for the nanotechnology discipline. In particular, the role of ZnO highly fascinates researchers in the variety of properties such as UV filter and photochemical, antifungal, high-catalyst, and antimicrobial activity (Kalpana & Devi Rajeswari, 2018). Upadhyay, Jain, Sharma, and Sharma (2020) discussed the synthesis and applications of ZnO NMs in biomedicine in great detail. Chauhan, Kant, Rai, Gupta, and Bhattacharya (2019) reported a synthesis method of ZnO nanoflower and graphene oxide (GO) composite grown uniformly over silicon substrate for decolorization of MB dye and industrial wastewater. The two methods for the biosynthesis of NMs are bottom-up and top-down (Firdhouse & Lalitha, 2015; Zhang et al., 2018). The most important reaction in case of a bottom-up solution is oxidation/reduction. Currently, the synthesis of NMs is an important field for research, aiming at an ecologically friendly green synthesis (Kalpana & Devi Rajeswari, 2018). The key materials in preparing the NMs are (i) the solvent medium for synthesis, (ii) the environmentally friendly reducing agent, and (iii) the nontoxic substrate for NMs stabilization. Because of their extraordinary properties (optical, mechanical, chemical, etc.), the synthesis of inorganic metal oxides NMs using biological elements is extremely significant.

16.4.1 Physical methods

ZnO NMs synthesis using physical methods include plasma blast, thermal evaporation, absorption of the liquid vapor, laser ablation, and ultrasonic radiation. These methods are chemically pure and technically basic, making them suitable for industrial processes at high rates of production (Gupta, Gangopadhyay, Gangopadhyay, & Bhattacharya, 2016; Kołodziejczak-Radzimska & Jesionowski, 2014; Naveed Ul Haq et al., 2017). Peng and coworkers (Peng, Fangli, Liuyang, Jinlin, & Yunfa, 2007) prepared high-performance ZnO NRs with a using a thermal plasma radio frequency (RF) device in one step, continuous and versatile manner. By heating the different Zn compounds and by changing the feeding rate of starting materials and the oxygen flow rate in the process, the aspect ratio of the synthesized NRs can be easily controlled (Peng et al., 2007). The findings demonstrate that a wall-free two-directional system controls the growth process of ZnO NRs in the RF plasma reactor. Plasma synthesis is a simple way to synthesize ZnO and other materials' 1D nanostructures. Thin ZnO films on silicon substrates approach thermal, and electrical vapor deposition was prepared by Fouad, Ismail, Zaki, and Mohamed (2006). Azoreactive dye decomposition on ZnO films was examined photocatalytically. The findings show that the efficiency of coloring decomposes decreases with a declining pH. Zhang et al. (2005) synthesized ZnO NRs and arrays on a cylindrical substratum by a simple thermal evaporation approach. In a well-spaced series, the diameter and length of each ZnO rod varied from 60 nm to 1.2 μm, and from 4 μm to 6 μm, respectively. Two key points for regulating rod diameters are heating temperature and deposition location. An ideal model for investigating the physical and chemical properties of 1D materials for their sizing effect is the perfect crystalline ZnO rods with various scales from nanometers to few micrometers. Well-aligned single-crystalline, ZnO nanowire array on an Al_2O_3 substructure was fabricated by Lyu, Zhang, Lee, Ruh, and Lee (2003) using a simple, low-temperature physical vapor deposition process of 450°C. They proved that ZnO nanowires adopted the self-catalyzed nuclei growth process. A curious hierarchically ordered ZnO structure was also observed often in addition to high-quality ZnO nanowires. Yadav, Mishra, and Pandey (2010) documented the tunability of ultrasonic irradiation of ZnO NMs' different sizes. They observed an increase in the amount of formed ZnO NMs as the time of ultrasonic irradiation increases and red change in the absorption edge is also observed. This supports the tunability of bandgap with ultrasound irradiation of histidine-capped ZnO NMs. Thareja and Shukla (2007) documented the colloidal formation of ZnO NMs in various liquid environments with a pulsed laser removal of Zn metal target at room temperature.

16.4.2 Chemical methods

Microemulsion, sol-gel, freezing, hydrothermal, solvothermal, and water vapor deposition are some of the chemical methods used to prepare ZnO NMs. Wet chemical synthesis is the most common process used for ZnO NMs manufacture based on the physical states of solid and liquid phases (Brayner et al., 2010; Jin & Jin, 2019; Król, Pomastowski, Rafińska, Railean-Plugaru, & Buszewski, 2017). Fricke, Voigt, Veit, and Sundmacher (2015) developed a microemulsion approach to generate NMs from crystalline ZnO with closely spaced sizes and well-defined face and form. Platelike NMs were obtained at low concentrations of Zn salt. If the concentration of Zn ions is increased, the average particle size is reduced. Oliveira, Hochepied, Grillon, and Berger (2003) precipitated Zn nitrate with NaOH solution at room temperature, in double-jet conditions. It led to the formation of the ellipse or starlike micrometric Zn particles isolated by pH variation. The existence of these additives has provided important signs of particulate production, indicating that nanocrystal-based aggregate is the product of submicronic particles. Demir, Muñoz-Espí, Lieberwirth, and Wegner (2006) proposed a wet-chemical approach to the manufacturing of ZnO monodispersed nanocrystals (NCs) (diameter range 20–80 nm). In the presence of 1-pentanol in m-xylene at 130°C, the synthesis begins from zinc acetate dihyde which is converted to ZnO. The p-toluene sulfonic acid monohydrate catalysis of this reaction (P-TSA), allows shorter reaction time and improves the reproductiveness of both the particle sizes distribution and the particle crystallinity. Aneesh, Vanaja, and Jayaraj (2007) synthesized continuous and hydrothermal OH-free ZnO NMs, by increasing the growth temperature and precursor concentration. In combination with a solid-state electrolyte centered on an aquatic dispersion of cellulose derivatives of stable vinyl acetate, acrylic acid ester in styrene, and lithium perchlorate, Santos et al. (2015) recorded electrolytic-gated transistors utilizing gallium dioxide NMs generated with solvothermal synthesis as channel substrate. Noothongkaew, Pukird, Sukkabot, and An (2014) synthesized ZnO NMs using chemical vapor deposition on silicon substrates. For gas sensors and solar cells, this synthesis may prove to be important.

16.4.3 Biological/green methods

Due to the simplest, most efficient, reproduced, and environmentally responsible options, biological synthesis otherwise called a "green synthesis" is desirable.

ZnO NMs can be synthesized with specific enzymes and biochemical pathways in suitable microorganisms (Jin & Jin, 2019). The synthesis of ZnO NMs with endophytic bacterial isolate culture *Sphingobacterium thalpophilum* and its antibacterial efficiency to bacterium pathogens were described by Rajabairavi et al. (2017). Two bacterial pathogens *Pseudomonas aeruginosa* and *Enterobacter aerogenes* were tested with disc diffusion methods for antibacterial efficacy of ZnO NMs. In both tested strains, this ZnO NMs showed improved antifungal activity. Gharagozlou, Baradaran, and Bayati (2015) have reported a new method to synthesize ZnO NMs from nonpoisonous and biocompatible chemicals where no toxins or fuel-free side-products are made. A binary complex Zn(II) was obtained in this method from alanine, where water is used as a solvent, and an amino acid that is biologically compatible instead of toxic amines is used as a source of nitrogen. Ambika and Sundrarajan (2015) synthesized ZnO NMs via greener method using Vitex negundo L. extract. The ZnO-coated protein in the egg albumin medium was prepared by Bhunia, Kamilya, and Saha (2016) by the green synthesis process. Such NMs showed potential anti-gram-negative activity in bacteria. ZnO NMs were synthesized by Ogunyemi et al. (2019), using chamomile floral plant extract (Matricaria chamomilla L.), olive-leaf (Olea europaea) and red tomato fruit. The ZnO NMs' antibacterial effect on *Xanthomonas oryzae* pv synthesized by Olea europaea was shown. Overall, the ZnO NMs promise biocontrol agents to combat rice diseases in bacterial leaf burnt. The single-step synthesis of polymer/ Zn NMs of nanocomposite thin films was recorded by Al-Jumaili et al. (2019) through the joint deposition of renewable geranium-specific polymers extracted from essential oil and ZnO NMs generated by Zn acetylacetonate thermal decomposition. Such ecofriendly nanocomposite films can be used to shield the surfaces of medical equipment from microbial attachment and invasion as encapsulation coatings. Some of the synthesis techniques, advantages and disadvantages are mentioned in the Table 16.1.

Table 16.1 ZnO nanomaterials (NMs) synthesis techniques, their advantages and disadvantages.

Synthesis techniques	Advantages	Disadvantages
Physical methods	Simple, cheap, catalyst-free, large-scale production	Parameter control
Chemical methods	Cheap, easy-to-handle chemicals, uncomplicated equipment, low-energy iNMut, easy parameter tailoring, large-scale production	Surfactant use, high cost of precursors
Biological/ green methods	Safe, ecofriendly, cheap organic solvents	NMs stability, unclear mechanism

16.5 Biomedical applications

ZnO NMs have attracted considerable interest in various biomedical fields such as anticancer, antibacterial, antibiotic, antidiabetic, antiinflammatory, and drug delivery and bioimaging applications, as a modern form for low-cost and low-toxic substances (Barui, Kotcherlakota, & Patra, 2018; Ng, Baeg, Yu, Ong, & Bay, 2017; Shivaramakrishnan, Gurumurthy, & Balasubramanian, 2017; Vishnukumar et al., 2018). We have summarized here the recent advances in the biomedicine use of ZnO NMs. ZnO NMs under 100 nm are known as fairly biocompatible that help their biomedical applications and constitute a good proprietary position in the promotion of biomedical research. Bhatt and Bhattacharya (2019) proposed a platform with noval dipstick biosensor, hybrid nanocomposite of ZnO, and gold (Au) particle for biomedical sensing applications. Flexible sensors and flexible sensing platforms in the field of biosensing application is emerging nowadays (Gupta & Pal, 2018; Jena, Gupta, Pippara, Pal, & Adit, 2019). The advantages of ZnO nanostructures include high surface area to volume ratio, nontoxicity, chemical stability, electrochemical activity, and highly conducting (Gupta, Patel, et al., 2016; Shivaramakrishnan et al., 2017). Fig. 16.4 represents the major applications in biomedicine field.

16.5.1 Imaging agent

ZnO is also known to be one of the well-known semiconductors that has good potential to replace traditional, biologically, and optically oriented species of Cd. ZnO NMs display efficient blue emissions and almost UV emissions that are green or yellow with oxygen vacancies, thus expanding their application into a field of bioimaging (Jiang et al., 2018; Xiong, 2013). ZnO@polymer core-shell NMs with tunable photoluminescence were made by Xiong, Xu, Ren, and Xia (2008) on a single sol-gel path. These ZnO@polymer core-shell NMs are nontoxic to human cells at concentrations of less than 0.2 mg/mL. The luminescent cells have sufficient life under UV light for microscopic imaging after using these ZnO@polymer NMs. Hence ZnO@polymer core-shell NMs are suitable for bioimaging applications. Under moderate conditions, that is, at a favorable and low pH, Jiang, Wang, and Wang (2011) prepared high-quality ZnO nanosheets. The study led to an efficient synthesis of fluorescent ZnO with a strong emission of approximately 560 nm attributed to ZnO nanostructure defect emissions. Tang, Choo, Li, Ding, and Xue (2010) synthesized ZnO NMs, including blue, green, yellow, and oranges, using a method for the precipitation of ethanol in various emission colors. By changing the pH value of the precipitation solution, the emission color of ZnO NMs could be adjusted. There was low cytotoxicity in the resulting ZnO@silica core-shell NMs, and they were successful in an application for cell labeling. Zhang and Liu (2010) fabricated low cytotoxic ZnO quantum dots (QDs) made with poly(2-(dimethylamino)ethyl methacrylate) (PDMAEMA), which was synthesized in situ by radical polymerization, with a dual purpose to supply DNA plasmid and labeling cells. The

Figure 16.4 Major applications of ZnO in biomedicine.

ZnO QDs, modified by polycation, were able to condense plasmid DNA to nanocomplexes filled with ZnO QDs, which emit heavy yellow light under UV light. The monodispersed ZnO silica-coated NMs onto the surfaces of hydrophilic amino groupings were synthesized and modified by Zhang, Xiong, Ren, Xia, and Kong (2012) by adjusting the reactant ratio and the reaction time. Both of these ZnO@silica NMs were verified by cytotoxicity data and cell phase kinetics studies under visible light, whereas yellow ZnO@silica was fatal to Hela cells under UV light at a distance of 365 nm because of the fact that a significant amount of ROS was released. Cho et al. (2011) demonstrated that FeO–ZnO NMs core injected in carcinoembryonic antigen into dendritic cells function as an imaging agent. The dendritic cells rapidly accumulate the NMs–antigen complex within an hour and can be observed in vitro by confocal microscopy and in vivo by magnetic resonance imaging. ZnO–NMs QDs were synthesized by Matsuyama et al. (2013) as a fluorescent agent for biopower use using the sol-gel technique. To enhance the water stability of ZnO NMs, they were treated with silica. They demonstrated the use of silicone-coated ZnO QDs with biotin in cell-labeling applications and the binding to the nerve cell and actin filaments of the silicon-coated ZnO–NMs QDs with biotin. In the similar way Gupta's research group developed an ultra-dense high aspect ZnO forest like nanostructure for detecting organic dyes from polluted water (Gupta, Mondal, Sharma, & Bhattacharya, 2015). Hong et al. (2011) showed that ZnO NWs for the molecular imaging of cancer cells could be implemented by using intrinsically fluorescent ZnO to return water solubilities, biocompatibilities, and low cell toxicity. Urban et al. (2012), Urban, Neogi, Butler, Fujita, and Neogi (2011) synthesized ZnO NMs for bioimaging extremely nonlinear visually adaptive. For nonlinearity in vitro imaging of living cells, ZnO NMs with enhanced second- and third-order nonlinearity were used. The second harmonic signal was used in NMs with extremely low iNMut light intensities to map and pictorially image live cells and ZnO-infiltrated thrombocytes in highly efficient nonlinear processes in NMs with noncentrosymmetric ZnO. Kachynski, Kuzmin, Nyk, Roy, and Prasad (2008) showed that ZnO NCs are biocompatible with a noncentrosymmetric structure can be used as nonresonant nonlinear optical samples in vitro for bioimaging applications. ZnO NCs were synthesized by a nonhydrolytic sol-gel process in organic media. Then, the ZnO NCs were dispersed into aqueously using phospholipid micelles. These nonresonant processes offer benefits beyond traditional two-photon bioimaging processes. The preparation, photophysical, and structural features of CdSe (S) alloys, which are protected by the ZnO layer, were reported by Aldeek et al. (2011). Water-dispersible CdSe(S) QDs had a shell ZnO obtained by $Zn(OAc)_2$ fundamental hydrolysis. Their results show that CDSe(S)/ZnO QDs are photostable higher than CdSe(S) cores. CdSe(S)/ZnO QDs hold potential uses for biological as fluorescent labeling and have been used in imaging the bacterial biofilms of *Shewanella oneidensis*.

16.5.2 Biosensor

Metal oxide-modified biosensors based on the ZnO nanostructure have been emphasized in order to improve highly sensitive biosensors (Hahn, Ahmad, & Tripathy, 2012; Tripathy & Kim, 2018). Ahmad and coworkers have worked on this aspect in various articles (Ahmad, Ahn, & Hahn, 2017; Ahmad, Tripathy, Ahn, & Hahn, 2017; Ahmad, Tripathy, & Hahn, 2013, 2014; Ahmad, Tripathy, Ahn, Bhat, et al., 2017; Ahmad, Tripathy, Jang, Khang, & Hahn, 2015; Ahmad, Tripathy, Kim, et al., 2014; Ahmad, Tripathy, Park, & Hahn, 2015). They made a high-performance cholesterol sensor with ZnO NRs grown directly on Ag/glass electrodes based on a solution-gated field-effect transistor. An all-in-one biosensor with ZnO FET array for selective glucose, cholesterol, and urea process synthesis of high-level ZnO NRs on electrode surfaces to detect reactive uric acid were also prepared generated with vertically directed ZnO NRs modified by Fe_2O_3 for nonenzymatic glucose field-effect transistors (Gupta & Bhattacharya, 2018). With electrospinning and subsequent thermal treatment, Wu and Yin (2013) have developed the best sensing properties in a nonenzymatic glucose sensor based on CuO–ZnO composite nanofibers. CuO–ZnO/Pt was reported to have electromechanical activity, high sensitivity, and fast amperometric sensing for glucose oxidation. The sensor was also selective, long lasting, and highly reproductive. Zhou et al. (2014) fabricated 3D transparent ZnO–CuO, nonenzymatic nanocomposites (HNCs) with different spin thicknesses relative to combined, flat CuO, ZnO/CuO nanowires. They demonstrated excellent long-term reliability, reproductivity, and excellent selectivity for these sensors. The development of special hierarchical heterojunction and the well-designed 3D structure were the key reasons for improved nonenzymatic biologic performance. The composite arrays of CuO nanoflowers/ZnO NRs have been demonstrated in a low temperature and simple single-step synthesis using an alkaline steam oxidation cycle (Soejima, Takada, & Ito, 2013). The nanoarrays CuO/ZnO are strongly electrocatalytic of glucose oxidation. SoYoon, Ramadoss, Saravanakumar, and Kim (2014) developed a new hierarchical architecture CuO nanoleaf/ZnO NRs for nonenzymatic sensor applications. There was a greater electrocatalytic activity for glucose oxidation in the CuO nanoleaf/ZnO NRs structure. The highly improved catalytic electrochemical reactivity is primarily due to the lower potential and broader electroactive surface area and the 1D structures of the ZnO NRs. ZnO–CuO NRs and nanospheres as a sensor were prepared by Karuppiah et al. (2015) using a simple hydrothermal process. The polythene, a stabilizer and a lowering agent of the growth of nanocomposites, was used. The sensor achieved substantial repetitiveness, reproducibility, and long-term reliability (up to 38 days) for detecting glucose. Cai et al. (2014) synthesized ZnO–CuO porous spheres with core-shell structure through a simple top-down controllable path. With its good electron transfer property and heterojunction between them, CuO can catalyze glucose in alkaline solutions, nonenzymatic glucose sensors. ZnO–CuO porous composite spheres display good performance in a wide linear range, high sensitivity, and low detection limit, which indicate a good detection behavior of glucose. Marie, Manoharan, Kuchuk, Ang, and Manasreh (2018) have developed a glucose enzyme-free sensor based on ferric oxide-functional vertical ZnO NRs (Fe_2O_3).

Similarly, Chauhan et al. fabricated a vertically grown rod nanostructure of ZnO on silicon substrate for industrial dye degradation (Chauhan, Rai, Gupta, & Bhattacharya, 2017). A high sensitivity, low detection level, quick and rapid response time, and linear reaction to changes in glucose concentrations were seen in the proposed nonenzymatic and modified glucose sensor. In order to detect nonenzymatic glucose with low cost, Strano and Mirabella (2017) developed an effective electrode. In terms of electrical stabilization, and sensitivity in comparison to the $Ni(OH)_2$/Cu, the presence of ZnO NRs was shown to increase the glucose sensor performance. The report showed a simple, scalable, and inexpensive manufacturing approach to successful glucose sensing systems within an urban mining environment. Jung, Ahmad, and Hahn (2018) fabricate nickel oxide (NiO) QDs with modified ZnO NRs. In order to boost the electrocatalytic properties and the surface of ZnO NRs, the ZNO NRs surfaces were filled with NiO QDs with RF magnetron sputtering. An analysis of whole blood and serum samples in humans showed that a nonenzymatic glucose sensor based on f-FET is able to effectively calculate the amount of glucose in the presence of intervening organisms and can thus be provided as a promising tool in more clinical and nonclinical fields.

16.5.3 Drug delivery

Drug delivery has emerged as a powerful tool in the treatment of various diseases among various applications for metal oxides (Martínez-Carmona, Gun'ko, & Vallet-Regí, 2018; Rasmussen, Martinez, Louka, & Wingett, 2010). The new approach to integrating ZnO QDs technology with cancer drug therapy was identified by Yuan, Hein, and Misra (2010). The findings indicate a possible medium for delivery of tumor-targeted drinks and recording the distribution process if necessary for the proposed QD generation loaded with anticancer agents and encapsulated with biocompatible polymers. In a simple, fast soft-chemical approach, Barick, Nigam, and Bahadur (2010) produced highly mesoporous 3D ZnO nanoassemblies. The research indicated that the release of medicines is dictated by the pH of the medium and externally applied ultrasound (continuous or pulsated), as well as by the existence of the materials that encapsulate the drug. In multiple signal conditions, Hanley et al. (2008) studied the reaction of normal human cells to ZnO NMs and compared them to the response of cancer cells. They showed ZnO NMs have a clear affinity for attacking cancer T cells in contrast with normal cells. Theoretical effectiveness of ZnO NMs in treating cancer and/or autoimmunity was shown in cell-specific toxicity results of potential cell diseases. ZnO NRs for the treatment of breast carcinoma cells (T47D) were proposed by Kishwar, Asif, Nur, Willander, and Larsson (2010). They developed ZnO NRs at the tips of borosilicate glass capillaries (0.5 μm diameter) with aqueous chemical growth technique in order to modify the sample for intracellular experiments. The novel effects of cell-localized toxicity indicated that PPDME-conjugated ZnO NRs might be added in a few minutes to breast cancer cell necrosis. In ZnO QDs with the latest ligand-swap free technique, Muhammad et al. (2011) reported ZnO QDs as a platform for a targeted intracellular pH-responsive delivery of an anticancer drug in addition to bioimaging. Antoine et al. (2012) reported that tetrapod micro–nano structures synthesized by flame transport considerably block HSV-2 entry into target cells and furthermore show the probability of preventing virus spread through already infected cells. A hybrid technique for ZnO/PEG nanospheres with anticancer drugs (DOX) in photodynamic therapies (PDTs) was investigated by Hariharan, Senthilkumar, Suganthi, and Rajarajan (2012). Within 26 h of the conjugate incubation in vitro in an acidic atmosphere, almost 91% of loaded medication was released. It indicates that DOX from DOX-ZnO/PEG nanocomposite is released efficiently by medication. DOX on NMs filled with ZnO/PEG showed greater antibacterial activity with gram-positive than gram-negative bacteria in visible light. This showed that ZnO/ PEG NMs for the medication delivery system of PDT could be used as a nanocarrier. Zhang et al. (2013) synthesized NMs with DOX in the center of ZnO@polymer for the treatment of human brain cells. A sampling of the ZnO@polymer-DOX nanocomposites by confocal laser scan microscopy revealed that they were decomposed at lysosomes to unleash DOX molecules that penetrated into the nucleus and eventually destroyed the cells. Chen et al. (2013) synthesized multifunctional nanocarriers formed in the center and amphiphilic hyperbranched layer of ZnO QDs coupled with gold nanomaterials (Au NMs) as a shell for selective anticancer drug delivery. The findings showed the tremendous potential of ZnO nanocarriers not only as a drug carrier that attacked tumors but also as an assistant to cancer care. Tan, Liu, Zhou, Wei, and Peng (2014) prepared a smart hybrid ZnO@ poly (N-isopropylacrylamide) (PNIPAM) for ZnO NMs by atomic transfer radical transmission (ATRP) and grafting thermal responsive PNIPAM. DOX was a model drug for anticancer and was loaded onto the hybrid NMs and an in vitro drug release study showed that ZnO@PNIPAM could be used as a thermal reaction device. Nanocomposites of nanostructured ZnO were synthesized by Upadhyaya et al. (2015) using an ex situ grafting process for the production of the cancer drug curcumin (Cr). The effectiveness of drugs trap was 74%, although a gradual and controlled release in the initial phase, and maintained in later period, analysis of drug release at 37°C at a pH of 4.5 and 7.4.

16.5.4 Anticancer activity

In recent decades, chemotherapy, radiation treatment, and operation are typically used to treat cancer, an uncontrolled malignant cell proliferation condition. While both of these treatments tend to be highly successful potentially in the destruction of cancer cells, they do have a range of significant side effects (Jiang et al., 2018; Sharma, Kumar, Choudhary, Mishra, & Vaidya, 2016). ZnO NMs have emerged as an anticancer drug and has been tried by various researchers. Tripathy, Ahmad, Ko, Khang, and Hahn (2015) investigated the use of Zn NMs, immune to acid, to modulate the liposome reaction characteristics. The integrated ZnO NMs were multimodal because they resulted not only in a pH-response system for drug supplies but also synergistic chemophotodynamic anticancer. This is a big step toward multimodal liposome structure growth. The cytotoxicity of ZnO NMs in the human cervical cell has been investigated by Pandurangan, Enkhtaivan, and Kim (2016). Viability of cells has been determined, and ZnO NMs have shown a possible cytotoxic effect. In cervical carcinoma cells, ZnO NMs showed a complex cytotoxic impact. ZnO NMs were concluded to exercise cytotoxicity on HeLa cell by apoptotic means the possible use of ZnO NMs in treating and therapy in cancer treatment. The bioavailability of curcumin has been ultimately improved by developing curcumin-loaded PMMA-PEG/ZnO nanocomposite using insoluble curcumin and ZnO NMs, which have a poor solubility (Dhivya et al., 2017). It has been shown that the therapeutic ability of the PMMA-PEG/ZnO-loaded curcumin can trigger cancer cell apoptosis via the apoptosis corridor of the cell cycle, which increases the probability of gastric cancer cell cures. A green ZnO biosynthesis of nanocomposite with seaweed Sargassum muticum water extract, and hyaluronic biopolymer, has been identified by Namvar et al. (2016). Moghimipour et al. (2018) prepared transferrin (Tf) targeting liposomal 5-fluorouracil (5FU) for enhancing medication safety and efficacy. The thin-layer system was used for processing liposomes. In contrast to free drugs and nontarget liposomes, the MTT study in cells of HT-29 demonstrated an improvement in the cytotoxic activity of target liposomes. Targeted liposomes, however, had no cytotoxic effect on human cells in contrast with cancer cells. In addition, activated liposomes induced apoptosis, as seen by decreased mitochondrial membrane potential and cytochrome c release, through triggering mitochondrial apoptosis pathways. ZnO NMs against cocultured C2C12 myoblastoma cancer and 3T3-L1 adipocytes were examined by Chandrasekaran and Pandurangan (2015). Their findings revealed that ZnO NMs had been reported that C2C12 myoblastoma cancer cells have apoptosis extreme mediated than 3T3-L1 cells. Wang, Lee, Kim, and Zhu (2017) demonstrated that NMs with ZnO is able to attack different cancer cell types, including skin cells, CSC, and macrophages, and at the same time perform numerous key functions, including skin inhalation, drug-resistant cancer, cancer recurrence prevention and metastasis, and the resuscitation of carcinogenesis. Ghaffari, Sarrafzadeh, Fakhroueian, Shahriari, and Khorramizadeh (2017) developed a basic surface functionalization method for the supply of water-soluble curcumin by ZnO NMs of 3-mercaptopropionic acid (MPA). The findings showed that nanoconjugation complex has an improved cancer-inhibiting capacity. In addition, ZnO-MPA-curcumin increased solubility and high flexibility and may be considered for new therapeutic activities. The toxicity of ZnO NMs to prokaryotic and eukaryotic cells was investigated by Premanathan, Karthikeyan, Jeyasubramanian, and Manivannan (2011). Myeloblastic leukemia (HL60) of human and normal peripheral blood mononuclear cells (PBMCs) were tested for the cytostatic effect of ZnO on the mammalian cells.

16.5.5 Antibacterial agent

Due to its benefits, such as high specific surface areas and high efficiency, ZnO NMs can be used as an antibacterial agent to kill a wide range of pathogens. Previously, however, ZnO NMs were not recognized for their antibacterial function. ZnO NMs were mainly toxic to antibacterial mechanisms based on their ability to induce excess ROS production, including superoxide anion, hydroxyl radicals, and production of hydrogen peroxides (Zhang & Xiong, 2015). Singh, Rawat, Khan, Naqvi, and Singh (2014) reported a biological route for synthesis with *P. aeruginosa* (rhamnolipids) NMs and their antioxidant properties of the ZnO NMs. The findings demonstrate the relevance for the production of surface defense ZnO NMs that can be used as promising antioxidants in the biological system, of *P. aeruginosa* RL as efficient stabilizing agents. Ishwarya et al. (2018) used the Ulva Lactuca seaweed extract as a reducing and encapsulating agent to synthesize Zn NMs by using a new, effortless green chemical method. The results of ZnO NMs on the morphology and histology of mosquito larvae were tracked. The periplasmic domain of the Vibrio cholera protein ZnO NMs with a diameter of 2.5 nm has been reported by Chatterjee et al. (2010). The protein plays a crucial function in regulating various virulence factors in cholera pathogenesis. Matai et al. (2014) synthesized nanocomposites of Ag–ZnO, which is antibacterial in the wide range against gram-positive and gram-negative bacteria. Ag–ZnO nanocomposites with various molar ratios were synthesized in the absence with surfactants by simple microwave-assisted reactions. Ghule, Ghule, Chen, and Ling (2006) reported a simple approach for ZnO nanoparticulate (alternative to 20 nm) ultrasound-assisted coating of paper with no binder assistance. Antibacterial resistance against *Escherichia coli* 11634 was detected in coated paper ZnO NMs. Shaban, Mohamed, and Abdallah (2018) prepared NMs of ZnO by sol-gel coated on fabrics of cotton by the process of spin-coating to increase the antimicrobial properties and repellents of soil. Several gram-positive and gram-negative bacteria such as *Salmonella typhimurium*, *Klebsiella pneumoniae*, *E. coli*, and *Bacillus subtilis* have been investigated for the bacterial presence of ZnO-coated clothing in their study. Bellanger, Billard, Schneider, Balan, and Merlin (2015) have investigated ZnO QDs using a dialysis device to separate QDs from leaked Zn^{2+} ions by using an inductively combined ICP-OES plasma emission spectrometer. The findings indicate that the design and usage of whole-cell biosensors and the metal release control from QDs were insufficient. Soren et al. (2018) reported the synthesis, characterization, and evaluation of mono-dispersed ZnO NMs synthesized by the aqueous phase and by the polyol method, the antimicrobial and antioxidants potential. In contrast to the particles synthesized by the aqueous precipitation process, ZnO NMs synthesized by polyol method indicate improved MIC values for both gram-positive and gram-negative bacteria. The current study shows the successful synthesis of ZnO NMs with the antioxidant property of several clinical bacterial pathogens, as well as significant broad-spectrum antibacterial activity. Sarwar, Ali, Pal, and Chakrabarti (2017) observed that ZnO NMs decrease fluid aggregation by cholera toxin-inducing protein in the mouse ileum. The findings indicate that ZnO NMs inhibit the translocalization into intestinal epithelial cells of the cholera toxin for the first time, without any detectable toxic effects on HEK293 and HT-29.

16.5.6 Antiinflammatory activity

Due to its benefits, such as high specific surface areas and high efficiency, ZnO NMs can be used as an antibacterial agent to kill a wide range of pathogens. Previously, however, ZnO NMs have not been recognized for their antibacterial function (Chen et al., 2018). ZnO NMs were mainly toxic to antibacterial mechanisms based on their ability to induce excess ROS production, including superoxide anion, hydroxyl radicals, and production of hydrogen peroxides. Wollina, Wiegand, Strehle, and Hipler (2013) investigated in vitro and in vivo the role of ZnO-functionalized textile fibers in the control of oxidative stress. The ZnO textile was evaluated in vitro to evaluate the antibacterial effect and biocompatibility. When patients with the AD wear ZnO textiles overnight for 3 consecutive days, the rapid development of AD frequency, pruritus, and subjective sleep consistency was noted. This was attributed to the high antioxidant potential and the distribution of heavy antibacking action of the ZnO fiber. In addition, the ZnO textiles have been shown to exhibit excellent biocompatibility and to be well tolerated by patients with AD. On the basis of regular topical application of ZnO particles, Ilves et al. (2014) investigated whether specific sized ZnO NMs can penetrate the damaged skin and allergic skin of the mouse model for atopic dermatitis. The findings showed that the deep layers of the allergic skin could only be covered with ZnO, whereas the high layers of both affected and allergic skin contain bulk sized ZnO (bZtnO).

Therefore, the local skin inflammation of the AD-mouse model reduces in all forms of particles, but ZnO has an increased capacity for systemic effects to be reversed. Thatoi et al. (2016) synthesized Ag and ZnO NMs in photocondition with water extracts of two mangrove plants, namely Heritiera fomes and Sonneratia apetala. ZnO NMs were found to have a relatively greater antiinflammatory capacity than Ag NMs.

16.5.7 Antidiabetic activity

Diabetes mellitus is due to an inability of the body to produce insulin or the ineffective use of insulin (Kharroubi, 2015; Khawandanah, 2019). The WHO has estimated that over 400 million diabetes adults around the world in 2014 were living with diabetes in the world (Piero, 2015; Roglic, 2016). In all human tissues and fabrics, Zn is a trace element that is found abundantly. The structural integrity of insulin is well known to keep Zn and is active in the secretion of insulin from pancreatic cells. It also includes synthesis, storage, and secretion of insulin (Li et al., 2013). Thus, ZnO NMs were developed and tested for their antidiabetic ability as a novel agent for the delivery of Zn. In combination with ZnO NMs, Kitture et al. (2015) investigated red sandalwood (RSW) natural extract as a potent antidiabetic agent. The conjugate ZnO−RSW NMs were found to show excellent activity in comparison with the individual ZnO NMs and RSW extract against crudely Moorish pancreas glucosidase. Hussein, El-Banna, Razik, and El-Naggar (2018) investigated the effect and application of chemically synthesized ZnO NMs as a special marker for endothelial dysfunction of streptozotocin-induced streptozole rats, and the protection of the degree of asymmetric dimethylarginine (ADMA). ZnO NMs were chemically synthesized in the presence of potassium hydroxide using natural, biological-benign biodegradable hydroxyl ethyl (HES) as a stabilizing and guiding agent.

16.6 Mechanism

Centered on their anticancer, antibacterial, antidiabetic, antiinflammatory, medication supply, and biomedical activity, the ZnO NMs have been promising for biomedical applications. Due to its intrinsic toxicity, the ZnO NMs are highly influenced by induced intracellular ROS generation and activated apoptotic signaling cascade that renders the ZnO NMs a possible anticancer and antibacterial nominee. However, ZnO NMs have proved well known to facilitate therapeutical or biomolecular bioavailability while acting as drug carriers in order to increase the performance of therapy (Kumar et al., 2013). In addition, ZnO NMs have demonstrated positive promise in the treatment of diabetes and attenuating its symptoms with the capacity to reduce blood glucose and insulin rates. The majority of biomedical uses in ZnO NMs are due to their ability to produce ROS, which results in cell death if the cell's antioxidant capability is exceeded (Ryter et al., 2007). ZnO semiconducting properties affect its ROS generation performance. Within semiconductors, the electrons (e^-) carry energy within such bands. For crystalline ZnO, an open area extending from the top of the filled valence band into the bottom is around 3.3 eV. UV rays provide enough energy to help the band and leave holes behind (h^+). e^- and h^+ move to the surface of the NMs and react respectively with oxygen and hydroxylic ions. Superoxide and hydroxyl radicals are thus formed (Mishra, Mishra, Ekielski, Talegaonkar, & Vaidya, 2017). For ZnO NMs, even in the absence of UV light a vast deal of the valence-strips h^+ and/or leading strip e^- are present because of nanosized material crystal defects. The numerous ROS are then released and cause irreparable damage to the redox cycle cascades in the cells. Necrosis and apoptosis are other mechanisms for the biomedical activity of ZnO NMs (Wilhelmi et al., 2013; Yang et al., 2015). DNA damage is caused by ROS triggers apoptotic pathways by removing apoptogenic factors from mitochondrial space. This contributes to apoptosome formation, which stimulates enzymes of the executioner, which results in apoptosis and cell death after their particular substrate have been broken. There was an alternative mechanism also suggested, which induces apoptosis by rapid dissolution within the macrophage acidic lysosomes after ZnO is taken into particulate agglomerated form (Wilhelmi et al., 2013). Fig. 16.5 shows the mechanism of action of ZnO NMs. ZnO NMs cause ROS to be generated. DNA damage to the ROS, which in turn triggers the activation of mitochondrial membrane apoptogenic factors. Apoptogenic conditions contribute to apoptosome development, contributing eventually to apoptosis. Because of partially

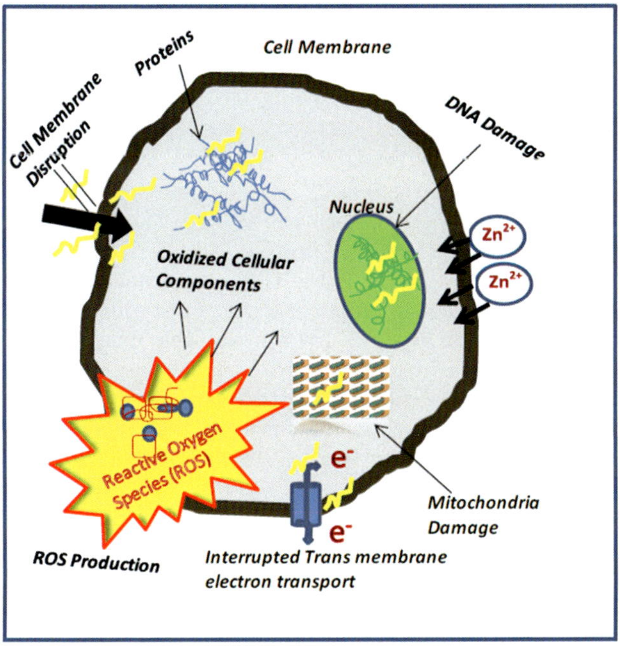

Figure 16.5 Mechanism of action of ZnO nanomaterials.

bonds of oxygen atoms, ZnO NMs carry a negative charge on their surface. During low pH, the protons are transferred by means of a positive charged surface ($ZNOH^{2+}$) from the atmosphere to the particle surface. Such positive particles interact on the outer membrane of the cell with negative phospholipids that cause their absorption by the cell. ZnO NMs absorb and release Zn^+ in acid lysosomes, which prevent the respiratory enzyme's activity causing cell death.

16.7 Conclusion

It can be concluded that the ZnO nanostructured materials have been immensely researched all across the globe and has come out as potential functional material for variegated applications viz., sensors, nanogenerators, biomedical, etc. Albeit, some significant issues in ZnO NMs need to be examined, such as ZnO NMs' toxicity to biological systems which remain a major problem in recent research. As a concluding remark, it can be deliberated that the production of medicines should be greatly facilitated by this class of semiconducting NMs and more exciting contributions from the same can be expected in future.

References

Ahmad, R., Ahn, M.-S., & Hahn, Y.-B. (2017). A highly sensitive nonenzymatic sensor based on Fe_2O_3 nanoparticle coated ZnO nanorods for electrochemical detection of nitrite. *Advanced Materials Interfaces*, *4*(22), 1700691. Available from https://doi.org/10.1002/admi.201700691.

Ahmad, R., Tripathy, N., Ahn, M.-S., Bhat, K. S., Mahmoudi, T., Wang, Y., ... Hahn, Y.-B. (2017). Highly efficient non-enzymatic glucose sensor based on CuO modified vertically-grown ZnO nanorods on electrode. *Scientific Reports*, *7*(1), 5715. Available from https://doi.org/10.1038/s41598-017-06064-8.

Ahmad, R., Tripathy, N., Ahn, M.-S., & Hahn, Y.-B. (2017). Solution process synthesis of high aspect ratio ZnO nanorods on electrode surface for sensitive electrochemical detection of uric acid. *Scientific Reports*, *7*, 46475. Available from https://doi.org/10.1038/srep46475.

Ahmad, R., Tripathy, N., & Hahn, Y.-B. (2013). High-performance cholesterol sensor based on the solution-gated field effect transistor fabricated with ZnO nanorods. *Biosensors and Bioelectronics*, *45*, 281–286. Available from https://doi.org/10.1016/j.bios.2013.01.021.

Ahmad, R., Tripathy, N., & Hahn, Y.-B. (2014). Highly stable urea sensor based on ZnO nanorods directly grown on Ag/glass electrodes. *Sensors and Actuators B: Chemical*, *194*, 290–295. Available from https://doi.org/10.1016/j.snb.2013.12.098.

Ahmad, R., Tripathy, N., Jang, N. K., Khang, G., & Hahn, Y.-B. (2015). Fabrication of highly sensitive uric acid biosensor based on directly grown ZnO nanosheets on electrode surface. *Sensors and Actuators B: Chemical*, *206*, 146–151. Available from https://doi.org/10.1016/j.snb.2014.09.026.

Ahmad, R., Tripathy, N., Kim, S. H., Umar, A., Al-Hajry, A., & Hahn, Y.-B. (2014). High performance cholesterol sensor based on ZnO nanotubes grown on Si/Ag electrodes. *Electrochemistry Communications*, *38*, 4–7. Available from https://doi.org/10.1016/j.elecom.2013.10.028.

Ahmad, R., Tripathy, N., Park, J.-H., & Hahn, Y.-B. (2015). A comprehensive biosensor integrated with a ZnO nanorod FET array for selective detection of glucose, cholesterol and urea. *Chemical Communications*, *51*(60), 11968–11971. Available from https://doi.org/10.1039/c5cc03656a.

Aldeek, F., Mustin, C., Balan, L., Medjahdi, G., Roques-Carmes, T., Arnoux, P., & Schneider, R. (2011). Enhanced photostability from CdSe(S)/ZnO core/shell quantum dots and their use in biolabeling. *European Journal of Inorganic Chemistry*, *2011*(6), 794–801. Available from https://doi.org/10.1002/ejic.201000790.

Al-Jumaili, A., Mulvey, P., Kumar, A., Prasad, K., Bazaka, K., Warner, J., & Jacob, M. V. (2019). Eco-friendly nanocomposites derived from geranium oil and zinc oxide in one step approach. *Scientific Reports*, *9*(1), 5973. Available from https://doi.org/10.1038/s41598-019-42211-z.

Ambika, S., & Sundrarajan, M. (2015). Green biosynthesis of ZnO nanomaterials using Vitex negundo L. extract: Spectroscopic investigation of interaction between ZnO nanomaterials and human serum albumin. *Journal of Photochemistry and Photobiology B: Biology*, *149*, 143–148. Available from https://doi.org/10.1016/j.jphotobiol.2015.05.004.

Aneesh, P. M., Vanaja, K. A., & Jayaraj, M. K. (2007). Synthesis of ZnO nanomaterials by hydrothermal method. In Z. Gaburro & S. Cabrini (Eds.), Proceedings of SPIE 6639, Nanophotonic Materials IV (Vol. 6639, p. 66390J). https://doi.org/10.1117/12.730364

Antoine, T. E., Mishra, Y. K., Trigilio, J., Tiwari, V., Adelung, R., & Shukla, D. (2012). Prophylactic, therapeutic and neutralizing effects of zinc oxide tetrapod structures against herpes simplex virus type-2 infection. *Antiviral Research*, *96*(3), 363–375. Available from https://doi.org/10.1016/j.antiviral.2012.09.020.

Banerjee, B. (2017). Recent developments on nano-ZnO catalyzed synthesis of bioactive heterocycles. *Journal of Nanostructure in Chemistry*, *7*(4), 389–413. Available from https://doi.org/10.1007/s40097-017-0247-0.

Barick, K. C., Nigam, S., & Bahadur, D. (2010). Nanoscale assembly of mesoporous ZnO: A potential drug carrier. *Journal of Materials Chemistry*, *20*(31), 6446. Available from https://doi.org/10.1039/c0jm00022a.

Barui, A. K., Kotcherlakota, R., & Patra, C. R. (2018). *Biomedical applications of zinc oxide nanomaterials*, . *Inorganic frameworks as smart nanomedicines* (Vol. 20, pp. 239–278). Elsevier Inc Issue 4. Available from https://doi.org/10.1016/B978-0-12-813661-4.00006-7.

Bellanger, X., Billard, P., Schneider, R., Balan, L., & Merlin, C. (2015). Stability and toxicity of ZnO quantum dots: Interplay between nanomaterials and bacteria. *Journal of Hazardous Materials*, *283*, 110–116. Available from https://doi.org/10.1016/j.jhazmat.2014.09.017.

Bhatia, S., & Verma, N. (2017). Photocatalytic activity of ZnO nanomaterials with optimization of defects. *Materials Research Bulletin*, *95*, 468–476. Available from https://doi.org/10.1016/j.materresbull.2017.08.019.

Bhatt, G., & Bhattacharya, S. (2019). Novel dipstick model for portable bio-sensing application. *Journal of Energy and Environmental Sustainability*, *7*, 32–35. Available from https://www.researchgate.net/publication/334762489.

Bhunia, A., Kamilya, T., & Saha, S. (2016). Optical and structural properties of protein capped ZnO nanomaterials and its antimicrobial activity. *Journal of Advances in Biology & Biotechnology*, *10*(1), 1–9. Available from https://doi.org/10.9734/jabb/2016/29626.

Brayner, R., Dahoumane, S. A., Yépremian, C., Djediat, C., Meyer, M., Couté., ... Fiévet, F. (2010). ZnO nanomaterials: Synthesis, characterization, and ecotoxicological studies. *Langmuir: The ACS Journal of Surfaces and Colloids*, *26*(9), 6522–6528. Available from https://doi.org/10.1021/la100293s.

Cai, B., Zhou, Y., Zhao, M., Cai, H., Ye, Z., Wang, L., & Huang, J. (2014). Synthesis of ZnO–CuO porous core–shell spheres and their application for non-enzymatic glucose sensor. *Applied Physics A*, *118*(3), 989–996. Available from https://doi.org/10.1007/s00339-014-8855-8.

Chandra, S., Barick, K. C., & Bahadur, D. (2011). Oxide and hybrid nanostructures for therapeutic applications. *Advanced Drug Delivery Reviews*, *63*(14–15), 1267–1281. Available from https://doi.org/10.1016/j.addr.2011.06.003.

Chandrasekaran, M., & Pandurangan, M. (2015). In vitro selective anti-proliferative effect of zinc oxide nanomaterials against co-cultured C2C12 myoblastoma cancer and 3T3-L1 normal cells. *Biological Trace Element Research*, *172*(1), 148–154. Available from https://doi.org/10.1007/s12011-015-0562-6.

Chatterjee, T., Chakraborti, S., Joshi, P., Singh, S. P., Gupta, V., & Chakrabarti, P. (2010). The effect of zinc oxide nanomaterials on the structure of the periplasmic domain of the Vibrio cholerae ToxR protein. *FEBS Journal*, *277*(20), 4184–4194. Available from https://doi.org/10.1111/j.1742-4658.2010.07807.x.

Chauhan, P. S., Kant, R., Rai, A., Gupta, A., & Bhattacharya, S. (2019). Facile synthesis of ZnO/GO nanoflowers over Si substrate for improved photocatalytic decolorization of MB dye and industrial wastewater under solar irradiation. *Materials Science in Semiconductor Processing*, *89*, 6–17. Available from https://doi.org/10.1016/j.mssp.2018.08.022, August 2018.

Chauhan, P. S., Rai, A., Gupta, A., & Bhattacharya, S. (2017). Enhanced photocatalytic performance of vertically grown ZnO nanorods decorated with metals (Al, Ag, Au, and Au-Pd) for degradation of industrial dye. *Materials Research Express*, *4*(5). Available from https://doi.org/10.1088/2053-1591/aa6d31.

Chen, L., Deng, H., Cui, H., Fang, J., Zuo, Z., Deng, J., ... Zhao, L. (2018). Inflammatory responses and inflammation-associated diseases in organs. *Oncotarget*, *9*(6), 7204–7218. Available from https://doi.org/10.18632/oncotarget.23208.

Chen, L., Li, X., Qu, L., Gao, C., Wang, Y., Teng, F., ... Xie, E. (2014). Facile and fast one-pot synthesis of ultra-long porous ZnO nanowire arrays for efficient dye-sensitized solar cells. *Journal of Alloys and Compounds*, *586*, 766–772. Available from https://doi.org/10.1016/j.jallcom.2013.10.118.

Chen, T., Zhao, T., Wei, D., Wei, Y., Li, Y., & Zhang, H. (2013). Core–shell nanocarriers with ZnO quantum dots-conjugated Au nanoparticle for tumor-targeted drug delivery. *Carbohydrate Polymers*, *92*(2), 1124–1132. Available from https://doi.org/10.1016/j.carbpol.2012.10.022.

Cho, N.-H., Cheong, T.-C., Min, J. H., Wu, J. H., Lee, S. J., Kim, D., ... Seong, S.-Y. (2011). A multifunctional core–shell nanoparticle for dendritic cell-based cancer immunotherapy. *Nature Nanotechnology*, *6*(10), 675–682. Available from https://doi.org/10.1038/nnano.2011.149.

Demir, M. M., Muñoz-Espí, R., Lieberwirth, I., & Wegner, G. (2006). Precipitation of monodisperse ZnO nanocrystals via acid-catalyzed esterification of zinc acetate. *Journal of Materials Chemistry*, *16*(28), 2940–2947. Available from https://doi.org/10.1039/B601451H.

Dhivya, R., Ranjani, J., Bowen, P. K., Rajendhran, J., Mayandi, J., & Annaraj, J. (2017). Biocompatible curcumin loaded PMMA-PEG/ZnO nanocomposite induce apoptosis and cytotoxicity in human gastric cancer cells. *Materials Science and Engineering: C, 80*, 59–68. Available from https://doi.org/10.1016/j.msec.2017.05.128.

Dong, Z., Han, B., Qian, S., & Chen, D. (2012). Fluorescent properties of ZnO nanostructures fabricated by hydrothermal method. *Journal of Nanomaterials, 2012*, 1–5. Available from https://doi.org/10.1155/2012/251276.

Firdhouse, M. J., & Lalitha, P. (2015). Biosynthesis of silver nanomaterials and its applications. *Journal of Nanotechnology, 2015*, 1–18. Available from https://doi.org/10.1155/2015/829526.

Fouad, O., Ismail, A., Zaki, Z., & Mohamed, R. (2006). Zinc oxide thin films prepared by thermal evaporation deposition and its photocatalytic activity. *Applied Catalysis B: Environmental, 62*(1–2), 144–149. Available from https://doi.org/10.1016/j.apcatb.2005.07.006.

Fricke, M., Voigt, A., Veit, P., & Sundmacher, K. (2015). Miniemulsion-based process for controlling the size and shape of zinc oxide nanomaterials. *Industrial & Engineering Chemistry Research, 54*(42), 10293–10300. Available from https://doi.org/10.1021/acs.iecr.5b01149.

Ghaffari, S.-B., Sarrafzadeh, M.-H., Fakhroueian, Z., Shahriari, S., & Khorramizadeh, M. R. (2017). Functionalization of ZnO nanomaterials by 3-mercaptopropionic acid for aqueous curcumin delivery: Synthesis, characterization, and anticancer assessment. *Materials Science and Engineering: C, 79*, 465–472. Available from https://doi.org/10.1016/j.msec.2017.05.065.

Gharagozlou, M., Baradaran, Z., & Bayati, R. (2015). A green chemical method for synthesis of ZnO nanomaterials from solid-state decomposition of Schiff-bases derived from amino acid alanine complexes. *Ceramics International, 41*(7), 8382–8387. Available from https://doi.org/10.1016/j.ceramint.2015.03.029.

Ghule, K., Ghule, A. V., Chen, B.-J., & Ling, Y.-C. (2006). Preparation and characterization of ZnO nanomaterials coated paper and its antibacterial activity study. *Green Chemistry, 8*(12), 1034. Available from https://doi.org/10.1039/b605623g.

Gupta, A., & Bhattacharya, S. (2018). On the growth mechanism of ZnO nano structure via aqueous chemical synthesis. *Applied Nanoscience (Switzerland), 8*(3), 499–509. Available from https://doi.org/10.1007/s13204-018-0782-0.

Gupta, A., Gangopadhyay, S., Gangopadhyay, K., & Bhattacharya, S. (2016). Palladium-functionalized nanostructured platforms for enhanced hydrogen sensing. *Nanomaterials and Nanotechnology, 6*. Available from https://doi.org/10.5772/63987.

Gupta, A., Mondal, K., Sharma, A., & Bhattacharya, S. (2015). Superhydrophobic polymethylsilsesquioxane pinned one dimensional ZnO nanostructures for water remediation through photo-catalysis. *RSC Advances, 5*(57), 45897–45907. Available from https://doi.org/10.1039/c5ra02938d.

Gupta, A., Nayak, M., Singh, D., & Bhattacharya, S. (2014). Antibody immobilization for ZnO nanowire based biosensor application. *Materials Research Society Symposium Proceedings, 1675*, 33–39. Available from https://doi.org/10.1557/opl.2014.848.

Gupta, A., & Pal, P. (2018). *Flexible sensors for biomedical application* (pp. 287–314). Singapore: Springer. Available from https://doi.org/10.1007/978-981-10-7751-7_13.

Gupta, A., Pandey, S. S., Nayak, M., Maity, A., Majumder, S. B., & Bhattacharya, S. (2014). Hydrogen sensing based on nanoporous silica-embedded ultra dense ZnO nanobundles. *RSC Advances, 4*(15), 7476–7482. Available from https://doi.org/10.1039/c3ra45316b.

Gupta, A., Parida, P. K., & Pal, P. (2019). *Functional films for gas sensing applications: A review*. Singapore: Springer. Available from https://doi.org/10.1007/978-981-13-3290-6_2.

Gupta, A., Patel, V. K., Kant, R., & Bhattacharya, S. (2016). Surface modification strategies for fabrication of nano-biodevices: A critical review. *Reviews of Adhesion and Adhesives*, *4*(2), 166–191. Available from https://doi.org/10.7569/RAA.2016.097307.

Gupta, A., Saurav, J. R., & Bhattacharya, S. (2015). Solar light based degradation of organic pollutants using ZnO nanobrushes for water filtration. *RSC Advances*, *5*(87), 71472–71481. Available from https://doi.org/10.1039/c5ra10456d.

Gupta, A., Singh, D., Raj, P., Gupta, H., Verma, S., & Bhattacharya, S. (2015). Investigation of ZnO-hydroxyapatite nanocomposite incorporated in restorative glass ionomer cement to enhance its mechanical and antimicrobial properties. *Journal of Bionanoscience*, *9*(3), 190–196. Available from https://doi.org/10.1166/jbns.2015.1299.

Gupta, A., Srivastava, A., Mathai, C. J., Gangopadhyay, K., Gangopadhyay, S., & Bhattacharya, S. (2014). Nano porous palladium sensor for sensitive and rapid detection of hydrogen. *Sensor Letters*, *12*(8), 1279–1285. Available from https://doi.org/10.1166/sl.2014.3307.

Gupta, A., Sundriyal, P., Basu, A., Manoharan, K., Kant, R., & Bhattacharya, S. (2020). Nano-finishing of MEMS-based platforms for optimum optical sensing. *Journal of Micromanufacturing*, *3*(1), 39–53. Available from https://doi.org/10.1177/2516598419862676.

Hahn, Y.-B., Ahmad, R., & Tripathy, N. (2012). Chemical and biological sensors based on metal oxide nanostructures. *Chemical Communications*, *48*(84), 10369. Available from https://doi.org/10.1039/c2cc34706g.

Hanley, C., Layne, J., Punnoose, A., Reddy, K. M., Coombs, I., Coombs, A., ... Wingett, D. (2008). Preferential killing of cancer cells and activated human T cells using ZnO nanomaterials. *Nanotechnology*, *19*(29), 295103. Available from https://doi.org/10.1088/0957-4484/19/29/295103.

Hariharan, R., Senthilkumar, S., Suganthi, A., & Rajarajan, M. (2012). Synthesis and characterization of doxorubicin modified ZnO/PEG nanomaterials and its photodynamic action. *Journal of Photochemistry and Photobiology B: Biology*, *116*, 56–65. Available from https://doi.org/10.1016/j.jphotobiol.2012.08.008.

Hong, H., Shi, J., Yang, Y., Zhang, Y., Engle, J. W., Nickles, R. J., ... Cai, W. (2011). Cancer-targeted optical imaging with fluorescent zinc oxide nanowires. *Nano Letters*, *11*(9), 3744–3750. Available from https://doi.org/10.1021/nl201782m.

Hussein, J., El-Banna, M., Razik, T. A., & El-Naggar, M. E. (2018). Biocompatible zinc oxide nanocrystals stabilized via hydroxyethyl cellulose for mitigation of diabetic complications. *International Journal of Biological Macromolecules*, *107*, 748–754. Available from https://doi.org/10.1016/j.ijbiomac.2017.09.056.

Ilves, M., Palomäki, J., Vippola, M., Lehto, M., Savolainen, K., Savinko, T., & Alenius, H. (2014). Topically applied ZnO nanomaterials suppress allergen induced skin inflammation but induce vigorous IgE production in the atopic dermatitis mouse model. *Particle and Fibre Toxicology*, *11*(1), 38. Available from https://doi.org/10.1186/s12989-014-0038-4.

Ishwarya, R., Vaseeharan, B., Kalyani, S., Banumathi, B., Govindarajan, M., Alharbi, N. S., ... Benelli, G. (2018). Facile green synthesis of zinc oxide nanomaterials using Ulva lactuca seaweed extract and evaluation of their photocatalytic, antibiofilm and insecticidal activity. *Journal of Photochemistry and Photobiology B: Biology*, *178*, 249–258. Available from https://doi.org/10.1016/j.jphotobiol.2017.11.006.

Jena, S., Gupta, A., Pippara, R. K., Pal, P., & Adit. (2019). *Wireless sensing systems: A review*. Singapore: Springer. Available from https://doi.org/10.1007/978-981-13-3290-6_9.

Jiang, J., Pi, J., & Cai, J. (2018). The advancing of zinc oxide nanomaterials for biomedical applications. *Bioinorganic Chemistry and Applications*, 2018. Available from https://doi.org/10.1155/2018/1062562.

Jiang, H., Wang, H., & Wang, X. (2011). Facile and mild preparation of fluorescent ZnO nanosheets and their bioimaging applications. *Applied Surface Science*, *257*(15), 6991–6995. Available from https://doi.org/10.1016/j.apsusc.2011.03.053.

Jin, S. E., & Jin, H. E. (2019). Synthesis, characterization, and three-dimensional structure generation of zinc oxide-based nanomedicine for biomedical applications. *Pharmaceutics*, *11*(11). Available from https://doi.org/10.3390/pharmaceutics11110575.

Jood, P., Mehta, R. J., Zhang, Y., Peleckis, G., Wang, X., Siegel, R. W., ... Ramanath, G. (2011). Al-doped zinc oxide nanocomposites with enhanced thermoelectric properties. *Nano Letters*, *11*(10), 4337–4342. Available from https://doi.org/10.1021/nl202439h.

Jung, D.-U.-J., Ahmad, R., & Hahn, Y.-B. (2018). Nonenzymatic flexible field-effect transistor based glucose sensor fabricated using NiO quantum dots modified ZnO nanorods. *Journal of Colloid and Interface Science*, *512*, 21–28. Available from https://doi.org/10.1016/j.jcis.2017.10.037.

Kachynski, A. V., Kuzmin, A. N., Nyk, M., Roy, I., & Prasad, P. N. (2008). Zinc oxide nanocrystals for non-resonant nonlinear optical microscopy in biology and medicine. *The Journal of Physical Chemistry. C, Nanomaterials and Interfaces*, *112*(29), 10721–10724. Available from https://doi.org/10.1021/jp801684j.

Kalpana, V. N., & Devi Rajeswari, V. (2018). A review on green synthesis, biomedical applications, and toxicity studies of ZnO NMs. *Bioinorganic Chemistry and Applications*, *2018*. Available from https://doi.org/10.1155/2018/3569758.

Karuppiah, C., Velmurugan, M., Chen, S.-M., Tsai, S.-H., Lou, B.-S., Ajmal Ali, M., & Al-Hemaid, F. M. A. (2015). A simple hydrothermal synthesis and fabrication of zinc oxide–copper oxide heterostructure for the sensitive determination of nonenzymatic glucose biosensor. *Sensors and Actuators B: Chemical*, *221*, 1299–1306. Available from https://doi.org/10.1016/j.snb.2015.07.075.

Khalafi, T., Buazar, F., & Ghanemi, K. (2019). Phycosynthesis and enhanced photocatalytic activity of zinc oxide nanomaterials toward organosulfur pollutants. *Scientific Reports*, *9* (1), 6866. Available from https://doi.org/10.1038/s41598-019-43368-3.

Kharroubi, A. T. (2015). Diabetes mellitus: The epidemic of the century. *World Journal of Diabetes*, *6*(6), 850. Available from https://doi.org/10.4239/wjd.v6.i6.850.

Khawandanah, J. (2019). Double or hybrid diabetes: A systematic review on disease prevalence, characteristics and risk factors. *Nutrition & Diabetes*, *9*(1), 33. Available from https://doi.org/10.1038/s41387-019-0101-1.

Kishwar, S., Asif, M. H., Nur, O., Willander, M., & Larsson, P.-O. (2010). Intracellular ZnO nanorods conjugated with protoporphyrin for local mediated photochemistry and efficient treatment of single cancer cell. *Nanoscale Research Letters*, *5*(10), 1669–1674. Available from https://doi.org/10.1007/s11671-010-9693-z.

Kitture, R., Chordiya, K., Gaware, S., Ghosh, S., More, P. A., Kulkarni, P., ... Kale, S. N. (2015). ZnO nanomaterials-red sandalwood conjugate: A promising anti-diabetic agent. *Journal of Nanoscience and Nanotechnology*, *15*(6), 4046–4051. Available from https://doi.org/10.1166/jnn.2015.10323.

Kołodziejczak-Radzimska, A., & Jesionowski, T. (2014). Zinc oxide—From synthesis to application: A review. *Materials*, *7*(4), 2833–2881. Available from https://doi.org/10.3390/ma7042833.

Kong, X. Y., & Wang, Z. L. (2004). Polar-surface dominated ZnO nanobelts and the electrostatic energy induced nanohelixes, nanosprings, and nanospirals. *Applied Physics Letters*, *84*(6), 975–977. Available from https://doi.org/10.1063/1.1646453.

Król, A., Pomastowski, P., Rafińska, K., Railean-Plugaru, V., & Buszewski, B. (2017). Zinc oxide nanomaterials: Synthesis, antiseptic activity and toxicity mechanism. *Advances in Colloid and Interface Science*, *249*, 37–52. Available from https://doi.org/10.1016/j.cis.2017.07.033.

Kumar, A., Gupta, A., Kant, R., Akhtar, S. N., Tiwari, N., Ramkumar, J., & Bhattacharya, S. (2013). Optimization of laser machining process for the preparation of photomasks, and its application to microsystems fabrication. *Journal of Micro/Nanolithography, MEMS, and MOEMS*, *12*(4), 041203. Available from https://doi.org/10.1117/1.jmm.12.4.041203.

Kumar, R., Umar, A., Kumar, G., & Nalwa, H. S. (2017). Antimicrobial properties of ZnO nanomaterials: A review. *Ceramics International*, *43*(5), 3940–3961. Available from https://doi.org/10.1016/j.ceramint.2016.12.062.

Li, B., Tan, Y., Sun, W., Fu, Y., Miao, L., & Cai, L. (2013). The role of zinc in the prevention of diabetic cardiomyopathy and nephropathy. *Toxicology Mechanisms and Methods*, *23*(1), 27–33. Available from https://doi.org/10.3109/15376516.2012.735277.

Lu, P.-J., Huang, S.-C., Chen, Y.-P., Chiueh, L.-C., & Shih, D. Y.-C. (2015). Analysis of titanium dioxide and zinc oxide nanomaterials in cosmetics. *Journal of Food and Drug Analysis*, *23*(3), 587–594. Available from https://doi.org/10.1016/j.jfda.2015.02.009.

Lyu, S. C., Zhang, Y., Lee, C. J., Ruh, H., & Lee, H. J. (2003). Low-temperature growth of ZnO nanowire array by a simple physical vapor-deposition method. *Chemistry of Materials*, *15*(17), 3294–3299. Available from https://doi.org/10.1021/cm020465j.

Marie, M., Manoharan, A., Kuchuk, A., Ang, S., & Manasreh, M. O. (2018). Vertically grown zinc oxide nanorods functionalized with ferric oxide for in vivo and non-enzymatic glucose detection. *Nanotechnology*, *29*(11), 115501. Available from https://doi.org/10.1088/1361-6528/aaa682.

Martínez-Carmona, M., Gun'ko, Y., & Vallet-Regí, M. (2018). ZnO nanostructures for drug delivery and theranostic applications. *Nanomaterials*, *8*(4), 268. Available from https://doi.org/10.3390/nano8040268.

Matai, I., Sachdev, A., Dubey, P., Uday Kumar, S., Bhushan, B., & Gopinath, P. (2014). Antibacterial activity and mechanism of Ag–ZnO nanocomposite on *S. aureus* and GFP-expressing antibiotic resistant *E. coli*. *Colloids and Surfaces B: Biointerfaces*, *115*, 359–367. Available from https://doi.org/10.1016/j.colsurfb.2013.12.005.

Matsuyama, K., Ihsan, N., Irie, K., Mishima, K., Okuyama, T., & Muto, H. (2013). Bioimaging application of highly luminescent silica-coated ZnO-nanoparticle quantum dots with biotin. *Journal of Colloid and Interface Science*, *399*, 19–25. Available from https://doi.org/10.1016/j.jcis.2013.02.047.

Mishra, P. K., Mishra, H., Ekielski, A., Talegaonkar, S., & Vaidya, B. (2017). Zinc oxide nanomaterials: A promising nanomaterial for biomedical applications. *Drug Discovery Today*, *22*(12), 1825–1834. Available from https://doi.org/10.1016/j.drudis.2017.08.006.

Moghimipour, E., Rezaei, M., Ramezani, Z., Kouchak, M., Amini, M., Angali, K. A., … Handali, S. (2018). Transferrin targeted liposomal 5-fluorouracil induced apoptosis via mitochondria signaling pathway in cancer cells. *Life Sciences*, *194*, 104–110. Available from https://doi.org/10.1016/j.lfs.2017.12.026.

Muhammad, F., Guo, M., Guo, Y., Qi, W., Qu, F., Sun, F., … Zhu, G. (2011). Acid degradable ZnO quantum dots as a platform for targeted delivery of an anticancer drug. *Journal of Materials Chemistry*, *21*(35), 13406. Available from https://doi.org/10.1039/c1jm12119g.

Namvar, F., Azizi, S., Rahman, H. S., Mohamad, R., Rasedee, A., Soltani, M., & Rahim, R. A. (2016). Green synthesis, characterization, and anticancer activity of hyaluronan/

zinc oxide nanocomposite. *OncoTargets and Therapy*, *9*, 4549–4559. Available from https://doi.org/10.2147/OTT.S95962.

Naveed Ul Haq, A., Nadhman, A., Ullah, I., Mustafa, G., Yasinzai, M., & Khan, I. (2017). Synthesis approaches of zinc oxide nanomaterials: The dilemma of ecotoxicity. *Journal of Nanomaterials*, *2017*, 1–14. Available from https://doi.org/10.1155/2017/8510342.

Ng, C., Baeg, G.-H., Yu, L., Ong, C., & Bay, B.-H. (2017). Biomedical applications of nanomaterials as therapeutics. *Current Medicinal Chemistry*, *24*(999), 1. Available from https://doi.org/10.2174/0929867324666170331120328.

Noothongkaew, S., Pukird, S., Sukkabot, W., & An, K. S. (2014). Zinc oxide nano walls synthesized by chemical vapor deposition. *Key Engineering Materials*, *608*, 127–131. Available from https://doi.org/10.4028/http://www.scientific.net/KEM.608.127.

Ogunyemi, S. O., Abdallah, Y., Zhang, M., Fouad, H., Hong, X., Ibrahim, E., ... Li, B. (2019). Green synthesis of zinc oxide nanomaterials using different plant extracts and their antibacterial activity against *Xanthomonas oryzae* pv. oryzae. *Artificial Cells, Nanomedicine, and Biotechnology*, *47*(1), 341–352. Available from https://doi.org/10.1080/21691401.2018.1557671.

Oliveira, A. P. A., Hochepied, J.-F., Grillon, F., & Berger, M.-H. (2003). Controlled precipitation of zinc oxide particles at room temperature. *Chemistry of Materials*, *15*(16), 3202–3207. Available from https://doi.org/10.1021/cm0213725.

Ong, C. B., Ng, L. Y., & Mohammad, A. W. (2018). A review of ZnO nanomaterials as solar photocatalysts: Synthesis, mechanisms and applications. *Renewable and Sustainable Energy Reviews*, *81*, 536–551. Available from https://doi.org/10.1016/j.rser.2017.08.020.

Özgür, Ü., Alivov, Y. I., Liu, C., Teke, A., Reshchikov, M. A., Doğan, S., ... Morkoç, H. (2005). A comprehensive review of ZnO materials and devices. *Journal of Applied Physics*, *98*(4), 041301. Available from https://doi.org/10.1063/1.1992666.

Pan, Z. W. (2001). Nanobelts of semiconducting oxides. *Science (New York, N.Y.)*, *291* (5510), 1947–1949. Available from https://doi.org/10.1126/science.1058120.

Pandurangan, M., Enkhtaivan, G., & Kim, D. H. (2016). Anticancer studies of synthesized ZnO nanomaterials against human cervical carcinoma cells. *Journal of Photochemistry and Photobiology B: Biology*, *158*, 206–211. Available from https://doi.org/10.1016/j.jphotobiol.2016.03.002.

Peng, H., Fangli, Y., Liuyang, B., Jinlin, L., & Yunfa, C. (2007). Plasma synthesis of large quantities of zinc oxide nanorods. *The Journal of Physical Chemistry C*, *111*(1), 194–200. Available from https://doi.org/10.1021/jp065390b.

Piero, M. N. (2015). Diabetes mellitus – a devastating metabolic disorder. *Asian Journal of Biomedical and Pharmaceutical Sciences*, *4*(40), 1–7. Available from https://doi.org/10.15272/ajbps.v4i40.645.

Premanathan, M., Karthikeyan, K., Jeyasubramanian, K., & Manivannan, G. (2011). Selective toxicity of ZnO nanomaterials toward gram-positive bacteria and cancer cells by apoptosis through lipid peroxidation. *Nanomedicine: Nanotechnology, Biology and Medicine*, *7*(2), 184–192. Available from https://doi.org/10.1016/j.nano.2010.10.001.

Rajabairavi, N., Raju, C. S., Karthikeyan, C., Varutharaju, K., Nethaji, S., Hameed, A. S. H., & Shajahan, A. (2017). Biosynthesis of novel zinc oxide nanomaterials (ZnO NMs) using endophytic bacteria *Sphingobacterium thalpophilum*. In *Springer proceedings in physics* (pp. 245–254). Springer International Publishing. https://doi.org/10.1007/978–3–319–44890-9_23

Rajalakshmi, M., Sohila, S., Ramya, S., Divakar, R., Ghosh, C., & Kalavathi, S. (2012). Blue green and UV emitting ZnO nanomaterials synthesized through a non-aqueous route.

Optical Materials, *34*(8), 1241–1245. Available from https://doi.org/10.1016/j.optmat.2012.01.021.

Ramos, A. P., Cruz, M. A. E., Tovani, C. B., & Ciancaglini, P. (2017). Biomedical applications of nanotechnology. *Biophysical Reviews*, *9*(2), 79–89. Available from https://doi.org/10.1007/s12551-016-0246-2.

Rasmussen, J. W., Martinez, E., Louka, P., & Wingett, D. G. (2010). Zinc oxide nanomaterials for selective destruction of tumor cells and potential for drug delivery applications. *Expert Opinion on Drug Delivery*, *7*(9), 1063–1077. Available from https://doi.org/10.1517/17425247.2010.502560.

Roglic, G. (2016). WHO global report on diabetes: A summary. *International Journal of Noncommunicable Diseases*, *1*(1), 3. Available from https://doi.org/10.4103/2468-8827.184853.

Ryter, S. W., Kim, H. P., Hoetzel, A., Park, J. W., Nakahira, K., Wang, X., & Choi, A. M. K. (2007). Mechanisms of cell death in oxidative stress. *Antioxidants & Redox Signaling*, *9*(1), 49–89. Available from https://doi.org/10.1089/ars.2007.9.49.

Saji, V. S., Choe, H. C., & Yeung, K. W. K. (2010). Nanotechnology in biomedical applications: A review. *International Journal of Nano and Biomaterials*, *3*(2), 119–139. Available from https://doi.org/10.1504/IJNBM.2010.037801.

Santos, L., Nunes, D., Calmeiro, T., Branquinho, R., Salgueiro, D., Barquinha, P., ... Fortunato, E. (2015). Solvothermal synthesis of gallium–indium-zinc-oxide nanomaterials for electrolyte-gated transistors. *ACS Applied Materials & Interfaces*, *7*(1), 638–646. Available from https://doi.org/10.1021/am506814t.

Sarwar, S., Ali, A., Pal, M., & Chakrabarti, P. (2017). Zinc oxide nanomaterials provide anticholera activity by disrupting the interaction of cholera toxin with the human GM1 receptor. *The Journal of Biological Chemistry*, *292*(44), 18303–18311. Available from https://doi.org/10.1074/jbc.M117.793240.

Shaban, M., Mohamed, F., & Abdallah, S. (2018). Production and characterization of superhydrophobic and antibacterial coated fabrics utilizing ZnO nanocatalyst. *Scientific Reports*, *8*(1), 3925. Available from https://doi.org/10.1038/s41598-018-22324-7.

Sharma, H., Kumar, K., Choudhary, C., Mishra, P. K., & Vaidya, B. (2016). Development and characterization of metal oxide nanomaterials for the delivery of anticancer drug. *Artificial Cells, Nanomedicine, and Biotechnology*, *44*(2), 672–679. Available from https://doi.org/10.3109/21691401.2014.978980.

Shivaramakrishnan, B., Gurumurthy, B., & Balasubramanian, A. (2017). Potential biomedical applications of metallic nanobiomaterials: A review. *International Journal of Pharmaceutical Sciences and Research*, *8*(3), 985–1000. Available from https://doi.org/10.13040/IJPSR.0975-8232.8(3)0.985-00.

Singh, B. R. B. N., Rawat, A. K. S., Khan, W., Naqvi, A. H., & Singh, B. R. B. N. (2014). Biosynthesis of stable antioxidant ZnO nanomaterials by Pseudomonas aeruginosa rhamnolipids. *PLOS ONE*, *9*(9), e106937. Available from https://doi.org/10.1371/journal.pone.0106937, e106937.

Soejima, T., Takada, K., & Ito, S. (2013). Alkaline vapor oxidation synthesis and electrocatalytic activity toward glucose oxidation of CuO/ZnO composite nanoarrays. *Applied Surface Science*, *277*, 192–200. Available from https://doi.org/10.1016/j.apsusc.2013.04.024.

Soren, S., Kumar, S., Mishra, S., Jena, P. K., Verma, S. K., & Parhi, P. (2018). Evaluation of antibacterial and antioxidant potential of the zinc oxide nanomaterials synthesized by aqueous and polyol method. *Microbial Pathogenesis*, *119*, 145–151. Available from https://doi.org/10.1016/j.micpath.2018.03.048.

SoYoon, S., Ramadoss, A., Saravanakumar, B., & Kim, S. J. (2014). Novel Cu/CuO/ZnO hybrid hierarchical nanostructures for non-enzymatic glucose sensor application. *Journal of Electroanalytical Chemistry*, *717–718*, 90–95. Available from https://doi.org/10.1016/j.jelechem.2014.01.012.

Sridar, R., Ramanane, U. U., & Rajasimman, M. (2018). ZnO nanomaterials – Synthesis, characterization and its application for phenol removal from synthetic and pharmaceutical industry wastewater. *Environmental Nanotechnology, Monitoring & Management*, *10*, 388–393. Available from https://doi.org/10.1016/j.enmm.2018.09.003.

Strano, V., & Mirabella, S. (2017). Low-cost and facile synthesis of $Ni(OH)_2$/ZnO nanostructures for high-sensitivity glucose detection. *Nanotechnology*, *29*(1), 15502. Available from https://doi.org/10.1088/1361-6528/aa98ec.

Tan, L., Liu, J., Zhou, W., Wei, J., & Peng, Z. (2014). A novel thermal and pH responsive drug delivery system based on ZnO@PNIPAM hybrid nanomaterials. *Materials Science and Engineering: C*, *45*, 524–529. Available from https://doi.org/10.1016/j.msec.2014.09.031.

Tang, X., Choo, E. S. G., Li, L., Ding, J., & Xue, J. (2010). Synthesis of ZnO nanomaterials with tunable emission colors and their cell labeling applications. *Chemistry of Materials*, *22*(11), 3383–3388. Available from https://doi.org/10.1021/cm903869r.

Thareja, R. K., & Shukla, S. (2007). Synthesis and characterization of zinc oxide nanomaterials by laser ablation of zinc in liquid. *Applied Surface Science*, *253*(22), 8889–8895. Available from https://doi.org/10.1016/j.apsusc.2007.04.088.

Thatoi, P., Kerry, R. G., Gouda, S., Das, G., Pramanik, K., Thatoi, H., & Patra, J. K. (2016). Photo-mediated green synthesis of silver and zinc oxide nanomaterials using aqueous extracts of two mangrove plant species, Heritiera fomes and Sonneratia apetala and investigation of their biomedical applications. *Journal of Photochemistry and Photobiology B: Biology*, *163*, 311–318. Available from https://doi.org/10.1016/j.jphotobiol.2016.07.029.

Tripathy, N., Ahmad, R., Ko, H. A., Khang, G., & Hahn, Y.-B. (2015). Enhanced anticancer potency using an acid-responsive ZnO-incorporated liposomal drug-delivery system. *Nanoscale*, *7*(9), 4088–4096. Available from https://doi.org/10.1039/c4nr06979j.

Tripathy, N., & Kim, D.-H. H. (2018). Metal oxide modified ZnO nanomaterials for biosensor applications. *Nano Convergence*, *5*(1), 27. Available from https://doi.org/10.1186/s40580-018-0159-9.

Upadhyay, P. K., Jain, V. K., Sharma, K., & Sharma, R. (2020). Synthesis and applications of ZnO nanomaterials in biomedicine. *Research Journal of Pharmacy and Technology*, *13*(4), 1636. Available from https://doi.org/10.5958/0974-360X.2020.00297.8.

Upadhyaya, L., Singh, J., Agarwal, V., Pandey, A. C., Verma, S. P., Das, P., & Tewari, R. P. (2015). Efficient water soluble nanostructured ZnO grafted O-carboxymethyl chitosan/curcumin-nanocomposite for cancer therapy. *Process Biochemistry*, *50*(4), 678–688. Available from https://doi.org/10.1016/j.procbio.2014.12.029.

Urban, B. E., Neogi, P. B., Butler, S. J., Fujita, Y., & Neogi, A. (2011). Second harmonic imaging of plants tissues and cell implosion using two-photon process in ZnO nanomaterials. *Journal of Biophotonics*, *5*(3), 283–291. Available from https://doi.org/10.1002/jbio.201100076.

Urban, B. E., Neogi, P., Senthilkumar, K., Rajpurohit, S. K., Jagadeeshwaran, P., Kim, S., . . . Neogi, A. (2012). Bioimaging using the optimized nonlinear optical properties of ZnO nanomaterials. *IEEE Journal of Selected Topics in Quantum Electronics*, *18*(4), 1451–1456. Available from https://doi.org/10.1109/jstqe.2012.2184793.

Vishnukumar, P., Vivekanandhan, S., Misra, M., & Mohanty, A. K. (2018). Recent advances and emerging opportunities in phytochemical synthesis of ZnO nanostructures.

Materials Science in Semiconductor Processing, *80*, 143–161. Available from https://doi.org/10.1016/j.mssp.2018.01.026.

Wang, Z. L. (2004a). Zinc oxide nanostructures: Growth, properties and applications. *Journal of Physics Condensed Matter*, *16*(25). Available from https://doi.org/10.1088/0953-8984/16/25/R01.

Wang, J., Lee, J. S., Kim, D., & Zhu, L. (2017). Exploration of zinc oxide nanomaterials as a multitarget and multifunctional anticancer nanomedicine. *ACS Applied Materials & Interfaces*, *9*(46), 39971–39984. Available from https://doi.org/10.1021/acsami.7b11219.

Wang, X., Liu, J., Song, J., & Wang, Z. L. (2007). Integrated nanogenerators in biofluid. *Nano Letters*, *7*(8), 2475–2479. Available from https://doi.org/10.1021/nl0712567.

Wilhelmi, V., Fischer, U., Weighardt, H., Schulze-Osthoff, K., Nickel, C., Stahlmecke, B., ... Albrecht, C. (2013). Zinc oxide nanomaterials induce necrosis and apoptosis in macrophages in a p47phox- and Nrf2-independent manner. *PLOS ONE*, *8*(6), e65704. Available from https://doi.org/10.1371/journal.pone.0065704.

Wollina, U., Wiegand, Boldt, Strehle., & Hipler, C. (2013). Skin-protective effects of a zinc oxide-functionalized textile and its relevance for atopic dermatitis. Clinical, cosmetic and investigational. *Dermatology (Basel, Switzerland)*, 115. Available from https://doi.org/10.2147/CCID.S44865.

Wu, J., & Yin, F. (2013). Easy fabrication of a sensitive non-enzymatic glucose sensor based on electrospinning CuO-ZnO nanocomposites. *Integrated Ferroelectrics*, *147*(1), 47–58. Available from https://doi.org/10.1080/10584587.2013.790695.

Xiong, H.-M. (2013). ZnO nanomaterials applied to bioimaging and drug delivery. *Advanced Materials*, *25*(37), 5329–5335. Available from https://doi.org/10.1002/adma.201301732.

Xiong, H.-M., Xu, Y., Ren, Q.-G., & Xia, Y.-Y. (2008). Stable aqueous ZnO@polymer core – shell nanomaterials with tunable photoluminescence and their application in cell imaging. *Journal of the American Chemical Society*, *130*(24), 7522–7523. Available from https://doi.org/10.1021/ja800999u.

Yadav, R. S., Mishra, P., & Pandey, A. C. (2010). Tuning the band gap of ZnO nanomaterials by ultrasonic irradiation. *Inorganic Materials*, *46*(2), 163–167. Available from https://doi.org/10.1134/S0020168510020135.

Yakimova, R. (2012). ZnO materials and surface tailoring for biosensing. *Frontiers in Bioscience*, *E4*(1), 254–278. Available from https://doi.org/10.2741/e374.

Yang, Y., Guo, W., Zhang, Y., Ding, Y., Wang, X., & Wang, Z. L. (2011). Piezotronic effect on the output voltage of P3HT/ZnO micro/nanowire heterojunction solar cells. *Nano Letters*, *11*(11), 4812–4817. Available from https://doi.org/10.1021/nl202648p.

Yang, X., Shao, H., Liu, W., Gu, W., Shu, X., Mo, Y., ... Jiang, M. (2015). Endoplasmic reticulum stress and oxidative stress are involved in ZnO nanoparticle-induced hepatotoxicity. *Toxicology Letters*, *234*(1), 40–49. Available from https://doi.org/10.1016/j.toxlet.2015.02.004.

Yuan, Q., Hein, S., & Misra, R. D. K. (2010). New generation of chitosan-encapsulated ZnO quantum dots loaded with drug: Synthesis, characterization and in vitro drug delivery response. *Acta Biomaterialia*, *6*(7), 2732–2739. Available from https://doi.org/10.1016/j.actbio.2010.01.025.

Zhang, P., & Liu, W. (2010). ZnO QD@PMAA-co-PDMAEMA nonviral vector for plasmid DNA delivery and bioimaging. *Biomaterials*, *31*(11), 3087–3094. Available from https://doi.org/10.1016/j.biomaterials.2010.01.007.

Zhang, Y., Wang, L., Liu, X., Yan, Y., Chen, C., & Zhu, J. (2005). Synthesis of nano/micro zinc oxide rods and arrays by thermal evaporation approach on cylindrical shape substrate. *The Journal of Physical Chemistry. B*, *109*(27), 13091–13093. Available from https://doi.org/10.1021/jp050851z.

Zhang, Z.-Y., & Xiong, H.-M. (2015). Photoluminescent ZnO nanomaterials and their biological applications. *Materials*, *8*(6), 3101–3127. Available from https://doi.org/10.3390/ma8063101.

Zhang, H.-J., Xiong, H.-M., Ren, Q.-G., Xia, Y.-Y., & Kong, J.-L. (2012). ZnO@silica core–shell nanomaterials with remarkable luminescence and stability in cell imaging. *Journal of Materials Chemistry*, *22*(26), 13159. Available from https://doi.org/10.1039/c2jm30855j.

Zhang, Z.-Y., Xu, Y.-D., Ma, Y.-Y., Qiu, L.-L., Wang, Y., Kong, J.-L., & Xiong, H.-M. (2013). Biodegradable ZnO@polymer core-shell nanocarriers: pH-triggered release of doxorubicin in vitro. *Angewandte Chemie International Edition*, *52*(15), 4127–4131. Available from https://doi.org/10.1002/anie.201300431.

Zhang, Z., Shen, W., Xue, J., Liu, Y., Liu, Y., Yan, P., ... Tang, J. (2018). Recent advances in synthetic methods and applications of silver nanostructures. *Nanoscale Research Letters*, *13*(1), 54. Available from https://doi.org/10.1186/s11671-018-2450-4.

Zhou, C., Xu, L., Song, J., Xing, R., Xu, S., Liu, D., & Song, H. (2014). Ultrasensitive nonenzymatic glucose sensor based on three-dimensional network of ZnO-CuO hierarchical nanocomposites by electrospinning. *Scientific Reports*, *4*, 7382. Available from https://doi.org/10.1038/srep07382.

Zhou, J., Xu, N. S., & Wang, Z. L. (2006). Dissolving behavior and stability of ZnO wires in biofluids: A study on biodegradability and biocompatibility of ZnO nanostructures. *Advanced Materials*, *18*(18), 2432–2435. Available from https://doi.org/10.1002/adma.200600200.

Recent progress on titanium oxide nanostructures for biosensing applications

17

Monsur Islam, Ahsana Sadaf, Dario Mager and Jan G. Korvink
Institute of Microstructure Technology, Karlsruhe Institute of Technology, Karlsruhe, Germany

17.1 Introduction

A biosensor is an analytical device which recognizes a biological substance and converts the recognition into a measurable or readable signal (Mehrotra, 2016; Su, Jia, Hou, & Lei, 2011; Wang, 1999). Since the first development in 1962 by Leland C. Clark (Mohanty & Kougianos, 2006), biosensors have been a hot topic in research communities. Biosensors have a huge impact on healthcare diagnostics, as well as on the monitoring and pharmaceutical industries. Rapid detection, portability, and the low cost of biosensors have been proved an essential step in the decentralization of clinical applications, by bringing emergency-room screening to self-testing at home.

A typical biosensor system consists of an analyte recognition element, a transducer, and a signal readout or measurement system. A schematic of a typical biosensor is shown in Fig. 17.1 (Solanki et al., 2011). Typical requirements for a biosensor include high sensitivity, a fast response time, and high selectivity. Low fabrication cost and mass-scale fabrication capabilities are also desirable for biosensors, toward rendering them affordable to users. Such qualities of the biosensor depend mainly on the different parts of the biosensor. With tremendous advances in nanotechnology and nanoscience, a significant progress has been observed in using nanomaterials as different components in a biosensor assembly (George, Antony, & Mathew, 2018; Lan, Yao, Ping, & Ying, 2017; Solanki et al., 2011; Zhu, Yang, Li, Du, & Lin, 2015). Nanomaterials offer high surface-to-volume ratio, which results in remarkably high sensitivity compared to traditional sensing materials. Furthermore, the small sizes of the nanomaterial patches help reducing the footprint of the biosensor, and excellent electronic properties decrease the power requirement. Such properties have enabled easier fabrication of portable, even wearable, devices (Kalambate et al., 2019; Kim, Kumar, Bandodkar, & Wang, 2017).

Among the many nanomaterials, titanium dioxide (TiO_2) nanomaterials have gathered significant attention among researchers. Over the past decades, TiO_2 nanomaterials have been widely researched for biosensing applications, toward the detection of numerous bioanalytes. Apart from the inherent properties of

Metal Oxides for Biomedical and Biosensor Applications. DOI: https://doi.org/10.1016/B978-0-12-823033-6.00015-6

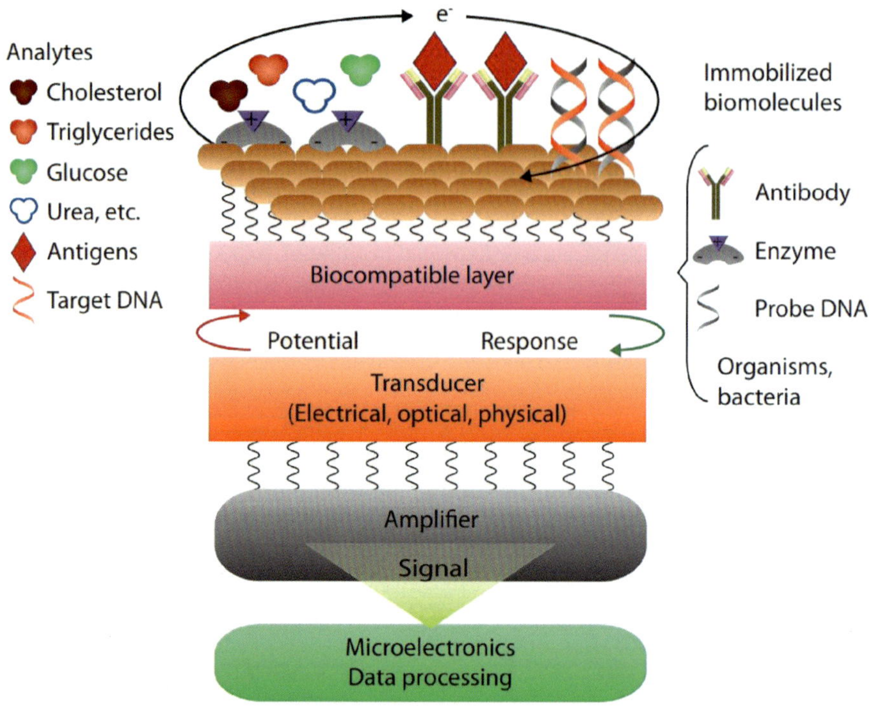

Figure 17.1 Schematic representation of a biosensor.
Source: Reprinted with permission from Solanki, P. R., Kaushik, A., Agrawal, V. V., & Malhotra, B. D. (2011). Nanostructured metal oxide-based biosensors. NPG Asia Materials 3, 17–24. https://doi.org/10.1038/asiamat.2010.137. Copyright 2011 Nature Publishing Group.

nanomaterials, TiO_2 nanomaterials further offer excellent biocompatibility, nontoxicity, photocatalytic activity, and inexpensive synthesis capabilities (Bai & Zhou, 2014). Furthermore, TiO_2 nanomaterials can be synthesized in various morphologies, which allow for a wider range of surface area. Such properties have enabled TiO_2 nanomaterials as biosensing material for the detection of various bioanalytes. In 2014 Bai and Zhou published an extensive review of the TiO_2-based biosensors up to the date of publication (Bai & Zhou, 2014). Therefore in this review, we focus on TiO_2-based biosensors reported over the last decade (2011–2020).

17.2 Properties of TiO_2

TiO_2 belongs to the family of transition metal oxides. In nature, it can exist in four different crystal groups: anatase (I41/amd), rutile (P42/mnm), brookite (Pbca), and TiO_2(B) (C2/m). The crystal structures of these phases are shown in Fig. 17.2 (Haggerty et al., 2017; Dylla et al. 2013) and their crystal properties are summarized in Table 17.1. In general, all the phases of TiO_2 are made of distorted

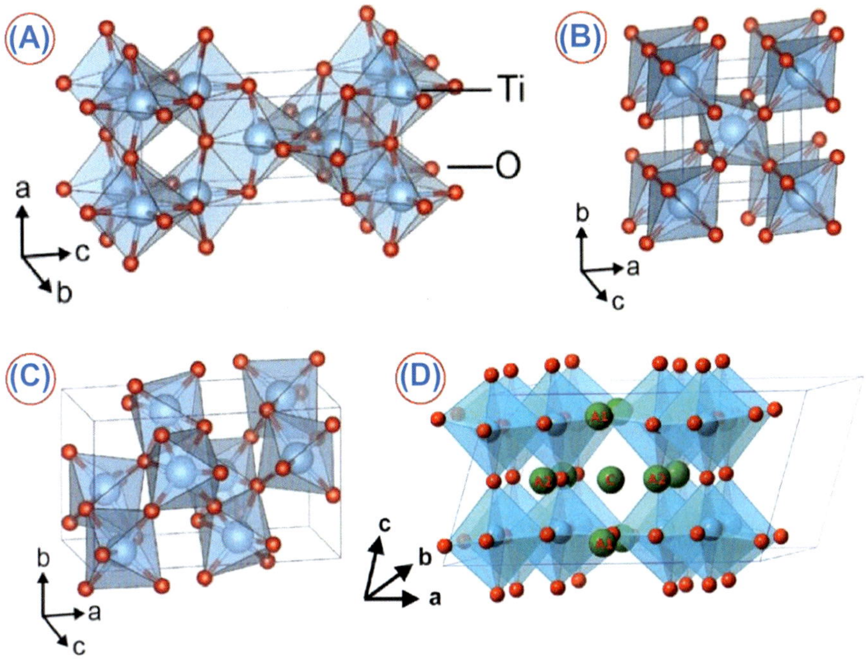

Figure 17.2 Crystal structures of TiO_2. (A) Anatase (tetragonal, I41/amd). (B) Rutile (tetragonal, P42/mnm). (C) Brookite (orthorhombic, Pbca). (D) Crystal structure of TiO_2(B) phase polyhydra.
Source: (C) Reprinted with permission from Haggerty, J. E. S., Schelhas, L. T., Kitchaev, D. A., Mangum, J. S., Garten, L. M., Sun, ... Tate, J. (2017). High-fraction brookite films from amorphous precursors. *Scientific Reports*, 7, 15232. https://doi.org/10.1038/s41598-017-15364-y. Copyright 2017 Nature Publishing Group; (D) Reprinted with permission from Dylla, A. G., Henkelman, G., & Stevenson, K. J. (2013). Lithium insertion in nanostructured TiO_2 (B) architectures. *Accounts of Chemical Research*, 46, 1104–1112. https://doi.org/10.1021/ar300176y. Copyright 2013 American Chemical Society.

Table 17.1 Lattice parameters and band gap of different phases of TiO_2.

Phase	Crystal structure	Density (g/cm^3)	a (Å)	b (Å)	c (Å)	Band gap (eV)
Rutile	Tetragonal	4.24	4.5937	4.5937	2.9581	3.00
Anatase	Tetragonal	3.83	3.7842	3.7842	9.5146	3.20
Brookite	Orthorhombic	3.17	9.16	5.43	5.13	3.26
TiO_2 (B)	Monoclinic	3.64	12.16	3.74	6.51	

Source: Reprinted with permission from Rahimi, N., Pax, R. A., & Gray, E. M. (2016). Review of functional titanium oxides. I: TiO_2 and its modifications. *Progress in Solid State Chemistry*, 44, 86–105. https://doi.org/10.1016/j.progsolidstchem.2016.07.002. Copyright 2016 Elsevier.

octahedra, where one central Ti^{4+} coordinates with six O^{2-} forming a TiO_6 unit. Each phase of the TiO_2 features different octahedral assembly to form the TiO_6 unit. Anatase and rutile phases form the TiO_6 octahedra by connecting their vertices and edges, forming a tetragonal structure. Such formation makes rutile and anatase the more stable forms than the brookite and TiO_2(B) phases (Rahimi, Pax, & Gray, 2016), which also leads to easy synthesis of rutile and anatase phases. Due to such properties, anatase and rutile are the most exploited forms in biosensing application. Furthermore, rutile exhibits higher stability at ambient temperature and pressure in macroforms (particle size >30 nm) due to its more compact crystal structure, whereas anatase becomes the most stable phase when particle size becomes lower than 5 nm (Rahimi et al., 2016; Zhu, Zhang, Hong, & Yin, 2005).

TiO_2 is known for its excellent photocatalytic activities. Due to the deep valence band (VB) of TiO_2, holes are generated on the surface of TiO_2 upon exposure to photons (Guo, Zhou, Ma, & Yang, 2019). Such surface holes allow to harvest free electrons produced by an outside reaction. For photocatalytic activities, the anatase phase of TiO_2 offers better performance than the rutile phase (Guo et al., 2019; Luttrell et al., 2015; Zhang, Zhou, Liu, & Yu, 2014). Due to nanocrystalline sizes, anatase offers a higher surface area, thereby resulting in more active sites. Furthermore, the higher concentration of oxygen vacancies and larger bandgap in anatase result in a higher charge separation efficiency and redox activity than for rutile nanoparticles (NPs).

The most important property of TiO_2 toward biosensing is its excellent biocompatibility and nontoxicity. Furthermore, titanium is known to form coordination bonds with amine and carboxyl groups (Chen & Dong 2003; He et al., 2015), which make TiO_2 surfaces a preferable interface for the immobilization of various biomolecules such as enzymes. Upon immobilization, TiO_2 interfaces maintain the enzyme's activity for long-term applications. Availability of various nanostructures of TiO_2 further enables immobilization onto a larger surface area, allowing for higher surface reactivity of a biosensor assembly. Furthermore, TiO_2 acts as an n-type semiconductor, which is known for its electron-accepting characteristics (Anitha, Banerjee, & Joo, 2015). Such characteristic makes it possible to harvest the electron produced by the reaction between the enzyme and the targeted bioanalyte by the TiO_2 surfaces. In addition to that, the photocatalytic properties, mentioned above, also enable TiO_2 nanomaterials as a transducer material for photocatalytic biosensor devices.

17.3 Synthesis of TiO_2 nanostructures for biosensors

Several methodologies have been developed over the years for the synthesis of TiO_2 nanostructures. While categorizing the synthesis methodologies adapted for biosensing applications, a number synthesis routes found in the literature have attracted disproportionate attention in the last decade. Among these methodologies, hydrothermal synthesis, and anodization, have been the most popular used for the fabrication of biosensors. The following sections present a brief description of these two methodologies along with some lesser used synthesis routes used in this field of research.

17.3.1 Hydrothermal method

The hydrothermal method is in the ceramic industry for the synthesis of nanostructures. It typically uses an aqueous or nonaqueous solution of the precursor as the starting material and is conducted in an autoclave at an elevated temperature and pressure (Lee, Mazare, & Schmuki, 2014). The synthesis mechanism includes formation of a supercritical fluid, which allows for high solubility of the solid precursor and precipitation of the nanostructures of inorganic materials. Due to its facile synthesis mechanism and control over the nanostructures, the hydrothermal method has been popular to grow 1-D nanostructures of TiO_2 for biosensing applications. Various nanostructures of TiO_2 such as nanorods (NRs), nanowires (NWs), and nanotubes (NTs) were synthesized. For example, Liu and Aydil synthesized TiO_2 nanorods (TNRs) on transparent conductive fluorine-doped tin oxide substrates by autoclaving a solution of titanium butoxide at 80°C–200°C for 1–24 h (Liu & Aydil, 2009).The synthesized NRs featured an average diameter of 90 nm and the length ranged from 600 to 2 μm (Fig. 17.3A). The hydrothermally synthesized TiO_2 nanostructures also allow to combine with different materials (e.g., polymer, metal, and metal oxide NPs) to yield a composite nanostructure. For example, Zheng et al. synthesized a poly (3,4-ethylenedioxythiophene) (PEDOT)/TiO_2 NW composite to construct a PEC biosensor (Zheng, Zhang, Liu, Zhou, & Alwarappan, 2020). To synthesis the composite, the authors first synthesized the TiO_2 NWs with a diameter of 110–200 nm and a length of 1.5–2 μM using the hydrothermal method (Fig. 17.3B and D), and then deposited PEDOT over the TiO_2 NWs using a second hydrothermal method (Fig. 17.3C and E).

17.3.2 Anodization method

Anodization is performed in an electrolytic cell, where titanium is used as the anode material. An acidic solution is used as the electrolyte in the cell. When current passes through the electrolyte, hydrogen is released at the cathode and oxygen is formed at the anode. The oxygen reacts with the titanium and forms a coating of TiO_2 on the surface of the titanium (Indira, Mudali, Nishimura, & Rajendran, 2015). The anodization route generally results in the formation of TiO_2 NT structures. An example of such TiO_2 NTs is shown in Fig. 17.3F, in which the NTs feature an inner diameter of 100 nm, a wall thickness of 15 nm, and a length up to 10 μM (Khaliq et al., 2020). The anodization-derived TiO_2 NTs further allow doping with several elements such as metals, metal oxides, and quantum dots. For example, Khaliq et al. deposited Cu_2O NPs on the TiO_2 NTs using a sonication-assisted chemical bath deposition method (Fig. 17.3G) (Khaliq et al., 2020). In another example, Grochowska et al. used magnetron sputtering to deposit Au

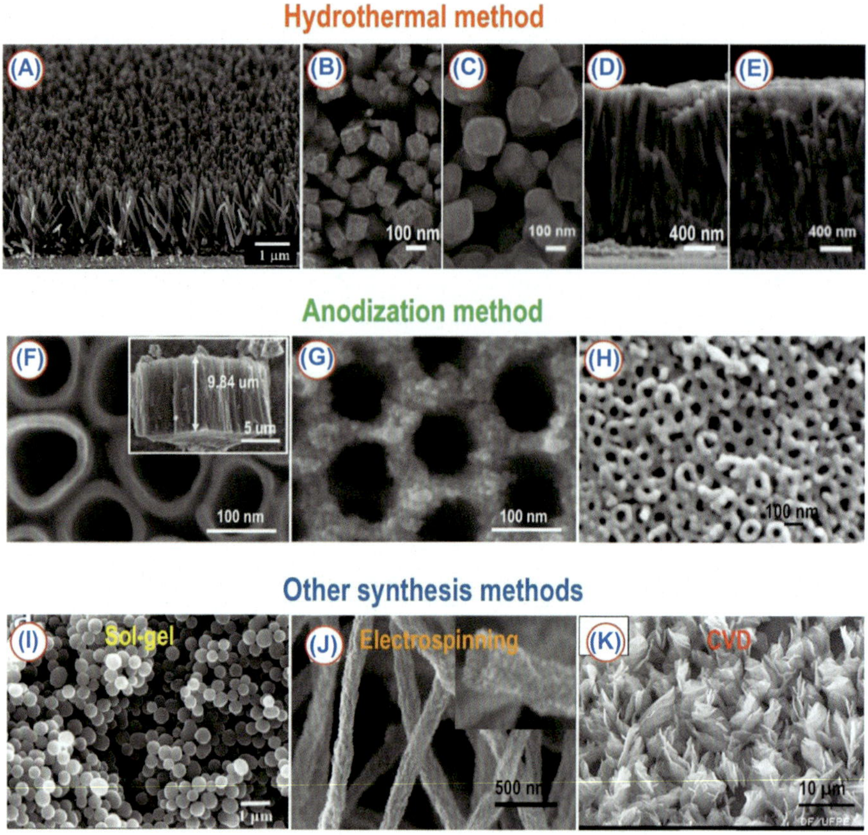

Figure 17.3 Examples of TiO_2 nanostructures synthesized using different synthesis methods for biosensing applications. *Hydrothermal method*: (A) TiO_2 nanorods (NRs) grown on a fluorinated-tin oxide (FTO) substrate. (B–E) TiO_2 nanowires (NWs). (B) and (D) show a top and side view of TiO_2 NWs, respectively, whereas (C) and (E) show TiO_2 NWs coated with poly(3,4-ethylenedioxythiophene) in top and side view, respectively. *Anodization method*: (F) TiO_2 nanotubes (NTs); the inset shows the length of the NTs. (G) Cu_2O-decorated TiO_2 NTs. (H) Au-NP-decorated TiO_2 NTs. Other synthesis methods: (I) TiO_2 nanoparticles synthesized using the sol-gel method. (J) TiO_2 nanofibers fabricated using electrospinning. (K) TiO_2 membrane nanostructure fabricated using the chemical vapor deposition (CVD) method. Source: (A) Liu, B., Aydil, E. S. (2009). Growth of oriented single-crystalline rutile TiO_2 nanorods on transparent conducting substrates for dye-sensitized solar cells. *Journal of the American Chemical Society*, 131, 3985–3990. https://doi.org/10.1021/ja8078972. Copyright 2009 American Chemical Society; (B)–(E) Reprinted with the permission from Zheng, H., Zhang, S., Liu, X., Zhou, Y., & Alwarappan, S. (2020). Synthesis of a PEDOT—TiO_2 heterostructure as a dual biosensing platform operating via photoelectrochemical (PEC) and electrochemical transduction mode. *Biosensors & Bioelectronics*, 162, 112234. https://doi.org/10.1016/j.bios.2020.112234. Copyright 2020 Elsevier; (G) Reprinted with permission from Khaliq, N., Rasheed, M. A., Cha, G., Khan, M., Karim, S., Schmuki, P., & Ali, G. (2020). Development of nonenzymatic cholesterol biosensor based on TiO_2 nanotubes decorated with Cu_2O nanoparticles, *Sensors & Actuators, B: Chemical*, 302, 127200. https://doi.org/10.1016/j.snb.2019.127200. Copyright 2020 Elsevier; (H) Reprinted with permission

(*Continued*)

nanoparticle on the TiO_2 NTs up to a thickness of 100 nm (Fig. 17.3H) (Grochowska et al., 2019).

17.3.3 Other synthesis methods

Apart from hydrothermal and anodization methods, the methods used for the synthesis of TiO_2 nanostructures for biosensors include sol-gel, electrospinning, and vapor deposition. The sol-gel method was the most popular approach for fabricating the first generation of TiO_2-based biosensors. It relies on controlled hydrolysis of a titanium alkoxide, followed by condensation and precipitation of TiO_2 nanostructures (Macwan, Dave, & Chaturvedi, 2011). The sol-gel method mostly yields formation of thin films (porous and nonporous) and NPs of TiO_2. For example, Liu et al. synthesized TiO_2 NPs featuring an average diameter of 400 nm (Fig. 17.3I) through sol-gel derivation from titanium isopropoxide, which they further used to fabricate a TiO_2-MoS_2-gold nanoparticle-based aptasensor (Liu, Liu et al., 2018). In another example, Zhang et al. used the sol-gel procedure on titanium isopropoxide to obtain a porous TiO_2 film, which was used for molecularly imprinting of polypyrrole for the fabrication of a PEC sensor for bilirubin (Zhang, Bai, & Yang, 2016).

Electrospinning results in the formation of nanofibers due to the interaction of a liquid droplet and a high electric field (Mondal & Sharma 2016). For the synthesis of TiO_2 nanofibers using electrospinning, typically a titanium alkoxide sol is used as the precursor material. Upon fabrication of the sol nanofibers, high temperature sintering is employed to transform the alkoxide to TiO_2 nanofibers. Such a methodology has been used by Ali et al. (2016), Arvand, Ghodsi, and Zanjanchi (2016), and Guo, Liu et al. (2017) to develop TiO_2 nanofiber-based biosensors. An example of electrospun TiO_2 nanofibers is shown in Fig. 17.3J.

Physical and chemical vapor deposition (PVD and CVD) are the most popular methods for the synthesis of TiO_2 nanostructures. However, these methods have not been explored much for biosensors, such as CVD grown TiO_2 films for biosensor

◀from Grochowska, K., Ryl, J., Karczewski, J., Śliwiński, G., Cenian, A., & Siuzdak, K. (2019). Nonenzymatic flexible glucose sensing platform based on nanostructured TiO_2—Au composite. *Journal of Electroanalytical Chemistry*, 837, 230–239. https://doi.org/10.1016/j.jelechem.2019.02.040. Copyright 2019 Elsevier; (I) Reprinted with permission from Liu, X., Liu, P., Tang, Y., Yang, L., Li, L., Qi, Z. ... Wong, D. K. Y. (2018). A photoelectrochemical aptasensor based on a 3D flower-like TiO_2-MoS_2-gold nanoparticle heterostructure for detection of kanamycin. *Biosensors & Bioelectronics*, 112, 193–201. https://doi.org/10.1016/j.bios.2018.04.041. Copyright 2018 Elsevier; (J) Guo, Q., Liu, L., Zhang, M., Hou, H., Song, Y., Wang, H. ... Wang, L. (2017). Hierarchically mesostructured porous TiO2 hollow nanofibers for high performance glucose biosensing. *Biosensors & Bioelectronics*, 92, 654–660. https://doi.org/10.1016/j.bios.2016.10.036. Copyright 2017 Elsevier; (K) Reprinted with permission from Zanghelini, F., Frías, I. A. M., Rêgo, M. J. B. M., Pitta, M. G. R., Sacilloti, M., Oliveira, M. D. L., & Andrade, C. A. S. (2017). Biosensing breast cancer cells based on a three-dimensional TiO_2 nanomembrane transducer. *Biosensors & Bioelectronics*, 92, 313–320. https://doi.org/10.1016/j.bios.2016.11.006. Copyright 2017 Elsevier.

applications (Dominik et al., 2017; Zanghelini et al., 2017). The methods mostly result in the fabrication of TiO_2 thin films. However, 3D membrane structures can also be obtained using these methods. For example, Zanghelini et al. synthesized a TiO_2 membrane nanostructure using a CVD method, which featured 3D butterfly-like nanostructures (Fig. 17.3K) (Zanghelini et al., 2017).

17.4 Working principle of TiO_2 biosensors

The working principle of biosensors depends on sensing mechanism of the biosensor. Different sensing mechanisms have been successfully employed using TiO_2 nanostructures. In the last decade, the most common types of biosensors fabricated using TiO_2 nanostructures are mainly two types: (A) amperometric biosensor and (B) PEC biosensor. Other methodologies such as potentiometric and optical methods have been also adapted by several researchers. However, usage of these methods has been found limited in terms of TiO_2-based biosensors when compared to amperometric and PEC biosensors. Therefore in the following subsections, we only present the mechanism for amperometric and PEC biosensors.

17.4.1 Amperometric biosensor

Schematic of a typical amperometric biosensor is shown in Fig. 17.4A. Amperometric biosensors rely on the (bio-)electrochemical reaction of biological species on the surface of the electrode, which can be measured by the current produced during the electrochemical reactions (Zhu et al., 2019). Change in the measurement current is translated to the concentration of the biological species. The electrochemical reaction is typically carried out in a three-electrode system, where the TiO_2 nanostructures are used as the working electrode. The biological species in interest are dispersed in the electrolyte solution. The output signal can be measured in two methods here, amperometry and voltammetry (Apetrei & Ghasemi-Varnamkhasti, 2013). In amperometry, a constant voltage is applied to the electrochemical cell and oxidation and reduction reactions of the electroactive species are created on the surface of the biosensor. The current produced by the oxidation–reduction reaction is collected using a typical electrochemical station. In contrary, voltametric biosensors operate in linear or cyclic voltametric mode, where variation of potential is applied between the working electrode and the counter electrode (Bai & Zhou, 2014).

17.4.2 Photoelectrochemical biosensor

In recent years, TiO_2 nanostructures have been extensively researched for the development of PEC biosensors due to excellent photocatalytic properties of TiO_2. In a typical PEC set up, a light harvesting semiconductor (TiO_2 in this case) is assembled on a transparent oxide working electrode. The sensing mechanism is

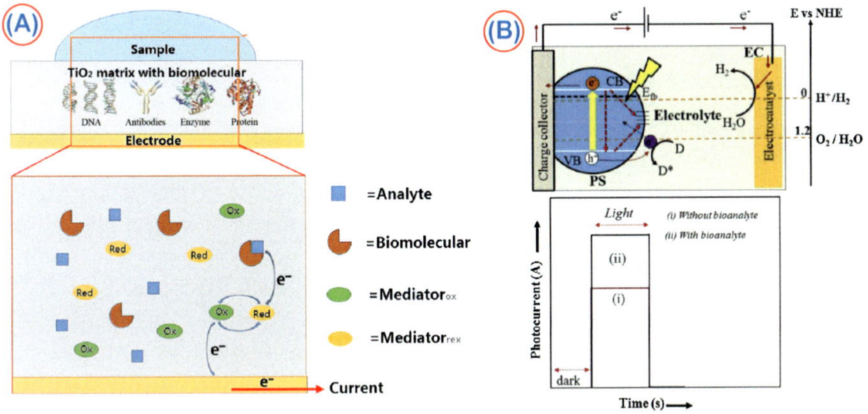

Figure 17.4 Schematics of popular TiO_2-based biosensors. (A) Amperometric biosensor. (B) Schematic of the sensing mechanism of a photoelectrochemical biosensor showing the photooxidation of the bioanalyte represented by D in the figure (top). The detection of the bioanalyte is realized by the enhancement in the photocurrent in the presence of the bioanalyte (down).
Source: (A) Reprinted with the permission from Bai, J., & Zhou, B. (2014). Titanium dioxide nanomaterials for sensor applications. *Chemical Reviews*, 114, 10131–10176. https://doi.org/10.1021/cr400625j. Copyright 2014 American Chemical Society; (B) Reprinted with the permission from Devadoss, A., Sudhagar, P., Terashima, C., Nakata, K., & Fujishima, A. (2015). Photoelectrochemical biosensors: New insights into promising photoelectrodes and signal amplification strategies. *Journal of Photochemistry and Photobiology C: Photochemistry Reviews*, 24, 43–63. https://doi.org/10.1016/j.jphotochemrev.2015.06.002. Copyright 2015 Elsevier.

schematically shown in Fig. 17.4B. Upon light irradiation, photon-excited charge carriers [holes (h^+) and electrons (e^-)] were generated on the TiO_2 surface (Devadoss et al., 2015; Zang, Lei, & Ju, 2017). The photoelectrons subsequently transport from the conduction band (CB) of the TiO_2 to the charge collector. In the absence of bioanalytes, the photogenerated holes in the VB of TiO_2 oxidize the water molecules present in the electrolytes Eq. (17.1) to generate a detectable photocurrent.

$$2H_2O + 4h^+ \leftrightarrow 4H^+ + O_2 \quad (17.1)$$

Upon exposure to bioanalytes, the biomolecules scavenge the holes to get oxidized which enable generation of further electrons generating further current. Hence, the detection of the bioanalyte is realized by the enhancement of the photocurrent, which is proportional to the amount of the bioanalyte. Such methodology is capable for the detection of small amount of bioanalyte, which is not possible in other sensing methods.

17.5 TiO_2 biosensors

17.5.1 Glucose sensor

Glucose is one of the most important biological substances that biosensors target for. Maintaining a proper concentration of glucose in the blood and homeostatic system in human body is extreme importance. Imbalance in glucose concentration triggers many deleterious consequences. Diabetes mellitus is the most concerning disease regarding such consequence. High concentration of glucose in blood can result in heart diseases, blindness, and kidney failure (Lee, Hong, Baik, Hyeon, & Kim, 2018). Therefore it is crucial to monitor the glucose level in blood. For such purpose, the development of glucose biosensors targeting high sensitivity, low cost, and high selectivity has been a hot topic in the medical science as well as food industries (Newman & Turner, 2005; Rahman, Ahammad, Jin, Ahn, & Lee, 2010). TiO_2 nanostructures have been successfully used by several researchers for successful detection of the glucose. Examples of such TiO_2 nanostructures are summarized in Table 17.2.

Toward development of TiO_2 nanostructures-based highly sensitive glucose biosensors, an enzymatic approach has been utilized by many researchers, where an electroactive enzyme is immobilized on the TiO_2 nanostructures and the enzyme promotes catalytic oxidation of the glucose. The most popular electroactive enzyme for glucose detection is glucose oxidase (GOD). During electrochemical characterization, TiO_2 and GOD exhibit typical square-shaped cyclic voltammetry (CV) curves, which are results of double-layer capacitance as shown in Fig. 17.5A. In contrast, GOD-immobilized TiO_2 electrode resulted in two well-defined redox peaks (Fig. 17.5A), which is a result of a quasireversible reaction between the flavin group in the enzyme [GOD(FAD)] and the TiO_2 nanomaterial, as mentioned in Eq. (17.2) (Guo, Liu et al., 2017).

$$\mathrm{GOD(FADH)} + 2\mathrm{H}^+ + 2\mathrm{e}^- \overset{\mathrm{TiO_2}}{\leftrightarrow} \mathrm{GOD(FADH_2)} \tag{17.2}$$

Upon exposure to glucose molecules, the redox peaks decrease due to formation of $GOD(FADH_2)$ as shown in Eq. (17.3) (Guo, Liu et al., 2017; Si, Ding, Yuan, David Lou, & Kim, 2011), and such decrease is proportional to the concentration of glucose as shown in Fig. 17.5B and C.

$$\mathrm{GOD(FAD)} + \mathrm{glucose} \rightarrow \mathrm{GOD(FADH_2)} + \mathrm{gluconic\ acid} \tag{17.3}$$

Such phenomenon leads to successful detection of glucose by the GOD-immobilized TiO_2 nanostructures without need for an oxygen reduction reaction as needed in the first-generation glucose sensors (Ronkainen, Halsall, & Heineman, 2010). In an oxidative condition, the glucose sensing is relied on the oxidation reaction of the glucose molecules by GOD(FAD), as shown in Eqs. (17.4) and (17.5) (Wooten, Karra, Zhang, & Gorski, 2014).

$$\mathrm{GOD(FAD)} + \mathrm{glucose} \rightarrow \mathrm{GOD(FADH_2)} + \mathrm{glucono} - 1,5 - \mathrm{lactone} \tag{17.4}$$

Table 17.2 Examples of TiO_2 nanostructure-based glucose sensors.

TiO_2 nanostructures	TiO_2 synthesis method	Sensing method	Sensitivity (μA/mM/cm^2)/LOD (μM)	Reference
TiO_2 nanofibers	Sol-gel/ Electrospinning	Amperometric	32.6/0.8	Guo, Liu et al. (2017)
TiO_2 NRs	Hydrothermal	Amperometric	139.69/28.67	Liu and Fu (2017)
CuO/TiO_2	Electrospinning	Amperometric	1321.0/0.39	Chen et al. (2012)
Ni-NPs/TiO_2 NTs	Anodization	Amperometric	700.2/ 2	Yu et al. (2012)
Au/TiO_2	Anodization	Amperometric	540.0/ 10	Grochowska et al. (2019)
Ag/TiO_2	Hydrothermal	Amperometric	1968.7/0.19	Dayakar et al. (2018)
Cu-Ni/TiO_2 NTs	Anodization	Amperometric	1590.9/ 5	Li, Yao et al. (2013)
Ni/CdS- Ti@TiO_2 core-shell NWs	Hydrothermal	Amperometric	1136.7/ 0.35	Guo, Huo, Han, Xu, and Li (2014)
Co-Cu NPs/TiO_2 NTs	Anodization	Amperometric	4651.0/0.6	Suneesh, Sara Vargis, Ramachandran, Nair, and Satheesh Babu (2015)
TiO_2 NRs	Hydrothermal	PEC	201.5/ 0.01	He et al. (2018)
Ag/TiO_2	Anodization	PEC	194.0/ 0.53	Xu et al. (2014)
Pt/TiO_2	Anodization	PEC	76.0/ 13.5	Xu et al. (2014)
Au-NRs/TiO_2	Anodization	PEC	812.0/ 1	Guo, Li et al. (2017)
$Ni/CdS/TiO_2$	Anodization	PEC	317.6/ 7.9	Huo, Xu, Zhang, and Xu (2015)
$Ni(OH)_2/TiO_2$	Sol-gel	PEC	163.5/0.19	Yang, Yan, and Zhang (2017)

LOD, limit of detection; *NP*, nanoparticle; *NR*, nanorod; *NT*, nanotube; *NW*, nanowire; *PEC*, photoelectrochemical.

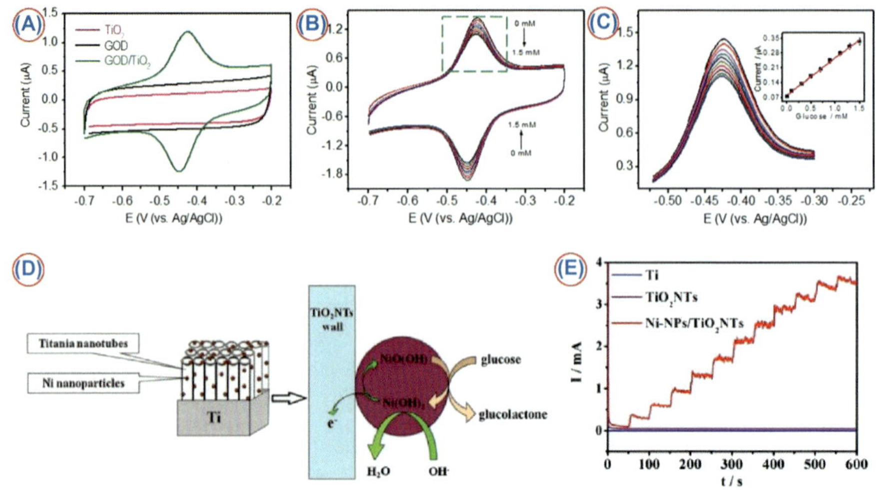

Figure 17.5 (A) Cyclic voltammetry (CV) curves of TiO_2, glucose oxidase (GOD) and GOD-immobilized TiO_2. (B) Cyclic voltammetry curves of GOD-immobilized TiO_2 electrodes in 0.1 M O_2-free PBS (pH 7.0) solution with different concentrations of glucose. (C) The enlarged view of the curves indicated by the rectangle in (B). Inset shows the calibration curve. This characterization was obtained for hierarchically porous TiO_2 nanofibers (D) Schematic of the Ni-nanoparticles/TiO_2-nanotubes (Ni-NPs/TiO_2-NTs) and their glucose sensing mechanism; (E) Amperometric response of the Ni-NPs/TiO_2-NTs for successive addition of 0.5 mM glucose in 0.1 M NaOH solution at 0.6 V and their comparison with the bare Ti and TiO_2 nanotubes.
Source: (C) Reprinted with permission from Guo, Q., Liu, L., Zhang, M., Hou, H., Song, Y., Wang, H. . . . Wang, L. (2017). Hierarchically mesostructured porous TiO_2 hollow nanofibers for high performance glucose biosensing. *Biosensors & Bioelectronics*, 92, 654–660. https://doi.org/10.1016/j.bios.2016.10.036. Copyright 2017 Elsevier; (E) Reprinted with permission from Yu, S., Peng, X., Cao, G., Zhou, M., Qiao, L., Yao, J., & He, H. (2012). Ni nanoparticles decorated titania nanotube arrays as efficient nonenzymatic glucose sensor. *Electrochimica Acta*, 76, 512–517. https://doi.org/10.1016/j.electacta.2012.05.079. Copyright 2012 Elsevier.

$$GOD(FADH_2) + O_2 \rightarrow GOD(FAD) + H_2O_2 \tag{17.5}$$

Using these methodologies, several TiO_2 nanostructures have been utilized for fabrication of glucose biosensors with high sensitivity, long stability, and high selectivity. For example, Guo et al. fabricated hierarchically porous TiO_2 nanofibers using an electrospun nanofiber film template-assisted sol-gel method (Guo, Liu et al., 2017), where the TiO_2 nanofibers featured a diameter of 100–150 nm with tens of micrometer in length and a surface area of 165.2 m^2/g^2. Such high surface area of the porous TiO_2 nanofibers facilitated immobilization of the GOD enzyme and retained the electroactive sites of the enzyme. This electrode exhibited glucose sensing within a linear range of 0.006–1.5 mM ($R = 0.997$) with a sensitivity of

22.5 μA/mM/cm^2, and a detection limit (LOD) of 2 μM in a mediator-free condition, and an oxidative condition resulted in sensing in a linear range of 0.002–3.17 mM with a sensitivity of 32.6 μA/mM/cm^2 and an LOD of 0.8 μM. In both conditions, the electrodes exhibited good stability and selectivity.

Even though the enzymatic biosensors exhibit high sensitivity toward the detection of glucose, they suffer from stability over long time, and complex and expensive storage facility (Si, Huang, Wang, & Ma, 2013). Therefore there has been a major push in the development of nonenzymatic glucose biosensors that does not require immobilization of any enzyme and still can produce high sensitivity, selectivity, and stability over long duration. In such effort, in recent years, several TiO_2-based composite materials have been emerged as nonenzymatic glucose sensing materials. For majority of the composite materials, the TiO_2 nanostructures serve as the substrate to the active electrocatalytic materials such as Ni, Cu, Pt, Au, and CuO (Guo, Li et al., 2017; Li, Yao et al., 2013; Tian, Prestgard, & Tiwari, 2014; Wang et al., 2013; Yu et al., 2012). These composite materials feature high specific surface area, high electron transfer rate, and excellent bioaffinity, which make them suitable candidate for highly sensitive biosensors. For example, Yu et al. reported the development of a highly sensitive nonenzymatic glucose sensor using a Ni nanoparticle/TiO_2 nanotube (Ni-NP/TiO_2-NT) array (Yu et al., 2012). They synthesized the sensing material by anodic growth of TiO_2 NTs on a Ti foil followed by pulse electrodeposition of Ni NPs onto the TiO_2 NTs. When exposed to glucose, the Ni-NPs enabled electrocatalytic oxidation of the glucose as shown in Fig. 17.5D and described in Eq. (17.6).

$$\mathrm{Ni(OH)} + \mathrm{glucose} \rightarrow \mathrm{Ni(OH)_2} + \mathrm{glucolactone} \tag{17.6}$$

Due to such electrocatalytic effect, the Ni-NPs/ TiO_2-NTs electrode displayed significantly higher response to the change in glucose concentration compared to the bare TiO_2-NTs electrode, as shown in Fig. 17.5E. These Ni-NPs/ TiO_2-NTs electrode exhibited a linear calibration curve for a glucose concentration range of 0.004–4.8 mM with a high sensitivity of 700.2 μA/mM/cm^2 and a low detection limit of 2 μM for a signal-to-noise ratio of 3. Toward achieving higher sensitivity, several researchers employed multiple electrocatalysts over a TiO_2 nanostructure. For example, Li et al. electrodeposited Ni and Cu NPs over a TiO_2-NTs array (Li, Yao et al., 2013). Due to the simultaneous electrocatalytical activity of the Ni and Cu NPs, the Ni-Cu/TiO_2-NTs composite electrode resulted in a high sensitivity of 1590.9 μA/mM/cm^2 toward the detection of glucose. In a similar approach, Suneesh et al. reported fabrication of a Co-Cu NPs/ TiO_2-NTs electrode and it exhibited a sensitivity of 4651 μA/mM/cm^2, which the authors claimed to be higher than any TiO_2-based glucose biosensor (Suneesh et al., 2015).

Majority of the TiO_2-based glucose biosensors developed till date have employed amperometric detection methodology. However, in recent years, PEC biosensors have been emerging to be a popular method for the detection of the glucose molecules, due to the strong photocatalytic performance, nontoxicity, and long-term photostability of TiO_2 nanomaterials. As discussed earlier, when TiO_2

nanostructures are illuminated by UV light with a wavelength higher than 3.2 eV, a photocurrent is generated from photocatalytic water oxidation by the photogenerated electron–holes on the surface of TiO_2 nanostructures. Upon exposure to glucose molecules, the photocurrent increases due to photooxidation of the glucose molecules on the TiO_2 nanostructure surface. Such increase in photocurrent enables detection of the glucose molecules. Utilizing such methodology, He et al. utilized TNR arrays for fabrication of a PEC glucose sensor, which exhibited a sensitivity of 201.5 $\mu A/mM/cm^2$ in the glucose concentration range of 0.01–0.2 mM (He et al., 2018). To enhance the PEC response for detection of glucose, several researchers have modified the TiO_2 nanostructures with narrow band gap semiconductors and noble metals (Guo, Li et al., 2017; Huo et al., 2015; Xu et al., 2014; Yang et al., 2017). Such modification promotes separation of the photogenerated electron–hole pairs and accelerates the transfer of photogenerated electrons, which consequently yield enhanced photocatalytic performance. For example, Guo et al. fabricated an Au nanorod/ TiO_2 (Au-NR/ TiO_2) heterostructure-based photochemical sensor for glucose sensing, which resulted in a sensitivity of 812 $\mu A/mM/cm^2$ and a limit of detection (LOD) of 2 μM (Guo, Li et al., 2017). Such high sensitivity was also a result of the localized surface plasmon resonance generated at the interfaces of the Au and the TiO_2. The authors further demonstrated a high selectivity and excellent stability of the sensor materials over more than 8 weeks.

17.5.2 Hydrogen peroxide (H_2O_2) sensor

H_2O_2 is an important chemical substance in bioanalysis, as concentration of H_2O_2 can be used as a measure to several biological conditions such as diabetes, atherosclerosis, immune cell activation, vascular remodeling, apoptosis, stomatal closure, and root growth (Chen, Cai, Ren, Wen, & Zhao, 2012; Halliwell, Clement, & Long, 2000; Laloi, Apel, & Danon, 2004; Narwal, Yadav, Thakur, & Pundir, 2017). Therefore detection of H_2O_2 is of practical significance both in academic and industrial research. Like glucose sensors, enzymatic approach is the most popular approach for the detection of H_2O_2, especially for electrochemical detection. Many redox proteins including horseradish peroxide (HRP), cytochrome C, myoglobin, and hemoglobin have been popularly used as the catalytic enzymes, which are immobilized on an electrode surface and reduces the H_2O_2 molecules on the electrode surface for detection of H_2O_2 (Liu, Weng, & Yang, 2017). For example, Eqs. (17.7) and (17.8) show the redox reactions of H_2O_2 in the presence of HRP (Li, Cheng, Weng, Du, & Han, 2013).

$$HRP_{reduced} + H_2O_2 \rightarrow HRP_{oxidized} + H_2O \quad (17.7)$$

$$HRP_{oxidized} + 2e^- + 2H^+ \rightarrow HRP_{reduced} + H_2O \quad (17.8)$$

The sensing process relies on the electron transfer generated in the redox reaction from the enzyme to the electrode. As mentioned in the previous section, TiO_2

nanostructures exhibit excellent capability of enzyme immobilization and excellent electron transfer capability toward such detection of bioanalytes. Therefore TiO_2 nanostructures have been also used for the development of enzymatic H_2O_2 biosensors by many researchers. For example, Li et al. fabricated an H_2O_2 biosensor by immobilizing HRP onto TNRs hydrothermally synthesized on a Ti foil, followed by sealing the assembly with Nafion for better packaging of the sensor, as schematically shown in Fig. 17.6A (Li, Cheng et al., 2013). The authors showed that absence of TiO_2 nanostructures in the electrode system resulted in no redox peaks upon exposure to H_2O_2 molecules, whereas a significantly high redox peak was observed in the presence of TNRs in the electrode system, as shown in Fig. 17.6B. This proves that it was the TiO_2 nanostructures which yielded strong electrostatic interaction to promote electron transfer between the enzyme and the electrode in the presence of H_2O_2. The Nafion/HRP/TNR/Ti electrode exhibited sensing capability within a linear range of H_2O_2 concentration ranging from 25 nM to 460 μM (Fig. 17.6C), with a high sensitivity of 416.9 μA/mM and a low LOD of 0.012 μM. Different other nanostructures of TiO_2 such as NTs and spheres have been also used in a similar enzymatic approach for the detection of H_2O_2 (Kafi, Wu, Benvenuto, & Chen, 2011; Xie et al., 2011).

In the last decade, nonenzymatic H_2O_2 biosensors have been extensively researched in comparison to the enzymatic biosensors. In these nonenzymatic biosensors, different combinations of metals (e.g., Ag, Au, and Pt) and metal oxides (e.g., MnO_2, Cu_2O, $BiVO_4$) with TiO_2 nanostructures were used toward achieving a high sensitivity for the detection of H_2O_2 in biological samples. Performance of these composite materials has been summarized in Table 17.3. Among metal-based TiO_2 nanocomposites, Khan et al. reported fabrication of Ag/ TiO_2 nanocomposite electrode using an electrochemically active biofilm approach (Khan et al., 2013). While using an Ag/ TiO_2 nanocomposite-modified glassy carbon electrode (GCE) for H_2O_2 sensing, electron transfer occurs from GCE to TiO_2, which is stored in the CB of TiO_2. These electrons further transfer to the H_2O_2 molecules through Ag for reduction of the H_2O_2 molecules (Fig. 17.6D), enabling the detection of the H_2O_2 molecules. In the absence of Ag, the electron transfer process becomes slower due to the absence of catalytic activity of the Ag, which produces significantly lower redox current during the reaction with the H_2O_2 molecules, as shown in the CV curves in Fig. 17.6E. The faster electron transfer capability of the Ag/ TiO_2 nanocomposite-modified GCE was further realized by the Nyquist diagram of electrochemical impedance spectrum (Fig. 17.6F), which shows Ag/ TiO_2 /GCE exhibited significantly lower electron transfer resistance compared to the TiO_2 /GCE. Such electrode exhibited an excellent response for H_2O_2 detection within a linear range of 0.8–43.3 μM, with a sensitivity of 65,232.8 $\mu A/mM/cm^2$ and an LOD of 0.83 μM. This sensitivity is the highest, till date, for any TiO_2-based H_2O_2 biosensor.

Among different metal oxides, Cu_2O-based TiO_2 composites exhibited high sensitivities in H_2O_2 detection. Wen et al. used flake such as Cu_2O on TiO_2 NTs for fabrication of nonenzymatic potentiometric H_2O_2 biosensors (Wen et al., 2017). The Cu_2O/ TiO_2 electrode exhibited good catalytic performance toward reduction

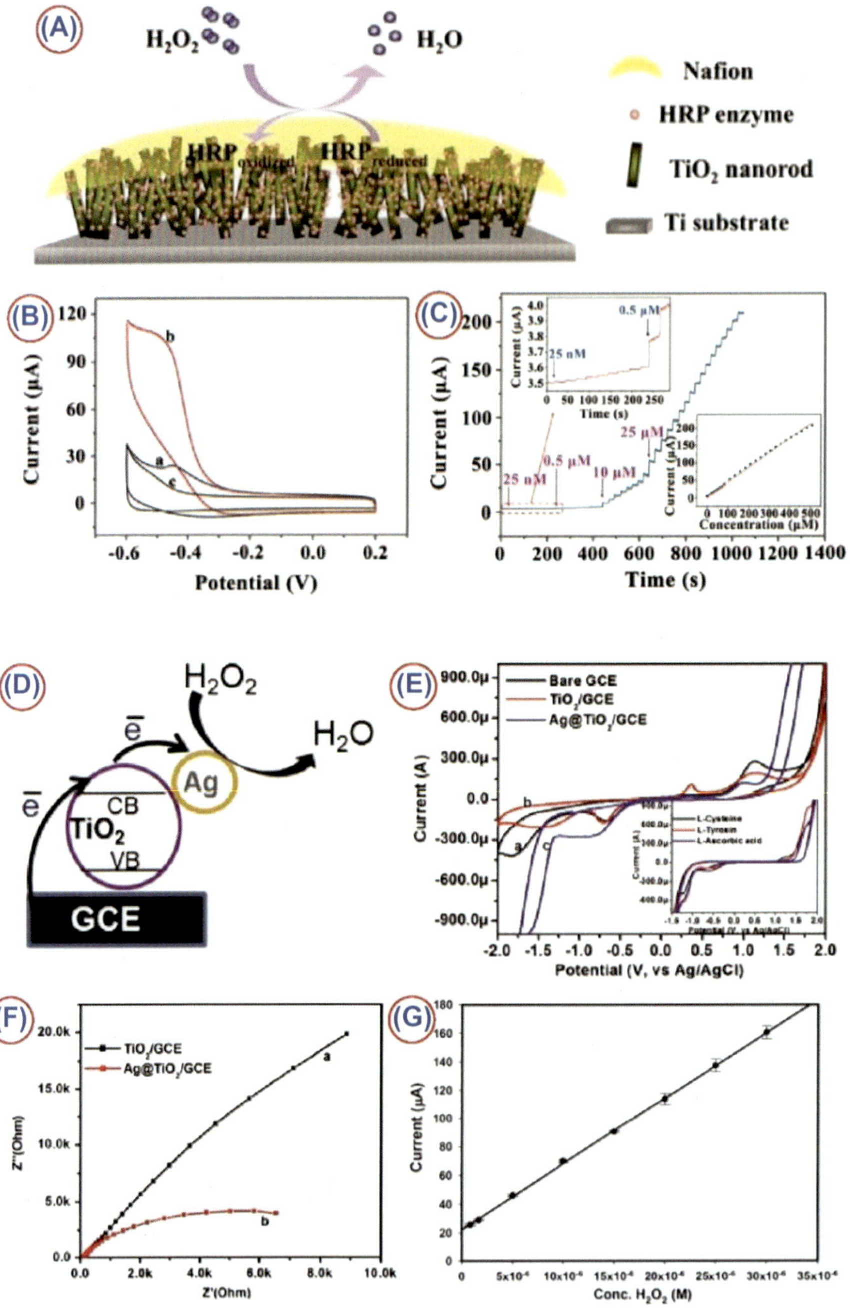

Figure 17.6 (A) Schematic representation of enzymatic H_2O_2 biosensor based on TiO_2 nanorod (TNR) electrodes. (B) Cyclic voltammetries (CVs) of Nafion/HRP/TNR/Ti in 0.1 m PBS solution in the absence (A) and presence (B) of H_2O_2, showing sensing capabilities of the electrode. However, the absence of TiO_2 in the electrode did not show any sensing

(Continued)

of H_2O_2 with a response time of 4 s. The sensitivity and LOD of these electrodes were 4412.1 μA/mM/cm^2 and 90.5 μM, respectively. Li et al. further improved the sensitivity and LOD of TiO_2 / Cu_2O for H_2O_2 sensing by using a different morphology of the composite (Li et al., 2015). They synthesized a quasi-core-shell structure of TiO_2 / Cu_2O by depositing sol-gel-derived TiO_2 layer around electrodeposited Cu_2O particles and utilized them for PEC sensor for H_2O_2 detection from living tumorigenic cells, which can be potentially used for cancel detection probe. They achieved a sensitivity in the range of 181.1–2229.1 μA/mM/cm^2 and a detection limit of 0.15 μM.

17.5.3 Urea sensor

Urea is the main end product of protein metabolism in human physiology. A typical urea level in blood serum is in the range from 2.5 to 7.5 mM/L, depending on the physiological features and health of the body (Dhawan, Sumana, & Malhotra, 2009). A higher urea level can indicate occurrence of renal dysfunction, urinary tract obstruction, dehydration, shock, and gastrointestinal bleeding, whereas a lower concentration of urea can be result of liver disease or insufficient protein intake. Hence, the health of liver and kidney in human body can be determined by monitoring the urea level in bloodstream. Although use of TiO_2 nanostructures has not been explored widely for the detection of urea as compared to glucose and H_2O_2, TiO_2 nanostructures have exhibited good potential as urea biosensor material.

Detection of urea in majority of the biosensor has been realized through immobilization of the enzyme urease onto the sensing material. Upon exposure to urea, an enzymatic catalytic hydrolysis reaction occurs Eq. (17.9), which generates NH^{4+}

◀capability (C), which proved that TiO_2 was the main sensor material in the electrode. (C) Amperometric response of Nafion/HRP/TNR/Ti electrode with successive addition of H_2O_2 The upper inset shows the amperometric response at low concentrations of H_2O_2. The lower inset shows the calibration curve of the response with different concentration of H_2O_2. (D) Nonenzymatic H_2O_2 sensing mechanism of Ag/TiO_2 nanocomposite. (E) Comparison of CV of Ag/TiO_2 nanocomposite compared to TiO_2/glassy carbon electrode (GCE) and bare GCE electrode in the presence of H_2O_2. (F) Impedance spectra of Ag/TiO_2 nanocomposite and TiO_2 in the presence of H_2O_2 for comparing the electron transfer capability. (G) Calibration curve of Ag/TiO_2 nanocomposite electrode for sensing of H_2O_2.
Source: (C) Reprinted with the permission from Li, Q., Cheng, K., Weng, W., Du, P., & Han, G. (2013). Titanium dioxide nanorod-based amperometric sensor for highly sensitive enzymatic detection of hydrogen peroxide. *Microchimica Acta*, 180, 1487–1493. https://doi.org/10.1007/s00604-013-1077-5. Copyright 2013 SpringerLink; (G) Reprinted with permission from X. Li, J. Yao, F. Liu, H. He, M. Zhou, N. Mao, ... Y. Zhang, Nickel/copper nanoparticles modified TiO_2 nanotubes for non-enzymatic glucose biosensors, *Sensors & Actuators, B: Chemical* 181 (2013) 501–508. https://doi.org/10.1016/j.snb.2013.02.035; M.M. Khan, S.A. Ansari, J. Lee, M.H. Cho, Novel Ag@TiO_2 nanocomposite synthesized by electrochemically active biofilm for nonenzymatic hydrogen peroxide sensor, *Materials Science and Engineering C*. 33 (2013) 4692–4699. https://doi.org/10.1016/j.msec.2013.07.028. Copyright 2013 Elsevier.

Table 17.3 Examples of TiO_2 nanostructure-based H_2O_2 sensors.

TiO_2 nanostructures	TiO_2 synthesis method	Sensing method	Sensitivity (μA/mM)/LOD (μM)	Reference
TiO_2 NTs	Anodization	Amperometric	− / 0.08	Kafi et al. (2011)
TiO_2 NRs	Hydrothermal	Amperometric	416.9/0.012	Li, Cheng et al. (2013)
TiO_2 microspheres	Hydrothermal	Amperometric	− / 0.05	Xie et al. (2011)
Ag/TiO_2	Hydrothermal	Amperometric	8.3/1.2	Yu et al. (2015)
Au/TiO_2/Glassy carbon	−	Amperometric	664.0/0.1	Yin, Guo, Xia, Huang, and Li (2014)
Ag/TiO_2	Sol-gel	Amperometric	65230.0/0.83	Khan, Ansari, Lee, and Cho (2013)
Pt/TiO_2/SWCNT	Electrochemical deposition	Amperometric	571.7/0.73	Han et al. (2012)
MnO_2/TiO_2	Hydrothermal	Amperometric	8.02/4.5	Ko et al. (2019)
Cu_2O/TiO_2	Anodization	Potentiometric	412.1/ 90.5	Wen, Long, and Tang (2017)
Cu_2O/TiO_2	Sol-gel	PEC	2229.1/0.15	Li, Xin, and Zhang (2015)
$BiVO_4/TiO_2$	Hydrothermal	Amperometric	3014.0/5	Derbali, Othmani, Kouass, Touati, and Dhaouadi (2020)
TiO_2 NPs	Hydrothermal	Colorimetric	− / 0.351	Gökdere, Üzer, Durmazel, Erçağ, and Apak (2019)

and HCO^{3-} and OH^{-} ions at the electrode interface (Dhawan et al., 2009). Generation of these ions increases the electronic flow at the electrode surface, which enables sensing of urea.

$$NH_2CONH_2 + 3H_2O \overset{\text{urease}}{\leftrightarrow} 2NH_4^+ + HCO_3^- + OH^- \tag{17.9}$$

Using this methodology, Srivastava et al. demonstrated detection of urea using a mediator-free microfluidics device based on titania-zirconia (TiO_2 /ZrO_2) nanocomposite (Srivastava et al., 2013). In their work, the authors mixed glutamate dehydrogenase (GLDH) with the urease and immobilized the coenzymes on the electrode surface. The electrochemical reactions with the analyte and the enzymes were performed in the presence of NADH and α-KG. The TiO_2 / ZrO_2 nanocomposite proved to create a favorable microenvironment for the immobilized enzymes, as they can reduce the tunneling distance between the active sites of enzymes and the electrode surface. Such phenomena resulted in enhanced electrochemical signal during cyclic voltammetry (Fig. 17.7A). Furthermore, addition of GLDH with urease facilitated generation of higher number of free electrons on the electrode surface through catalytic reaction among $NH4^+$ ions, α-KG and NADH, as shown in Fig. 17.7B, which further enhanced the sensing capabilities. The electrodes exhibited good sensing capabilities within a urea range of 5–100 mg/dL, with an LOD of 0.07 mg/dL and a sensitivity of 2.74 μA/mM/cm^2. In another example, Rahmanian et al. fabricated a 3D hierarchical nano-ZnO/ TiO_2 electrode on conductive fluorinated-tin oxide (FTO) for enzymatic detection of urea (Rahmanian, Mozaffari, Amoli, & Abedi, 2018). Although, the main sensor material in their device was ZnO, the presence of TiO_2 enhanced the sensing performance of their devices significantly through formation of heterojunction with ZnO. The TiO_2 layer further promoted the electron transfer between ZnO to FTO substrate. Furthermore, it enabled high electronic density to the biosensor surface as an electrostatic repulsion layer, which yielded reduction of anionic interferents at the biological media. Such characteristics resulted in a faster detection kinetics upon catalytic reaction between surface immobilized urease and urea. The nano-ZnO/ TiO_2 electrode exhibited a response time of 4 s and a detection limit of 2 mg/dL, which were comparable or even superior to other urea sensors.

TiO_2 nanostructures also demonstrated the capability of nonenzymatic detection of urea. Yang et al. reported a novel piezoelectric biosensor for urea detection by using the molecularly imprinted TiO_2 (MIT) thin film onto a transparent TiO_2 NT array (Yang et al., 2015). During piezoelectric measurements, the MIT electrodes exhibited a frequency response 4.6 times greater than the TiO_2 NT electrodes. The MIT electrodes facilitated rebinding and accumulating higher number of urea molecules compared to the TiO_2 NT electrode, resulting in higher photocatalytic degradation of urea-generating NO_3^- and CO_3^{2-} ions. The sensing mechanism is schematically illustrated in Fig. 17.7C. Generation of such ions increased the electronic transfer on the electrode surface, which yielded high sensitivity of the electrode. The MIT electrode exhibited a detection limit of 0.01 μM and a response time of 2 s.

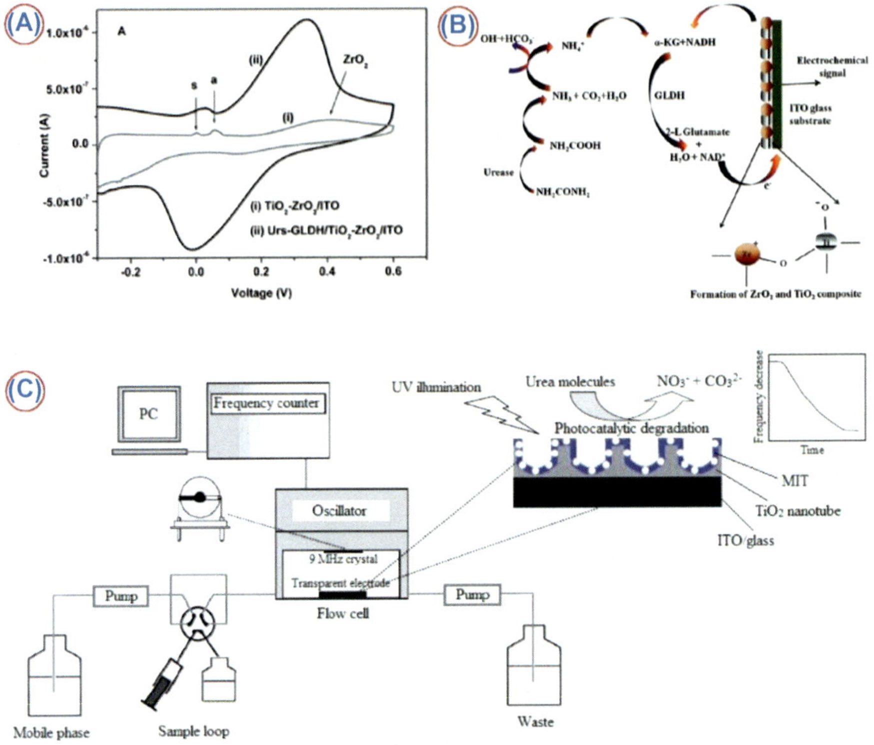

Figure 17.7 (A) Cyclic voltammetry (CV) of TiO_2/ZrO_2 nanocomposite electrode, showing higher CV response for the enzyme-immobilized electrodes and (B) schematic sensing mechanism of the TiO_2/ZrO_2 nanocomposite electrode toward detection of urea in the presence of NADH and α-KG. (C) Sensing mechanism of the molecularly imprinted transparent TiO_2 film for the detection of urea.
Source: (A) and (B): Reprinted with permission from Srivastava, S., Ali, M. A., Solanki, P. R., Chavhan, P. M., Pandey, M. K., Mulchandani, A., Malhotra, B. D. (2013). Mediator-free microfluidics biosensor based on titania–zirconia nanocomposite for urea detection. *RSC Advances*, 3, 228–235. https://doi.org/10.1039/C2RA21461J. Copyright 2013 The Royal Society of Chemistry; (C) Reprinted with permission from Yang, Z., Liu, X., Zhang, C., & Liu, B. (2015). A high-performance nonenzymatic piezoelectric sensor based on molecularly imprinted transparent TiO_2 film for detection of urea. *Biosensors & Bioelectronics*, 74, 85–90. https://doi.org/10.1016/j.bios.2015.06.022. Copyright 2015 Elsevier.

17.5.4 TiO_2 biosensors in cancer research

TiO_2 nanomaterials have been investigated by several researchers toward development of a biosensor for diagnosis of human cancer. Although a direct sensor for cancer cell is still yet to be developed, various TiO_2 nanomaterials have successfully demonstrated detection of cancer-specific indicators such as tumor markers, cells, and mutated DNA. Such studies have proved to be an important step toward early diagnosis of cancer. In 2018 Mavrič et al. published an extensive review on the development of these TiO_2-based biosensors toward human cancer detection and their impact on the research on cancer diagnosis (Mavrič et al., 2018). Therefore in this section, we mostly focus on the studies reported in 2018 and after to complement the review by Mavrič et al. The TiO_2-based biosensors used for cancer research reported in the mentioned period are summarized in Table 17.4.

While going through the works on TiO_2-based biosensors in cancer research, we noticed that researchers preferred to use a TiO_2-based composite material than a pure TiO_2 material as the sensor material. However, few researchers still solely relied on the intrinsic properties of TiO_2 nanostructures and used them in cancer research. For example, Wang et al. used a yeast templated mesoporous TiO_2 electrode for detection of pancreatic cancer microRNAs (miRNAs), due to its high specific surface and good electrocatalytic ability of oxygen reduction (Wang et al., 2020). The mesoporous TiO_2 electrode demonstrated good electrocatalysis of the oxidation reaction of the pancreatic cancer miRNAs and showed a high selectivity toward detection of a perfect match and mismatch pancreatic cancer miRNAs with single-nucleotide discrimination. A sensitivity of 63.8 nA/μM was achieved by the TiO_2 electrode for the detection of the miRNAs.

For development of TiO_2-based composite biosensors, graphene, and its derivatives [e.g., graphene oxide (GO) and reduced graphene oxide (rGO)] have been of interest for many researchers. The graphene-based materials promote surface immobilization of biomolecules due to their $\pi-\pi$ interaction on the graphene nanomaterials (Szunerits & Boukherroub 2018). Integrating graphene nanomaterials with TiO_2 nanostructures have resulted in development of high-performance biosensors toward detection of various cancer cells and tumor biomarkers. For example, Safavipour et al. demonstrated direct detection of breast cancer cell MCF-7 by developing an aptasensor based on TiO_2 NT/rGO composite (Safavipour et al., 2020). The authors immobilized human mucin-1 (MUC1) aptamer, a common biomarker for breast cancer, on the electrode surface. The TiO_2-rGO aptasensor exhibited high selectivity and sensitivity for MCF-7 cell detection with an LOD of 40 cells/mL within a detection range of 103−107 cells/mL. Furthermore, when exposed to NIR laser, the hybrid sensor exhibited high photothermal effect, reaching to a maximum temperature of 64.8°C within 10 min. Such photothermal effect can potentially be used for photothermal therapy in cancer treatment. In another example, Tian et al. developed an electrochemiluminescence (ECL) immunosensor based on graphene quantum dots (GQDs) infilled TiO_2 NTs for detection of prostate-specific antigen (PSA) toward early diagnosis of prostate cancer (Tian et al., 2019). The authors showed that the GQDs/ TiO_2 nanocomposite exhibited an ECL intensity six times stronger than that of a pure TiO_2 NT. In the immunosensor, the amount of the PSA was quantified by the decrease in the ECL intensity of the nanocomposite electrode, which occurred due to the increased electron transfer resistance induced by the antigen protein. Such ECL signal reduction was further enhanced by introducing CdTe NPs modified with Fe_3O_4 magnetic nanoparticles (CdTe/MNPs) as the quencher on the electrode surface, as schematically presented

Table 17.4 Examples of TiO_2 nanostructure-based biosensors used for cancer diagnosis.

TiO_2 nanostructures	TiO_2 synthesis method	Analyte	Sensing method	Sensitivity/LOD	Reference
Mesoporous TiO_2	Yeast template	Pancreatic cancer miRNAs	Amperometric	63.8 nA/μM / –	Wang et al. (2020)
TiO_2 nanomembrane	PVD	Breast cancer cell T47D	Impedimetric	– / 10 cells/mL	Zanghelini et al. (2017)
TiO_2 / rGO	Anodization	Breast cancer cell MCF-7	Amperometric	– / 40 cells/mL	Safavipour, Kharaziha, Amjadi, Karimzadeh, and Allafchian (2020)
rGO/TiO_2	Hydrothermal	EpCAM	Amperometric	3.24 μA/mM/cm^2/ 6.5 pg/mL	Jalil, Pandey, and Kumar (2020)
GQDs/TiO_2 NTs	Anodization	Prostate protein antigen	ECL	– / 1 fg/mL	Tian, Wang, Luan, and Zhuang (2019)
Pt/TiO_2/N-CDs	–	Carcinoembryonic antigen	PEC	43.77 μA · mL/ng / 0.48 pg/mL	Li et al. (2018)
Au/ TiO_2/ graphene	–	Carcinoembryonic antigen	Amperometric	– / 3.33 pg/mL	Chen et al. (2018)

CD, carbon dots; *ECL*, electrochemiluminescence; *EpCAM, epithelial cellular adhesion molecule*; *GQD*, graphene quantum dot; *LOD*, limit of detection; *PEC*, photoelectrochemical; *PVD*, physical vapor deposition; *rGO*, reduced graphene oxide.

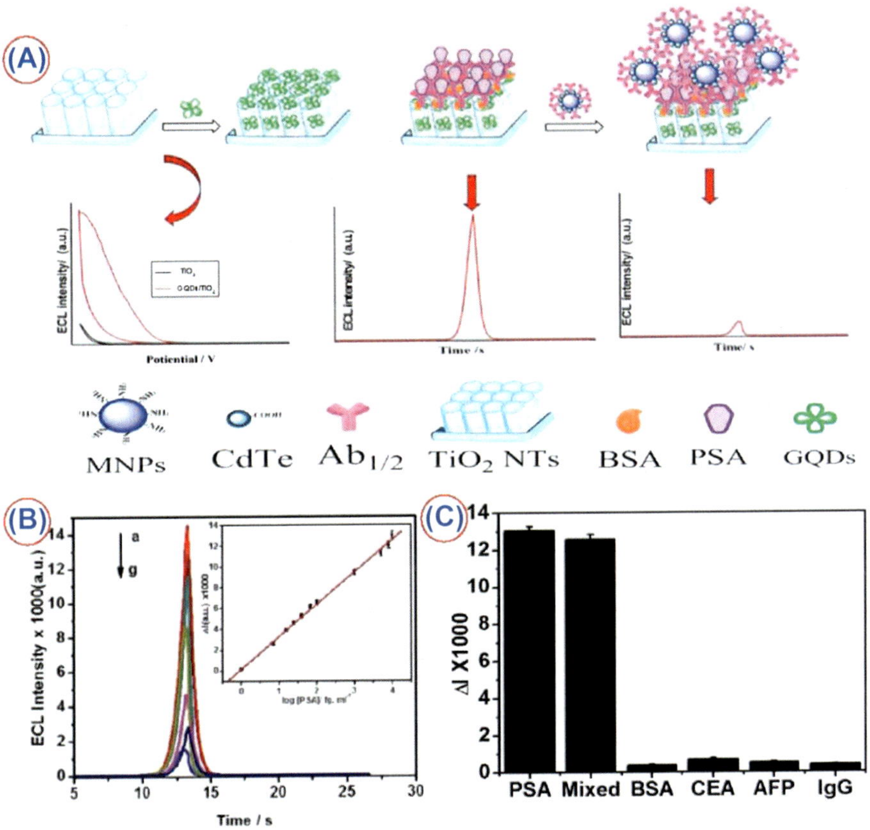

Figure 17.8 (A) Schematic illustration of the electrochemiluminescence (ECL) sensing mechanism of graphene quantum dot-filled TiO_2 nanotube electrode for detection of prostate-specific antigen (PSA) toward early diagnosis of prostate cancer. CdTe/magnetic nanoparticles (MNPs) composite was used here as electrochemiluminescence signal quencher to enhance the sensitivity. (B) The ECL intensity of the fabricated electrode in response to different concentration of PSA (fM: $a = 0$, $b = 1$, $c = 50$, $d = 100$, $e = 1000$, $f = 5000$, and $g = 10{,}000$). Inset shows the calibration curve of the ECL intensity response. (C) ECL intensity response of the ECL immunoassay in the presence of different interfering proteins at 1.0 pg/mL, showing high selectivity of the immunoassay toward PSA at 100 fg/mL. Source: (C) Reprinted with permission from Tian, C., Wang, L., Luan, F., & Zhuang, X. (2019). An electrochemiluminescence sensor for the detection of prostate protein antigen based on the graphene quantum dots infilled TiO_2 nanotube arrays. *Talanta*, 191, 103–108. https://doi.org/10.1016/j.talanta.2018.08.050. Copyright 2019 Elsevier.

in Fig. 17.8. The quencher/antigen/nanocomposite system resulted in ultra-high sensitivity and selectivity within a PSA concentration range of 1×10^{-3}–1 pg/mL (Fig. 17.8B and C). The authors further demonstrated detection of PSA in a blood serum sample to exhibit its potential to clinical use for prostate cancer detection.

17.5.5 Biosensors for different other analytes

Apart from the bioanalytes discussed in the previous sections, TiO_2 nanostructures have been used for various other bioanalytes, ranging from nucleic acids, proteins, fatty acids, and hormones (Table 17.5), which are essential for maintenance and regulation of a healthy state of a human body. Such detection often leads to early diagnosis of diseases so that an early treatment can be initiated toward fast recovery or prevention of a disease. For example, cholesterol, an important lipoprotein for biosynthesis of steroid hormones, bile acid, and vitamin D (Sulimovici & Boyd, 1970), plays an important role in diagnosis and treatment of several cardiovascular diseases, liver diseases, diabetes, and nephrotic syndrome (Narwal et al., 2019). Toward detection of cholesterol in human blood serum, Khaliq et al. fabricated an amperometric nonenzymatic biosensor by using a Cu_2O nanoparticle-decorated TiO_2 NT electrode (Khaliq et al., 2020). The TiO_2-NTs allowed for a higher electroactive site, which further increased to fivefold by the incorporation of the CuO_2 NPs on the TiO_2-NT surface. Such combination resulted in an enhanced electron transfer upon oxidation of the cholesterol molecules by the Cu_2O NPs. The fabricated electrode exhibited a wide linear range of detection (24.4–622 μM) with a sensitivity of 6034.04 $\mu A/mM/cm^2$, a low detection limit of 0.05 μM and a response time less than 3 s, which the authors claimed superior performance than previously reported cholesterol biosensors.

Another important biomolecule group is nucleic acid, which includes deoxyribonucleic acid (DNA) and ribonucleic acid (RNA). These nucleic acids have been widely used in several healthcare applications including molecular diagnostics, gene therapy, pathogen detection, and early detection of diseases (Gootenberg et al., 2017). Few efforts have been made as well toward using TiO_2 nanostructures toward detection of the nucleic acids (Gao, Sun, Lin, & Wang, 2012; Liu, Chen et al., 2018; Sakib, Pandey, Soleymani, & Zhitomirsky, 2020; Wang, Yin et al., 2019). For example, Liu et al. fabricated a PEC biosensor using an Au-NP/TiO_2-NP/FTO electrode (Liu, Chen et al., 2018). To enhance the photocurrent of the electrode, CdSe–COOH quantum dots were adsorbed on the Au/TiO_2 electrodes, which increased the electron transfer by 1.5-folds. The fabricated electrode exhibited detection of DNA molecules within a wider range of 10 fM to 0.1 μM with a low detection limit of 3 fM.

TiO_2 nanostructures further exhibited good capability in detection of cells as well. Viter et al. demonstrated use of TiO_2-NPs for developing a photoluminescence (PL) biosensor for detection of *Salmonella typhimurium* cells (Viter et al., 2017), which is one of most screened pathogens in food and agricultural industry in the European Union (Burris & Stewart, 2012). To develop the biosensor, the authors deposited TiO_2-NPs on a glass substrate followed by immobilization of anti-S-Ab antibody, which is a recognition antibody for *Salmonella* antigens. After incubating the optical surface along with *Salmonella*-antigens, a significant drop and a small left shift in the PL intensity were observed, which were proportional to the concentration of the *Salmonella*. The authors hypothesized that such detection of the *Salmonella* was attributed to the passivation of surface states, charge transfer,

Table 17.5 Examples of TiO_2 nanostructure-based biosensors of different bioanalytes.

TiO_2 nanostructures	TiO_2 synthesis method	Analyte	Sensing method	Sensitivity/LOD	Reference
Au/TiO_2 NPs	–	DNA	PEC	– / 3fM	Liu et al. (2018)
Ag / TiO_2 / rGO	Hydrothermal	8-hydroxy-2'-deoxy-guanosine	Amperometric	59 mV/10 nM	Jirjees Dhulkefl, Atacan, Bas, and Ozmen (2020)
TiO_2 nanobrush	Hydrothermal	-amyloid Peptides	PEC	114.8 A·ng/mL / 26.3 ng/mL	Lu et al. (2020)
Cu_2O/TiO_2 NTs	Anodization	Cholesterol	Amperometric	6034.0 A/mM/cm^2 / 0.05 pg/mL	Khaliq et al. (2020)
PEDOT/TiO_2 NWs	Hydrothermal	Lactate	PEC	0.1386 A/M / 0.08 M	Zheng et al. (2020)
N-doped carbon Coated TiO_2-NTs	Anodization	Ascorbic acid Dopamine Uric acid	Amperometric	– / 1.8 M – / 0.015 M – / 0.11 M	Wang, Zheng et al. (2020)
TiO_2-NGO/ Au/ Pd HSs	–	Human epididymis specific protein 4 antigen	Amperometric	9.06 A/pM / 1.48 nM	Yan et al. (2019)
$MgIn_2S_4$-TiO_2 nanoarray	Hydrothermal	Adenosine triphosphate (ATP)	PEC	5.1 A/pM / 9.2 pM	Yang et al. (2019)
CdS/$CuInS_2$/ Au/ TiO_2 NTs	Anodization	Cytochrome c	PEC	– / –	Wang, Gu et al. (2019)
TiO_2/ MoS_2/ Au	Sol-gel	Kanamycin	PEC	0.011 A/pM / 0.05 nM	Liu, Liu et al. (2018)
TiO_2 NPs	–	*Salmonella*	Photoluminescence	110^3 cL/mL–110^3 cL/mL / –	Viter et al. (2017)

LOD, limit of detection; *NP*, nanoparticle; *NT*, nanotube; *NW*, nanowire; *PEC*, photoelectrochemical; *PEDOT*, poly(3,4-ethylenedioxythiophene); *rGO*, reduced graphene oxide.

electrostatic interaction, and Förster energy transfer between the TiO_2 surface and the biological compounds.

17.6 Conclusion

TiO_2 nanostructures have been proved to be an excellent candidate for biosensor applications and extensively used for detection of numerous biological analytes. Among various nanostructures, TiO_2-NTs and -NWs have found the primary interest among the researchers, due to their high surface area to volume ratio, easy synthesis by hydrothermal and anodization route, and ability to be doped by several other functional electroactive materials. Although amperometric biosensors have got majority of the attention due to its high sensitivity and accuracy in measurement, a significant interest has been growing toward using PEC sensors utilizing the excellent photocatalytic activities of TiO_2 nanostructures. Furthermore, in the last decade, majority of the focus has been on using a TiO_2-based composite material instead of pure TiO_2 nanostructures for biosensor development, as doping or integrating other electroactive material with TiO_2 has shown significantly enhanced performance.

Toward developing biosensors with higher sensitivity and a lower detection limit, it is necessary to investigate further nanostructural schemes of TiO_2. Novel nanostructures and synthesis methods should be explored toward construction of a robust and highly efficient biosensor. In addition to these, current world still needs biosensors which are affordable to poor to middle-class society, free of clinical settings, and functional for extreme point-of-care conditions. Toward achieving these criteria, more simple and rapid synthesis routes need to be explored which can reduce the fabrication cost drastically and make the device integration faster. Furthermore, biosensing performance of the TiO_2 nanostructures should be investigated extensively in extreme point-of-care settings. If required, additional modification of the nanostructures should be investigated so that they can retain or even improve their performance in such conditions.

References

Ali, M. A., Mondal, K., Jiao, Y., Oren, S., Xu, Z., Sharma, A., ... Dong, L. (2016). Microfluidic immuno-biochip for detection of breast cancer biomarkers using hierarchical composite of porous graphene and titanium dioxide nanofibers. *ACS Applied Materials & Interfaces*, *8*, 20570–20582. Available from https://doi.org/10.1021/acsami.6b05648.

Anitha, V. C., Banerjee, A. N., & Joo, S. W. (2015). Recent developments in TiO_2 as n- and p-type transparent semiconductors: Synthesis, modification, properties, and energy-related applications. *Journal of Materials Science*, *50*, 7495–7536. Available from https://doi.org/10.1007/s10853-015-9303-7.

Apetrei, C., & Ghasemi-Varnamkhasti, M. (2013). Biosensors in food PDO authentication. *Comprehensive Analytical Chemistry*, 279–297. Available from https://doi.org/10.1016/B978-0-444-59562-1.00011-6.

Arvand, M., Ghodsi, N., & Zanjanchi, M. A. (2016). A new microplatform based on titanium dioxide nanofibers/graphene oxide nanosheets nanocomposite modified screen printed carbon electrode for electrochemical determination of adenine in the presence of guanine. *Biosensors & Bioelectronics*, *77*, 837–844. Available from https://doi.org/10.1016/j.bios.2015.10.055.

Bai, J., & Zhou, B. (2014). Titanium dioxide nanomaterials for sensor applications. *Chemical Reviews*, *114*, 10131–10176. Available from https://doi.org/10.1021/cr400625j.

Burris, K. P., & Stewart, C. N. (2012). Fluorescent nanoparticles: Sensing pathogens and toxins in foods and crops. *Trends in Food Science and Technology*, *28*, 143–152. Available from https://doi.org/10.1016/j.tifs.2012.06.013.

Chen, W., Cai, S., Ren, Q.-Q., Wen, W., & Zhao, Y.-D. (2012). Recent advances in electrochemical sensing for hydrogen peroxide: A review. *Analyst*, *137*, 49–58. Available from https://doi.org/10.1039/C1AN15738H.

Chen, X., & Dong, S. (2003). Sol-gel-derived titanium oxide/copolymer composite based glucose biosensor. *Biosensors & Bioelectronics*, *18*, 999–1004. Available from https://doi.org/10.1016/S0956-5663(02)00221-X.

Chen, Y., Li, Y., Deng, D., He, H., Yan, X., Wang, Z., ... Luo, L. (2018). Effective immobilization of Au nanoparticles on TiO_2 loaded graphene for a novel sandwich-type immunosensor. *Biosensors & Bioelectronics*, *102*, 301–306. Available from https://doi.org/10.1016/j.bios.2017.11.009.

Chen, J., Xu, L., Xing, R., Song, J., Song, H., Liu, D., ... Zhou, J. (2012). Electrospun three-dimensional porous CuO/TiO_2 hierarchical nanocomposites electrode for nonenzymatic glucose biosensing. *Electrochemistry Communications*, *20*, 75–78. Available from https://doi.org/10.1016/j.elecom.2012.01.032.

Dayakar, T., Venkateswara Rao, K., Vinod Kumar, M., Bikshalu, K., Chakradhar, B., & Ramachandra Rao, K. (2018). Novel synthesis and characterization of Ag@ TiO_2 core shell nanostructure for non-enzymatic glucose sensor. *Applied Surface Science*, *435*, 216–224. Available from https://doi.org/10.1016/j.apsusc.2017.11.077.

Derbali, M., Othmani, A., Kouass, S., Touati, F., & Dhaouadi, H. (2020). $BiVO_4/TiO_2$ nanocomposite: Electrochemical sensor for hydrogen peroxide. *Materials Research Bulletin*, *125*, 110771. Available from https://doi.org/10.1016/j.materresbull.2020.110771.

Devadoss, A., Sudhagar, P., Terashima, C., Nakata, K., & Fujishima, A. (2015). Photoelectrochemical biosensors: New insights into promising photoelectrodes and signal amplification strategies. *Journal of Photochemistry and Photobiology C: Photochemistry Reviews*, *24*, 43–63. Available from https://doi.org/10.1016/j.jphotochemrev.2015.06.002.

Dhawan, G., Sumana, G., & Malhotra, B. D. (2009). Recent developments in urea biosensors. *Biochemical Engineering Journal*, *44*, 42–52. Available from https://doi.org/10.1016/j.bej.2008.07.004.

Dominik, M., Leśniewski, A., Janczuk, M., Niedziółka-Jönsson, J., Hołdyński, M., Wachnicki, Ł., ... Śmietana, M. (2017). Titanium oxide thin films obtained with physical and chemical vapour deposition methods for optical biosensing purposes. *Biosensors & Bioelectronics*, *93*, 102–109. Available from https://doi.org/10.1016/j.bios.2016.09.079.

Dylla, A. G., Henkelman, G., & Stevenson, K. J. (2013). Lithium insertion in nanostructured TiO_2 (B) architectures. *Accounts of Chemical Research*, *46*, 1104–1112. Available from https://doi.org/10.1021/ar300176y.

Gao, H., Sun, M., Lin, C., & Wang, S. (2012). Electrochemical DNA biosensor based on graphene and TiO_2 nanorods composite film for the detection of transgenic soybean gene sequence of MON89788. *Electroanalysis*, *24*, 2283–2290. Available from https://doi.org/10.1002/elan.201200403.

George, J. M., Antony, A., & Mathew, B. (2018). Metal oxide nanoparticles in electrochemical sensing and biosensing: A review. *Microchimica Acta*, *185*, 358. Available from https://doi.org/10.1007/s00604-018-2894-3.

Gökdere, B., Üzer, A., Durmazel, S., Erçağ, E., & Apak, R. (2019). Titanium dioxide nanoparticles–based colorimetric sensors for determination of hydrogen peroxide and triacetone triperoxide (TATP). *Talanta*, *202*, 402–410. Available from https://doi.org/10.1016/j.talanta.2019.04.071.

Gootenberg, J. S., Abudayyeh, O. O., Lee, J. W., Essletzbichler, P., Dy, A. J., Joung, J., ... Zhang, F. (2017). Nucleic acid detection with CRISPR-Cas13a/C2c2. *Science (New York, N.Y.)*, *356*(80-), 438–442. Available from https://doi.org/10.1126/science.aam9321.

Grochowska, K., Ryl, J., Karczewski, J., Śliwiński, G., Cenian, A., & Siuzdak, K. (2019). Non-enzymatic flexible glucose sensing platform based on nanostructured TiO_2 – Au composite. *Journal of Electroanalytical Chemistry*, *837*, 230–239. Available from https://doi.org/10.1016/j.jelechem.2019.02.040.

Guo, C., Huo, H., Han, X., Xu, C., & Li, H. (2014). Ni/CdS bifunctional Ti@TiO_2 core–shell nanowire electrode for high-performance nonenzymatic glucose sensing. *Analytical Chemistry*, *86*, 876–883. Available from https://doi.org/10.1021/ac4034467.

Guo, L., Li, Z., Marcus, K., Navarro, S., Liang, K., Zhou, L., ... Yang, Y. (2017). Periodically patterned Au-TiO_2 heterostructures for photoelectrochemical sensor. *ACS Sensors*, *2*, 621–625. Available from https://doi.org/10.1021/acssensors.7b00251.

Guo, Q., Liu, L., Zhang, M., Hou, H., Song, Y., Wang, H., ... Wang, L. (2017). Hierarchically mesostructured porous TiO_2 hollow nanofibers for high performance glucose biosensing. *Biosensors & Bioelectronics*, *92*, 654–660. Available from https://doi.org/10.1016/j.bios.2016.10.036.

Guo, Q., Zhou, C., Ma, Z., & Yang, X. (2019). Fundamentals of TiO_2 photocatalysis: Concepts, mechanisms, and challenges. *Advanced Materials*, *31*, 1901997. Available from https://doi.org/10.1002/adma.201901997.

Haggerty, J. E. S., Schelhas, L. T., Kitchaev, D. A., Mangum, J. S., Garten, L. M., Sun, W., ... Tate, J. (2017). High-fraction brookite films from amorphous precursors. *Scientific Reports*, *7*, 15232. Available from https://doi.org/10.1038/s41598-017-15364-y.

Halliwell, B., Clement, M. V., & Long, L. H. (2000). Hydrogen peroxide in the human body. *FEBS Letters*, *486*, 10–13. Available from https://doi.org/10.1016/S0014-5793(00)02197-9.

Han, K. N., Li, C. A., Bui, M.-P. N., Pham, X.-H., Kim, B. S., Choa, Y. H., & Seong, G. H. (2012). Development of Pt/TiO_2 nanohybrids-modified SWCNT electrode for sensitive hydrogen peroxide detection. *Sensors & Actuators, B: Chemical*, *174*, 406–413. Available from https://doi.org/10.1016/j.snb.2012.08.066.

He, X.-M., Chen, X., Zhu, G.-T., Wang, Q., Yuan, B.-F., & Feng, Y.-Q. (2015). Hydrophilic carboxyl cotton chelator for titanium (IV) immobilization and its application as novel fibrous sorbent for rapid enrichment of phosphopeptides. *ACS Applied Materials & Interfaces.*, *7*, 17356–17362. Available from https://doi.org/10.1021/acsami.5b04572.

He, L., Liu, Q., Zhang, S., Zhang, X., Gong, C., Shu, H., ... Zhang, B. (2018). High sensitivity of TiO_2 nanorod array electrode for photoelectrochemical glucose sensor and its photo fuel cell application. *Electrochemistry Communications*, *94*, 18–22. Available from https://doi.org/10.1016/j.elecom.2018.07.021.

Huo, H., Xu, Z., Zhang, T., & Xu, C. (2015). Ni/CdS/TiO_2 nanotube array heterostructures for high performance photoelectrochemical biosensing. *Journal of Materials Chemistry A.*, *3*, 5882–5888. Available from https://doi.org/10.1039/C4TA07190E.

Indira, K., Mudali, U. K., Nishimura, T., & Rajendran, N. (2015). A review on TiO_2 nanotubes: Influence of anodization parameters, formation mechanism, properties, corrosion behavior, and biomedical applications. *Journal of Bio- Tribo-Corrosion*, *1*, 28. Available from https://doi.org/10.1007/s40735-015-0024-x.

Jalil, O., Pandey, C. M., & Kumar, D. (2020). Electrochemical biosensor for the epithelial cancer biomarker EpCAM based on reduced graphene oxide modified with nanostructured titanium dioxide. *Microchimica Acta.*, *187*, 275. Available from https://doi.org/10.1007/s00604-020-04233-7.

Jirjees Dhulkefl, A., Atacan, K., Bas, S. Z., & Ozmen, M. (2020). An Ag–TiO_2 –reduced graphene oxide hybrid film for electrochemical detection of 8-hydroxy-2′-deoxyguanosine as an oxidative DNA damage biomarker. *Analytical Methods*, *12*, 499–506. Available from https://doi.org/10.1039/C9AY02175B.

Kafi, A. K. M., Wu, G., Benvenuto, P., & Chen, A. (2011). Highly sensitive amperometric H_2O_2 biosensor based on hemoglobin modified TiO_2 nanotubes. *Journal of Electroanalytical Chemistry*, *662*, 64–69. Available from https://doi.org/10.1016/j.jelechem.2011.03.021.

Kalambate, P. K., Dhanjai., Huang, Z., Li, Y., Shen, Y., Xie, M., ... Srivastava, A. K. (2019). Core@shell nanomaterials based sensing devices: A review. *TrAC Trends in Analytical Chemistry*, *115*, 147–161. Available from https://doi.org/10.1016/j.trac.2019.04.002.

Khaliq, N., Rasheed, M. A., Cha, G., Khan, M., Karim, S., Schmuki, P., & Ali, G. (2020). Development of non-enzymatic cholesterol bio-sensor based on TiO_2 nanotubes decorated with Cu_2O nanoparticles. *Sensors & Actuators, B: Chemical*, *302*, 127200. Available from https://doi.org/10.1016/j.snb.2019.127200.

Khan, M. M., Ansari, S. A., Lee, J., & Cho, M. H. (2013). Novel Ag@TiO_2 nanocomposite synthesized by electrochemically active biofilm for nonenzymatic hydrogen peroxide sensor. *Materials Science and Engineering C*, *33*, 4692–4699. Available from https://doi.org/10.1016/j.msec.2013.07.028.

Kim, J., Kumar, R., Bandodkar, A. J., & Wang, J. (2017). Advanced materials for printed wearable electrochemical devices: A review. *Advanced Electronic Materials*, *3*, 1600260. Available from https://doi.org/10.1002/aelm.201600260.

Ko, W., Chen, Y., Li, M., Lai, J., Lin, K., & Novel, A. (2019). Hydrogen peroxide amperometric sensor based on hierarchical 3D porous MnO_2 $-TiO_2$ composites. *Electroanalysis*, *31*, 797–804. Available from https://doi.org/10.1002/elan.201800783.

Laloi, C., Apel, K., & Danon, A. (2004). Reactive oxygen signalling: The latest news. *Current Opinion in Plant Biology*, *7*, 323–328. Available from https://doi.org/10.1016/j.pbi.2004.03.005.

Lan, L., Yao, Y., Ping, J., & Ying, Y. (2017). Recent advances in nanomaterial-based biosensors for antibiotics detection. *Biosensors & Bioelectronics*, *91*, 504–514. Available from https://doi.org/10.1016/j.bios.2017.01.007.

Lee, H., Hong, Y. J., Baik, S., Hyeon, T., & Kim, D. (2018). Enzyme-based glucose sensor: From invasive to wearable device. *Advanced Healthcare Materials*, *7*, 1701150. Available from https://doi.org/10.1002/adhm.201701150.

Lee, K., Mazare, A., & Schmuki, P. (2014). One-dimensional titanium dioxide nanomaterials: Nanotubes. *Chemical Reviews*, *114*, 9385–9454. Available from https://doi.org/10.1021/cr500061m.

Li, Q., Cheng, K., Weng, W., Du, P., & Han, G. (2013). Titanium dioxide nanorod-based amperometric sensor for highly sensitive enzymatic detection of hydrogen peroxide. *Microchimica Acta*, *180*, 1487–1493. Available from https://doi.org/10.1007/s00604-013-1077-5.

Li, Z., Xin, Y., & Zhang, Z. (2015). New photocathodic analysis platform with quasi-core/shell-structured $TiO_2@Cu_2O$ for sensitive detection of H_2O_2 release from living cells. *Analytical Chemistry*, *87*, 10491–10497. Available from https://doi.org/10.1021/acs.analchem.5b02644.

Li, X., Yao, J., Liu, F., He, H., Zhou, M., Mao, N., ... Zhang, Y. (2013). Nickel/copper nanoparticles modified TiO_2 nanotubes for non-enzymatic glucose biosensors. *Sensors & Actuators, B: Chemical*, *181*, 501–508. Available from https://doi.org/10.1016/j.snb.2013.02.035.

Li, L., Zheng, X., Huang, Y., Zhang, L., Cui, K., Zhang, Y., & Yu, J. (2018). Addressable TiO_2 nanotubes functionalized paper-based cyto-sensor with photocontrollable switch for highly-efficient evaluating surface protein expressions of cancer cells. *Analytical Chemistry*, *90*, 13882–13890. Available from https://doi.org/10.1021/acs.analchem.8b02849.

Liu, B., & Aydil, E. S. (2009). Growth of oriented single-crystalline rutile TiO_2 nanorods on transparent conducting substrates for dye-sensitized solar cells. *Journal of the American Chemical Society*, *131*, 3985–3990. Available from https://doi.org/10.1021/ja8078972.

Liu, S., & Fu, D. (2017). *Glucose biosensor based on glucose oxidase immobilized on branched TiO_2 nanorod arrays. Advanced energy and environment research* (pp. 195–198). Boca Raton, FL: CRC Press, Taylor & Francis Group. Available from https://doi.org/10.1201/9781315212876-39.

Liu, X.-P., Chen, J.-S., Mao, C., Niu, H.-L., Song, J.-M., & Jin, B.-K. (2018). Enhanced photoelectrochemical DNA sensor based on TiO_2/Au hybrid structure. *Biosensors & Bioelectronics*, *116*, 23–29. Available from https://doi.org/10.1016/j.bios.2018.05.036.

Liu, X., Liu, P., Tang, Y., Yang, L., Li, L., Qi, Z., ... Wong, D. K. Y. (2018). A photoelectrochemical aptasensor based on a 3D flower-like TiO_2-MoS_2-gold nanoparticle heterostructure for detection of kanamycin. *Biosensors & Bioelectronics*, *112*, 193–201. Available from https://doi.org/10.1016/j.bios.2018.04.041.

Liu, H., Weng, L., & Yang, C. (2017). A review on nanomaterial-based electrochemical sensors for H_2O_2, H_2S and NO inside cells or released by cells. *Microchimica Acta*, *184*, 1267–1283. Available from https://doi.org/10.1007/s00604-017-2179-2.

Lu, Y.-J., Purwidyantri, A., Liu, H.-L., Wang, L.-W., Shih, C.-Y., Pijanowska, D. G., & Yang, C.-M. (2020). Photoelectrochemical detection of β-amyloid peptides by a TiO_2 nanobrush biosensor. *IEEE Sensors Journal*, *20*, 6248–6255. Available from https://doi.org/10.1109/JSEN.2020.2976561.

Luttrell, T., Halpegamage, S., Tao, J., Kramer, A., Sutter, E., & Batzill, M. (2015). Why is anatase a better photocatalyst than rutile? - Model studies on epitaxial TiO_2 films. *Scientific Reports*, *4*, 4043. Available from https://doi.org/10.1038/srep04043.

Macwan, D. P., Dave, P. N., & Chaturvedi, S. (2011). A review on nano-TiO_2 sol–gel type syntheses and its applications. *Journal of Materials & Science*, *46*, 3669–3686. Available from https://doi.org/10.1007/s10853-011-5378-y.

Mavrič, T., Benčina, M., Imani, R., Junkar, I., Valant, M., Kralj-Iglič, V., & Iglič, A. (2018). Electrochemical biosensor based on TiO_2 nanomaterials for cancer diagnostics. *Advanced biomembrane lipid self-assembly*, 63–105. Available from https://doi.org/10.1016/bs.abl.2017.12.003.

Mehrotra, P. (2016). Biosensors and their applications – A review. *Journal of Oral Biology and Craniofacial Research*, *6*, 153–159. Available from https://doi.org/10.1016/j.jobcr.2015.12.002.

Mohanty, S. P., & Kougianos, E. (2006). Biosensors: A tutorial review. *IEEE Potentials*, *25*, 35–40. Available from https://doi.org/10.1109/MP.2006.1649009.

Mondal, K., & Sharma, A. (2016). Recent advances in electrospun metal-oxide nanofiber based interfaces for electrochemical biosensing. *RSC Advances*, *6*, 94595–94616. Available from https://doi.org/10.1039/C6RA21477K.

Narwal, V., Deswal, R., Batra, B., Kalra, V., Hooda, R., Sharma, M., & Rana, J. S. (2019). Cholesterol biosensors: A review. *Steroids*, *143*, 6–17. Available from https://doi.org/10.1016/j.steroids.2018.12.003.

Narwal, V., Yadav, N., Thakur, M., & Pundir, C. S. (2017). An amperometric H_2O_2 biosensor based on hemoglobin nanoparticles immobilized on to a gold electrode. *Bioscience Reports*, *37*. Available from https://doi.org/10.1042/BSR20170194.

Newman, J. D., & Turner, A. P. F. (2005). Home blood glucose biosensors: A commercial perspective. *Biosensors & Bioelectronics*, *20*, 2435–2453. Available from https://doi.org/10.1016/j.bios.2004.11.012.

Rahimi, N., Pax, R. A., & Gray, E. M. (2016). Review of functional titanium oxides. I: TiO_2 and its modifications. *Progress in Solid State Chemistry*, *44*, 86–105. Available from https://doi.org/10.1016/j.progsolidstchem.2016.07.002.

Rahman, M. M., Ahammad, A. J. S., Jin, J.-H., Ahn, S. J., & Lee, J.-J. (2010). A comprehensive review of glucose biosensors based on nanostructured metal-oxides. *Sensors*, *10*, 4855–4886. Available from https://doi.org/10.3390/s100504855.

Rahmanian, R., Mozaffari, S. A., Amoli, H. S., & Abedi, M. (2018). Development of sensitive impedimetric urea biosensor using DC sputtered Nano-ZnO on TiO_2 thin film as a novel hierarchical nanostructure transducer. *Sensors & Actuators, B: Chemical*, *256*, 760–774. Available from https://doi.org/10.1016/j.snb.2017.10.009.

Ronkainen, N. J., Halsall, H. B., & Heineman, W. R. (2010). Electrochemical biosensors. *Chemical Society Reviews*, *39*, 1747. Available from https://doi.org/10.1039/b714449k.

Safavipour, M., Kharaziha, M., Amjadi, E., Karimzadeh, F., & Allafchian, A. (2020). TiO_2 nanotubes/reduced GO nanoparticles for sensitive detection of breast cancer cells and photothermal performance. *Talanta*, *208*, 120369. Available from https://doi.org/10.1016/j.talanta.2019.120369.

Sakib, S., Pandey, R., Soleymani, L., & Zhitomirsky, I. (2020). Surface modification of TiO_2 for photoelectrochemical DNA biosensors. *Medical Devices & Sensors.*, *3*, 10066. Available from https://doi.org/10.1002/mds3.10066.

Si, P., Ding, S., Yuan, J., (David) Lou, X. W., & Kim, D.-H. (2011). Hierarchically structured one-dimensional TiO_2 for protein immobilization, direct electrochemistry, and mediator-free glucose sensing. *ACS Nano*, *5*, 7617–7626. Available from https://doi.org/10.1021/nn202714c.

Si, P., Huang, Y., Wang, T., & Ma, J. (2013). Nanomaterials for electrochemical non-enzymatic glucose biosensors. *RSC Advances*, *3*, 3487. Available from https://doi.org/10.1039/c2ra22360k.

Solanki, P. R., Kaushik, A., Agrawal, V. V., & Malhotra, B. D. (2011). Nanostructured metal oxide-based biosensors. *NPG Asia Materials*, *3*, 17–24. Available from https://doi.org/10.1038/asiamat.2010.137.

Srivastava, S., Ali, M. A., Solanki, P. R., Chavhan, P. M., Pandey, M. K., Mulchandani, A., . . . Malhotra, B. D. (2013). Mediator-free microfluidics biosensor based on titania–zirconia nanocomposite for urea detection. *RSC Advances*, *3*, 228−235. Available from https://doi.org/10.1039/C2RA21461J.

Su, L., Jia, W., Hou, C., & Lei, Y. (2011). Microbial biosensors: A review. *Biosensors & Bioelectronics*, *26*, 1788−1799. Available from https://doi.org/10.1016/j.bios.2010.09.005.

Sulimovici, S. I., & Boyd, G. S. (1970). The cholesterol side-chain cleavage enzymes in steroid hormone-producing tissues. *Vitamins and Hormones*, *27*, 199−234. Available from https://doi.org/10.1016/S0083-6729(08)61127-9.

Suneesh, P. V., Sara Vargis, V., Ramachandran, T., Nair, B. G., & Satheesh Babu, T. G. (2015). Co−Cu alloy nanoparticles decorated TiO_2 nanotube arrays for highly sensitive and selective nonenzymatic sensing of glucose. *Sensors & Actuators, B: Chemical*, *215*, 337−344. Available from https://doi.org/10.1016/j.snb.2015.03.073.

Szunerits, S., & Boukherroub, R. (2018). Graphene-based biosensors. *Interface Focus*, *8*, 20160132. Available from https://doi.org/10.1098/rsfs.2016.0132.

Tian, K., Prestgard, M., & Tiwari, A. (2014). A review of recent advances in nonenzymatic glucose sensors. *Materials Science and Engineering C*, *41*, 100−118. Available from https://doi.org/10.1016/j.msec.2014.04.013.

Tian, C., Wang, L., Luan, F., & Zhuang, X. (2019). An electrochemiluminescence sensor for the detection of prostate protein antigen based on the graphene quantum dots infilled TiO_2 nanotube arrays. *Talanta*, *191*, 103−108. Available from https://doi.org/10.1016/j.talanta.2018.08.050.

Viter, R., Tereshchenko, A., Smyntyna, V., Ogorodniichuk, J., Starodub, N., Yakimova, R., ... Ramanavicius, A. (2017). Toward development of optical biosensors based on photoluminescence of TiO_2 nanoparticles for the detection of Salmonella. *Sensors & Actuators, B: Chemical*, *252*, 95−102. Available from https://doi.org/10.1016/j.snb.2017.05.139.

Wang, J. (1999). Amperometric biosensors for clinical and therapeutic drug monitoring: A review. *Journal of Pharmaceutical and Biomedical Analysis*, *19*, 47−53. Available from https://doi.org/10.1016/S0731-7085(98)00056-9.

Wang, X., Cui, J., Chen, S., Yang, Y., Gao, L., & He, Q. (2020). Electrochemical sensing of pancreatic cancer miR-1290 based on yeast-templated mesoporous TiO_2 modified electrodes. *Analytica Chimica Acta*, *1105*, 82−86. Available from https://doi.org/10.1016/j.aca.2020.01.030.

Wang, G., He, X., Wang, L., Gu, A., Huang, Y., Fang, B., ... Zhang, X. (2013). Non-enzymatic electrochemical sensing of glucose. *Microchimica Acta.*, *180*, 161−186. Available from https://doi.org/10.1007/s00604-012-0923-1.

Wang, L., Gu, W., Sheng, P., Zhang, Z., Zhang, B., & Cai, Q. (2019). A label-free cytochrome c photoelectrochemical aptasensor based on $CdS/CuInS_2/Au/TiO_2$ nanotubes. *Sensors & Actuators, B: Chemical*, *281*, 1088−1096. Available from https://doi.org/10.1016/j.snb.2018.11.004.

Wang, M., Yin, H., Zhou, Y., Sui, C., Wang, Y., Meng, X., ... Ai, S. (2019). Photoelectrochemical biosensor for microRNA detection based on a $MoS_2/g\text{-}C_3N_4$/black TiO_2 heterojunction with Histostar@AuNPs for signal amplification. *Biosensors & Bioelectronics*, *128*, 137−143. Available from https://doi.org/10.1016/j.bios.2018.12.048.

Wang, J., Zeng, Y., Wan, L., Zhao, J., Yang, J., Hu, J., ... Liang, F. (2020). Catalyst-free fabrication of one-dimensional N-doped carbon coated TiO_2 nanotube arrays by template carbonization of polydopamine for high performance electrochemical sensors. *Applied Surface Science*, *509*, 145301. Available from https://doi.org/10.1016/j.apsusc.2020.145301.

Wen, X., Long, M., & Tang, A. (2017). Flake-like Cu_2O on TiO_2 nanotubes array as an efficient nonenzymatic H_2O_2 biosensor. *Journal of Electroanalytical Chemistry, 785*, 33–39. Available from https://doi.org/10.1016/j.jelechem.2016.12.018.

Wooten, M., Karra, S., Zhang, M., & Gorski, W. (2014). On the direct electron transfer, sensing, and enzyme activity in the glucose oxidase/carbon nanotubes system. *Analytical Chemistry, 86*, 752–757. Available from https://doi.org/10.1021/ac403250w.

Xie, Q., Zhao, Y., Chen, X., Liu, H., Evans, D. G., & Yang, W. (2011). Nanosheet-based titania microspheres with hollow core-shell structure encapsulating horseradish peroxidase for a mediator-free biosensor. *Biomaterials, 32*, 6588–6594. Available from https://doi.org/10.1016/j.biomaterials.2011.05.055.

Xu, G., Liu, H., Wang, J., Lv, J., Zheng, Z., & Wu, Y. (2014). Photoelectrochemical performances and potential applications of TiO_2 nanotube arrays modified with Ag and Pt nanoparticles. *Electrochimica Acta, 121*, 194–202. Available from https://doi.org/10.1016/j.electacta.2013.12.154.

Yan, Q., Cao, L., Dong, H., Tan, Z., Hu, Y., Liu, Q., ... Dong, Y. (2019). Label-free immunosensors based on a novel multi-amplification signal strategy of TiO_2-NGO/Au@Pd hetero-nanostructures. *Biosensors & Bioelectronics, 127*, 174–180. Available from https://doi.org/10.1016/j.bios.2018.12.038.

Yang, L., Liu, X., Li, L., Zhang, S., Zheng, H., Tang, Y., & Ju, H. (2019). A visible light photoelectrochemical sandwich aptasensor for adenosine triphosphate based on $MgIn_2S_4$-TiO_2 nanoarray heterojunction. *Biosensors & Bioelectronics, 142*, 111487. Available from https://doi.org/10.1016/j.bios.2019.111487.

Yang, Z., Liu, X., Zhang, C., & Liu, B. (2015). A high-performance nonenzymatic piezoelectric sensor based on molecularly imprinted transparent TiO_2 film for detection of urea. *Biosensors & Bioelectronics, 74*, 85–90. Available from https://doi.org/10.1016/j.bios.2015.06.022.

Yang, Y., Yan, K., & Zhang, J. (2017). Dual non-enzymatic glucose sensing on $Ni(OH)_2$/TiO_2 photoanode under visible light illumination. *Electrochimica Acta, 228*, 28–35. Available from https://doi.org/10.1016/j.electacta.2017.01.050.

Yin, X., Guo, M., Xia, Y., Huang, W., & Li, Z. (2014). Amperometric sensing of hydrogen peroxide on a modified electrode with layered Au/TiO_2 nanofilms from self-assembly at air/water interface. *Journal of Electroanalytical Chemistry, 720–721*, 19–23. Available from https://doi.org/10.1016/j.jelechem.2014.02.017.

Yu, S., Peng, X., Cao, G., Zhou, M., Qiao, L., Yao, J., & He, H. (2012). Ni nanoparticles decorated titania nanotube arrays as efficient nonenzymatic glucose sensor. *Electrochimica Acta, 76*, 512–517. Available from https://doi.org/10.1016/j.electacta.2012.05.079.

Yu, L., Zhang, Y., Zhi, Q., Wang, Q., Gittleson, F. S., Li, J., & Taylor, A. D. (2015). Enhanced photoelectrochemical and sensing performance of novel TiO_2 arrays to H_2O_2 detection. *Sensors & Actuators, B: Chemical, 211*, 111–115. Available from https://doi.org/10.1016/j.snb.2015.01.060.

Zang, Y., Lei, J., & Ju, H. (2017). Principles and applications of photoelectrochemical sensing strategies based on biofunctionalized nanostructures. *Biosensors & Bioelectronics, 96*, 8–16. Available from https://doi.org/10.1016/j.bios.2017.04.030.

Zanghelini, F., Frías, I. A. M., Rêgo, M. J. B. M., Pitta, M. G. R., Sacilloti, M., Oliveira, M. D. L., & Andrade, C. A. S. (2017). Biosensing breast cancer cells based on a three-dimensional TiO_2 nanomembrane transducer. *Biosensors & Bioelectronics, 92*, 313–320. Available from https://doi.org/10.1016/j.bios.2016.11.006.

Zhang, C., Bai, W., & Yang, Z. (2016). A novel photoelectrochemical sensor for bilirubin based on porous transparent TiO_2 and molecularly imprinted polypyrrole. *Electrochimica Acta*, *187*, 451–456. Available from https://doi.org/10.1016/j.electacta.2015.11.098.

Zhang, J., Zhou, P., Liu, J., & Yu, J. (2014). New understanding of the difference of photocatalytic activity among anatase, rutile and brookite TiO_2. *Physical Chemistry Chemical Physics: PCCP*, *16*, 20382–20386. Available from https://doi.org/10.1039/C4CP02201G.

Zheng, H., Zhang, S., Liu, X., Zhou, Y., & Alwarappan, S. (2020). Synthesis of a PEDOT-TiO_2 heterostructure as a dual biosensing platform operating via photoelectrochemical and electrochemical transduction mode. *Biosensors & Bioelectronics*, *162*, 112234. Available from https://doi.org/10.1016/j.bios.2020.112234.

Zhu, Y.-C., Mei, L.-P., Ruan, Y.-F., Zhang, N., Zhao, W.-W., Xu, J.-J., & Chen, H.-Y. (2019). *Enzyme-based biosensors and their applications. Advanced Enzyme Technologies* (pp. 201–223). Elsevier. Available from https://doi.org/10.1016/B978-0-444-64114-4.00008-X.

Zhu, C., Yang, G., Li, H., Du, D., & Lin, Y. (2015). Electrochemical sensors and biosensors based on nanomaterials and nanostructures. *Analytical Chemistry*, *87*, 230–249. Available from https://doi.org/10.1021/ac5039863.

Zhu, K.-R., Zhang, M.-S., Hong, J.-M., & Yin, Z. (2005). Size effect on phase transition sequence of TiO_2 nanocrystal. *Materials Science and Engineering A*, *403*, 87–93. Available from https://doi.org/10.1016/j.msea.2005.04.029.

ZrO_2 in biomedical applications

18

Shweta J. Malode and Nagaraj P. Shetti
Department of Chemistry, School of Advanced Sciences, KLE Technological University, Hubballi, India

18.1 Introduction

Zirconium metal (Zr) was discovered by Martin Klaproth, a researcher, in 1789. Zr possesses a density of 6.49 g/cm^3 with hexagonal crystal structure, boiling point of 3580°C, and a melting point of 1852°C. The Zr in its oxide composition, that is, ZrO_2, possesses outstanding mechanical feasibility, nontoxicity, chemical stableness, and high affinity to oxygen-including groups. Zirconium is not found as a pure oxide in nature. The minerals from which Zr is chemically extracted are baddeleyite (ZrO_2) and zirconate (ZrO_2-SiO_2, $ZrSiO_4$). Amongst these, zirconate is abundant, yet it is less pure and requires treatment to obtain Zr (Piconi & Maccauro, 1999). Baddeleyite comprises of 96.5%–98.5% zirconia. Subsequently, baddeleyite is reported as a source of the utmost purity to obtain Zr metal and its compounds. Zirconium dioxide (ZrO_2), also named as zirconia, is acquired from baddeleyite, a monoclinic crystal structure at ambient temperature. At high temperatures, powder can be purified and refined to obtain cubic-structured Zr. The resultant material is strong, optically pure and semitransparent, and is generally used as valuable stones or in gas sensors (Koutayas, Vagkopoulou, Pelekanos, Koidis & Strub, 2009).

Zirconium exerts mechanical stability, anticorrosion quality, and biocompatibility, and is used in biomedical applications such as implantology (Kubasiewicz-Ross, Dominiak, Gedrange & Botzenhart, 2017). Zr hybrids have been prepared using various techniques to enhance the Zr attributes. The sol–gel method was utilized to prepare bioactive glass/zirconium titanate thin film for bone and dental implants (Mozafari et al., 2013; Salahinejad et al., 2012). In addition, significant information about zirconium is required to apply it in biomedical applications. This chapter sheds light on a better understanding of zirconium and its hybrid applications. Zirconia is a fascinating ceramic for biomedical applications due to its appropriate bioactivity and biocompatibility (Zarrintaj et al., 2017). Its applications incorporate energy units, thermal barrier covering, and assurance against degradation by nuclear fuels. In addition, Zr is utilized in dental recovery, fuel cells, and biomedical hip implants (Chroneos, Yildiz, Tarancon, Parfitt & Kilner, 2011; Schulz et al., 2003). In the past, Zr was utilized in the semiconductor industry as a reinforcement in metal oxide semiconductor field effect transistor (MOSFET) gate

Metal Oxides for Biomedical and Biosensor Applications. DOI: https://doi.org/10.1016/B978-0-12-823033-6.00016-8

dielectrics for SiO_2 (Robertson, Xiong & Falabretti, 2005; Samanta, Lenosky & Li, 2010). With the conspicuous status of environmental gas in increasing numbers of locations, many investigations have taken place on the advancement of reasonable gas-detection materials. ZrO_2 thin films have been manufactured effectively utilizing different applications, such as spin coating sol–gel technique (Salari & Ghodsi, 2017; Sooa et al., 2012), solution combustion (Ravichandran et al., 2014), atomic layer (Kukli et al., 2017; Lima et al., 2016), pulse laser deposition (Dikovska, Atanasova, Avdeev & Strijkova, 2016), radio frequency magnetron (Haque, Tratipai, Jha, Bhattacharyya & Sahoo, 2016), dip surfacing (Abd El-Lateef & Khalaf, 2015; Khalaf & Abd El-Lateef, 2016), and ultrasonic spray pyrolysis (Perednisa, Wilhelmb, Pratsinisb & Gauckler, 2005; Zhang, Li & Tanji, 2014). Sol–gel and chemical vapor deposition must be executed in a vacuum as these involve multiple steps. Spray pyrolysis offers enormous opportunities for producing thin films and advanced ceramic powders due to its chemical flexibility (Vergnieres, Odier, Weiss, Bruzek & Saugrain, 2005).

18.1.1 Phases of zirconia

The Zr atoms' spacial arrangement is characterized by discrete crystallographic structures, presenting a measurement also called polymorphism. The structures are described by explicit terms such as monoclinic, tetragonal, and cubic (Fig. 18.1). The pure Zr monoclinic phase is steady up to 1170°C. Between 1170°C and 2370°C, tetragonal-shaped Zr is obtained, and when the temperature is greater than 2370°C, cubic Zr is established. Further dependent on the cooling process, the tetragonal phase turns into monoclinic at around 970°C. By virtue of polymorphism, unmodified Zr cannot be utilized at rarified temperatures because of the enormous volume change on cooling (3%–5%) to the monoclinic stage. This alteration can exceed the flexible and crack limits, leading to breaks and imperfections in ceramics production (Denry & Kelly, 2008). The transformation from tetragonal to monoclinic phases is applied to better the mechanical properties of Zr, peculiarly its tenacity. This transformation is martensitic in nature, that is, the change in atomic position takes place abruptly at a speed close to the speed of sound proliferation in solids. The opposite transition, a monoclinic > tetragonal change, takes place at around 1170°C, though the tetragonal > monoclinic change happens on cooling and is found in the range of 850°C–1000°C. Subsequently, the production of pure Zr components is not feasible because of spontaneous failure. The addition of stabilizing oxides allows the tetragonal form to be sustained at ambient temperature (Terki, Bertrand, Aouraf & Coddet, 2006). To stabilize Zr, assorted oxides, for example, CaO or MgO, and yttrium oxide (Y_2O_3), can be incorporated to permit the tetragonal structure to exist at normal temperature after production. A completely stabilized Zr is created by including adequate amounts of stabilizing oxides, for example, limestone (CaO) 16 mol%, magnesia (MgO) 16 mol%, or yttria (Y_2O_3) 8 mol%. Limited stabilization of Zr is found with similar oxides, however in smaller amounts of 2–3 mol% yttria, a polyphase development is made, which involves tetragonal and cubic Zr in mass/monoclinic stage precipitated in

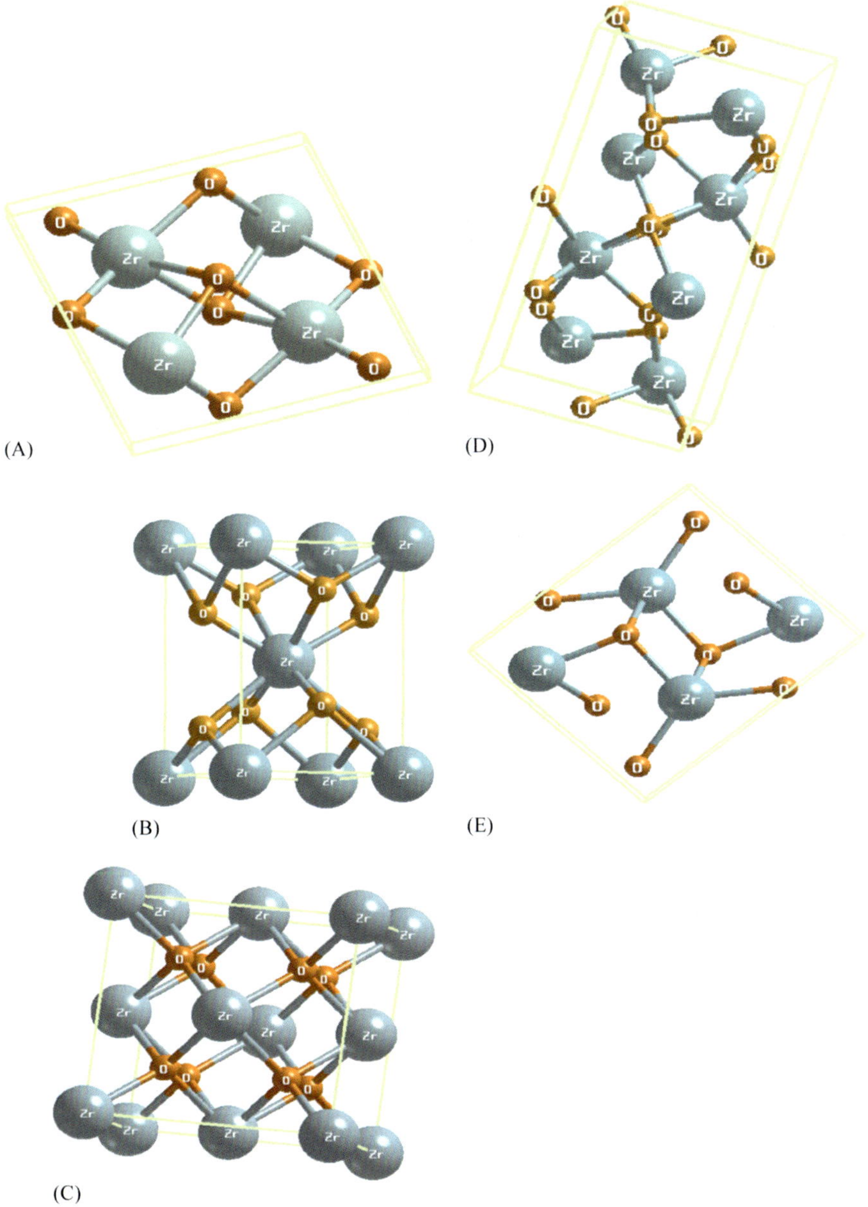

Figure 18.1 Zr polymorph crystal structures: (A) monoclinic, (B) tetragonal, (C) cubic, (D) brookite, and (E) cotunnite.

Source: Reprinted with permission from Terki, R., Bertrand, G., Aouraf, H, & Coddet, C. (2006). Structural and electronic properties of zirconia phases: A FP-LAPW investigations. *Materials Science in Semiconductor Processing*, *9*, 1006–1013.

small quantities (Piconi & Maccauro, 1999). The transformation of tetragonal Zr into the monoclinic stage is dependent on the temperature, size of particles, vapor, material structure, and stabilizing oxides concentration. The partially stabilized Zr molecule size to be held in the tetragonal structure is 0.2–1 μm, as, under 0.2 μm, the change to the monoclinic stage is not possible at room temperature (Kelly & Denry, 2008).

18.1.1.1 Monoclinic zirconia

Baddeleyite normally consists of 2% HfO_2 (hafnium oxide), which has similar shape and chemical properties to Zr. Zr^{4+} particles have the coordination number 7, for the oxygen particles filling the tetrahedral body structure. Regular separation between three of the seven oxygen particles and the zirconia particle is 2.07 Å. Regular separation between four oxygen ions and the zirconium particle is 2.21 Å. One of the edges (134.3 degrees) diverges extensively with the tetrahedral value (109.5 degrees). As an outcome, the shape of the oxygen ion is not planar and a curve occurs at the level of the four oxygens, and the planes of three oxygens are completely unreliable (Mamivand, Zaeem, Kadiri & Chen, 2013).

18.1.1.2 Tetragonal Zr

The Zr tetragonal phase has an unbent prism with rectangular faces. The shape of Zr^{4+} looks to be distorted with a coordination number of 8, on account of the placement of four oxygen ions at a length of 2.065 Å in the tetrahedron plane, and the other four at a space of 2.455 Å in a tetrahedron which is stretched and turned out at 90 degrees (Vagkopoulou, Koutayas, Koidis & Strub, 2009).

18.1.1.3 Cubic Zr

The cubic Zr structure comprises eight oxygen ions, surrounded by cubic arrangement, known as fluorite. These ions take the tetrahedral body structure of a cubic lattice of cations (CFC) (Vagkopoulou et al., 2009).

18.1.2 Zirconia-based coating

Zr has been used in different manufacturing applications because of its properties like biocompatibility, wear resistance, and so forth. The polyethylene coat for total knee replacement (TKR) when was covered by Zr, the wear resistance of Zr-2.5 Nb was seen as 90%, with suitable scratch and wear resistance which was higher than for a cobalt-chromium compound (Ries, Salehi, Widding & Hunter, 2002). In light of the excellent execution of Zr, for the related segment a joint operation was planned utilizing Zr-2.5 Nb (Lazennec & Dietrich, 2004). The oxidized Zr components had better bacterial resistance than TiAl and CoCr (Tokash, Stojilovic, Ramsier, Kovacik & Mostardi, 2005). The sodium hydroxide treated Zr metal increased the apatite practice (Kamitakahara, Ohtsuki & Miyazaki, 2007). Kumar and Kaliaraj (2018), prepared a coating based on zirconium nitrate/copper (ZrN/Cu) polylayer with an appropriate wear and corrosion resistance. This possessed the bactericide features due to ion releasing nature.

Zirconia-based coatings increase the thermal insulation. There was a low estimation of heat conduction of such coverings left in an increasing metallic substrate temperature (Song et al., 2017). ZrO_2 was coated using a plasma sprayer as a thermal barrier, which introduced cyclic temperature deviation. The covering showed strain-resistant measurements. The oxide component was formed because high temperature resulted in better protection (Imran, Alam, Irfan & Farooq, 2015). Zr has a drawback of presenting most extreme thermal enlargement and shrinkage because of phase change among monoclinic and tetragonal stages. Henceforth, the utilization of Zr as a refractory, thermal insulator should be supported to prevent this phase transition (Rendtorff, Suarez, Sakka & Aglietti, 2012).

Sol-gel process has been applied for Zr coating owing to accomplishing exceptional structural development. Tetragonal crystal and consistent molecules were yielded dominantly; besides, heat drying resulted in decrement in crystal size (Tyagi, Sidhpuria, Shaik & Jasra, 2006). Zr titanate multilayer was prepared with a 50 nm diameter spherical nanoparticle, furthermore, carboxymethyl cellulose utilization as a dispersant reduced size of molecule to 20 nm. The surface examinations on coatings arranged by sol gel method demonstrated the bearing of the hydroxyl group on the surface which as needs be elevates the validation of calcium phosphate and later encourages cell adherence on material implantation (Gupta &Kumar, 2008; Kalantari, Naghib, Naimi-Jamal & Mozafari, 2017; Salahinejad et al., 2013). However, the sol gel practice has generally shortcomings. Sol gel brings about major porous coating done with unsatisfactory mechanical property and completeness; Furthermore, achieving a thick development includes high temperature forming. High temperature molding leads to breaking and separation of coating attributable to huge divergence among thermal elaboration constants of substrate and coating (Gupta & Kumar, 2008). The bio and organic elements cannot withstand high temperature, in this way, hybrid applications have been utilized for making up such limitations.

For bone tissue design, the advancement of bone-like biomaterials to accomplish appropriate biomimetic dimensions was necessary. Numerous intercrossed organic-inorganic materials have been come out to emulate the natural bone ordered of inorganic nanocrystallites of hydroxyapatite ($Ca_{10}(PO_4)_6(OH)_2$) and organic collagen fibrils (Brun et al., 2014; Depan, Venkat Surya, Girase & Misra, 2011). In view of their biocompatibility with the human body, hydroxyapatite (HA) and chitosan (CS) has been chosen generally (Wang, 2003). Bhowmick et al. (2017), incorporated ZrO_2 NPs into CS–PEG–HA combination for increment in the mechanical durability, retention of water, antimicrobial attributes, by moderating props required for effective bone regeneration, that is, porosity and cytocompatible characters.

The metal oxide coatings of ZrO_2, TiO_2 and Al_2O_3 expose increasingly helpful optical and mechanical possessions by biocompatibility (Choi & Suresh, 2002; Duan et al., 2013; Vinci & Vlassak, 1996; Volinsky, Vella & Gerberich, 2002). Bukhari, Imran, Bashir, Riaz and Naseem (2018) applied a sol–gel strategy to examine the impact of titania doping fixation with various substance of TiO_2 extending from 1–5 wt.%, TiO_2 doped ZrO_2 and Un-doped thin films. Photocatalytic responses sensified by ZrO_2 have attracted wide contributions for solving energy and ecological issues. Gurushantha et al. (2017) inquired about the photoluminescence properties of 1:1, 1:2 and 2:1 proportions of ZrO_2 and ZnO NPs, and ZrO_2:ZnO nanocomposites (NCs). The combination of ZnO composite material brought about reducing the energy gap of ZrO_2 band from 5.3−3.1 eV. The photoluminescence extinguishing is found later the promotion of ZnO NPs is hindered as an unadulterated concealing impact linked the method of sample preparation. The NC of ZrO_2: ZnO showed an expanded 93% degradation of acid orange 8 dye photocatalytic capacity neither UV light equated to different complexes (Vidya et al., 2015).

Different implant materials like titanium and its composites are possible in the market; however 316L SS have appreciable attention because of its easy preparation, more affordable and improved corrosive resistant qualities (Kaliaraj, Ramadoss, Sundaram, Balasubramanian & Muthirulandi, 2014). Because of aesthetic characteristics of a natural tooth, prominent resistance to corrosion, and mechanical strength, ceramics production coatings are incredibly applied for dental implant durability (Guess, Att & Strub, 2012). These are the better options in contrast to the metallic embeds for dental prosthesis remodels. Toward the start, alumina was utilized as dental embed material (Thomas et al., 2016). Nevertheless, alumina showed low crack endurance and twisting toughness rather than ZrO_2 ceramics. Along these lines in the present premises, Zr based ceramics productions were utilized to create implant abutments, copings, synthetic teeth, oral cavity regions (Papaspyridakos & Lal, 2013). Kaliaraj et al. (2018) created ZrO_2 film by electron beam physical vapor deposition (EBPVD) method. Hydrophilic nature was distinguished on ZrO_2 surface film and to the most reduced degree hydrophilic conduct was encountered on bare 316L SS substrate. The adhesiveness of bacteria was executed versus Streptococcus mutans. The ZrO_2 film defended the connection of bacteria on its coating. Electrochemical studies utilizing 0.2% and NaF of 2% disclosed that ZrO_2 surfaced 316L SS designated brilliant corrosion resistance and can be counted a potential bioimplant for dental diligences.

18.1.3 Synthesis of zirconia

Zirconium Oxide (ZrO_2), n-type semiconductors possess a wide 5.0 eV band gap and large energy holding capacity. Zirconium (II) oxide is chemically stable relying on temperature subsists in various forms. Monoclinic phase is static under 1000°C, tetragonal stage subsists over 1170°C and transforms to cubic phase after 2370°C. The chemical reactions can be stated as (Jaenicke, Chuah, Raju & Nie, 2008):

Stefanic et al. have studied the dependency parameters such as pH value, reaction temperature and type of anion on the properties of hydrous Zr. They blended Zr n-propoxide in isopropanol with dissimilar pH solutions and ion types. When they are calculated at 400°C, they get tetragonal Zr as the first stage (Stefanic, Stefanic & Music, 2000). Stocker and Baiker have created aerogel of Zr by sol-gel technique through acid catalysis of tetra-n-butoxy zirconium (IV) (Stocker & Baiker, 1998). Beena et al. prepared Zr NPs applying sol-gel and precipitation procedures. Towards sol-gel amalgamation, the Zr propoxide was diluted to 30 wt.%

by contributing propanol, and then hydrolyzed with dropwise NH_3 solution till pH 10–10.5 with consistent mixing. For the precipitation method, zirconium oxychloride was disintegrated in distilled water followed by addition of ammonia solution drop wise till pH 10–10.5 with constant stirring. The precipitate was filtered, water-washed to get rid of chlorine ions. The samples were dried out in an oven, beneath vacuum in rotavapor and finally heated at varying temperatures from 400°C to 700°C. Tetragonal stage stabilization at higher temperature was found by sol-gel method of synthesis. The crystallite size was 13–28 nm by sol gel synthesis and for precipitation synthesis 11–37 nm (Tyagi et al., 2006). Ward & Ko synthesized ZrO_2 by sol-gel method. Zr n-propoxide combined with HNO_3 and n-propanol is aggregated with some other solution comprising n-propanol and H_2O with continuous shaking. Before drying, the resulting gel is aged for 2 h. The gel time increased with rising HNO_3 addition. The condensation reaction slows down with rising gel time, hence creating a greater surface area of 134 m^2/g and 0.37 cm^3/g as pore volume of product (Ward & Ko, 1995). Heshmatpour and cowworkers used glucose and fructose to synthesize ZrNPs through sol-gel technique. Propanol is poured to dilute Zr n-propoxide to 30 wt.% followed by drop wise addition of ammonia solution till pH 9–10. The gel obtained is dried for 12 h at 110°C in an oven. Further, calcinations were done around 300°C–700°C. They Zr NPs had crystallite size within 10–30 nm and the organic linear is encouraged to steady tetragonal form at 700°C (Heshmatpour & Aghakhanpour, 2011).

18.1.4 Characterization of sample

To view the corresponding phase using an X-shaft diffraction (XRD) scheme, J.M.E. (Matos et al., 2009) worked on reflux amalgam and aqueous preparation of ZrO_2 NPs under various conditions. In Fig. 18.2A, the peaks are extremely broad owing to the nature of the nanocrystalline structure and the modest transparency size. The size of the stone is less than 10 nm. Zirconia produced without hydrogen peroxide gives a distinctive monoclinic and enormous crystals. Fig. 18.2B evidences the FT-Raman spectrum of ZrO_2 NPs placed in an aqueous system for 72 h at 120°C. Dynamic Raman modes can be distinguished in Fig. 18.2B. As indicated in the text (Matos et al., 2009), these groups were designated by a monoclinic design. Fig. 18.2C shows the measurement of the decomposition of ZrO_2 found by distinct A, B, C and D methods in gravity test. Fig. 18.2C shows a total weight loss of about 18% between 25°C and 100°C for all examples sorted without NH_4OH. Bulk at 100°C can be demonstrated by rinsing with water that appears outside the ZrO_2 particles. Around 200°C could be due to synthetic and actually improved water leaking from outside ZrO_2, while the third major misfortune could be due to the phase transformation of ZrO_2. In Fig. 18.2D shows FT-IR spectra between 2000 and 400 cm^{-1} for ZrO_2 obtained by various methods. The hydroxide (Zr–OH) is observed at 1401 cm^{-1}, peroxide accumulation ($O–O^-$) is observed between 900 cm^{-1}, the metal–oxygen bond (Zr–O) is located at 583 cm^{-1} and the metal, oxygen and metal bond capacity is low (Zr–O–Zr) can be found at 506 cm^{-1}. Similarly, the band around 1400 cm^{-1} was assigned to prolonged N and O bond vibrations in NH_4^+, owing to the use of

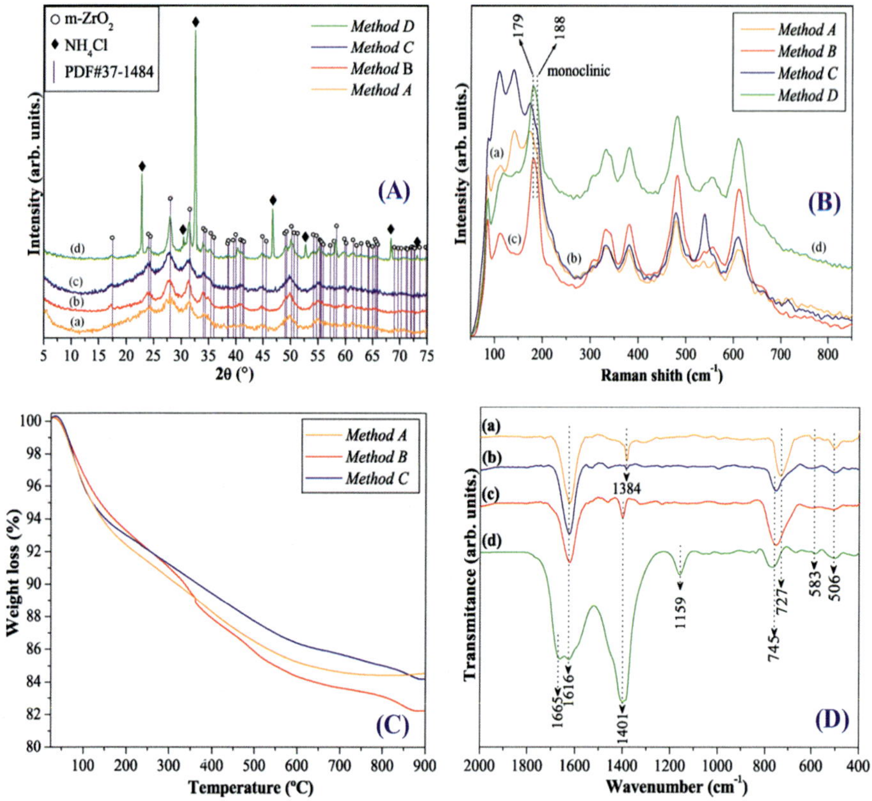

Figure 18.2 (A) XRD patterns for the ZrO_2 nanopowders: (a,b,c,d) Method A, B, C, D. (B) FT-Raman spectra for the ZrO_2 NPs: (a,b,c,d) Method A, B, C, D. (C) TGA curves for the ZrO_2 nanopowders obtained by different methods: Method A B, C. (D) FT-IR spectra for the ZrO_2 nanopowders: (a,b,c,d) Method A, B, C and D.

Source: Reprinted with permission from Matos, J. M. E., Anjos Jr., F. M., Cavalcante, L. S., Santos, V., Leal, S. H., Santosh Jr., L. S., . . . Longo, E. (2009). Reflux synthesis and hydrothermal processing of ZrO2 nanopowders at low temperature. *Materials Chemistry and Physics*, *117*, 455–459.

CH_4N_2O in Method D. Subsequent studies of FTIR spectra showed that the characteristics of the acidic main point of the surface of the ZrO_2 are unique. This conduct is due to the presence of H_2O_2 or NH_4OH within the techniques used that brings a couple of alkaline and acidic nature of the ZrO_2 coat. Table 18.1 shows the various routes and techniques followed to obtain ZrO_2 powders.

18.1.5 Properties of zirconia

In modern dentistry, Zr finds several applications on account of its biocompatibility, mechanical properties, color and so on. Even so, some of the factors such as ageing,

Table 18.1 Preparing route and synthesis technique used to obtain ZrO_2 powders (Somiya & Akiba, 1999).

Processing route	Synthesis method
1. Thermal breakdown	a. Heating b. Spray dying c. Flame spraying d. Plasma spraying e. Vapor phase f. Freeze drying g. Hot kerosene drying h. Hot petroleum drying
2. Precipitation or hydrolysis	a. Neutralization and precipitation b. Homogeneous precipitation c. Co-precipitation d. Salts solution e. Alkoxides f. Sol-gel
3. Hydrothermal	a. Precipitation b. Crystallization c. Decomposition d. Oxidation e. Synthesis f. Electrochemical g. Mechanochemical h. RESA (reactive submerged arc) i. Hydrothermal + microwave j. Hydrothermal + Ultrasonic

ceramic bonding, and light transmission need to be evaluated to have Zr as a positive renewing material. Processing Zr to full contour may be a substitute to customarily secured redesigns.

18.1.5.1 Mechanical attributes and aging of zirconia

The Zr mechanical properties are comparable to stainless steel. It possesses 900–1200 MPa resistance to traction and resistance to compression about 2000 MPa (Piconi & Maccauro, 1999). The Zr can also endure repeating load stresses. Cales and Stefani found that with an intermittent power of 28 kN to Zr substrata, around 50 billion cycles were expected to split it, though with a power more than 90 kN, after just 15 cycles the structural failure happened (Cales & Stefani, 1994). The properties of Zr can be commuted by surface treatments. The low-temperature degradation or "aging" has not been well contemplated. Non-aqueous solvents and water can cause the establishment of zirconia hydroxides on a break resulting in the elaboration of the crack. This results in reduced strength, toughness, and density (Swab, 1991). Surface abrasion canas well reduce durability (Işeri, Ozkurt, Yalnız & Kazazoglu, 2012).

Zirconia is characterized by a property called transformation toughening exhibiting eminent bending strength and transformation toughness (Peláez, Cogolludo, Serrano, Lozano & Suárez, 2012). The occurrence of structure crack was straightforwardly identified with the planning of the FPD, wherever inlay retained FPDs (IRFPD) indicated the foremost noteworthy unsuccessful rate (Al-Amleh, Lyons & Swain, 2010). The foremost well-known inconvenience noticed in Zr-based renovations was cracks of the veneer ceramic ware, showing as knapping breaks of the veneer terminated exposing the essential Y-TZP model. This may bring about an adjustment in the temperature development coefficients. Different components incorporate the diverse surface medicines of the systems and the attachment durability between the veneering ceramics production and Zr structures (Fischer, Stawarczyk & Hammerle, 2008; Hisbergues, Vendeville & Vendeville, 2009). Forming a CAD/CAM-prepared lithium disilicate forming exterior top onto the Zr adjusting has in a general sense extended the mechanical nature of crown reclamations and speaks to a savvy method of creating every single artistic rebuilding (Beuer et al., 2009). Processing of new propagation full-form zirconia may be an elective way to deal with conquering chipping breaks of covered zirconia regainings. Manufacturing mono-block rebuilding efforts from pure zirconia could increment mechanical stability and extend the scope of denotations (Guess et al., 2012).

18.1.5.2 Biocompatibility of zirconia

Eminent bio compatibilities of Y-TZP on the utilization of pure Zr powders that have been purged of their radioactive substance have been confirmed (Covacci et al., 1999; Lohmann et al., 2002; Stanford, Oates & Beirne, 2006). Advancing evaluations have shown that few bacteria aggregate around Y-TZP than titanium (Scarano, Piattelli, Caputi, Favero & Piattelli, 2004; Welander, Abrahamsson & Berglundh, 2008). This might be explicated by not at all like protein surface osmosis characteristics. Regarding periodontal health, none of the evaluations point by point any separation or any adjustments in the characteristic adequacy of the tissues across the Zr-based renovations. However two or three information evaluated and investigated deviations in the biocompatibility, no gingival worsening or periodontitis could be delineated (Raigrodski, Hillstead, Meng & Chung, 2012).

The Zr was presented as supporting material in 1996 (Wohlwend, Studer & Scharer Das, 1996). A randomized checked objective clinical test equating Zr and Ti projections upheld by 40 single implants was reported (Zembic, Sailer, Jung & Hämmerle, 2009). Subsequent to being in work for a long time, 18 Zr and 10 Ti projections were followed-up. Both projection materials showed endurance paces of 100%, just as comparative biological and esthetic results. In a creature field, the collagen fiber direction was comparable throughout Zr and Ti implant contracts. For the two materials, the fibers orientation was equal slanted and corresponding to the implant surface (Tetè, Mastrangelo, Bianchi, Zizzari & Scarano, 2009). A comparable level of plaque collection was ground at Zr and Ti projections at 3 years. In a similar report, when Zr projections are utilized as regaining endure, there were no noteworthy contrasts in bone layers among Zr and Ti supports following three time development.

18.1.5.3 Aesthetic dimensions and light transmittance of Zr

Every ceramic material agreeably turns to the interest for esthetic regaining than metal-ceramics regaining efforts with hard cores (McLaren, 1998). In any case, the translucence of Zr-based ceramic caps is accounted for to be not as much as that of lithium disinclined glass ceramics, due to magnificent artistic outcomes are recorded (Culp & McLaren, 2010; Koutayas et al., 2009). In-Ceram Zr (VITA Zahnfabrik, Bad Sackingen, Germany), an Al_2O_3-grounded ceramics of 35% ZrO_2, bears low semi transparency (Heffernan et al., 2002; Narcisi, 1999; Odman & Andersson, 2001). This could be a hindrance to accomplishing an artistic satisfactory rebuilding. Alumina and glass-ceramic have translucence and mechanical dimensions smaller than ZrO_2 ceramics (Baldissara, Llukacej, Ciocca, Valandro & Scotti, 2010; Stappert, Guess, Chitmongkolsuk, Gerds & Strub, 2007).

In light of this, the utilization of Zr ceramics production with various chemical compositions might be noteworthy for practitioners. Also, estimating the level of change of various resin lute agents underneath Zr clay materials may deliver improved objective results (Cekic-Nagas, Egilmez & Ergun, 2012). Next investigations ought to be extended to incorporate new-age full-form zirconia (Beuer, Stimmelmayr, Gueth, Edelhoff & Naumann, 2012). Full-shape zirconia processing spaces are made through an interesting patent-pending procedure. In a procedure the ZrO_2 powders are set up to likewise decrease the molecule size of ZrO_2 and blended in with a reasonable spread to broaden the compression and density of the green state and remove the blocked porousness. The makers guarantee that dissimilar to standard high-pressure milling blank manufacturing, this dealing contributes full-shape Zr mended light transmittance, giving a lower, more biological tint measure (BruxZir Solid Zirconia crowns & bridges, 2010).

18.1.5.4 Bonding to zirconia

The life span of an indirect regaining is firmly identified with the wholeness of the concrete at the edge (Valandro et al., 2006). Regardless of the way that the usage of Zr ceramics production for dental coatings is continuous, the good technique to accomplish a solid alliance between the ceramic and the tooth construction is as yet obscure. The main agreement noticed in the writing is that hydrofluoric acid engraving and normal silane factors are unsuccessful with Zr ceramics production (Kern & Wegner, 1998; Yoshida, Tsuo & Atsuta, 2006). A few examinations have explored the bond quality and the solidness of different holding techniques used to shape high-quality Zr ceramics production. Air abrasion with Al_2O_3 molecules is executed to evacuate contaminants layers, in this manner expanding micromechanical maintenance between the resin cement and the regaining (Foxton et al., 2011;

Yang, Barloi & Kern, 2010). These particles could possibly be silica-covered (with tribochemical treatment) (Chen, Suh, Kim & Tay, 2011).

In different examinations a few covering agents were utilized to upgrade the development of substance holding with Zr however just those agents that comprise a monomer of phosphate operator were successful in setting up a solid adhesiveness with Zr (Wegner & Kern, 2000). An ongoing report concentrating on the long-term stability of Zr resin holding exhibits legitimately identified with the science of the stuff utilized, admitting preliminaries. The authors recommend that an increasingly hydrophobic compound is compulsory to all the more likely oppose the impending impact of hydrolysis so as to increase full profit by the preliminaries (Aboushelib, Matinlinna, Salameh & Ounsi, 2008; Mirmohammadi, Aboushelib, Salameh, Feilzer & Kleverlaan, 2010). In a novel way to deal with improving zirconia resin bond quality; particular penetration engraving of Zr-based stuff has been attempted. This makes a tenacious coat whereas the resin can infiltrate and enlace so as to set up a firm bond with Zr (Aboushelib, 2011; Aboushelib, Feilzer & Kleverlaan, 2010).

18.2 Applications of zirconia

18.2.1 Engineering applications

Doufar et al. (2020) utilized daylight for the total degradation of inorganic toxin (Cr(VI)). To accomplish this objective, the solution is to utilize hetero-intersection by presenting different materials, for example, TiO_2. The p-n hetero-intersection Fe-ZrO_2/TiO_2 is worked so as to lessen the combining of electron/gap (e^-/h^+) sets and to amend the photocatalytic execution. N. Doufar et al., examined the physico-chemical attributes of the as-arranged nanoparticles by coprecipitation of $Zr_{(1-x)}FexO_{(2-x/2)}$ studied by physical procedures. The utilization of p-n hetero-intersection improves the photoactivity by upgrading the partition of couples and diminishing the combining procedure. Gurushantha et al. (2017) exhibited a low temperature flexible technique for synthesizing ZrO_2:ZnO NCs. These NCs were very much described and their PL and photocatalytic considerations were talked about their possible utilization in lighting and waste water treatment applications. The photoluminescence extinguishing seen behind the expansion of ZnO NPs is hindered as an unadulterated hiding impact legitimately identified with the method of test sample preparation.

The vicinity of Zr grains in an alumina cross-section as a specific II-phase connects with the Zr to carry on in an intrinsic way that is to experience tetragonal to monoclinic stage change or to be held in a metastable tetragonal phase structure on cooling of the composites after molding. It is the volume growth and shear distort identified with the t−m stage transformation that results in different strengthening structures in these composites, yielding weight influenced change continuing, microcrack sustaining, compressive surface loads, and crack redirection. Mondal (2005) synthesized the Zr components of toughened alumina (ZTA) by EDS and coprecipitation structures are reasonable for wear limitation coatings. ZTA with its improved hardness and break quality is considerably more regularly fitting and logically beneficial and significant in metal carving applications. ZTA instruments can be utilized in completing tasks even at ordinarily controlled curving rates in the range 150–280 m/min. The adjustment in reinforcing sway as a result of the extension of Y-PSZ accepts an enormous valuable activity concerning hardness similarly as crack solidness for making cutting gadget implants.

18.2.2 Biomedical applications

As exhibited by Bhowmick et al. (2017) ZrO_2 NPs were incorporated in organic-inorganic blend composites including poly(ethylene glycol), chitosan, and nano-hydroxyapatiteto make nanocomposites for bone tissue covering. In the FT-IR spectra (Fig. 18.3A) of BNCI-III, wide band focused at 3305 cm^{-1} was noticed for O–H widening frequency of CS and PEG. Fig. 18.3B addresses the XRD patterns of the prepared nanocomposites. The addition of tetragonal ZrO_2 NPs into these nanocomposites was supported by watching featured peaks at 30.06 and 51.17 degrees; chosento (101) and (112) significant crystal planes (Saha, Payra & Banerjee, 2015).

SEM pictures (Fig. 18.3C) shows the BNC I-III porous structure. The BNC I-III antimicrobial holding was more prominent if there should be an occasion of gram-negative bacteria *Escherichia coli* XL1B stand apart from gram-positive bacteria Bacillus cereus and Lysinibacillus fusiformis, as appeared in Table 18.2. Most great mechanical quality, most necessary porosity, most potent antimicrobial property and most imperative MG-63 cell proliferation by using suited water soaking up capacity and similitude with the human body make BNC I becoming bone-like nanocomposite for productive bone regeneration (Feng et al., 2000; Haefili, Franklin & Hardy,1984; Jangra et al., 2012; Kundu, Mandal, Ghosh & Pal, 2004; Ragab et al., 2014; Yamanaka, Hara & Kudo, 2005).

Bukhari et al. (2018) used different content of TiO_2 substance 1–5 wt.% and sorted out undoped and TiO_2 doped ZrO_2 thin films by sol–gel method. Crystallite estimation ~13 nm for unadulterated t-ZrO_2 phase with reduced unit portable volume 136.7 $Å^3$ was to be used. By and large more prominent X-pillar thickness and low estimation of porosity of 6% was once watched in citing to some degree exorbitant estimation of hardness at 3 wt.% of TiO_2 content. For the most part more noteworthy hardness regard has been looked for 3 wt.% of TiO_2 doped ZrO_2 unstable film because of the reality of substance of single stage (t-ZrO_2). Table 18.3 proposes the range in Vickers hardness with TiO_2 concentration close by type sin ZrO_2 stages and the range fluctuation in crack ruggedness with phase translation.

Kaliaraj et al. (2018) analyzed the deposition of Zr film onto 316L SS substrate through EBPVD method. Hydrophilic nature used to be considered on ZrO_2 surfaced film and hydrophilic conduct that used to be seen on bare 316L SS substrate. The XRD pattern of ZrO_2 film is shown in Fig. 18.4A. The polycrystalline existence of ZrO_2 show monoclinic stage with the reflections of (011), (111), (111),

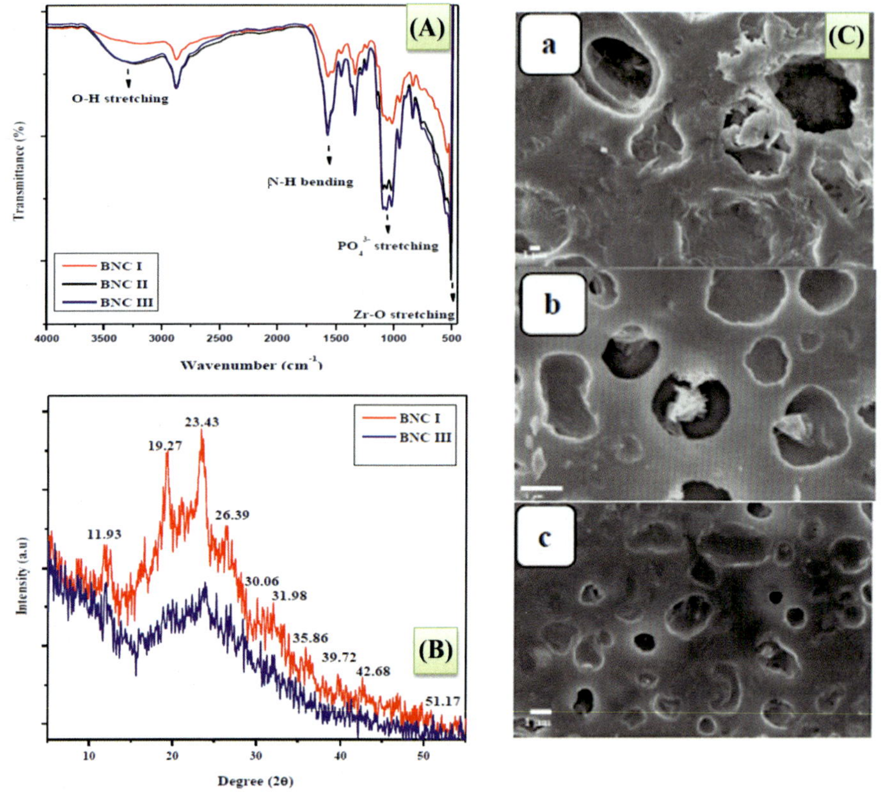

Figure 18.3 (A) FT-IR spectra of BNCI-III. (B) Powder XRD spectra of BNC I and III. (C) SEM images of (a, b, c) BNC I, II and III.
Source: Reprinted with permission from Bhowmick, A., Pramanik, N., Jana, P., Mitra, T., Gnanamani, A., Das, M., ... Kundu, P.P. (2017). Development of bone-like zirconium oxide nanoceramic modified chitosan based porous nanocomposites for biomedical application. *International Journal of Biological Macromolecules*, *95*, 348–356.

(020), (120), (201), (211), (022), (031), (131), (131), (113), (023) and (140) and their referring to d-dispersing is definitely coordinated with monoclinic ZrO_2 JCPSD data. The received XRD diagram regarded (111) aircraft as the most appreciated improvement direction. Surface geology and quality of the ZrO_2 films have been estimated via AFM. Fig. 18.4B (a) and (b) exhibit 2D, 3D images of bare 316L SS substrata and greater surface roughness established (Ra = 10 nm) as abnormalities on a superficial level. Fig. 18.4B (c) and (d) exhibit 2D and 3D pictures of ZrO_2 films. 2D and 3D topography of the ZrO_2 films find unvarying dispersion, defect-free ZrO_2 films with spherical molecules. (Fig. 18.4C) indicates the surface structure of ZrO_2 coating and shows thick; crack free and slightly clustered molecules. The mild assemblage used to be due to the fact of the greater portability of adatoms on the metallic surface at a greater temperature of substrate which improvements the aggregation of the particles.

Table 18.2 BNC I-III antimicrobial properties (Bhowmick et al., 2017).

	Antimicrobial zone of inhibition [(Mean ± SD) mm] of BNC I-III against bacterial strains		
	Gram-positive bacterial strains		**Gram-negative bacterial strains**
	Bacillus cereus	***Lysinibacillus fusiformis***	***Escherichia coli XL1B***
BNC I	12.2 ± 0.62	11.8 ± 0.56	12.5 ± 0.58
BNC II	12.5 ± 0.55	12.1 ± 0.51	12.8 ± 0.61
BNC III	12.7 ± 0.51	12.4 ± 0.68	13.2 ± 0.57

*Significant values $P < .05$.

Table 18.3 TiO_2 doped ZrO_2 Hardness Vickers (Bukhari et al., 2018).

[TiO_2] (wt.%)	**Hardness Vickers (HV) at constant load* and time****	**Porosity (%)**	**Crack toughness (MPa $m^{-1/2}$)**	**Zirconia phase**
0	645 ± 5	14.4	13.48	Monoclinic-tetragonal
1	700 ± 5	12.8	13.55	Monoclinic-tetragonal
2	782 ± 5	11.8	13.96	Monoclinic-tetragonal
3	1363 ± 5	5.7	16.78	Tetragonal
4	653 ± 5	13.9	12.12	Monoclinic-tetragonal
5	623 ± 5	14.8	11.56	Monoclinic-tetragonal

Liu, Zou, Xing and Huang (2019) incorporated the UV treatable ZrO_2-Al_2O_3 composite ceramics paste dependent on SLA-3D printing innovation, and the comparing ceramic green bodies were 3D printed, lastly, the $ZrO_2-Al_2O_3$ composite were produced by an ensuing de binding and forging action. Fig. 18.5 shows the schematic representation to the manufacture procedure of ZTA parts by means of SLA innovation.

The ZTA composite ceramics fracture morphology sintered at 1500°C is showed up in Fig. 18.6. With increase in holding time, Zr and alumina grain size increases and reduction in porosity in ceramic parts was observed. The nearness of transgranular crack builds the ceramic model need increasingly split imperativeness when it discloses, which unquestionably influences the mechanical properties of the ceased parts.

The morphology of the crack of the production of composite ceramics sintered in various retention times, while the sintering temperature is 1500°C has been shown in Fig. 18.7. With an extent of the retention time, the grain size densifies and the ceramic porousness in the parts decreases. Coupled with the conservative,

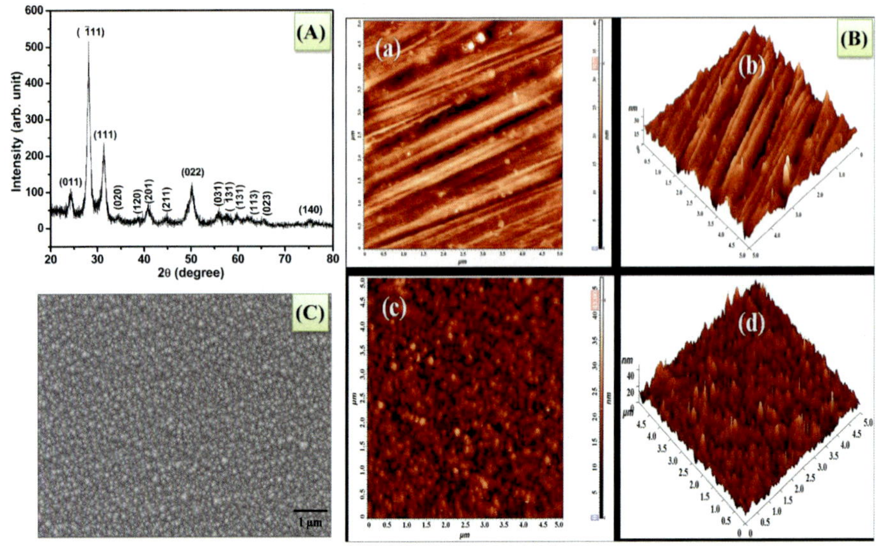

Figure 18.4 (A) XRD example of ZrO_2 covering. (B) Topographical examination of ZrO_2 covered and exposed 316L SS substrate by AFM investigation. (a) and (b): 2D and 3D pictures of ZrO_2 covering; (c) and (d): 2D and 3D pictures of exposed 316L SS substrate. (C) SEM morphology of ZrO_2 covering.
Source: Reprinted with permission from Kaliaraj, G. S., Vishwakarma, V., Kirubaharan, K., Dharini, T., Ramachandran, D., & Muthaiah, B. (2018). Corrosion and biocompatibility behaviour of zirconia coating by EBPVD for biomedical applications. *Surface & Coatings Technology*, *334*, 336–343.

the crack surface rupture method changes from the intergranular crack in Fig. 18.7A and B retention time of 0 and 30 minutes to the interim crack and the interim crack in Fig. 18.7C and D retention time of 60 and 90 minutes. At the time when the retention time is 60 minutes, the zircon grains scatter at the grain edges of the alumina grains in Fig. 18.7D, and the inter-grain retention is minimized, and there are crack conspicuous transgresses on the crack surface. Be that as it may, when the retention time is prolonged to 90 minutes, despite the fact that there is a transgranular crack despite everything, and the transport of the grain sizing is not unvarying, which influences the ceramics mechanical properties.

(1) The actual thickness of ceramics components increments step by step till they arrive at the maximum of 3.78 g/cm^3 at 1550°C. On further rising of sintering temperature, the real thickness of ceramic components dropped. With the dragging out of sintering retaining time, the actual thickness of ceramic components increments slowly, arriving at the most severe thickness of 3.79 g/cm^3 at 90 minutes, then again the conserving time surpasses upon 60 minutes, the enlargement of true thickness continues to be secure fundamentally. (2) On increasing temperature, the hardness of ceramics components accelerated first and later on diminished, arriving at the best hardness of 14.1 GPa at 1500°C. The stretching out of controlling time

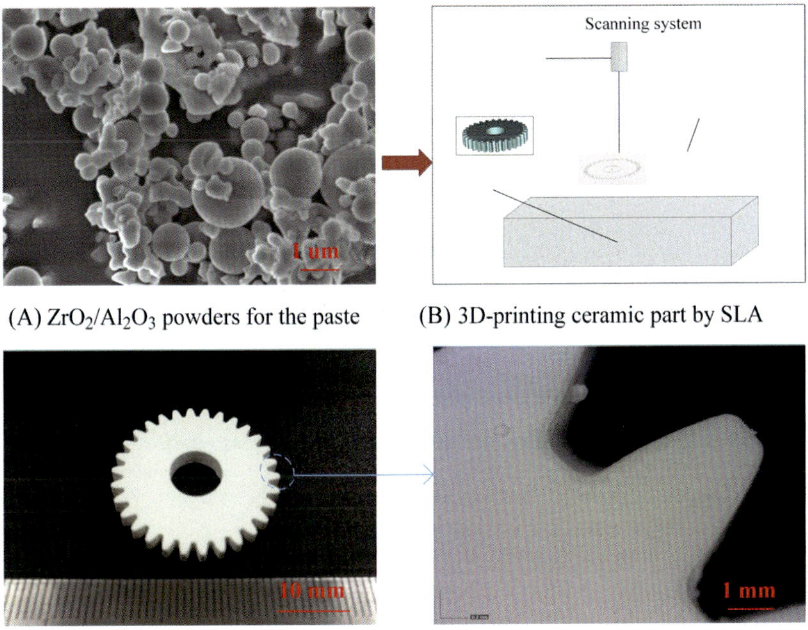

(A) ZrO_2/Al_2O_3 powders for the paste (B) 3D-printing ceramic part by SLA

(C) Sintered ZrO_2/Al_2O_3 ceramic gear and its microscopic structure

Figure 18.5 Schematic representation of the material ion of ZTA constituents through SLA technology.
Source: Reprinted with permission from Liu, X., Zou, B., Xing, H., & Huang, C. (2019). The preparation of ZrO2-Al2O3 composite ceramic by SLA-3D printing and sintering processing. *Ceramics International*, *46*, 937–944.

builds the stiffness of ceramics and later on diminishes, arriving at the restriction of 13.3 GPa when the conserving time is 60 minutes. (3) The ZrO_2/Al_2O_3 ceramics paste based on SLA-3D printing was synthesized.

Maham, Nasrollahzadeh and Sajadi (2020) developed a green methodology to draw Ag/ZrO_2 nanocomposite by utilizing extract of Ageratum conyzoides L leaf. The characterization of the as-composed heterogeneous catalyst was done utilizing various techniques. The immobilization of AgNPs on the ZrO_2 surface incited higher reactant execution because of the reality of the synergistic participation between AgNPs and substantially low propensity of AgNPs to accumulate all through the reaction. Furthermore, the as-incorporated nanocomposite should be reprocessed in 5 successive cycles. This method has the advantages of simple removal of extravagant and harmful mixtures, the ejection of homogeneous catalysts and eminent yield.

18.2.3 Sensor applications

Keerthana, Sakthivel and Prabha (2019) examined the manufactured method for getting an oval kind of $MgO-ZrO_2$ that could be accomplished utilizing different

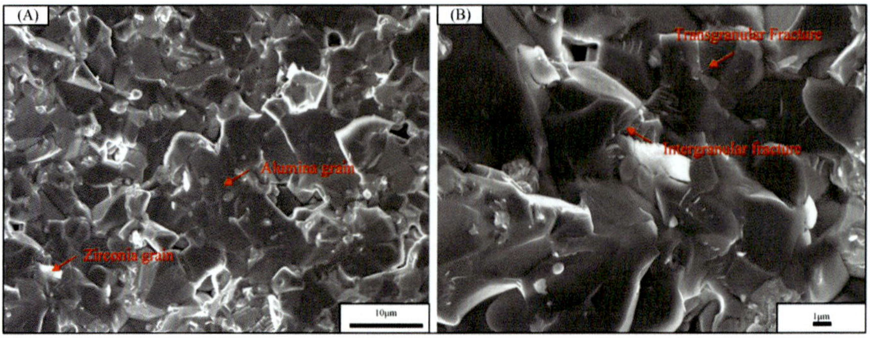

Figure 18.6 SEM micrographs of fracture surfaces of ZTA composite ceramic at 1500°C at 60 min conserving time.
Source: Reprinted with permission from Liu, X., Zou, B., Xing, H., & Huang, C. (2019). The preparation of ZrO_2-Al_2O_3 composite ceramic by SLA-3D printing and sintering processing. *Ceramics International*, *46*, 937–944.

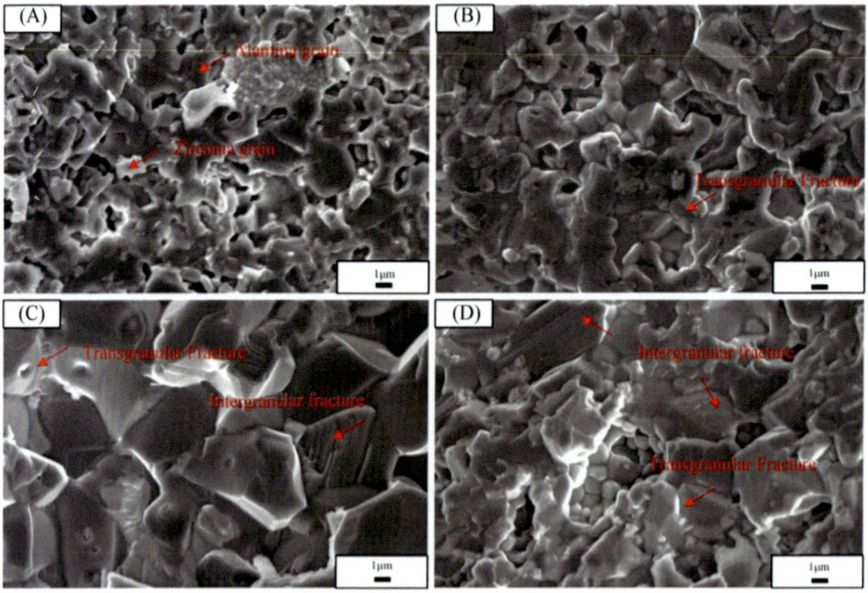

Figure 18.7 SEM micrographs of crack surfaces of ZTA composite ceramic sintered at 1500°C with holding time: (A) 0 min, (B) 30 min, (C) 60 min, (D) 90 min.
Source: Reprinted with permission from Liu, X., Zou, B., Xing, H., & Huang, C. (2019). The preparation of ZrO_2-Al_2O_3 composite ceramic by SLA-3D printing and sintering processing. *Ceramics International*, *46*, 937–944.

methods, amid sol-gel technique. A special element of this methodology is to get particles in a profoundly crystalline structure straightforwardly after blend, taking out the requirement for any resulting calcination step to instigate crystallization. The response frameworks are straight forward, and the metal oxide fore runners may be utilized in high focuses. In recent years, a novel, exhaustively material non-aqueous approach to manage NPs of metal oxide engages the amalgamation of a collection of two to threefold metallic oxides that have made. The warm strengthening methodology can assemble compressibility, light thickness, conductivity, and so on (Honakeri, Malode, Kulkarni & Shetti, 2020; Kulkarni et al., 2020; Malode, Keerthi, Shetti & Kulkarni, 2020; Shetti, Malode, Nayak, Aminabhavi & Reddy, 2019; Shetti, Nayak, Malode, et al., 2019; Shetti, Malode, Malladi et al., 2019; Shetti, Nayak, Malode, Reddy et al., 2019; Shetti, Malode, Ilager et al., 2019; Shetti et al., 2020; Shetti, Malode, Nayak, Reddy & Reddy, 2019; Shetti, Shanbhag, Malode, Srivastava & Reddy, 2020; Shetti et al., 2019; Shetti et al., 2020a). The surface zone can be reduced when the aggregation is kept away from. Eminent conduction, improved quality, and flexibility can be extended with fundamentally less porosity essentially as refreshed mechanical property by methods for basic adsorption segments (Malode, Keerthi, & Shetti, 2020; Shetti, Malode, Vernekar et al., 2019; Shetti, Malode, Nayak, Bagihalli et al., 2019; Shetti, Malode, Nayak .& Reddy, 2019; Shetti, Malode, Bukkitgar et al., 2019; Shetti, Ilager et al.,2020; Shetti, Malode, Nayak, Naik et al., 2020; Shetti, Malode & Nandibewoor, 2015; Shetti, Nayak & Malode et al., 2019; Shetti et al., 2019; Shetti, Nayak & Malode, 2018; Shetti, Nayak, Malode & Kulkarni, 2017; Shetti, Nayak, Malode & Kulkarni, 2018; Vernekar et al., 2020; Shetti et al., 2019a). The citrate gel approach passed on about a round-molded equilateral composite. Tetragonal degrees are viewed as overwhelming the monoclinic phase when magnesia-counter balance zirconia used to be set up by techniques for the expanding course. Tetragonal zirconia has a tremendous degree of use in dentistry. Agglomeration reduces the photocatalytic productiveness and leftovers the photodegradation, and in this manner, heterogeneous structures improve the photocatalytic development. Perovskites constructions have been set up over high stability and utilized in the energy storage devices, for example, solar cells. The cubic term of zirconia should be settled at high temperature utilizing suitable metallic oxides. Particular morphology of mixed metallic oxides should be performed with the guide of varying the fuel-oxidant extent.

The best possible way to implant biomaterial is a key feature for the drawn-out achievement of additions. Renganathan, Tanneru and Madurai (2018) analyzed an install containing Ti-based alloys that had a flexible profitable introduction owing to physical, mechanical, and biological holdings. Contempt to the nearness of a lot of Ti-based alloys, modest modulus β-type Ti-founded blends have not been able to be executed in assessment with one of a kindsorts. Giving incredibly great biological fixation by means of bone tissue ingrowth into the porous contraption is its main feature. Like titanium, the variables having a region with bundle VI of the Periodic table are considered as suitable for the upgrade of implants. To face the more prominent drawn out presence of an implant in a biological medium, it has over the top utilization restriction essentially as tissue in growth. The additions created the utilization of Ti and the Zr compound concedes higher plans. For this changed alloy plan, improvement of solidarity is standard by methods for combining steel as an option of character metal.

Mekala, Deepa and Rajendran (2018) synthesized undoped and rare earth, Dysprosium (Dy) and Cerium (Ce) doped ZrO_2 NPs through Co-precipitation. The as synthesized products characterized with the help of XRD, SEM, EDX, and UV vis- Drs examination. The nature of dopants impacts the shape of the nanoparticles revealed by XRD. EDX finds out about the proximity of Dy, Zr, O, and Ce and a relevant addition in the Dy and Ce degree in a definitive way. Zirconia (ZrO_2) and Alumina (Al_2O_3) are familiar bioinert ceramic assemblies inferable from mechanical living arrangements. Vasanthavel, Meenakshi, Nivedha, Ballamurugan and Kannan (2018) inspected the impact of yttrium (Y^{3+}) fuses in ZrO_2-SiO_2 combined system (YSZ) on its essential features. The sol-gel technique was used to mix the powders and has been characterized to look into the mechanical and structural features influenced by the method of Y^{3+} increase. Depiction forms confirmed the basic employment of Y^{3+} on the resultant structural and thermal adequacy of the YZS scheme. Instrumented indentation gathered the updated mechanical habitations displayed through the YZS system in assessment with an unadulterated ZrO_2-SiO_2 twofold structure (ZS).

Yao et al. (2018) compared at improving impacts of ZrO_2 and TiO_2 control hydroxyapatite (HA), HA/ZrO_2 and HA/TiO_2 composite coatings particles deposited on 316L stainless metal substrates by high-velocity suspension flame spray (HVSFS) method via combination of the sum total of coatings have been separating through XRD and FTIR. The microstructures and surface morphology was depicted by utilizing SEM (Fig. 18.8). The composite coatings affirmed for all intents and purposes equivalent and improved utilization contradict displays in Hanks'copybody-fluids appeared differently in relation to the revealed substrate. Negahdary, Habibi-Tamijani, Asadi and Ayati (2013) investigated amalgamation of Zr Nps and their usages as brought segments of solid structures. They used far reaching Portland concrete in related preliminary Concrete Structures. The preliminary or E affiliation (E1–E4) mixes had been set up with various proportions of ZrO_2 Nps with anormalatom estimation of 20 nm. The altered cement with ZrO_2 used to be focused with the separation versatility, flexural quality, and putting time methodologies.

Shevchenko et al. (2002) delivered 80 mass% $ZrO_2 - 20$ mass% Al_2O_3 powder utilizing an unpredictable strategy which coordinates sol-gel innovation and aqueous union. The particular surface zones of the powder changed from 39 to 5.3 m^2/g relying upon the warm treatment conditions. Metastable F-ZrO_2 framed after powder toughening at 400°C. The stage change F-$ZrO_2 \rightarrow$ T-ZrO_2 (hints of M-ZrO_2) happened under powder warm treatment from 700°C to 1000°C. Just Θ-Al_2O_3 was identified under trial conditions. The powder was portrayed by a sintering movement. Working the procedures under powder warm treatment in the $ZrO_2 - Y_2O_3 - CeO_2 - Al_2O_3$ framework will permit one to deliver an assortment of artistic microstructures from fine-grained to "self-strengthened." These powders can be utilized in assembling surgical cutting instruments just as in ceramic passive bioimplants and strong electrolytes for fuel components.

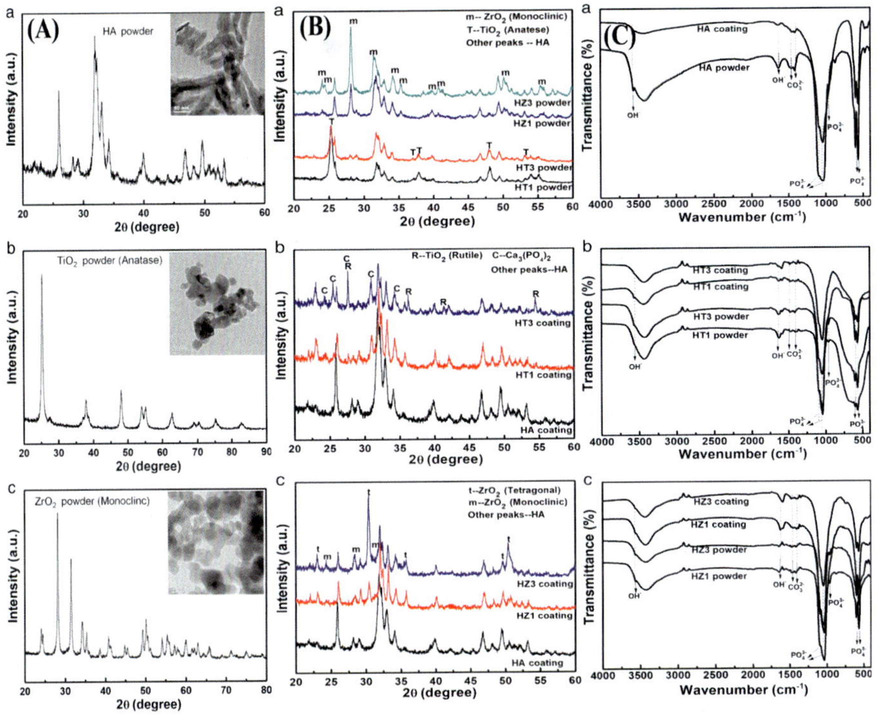

Figure 18.8 (A) XRD patterns of (a) HA, (b) TiO_2 and (c) ZrO_2 powders. (B) XRD examples of the (a) blended powders, (b) HA/TiO_2 composite coatings and (c) HA/ZrO_2 composite coverings. (C) FTIR spectra of the blended powders and composite coatings for (a) HA, (b) HA/TiO_2 and (c) HA/ZrO_2.
Source: Reprinted with permission from Yao, H.-L., Hu, X.-Z., Bai, X.-B., Wang, H.-T., Chen, Q.-Y., & Ji, G.-C. (2018). Comparative study of HA/TiO2 and HA/ZrO2 composite coatings deposited by high-velocity suspension flame spray (HVSFS). *Surface & Coatings Technology*, *351*, 177–187.

18.2.4 Limitations and challenges of zirconia

The zirconia-based ceramics productions have been utilized in an assortment of studies, in vitro, in vivo, or in silico. The Zr has numerous points of interest, a few disadvantages and hence a few perspectives to be corrected. The primary obstacle to defeat is to develop materials that can substitute dental tissues. In the logical site, Zr gives stand-out snags of chipping/delamination of veneering porcelain, and the trouble in connection, which takes mechanically retentive backings. In research facility preliminaries, the establishment of disappointments, the over the top burdens, and the time required for crack hinder the recreation of the logical circumstance. The breaking off and delamination recurrence of veneering porcelain of Zr reclamations in clinical audits has driven the creators to advance Zr for solid recapturing, which may furthermore be varnished for a more prominent suggested stylish meat. Polychromatic CAD/CAM blocks are reachable in the commercial centers and reenact the shading differences from cervical to incision thirds of teeth. The Zr squares translucency has furthermore been upgraded, trying to copy the esthetic effect of glass involving ceramics. Be that as it may, clinically they are suitable for the posterior district.

The Zr ceramics notwithstanding shiny segment or silica content material stretches out to a texture with repaired mechanical highlights, aside from a synthetically inactive material. The non-communication of the artistic floor hampers the holding measures. Sand blasting can likewise help grip, nonetheless, the mischief welcomed on to the texture is something to consider in the drawn out logical supplier of rebuilding efforts. The utility of exceptionally muddled floor medicines, similar to plasma affidavit, does not now help with conquering this trouble either. Right now, zirconia reclamations are by certain methods set up on mechanical maintenance, restricting the pointers of the material. The blast of glass-penetrated Zr is a guaranteeing final product for attachment. The paste measures are eased, the look and the facade restricting addition when the glass is acquainted with the cementation surface. The most perfectly awesome vitality was once said in glass-penetrated Zr. The various techniques had been as of now presented in the writing for glass penetration into Zr grains, notwithstanding, there are no models conveniently available.

Concerning tests, the gold popular smaller scale malleable bond vitality investigates is hard to execute because of the exceptionally high effectiveness and solidness of sintered Zr, which vitiates the cutting of inconceivably little examples. To consider holding fast to the Zr surface, the most drawn in looks are the shear and miniaturized scale sheer bond proficiency tests. The limits, for example, deficient pressure scattering, the expansion of compaction and pliable worries at the associated interface, and ordinary screw-ups of the substrate, can likewise control the results anyway positively are assented in writing. The Zr partitions outwear mannequin is the best suggestion for the provision of the general execution of the material in ailments. Attributable to the huge presentation, over the top transfers, and quite a while are indispensable to finding disappointment through weariness. The lab tests need to be devastated eating more prominent time. Along these lines, exceptionally over the top masses use may likewise underestimate the lifetime of Zr tests. Over 10 Hz of unreasonable frequencies of burden, constancy may likewise also surpass the research facility check from the clinical position. Besides, more noteworthy unrivaled rigging is needed when exorbitant masses are utilized. Henceforth, the bringing about of Zr tests weakness fiascos is currently not a minor occupation. At the point when over the top masses are upheld, for load-to-break keeps an eye on Zr pieces, it might likewise battle cracks at the heap utility spot, which is not, at this point, recouped clinically. As the lab checks mean to yield clinically relevant data, the flexibly settled in dislike site minifies the ramifications of the investigation. Regardless of the confinements, Zr is on the other hand the most famous artistic usable for dental applications, to be created in an all-clay gather. The cement cementation with analyzed convention should be what's more asked. The limitations on the utilization of solid Zr essentially dependent on veneering are to be toiled out and the reenactment of logical adjusting should also be altered. Notwithstanding, going with the meanings, Zr is a right and inescapable material, with prosperous logical achievement esteems.

References

Abd El-Lateef, H., & Khalaf, M. (2015). Corrosion resistance of ZrO_2-TiO_2 nanocomposite multilayer thin films coated on carbon steel in hydrochloric acid solution. *Materials Characterization*, *108*, 29–41.

Aboushelib, M. N. (2011). Evaluation of zirconia/resin bond strength and interface quality using a new technique. *Journal of Adhesive Dentistry*, *13*, 255–260.

Aboushelib, M. N., Matinlinna, J. P., Salameh, Z., & Ounsi, H. (2008). Innovations in bonding to zirconia-based materials: Part I. *Dental Materials Journal*, *24*, 1268–1272.

Aboushelib, M. N., Feilzer, A. J., & Kleverlaan, C. J. (2010). Bonding to zirconia using a new surface treatment. *Journal of Prosthodontics*, *19*, 340–346.

Al-Amleh, B., Lyons, K., & Swain, M. (2010). Clinical trials in zirconia: A systematic review. *Journal of Oral Rehabiliation*, *37*, 641–652.

Baldissara, P., Llukacej, A., Ciocca, L., Valandro, F. L., & Scotti, R. (2010). Translucency of zirconia copings made with different CAD/CAM systems. *Journal of Prosthetic Dentistry*, *104*, 6–12.

Beuer, F., Schweiger, J., Eichberger, M., Kappert, H. F., Gernet, W., & Edelhoff, D. (2009). High-strength CAD/CAM-materialated veneering material sintered to zirconia copings—a new materialation mode for all-ceramic restorations. *Dental Materials Journal*, *25*, 121–128.

Beuer, F., Stimmelmayr, M., Gueth, J. F., Edelhoff, D., & Naumann, M. (2012). In vitro performance of full-contour zirconia single crowns. *Dental Materials Journal*, *28*, 449–456.

Bhowmick, A., Pramanik, N., Jana, P., Mitra, T., Gnanamani, A., Das, M., ... Kundu, P. P. (2017). Development of bone-like zirconium oxide nanoceramic modified chitosan based porous nanocomposites for biomedical application. *International Journal of Biological Macromolecules*, *95*, 348–356.

Brun, V., Guillaume, C., Mechiche Alami, S., Josse, J., Jing, J., Draux, F., ... Velard, F. (2014). *Biomedical Materials and Engineering*, *24*, 63–73.

BruxZir Solid Zirconia crowns & bridges. (2010). Clinical Solution Series, Scientific Validation Document: 1–8. Available from: http://www.bruxzir.com/downloads-bruxzirzirconia-dental-crown/bruxzir-solid-zirconia-businessintegration-program.pdf.

Bukhari, S. B., Imran, M., Bashir, M., Riaz, S., & Naseem, S. (2018). Room temperature stabilized TiO_2 doped ZrO_2 thin films for teeth coatings—A sol-gel Approach. *Journal of Alloys and Compounds*, *767*, 1238–1252.

Cales, B., & Stefani, Y. (1994). Mechanical properties and surface analy sis of retrieved zirconia femoral hip joint heads after an implantation time of two to three years. *Journal of Materials Science: Materials in Medicine*, *5*, 376–380.

Cekic-Nagas, I., Egilmez, F., & Ergun, G. (2012). Comparison of light transmittance in different thicknesses of zirconia under various light curing units. *Journal of Advanced Prosthodontics*, *4*, 93–96.

Chen, L., Suh, B. I., Kim, J., & Tay, F. R. (2011). Evaluation of silica-coating techniques for zirconia bonding. *American Journal of Dentistry*, *24*, 79–84.

Choi, Y., & Suresh, S. (2002). Size effects on the mechanical properties of thin polycrystalline metal films on substrates. *Acta Materialia*, *50*, 1881–1893.

Chroneos, A., Yildiz, B., Tarancon, A., Parfitt, D., & Kilner, J. A. (2011). Oxygen diffusion in solid oxide fuel cell cathode and electrolyte materials: Mechanisticinsights from atomistic simulations. *Energy and Environmental Sciences*, *4*, 2774–2789.

Covacci, V., Bruzzese, N., Maccauro, G., Andreassi, C., Ricci, G. A., Piconi, C., ... Cittadini, A. (1999). In vitro evaluation of the mutagenic and carcinogenic power of high purity zirconia ceramic. *Biomaterials*, *20*, 371–376.

Culp, L., & McLaren, E. A. (2010). Lithium disilicate: The restorative material of multiple options. *Compendium Continuing Education in Dentistry*, *31*, 716–720, 722, 724–725.

Denry, I., & Kelly, J. R. (2008). State of the art of zirconia for dental applications. *Dental Materials*, *24*, 299–307.

Depan, D., Venkat Surya, P. K. C., Girase, B., & Misra, R. D. K. (2011). Organic/inorganic hybrid network structure nanocomposite scaffolds based on grafted chitosan for tissue engineering. *Acta Biomaterialia*, *7*, 2163–2175.

Dikovska, A., Atanasova, G., Avdeev, G., & Strijkova, V. (2016). Thin nanocrystalline zirconia films prepared by pulsed laser deposition. *Journal of Physics Conference Series*, *700*, 012024.

Doufar, N., Benamira, M., Lahmar, H., Trari, M., Avramova, I., & Caldes, M. T. (2020). Structural and photochemical properties of Fe-doped ZrO_2 and their application as photocatalysts with TiO_2 for chromate reduction. *Journal of Photochemistry & Photobiology, A: Chemistry*, *386*, 112105.

Duan, N., Lin, H., Li, L., Hu, J., Bi, L., Lu, H., ... Deng, L. (2013). ZrO_2-TiO_2 thin films: A new material system for mid-infrared integrated photonics. *Optical Materials Express*, *3*, 1537–1545.

Feng, Q. L., Wu, J., Chen, G. Q., Cui, F. Z., Kim, T. N., & Kim, J. O. (2000). A mechanistic study of the antibacterial effect of silver ions on *Escherichia coli* and *Staphylococcus aureus*. *Journal of Biomedical Materials Research*, *52*, 662–668.

Fischer, J., Stawarczyk, B., & Hammerle, C. H. (2008). Flexural strength of veneering ceramics for zirconia. *Journal of Dentistry*, *36*, 316–321.

Foxton, R. M., Cavalcanti, A. N., Nakajima, M., Pilecki, P., Sherriff, M., Melo, L., ... Watson, T. F. (2011). Durability of resin cement bond to aluminium oxide and zirconia ceramics after air abrasion and laser treatment. *Journal of Prosthodontics*, *20*, 84–92.

Guess, P. C., Att, W., & Strub, J. R. (2012). Zirconia in fixed implant prosthodontics. *Clinical Implant Dentistry and Related Research*, *14*, 633–645.

Gupta, R., & Kumar, A. (2008). Bioactive materials for biomedical applications using sol–gel technology. *Biomedical Materials*, *3*, 034005.

Gurushantha, K., Renuka, L., Anantharaju, K. S., Vidya, Y. S., Nagaswarupa, H. P., Prashantha, S. C., ... Nagabhushana, H. (2017). Photocatalytic and Photoluminescence studies of ZrO_2/ZnO nanocomposite for LED and Waste water treatment applications. *Materials Today: Proceedings*, *4*, 11747–11755.

Haefili, C., Franklin, C., & Hardy, K. (1984). Plasmid-determined silver resistance in *Pseudomonas stutzeri* isolated from a silver mine. *Journal of Bacteriology*, *158*, 389–392.

Haque, S., Tratipai, S., Jha, S., Bhattacharyya, D., & Sahoo, N. (2016). EXAFS studies on Gd-doped ZrO_2 thin films deposited by RF magnetron sputtering. *Applied Optics*, *55* (26), 7355–7564.

Heffernan, M. J., Aquilino, S. A., Diaz-Arnold, A. M., Haselton, D. R., Stanford, C. M., & Vargas, M. A. (2002). Relative translucency of six allceramic systems. Part I: Core materials. *Journal of Prosthetic Dentistry*, *88*, 4–9.

Heshmatpour, F., & Aghakhanpour, R. B. (2011). Synthesis and characterization of nanocrystalline zirconia powder by simple sol–gel method with glucose and fructose as organic additives. *Powder Technology*, *205*, 193–200.

Hisbergues, M., Vendeville, S., & Vendeville, P. (2009). Zirconia: Established facts and perspectives for a biomaterial in dental implantology. *Journal of Biomedical Materials Research—Part B Applied Biomaterials*, *88*, 519–529.

Honakeri, N. C., Malode, S. J., Kulkarni, R. M., & Shetti, N. P. (2020). Electrochemical behavior of diclofenac sodium at coreshell nanostructure modified electrode and its analysis in human urine and pharmaceutical samples. *Sensors International*, *1*, 100002.

Işeri, U., Ozkurt, Z., Yalnız, A., & Kazazoglu, E. (2012). Comparison of different grinding procedures on the flexural strength of zirconia. *Journal of Prosthetic Dentistry*, *107*, 309–315.

Imran, A., Alam, S., Irfan, M., & Farooq, M. (2015). Micro structural study of plasma-sprayed zirconia-CaO thermal barrier coatings. *Materials Today: Proceedings*, *2*, 5318–5323.

Jaenicke, S., Chuah, G. K., Raju, V., & Nie, V. T. (2008). *Catalysis Surveys from Asia*, *12*, 153–169.

Jangra, S. L., Stalin, K., Dilbaghi, N., Kumar, S., Tawale, J., Singh, S. P., … Pasricha, R. (2012). Antimicrobial activity of zirconia (ZrO_2) nanoparticles and zirconium complexes. *Journal of Nanoscience and Nanotechnology*, *12*, 7105–7112.

Kalantari, E., Naghib, S. M., Naimi-Jamal, M. R., & Mozafari, M. (2017). Green solvent-based sol–gel synthesis of monticellite nanoparticles: A rapid and efficient approach. *Journal of Sol-Gel Science and Technology*, *84*, 87–95.

Kaliaraj, G. S., Ramadoss, A., Sundaram, M., Balasubramanian, S., & Muthirulandi, J. (2014). Studies of calcium-precipitating oral bacterial adhesion on TiN, TiO_2 single layer, and TiN/TiO_2 multilayer- coated 316L SS. *Journal of Materials Science*, *49*(20), 7172–7180.

Kaliaraj, G. S., Vishwakarma, V., Kirubaharan, K., Dharini, T., Ramachandran, D., & Muthaiah, B. (2018). Corrosion and biocompatibility behaviour of zirconia coating by EBPVD for biomedical applications. *Surface & Coatings Technology*, *334*, 336–343.

Kamitakahara, M., Ohtsuki, C., & Miyazaki, T. (2007). Coating of bone-like apatite for development of bioactive materials for bone reconstruction. *Biomedical Materials*, *2*, R17.

Keerthana, L., Sakthivel, C., & Prabha, I. (2019). MgO-ZrO_2 mixed nanocomposites: Materialation methods and Applications. *Materials TodaySustainability*, *3–4*, 100007.

Kelly, J. R., & Denry, I. (2008). Stabilized zirconia as a structural ceramic: An overview. *Dental Materials*, *24*, 289–298.

Kern, M., & Wegner, S. M. (1998). Bonding to zirconia ceramic: Adhesion methods and their durability. *Dental Materials Journal*, *14*, 64–71.

Khalaf, M., & Abd El-Lateef, H. (2016). Corrosion protection of mild steel by coating with TiO_2 thin films codoped with NiO and ZrO_2 in acidic chloride environments. *Materials Chemistry and Physice*, *177*, 250–265.

Koutayas, S. O., Vagkopoulou, T., Pelekanos, S., Koidis, P., & Strub, J. R. (2009). Zirconia in Dentistry: Part 2. Evidence-based clinical breakthrough. *European Journal of Esthetic Dentistry*, *4*, 348–380.

Kubasiewicz-Ross, P., Dominiak, M., Gedrange, T., & Botzenhart, U. U. (2017). Zirconium: The material of the future in modern implantology. *Advances in Clinical and Experimental Medicine*, *26*, 533.

Kukli, K., Kemell, M., Vehkamaki, M., Heikkil, M., Mizohata, K., Kalam, K., . . . Frohlich, K. (2017). Atomic layer deposition andproperties of mixed Ta_2O_5 and ZrO_2 films. *AIP Advances*, *7*, 025001.

Kulkarni, D. R., Malode, S. J., Prabhu, K. K., Ayachit, N. H., Kulkarni, R. M., & Shetti, N. P. (2020). Development of a novel nanosensor using Ca-doped ZnO for antihistamine drug. *Materials Chemistry and Physics*, *246*, 122791.

Kumar, D. D., & Kaliaraj, G. S. (2018). Multifunctional zirconium nitride/copper multilayer coatings on medical grade 316LSS and titanium substrates for biomedical applications. *Journal of the Mechanical Behavior of Biomedical Materials*, *77*, 106−115.

Kundu, S., Mandal, M., Ghosh, S. K., & Pal, T. (2004). *Journal of Colloid and Interface Science*, *272*, 134−144.

Lazennec, J. Y., & Dietrich, M. (2004). *Bioceramics in joint arthroplasty: 9th BIOLOX symposium.* Proceedings: Springer Science & Business Media.

Lima, W., Quaha, H., Luc, Q., Muc, Y., Azli, W., Ismaild, W., . . . Cheonga, K. (2016). Effects of rapid thermal annealingon structural, chemical, and electrical characteristics of atomic-layer deposited lanthanum doped zirconium dioxide thin film on 4H-SiC substrate. *Applied Surface Science*, *365*, 296−305.

Liu, X., Zou, B., Xing, H., & Huang, C. (2019). The preparation of ZrO_2-Al_2O_3 composite ceramic by SLA-3D printing and sintering processing. *Ceramics International*, *46*, 937−944.

Lohmann, C. H., Dean, D. D., Koster, G., Casasola, D., Buchhorn, G. H., Fink, U., . . . Boyan, B. D. (2002). Ceramic and PMMA particles differentially affect osteoblast phenotype. *Biomaterials*, *23*, 1855−1863.

Maham, M., Nasrollahzadeh, M., & Sajadi, S. M. (2020). Facile synthesis of Ag/ZrO_2 nanocomposite as a recyclable catalyst for the treatment of environmental pollutants. *Composites Part B*, *185*, 107783.

Malode, S. J., Keerthi, P. K., & Shetti, N. P. (2020). Electrocatalytic behavior of a heterostructured nanocomposite sensor for aminotriazole. *New Journal of Chemistry*, *44*, 19376.

Malode, S. J., Keerthi, P. K., Shetti, N. P., & Kulkarni, R. M. (2020). Electroanalysis of carbendazim using MWCNT/Ca-ZnO modified electrode. *Electroanalysis*, *32*(7), 1590−1599.

Mamivand, M., Zaeem, M. A., Kadiri, H. E., & Chen, L. Q. (2013). Phase field modeling of the tetragonal-to-monoclinic phase transformation in zirconia. *Acta Materialia*, *61*, 5223−5235.

Matos, J. M. E., Anjos Jr, F. M., Cavalcante, L. S., Santos, V., Leal, S. H., Santosh Jr, L. S., . . . Longo, E. (2009). Reflux synthesis and hydrothermal processing of ZrO_2 nanopowders at low temperature. *Materials Chemistry and Physics*, *117*, 455−459.

McLaren, E. A. (1998). All-ceramic alternatives to conventional metalceramic restorations. *Compendium Continuing Education in Dentistry*, *19*, 307−308, 310, 312 passim; quiz 326.

Mekala, R., Deepa, B., & Rajendran, V. (2018). Preparation, characterization and antibacterial property of rare earth (Dy and Ce) doping on ZrO_2 nanoparticles prepared by coprecipitation method. *Materials Today: Proceedings*, *5*, 8837−8843.

Mirmohammadi, H., Aboushelib, M. N., Salameh, Z., Feilzer, A. J., & Kleverlaan, C. J. (2010). Innovations in bonding to zirconia based ceramics: Part III. Phosphate monomer resin cements. *Dental Materials Journal*, *26*, 786−792.

Mondal, B. (2005). Zirconia toughened alumina for wear resistant engineering and machinability of steel application. *Advances in Applied Ceramics*, *104*(5), 256−260.

Mozafari, M., Salahinejad, E., Shabafrooz, V., Yazdimamaghani, M., Vashaee, D., & Tayebi, L. (2013). Multilayer bioactive glass/zirconium titanate thin films in bone tissue engineering and regenerative dentistry. *International Journal of Nanomedicine*, *8*, 1665.

Narcisi, E. M. (1999). Three-unit bridge construction in anterior single-pontic areas using a metal-free restorative. *Compendium of Continuing Education in Dentistry*, *20*, 109–112, 114, 116–119.

Negahdary, M., Habibi-Tamijani, A., Asadi, A., & Ayati, S. (2013). Synthesis of zirconia nanoparticles and their ameliorative roles as additives concrete structures. *Journal of Chemistry*, *2013*, Article ID 314862.

Odman, P., & Andersson, B. (2001). Procera AllCeram crowns followed for 5 to 10.5 years: A prospective clinical study. *International Journal of Prosthodontics*, *14*, 504–509.

Papaspyridakos, P., & Lal, K. (2013). Computer-assisted design/computer-assisted manufacturing zirconia implant fixed complete prostheses: Clinical results and technical complications up to 4 years of function. *Clinical Oral Implant Research*, *24*, 659–665.

Peláez, J., Cogolludo, P. G., Serrano, B., Lozano, J. F., & Suárez, M. J. (2012). A prospective evaluation of zirconia posterior fixed dental prostheses: Three-year clinical results. *Journal of Prosthetic Dentistry*, *107*, 373–379.

Perednisa, D., Wilhelmb, O., Pratsinisb, S., & Gauckler, L. (2005). Morphology and deposition of thin yttria-stabilized zirconia, films using spray pyrolysis. *Thin Solid Films*, *474*, 84–95.

Piconi, C., & Maccauro, G. (1999). Zirconia as a ceramic biomaterial. *Biomaterials*, *20*(1), 1–25.

Ragab, H. S., Ibrahim, F. A., Abdallah, F., Al-Ghamdi, A. A., El-Tantawy, F., & Yakuphanoglu, F. (2014). Synthesis and in vitro antibacterial properties of hydroxyapatite nanoparticles. *IOSR Journal of Pharmacy and Biological Sciences*, *9*, 77–85.

Raigrodski, A. J., Hillstead, M. B., Meng, G. K., & Chung, K. H. (2012). Survival and complications of zirconia-based fixed dental prostheses: A systematic review. *Journal of Prosthetic Dentistry*, *107*, 170–177.

Ravichandran, A., Pushpa, K., Ravichandran, K., Karthika, K., Nagabhushana, B., Mantha, S., . . . Swaminathan, K. (2014). Effect of Al doping on the structural andoptical properties of ZrO_2 nanopowders synthesized using solution combustion method. *Superlattices and Microstructures*, *75*, 533–542.

Rendtorff, N. M., Suarez, G., Sakka, Y., & Aglietti, E. F. (2012). Influence of the zirconia transformation on the thermal behavior of zircon–zirconia composites. *Journal of Thermal Analysis and Calorimetry*, *110*, 695–705.

Renganathan, G., Tanneru, N., & Madurai, S. L. (2018). Orthopedical and biomedical applications of titanium and zirconium metals. *Fundamental Biomaterials: Metals*, 211–241.

Ries, M. D., Salehi, A., Widding, K., & Hunter, G. (2002). Polyethylene wear performance of oxidized zirconium and cobalt-chromium knee components under abrasive conditions. *Journal of Bone and Joint Surgery*, *84*, S129–S135.

Robertson, J., Xiong, K., & Falabretti, B. (2005). Point defects in ZrO_2 high-k gate oxide. *IEEE Transactions on Device and Materials Reliability*, *5*, 84–89.

Saha, A., Payra, S., & Banerjee, S. (2015). One-pot multicomponent synthesis of highly functionalized bio-active pyrano[2,3-c]pyrazole and benzylpyrazolyl coumarin derivatives using ZrO_2 nanoparticles as a reusable catalyst. *Green Chemistry*, *17*, 2859–2866.

Salahinejad, E., Hadianfard, M., Macdonald, D., Mozafari, M., Vashaee, D., & Tayebi, L. (2012). Zirconium titanate thin film prepared by an aqueous particulate sol–gel spin coating process using carboxymethyl cellulose as dispersant. *Materials Letters*, *88*, 5–8.

Salahinejad, E., Hadianfard, M., Macdonald, D., Mozafari, M., Walker, K., Rad, A. T., et al. (2013). Surface modification of stainless steel orthopedic implants by sol–gel $ZrTiO_4$ and $ZrTiO_4$–PMMA coatings. *Journal of Biomedical Nanotechnology*, *9*, 1327–1335.

Salari, S., & Ghodsi, F. (2017). A significant enhancement in the photoluminescence emission of the Mg doped ZrO_2 thin films by tailoring the effect of oxygen vacancy. *Journal of Luminescence*, *182*, 289–299.

Samanta, A., Lenosky, T., & Li, J. (2010). Thermodynamic stability of oxygen point defects in cubic zirconia, *arXiv:1009.5567* [cond-mat.mtrl-sci].

Scarano, A., Piattelli, M., Caputi, S., Favero, G. A., & Piattelli, A. (2004). Bacterial adhesion on commercially unadulterated titanium and zirconium oxide disks: An in vivo human study. *Journal of Periodontology*, *75*, 292–296.

Schulz, U., Leyens, C., Fritscher, K., Peters, M., Saruhan-Brings, B., Lavigne, O., … Caliez, M. (2003). Some recent trends inresearch and technology of advanced thermal barrier coatings. *Aerospace Science and Technology*, *7*, 73–80.

Shetti, N. P., Malode, S. J., & Nandibewoor, S. T. (2015). Electro-oxidation of captopril at a gold electrode and its determination in pharmaceuticals and human fluids. *Analytical Methods*, *7*(20), 8673–8682.

Shetti, N. P., Nayak, D. S., Malode, S. J., & Kulkarni, R. M. (2017). An electrochemical sensor for clozapine at ruthenium doped TiO_2 nanoparticles modified electrode. *Sensors and Actuators B: Chemical*, *247*, 858–867.

Shetti, N. P., Nayak, D. S., Malode, S. J., & Kulkarni, R. M. (2018). Fabrication of MWCNTs and Ru doped TiO_2 nanoparticles composite carbon sensor for biomedical application. *ECS Journal of Solid State Science and Technology*, *7*(7), Q3070–Q3078.

Shetti, N. P., Nayak, D. S., & Malode, S. J. (2018). Electrochemical behavior of azo food dye at nanoclay modified carbon electrode—A nanomolar determination. *Vacuum*, *155*, 524–530.

Shetti, N. P., Malode, S. J., Nayak, D. S., Bagihalli, G. B., Kalanur, S. S., Malladi, R. S., … Reddy, K. R. (2019a). Fabrication of ZnO nanoparticles modified sensor for electrochemical oxidation of methdilazineApplied. *Surface Science*, *496*, 143656.

Shetti, N. P., Malode, S. J., Nayak, D. S., Reddy, K. R., Reddy, C. V., & Ravindranadh, K. (2019a). Silica gel-modified electrode as an electrochemical sensor for the detection of acetaminophen. *Microchemical Journal*, *150*, 104206.

Shetti, N. P., Malode, S. J., Nayak, D. S., Bagihalli, G. B., Reddy, K. R., Ravindranadh, K., … Reddy, C. V. (2019b). A novel biosensor based on graphene oxide-nanoclay hybrid electrode for the detection of Theophylline for healthcare applications. *Microchemical Journal*, *149*, 103985.

Shetti, N. P., Nayak, D. S., Malode, S. J., Reddy, K. R., Shukla, S. S., & Aminabhavi, T. M. (2019). Electrochemical behavior of flufenamic acid at amberlite XAD-4 resin and silver-doped titanium dioxide/amberlite XAD-4 resin modified carbon electrodes. *Colloids and Surfaces B: Biointerfaces*, *177*, 407–415.

Shetti, N. P., Malode, S. J., Nayak, D. S., Bagihalli, G. B., Kalanur, S. S., Malladi, R. S., … Reddy, K. R. (2019). Fabrication of ZnO nanoparticles modified sensor for electrochemical oxidation of methdilazine. *Applied Surface Science*, *496*, 143656.

Shetti, N. P., Malode, S. J., Nayak, D. S., Reddy, K. R., Reddy, C. V., & Ravindranadh, K. (2019). Silica gel-modified electrode as an electrochemical sensor for the detection of acetaminophen. *Microchemical Journal*, *150*, 104206.

Shetti, N. P., Malode, S. J., Bukkitgar, S. D., Bagihalli, G. B., Kulkarni, R. M., Pujari, S. B., … Reddy, K. R. (2019). Electro-oxidation and determination of nimesulide at nanosilica modified sensor. *Materials Science for Energy Technologies*, *2*(3), 396–400.

Shetti, N. P., Malode, S. J., Ilager, D., Raghava Reddy, K., Shukla, S. S., & Aminabhavi, T. M. (2019). A novel electrochemical sensor for detection of molinate using ZnO nanoparticles loaded carbon electrode. *Electroanalysis*, *31*(6), 1040–1049.

Shetti, N. P., Malode, S. J., Malladi, R. S., Nargund, S. L., Shukla, S. S., & Aminabhavi, T. M. (2019). Electrochemical detection and degradation of textile dye Congo red at graphene oxide modified electrode. *Microchemical Journal*, *146*, 387–392.

Shetti, N. P., Malode, S. J., Nayak, D. S., Aminabhavi, T. M., & Reddy, K. R. (2019). Nanostructured silver doped TiO_2/CNTs hybrid as an efficient electrochemical sensor for detection of anti-inflammatory drug, cetirizine. *Microchemical Journal*, *150*, 104124.

Shetti, N. P., Malode, S. J., Nayak, D. S., Reddy, C. V., & Reddy, K. R. (2019). Novel biosensor for efficient electrochemical detection of methdilazine using carbon nanotubes-modified electrodes. *Materials Research Express*, *6*, 116308.

Shetti, N. P., Malode, S. J., Nayak, D. S., & Reddy, K. R. (2019). Novel heterostructured Ru-doped TiO_2/CNTs hybrids with enhanced electrochemical sensing performance for Cetirizine. *Materials Research Express*, *6*(11), 115085.

Shetti, N. P., Malode, S. J., Vernekar, P. R., Nayak, D. S., Shetty, N. S., Reddy, K. R., ... Aminabhavi, T. M. (2019). Electro-sensing base for herbicide aclonifen at graphitic carbon nitride modified carbon electrode – Water and soil sample analysis. *Microchemical Journal*, *149*, 103976.

Shetti, N. P., Nayak, D. S., Malode, S. J., Kakarla, R. R., Shukla, S. S., & Aminabhavi, T. M. (2019). Sensors based on ruthenium-doped TiO_2 nanoparticles loaded into multiwalled carbon nanotubes for the detection of flufenamic acid and mefenamic acid. *Analytica Chimica Acta*, *1051*, 58–72.

Shetti, N. P., Shanbhag, M. M., Malode, S. J., Srivastava, R. K., & Reddy, K. R. (2020). Amberlite XAD-4 modified electrodes for highly sensitive electrochemical determination of nimesulide in human urine. *Microchemical Journal*, *153*, 104389.

Shetti, N. P., Malode, S. J., Nayak, D. S., Bukkitgar, S. D., Bagihalli, G. B., Kulkarni, R. M., ... Reddy, K. R. (2020). Novel nanoclay-based electrochemical sensor for highly efficient electrochemical sensing nimesulide. *Journal of Physics and Chemistry of Solids*, *137*, 109210.

Shetti, N. P., Malode, S. J., Nayak, D. S., Bukkitgar, S. D., Bagihalli, G. B., Kulkarni, R. M., & ...Reddy, K. R. (2020). Novel nanoclay-based electrochemical sensor for highly efficient electrochemical sensing nimesulide. *Journal of Physics and Chemistry of Solids137,109-210.*

Shetti, N. P., Ilager, D., Malode, S. J., Monga, D., Basu, S., & Reddy, K. R. (2020). Poly (eriochrome black T) modified electrode for electrosensing of methdilazine. *Materials Science in Semiconductor Processing*, *120*, 105261.

Shetti, N. P., Malode, S. J., Nayak, D. S., Naik, R. R., Kuchinad, G. T., Reddy, K. R., ... Aminabhavi, T. M. (2020). Hetero-nanostructured iron oxide and bentonite clay composite assembly for the determination of an antiviral drug acyclovir. *Microchemical Journal*, *155*, 104727.

Shevchenko, A. V., Dudnik, E. V., Ruban, A. K., Red'ko, V. P., Vereschaka, V. M., & Lopato, L. M. (2002). Nanocrystalline powders based on ZrO_2 for biomedical applications and power engineering. *Powder Metallurgy and Metal Ceramics*, *41*, 11–12.

Somiya, T., & Akiba, T. (1999). Hydrothermal zirconia powders: A bibliography. *Journal of European Ceramic Society*, *19*, 81–87.

Song, X., Liu, Z., Kong, M., Lin, C., Huang, L., Zheng, X., et al. (2017). Thermal stability of yttria-stabilized zirconia (YSZ) and YSZ-Al_2O_3 coatings. *Ceramics International*, *43*, 14321–14325.

Sooa, M., Prastomo, N., Matsuda, A., Kawamura, G., Muto, H., Fauzi, A., ... Cheong, K. (2012). Elaboration and characterization of sol–gel derived ZrO_2 thin films treated with hot water. *Applied Surface Science*, *258*, 5250–5258.

Stanford, C., Oates, T., & Beirne, R. (2006). Zirconia as an implant and restorative biomaterial. *The International Journal of Oral & Maxillofacial Implants*, *21*, 841–844.

Stappert, C. F., Guess, P. C., Chitmongkolsuk, S., Gerds, T., & Strub, J. R. (2007). All-ceramic partial coverage restorations on natural molars. Masticatory fatigue loading and crack resistance. *American Journal of Dentistry*, *20*, 21–26.

Stefanic, G., Stefanic, I. I., & Music, S. (2000). *Materials Chemistry and Physics*, *65*, 197–207.

Stocker, C., & Baiker, A. (1998). *Journal of Non-Crystalline Solids*, *223*, 165–178.

Swab, J. J. (1991). Low temperature degradation of Y-TZP materials. *Journal of Materials Science*, *26*, 6706–6714.

Terki, R., Bertrand, G., Aouraf, H., & Coddet, C. (2006). Structural and electronic properties of zirconia phases: A FP-LAPW investigations. *Materials Science in Semiconductor Processing*, *9*, 1006–1013.

Tetè, S., Mastrangelo, F., Bianchi, A., Zizzari, V., & Scarano, A. (2009). Collagen fiber orientation around machined titanium and zirconia dental implant necks: An animal study. *International Journal of Oral & Maxillofac Implants*, *24*, 52–58.

Thomas, A., Sridhar, S., Aghyarian, S., Watkins-curry, P., Chan, J. Y., Pozzi, A., ... Rodrigues, D. C. (2016). Corrosion behavior of zirconia in acidulated phosphate fluoride. *Journal of Applied Oral Science*, *24*(1), 52–60.

Tokash, J., Stojilovic, N., Ramsier, R., Kovacik, M., & Mostardi, R. (2005). Surface analysis of prosthetic wear debris. *Surface and Interface Analysis*, *37*, 379–384.

Tyagi, B., Sidhpuria, K., Shaik, B., & Jasra, R. V. (2006). Synthesis of nanocrystalline zirconia using sol – gel and precipitation techniques. *Industrial & Engineering Chemistry Research*, *45*, 8643–8650.

Vagkopoulou, T., Koutayas, S. O., Koidis, P., & Strub, J. R. (2009). Zirconia in dentistry: Part 1. Discovering the nature of an upcoming bioceramic. *European Journal of Esthetic Dentistry*, *4*, 130–151.

Valandro, L. F., Ozcan, M., Bottino, M. C., Bottino, M. A., Scotti, R., & Bona, A. D. (2006). Bond strength of a resin cement to high-alumina and zirconia-reinforced ceramics: The effect of surface conditioning. *Journal of Adhesive Dentistry*, *8*, 175–181.

Vasanthavel, S., Meenakshi, K., Nivedha, V., Ballamurugan, A. M., & Kannan, S. (2018). Tuning the structural and mechanical properties in ZrO_2-SiO_2 binary system through Y^{3+} inclusions. *Materials Science & Engineering*, *C84*, 230–235.

Vergnieres, L., Odier, P., Weiss, F., Bruzek, C., & Saugrain, J. (2005). Epitaxial thick films by spray pyrolysis for coated conductors. *Journal of European Ceramic Society*, *25*(12), 2951–2954.

Vernekar, P. R., Shetti, N. P., Shanbhag, M. M., Malode, S. J., Malladi, R. S., & Malladi, K. R. (2020). Novel layered structured bentonite clay-based electrodes for electrochemical sensor applications. *Microchemical Journal*, 105441.

Vidya, Y. S., Anantharaju, K. S., Nagabhushana, H., Sharma, S. C., Nagaswarupa, H. P., Prashantha, S. C., ... Kumar, Danith (2015). *Spectrochimica Acta A*, *135*, 241–251.

Vinci, R. P., & Vlassak, J. J. (1996). Mechanical behavior of thin films. *Annual Review of Materials Science*, *26*, 431–462.

Volinsky, A. A., Vella, J. B., & Gerberich, W. W. (2002). Interfacial toughness measurements for thin films on substrates. *Thin Solid Films*, *429*, 201–210.

Wang, M. (2003). Developing bioactive composite materials for tissue replacement. *Biomaterials*, *24*, 2133–2151.

Ward, D. A., & Ko, E. I. (1995). *Preprints of papers presented at the ACS National Meeting*, 40(2), 356–359.

Wegner, S. M., & Kern, M. (2000). Long-term resin bond strength to zirconia ceramic. *The Journal of Adhesive Dentistry*, *2*, 139–147.

Welander, M., Abrahamsson, I., & Berglundh, T. (2008). The mucosal barrier at implant abutments of different materials. *Clinical Oral Implants Research*, *19*, 635–641.

Wohlwend, A., Studer, S., & Scharer Das, P. (1996). Zirkonoxidabutment ein neues vollkeramisches konzept zur ästhetischen verbesserung der suprastruktur in der implantologie. *Quint Zahnt*, *22*, 364–381.

Yamanaka, M., Hara, K., & Kudo, J. (2005). Bactericidal actions of a silver ion solution on *Escherichia coli*, studied by energy-filtering transmission electron microscopy and proteomic analysis. *Applied and Environmental Microbiology*, *71*, 7589–7593.

Yang, B., Barloi, A., & Kern, M. (2010). Influence of air-abrasion on zirconia ceramic bonding using an adhesive composite resin. *Dental Materials Journal*, *26*, 44–50.

Yao, H.-L., Hu, X.-Z., Bai, X.-B., Wang, H.-T., Chen, Q.-Y., & Ji, G.-C. (2018). Comparative study of HA/TiO_2 and HA/ZrO_2 composite coatings deposited by high-velocity suspension flame spray (HVSFS). *Surface & Coatings Technology*, *351*, 177–187.

Yoshida, K., Tsuo, Y., & Atsuta, M. (2006). Bonding of dual-cured resin cement to zirconia ceramic using phosphate acid ester monomer and zirconate coupler. *Journal of Biomedical Materials Research*, *77*, 28–33.

Zarrintaj, P., Moghaddam, A. S., Manouchehri, S., Atoufi, Z., Amiri, A., Amirkhani, M. A., et al. (2017). Can regenerative medicine and nanotechnology combine to heal wounds? The search for the ideal wound dressing. *Nanomedicine: Nanotechnology, Biology, and Medicine*, *12*, 2403–2422.

Zembic, A., Sailer, I., Jung, R. E., & Hämmerle, C. H. (2009). Randomized controlled clinical trial of customized zirconia and titanium implant abutments for single-tooth implants in canine and posterior regions: 3-year results. *Clinical Oral Implants Research*, *20*, 802–808.

Zhang, J., Li, W., & Tanji, T. (2014). Synthesis of zirconia oxide (ZrO_2) nanofibers on zirconnia substrates by ultrasonic spray pyrolysis. *Materials Sciences and Applications*, *5*, 193–198.

Iron oxides and their prospects for biomedical applications

19

Bhuvaneshwari Balasubramaniam[1,*], *Bidipta Ghosh*[2], *Richa Chaturvedi*[1] *and Raju Kumar Gupta*[1]
[1]Department of Chemical Engineering, Indian Institute of Technology Kanpur, Kanpur, India, [2]Department of Chemical Engineering, National Institute of Technology Durgapur, Durgapur, India

19.1 Introduction: iron oxide in biomedical applications

Nanotechnology has advanced tremendously in the last few years and continues to be a significant field of research and innovation that fabricates efficacious materials with designed characteristics. The application of nanotechnology in the production of metal oxides has provided a variety of nano-sized materials with distinctive properties. Metal oxide nanoparticles (MONPs) have interesting redox and catalytic properties, unique structure, high surface area, good biocompatibility, and mechanical strength because they are of smaller size than their bulk-sized counterparts. Because of their distinct physicochemical properties, MONPs have gained significant attention for use in medical implants, biomedical therapeutics, biosensing, bioimaging, cancer diagnosis and therapy, neurochemical monitoring, etc. Titania (TiO_2), iron oxides (Fe_2O_3 and Fe_3O_4), zinc oxide (ZnO), and ceria (CeO_2) are some examples of these metal oxide nanoparticles (Murthy, Effiong, & Fei, 2020) (Fig. 19.1).

For instance, titania is extensively used in medical implants and exhibits good biocompatibility, which enables it to act as a suitable exterior for proliferation and cell attachment.

Zinc oxide nanoparticles, because of their biodegradability and low toxicity, are used in biosensing and drug delivery.

Ceria-based nanoparticles are widely known for their antioxidant, autocatalytic, and redox attributes.

There are numerous other metal oxides which are used in magnetic resonance imaging (MRI) as contrast agents, as transporters in targeted drug delivery, and in cell labeling and separation as gas-sensing nanoprobes (Andreescu, Ornatska, Erlichman, Estevez, & Leiter, 2012). The iron oxide nanoparticles (IONPs), owing to their nontoxic nature in biological systems, have immense utilization in the biomedical field. They have both semiconductor properties and magnetic behavior, which lead to their multifunctional biomedical applications (Sangaiya & Jayaprakash, 2018). IONPs have commanding applications, such as bioimaging,

* Equal first authors.

Metal Oxides for Biomedical and Biosensor Applications. DOI: https://doi.org/10.1016/B978-0-12-823033-6.00017-X

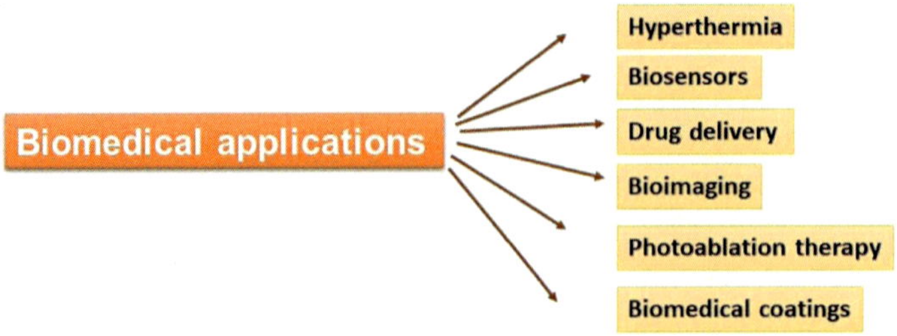

Figure 19.1 Biomedical applications of nanoparticles (McNamara & Tofail, 2017).

hyperthermia, cell labeling, drug delivery, and gene delivery due to their outstanding properties such as nontoxicity, biocompatibility, high saturation magnetization, chemical stability, and high magnetic susceptibility (McNamara & Tofail, 2017). When IONPs attain smaller dimensions ($\sim$10–20 nm), their superparamagnetic properties are evident, and therefore they show better efficacy for the abovementioned applications (Arias et al., 2018).

The following are a few IONP applications:

Antimicrobial activity: IONPs have vast applications in fungal infection and antimicrobial indication treatments. They have the potential to combat bacterial infections and can be used in combination with market-available antibiotics. The pure and various doped superparamagnetic iron oxide nanoparticles (SIONPs) are suitable for medical devices and antimicrobial applications due to their biocompatibility. Therefore IONPs have very good safety and excellent antimicrobial activity for mammalian cells. Hematite shows comparatively better antimicrobial activity than traditional magnetite nanoparticles (Sangaiya & Jayaprakash, 2018).

Anticancer activity: IONPs are used in cancer diagnosis and hyperthermia therapy, and also in iron deficiency anemia (Soetaert, Korangath, Serantes, Fiering, & Ivkov, 2020). Magnetic IONPs are more suitable for diagnosis of cancer and its treatment than semiconducting IONPs. The research into nanoparticles has become successful in recognizing numerous forms of breast cancer in mice (Sangaiya & Jayaprakash, 2018).

IONP treatment in human bone marrow cells: IONP treatment influences morphological and functional improvements. This may greatly affect the treatment of bone marrow cells (Urdzíková et al., 2006).

Special targeted drug delivery: Magnetic iron oxide nanoparticles constitute vigorous nanoplatforms which facilitates them to achieve high potential drug loading and targeting abilities due to their exceptional properties (biological and magnetic properties) (Vangijzegem, Stanicki, & Laurent, 2019).

19.2 Synthesis of iron oxide nanoparticles with respect to biomedical applications

The main three paths for the synthesis of IONPs used for biomedical applications are physical, chemical, and biological methods. These methods are mainly used to

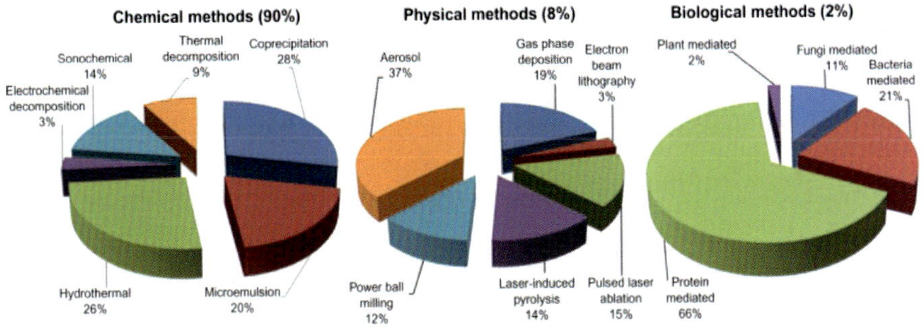

Figure 19.2 A comparison of the synthesis of superparamagnetic iron oxide nanoparticles (SPIONs) using three different routes (Ali et al., 2016).

synthesize more soluble, shape- and size-fine-tuned IONPs which are also biocompatible and stable (Wu, He, & Jiang, 2008). The following are some of the most common methods of synthesis (Fig. 19.2).

19.2.1 Physical methods

19.2.1.1 Pyrolysis method

Laser-aided pyrolysis in a continuous flow reactor was used for synthesizing IONPs (Martelli et al., 2000). In this process, gaseous organometallic-based precursors were subjected to laser radiation which further transmits energy according to the wavelength of the chemicals, thereby producing IONPs (Bomatí-Miguel, Zhao, Martelli, Di Nunzio, & Veintemillas-Verdaguer, 2010). Compared to other methods, laser-assisted pyrolysis is a potentially clean method which produces controllable and uniform particle size distributions (Martelli et al., 2000). This process operates at atmospheric pressure and therefore it is less expensive than other methods (Thongsuwan, Aukkaravittayapun, & Singjai, 2007). The impact of the major process parameters, such as gas flow speed, relative concentration of reactant gases, laser power, and reaction chamber pressure affect the viability of the process in terms of production yield, synthesized phases, magnetization, size, and morphology of particles (Martelli et al., 2000).

19.2.1.2 Laser ablation-based synthesis in solution

This is one of the green techniques for preparing metal nanoparticles which offers a distinctive tool for nanofabrication of IONPs. In this method, a high-power laser beam ablates the metal plate producing nanoparticles in the liquid. Energy, ablation time, wavelength, absorption of an aqueous solution, and repetition rate of laser are significant parameters in producing metal NPs. This is the simplest method for synthesizing metal nanoparticles without any chemical or surfactant addition. When the pulsed laser emission touches the target and the material is submerged in liquid solution, the synthesis is triggered. This process causes changes in the composition of the liquid solution and in the ablation target (Amendola & Meneghetti, 2013).

Recent experiments have shown that the laser ablation method was successful in the reactions involved in the bulk iron and phosphonates aqueous solution to produce a reduced FeOx crystal size with a few atom agglomerates (Fracasso et al., 2018).

19.2.2 Chemical methods

19.2.2.1 Microemulsion

The microemulsion-assisted synthesis process is extensively used to produce magnetic nanoparticles. Microemulsions are thermodynamically stable and transparent colloidal dispersions wherein two initially immiscible liquids (e.g., oil and water) coexist in a single phase because of the existence of a monolayer of the surfactant molecules (Danielsson & Lindman, 1981). The microemulsion can be represented as oil-swollen micelles that are dispersed in water or water-swollen micelles that are dispersed in oil, and these microemulsions are represented as o/w microemulsions or w/o microemulsions. These emulsions are differentiated based on the hydrophilic − lipophilic balance (HLB) of the surfactant along with the ratio of water and oil (Okoli, Boutonnet, Mariey, Järås, & Rajarao, 2011). The oil and water ratios with further surfactant adjustment facilitate controlled size of the microemulsion, which ultimately leads to the production of size-tuned IONPs (Deng, Wang, Yang, Fu, & Elaïssari, 2003; Wu, Wu, Yu, Jiang, & Kim, 2015). The concentration of Fe^{2+}/Fe^{3+} ions, nature of the surfactant, as well as the pH value and temperature greatly impact the NP size distribution and, therefore, their magnetization (Drmota, Drofenik, Koselj, & Žnidaršič, 2012). This method has certain drawbacks for biomedical applications such as low temperature requirements and high quantity of oil. This limits the high-scale production of IONPs. In addition, surfactants cling to iron oxide nanoparticles and makes them difficult to remove, -hence it needs further research focus (Malik, Wani, & Hashim, 2012).

19.2.2.2 Coprecipitation method

Aqueous coprecipitation is the widely used chemical synthesis method of IONPs (Assa et al., 2016). Ferrous chloride and ferric chloride were used as precursors and sodium hydroxide (NaOH) was used as a precipitation agent. First, $FeCl_2 \cdot 4H_2O$ and $FeCl_3 \cdot 6H_2O$ were dissolved in deionized water in a 1:2 ratio. Then, the solution was stirred and bubbled in an inert (argon) atmosphere where the temperature was kept constant at 80°C. Argon prevents excess oxidation in the solution. Subsequently, sodium hydroxide was added drop wise into the solution and precipitates were formed. Nanoparticles were separated using a magnet before removing the flask from the hot plate and washed two to three times with deionized water. The amount of NaOH was adjusted to control the pH, thereby nanoparticles of different sizes are formed. The following are the reactions for the synthesis of IONPs (Riaz, Bashir, & Naseem, 2014):

$$2FeCl_3 + FeCl_2 + 8NaOH\ Fe_3O_4 + 8NaCl + 4H_2O \tag{19.1}$$

$$Fe^{2+} + 2Fe^{3+} + 8OH^{-} Fe_3O_4 + 4H_2O \quad (19.2)$$

This is a low-cost and convenient technique that allows rapid and large-scale production of nanoparticles. Nevertheless, the resulting nanoparticles face large size distribution problems and aggregation, which are very common in the aqueous methods, apart from the other issues such as a tendency to oxidize and poor crystallinity, therefore compromising their magnetic properties.

It has been proven that the particle shape, size, magnetic properties, and composition of IONPs may vary with temperature, base concentration, Fe^{2+}/Fe^{3+} proportion, order of the reactants, ionic strength of the media, and the use of surfactants (Levy, Sahoo, Kim, Bergey, & Prasad, 2002). This can be achieved via a coprecipitation method with functionally upgraded IONPs. For instance, spherical iron oxide nanoparticles of different sizes were achieved by varying the amount of sodium hydroxide and the pH of the precipitates (Yan et al., 2012). Hence, it is understood that the linear relation between IONP diameter and pH helps in controlling nanoparticle aggregation (Arias et al., 2018).

19.2.2.3 Solvothermal synthesis

IONPs can be synthesized by a simple and quick one-pot solvothermal method using any solvent, for example, ethylene glycol, iron precursor ($FeCl_3 \cdot 6H_2O$), and nucleating and precipitating agents such as $(NH_4)_2CO_3$, NH_4Ac, aqueous NH_3, or NH_4HCO_3, respectively. This technique allows the tuning of particle size and crystallinity, which gives an upgraded magnetic property upon changing the temperature and nucleating agents. Most importantly, in this process, the conversion efficiency of raw material into the product is close to 100% (Kozakova et al., 2015).

19.2.2.4 Hydrothermal synthesis

In the hydrothermal method, IONPs were uniformly synthesized by oxidation of $FeCl_2 \cdot 4H_2O$ in basic aqueous solution under raised pressure and temperature (Ge et al., 2009). The mineralization of Fe^{3+} ions and controlled oxidation of Fe_3O_4 occur under these conditions (Wu et al., 2008). By varying the solvent composition and concentration of the reactants, the diameter varies between 15 and 30 nm. The resultant Fe_3O_4 nanoparticles are single crystals with high purity. As the particle diameter decreases, the magnetic property displays a transition from ferromagnetic to superparamagnetic behavior (Ge et al., 2009). Moreover, hydrothermal synthesis is one of the longer processes, from a few hours to days compared to the microemulsion method (Lu, Salabas, & Schüth, 2007).

19.2.2.5 Thermal decomposition

In this method, iron precursors and surfactant [e.g., oleic acid ($C_{18}H_{34}O_2$)] are subjected to thermal decomposition which results in the development of spherical and monodisperse IONPs. The key principle behind such a monodisperse synthesis method is a separation between the growth phase and burst nucleation. Here, the size of the population is adjusted mainly through the precursor-to-surfactant ratio (Lassenberger et al., 2017). This is a nonaqueous synthesis method, which is advantageous in terms of synthesizing high-quality nanoparticles with narrow particle size distribution and particle size control, which further enhances their degree of crystallinity and magnetism (Huang, Shieh, Yeh, Wu, & Cheng, 2014; Maity, Ding, & Xue, 2008). The particle size of the IONPs is directly proportional to the phase of inorganic cluster formed which occurs between the precursor decomposition and nucleation. Further, the concentration and size of the clusters depend on the heating rate and precursor-to-surfactant ratio (Lassenberger et al., 2017).

Nanocrystals of iron oxide with high yield, narrow size distributions, and negligible congregation were obtained using this method (Yu, Falkner, Yavuz, & Colvin, 2004). Nonetheless, the obtained aqueous insoluble nanoparticles were subjected to functionalization to make their surfaces hydrophilic and also to make them compatible for biological solutions (Assa et al., 2016). Further, shorter synthesis reactions and higher temperatures result in high reactivity and saturation magnetization (Belaïd, Stanicki, Vander Elst, Muller, & Laurent, 2018).

19.2.2.6 Sonochemical method

High-intensity ultrasonic radiation is used in this method to disperse the ferrous salt solution at room temperature, which can facilitate the formation of blisters in solution through physical effects which further supplies the energy required for the reaction (Raghunath & Perumal, 2017). There are certain process parameters such as temperature, ultrasonication power play, and sonication time, that are responsible for the size, shape, and morphology of the obtained products (Hassanjani-Roshan, Vaezi, Shokuhfar, & Rajabali, 2011). This method is a versatile and better substitute for materials synthesis involving a short reaction time and the absence of high process temperature and pressure (Sodipo & Aziz, 2018). Similarly, the ultrasonic frequency used in this method determines the resultant product, for example, in many cases, the quantity of the drug loaded is proportional to the ultrasound frequency (Dolores, Raquel, & Adianez, 2015).

19.2.2.7 Sol–gel reaction

In general, the sol–gel reaction is a wet method where salts (e.g., chlorides, acetates, and nitrates) and iron alkoxides experience a condensation and hydrolysis reaction (Pandey & Mishra, 2011). Sol–gel is the most suitable nanoparticle fabrication method because of the use of low temperature and low-cost precursors. It is also a relatively simple process. Several parameters, for example, the precursor solution's pH, temperature, viscosity of the solution, and post synthesis annealing determine the microstructural properties of NPs. This process gains a great deal of significance in the preparation of hematite, due to the yield of multiple iron oxide phases such as magnetite and maghemite along with the other phases

(Kayani, Arshad, Riaz, & Naseem, 2014). Further, it has been mentioned that although the precursors used in this processes are slightly expensive, the NPs produced through this process possess high permeability and low wear resistance (Reddy, Arias, Nicolas, & Couvreur, 2012).

19.2.2.8 Polyol method

A simple polyol synthesis method is used for synthesizing IONPs having low polydispersity with the synthetic conditions of high temperature and high pressure. In comparison with the other conventional processes, like coprecipitation and thermal decomposition, the polyol method yields nanoparticles with a narrow particle size distribution, which is a cost-effective, simple, and reproducible approach without the need for an inert atmosphere (Hachani et al., 2016). A reduction reaction is prominent in the polyol method, where polyol is subjected to heat to its boiling temperature, which in turn can act as a reducing agent and also controls the growth of the particles upon the reaction with iron precursors (Cai & Wan, 2007). The size of the nanoparticles is dependent on various parameters such as the type of polyol, reaction time, and concentration of the iron precursor. It is further understood from this process that the nanoparticle size is determined by the glycol length (Hachani et al., 2016).

19.2.2.9 Microwave-assisted synthesis

This is one of the most famous and simple synthesis methods for producing highly pure iron oxide products. In the microwave-assisted synthesis process, iron precursors are subjected to microwave electromagnetic radiation, which causes molecule reorientation and mixing homogeneously due to the high internal heating developed (Raghunath & Perumal, 2017; Wu et al., 2015). In this process, polyethylene glycol (PEG) and water solutions of chloride salts of iron with β-cyclodextrin (β-CD) along with the ammonia (NH_3) solutions were majorly involved in IONP synthesis (Aivazoglou, Metaxa, & Hristoforou, 2017). This method is very advantageous due to its low-cost, decreased reaction time, and specific control over the shape and size of the nanoparticles (Unsoy et al., 2015). This method is attractive due to its cost-effectiveness in commercial production and because it facilitates products which have good biocompatibility and solubility in clinical trials (Osborne et al., 2012).

19.2.3 Biological methods

In biological methods, the synthesis of iron oxide nanoparticles has been carried out using environment-friendly materials, for example, plant leaf extracts, due to its biocompatibility and eco-friendliness. These iron nanoparticles have been extensively used in biomedical applications due to the absence of toxic and harmful materials in the synthesis protocol (Awwad & Salem, 2012).

This biosynthesis process is a simple, eco-friendly, and low-cost method to produce IONPs (Fatemi, Mollania, Momeni-Moghaddam, & Sadeghifar, 2018).

During synthesis, the plant extracts or microbe-derived products possessing reduction capability interact with the iron-based precursor materials under vigorous stirring of the reaction mixture. The resultant nanoparticles also exhibit excellent biocompatibility (Raghunath & Perumal, 2017).

In this method, an iron-reducing bacteria like *Geobacter metallireducens* was used for synthesis of nanoparticles (Smith, Lovley, & Tremblay, 2013). For example, enzyme lumazine synthase (produced both by bacteria and fungi), a biological nanoreactor, nanoparticle is produced inside the small-diameter capsid-based templates using the specific bacteria and fungi(Shenton, Mann, Cölfen, Bacher, & Fischer, 2001). Despite a lower yield compared to other physical and chemical methods, this method has gained significant importance because of its nontoxic nature of green synthesis and easy functionalization to promote herbal properties (Sankaralingam & Kadirvelu, 2017).

19.3 Methods of physicochemical characterization

Surface characterization techniques are used to understand the surface characteristics of nanoparticles or nanocomposites, in general and for iron-based nanoparticles or nanocomposites, in particular. The characterization techniques form a basic set to understand the chemical composition, surface morphology, and spatial distribution of the functional groups, etc. (Hyeon, 2003). There are a number of characterization techniques used to investigate magnetic nanoparticles, namely scanning electron microscopy (SEM), X-ray diffraction (XRD), transmission electron microscopy (TEM), Fourier transform infrared spectroscopy (FTIR), thermogravimetric analysis (TGA), atomic force microscopy (AFM), X-ray photoelectron spectroscopy (XPS), electron paramagnetic resonance (EPR) or electron spin resonance (ESR) spectroscopy (EPR), nuclear magnetic resonance spectroscopy (NMR), and vibrating sample magnetometry, etc. (Xu et al., 2014). Other characterization techniques, such as ion–particle probe, nanoparticle tracking analysis, hydrophobic interaction chromatography, thermodynamics, field flow fractionation, isopycnic centrifugation, tilted laser microscopy, zeta-potential analysis, electrophoresis, and turbidimetry are also used to characterize iron-based nanomaterials (Sarkar, Singh, Mandal, Kumar, & Parmar, 2015; Sosa, Noguez, & Barrera, 2003). The chemical and physical properties, advantages, and limitations of each technique are summarized in Table 19.1.

19.4 Methods for functionalization of metal nanoparticles for biomedical applications

In general, long-chain hydrocarbons are used to coat monodisperse nanoparticles with controlled size and shape, helping to form a hydrophobic surface. Here, the surfaces of the nanoparticles are functionalized by means of surfactant exchange or surfactant addition to impede a biocompatible property, mainly in biological

Table 19.1 The analytical techniques used to assess the physicochemical characteristics of nanomaterials (Ali et al., 2016).

S. no.	Modalities	Physical and chemical properties	Advantages	Limitations
1	Dynamic light scattering (DLS)	Size distribution is made based on hydrodynamics	DLS is a constructive method of rapid and consistent measurement. Small to moderate expenses on equipment. Hydrodynamic sizes are exactly determined	If large particles are present in even small quantities, they may be accounted during the data analysis. Restricted size determination. Size restrictions
2	Fluorescence correlation spectroscopy (FCS)	Dimension and binding kinetics of hydrodynamics	Uptake of sample is low. High spatial and temporal magnification. To study concentration effect, chemical kinetics, molecular diffusion, and conformation dynamics are performed via fluorescent probes technique	Because of deficiency of proper methods, it causes inaccuracy, restriction in usage, and limitations in fluorophore species
3	Surface-enhanced Raman scattering (SERS)	Hydrodynamic size and size distribution. Conformational variations in structural, chemical, conjugate, and electronic characteristics	No need for sample preparation. Topological information. Capability of detecting tissue abnormality. Increased spatial resolution of the nanomaterials. Better Raman scattering signal	Compared to RS, there is a weak signal restricted spatial resolution, very minute cross-section
4	Zetapotential	Stability based on charge on the surface	Allows concurrent measurement of many particles	Electro-osmotic effect, inadequate accuracy, and repetitive measurement
5	Near-field scanning optical microscopy (NSOM)	Nanomaterial shapes and size	Close situation analysis. Instantaneous measurement of fluorescence and spectroscopy	Long scanning time. Problems to visualize soft materials. Analysis of tiny sample area. Incident light intensity is deficient to stimulate fine fluorescent molecules

(Continued)

Table 19.1 (Continued)

S. no.	Modalities	Physical and chemical properties	Advantages	Limitations
6	Circular dichroism (CD)	For biomolecules. Thermal constancy. Structural and conformational variations (like protein DNA)	Motivated and constructive method	CD signals are weak for nonchiral chromophores. Analysis of molecules consisting of multiple chiral chromophores faces challenges
7	Mass spectroscopy (MS)	Surface properties, molecular weight, structure, composition	High precision and accuracy in measurement. Very small amount of sample is required. High sensitivity to detection	Costly equipment. Deficiency of complete databases for molecular species identification
8	Infrared; ATR, attenuated total reflection (IR ATR-FTIR)	Surface properties such as conformation and structure. Bioconjugate	Cheap and fast measurement. Minimal sample preparation requirements. Enhanced reproducibility irrespective of the thickness of the sample	Sample preparation (IR) is complex. In nanoscale analysis sensitivity is low
9	Scanning electron microscopy (SEM); environmental SEM (ESEM)	Shape. Size and size distribution. Dispersion. Aggregation	Simultaneous measurement of the size navigation and shape of NMs. High deliration (below subnanometers) in natural state visualization of biomolecules supplied by the usage of ESEM technique	Requirement of dry samples. Conducting sample requirement or conductive coating materials. Heterogeneous sample requirement. Expensive apparatus. Cryogenic method is needed for various nanoparticle bioconjugates
10	Transmission electron microscopy (TEM)	Accumulation. Shape heterogeneity. Dispersion. Size and size navigation	Higher spatial resolution than SEM. Direct measurement of the shape and size of nanomaterials occurs. Used for the investigation of electronic structure and chemical composition	Ultrathin sample requirement. Samples required in nonphysiological states. Expensive equipment. Insufficient sampling

11	Scanning tunneling microscopy (STM)	Accumulation. Shape heterogeneity. Dispersion. Size and size navigation	High spatial resolution. Sudden measurement at atomic level	Requirement for conductive surfaces
12	Atomic force microscopy (AFM)	Shape heterogeneity. Dispersion. Size and size navigation	Direct measurement in dry conditions, ambient, or aqueous environment	Sampling is time consuming and poor
13	Nuclear magnetic resonance (NMR)	Indirect analysis of size. Structure. Concentration. purity	Constructive procedure. Less sample preparation needed	Low sensitivity. Time consuming. Large amount of sample requirement
14	X-ray diffraction analysis (XRD)	Used for crystalline materials, size, shape, and structure determination	High spatial resolution at atomic level. Well-organized modalities	Usage in crystalline materials has been reduced. Only conformation site; low accessibility
15	Small-angle X-ray scattering (SAXS)	Structure, shape, size, and size transportation	Constructive procedure, sample preparation is very simple. Accessibility of amorphous materials and sample in solution	Comparatively low resolution

applications. The surfactant addition is made through the adsorption of amphiphilic molecules containing both a hydrophilic segment and a hydrophobic component. Further, a double-layered structure is formed by the hydrophobic part of the hydrocarbon chain, wherein the hydrophilic part is attached to the outside of the nanoparticles, which imparts water solubility to the molecule.

In the case of the surfactant exchange method, the original surfactant is replaced with advanced bifunctional surfactant, which is responsible for binding the nanoparticle surface through a strong chemical bond. The other end of the nanoparticles with a second functional group has polar characteristics which help to disperse them in water or to impart functionalization characteristics.

There are several methods available for coating nanoparticles (Suh, Suslick, Stucky, & Suh, 2009). Polymer coating materials of proteins, dextran, dendrimers, gelatine, lipids, poly(ethylene-*co*-vinyl acetate), poly(vinyl alcohol) (PVA), poly ethylene glycol (PEG), chitosan, poly(vinylpyrrolidone) (PVP), and pullulan, are frequently chosen for this purpose (Jeong et al., 1999; Kellar et al., 2000; Massia, Stark, & Letbetter, 2000; Schwick & Heide, 1969; Suh et al., 2009; Zhao & Milton Harris, 1998). Some other special molecules, namely, bifunctional 2,3-dimercaptosuccinic acids (Chen et al., 2008), dopamine (Xie, Xu, Kohler, Hou, & Sun, 2007; Xu et al., 2004), and silanes (De Palma et al., 2007), are also used as functionalization agents. It has also been reported that dopamine is attached as a stable anchor on the Fe_3O_4 nanoparticle surface (Xie et al., 2006, 2007). On ferrite magnetic nanostructures, silanes are particularly used to exchange the hydrophobic ligands (De Palma et al., 2007). The end group of silanes, including acrylate, amino, isocyanine, and carboxylic thiol groups, offer considerable chemistry for the modification of the nanostructure. In the biological field, saline and dopamine are often combined with PEG or other polymers to provide stabilization of long-term nanoparticles (Na et al., 2007; Xie et al., 2006). Besides PEG, alginate (Xu, Shen, Xu, Xie, & Li, 2006), poly(acrylic acid) (PAA) (Lin, Lee, & Chiu, 2005), chitosan (Zhu, Yuan, & Dai, 2008), and dextran (Moore, Weissleder, & Bogdanov, 1997; Veintemillas-Verdaguer, Morales, & Serna, 1998) are also used for nanoparticle stabilization and offer deep-rooted biocompatibility and stability (Seo et al., 2006). The superparamagnetic carbon-coated FeCo nanoparticles display ultrahigh R1 and R2 relaxivities and high aqueous biocompatibility along with solubility (Fig. 19.3).

19.5 Biomedical applications

19.5.1 Magnetic resonance molecular imaging

The entry of nanoparticles into medical applications has gathered an exceptional amount of research, especially for applications related to molecular imaging. The main advantages of having nanoparticles in molecular imaging is their small size, generally less than 100 nm, that are helpful to conjugate with several molecular markers. It helps them to interact at cellular and molecular levels, resulting in an increased range of disease targets in molecular imaging. Further, it acts as an

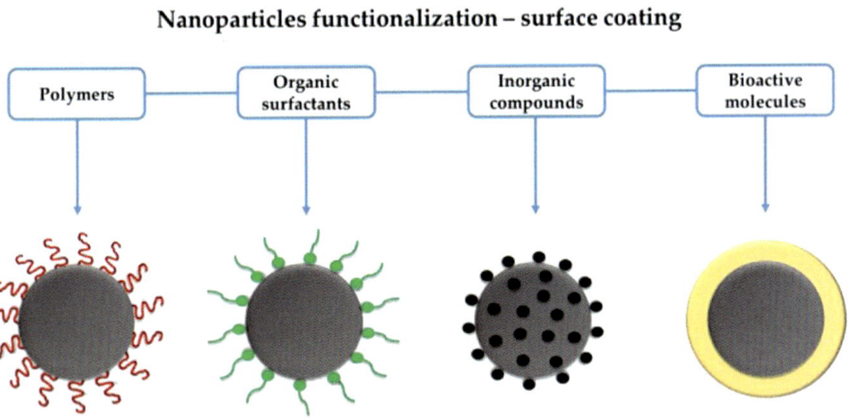

Figure 19.3 Schematic diagram of the main shells for iron oxide nanoparticle (IONP) functionalization. Cores of IONPs are represented by *gray* circles (Arias et al., 2018).

alternate to conventional imaging techniques. The demerits of conventional imaging modalities are their deficiency in combining high spatial resolution and sensitivity needed for molecular imaging. Despite its high resolution, the major drawback MRI faces is its lack of sensitivity to molecular signals. However, nuclear medicine modalities with high sensitivity, namely positron emission tomography (PET) and single photon emission computed tomography (SPECT), offer excellent sensitivity even at the reduced spatial resolution (Lee et al., 2007; Weissleder & Mahmood, 2001). MRI has high spatial resolution, as it uses nonionizing radiation, resulting in multiplanar tomographic capabilities that are noninvasive in nature (Ito, Shinkai, Honda, & Kobayashi, 2005). IONPs are widely researched for MRI, due to their superparamagnetic characteristics. There are various types of IONPs, namely γ-Fe_2O_3 (maghemite), Fe_3O_4 (magnetite), and α-Fe_2O_3 (hematite). Of the various IONPs, magnetite, Fe_3O_4, is considered very promising, due to its outstanding biocompatibility (Gupta & Gupta, 2005). The superparamagnetic iron oxide nanoparticles (SPIONs) act as a precursor for molecular imaging, hence they need to be magnetic, nontoxic, and biocompatible. Further, these precursor molecules should also be able to bind to a range of drugs, antibodies, enzymes, proteins, and various other molecular targets (Lodhia, Mandarano, Ferris, Eu, & Cowell, 2010).

19.5.2 Hyperthermia

SPIONs find potential applications in hyperthermia. In general, hyperthermia is described as a supplementary treatment to conventional radiotherapy, chemotherapy, and surgery, because all these conventional techniques fail due to a lack of specificity and have enormous side effects. Hyperthermia treatment is famous because of the SPIONs as it unveiled a changing magnetic field that produces heat due to magnetic hysteresis loss. Because of the superparamagnetic nature, these nanoparticles are subjected to an alternating magnetic field, and thereby the particles behave as powerful heat sources that are used to kill cancer cells. In comparison, cancer cells are more sensitive to temperatures above 41°C than normal cells. Here, the heat is generated mainly dependent upon the magnetic properties of SPIONs and their properties are mainly governed by their physicochemical characteristics, that is, their shape and size. Particles falling in the size range of 10–20 nm exhibit the best superparamagnetic properties. However, the major issue is their agglomeration, which hinders their dispersibility and stability. There are certain biocompatible coating materials available for superparamagnetic iron oxide nanoparticles, namely peptides, silica, and lipids, which simultaneously provide a good strategy to protect magnetic nanoparticles (Hedayatnasab, Abnisa, & Wan Daud, 2018; Kaushik et al., 2020).

19.5.3 Multimodal imaging

Multifunctional nanoparticles are used in multimodal imaging through combining two or more imaging modalities toward concurrent imaging and therapy. Multimodal imaging is considered as an alternate to traditional therapy and diagnosis, and is also called optimized therapy or "personalized medicine." Imaging modalities generally include MRI, ultrasound (US), PET, computed tomography (CT), optical imaging, and SPECT. Despite the advantages, all imaging modalities have drawbacks, insufficient spatial resolution or sensitivity. These issues hinder the ability to achieve reliable and accurate information at the disease area. To overcome this issue, a combination of imaging modalities, namely PET/CT or PET/MRI, has gained significant interest for improving the imaging instruments for diagnosis.

This type of multimodal imaging also provides highly accurate and reliable detection of disease sites. MRI and CT exhibit high-resolution images for anatomical information. On the other hand, high-sensitivity functional information about the disease can be obtained through PET images. In sequence, the combination of these various imaging modalities further accomplishes high resolution and enhanced sensitivity. Through those data, detailed biological or anatomical information about the target disease can be achieved (Lee et al., 2012). Nanoscale multimodal imaging probes can usually carry more than two imaging agents which helps to overcome the disadvantages of a single imaging modality. Elaborated information of the target site through targeted delivery is also possible with nanoscale multimodal imaging (Jennings & Long, 2009; Louie, 2010). IONPs also exhibit intrinsic imaging abilities for MR (Lee et al., 2012).

19.5.4 Cellular labeling

Although different strategies are used to endow cells with sufficient magnetization to be detectable by MRI, the easiest method is coincubation of cells with magnetic nanoparticles. Here, the cells are internalized with the spontaneous endocytosis pathway or phagocytosis. The surface functionalization of nanoparticles elevates the cellular uptake (Kolosnjaj-Tabi, Wilhelm, Clément, & Gazeau, 2013). Among all the iron nanoparticles, SPIONs have gathered a great deal of attention in cellular labeling due to their unique magnetic characteristics. Magnetic cell targeting is a safe, simple, and efficient delivery technique. Since SPIONs are biocompatible, biodegradable, and are actively transported into cells, they are responsive to magnetic fields. For example, the fabrication process consists of synthesizing magnetite NPs along with high-speed emulsification, resulting in the formation of a poly(lactic-*co*-glycolic acid) (PGLA) coating. These nanoparticles transmit enough magnetic mass to the cells which respond to the target in line with the magnetic fields (Tefft et al., 2015).

19.5.5 Magnetic particle imaging

Magnetic particle imaging is one of the emerging imaging modalities through which the direct and quantitative mapping of IONPs is visualized. In this method, the tailored IONPs are synthesized with the aim of achieving good spatial resolution and high sensitivity. The magnetic particle imaging (MPI) performances of IONP tracers are altered by the chemical composition and shape anisotropy variations effect. The selective doping of magnetite NPs with zinc moieties enhances the MPI signal by twofold. In this process, the target region of the test sample used to be selectively heated by saturating the IONPs outside of a field free region with the support of an external static field (Bauer, Situ, Griswold, & Samia, 2016).

19.5.6 Therapies and treatments

Cancer is a major cause of death worldwide, claiming many lives, often unexpectedly. Recently, nanotechnology approaches, for example, photodynamic therapy (PDT), chemotherapy, photothermal therapy (PTT), immunotherapy, and magnetic hyperthermia technique (MHT) have been used to treat cancer. However, the drawback to these methods lies in the nanoparticles that are used for therapies as they may not meet the required low biopersistence criteria and toxicity. Magnetic iron oxides nanoparticles are biocompatible, nontoxic, and highly dissolvable in the desired medium, which makes them a suitable candidate for MRI applications. In PDT and PTT techniques, high laser power may damage neighboring healthy tissues and this problem can be addressed by magnetically targeted delivery. Till now, iron oxide nanoparticles have been used for imaging-guided delivery as well for multimodal theranostics. Therefore with the applied external magnetic field, IONPs can be tuned to effectively deliver a photosensitizer to a tumor cell. Due to the enzyme-like activities and intrinsic properties of IONPs in a cellular environment, they are more effective at targeting and killing cancer cells (Saeed, Ren, & Wu, 2018). As far as drug nanocarriers are concerned, magnetic iron oxide nanoparticles show high drug loading capacity with focused targeting abilities. As is shown by the strong magnetic characteristics, it is understood that IONPs are activated by an external magnetic field to the targeted locations, mainly in vivo, which enhances the therapeutic compounds delivery to their place of action. By simultaneously vectorizing the IONPs with targeting agents, it is possible to achieve targeted drug delivery (Vangijzegem et al., 2019).

19.5.7 Nanocytotoxicity

Among all the various metal NPs, iron oxide nanoparticles have been found to be more potent in the cytotoxicity aspects (Hilger et al., 2003; Jeng & Swanson, 2006; Mahmoudi, Simchi, & Imani, 2009). Many studies have proven that IONPs at lower doses from 10 to 50 μg/mL show no measurable cytotoxic effect, however there is a substantial effect at higher levels of 100–250 μg/mL (Hussain, Hess, Gearhart, Geiss, & Schlager, 2005). Moreover, it is reported that, in general, metal oxide NPs show lower cytotoxicity compared to micron-sized particles, with equal masses of the same nominal composition (Veranth, Kaser, Veranth, Koch, & Yost, 2007). Although magnetic iron oxides are used extensively in biomedical applications, there is a lack of information at the cellular and molecular levels of the toxicity of these IONPs, which needs to be addressed to increase their potential use.

19.6 Conclusion

In this chapter, we have discussed the biomedical insights into iron oxide nanoparticles for their potential usage. The synthesis, characterization, and biomedical applications of IONPs, such as multimodal imaging, MR molecular imaging, hyperthermia, magnetic particle imaging (MPI), cellular labeling, therapies, and treatments were briefly presented. Report summarizes that it is essential to explore multiple and green synthetic routes to develop novel and cost-effective IONPs. Also, doping engineering of IONPs needs to be explored to expand their usage in various biomedical applications. Although many characterization studies have been reported to understand the physicochemical properties of IONPs, there remains a gray area on surface functionalization and its characterization, which needs further attention. For nano-cytotoxicity of IONPs, it is essential to explore further in-depth studies to enable them to be potential candidates for biomedical applications.

Acknowledgments

B.B. acknowledges support from the Department of Science and Technology (DST), India, in providing the women scientist project (Grant No: SR/WOS-A/CS-17/2017). RKG thanks DST India and Indo-US Science & Technology Forum (IUSSTF), India for funding the Indo-U.S. Virtual Networked Joint Center (Project No. IUSSTF/JC-025/2016).

References

Aivazoglou, E., Metaxa, E., & Hristoforou, E. (2017). Microwave-assisted synthesis of iron oxide nanoparticles in biocompatible organic environment. *AIP Advances*, *8*(4), 048201.

Ali, A., Zafar, H., Zia, M., Ul Haq, I., Phull, A. R., Ali, J. S., et al. (2016). Synthesis, characterization, applications, and challenges of iron oxide nanoparticles. *Nanotechnology, Science and Applications*, *9*, 49–67.

Amendola, V., & Meneghetti, M. (2013). What controls the composition and the structure of nanomaterials generated by laser ablation in liquid solution? *Physical Chemistry Chemical Physics*, *15*(9), 3027–3046.

Andreescu, S., Ornatska, M., Erlichman, J., Estevez, A., & Leiter, J. (2012). *Biomedical Applications of Metal Oxide Nanoparticles*, 57–100.

Arias, L. S., Pessan, J. P., Vieira, A. P. M., Lima, T. M. Td, Delbem, A. C. B., & Monteiro, D. R. (2018). Iron oxide nanoparticles for biomedical applications: A perspective on synthesis, drugs, antimicrobial activity, and toxicity. *Antibiotics (Basel)*, *7*(2), 46.

Assa, F., Jafarizadeh-Malmiri, H., Ajamein, H., Anarjan, N., Vaghari, H., Sayyar, Z., et al. (2016). A biotechnological perspective on the application of iron oxide nanoparticles. *Nano Research*, *9*(8), 2203–2225.

Awwad, A., & Salem, N. (2012). A green and facile approach for synthesis of magnetite nanoparticles. *Journal of Nanoscience and Nanotechnology*, *2*, 208–213.

Bauer, L. M., Situ, S. F., Griswold, M. A., & Samia, A. C. S. (2016). High-performance iron oxide nanoparticles for magnetic particle imaging–guided hyperthermia (hMPI). *Nanoscale*, *8*(24), 12162–12169.

Belaïd, S., Stanicki, D., Vander Elst, L., Muller, R. N., & Laurent, S. (2018). Influence of experimental parameters on iron oxide nanoparticle properties synthesized by thermal decomposition: Size and nuclear magnetic resonance studies. *Nanotechnology*, *29*(16), 165603.

Bomatí-Miguel, O., Zhao, X. Q., Martelli, S., Di Nunzio, P. E., & Veintemillas-Verdaguer, S. (2010). Modeling of the laser pyrolysis process by means of the aerosol theory: Case of iron nanoparticles. *Journal of Applied Physics*, *107*(1), 014906.

Cai, W., & Wan, J. (2007). Facile synthesis of superparamagnetic magnetite nanoparticles in liquid polyols. *Journal of Colloid and Interface Science*, *305*(2), 366–370.

Chen, Z. P., Zhang, Y., Zhang, S., Xia, J. G., Liu, J. W., Xu, K., et al. (2008). Preparation and characterization of water-soluble monodisperse magnetic iron oxide nanoparticles via surface double-exchange with DMSA. *Colloids and Surfaces A: Physicochemical and Engineering Aspects*, *316*(1), 210–216.

Deng, Y., Wang, L., Yang, W., Fu, S., & Elaïssari, A. (2003). Preparation of magnetic polymeric particles via inverse microemulsion polymerization process. *Journal of Magnetism and Magnetic Materials*, *257*, 69–78.

Danielsson, I., & Lindman, B. (1981). The definition of microemulsion. *Colloids and Surfaces*, *3*, 391–392.

De Palma, R., Peeters, S., Van Bael, M. J., Van den Rul, H., Bonroy, K., Laureyn, W., et al. (2007). Silane ligand exchange to make hydrophobic superparamagnetic nanoparticles water-dispersible. *Chemistry of Materials*, *19*(7), 1821–1831.

Dolores, R., Raquel, S., & Adianez, G. L. (2015). Sonochemical synthesis of iron oxide nanoparticles loaded with folate and cisplatin: Effect of ultrasonic frequency. *Ultrasonics Sonochemistry*, *23*, 391–398.

Drmota, A., Drofenik, M., Koselj, J., & Žnidaršič, A. (2012). Microemulsion method for synthesis of magnetic oxide nanoparticles. *Microemulsions—An Introduction to Properties and Applications*, *10*, 191–215.

Fatemi, M., Mollania, N., Momeni-Moghaddam, M., & Sadeghifar, F. (2018). Extracellular biosynthesis of magnetic iron oxide nanoparticles by *Bacillus cereus* strain HMH1: Characterization and in vitro cytotoxicity analysis on MCF-7 and 3T3 cell lines. *Journal of Biotechnology*, *270*, 1–11.

Fracasso, G., Ghigna, P., Nodari, L., Agnoli, S., Badocco, D., Pastore, P., et al. (2018). Nanoaggregates of iron poly-oxo-clusters obtained by laser ablation in aqueous solution of phosphonates. *Journal of Colloid and Interface Science*, *522*, 208–216.

Ge, S., Shi, X., Sun, K., Li, C., Uher, C., Baker, J. R., et al. (2009). Facile hydrothermal synthesis of iron oxide nanoparticles with tunable magnetic properties. *The Journal of Physical Chemistry C*, *113*(31), 13593–13599.

Gupta, A. K., & Gupta, M. (2005). Synthesis and surface engineering of iron oxide nanoparticles for biomedical applications. *Biomaterials*, *26*(18), 3995–4021.

Hachani, R., Lowdell, M., Birchall, M., Hervault, A., Mertz, D., Begin-Colin, S., et al. (2016). Polyol synthesis, functionalisation, and biocompatibility studies of superparamagnetic iron oxide nanoparticles as potential MRI contrast agents. *Nanoscale*, *8*(6), 3278–3287.

Hassanjani-Roshan, A., Vaezi, M. R., Shokuhfar, A., & Rajabali, Z. (2011). Synthesis of iron oxide nanoparticles via sonochemical method and their characterization. *Particuology*, *9* (1), 95–99.

Hedayatnasab, Z., Abnisa, F., & Wan Daud, W. M. A. (2018). Investigation properties of superparamagnetic nanoparticles and magnetic field-dependent hyperthermia therapy. *IOP Conference Series: Materials Science and Engineering*, *334*, 012042.

Hilger, I., Frühauf, S., Linß, W., Hiergeist, R., Andrä, W., Hergt, R., et al. (2003). Cytotoxicity of selected magnetic fluids on human adenocarcinoma cells. *Journal of Magnetism and Magnetic Materials*, *261*, 7–12.

Huang, K. S., Shieh, D. B., Yeh, C. S., Wu, P. C., & Cheng, F. Y. (2014). Antimicrobial applications of water-dispersible magnetic nanoparticles in biomedicine. *Current Medicinal Chemistry*, *21*(29), 3312–3322.

Hussain, S. M., Hess, K. L., Gearhart, J. M., Geiss, K. T., & Schlager, J. J. (2005). In vitro toxicity of nanoparticles in BRL 3A rat liver cells. *Toxicology In Vitro: An International Journal Published in Association With BIBRA*, *19*(7), 975–983.

Hyeon, T. (2003). Chemical synthesis of magnetic nanoparticles. *Chemical Communications*, *8*, 927–934.

Ito, A., Shinkai, M., Honda, H., & Kobayashi, T. (2005). Medical application of functionalized magnetic nanoparticles. *Journal of Bioscience and Bioengineering*, *100*(1), 1–11.

Jeng, H. A., & Swanson, J. (2006). Toxicity of metal oxide nanoparticles in mammalian cells. *Journal of Environmental Science and Health, Part A*, *41*(12), 2699–2711.

Jennings, L. E., & Long, N. J. (2009). 'Two is better than one'—Probes for dual-modality molecular imaging. *Chemical Communications*, *24*, 3511–3524.

Jeong, Y. I., Nah, J. W., Na, H. K., Na, K., Kim, I. S., Cho, C. S., et al. (1999). Self-assembling nanospheres of hydrophobized pullulans in water. *Drug Development and Industrial Pharmacy*, *25*(8), 917–927.

Kaushik, S., Thomas, J., Panwar, V., Ali, H., Chopra, V., Sharma, A., et al. (2020). In situ biosynthesized superparamagnetic iron oxide nanoparticles (SPIONS) induce efficient hyperthermia in cancer cells. *ACS Applied Bio Materials*, *3*(2), 779–788.

Kayani, Z. N., Arshad, S., Riaz, S., & Naseem, S. (2014). Synthesis of iron oxide nanoparticles by sol–gel technique and their characterization. *IEEE Transactions on Magnetics*, *50*(8), 1–4.

Kellar, K. E., Fujii, D. K., Gunther, W. H., Briley-Saebø, K., Bjørnerud, A., Spiller, M., et al. (2000). NC100150 injection, a preparation of optimized iron oxide nanoparticles for positive-contrast MR angiography. *Journal of Magnetic Resonance Imaging*, *11*(5), 488–494.

Kolosnjaj-Tabi, J., Wilhelm, C., Clément, O., & Gazeau, F. (2013). Cell labeling with magnetic nanoparticles: Opportunity for magnetic cell imaging and cell manipulation. *Journal of Nanobiotechnology*, *11*(1), S7.

Kozakova, Z., Kuritka, I., Kazantseva, N. E., Babayan, V., Pastorek, M., Machovsky, M., et al. (2015). The formation mechanism of iron oxide nanoparticles within the microwave-assisted solvothermal synthesis and its correlation with the structural and magnetic properties. *Dalton Transactions*, *44*(48), 21099–21108.

Lassenberger, A., Grünewald, T. A., van Oostrum, P. D. J., Rennhofer, H., Amenitsch, H., Zirbs, R., et al. (2017). Monodisperse iron oxide nanoparticles by thermal decomposition: Elucidating particle formation by second-resolved in situ small-angle X-ray scattering. *Chemistry of Materials*, *29*(10), 4511–4522.

Lee, D.-E., Koo, H., Sun, I.-C., Ryu, J. H., Kim, K., & Kwon, I. C. (2012). Multifunctional nanoparticles for multimodal imaging and theragnosis. *Chemical Society Reviews*, *41*(7), 2656–2672.

Lee, J. H., Huh, Y. M., Jun, Y. W., Seo, J. W., Jang, J. T., Song, H. T., et al. (2007). Artificially engineered magnetic nanoparticles for ultra-sensitive molecular imaging. *Nature Medicine*, *13*(1), 95–99.

Levy, L., Sahoo, Y., Kim, K.-S., Bergey, E. J., & Prasad, P. N. (2002). Nanochemistry: Synthesis and characterization of multifunctional nanoclinics for biological applications. *Chemistry of Materials*, *14*(9), 3715–3721.

Lin, C.-L., Lee, C.-F., & Chiu, W.-Y. (2005). Preparation and properties of poly (acrylic acid) oligomer stabilized superparamagnetic ferrofluid. *Journal of Colloid and Interface Science*, *291*(2), 411–420.

Lodhia, J., Mandarano, G., Ferris, N., Eu, P., & Cowell, S. (2010). Development and use of iron oxide nanoparticles (Part 1): Synthesis of iron oxide nanoparticles for MRI. *Biomedical Imaging and Intervention Journal*, *6*(2), e12.

Louie, A. (2010). Multimodality imaging probes: Design and challenges. *Chemical Reviews*, *110*(5), 3146–3195.

Lu, A.-H., Salabas, E. L., & Schüth, F. (2007). Magnetic nanoparticles: Synthesis, protection, functionalization, and application. *Angewandte Chemie International Edition*, *46*(8), 1222–1244.

Mahmoudi, M., Simchi, A., & Imani, M. (2009). Cytotoxicity of uncoated and polyvinyl alcohol coated superparamagnetic iron oxide nanoparticles. *The Journal of Physical Chemistry C*, *113*(22), 9573–9580.

Maity, D., Ding, J., & Xue, J.-M. (2008). Synthesis of magnetite nanoparticles by thermal decomposition: Time, temperature, surfactant and solvent effects. *Functional Materials Letters*, *1*(3), 189–193.

Malik, M. A., Wani, M. Y., & Hashim, M. A. (2012). Microemulsion method: A novel route to synthesize organic and inorganic nanomaterials: 1st nano update. *Arabian Journal of Chemistry*, *5*(4), 397–417.

Martelli, S., Mancini, A., Giorgi, R., Alexandrescu, R., Cojocaru, S., Crunteanu, A., et al. (2000). Production of iron-oxide nanoparticles by laser-induced pyrolysis of gaseous precursors. *Applied Surface Science*, *154–155*, 353–359.

Massia, S. P., Stark, J., & Letbetter, D. S. (2000). Surface-immobilized dextran limits cell adhesion and spreading. *Biomaterials*, *21*(22), 2253–2261.

McNamara, K., & Tofail, S. A. M. (2017). Nanoparticles in biomedical applications. *Advances in Physics: X*, *2*(1), 54–88.

Moore, A., Weissleder, R., & Bogdanov, A., Jr (1997). Uptake of dextran-coated monocrystalline iron oxides in tumor cells and macrophages. *Journal of Magnetic Resonance Imaging*, *7*(6), 1140–1145.

Murthy, S., Effiong, P., & Fei, C. C. (2020). Metal oxide nanoparticles in biomedical applications. In Y. Al-Douri (Ed.), *Metal oxide powder technologies* (pp. 233–251). Elsevier.

Na, H. B., Lee, I. S., Seo, H., Park, Y. I., Lee, J. H., Kim, S.-W., et al. (2007). Versatile PEG-derivatized phosphine oxide ligands for water-dispersible metal oxide nanocrystals. *Chemical Communications*, *48*, 5167–5169.

Okoli, C., Boutonnet, M., Mariey, L., Järås, S., & Rajarao, G. (2011). Application of magnetic iron oxide nanoparticles prepared from microemulsions for protein purification. *Journal of Chemical Technology & Biotechnology*, *86*(11), 1386–1393.

Osborne, E. A., Atkins, T. M., Gilbert, D. A., Kauzlarich, S. M., Liu, K., & Louie, A. Y. (2012). Rapid microwave-assisted synthesis of dextran-coated iron oxide nanoparticles for magnetic resonance imaging. *Nanotechnology*, *23*(21), 215602.

Pandey, S., & Mishra, S. B. (2011). Sol–gel derived organic–inorganic hybrid materials: Synthesis, characterizations and applications. *Journal of Sol-Gel Science and Technology*, *59*(1), 73–94.

Raghunath, A., & Perumal, E. (2017). Metal oxide nanoparticles as antimicrobial agents: A promise for the future. *International Journal of Antimicrobial Agents*, *49*(2), 137–152.

Reddy, L. H., Arias, J. L., Nicolas, J., & Couvreur, P. (2012). Magnetic nanoparticles: Design and characterization, toxicity and biocompatibility, pharmaceutical and biomedical applications. *Chemical Reviews*, *112*(11), 5818–5878.

Riaz, S., Bashir, M., & Naseem, S. (2014). Iron oxide nanoparticles prepared by modified co-precipitation method. *IEEE Transactions on Magnetics*, *50*(1), 1–4.

Saeed, M., Ren, W., & Wu, A. (2018). Therapeutic applications of iron oxide based nanoparticles in cancer: Basic concepts and recent advances. *Biomaterials Science*, *6*(4), 708–725.

Sangaiya, P., & Jayaprakash, R. (2018). A review on iron oxide nanoparticles and their biomedical applications. *Journal of Superconductivity and Novel Magnetism*, *31*(11), 3397–3413.

Sankaralingam, K., & Kadirvelu, K. (2017). Green synthesis of Iron oxide nanoparticles using *Lagenaria siceraria* and evaluation of its Antimicrobial activity. *Defence Life Science Journal*, *2*, 422.

Sarkar, D. J., Singh, A., Mandal, P., Kumar, A., & Parmar, B. S. (2015). Synthesis and characterization of poly (CMC-g-cl-PAam/Zeolite) superabsorbent composites for controlled delivery of zinc micronutrient: Swelling and release behavior. *Polymer-Plastics Technology and Engineering*, *54*(4), 357–367.

Schwick, H. G., & Heide, K. (1969). Immunochemistry and immunology of collagen and gelatin. *Bibliotheca Haematologica*, *33*, 111–125.

Seo, W. S., Lee, J. H., Sun, X., Suzuki, Y., Mann, D., Liu, Z., et al. (2006). FeCo/graphitic-shell nanocrystals as advanced magnetic-resonance-imaging and near-infrared agents. *Nature Materials*, *5*(12), 971–976.

Shenton, W., Mann, S., Cölfen, H., Bacher, A., & Fischer, M. (2001). Synthesis of nanophase iron oxide in lumazine synthase capsids, . This work was supported by the BBSRC (W. S.). We thank A. M. Seddon for help with transmission electron microscopy and analytical ultracentrifufation studies and G. D. Ruggiero for the generation of computer images *Angewandte Chemie (International Ed. in English)*, *40*(2), 442–445.

Smith, J. A., Lovley, D. R., & Tremblay, P.-L. (2013). Outer cell surface components essential for Fe(III) oxide reduction by *Geobacter metallireducens*. *Applied and Environmental Microbiology*, *79*(3), 901.

Sodipo, B. K., & Aziz, A. A. (2018). One minute synthesis of amino-silane functionalized superparamagnetic iron oxide nanoparticles by sonochemical method. *Ultrasonics Sonochemistry*, *40*, 837–840.

Soetaert, F., Korangath, P., Serantes, D., Fiering, S., & Ivkov, R. (2020). Cancer therapy with iron oxide nanoparticles: Agents of thermal and immune therapies. *Advanced Drug Delivery Reviews*, *163–164*, 65–83.

Sosa, I. O., Noguez, C., & Barrera, R. G. (2003). Optical properties of metal nanoparticles with arbitrary shapes. *Journal of Physical Chemistry B*, *107*(26), 6269–6275.

Suh, W. H., Suslick, K. S., Stucky, G. D., & Suh, Y.-H. (2009). Nanotechnology, nanotoxicology, and neuroscience. *Progress in Neurobiology*, *87*(3), 133–170.

Tefft, B. J., Uthamaraj, S., Harburn, J. J., Klabusay, M., Dragomir-Daescu, D., & Sandhu, G. S. (2015). Cell labeling and targeting with superparamagnetic iron oxide nanoparticles. *Journal of Visualized Experiments* (105), e53099.

Thongsuwan, W., Aukkaravittayapun, S., & Singjai, P. (2007). Preparation of iron oxide nanoparticles by a pyrosol technique. *Key Engineering Materials*, *353–358*, 2175–2178.

Unsoy, G., Gunduz, U., Oprea, O., Ficai, D., Sonmez, M., Radulescu, M., et al. (2015). Magnetite: From synthesis to applications. *Current Topics in Medicinal Chemistry*, *15* (16), 1622–1640.

Urdzíková, L., Jendelová, P., Glogarová, K., Burian, M., Hájek, M., & Syková, E. (2006). Transplantation of bone marrow stem cells as well as mobilization by granulocyte-colony stimulating factor promotes recovery after spinal cord injury in rats. *Journal of Neurotrauma*, *23*(9), 1379–1391.

Vangijzegem, T., Stanicki, D., & Laurent, S. (2019). Magnetic iron oxide nanoparticles for drug delivery: Applications and characteristics. *Expert Opinion on Drug Delivery*, *16*(1), 69–78.

Veintemillas-Verdaguer, S., Morales, M., & Serna, C. (1998). Continuous production of γ-Fe_2O_3 ultrafine powders by laser pyrolysis. *Materials Letters*, *35*(3–4), 227–231.

Veranth, J. M., Kaser, E. G., Veranth, M. M., Koch, M., & Yost, G. S. (2007). Cytokine responses of human lung cells (BEAS-2B) treated with micron-sized and nanoparticles of metal oxides compared to soil dusts. *Particle and Fibre Toxicology*, *4*, 2.

Weissleder, R., & Mahmood, U. (2001). Molecular imaging. *Radiology*, *219*(2), 316–333.

Wu, W., He, Q., & Jiang, C. (2008). Magnetic iron oxide nanoparticles: Synthesis and surface functionalization strategies. *Nanoscale Research Letters*, *3*(11), 397–415.

Wu, W., Wu, Z., Yu, T., Jiang, C., & Kim, W.-S. (2015). Recent progress on magnetic iron oxide nanoparticles: Synthesis, surface functional strategies and biomedical applications. *Science and Technology of Advanced Materials*, *16*(2), 023501.

Xie, J., Xu, C., Kohler, N., Hou, Y., & Sun, S. (2007). Controlled PEGylation of monodisperse Fe_3O_4 nanoparticles for reduced non-specific uptake by macrophage cells. *Advanced Materials*, *19*(20), 3163–3166.

Xie, J., Xu, C., Xu, Z., Hou, Y., Young, K. L., Wang, S. X., et al. (2006). Linking hydrophilic macromolecules to monodisperse magnetite (Fe_3O_4) nanoparticles via trichloro-s-triazine. *Chemistry of Materials*, *18*(23), 5401–5403.

Xu, C., Xu, K., Gu, H., Zheng, R., Liu, H., Zhang, X., et al. (2004). Dopamine as a robust anchor to immobilize functional molecules on the iron oxide shell of magnetic nanoparticles. *Journal of the American Chemical Society*, *126*(32), 9938–9939.

Xu, J., Sun, J., Wang, Y., Sheng, J., Wang, F., & Sun, M. (2014). Application of iron magnetic nanoparticles in protein immobilization. *Molecules (Basel, Switzerland)*, *19*(8), 11465–11486.

Xu, X., Shen, H., Xu, J., Xie, M., & Li, X. (2006). The colloidal stability and core-shell structure of magnetite nanoparticles coated with alginate. *Applied Surface Science*, *253* (4), 2158–2164.

Yan, A., Liu, Y., Liu, Y., Li, X., Lei, Z., & Liu, P. (2012). A NaAc-assisted large-scale coprecipitation synthesis and microwave absorption efficiency of Fe_3O_4 nanowires. *Materials Letters*, *68*, 402–405.

Yu, W. W., Falkner, J. C., Yavuz, C. T., & Colvin, V. L. (2004). Synthesis of monodisperse iron oxide nanocrystals by thermal decomposition of iron carboxylate salts. *Chemical Communications* (20), 2306–2307.

Zhao, X., & Milton Harris, J. (1998). Novel degradable poly (ethylene glycol) hydrogels for controlled release of protein. *Journal of Pharmaceutical Sciences*, *87*(11), 1450–1458.

Zhu, A., Yuan, L., & Dai, S. (2008). Preparation of well-dispersed superparamagnetic iron oxide nanoparticles in aqueous solution with biocompatible *N*-succinyl-*O*-carboxymethylchitosan. *The Journal of Physical Chemistry C*, *112*(14), 5432–5438.

Flexible and stretchable indium-fallium-zinc oxide-based electronic devices for sweat pH sensor application

20

Yogeenth Kumaresan[1], *Nirmal G. R.*[2] *and Praveen Kumar Poola*[3]
[1]Electronics & Nanoscale Engineering, School of Engineering, University of Glasgow, Glasgow, United Kingdom, [2]Graduate Institute of Biomedical Scienece, Chang Gung University, Taoyuan City, Taiwan, [3]Electronics and Communication Enhineering, K.L. University Hyderabad, Hyderabad, India

Wearable biosensors with the capability to attach conformably on any nonplanar surfaces, such as human skin, have attracted significant interest in human health and fitness monitoring (Rodrigues et al., 2020; Song et al., 2020). They have the capability to continuously monitor the key health parameters such as blood pressure, skin temperature, heart rate, diabetes, hydration/dehydration, etc. using biological body specimen such as sweat, blood, urine, hair, and saliva (Choi et al., 2018; Gao et al., 2016; Koh et al., 2016). In conventional medical approach, blood tests are utilized to analyze the human health condition through invasive diagnosis which contains a risk of infection (Nakata, Arie, Akita, & Takei, 2017). Therefore, readily accessible sweat, consists of 99% of water and 1% of electrolyte element, such as sodium, potassium, magnesium, and calcium ions useful for health monitoring, has been widely explored as noninvasive approach (Khan, Ali, & Bermak, 2019; Nakata et al., 2017). One way of identifying the chemical content of the sweat is through cyclic voltammetry (CV) measurement which is more complex (Cinti et al., 2018; He et al., 2019). Alternatively, ion-sensitive and gas-sensitive field effect transistors (FETs)-based approach has been utilized as a simple and direct measurement technique (Das et al., 2014; Jamasb, 2019), which is compactable with integrated circuits (ICs) for easy integration and data transfer (Nakata et al., 2017). In this regard, flexible and wearable FETs are utilized to monitor the chemical component of the sweat through chemically active, sensitive, and reactive layers (Han et al., 2017; Wang, Liu, & Zhang, 2017). In addition, monitoring the pH level in sweat is most important indicator for disease diagnosis (Nakata et al., 2017). Primarily, sweat is generated through two sweat glands, namely eccrine and apocrine, of human skin (Kaya et al., 2019). Eccrine sweat glands are in most region of the human skin with high density in hand, feet, and forehead; likewise, apocrine is located around the armpit, anus, and genital region. Therefore, sensors with the

Metal Oxides for Biomedical and Biosensor Applications. DOI: https://doi.org/10.1016/B978-0-12-823033-6.00018-1

potential to conformably attach on human skin are inevitable to enable real-time health monitoring and early diagnosis of the critical diseases.

In this chapter, we discuss recent progress of wearable sweat pH sensor using flexible and stretchable indium gallium zinc oxide (IGZO) FETs. This chapter is organized as follows: Section 20.1 presents an overview of metal oxide-based electronic devices that includes introduction of electronic devices and the basic working principle of oxide-based FETs. Section 20.2 presents the recent progress in flexible and stretchable inorganic electronic devices. Section 20.3 presents the sensing mechanism of transistor-based sweat pH sensors and finally, key points are summarized.

20.1 Overview of oxides-based electronic devices

20.1.1 Electronic devices: an introduction

Electronics devices such as transistors, solar cells, diodes, sensors, photodetectors, and so on, are widely used all over the world and nowadays it cannot be separated from our daily life (Johnston, Islam, Yodh, & Johnson, 2005; Oostinga, Heersche, Liu, Morpurgo, & Vandersypen, 2008; Pak et al., 2015, 2018). Their usage has made deep in road in various applications including home appliances, health care, medical diagnosis, automobiles, military, electronic skin, and robotics (Kumaresan, Kim, H., Pak, Y., Poola, P. K., Lee, R., & Lim, 2020; Schwartz et al., 2013; Scidà et al., 2018; Zhan et al., 2014). These devices consist of active materials namely semiconductors and/or metals in which the flow of electron is controlled for the purpose of information processing and system control. The brief history of electronics began in 1987 with the invention of first vacuum tube called vacuum diode invention by J.A. Fleming's (Dewar & Fleming, 1897) and sequentially, in 1948, the transistor era started with the invention of junction transistor from Bells laboratory by physicists John Bardeen, Walter Brattain, and William Shockley (Niehenke, 2015). Later, with the invention of ICs, many transistors and electronic components are accommodated in a single circuit that revolutionized the advancement in electronics and semiconductor industry. Since then, from 1965, the advancement in electronics follows the Moore's law, predicted from Gordan Moore: miniaturizing the transistor size and thereby doubling the number of components on a circuit (chip) ever year to accommodate more electronics component in a single circuit (Rosenbaum, 1969). As a result, the electronic components with the size of entire room were reduced to a tiny chip with increased ability to process, store, and communicate information. This tremendous transformation in giant industrial plant brings a revolution in semiconductor industry in 1980, which resulted in third industrial revolution (1969), with great achievement in computers and robots (Cannadine, 1984; Xu, David, & Kim, 2018). The fourth industrial revolution, the ongoing revolution (since 2000), is built on the third industrial revolution, with achievement in internet, 3D printer and genetic engineering (Xu et al., 2018).

The industrial revolution brought a positive transformation by helping the people to connect with the world and enable them to access unlimited knowledge. Further with every technological revolution, the capability will be multiplied in many different areas such as nanotechnology, material science, biotechnology, artificial intelligence, robotics, the Internet of Things, energy storage, autonomous vehicles, 3D printing, and quantum computing (Lu, Li, Chen, Kim, & Serikawa, 2018; Rich, Wood, & Majidi, 2018; Yu, Beam, & Kohane, 2018). Such a rapid transformation demands advancement in electronic devices with a portable property with consistent device performance upon being bent or rolled or crumpled, which can be utilized for the future technologies such as wearable computers, physical and chemical sensors, wearable health monitoring devices, and flexible and rollable displays (Han & Dong, 2018; Kumaresan et al., 2018; Linghu, Zhang, Wang, & Song, 2018). The upcoming sections in this chapter demonstrate the recent progress in the fabrication of flexible and stretchable electronic devices using inorganic semiconducting channel and their potential in sweat pH sensor application.

Electronic devices fabricated on flexible plastic substrates have drawn much attention and have become the main focus of research in large-area electronics due to their attractive advantages, such as being mechanically robust, light weight and potentially having simple roll-to-roll-based fabrication and mass production (Kumaresan et al., 2018; Kumaresan, Kim, H., Pak, Y., Poola, P. K., Lee, R., & Lim, 2020; Linghu, Zhang, Wang, & Song, 2018; Nakagawa et al., 2020). In general, the amorphous oxide semiconductors (AOSs), such as ZnO, indium tin oxide (ITO), zinc tin oxide (ZTO), and IGZO have attracted many researchers owing to their high mobility, large area uniformity, reproducibility, and low processing temperature (Nakagawa et al., 2020; Sim et al., 2019). Those properties made them an attractive channel material in FETs for flexible applications, such as wearable computers, sensors, photodetectors, and rollable displays (Billah, Hasan, & Jang, 2017; Kumaresan, Kim, H., Pak, Y., Poola, P. K., Lee, R., & Lim, 2020). In particular, IGZO has drawn considerable attention as an extremely promising alternative to amorphous silicon with 20–50 times increased electron mobility, which has often been used in liquid-crystal displays (LCDs) and e-papers. IGZO is an n-type semiconductor material with wide bandgap (~ 3 eV) and it is the only material which has high optical transparency and high carrier mobility with high yield and low cost (Kumaresan et al., 2017).

20.1.2 Basic principle of oxide-based field effect transistors

FET is a three-terminal switching device, made of an active semiconducting channel layer, most commonly used is either IGZO or ZnO, as well as a dielectric layer, such as SiO_2 or Al_2O_3, etc., and three metallic contact electrodes namely source, drain, and gate metal electrodes (Fu, Zhu, Zhang, Li, & Chen, 2016; Pešić et al., 2016; Yu, Marks, & Facchetti, 2016). According to the position of contact electrodes, the FETs are classified into four basic structures namely staggered top gate, staggered bottom gate, coplanar top gate, and coplanar bottom gate as shown in Fig. 20.1 (Saji, 2008).

In staggered structure, the source and drain electrodes and the dielectric layer are on the opposite side of the semiconducting channel but in coplanar structure,

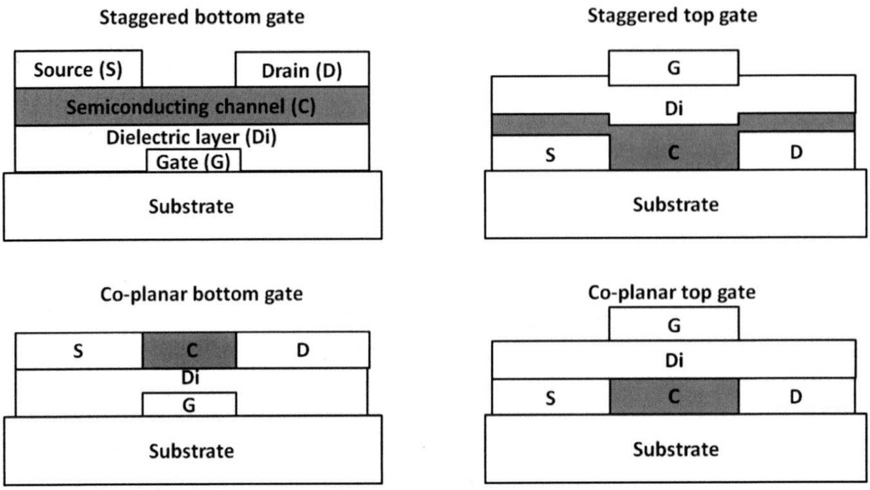

Figure 20.1 The schematic of different types of field effect transistor structures depending on the positioning of the source, drain, and gate electrodes.

the source and drain electrodes and the dielectric layer are located on the same side with respect to the semiconducting channel. In addition, both the staggered and coplanar structures are further classified into top gate and bottom gate based on the location of the gate electrode (Saji, 2008). In all the structures the semiconductor is placed between the source and drain electrode and the dielectric layer is sandwiched between the gate electrode and the semiconducting layer (Pachauri & Ingebrandt, 2016). The basic idea of this device is to control the current flow, by adjusting the free electrons in the semiconducting layer, between the source and drain electrodes by varying the applied gate bias; the applied gate voltage (V_{GS}) controls the on-state or off-state of FETs (Pachauri & Ingebrandt, 2016). The minimum gate voltage required to turn the device into on-state is defined as the threshold voltage (V_T). Further, the FETs operate on enhancement mode or depletion mode depending on whether it required gate voltage to obtain a conducting channel between the source and drain. In depletion mode, the gate voltage is required to deplete the free carriers in the channel, and turn-off the device. Therefore, depletion mode consumes power even in the off-state. In contrary, the enhancement mode FET will be in off-state under zero gate voltage.

The operation of n-type (semiconductor channel) FETs (IGZO FETs); under no gate bias (V_{GS}) or $0 < V_{GS} < V_T$, the free electron carriers in the n-type semiconducting channel are randomly located resulting in no channel formation between the source and drain electrode and there is no drain current (I_D), and FET is in off-state as shown in Fig. 20.2A. If $V_{GS} > V_T$, then the electrons on the semiconductor channel are accumulated near to the semiconductor channel/dielectric layer interface and further increase in V_{GS} will result in increase in the drain current (I_D). If $V_{DS} < V_{GS}$ - V_T, then the value of drain current linearly increases with respect to the applied drain voltage (V_{DS}) as shown in Fig. 20.2B, this is defined as linear mode and the value of I_D in the linear region is described in Eq. (20.1).

$$I_D = C_i \quad \mathbf{c}_{FE}\frac{W}{L}\left[\quad (V_{GS}V_T)V_{DS}\frac{1}{2}V_{DS}^2\right] \tag{20.1}$$

where C_i is the gate capacitance per unit area, $\mathbf{c}_{FE}$ is the field effect mobility, W is the width of the channel, and L is the length of the channel.

If $V_{DS} >> V_{GS}$ - V_T (Fig. 20.2C), then the value of drain current (I_D) is independent of the drain voltage (V_{DS}), this region is defined as saturation region (Fig. 20.2C) and the value of I_D in the saturation region is described in Eq. (20.2).

$$I_D = \frac{1}{2}C_i \quad \varsigma_{sat} \quad \frac{W}{L} \quad (V_{GS}V_T)^2 \tag{20.2}$$

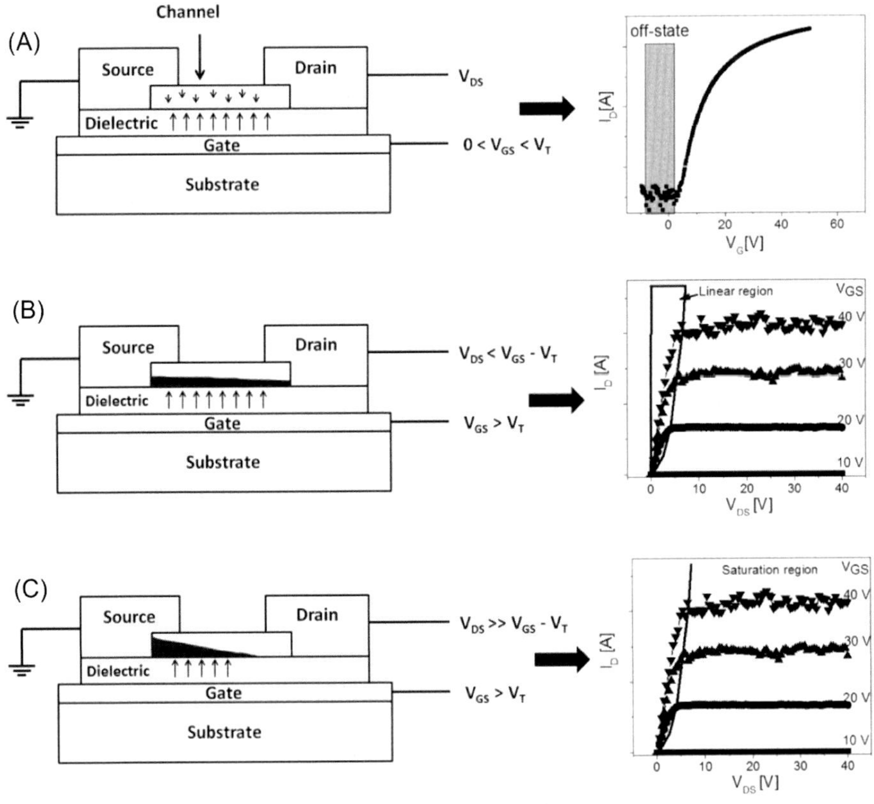

Figure 20.2 The schematic illustrates the working principle of n-type field effect transistor under different gate and drain voltage, and their respective IV characteristics; (A) $0 < V_{GS} < V_T$, (B) $V_{GS} > V_T$ and $V_{DS} < V_{GS} - V_T$, and (C) $V_{GS} > V_T$ and $V_{DS} >> V_{GS} - V_T$.

The basic characterization of the FET is the current–voltage (*IV*) measurement which provides the transfer and output characteristics of FETs as shown in Fig. 20.3A and B, respectively. Further the electrical parameters of the FETs, such as threshold voltage, subthreshold swing (SS), turn on voltage, on–off ration, and mobility, are extracted using the transfer characteristics as shown in Fig. 20.3A.

The on–off ratio is the ratio between the maximum on-current value and the minimum off-current value of the transfer curve, and the threshold voltage (V_T) is the minimum voltage at which the sufficient charge carries are accumulated in the conducting channel at the semiconductor/dielectric interface, which can be roughly calculated by plotting the linear curve as shown in the inset of Fig. 20.3A. Turn-on voltage (V_{on}) is the minimum gate voltage above which the drain current started increasing as shown in Fig. 20.3A. The SS value is the inverse of the subthreshold slope as given in Eq. (20.3).

$$\mathrm{SS} = \left[\frac{d\log(I_D)}{dV_G} \right]^{1} \tag{20.3}$$

In general, very small SS value is desirable because this value is related to the sharp transition between off- and on-state. In contrary, the large SS value is related to the poor interface quality between the semiconductor / dielectric layer and increase in the charge trap density (Kumaresan et al., 2016a).

$$Nss = \left[\frac{SS\tilde{n}\log(e)}{KT/q} 1 \right] \frac{C_{OX}}{\mathrm{q}} \tag{20.4}$$

The number of interface trap density (N_{SS}) can be calculated using Eq. (20.4), where k is Boltzmann coefficient, T is temperature in Kelvin, q is electron charge, and C_{OX} is gate oxide capacitance. Therefore, large SS value indicates large interface charge trapping density. The mobility values are calculated using different

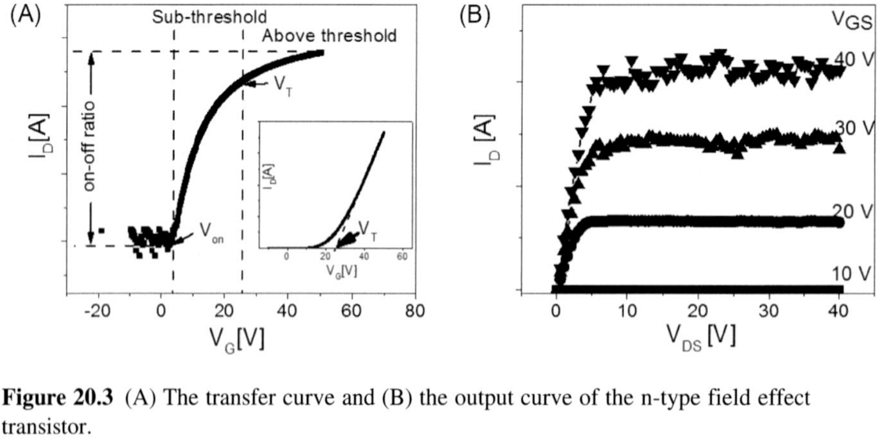

Figure 20.3 (A) The transfer curve and (B) the output curve of the n-type field effect transistor.

methodology and the most important techniques are highlighted below from Eqs. (20.5) to (20.9). Effective mobility ($_{eff}$) is calculated using the linear region drain conductance value (g_d) as given in Eqs. (20.5) and (20.7) (Saji, 2008).

Field effect mobility ($_{FE}$) is calculated using the linear region transconductance value (g_m) as given in Eqs. (20.6) and (20.8). The slope value for $_{FE}$ can be directly measured from the transfer curve as shown in Fig. 20.4A, from the slope value the $_{FE}$ can be calculated using Eq. (20.8). Saturation mobility ($_{sat}$) is calculated using the slope of transfer characteristic in the saturation region as shown in Fig. 20.4B,

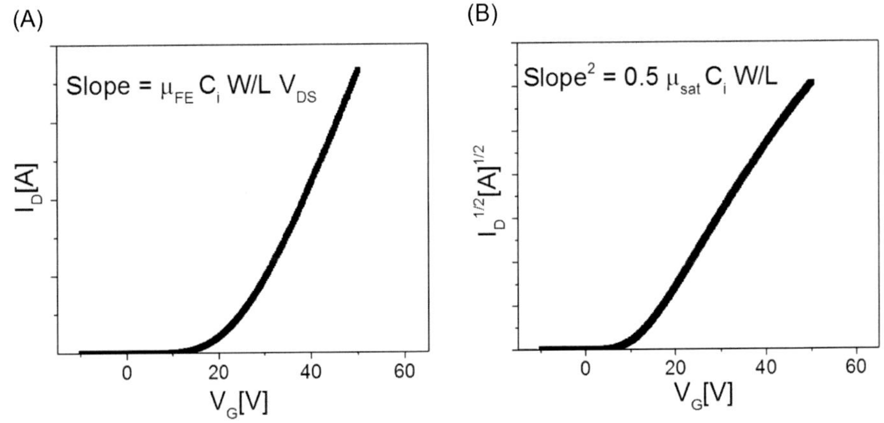

Figure 20.4 Determination of field effect mobility and saturation mobility from (A) the linear region of device operation and (B) the saturation region of the device operation, respectively.

$$g_d = \frac{dI_D}{dV_{DS}}\Bigg]_{V_{GS}=Conctant} \tag{20.5}$$

$$g_m = \frac{dI_D}{dV_{GS}}\Bigg]_{V_{DS}=Conctant} \tag{20.6}$$

$$c_{eff} = \frac{g_d}{C_i\left(\frac{W}{L}\right)(V_{GS}V_T)} \tag{20.7}$$

$$\varsigma_{FE} = \frac{g_m}{C_i\left(\frac{W}{L}\right)V_{DS}} = \frac{Slope}{C_i\left(\frac{W}{L}\right)V_{DS}} \tag{20.8}$$

$$c_{sat} = \frac{Slope^2}{\frac{1}{2}C_i\left(\frac{W}{L}\right)} = \frac{\left(d\sqrt{I_D}/dV_{GS}\right)^2}{\frac{1}{2}C_i\left(\frac{W}{L}\right)} \tag{20.9}$$

from that value the $_{sat}$ can be calculated using Eq. 20.9 (Saji, 2008). All these parameters are extremely useful to evaluate the FET performance.

20.2 Flexible and stretchable IGZO-based field effect transistors

Advancements in electronic devices for display and sensor applications demands portable properties with consistent device performance upon being bent, rolled, or crumpled (Bhatt, Kumar, & Panda, 2020; Knobelspies et al., 2018; Yoo, Sang, & Hwang, 2016). In other word, flexibility without degradation in the device performance is a key requirement (Kumaresan et al., 2018). Researchers have utilized semiconducting polymers as channel materials to enhance the flexibility of FETs (Sekitani, Zschieschang, Klauk, & Someya, 2010), but the device mobilities were low. In this regard, metal oxide semiconductors (MOS), especially IGZO, have received considerable attention because of their high mobility (>10 cm^2/Vs) (Kumaresan et al., 2016a, 2016b; Lee, Jeong, Mativenga, & Jang, 2017; Xia, Zhang, & Wang, 2018). However, the metal oxide materials cannot endure more than 1% strain due to the intrinsic limitation of its mechanical property (Sharma & Ahn, 2016). Therefore, two major approaches, (1) reducing the substrate thickness and (2) an unusual material structure—by patterning or wavy layout or bridge and island architecture, are broadly utilized to achieve flexible and stretchable properties in inorganic electronic devices (Han & Khang, 2015; Jang, Kim, & Yoon, 2019; Kumaresan et al., 2018; Kumaresan, Kim, H., Pak, Y., Poola, P. K., Lee, R., & Lim, 2020; Münzenrieder et al., 2015). As typical metal oxide-based electronic devices have an active device thickness of less than 1 m, the overall flexibility of the devices is determined by the substrate thickness (Sharma & Ahn, 2016). This is because the strain acting on the device while being bent is controlled by the substrate thickness which is calculated from the formula $S=d/2r$, whereas, S is strain, r is the bending radius, and d is the substrate thickness (Kumaresan et al., 2018). The physical strain acting on the device is directly proportional to the device thickness. In this regard, thinner substrates are generally preferable to enhance the flexibility. So far, polyethylene terephthalate (PET), polyethylene naphthalate, polyimide, polymethylmethacrylate (PMMA), and thin glasses have been utilized as the thin flexible substrates (Han & Khang, 2015; Hassan, Uddin, Ullah, Kim, & Chung, 2016). Han and Khang (2015) obtained a conductive ITO film on foldable glass substrate by selectively thinning the glass down to 5 m to achieve alternative thick and thin region that enabled reversible and repeated foldability at predesigned location. Similarly, IGZO transistor was fabricated on 1.5 m thick polyimide substrate to complement extreme bending and folding property as shown in Fig. 20.5A (Kim, Lee, Um, Mativenga, & Jang, 2016). The inorganic device on polyimide exhibited stable device performance even after 20,000 bending cycles. As handling of thin flexible substrates poses undesirable wrapping and bending during the device fabrication process, these thin flexible substrates are attached to a rigid substrate via a sacrificial layer. Ultimately, the ultrathin substrate, along with the devices, is released from the rigid substrate by selectively removing the underlying sacrificial layer via a solution process through intentionally made openings; however, easy separation and large area fabrication are still challenging issues. To address these issues, spin-coated PMMA/glass substrate was utilized and bottom-gated IGZO transistors was fabricated over PMMA. Sequentially, PMMA substrate along with the IGZO transistor was separated from the glass slide by engineering the surface energy at the interface between glass-PMMA (Kumaresan et al., 2018). Fig. 20.5B depicts the fabrication scheme of IGZO transistor on ultrathin PMMA substrate. The ultrathin PMMA substrate, along with the electronic devices, was separated from the rigid glass slide without the use of any sacrificial layers (Fig. 20.5B), and the thin PMMA substrate demonstrated excellent conformal contact with nonplanar surfaces such as wrinkled hand gloves and flexible PVC gas tubes. The bottom-gated IGZO transistors on the ultrathin PMMA substrate demonstrated highly bendable transistor performance even after harsh bending treatment at a bending radius of < 1 mm (Fig. 20.5D). As shown in Fig. 20.5E, the IGZO transistors revealed good transistor performance with an on/off-current ratio of $5 \quad 10^5$, a mobility of 10.7 cm^2/Vs, and a V_T of 7.2 V (Kumaresan et al., 2018).

Likewise, unusual material structures are utilized to obtain stretchable and wearable property in IGZO-based electronic devices. As shown in Fig. 20.5F, patterning technique was adopted to obtain flexible and stretchable property in IGZO film (Kumaresan, Kim, H., Pak, Y., Poola, P. K., Lee, R., & Lim, 2020). The honeycomb-patterned IGZO-based electronic devices on an elastomeric substrate assured omnidirectional stretchability up to 20% strain, with stable electrical performances. Alternatively, a stretchable wavy structured ZnO transistor was fabricated by transfer-printing the conventional device in to prestrained PDMS substrate (Park et al., 2010). After releasing the prestrain, wavy configuration on inorganic material was observed that enables the stretching up to 5% strain in ZnO transistors. In a similar way, an IGZO transistor on ultrathin polyimide/prestrained elastomer substrate was realized to obtain wavy configuration (Jang et al., 2019). The IGZO transistors were fabricated on ultrathin polyimide substrate and transfer printed on prestrained PDMS substrate. Furthermore, the wavy configuration was achieved after releasing the prestrain. The stretchable IGZO thin film transistors (TFTs) demonstrated outstanding device characteristics without any performance degradation up to 40% strain (Jang et al., 2019). These IGZO-based flexible and stretchable transistors have the potential to be comfortably attached to the human skin to detect the sweat pH for real-time continuous health monitoring.

20.3 Sweat pH sensor using flexible IGZO field effect transistors

Sweat, the native of abundant chemical information, could give us better knowledge about our body's deeper molecular state is generally being underutilized for the source of information it can provide in the field of health care and management

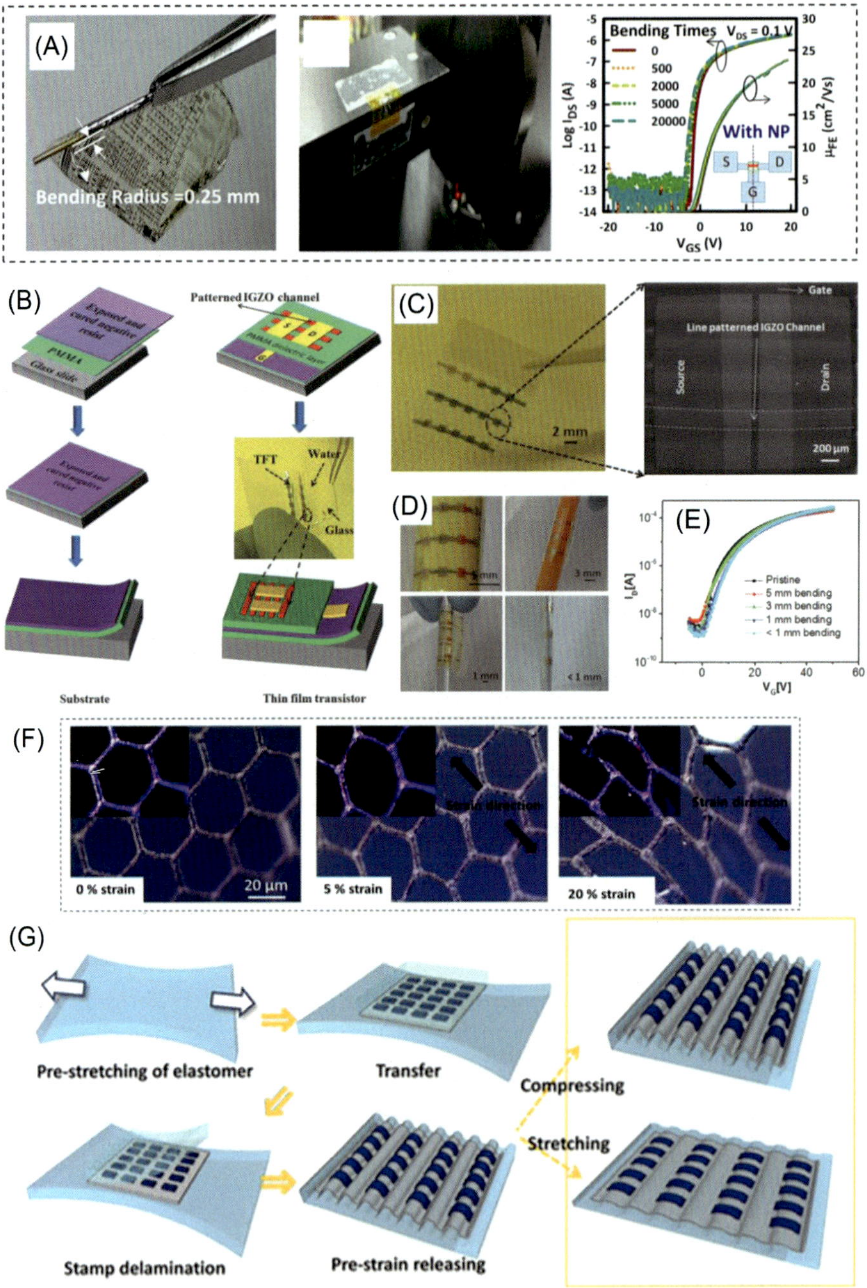

Figure 20.5 Flexible and stretchable inorganic electronic devices: (A) Optical image of indium gallium zinc oxide (IGZO) transistors on 1.5 m-thick polyimide substrate, wrapped

(*Continued*)

(Bariya, Nyein, & Javey, 2018). The existing conventional wearable sensors provide useful information about heart rate and other physical activities but fail to provide information at a deeper molecular level (Bandodkar & Wang, 2014). This void has ignited the biosensors' developers to study accessible biofluids to obtain deeper understanding about one's health through noninvasive methods. The pH level of human sweat is one of the major sources through which human health could be monitored. The pH level of a normal healthy man ranges between 4.5–6.5 and the abnormal pH level indicates the exposure of humans to certain diseases (Schmid-Wendtner & Korting, 2006; Strong, Smit, L. S., Turpin, S. V., Cole, J. L., Hon, C. T., & Markiewicz, 1991). Recent developments in flexible and stretchable electronic paved the way for developing wearable sweat sensors for real-time monitoring the human health through pH level of the sweat. In this regard, stretchable pH sensor with stretchable RFID antenna was fabricated for real-time monitoring of pH value in sweat equivalent solution (Dang et al., 2018). The photographic image of stretchable pH sensing patch using patterned inorganic material and their schematic representation of sensing components are given in Fig. 20.6A and B, respectively. A serpentine-patterned inorganic material has been utilized to enhance the stretchability required for wearable application. The sensor device demonstrated the sensitivity of 11.13 ◡5.8 mV/pH with a maximum response time of 8 s (Dang et al., 2018). The advancements in the pH sensors especially thin-film transistors-based pH sensors give a bright ray of hope in analyzing the pH level with greater sensitivity and maximized electrical performance (Pullano et al., 2018; Schmid, 2006;

◀ around 0.25 mm radius cylinder (right), placed on extreme bending machine (middle), and the transfer characteristics under extreme bending test up to 20,000 cycles (left). (B) A schematic representing the fabrication of bottom-gated IGZO transistors on an ultrathin polymethylmethacrylate (PMMA) substrate and the photographic image of IGZO transistor during separation form underlying glass slide is given in the middle. (C) The photographic image of IGZO transistor on ultrathin PMMA (right) and its scanning electron microscope image (left). (D) Digital image of IGZO transistor subjected to different bending radii ranging from 5 mm to $<$ 1 mm and (E) their respective transfer characteristics. (F) Optical microscope image of honeycomb-patterned inorganic material subjected to stretching strain ranges from 0% to 20% strain. (G) A schematic representing the fabrication of wavy dimensional IGZO transistor on elastomeric substrate and its respective compressive and stretching strain.
Source: (A) Reprinted from Kim, Y.-H., Lee, E., Um, J. G., Mativenga, M., & Jang, J. (2016). Highly robust neutral plane oxide TFTs withstanding 0.25 mm bending radius for stretchable electronics. Scientific Reports, *6*(1), 25734; (B)–(E) Reprinted from Kumaresan, Y., et al. (2018). Extremely flexible indium-gallium-zinc oxide (IGZO) based electronic devices placed on an ultrathin poly(methyl methacrylate) (PMMA) substrate. Advanced Electronic Materials, *4*(7), 1800167; (F) Reprinted from Kumaresan, Y., et al. (2020). Omnidirectional stretchable inorganic-material-based electronics with enhanced performance. Advanced Electronic Materials, *6*(7), 2000058; (G) Adapted with permission from Jang, H. -W., Kim, S., & Yoon, S. -M. (2019). Impact of polyimide film thickness for improving the mechanical robustness of stretchable InGaZnO thin-film transistors prepared on wavy-dimensional elastomer substrates. ACS Applied Materials & Interfaces, *11*(37), 34076–34083, Copyright (2019) American Chemical Society.

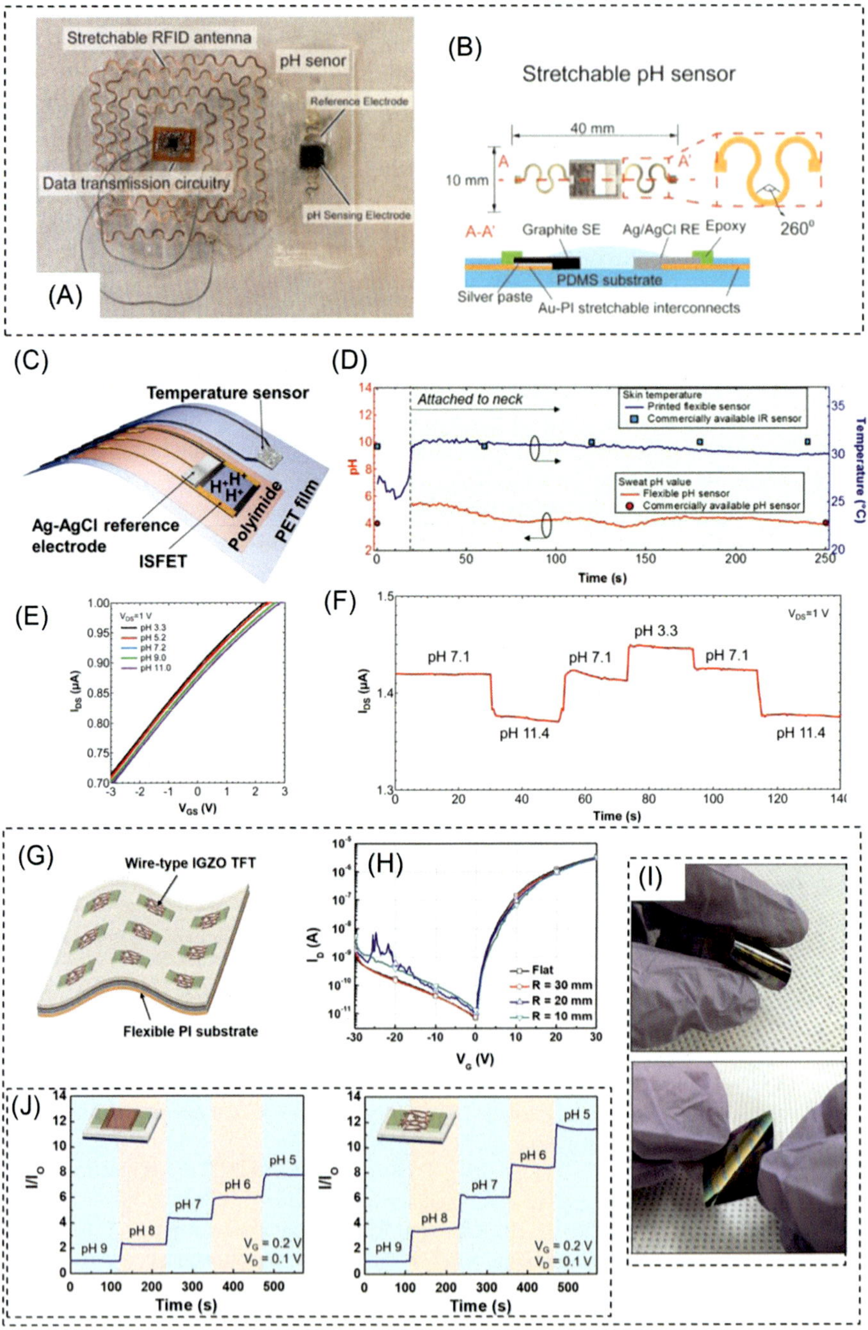
(A)
Stretchable RFID antenna
pH senor
Reference Electrode
Data transmission circuitry
pH Sensing Electrode
(B)
Stretchable pH sensor
40 mm
10 mm
A
A'
260°
A-A'
Graphite SE
Ag/AgCl RE
Epoxy
PDMS substrate
Silver paste
Au-PI stretchable interconnects
(C)
Temperature sensor
Ag-AgCl reference electrode
ISFET
Polyimide
PET film
(D)
Attached to neck
Skin temperature
Printed flexible sensor
Commercially available IR sensor
Sweat pH value
Flexible pH sensor
Commercially available pH sensor
pH
Temperature (°C)
Time (s)
(E)
V_DS=1 V
pH 3.3
pH 5.2
pH 7.2
pH 9.0
pH 11.0
I_DS (μA)
V_GS (V)
(F)
V_DS=1 V
pH 7.1
pH 11.4
pH 7.1
pH 3.3
pH 7.1
pH 11.4
I_DS (μA)
Time (s)
(G)
Wire-type IGZO TFT
Flexible PI substrate
(H)
Flat
R = 30 mm
R = 20 mm
R = 10 mm
I_D (A)
V_G (V)
(I)
(J)
pH 9
pH 8
pH 7
pH 6
pH 5
V_G = 0.2 V
V_D = 0.1 V
I/I_O
Time (s)

(Continued)

Torsi, Magliulo, Manoli, & Palazzo, 2013). In this regard, an IGZO-based thin-film transistor proves to be the number one contender. An improvement in pH sensitivity has been reported in the extended-gate field-effect transistor (EGFET) pH sensors. The amorphous IGZO-modified silicon nanowire (SiNW) pH sensors showcased an increased sensitivity compared to that of the pristine SiNW sensor. This improved sensitivity results from the adsorption of additional H^+ ions as the IGZO nanoparticles sputtered over the silicon nanowire provided more oxygen biding sites in the pH sensor device. The presence of IGZO nanoparticles with appropriate thickness raised the pH sensitivity with improved differentiation in various pH buffer solutions (Lin, Huang, & Yang, 2013). The pH sensitivities of 202 mV/pH were obtained from flexible extended gate field effect transistors (EGFETs) with tin oxide as the source and drain coated with amorphous IGZO (a-IGZO) fabricated on PET. The low-cost fully flexible EGFETs with a-IGZO as a semiconductor have been utilized for both pH sensing and the detection of a prostate-specific antigen. The a-IGZO was deposited over ITO at room temperature by radio frequency sputtering (Bhatt et al., 2020). Meanwhile, IGZO-MESFET (metal-semiconductor field effective transistor) biosensor showcased high sensitivity covering physiological values for glucose in the bloodstream as well as sweat, saliva, and tears (Kaczmarski, Jankowska-Śliwińska, & Borysiewicz, 2019).

Measuring the body temperature along with the various chemical state using pH sensor is an important indicator to detect the health parameters (Nakata et al., 2017). Based on this consideration, ion-sensitive FETs (ISFETs) was integrated with the temperature sensor as shown in Fig. 20.6C. The device was attached on neck and the

Figure 20.6 (A) Photo of stretchable wireless system for sweat pH monitoring; (B) Schematic diagram of stretchable pH sensor with graphite–polyurethane composite SE, Ag/AgCl-based RE and a pair of stretchable interconnects. (C) Schematic of a wearable device integrating flexible pH and temperature sensors, (D) real-time pH and skin temperature acquired by the device, (E) transfer characteristics of the ion-sensitive field effect transistor (ISFET) at V_{DS}=1 V in solutions of different pH, and (F) results from real-time measurement of solutions with different pH. (G) Schematic diagram of wire-type indium gallium zinc oxide (IGZO) TFTs on a flexible substrate. (H) Transfer characteristics of flexible wire-type IGZO TFTs measured in the bending state at various radii, (I) photograph of flexible wire-type IGZO TFTs, and (J) real-time response of the drain current of film-type (left) and wire-type (right) IGZO sensors.
Source: (A and B) Reprinted from Dang, W., Manjakkal, L., Navaraj, W.T., Lorenzelli, L., Vinciguerra, V., & Dahiya, R. (2018). Stretchable wireless system for sweat pH monitoring. Biosensors and Bioelectronics, *107*, 192–202, Copyright Elsevier; (C)–(F) Adapted with permission from Nakata, S., Arie, T., Akita, S., & Takei, K. (2017). Wearable, flexible, and multifunctional healthcare device with an ISFET chemical sensor for simultaneous sweat pH and skin temperature monitoring. ACS Sensors, *2*(3), 443–448, Copyright (2017) American Chemical Society; (J) Reprinted with permission from Kim, Y.-g., Tak, Y. J., Kim, H. J., Kim, W.-G., Yoo, H., & Kim, H. J. (2018). Facile fabrication of wire-type indium gallium zinc oxide thin-film transistors applicable to ultrasensitive flexible sensors. Scientific Reports, *8*(1), 5546.

real-time temperature and pH values are obtained over time as shown in Fig. 20.6D. The transfer characteristics of ISFETs and the real-time measurement of current with different pH concentration are given in Fig. 20.6E and F, respectively (Nakata et al., 2017). The pH sensitivity had been efficiently amplified through the development of an extended-gate in-plane-gate structure. Amorphous-IGZO TFT developed by introducing an in-plane-gate (IPG) structure showcased 99% or more for a small IPG area and an improved sensitivity could be noticed with further reduction in the area. The more effective amplification in the sensitivity is generated by the capacitive coupling effect (Pyo & Cho, 2018). A high pH sensitivity of 649.04 mV/pH was achieved by amorphous-IGZO separative extended gate ion-sensitive field-effect transistor (SEGISFET) under dual-gate (DG) sensing mode operation. The high-performance SEGISFET possesses tin oxide (SnO_2) SEG sensing part along with the double-gate structure amorphous IGZO thin-film transistor with tantalum pentoxide/silicon dioxide (Ta_2O_5 / SiO_2)-engineered gate oxide. The DG sensing mode yielded good linearity of 99.9% between pH 4 and pH 10 solutions (Pyo & Cho, 2017).

pH sensitivity beyond the Nernst limit was achieved through an amorphous-IGZO-based TiO_2 extended-gate field-effect transistor (EGFETs). The pH sensitivity obtained through the device was 129.1 mV/pH. This enhanced sensitivity results from the capacitive coupling between the front- and bottom-gate oxides. In addition to this, the double-gate operation exhibited remarkable stability than the single-gate operation which may reinforce the signal-to-noise ratio of the biosensor application. This high sensitivity reported aids in achieving an outstanding electrical characteristic of a-IGZO thin-film transistor (Jang, Gu, & Cho, 2013). Similarly, pH sensitivity of nearly four times as high as the Nernst limit was achieved through liquid–solid dual-gated (LSDG) IGZO thin-film transistor in the ISFET. The pH sensitivity of ~237 mV/pH was achieved. The device was fabricated with SiO_2 as the bottom solid-gated and the pH buffer solution as a top liquid gate. The theoretically estimated sensitivity of the IGZO–LSDG thin-film transistor pH sensor was 504 mV/pH at 298 K (Cai, Yang, Wang, Pei, & Wang, 2019).

In addition, the use of a-IGZO as the pH sensitive layer over an SiO_2/Si surface showed better pH sensitivity. The electrolyte–insulator–semiconductor (EIS) device was fabricated by depositing IGZO over SiO_2/Si followed by annealing of IGZO films. Annealing of the IGZO films has greater impact over the pH sensitivity of the EIS device. The EIS device with IGZO films annealed at 400°C in the presence of N_2 showed the better sensitivity of 53.3 mV/pH than that of O_2 ambience and the EIS device having IGZO film deposited without annealing. The annealing temperature also affects the pH sensitivity of the device. The pH sensitivity was decreased to 40.7 mV/pH when the annealing process takes place at 500°C in the same N_2 ambient (Kumar, Kumar, & Panda, 2015). Considerable improvement in the sensitivity has been witnessed in the wire-type IGZO thin-film transistor in comparison with the film-type device. The schematic representation of the device is shown in Fig. 20.6G. The wire-type IGZO TFTs fabricated simply and effectively on the flexible substrate possess a stable electrical characterization under mechanical stresses at a bending radius of 10 mm (Fig. 20.6H). The IGZO nanowires fabricated through self-formed cracked templates are in the shape of networked nanowires to improve the flexibility (Fig. 20.6I) and sensitivity. The IGZO nanowires in this device are employed as the active layer.

The electrical characteristics of the thin-film transistor can be controlled by varying width and density of the wires. The speed of the template solution coating through spin coating also contributes to control of the electrical characteristics of the thin-film transistor (Kim et al., 2018). The wire-type IGZO sensors demonstrated high response than the flat IGZO film as shown in Fig. 20.6J. The wire-type IGZO devices have flexibility with enhanced response suitable for wearable biosensors for real-time health monitoring.

20.4 Conclusion

Sensor is essentially a transducer, in which the surface reaction takes place, between the sensing channel and sensing molecules, resulting in electrical signal variation (physical or chemical properties variation in semiconducting channel) for practical application. Biosensors for health monitoring utilizes sweat, saliva, and tears as a biological sample. Among them, sweat is most abundant and readily available and monitoring the pH level of sweat is most useful for identifying critical diseases. In general, the extraction of pH value is achieved through measuring the concentration of hydrogen ions in the analyte. Especially FETs-based devices are effectively investigated due to its high sensitivity under low voltage. Therefore, flexible, and stretchable FETs are essential for real-time health monitoring. In this regard, the basic working principle of FETs and the recent progress in flexible and stretchable IGZO FETs have been discussed. The extraction of device parameters, such as on/off current ratio, threshold voltage, SS, and mobility from current–voltage curve for TFTs are briefly described. As the total device thickness is less than 1 m, the flexibility and stretchability in inorganic IGZO transistors are achieve either by reducing the substrate thickness or by patterning. Bottom-gated IGZO FETs on the ultrathin PMMA substrate demonstrated high flexibility even after harsh bending treatment at a bending radius of < 1 mm with consistent device performance. Similarly, IGZO FETs transferred on to prestrained PDMS substrate revealed wave configuration that enables the flexible and stretchable property in inorganic IGZO devices. Such devices open a door to fabricate wearable biosensor devices for real-time health monitoring. As compared to flat IGZO sensor, the wire-type IGZO sensor demonstrated enhanced sensitivity due to higher surface area and enhanced flexibility due to the patterned geometry. Further, the pH sensor device with IGZO FETs demonstrated enhanced pH sensitive of more than 500 mV/pH, that greatly improve the real-time detection and accurate monitoring of health condition.

References

Bandodkar, A. J., & Wang, J. (2014). Non-invasive wearable electrochemical sensors: Areview. *Trends in Biotechnology*, *32*(7), 363–371.

Bariya, M., Nyein, H. Y. Y., & Javey, A. (2018). Wearable sweat sensors. *Nature Electronics*, *1*(3), 160–171.

Bhatt, D., Kumar, S., & Panda, S. (2020). Amorphous IGZO field effect transistor based flexible chemical and biosensors for label free detection. *Flexible and Printed Electronics*, *5* (1), 014010.

Billah, M. M., Hasan, M. M., & Jang, J. (2017). Effect of tensile and compressive bending stress on electrical performance of flexible a-IGZO TFTs. *IEEE Electron Device Letters*, *38*(7), 890–893.

Cai, G., Yang, P., Wang, X., Pei, Y., & Wang, G. (2019). Investigation of pH sensor based on liquid-solid dual-gated IGZO thin-film transistor. *Materials Research Express*, *6*(9), 096305.

Cannadine, D. (1984). The present and the past in the English industrial revolution 1880–1980. *Past & Present*, *103*, 131–172.

Choi, D.-H., Thaxton, A., Jeong, I. c., Kim, K., Sosnay, P. R., Cutting, G. R., … Searson, P. C. (2018). Sweat test for cystic fibrosis: Wearable sweat sensor vs. standard laboratory test. *Journal of Cystic Fibrosis*, *17*(4), e35–e38, 2018/07/01.

Cinti, S., Fiore, L., Massoud, R., Cortese, C., Moscone, D., Palleschi, G., & Arduini, F. (2018). Low-cost and reagent-free paper-based device to detect chloride ions in serum and sweat. *Talanta*, *179*, 186–192.

Dang, W., Manjakkal, L., Navaraj, W. T., Lorenzelli, L., Vinciguerra, V., & Dahiya, R. (2018). Stretchable wireless system for sweat pH monitoring. *Biosensors and Bioelectronics*, *107*, 192–202.

Das, A., Ko, D. H., Chen, C.-H., Chang, L.-B., Lai, C. S., & Chu, F.-C. (2014). Highly sensitive palladium oxide thin film extended gate FETs as pH sensor. *Sensors and Actuators B: Chemical*, *205*, 199–205.

Dewar, J., & Fleming, J. A. (1897). On the electrical resistivity of pure mercury at the temperature of liquid air. *Proceedings of the Royal Society of London*, *60*(359–367), 76–81.

Fu, D., Zhu, C., Zhang, X., Li, C., & Chen, Y. (2016). Two-dimensional net-like SnO_2/ZnO heteronanostructures for high-performance H2S gas sensor. *Journal of Materials Chemistry A*, *4*(4), 1390–1398. Available from https://doi.org/10.1039/C5TA09190J.

Gao, W., Emaminejad, S., Nyein, H. Y. Y., Challa, S., Chen, K., & Peck, A. (2016). Fully integrated wearable sensor arrays for multiplexed in situ perspiration analysis. *Nature*, *529*(7587), 509–514.

Han, M. J., & Khang, D.-Y. (2015). Glass and plastics platforms for foldable electronics and displays. *Advanced Materials*, *27*(34), 4969–4974.

Han, S.-T., Peng, H., Sun, Q., Venkatesh, S., Chung, K.-S., & Lau, C. S. (2017). An overview of the development of flexible sensors. *Advanced Materials*, *29*(33), 1700375.

Han, Y., & Dong, J. (2018). Fabrication of self-recoverable flexible and stretchable electronic devices. *Journal of Manufacturing Systems*, *48*, 24–29.

Hassan, K., Uddin, A. S. M. I., Ullah, F., Kim, Y. S., & Chung, G.-S. (2016). Platinum/palladium bimetallic ultra-thin film decorated on a one-dimensional ZnO nanorods array for use as fast response flexible hydrogen sensor. *Materials Letters*, *176*, 232–236.

He, W., Wang, C., Wang, H., Jian, M., Lu, W., & Liang, X. (2019). Integrated textile sensor patch for real-time and multiplex sweat analysis. *Science Advances*, *5*(11), eaax0649.

Jamasb, S. (2019). Continuous monitoring of pH and blood gases using ion-sensitive and gas-sensitive field effect transistors operating in the amperometric mode in presence of drift. *Biosensors*, *9*(1).

Jang, H.-J., Gu, J.-G., & Cho, W.-J. (2013). Sensitivity enhancement of amorphous InGaZnO thin film transistor based extended gate field-effect transistors with dual-gate operation. *Sensors and Actuators B: Chemical, 181*, 880–884.

Jang, H.-W., Kim, S., & Yoon, S.-M. (2019). Impact of polyimide film thickness for improving the mechanical robustness of stretchable InGaZnO thin-film transistors prepared on wavy-dimensional elastomer substrates. *ACS Applied Materials & Interfaces, 11*(37), 34076–34083.

Johnston, D. E., Islam, M. F., Yodh, A. G., & Johnson, A. T. (2005). Electronic devices based on purified carbon nanotubes grown by high-pressure decomposition of carbon monoxide. *Nature Materials, 4*(8), 589–592.

Kaczmarski, J., Jankowska-Śliwińska, J., & Borysiewicz, M. A. (2019). IGZO MESFET with enzyme-modified Schottky gate electrode for glucose sensing. *Japanese Journal of Applied Physics, 58*(9), 090603.

Kaya, T., Liu, G., Ho, J., Yelamarthi, K., Miller, K., Edwards, J., & Stannard, A. (2019). Wearable sweat sensors: Background and current trends. *Electroanalysis, 31*(3), 411–421.

Khan, S., Ali, S., & Bermak, A. (2019). Recent developments in printing flexible and wearable sensing electronics for healthcare applications. *Sensors, 19*(5).

Kim, Y.-H., Lee, E., Um, J. G., Mativenga, M., & Jang, J. (2016). Highly robust neutral plane oxide TFTs withstanding 0.25 mm bending radius for stretchable electronics. *Scientific Reports, 6*(1), 25734.

Kim, Y.-g, Tak, Y. J., Kim, H. J., Kim, W.-G., Yoo, H., & Kim, H. J. (2018). Facile fabrication of wire-type indium gallium zinc oxide thin-film transistors applicable to ultrasensitive flexible sensors. *Scientific Reports, 8*(1), 5546.

Knobelspies, S., Bierer, B., Daus, A., Takabayashi, A., Salvatore, G. A., & Cantarella, G. (2018). Photo-induced room-temperature gas sensing with a-IGZO based thin-film transistors fabricated on flexible plastic foil. *Sensors, 18*(2).

Koh, A., Kang, D., Xue, Y., Lee, S., Pielak, R. M., & Kim, J. (2016). A soft, wearable microfluidic device for the capture, storage, and colorimetric sensing of sweat. *Science Translational Medicine, 8*(366), 366ra165.

Kumar, N., Kumar, J., & Panda, S. (2015). Low temperature annealed amorphous indium gallium zinc oxide (a-IGZO) as a pH sensitive layer for applications in field effect based sensors. *AIP Advances, 5*(6), 067123.

Kumaresan, Y., Pak, Y., Lim, N., kim, Y., Park, Y., & Yoon, S.-M (2016a). Highly bendable In-Ga-ZnO thin film transistors by using a thermally stable organic dielectric layer. *Scientific Reports, 6*(1), 37764.

Kumaresan, Y., Pak, Y., Lim, N., Lee, R., Song, H., & Kim, T. H (2016b). Effect of channel thickness, annealing temperature and channel length on nanoscale Ga_2O_3–In_2O_3–ZnO thin film transistor performance. *Journal of Nanoscience and Nanotechnology, 16*(6), 6364–6367.

Kumaresan, Y., Kim, H., Pak, Y., Poola, P. K., Lee, R., & Lim, N. (2020). Omnidirectional stretchable inorganic-material-based electronics with enhanced performance. *Advanced Electronic Materials, 6*(7), 2000058.

Kumaresan, Y., Kim, H., Jeong, Y., Pak, Y., Cho, S., & Lee, R. (2017). Ultra-high Sensitivity to low hydrogen gas concentration with Pd-decorated IGZO film. *IEEE Electron Device Letters, 38*(12), 1735–1738.

Kumaresan, Y., Lee, R., Lim, N., Pak, Y., Kim, H., Kim, W., & Jung, G. Y. (2018). Extremely flexible indium-gallium-zinc oxide (IGZO) based electronic devices placed on an ultrathin poly(methyl methacrylate) (PMMA) substrate. *Advanced Electronic Materials, 4*(7), 1800167.

Lee, S., Jeong, D., Mativenga, M., & Jang, J. (2017). Highly robust bendable oxide thin-film transistors on polyimide substrates via mesh and strip patterning of device layers. *Advanced Functional Materials*, *27*(29), 1700437.

Lin, J.-C., Huang, B.-R., & Yang, Y.-K. (2013). IGZO nanoparticle-modified silicon nanowires as extended-gate field-effect transistor pH sensors. *Sensors and Actuators B: Chemical*, *184*, 27–32.

Linghu, C., Zhang, S., Wang, C., & Song, J. (2018). Transfer printing techniques for flexible and stretchable inorganic electronics. *NPJ Flexible Electronics*, *2*(1), 26.

Lu, H., Li, Y., Chen, M., Kim, H., & Serikawa, S. (2018). Brain intelligence: Go beyond artificial intelligence. *Mobile Networks and Applications*, *23*(2), 368–375.

Münzenrieder, N., Cantarella, G., Vogt, C., Petti, L., Büthe, L., & Salvatore, G. A. (2015). Stretchable and conformable oxide thin-film electronics. *Advanced Electronics Materials*, *1*(3), 1400038.

Nakagawa, T., Negoro, Y., Yoneda, S., Shishido, H., Kobayashi, H., & Oota, M. (2020). Image sensor with In-pixel calculation using crystalline IGZO FET in display. *Japanese Journal of Applied Physics*, *59*(SG), SGGE01.

Nakata, S., Arie, T., Akita, S., & Takei, K. (2017). Wearable, flexible, and multifunctional healthcare device with an ISFET chemical sensor for simultaneous sweat pH and skin temperature monitoring. *ACS Sensors*, *2*(3), 443–448.

Niehenke, E. C. (2015). The evolution of transistors for power amplifiers: 1947 to today. In *2015 IEEE MTT-S international microwave symposium* (pp. 1–4).

Oostinga, J. B., Heersche, H. B., Liu, X., Morpurgo, A. F., & Vandersypen, L. M. K. (2008). Gate-induced insulating state in bilayer graphene devices. *Nature Materials*, *7*(2), 151–157.

Pachauri, V., & Ingebrandt, S. (2016). Biologically sensitive field-effect transistors: From ISFETs to NanoFETs. *Essays in Biochemistry*, *60*(1), 81–90.

Pak, Y., Lim, N., Kumaresan, Y., Lee, R., Kim, K., & Kim, T. H. (2015). Palladium nanoribbon array for fast hydrogen gas sensing with ultrahigh sensitivity. *Advanced Materials*, *27*(43), 6945–6952.

Pak, Y., Park, W., Mitra, S., Sasikala Devi, A. A., Loganathan, K., & Kumaresan, Y. (2018). Enhanced performance of MoS_2 photodetectors by inserting an ALD-processed TiO_2 interlayer. *Small (Weinheim an der Bergstrasse, Germany)*, *14*(5), 1703176.

Park, K., Lee, D.-K., Kim, B.-S., Jeon, H., Lee, N.-E., & Whang, D. (2010). Stretchable, transparent zinc oxidec thin film transistors. *Advanced Functional Materials*, *20*(20), 3577–3582.

Pešić, M., Fengler, F. P. G., Larcher, L., Padovani, A., Schenk, T., & Grimley, E. D. (2016). Physical mechanisms behind the field-cycling behavior of HfO_2-based ferroelectric capacitors. *Advanced Functional Materials*, *26*(25), 4601–4612.

Pullano, S. A., Critello, C. D., Mahbub, I., Tasneem, N. T., Shamsir, S., & Islam, S. K. (2018). EGFET-based sensors for bioanalytical applications: A review. *Sensors*, *18*(11).

Pyo, J.-Y., & Cho, W.-J. (2017). High-performance SEGISFET pH sensor using the structure of double-gate a-IGZO TFTs with engineered gate oxides. *Semiconductor Science and Technology*, *32*(3), 035015.

Pyo, J.-Y., & Cho, W.-J. (2018). In-plane-gate a-IGZO thin-film transistor for high-sensitivity pH sensor applications. *Sensors and Actuators B: Chemical*, *276*, 101–106.

Rich, S. I., Wood, R. J., & Majidi, C. (2018). Untethered soft robotics. *Nature Electronics*, *1*(2), 102–112.

Rodrigues, D., Barbosa, A. I., Rebelo, R., Kwon, I. K., Reis, R. L., & Correlo, V. M. (2020). Skin-integrated wearable systems and implantable biosensors: A comprehensive review. *Biosensors*, *10*(7).

Rosenbaum. (1969). G.E. Moore's the elements of ethics. *University of Toronto Quarterly*, *38*(3), 214–232.

Saji, K. J. (2008). Amorphous oxide transparent thin films: Growth, characterisation and application to thin film transistors. PhD thesis, Department of Physics, Cochin University of Science and Technology, Cochi, 2519.

Schmid, S. (2006). Ion sensitive field effect transistors (ISFETs) basics and applications. *Moscow Bavarian Joint Advanced Student School.*

Schmid-Wendtner, M. H., & Korting, H. C. (2006). The pH of the skin surface and its impact on the barrier function. *Skin Pharmacology and Physiology*, *19*(6), 296–302.

Schwartz, G., Tee, B. C. K., Mei, J., Appleton, A. L., Kim, D. H., Wang, H., & Bao, Z. (2013). Flexible polymer transistors with high pressure sensitivity for application in electronic skin and health monitoring. *Nature Communications*, *4*(1), 1859.

Scidà, A., Haque, S., Treossi, E., Robinson, A., Smerzi, S., & Ravesi, S. (2018). Application of graphene-based flexible antennas in consumer electronic devices. *Materials Today*, *21*(3), 223–230.

Sekitani, T., Zschieschang, U., Klauk, H., & Someya, T. (2010). Flexible organic transistors and circuits with extreme bending stability. *Nature Materials*, *9*(12), 1015–1022.

Sharma, B. K., & Ahn, J.-H. (2016). Flexible and stretchable oxide electronics. *Advanced Electronic Materials*, *2*(8), 1600105.

Sim, K., Rao, Z., Zou, Z., Ershad, F., Lei, J., & Thukral, A. (2019). Metal oxide semiconductor nanomembrane–based soft unnoticeable multifunctional electronics for wearable human-machine interfaces. *Science Advances*, *5*(8), eaav9653.

Song, Y., Min, J., Yu, Y., Wang, H., Yang, Y., Zhang, H., & Gao, W. (2020). Wireless battery-free wearable sweat sensor powered by human motion. *Science Advances*, *6*(40), eaay9842.

Strong, T. V., Smit, L. S., Turpin, S. V., Cole, J. L., Hon, C. T.., & Markiewicz, D. . . (1991). Cystic fibrosis gene mutation in two sisters with mild disease and normal sweat electrolyte levels. *New England Journal of Medicine*, *325*(23), 1630–1634.

Torsi, L., Magliulo, M., Manoli, K., & Palazzo, G. (2013). Organic field-effect transistor sensors: A tutorial review. *Chemical Society Reviews*, *42*(22), 8612–8628. Available from https://doi.org/10.1039/C3CS60127G.

Wang, X., Liu, Z., & Zhang, T. (2017). Flexible sensing electronics for wearable/attachable health monitoring. *Small (Weinheim an der Bergstrasse, Germany)*, *13*(25), 1602790.

Xia, G., Zhang, Q., & Wang, S. (2018). High-mobility IGZO TFTs by infrared radiation activated low-temperature solution process. *IEEE Electron Device Letters*, *39*(12), 1868–1871.

Xu, M., David, J. M., & Kim, S. H. (2018). The fourth industrial revolution: Opportunities and challenges. *International Journal of Financial Research*, *9*(2), 90–95.

Yoo, T.-H., Sang, B.-I., & Hwang, D. K. (2016). One-dimensional InGaZnO field-effect transistor on a polyimide wire substrate for an electronic textile. *Journal of the Korean Physical Society*, *68*(4), 599–603.

Yu, K.-H., Beam, A. L., & Kohane, I. S. (2018). Artificial intelligence in healthcare. *Nature Biomedical Engineering*, *2*(10), 719–731.

Yu, X., Marks, T. J., & Facchetti, A. (2016). Metal oxides for optoelectronic applications. *Nature Materials*, *15*(4), 383–396.

Zhan, B., Li, C., Yang, J., Jenkins, G., Huang, W., & Dong, X. (2014). Graphene field-effect transistor and its application for electronic sensing. *Small (Weinheim an der Bergstrasse, Germany)*, *10*(20), 4042–4065.

Layered metal oxides for biomedical applications

21

Uttam Gupta and Suchitra
Max Planck Institute for Chemical Physics of Solids, Dresden, Germany

21.1 Introduction

Layered materials have found various applications—due to their large surface to volume ratio and atomic level confinement along with the thickness—in areas such as optics, electronics, sensors, energy storage, catalysis, and biomedical applications (Chimene, Alge, & Gaharwar, 2015; Kurapati, Kostarelos, Prato, & Bianco, 2016). At present, a large number of layered materials are being discovered and are continually growing. After the discovery of graphene, other classes of layered materials such as single elements, phosphorene, silicene, germanene, arsenide, and other metals, and compounds such as hexagonal boron nitride, transition metal dichalcogenides, metal oxides, and MXenes gained new interests in the recent past (Miró, Audiffred, & Heine, 2014). Layered metal oxides (LMOs) are an ideal purposeful material due to their abundance, scalable synthesis, low biotoxicity with natural elimination.

LMOs have different oxidation states with tunable electronic and physicochemical properties and broad absorption spectra with fluorescence quenching ability for imaging and sensing (Ren, Wang, & Ou, 2020). The LMOs can be used for drug delivery and cancer therapy due to their large specific surface areas which acts as an excellent platform for high drug loading amounts. Owing to their strong oxidation and catalytic activity they can easily decompose in the presence of reducing substances leading to the release of loaded drugs or can be also used to detect the reducing substance in a biological sample (Chimene et al., 2015; Kalantar-zadeh et al., 2016; Kurapati et al., 2016; Ren et al., 2020). The properties of LMOs such as physiochemical and electronic properties can be easily tuned by ion intercalation or surface functionalization. These surface modifications result in an enhanced stability and biocompatibility in microenvironments. The ultra-doped LMOs show surface plasmon resonance in the visible region and hence can be used for the plasmonic biosensors' construction (Ren et al., 2020). These distinctive characteristics of LMOs have already given rise to lots of exciting applications and the dynamical reconfiguration of these materials paved the way for many exciting future development.

In this chapter, we overview developments of LMOs in the biomedical applications, synthesis, toxicity, and applications in biosystems. We have also touched on the various synthetic routes and the functionalization strategies. Our aim is to

Metal Oxides for Biomedical and Biosensor Applications. DOI: https://doi.org/10.1016/B978-0-12-823033-6.00019-3

provide a comprehensive insight into LMOs in existing and emerging recent advances in biomedical applications and an outlook for future research.

21.2 Structures and polymorphs of layered metal oxides

The metal oxides such as MoO_3, WO_3, MnO_2, and V_2O_5 naturally exist in the layered structure either in hydrate or anhydrate phases (Fig. 21.1) (Kalantar-zadeh et al., 2016; Miró et al., 2014; Ren et al., 2020). LMOs, especially MnO_2 and MoO_3, are widely studied for biomedical applications.

MoO_3 has three different polymorphs depending on the arrangement of MoO_6 octahedra: α-MoO_3, β-MoO_3, and h-MoO_3 (de Castro et al., 2017; Lou & Zeng, 2002). Out of these polymorphs, orthorhombic α-MoO_3 is thermodynamically stable, whereas β-MoO_3 and h-MoO_3 phases are metastable (Lou & Zeng, 2002). α-MoO_3 has a double-layered planar structure in which it has edge-sharing MoO_6 octahedra along the z-direction and is connected through their corners along the x-direction. It results in a double-layered planar structure, and these double-layered are stacked along y-direction bonded together by the weak van der Waal's forces (de Castro et al., 2017). The β-MoO_3 has a structure similar to ReO_3 and has corner-sharing MoO_6 octahedra leading to a monoclinic structure. The hexagonal h-MoO_3 consists of zig−zag chains of MoO_6 octahedra. The two-dimensional (2D) orthorhombic α-MoO_3 phase is thermodynamically more stable compared to other polymorphs (de Castro et al., 2017). MoO_2 exists in three phases: the hexagonal, tetragonal, and monoclinic phase; its monoclinic structure is thermodynamically more stable (Shi et al., 2009).

MnO_2 is another class of material widely studied for its biological applications due to its broad optical absorption, strong oxidation, and catalytic capability. It consists of the corner or edge-shared MoO_6 octahedral leading to the tunnel and layered structures. MnO_2 exists in the five different crystal structural forms: α-MnO_2, β-MnO_2, γ-MnO_2, δ-MnO_2, and λ-MnO_2 (Hayashi et al., 2019; Kijima, Yasuda, Sato, & Yoshimura, 2001). Among which δ-MnO_2 has a layered structure. The 2D single layer of MnO_2 consists of Mn layers sandwiched between two oxygen layers, and Mn is coordinated octahedrally to six oxygen atoms forming edge-sharing

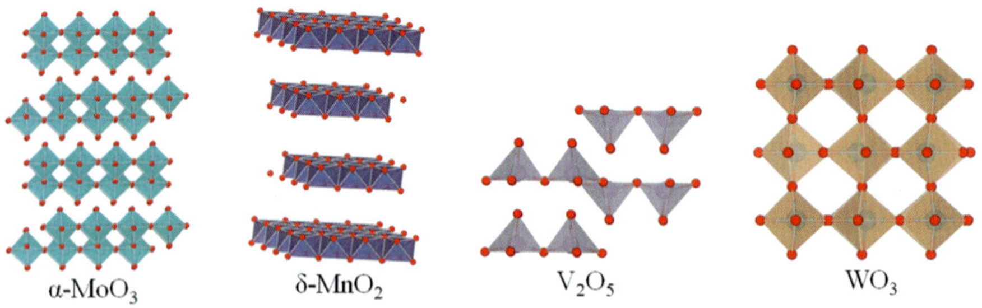

Figure 21.1 Crystal structures of layered metal oxides. *Gray*, *green*, *purple*, *cyan*, and *red* spheres are Mo, Mn, V, W, and O, respectively.

MnO_6 octahedra (Chen, Meng, et al., 2019). V_2O_5 is another important layered transition metal oxide.

LMOs also exist in perovskite structure with the typical chemical formula of ABO_3. The ABO_3 type structure consists of large-sized cations at A site, which occupies the center of the unit cell and is a 12-coordinated cation, and the small-sized cation is at the B site. Their 2D form can be easily synthesized using layer-by-layer deposition technique. The layered structure of ABO_3 comprises of an alternate thin layer with A and B cations of different sizes. The thickness of layers formed by A and B cations and the offsetting of layers with respect to another play a vital role in their characteristic properties. Tungsten trioxide exhibits a cubic perovskite structure and has a corner-sharing WO_6 octahedra. The structure consists of alternate layers of O and WO_2 planes stacked together. It has a tilted WO_6 octahedron, and W is displaced from its center leading to its lower symmetry. The stable form of WO_3 exists in a monoclinic structure with the space group of $P2_1/n$ (Wang, Di Valentin, & Pacchioni, 2011).

21.3 Synthesis

The naturally occurring LMOs such as MoO_3, WO_3, MnO_2, and V_2O_5 (both in hydrate or anhydrate phases) (Kalantar-zadeh et al., 2016; Miró et al., 2014; Ren et al., 2020) can be exfoliated down to nanosheets with O-terminated basal planes using liquid or gas phase techniques (Chimene et al., 2015; Gupta & Rao, 2019; Zavabeti et al., 2020). These metal oxides can be easily reduced to their thinnest most stable configuration, as their bulk form consists of one or two fundamental layers. They are either deposited by layer-by-layer techniques or can be exfoliated from their stable form. In case of nonlayered oxides such as ZnO and TiO_2, the known bulk structures of these materials transform to different 2D arrangements. For example, Wurtzite form gets transformed to planar structure during synthesis in layered ZnO. The LMOs are prepared either by reducing their dimensions from bulk to few layers (top-down) or synthesizing them from their reagents (bottom-up).

21.3.1 Top-down methods

The parent-layered bulk crystals with weak interlayer attractions allowing the facile creation of LMO nanosheets from the bulk or other high-dimensional systems. These synthesis methods rely on the parent-layered bulk crystals such as MoO_3, WoO_3, or V_2O_5 which exist as a binary-layered crystals naturally stacked via weak van der Waals interactions. In mechanical exfoliation, LMOs are thinned down from their stratified crystals using a sticky tape, as first demonstrated by Novoselov for graphene (Novoselov et al., 2004) (Fig. 21.2A). Although the synthesis method is low cost and yields high-quality crystals, it is not viable in biomedical systems due to low yield and intermittent substrates coverage. Liquid-assisted mechanical exfoliation is beneficial in generating large quantities of LMOs nanosheets in the solvent (such as water, NMP, DMSO) using external mechanical processes such as

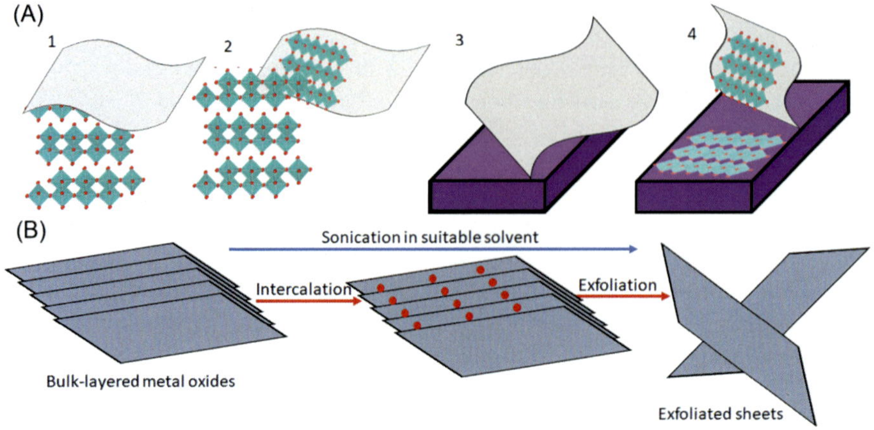

Figure 21.2 (A) Mechanical exfoliation of single crystals of layered materials using sticky tape and transferring it to the substrate. (B) Bulk compounds of layered materials' oxides are exfoliated using ultrasound sonication (bath or probe) in a suitable solvent to obtain layered nanosheets. Intercalation of ions such as Li^+ or H^+ and subsequently exfoliation yield a large number of nanosheets.

grinding, ball-milling, or sonication, which weakens van der Waals bonds between the adjacent layers (Fig. 21.2B) (Balendhran, Walia, et al., 2013; Kalantar-zadeh et al., 2016; Mannix, Kiraly, Hersam, & Guisinger, 2017). It has a relatively high yield, but exfoliation of the layered bulk material using this technique is a time-consuming process (a few days).

Exfoliation can be accelerated by intercalation and subsequent deintercalation by chemical or electrochemical process, which favors electrostatic repulsion of adjacent layers followed by liquid exfoliation. In liquid phase intercalation/exfoliation, a guest species such as H^+ or Li^+ or organic compounds such as polypyrrole, pyridine, and alkylamines, are inserted in the layered materials' gaps, subsequently separating them (Hu, Shin, Read, & Casiraghi, 2021; Nicolosi, Chhowalla, Kanatzidis, Strano, & Coleman, 2013; Shen et al., 2015; Zhang, Mei, Cao, Tang, & Zeng, 2020). Liquid phase intercalation/exfoliation method has been applied to prepare a wide variety of LMO nanosheets such as layered titania, manganese oxides, and tantalum oxides, ruthenium oxides, and their binary and ternary compounds (Kalantar-zadeh et al., 2016; Nicolosi et al., 2013; Ren et al., 2020). Further, some preform of the metal oxides such as layered chlorides, oxy chlorides, and some hydrates, which on rapid thermal expansion, exfoliate the bulk to thin oxides and release a large amount of gas (Zhao, Zhang, Si, & Wu, 2016).

21.3.2 Bottom-up methods

The uniform nanosheets of both layered and non-LMOs are synthesized in a large scale by various bottom-up approaches or chemical synthesis strategies (Zavabeti et al., 2017). Nanomaterials' growth processes assembled from atoms or molecules are facile, efficient, scalable, and versatile. Wet-chemical synthesis strategies include the electrodeposition, hydro-/solvo-thermal method, or sol-gel method. The applied voltages, high temperatures, high pressures, and reducing agents are the driving forces for the growth of LMOs in liquid phase with ionic precursors (Breedon et al., 2010; Yao et al., 2012). The advantage of the synthesis using wet chemical methods is that the products forms are of uniform size, shape, and thickness. Also, the yield is high and hence is suitable for scalable synthesis. But this method involves complex reaction steps, stringent requirements on reaction/solution preparation, expensive precursors, and it's difficult to obtain capping agent-free nanosheets (Fig. 21.3).

In vapor-phase synthesis, PVD and CVD techniques are used in which solid/gaseous precursors are evaporated by heating and deposited on a cool substrate. PVD/CVD techniques are expensive and complicated, but it can fabricate 2D materials of a large area with accurately controlled thickness, excellent quality, and charge mobility (Fig. 21.3A and B). The thermal evaporation, molecular beam epitaxy (MBE), pulsed-laser deposition (PLD), and direct current sputtering (DCS) are the common PVD synthesis techniques (Vila-Fungueiriño et al., 2015; Yang et al., 2019). Thermal evaporation techniques provide precise control over the deposition parameters by controlling the temperature and the gas flow (Vila-Fungueiriño et al., 2015). CVD technique is suitable for fabricating large domain-sized 2D thin-film nanomaterials of superior quality of uniform thickness with tunable orientation.

Space/surface confined synthesis is another important method to synthesize distinct LMOs. In this, 2D-layered materials such as graphene oxide acts as the host matrices and the guest molecules are deposited on it for the fabrication of well-

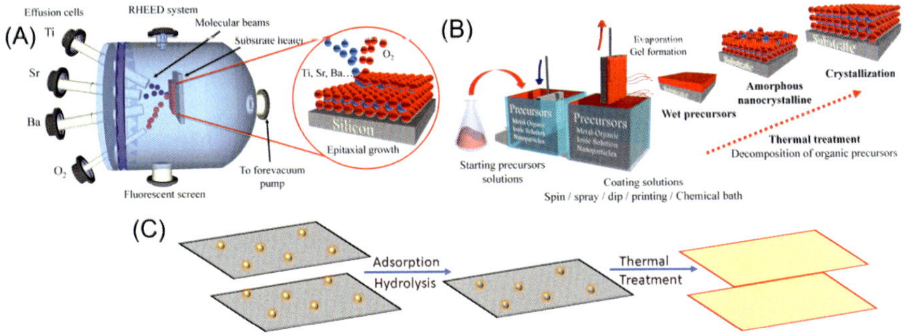

Figure 21.3 (A) Schematic of molecular beam epitaxy system for depositing layered metal oxides on a substrate. (B) Schematic of solution processing routes to deposit layered metal oxides on substrates. (C) Schematic of surface-confined growth of metal oxides using templates.
Source: (B) Adapted with permission from Vila-Fungueiriño, J. M., Bachelet, R., Saint-Girons, G., Gendry, M., Gich, M., Gazquez, J., et al. (2015). Integration of functional complex oxide nanomaterials on silicon. *Frontiers in Physics*, *3*, 38, Copyright 2015 Springer.

defined LMOs (Fig. 21.3C) (Li et al., 2020). The guest molecules are embedded in the interlayers' spaces of layered materials which leads to the confined growth of 2D materials in the spacing between two layers (Jia, Yang, Hao, Li, & Guo, 2020). Some of the examples of 2D metal oxides nanosheets grown by the surface confinement of graphene oxide are MgO, ZrO_2, Al_2O_3, TiO_2, SnO_2, and Sb_2O_5 (Jia et al., 2020; Zhao et al., 2017). The mechanism includes the adsorption of positively charged metal ions on the graphene oxide. The charged metal ions get attached via electrostatic attraction by conjugated pi bands. The next step involves the hydrolysis process which leads to the formation of ultrathin metal oxide coating by the nucleation process and development of the MO precursor on the graphene oxide surface. Finally, GO is removed by the application of thermal treatment and ultrathin LMO nanosheets are obtained. The surface properties, their shapes and crystal structure depend on the synthetic routes which also affect their functionalization behaviors.

21.4 Functionalization of layered metal oxides

LMOs are intrinsically biocompatible but are unstable in the physiological environment and can easily aggregate. However, the dispersity, biostability, and biocompatibility can be improved by surface modifications through the chemical and physical routes (Kalantar-zadeh et al., 2016). Functionalization also enhances their sensitivity as a diagnostic tool and makes it suitable to be used in DNA sensing, drug delivery, or contrast imaging. The tunable electronic structure and the large surface to volume ratio of LMOs make their functionalization feasible. The surface modifications by physical adsorption, oxygen vacancies, doping, and chemical bonding are discussed in the following subsections.

21.4.1 Intrinsic functionalization

The electronic and catalytic properties of LMOs are not only influenced by the cation oxidation state, but also defects and doping significantly affect their properties. Systematically introducing oxygen vacancies or doping LMOs modify the electronic band structure and adjust the Fermi level. Hence, they are essential modification strategies for the surface functionalization of LMOs (Grisolia et al., 2016; Kalantar-zadeh et al., 2016). The stoichiometric MoO_3 is catalytically inactive whereas substoichiometric MoO_3 has a large number of active sites due to the presence of oxygen vacancies and is catalytically active (Haque et al., 2019; Zhang et al., 2018). The oxygen-deficient LMOs instigate the intensive exposure of metal atoms, making them more accessible to biofunctional molecules such as cancer targeting molecules via conjugation reactions on the oxides surfaces. Doping of LMOs with ions or molecules readily tune the carrier types and intensity, which affects their photochemical properties (Xu, Wang, Zhao, & Chai, 2017). LMOs can be doped either by intercalation or surface modification with charged or polarized entities using wet-chemical or electrochemical methods (Stamenkovic et al., 2006). The presence of mechanical flexibility in some of the layered oxides makes the incorporation of size-mismatched dopants feasible (Nilius & Freund, 2015). The free charge carriers of dopants donate electron to the 2D oxide plane. As a result of which the plasmon resonance wavelength of LMOs shift to visible near-infrared (vis-NIR) wavelength and makes it suitable to be used in plasmonic biosensing. For example, the reversible doping of MoO_3 nano disks with H^+ ions modify its plasmon resonance features (Zhang et al., 2018). The introduction of dopants and oxygen vacancies in LMOs lead to a high carrier concentration ranging between $10^{20}–10^{21}$ cm^{-3}. The increased carrier concentration in LMOs shift the plasmonic absorption peaks in the NIR and paved the path for the efficient photothermal therapy (PTT) (Kalantar-zadeh et al., 2016; Yu et al., 2017).

21.4.2 Extrinsic functionalization

Surface functionalization by physisorption is suitable to keep the intrinsic properties intact. This is often desired when the absence/removal of functional groups is a part of the diagnostic/delivery mechanisms (Yang, Chen, & Shi, 2018). The extrinsic functionalization involves the physical adsorption of the desired probes or drug carriers onto the basal planes of the MOs. The relatively large surface area of basal planes of the MOs maximizes the loading and its interaction with the surrounding environment. The physisorption typically involves the noncovalent bonding achieved through electrostatic attraction, hydrophobic interactions, and van der Waals forces (Yang et al., 2018). The ultrathin thickness and the large surface area of LMOs provide large number of active sites where drugs/gene can get attached and hence it makes it a suitable candidate for drug/gene delivery systems.

The diverse guest agents such as upconversion nanoparticles (UCNPs) and single-stranded/double-stranded DNA (ssDNA/dsDNA) can easily be adsorbed on the LMOs nanosheets facilitating the endocytosis of guest agents (Deng, Xie, Vendrell, Chang, & Liu, 2011; Fan, Zhao, et al., 2015). The MO nanosheets with guest agents attached to it makes it suitable to be used in biomedical applications such as MOs disintegration or the drugs/genes delivery induced by pH, H_2O_2, or GSH in the physiological environment (Kalantar-zadeh et al., 2016; Ren et al., 2020; Solanki, Kaushik, Agrawal, & Malhotra, 2011; Tartaj, Morales, Gonzalez-Carreño, Veintemillas-Verdaguer, & Serna, 2011). The smart nanosystem can be also used for fluorescence imaging, gene silencing, and photodynamic therapy (PDT). MnO_2 nanosheets act as the nanocarrier for the DNAzyme suppressing the endogenous nuclease digestion (Fan et al., 2015; Meng et al., 2015; Yan et al., 2017; Zhao et al., 2014). The intracellular GSH reduces the MnO_2 nanosheets, Mn^{2+} ions are efficient cofactors of DNAzyme for gene silencing, providing activatable magnetic resonance and recover fluorescence signals simultaneously for monitoring the efficacy of delivery (Fan, Zhao, et al., 2015).

To facilitate the biological applications to a greater extent both methodologies chemical and physical surface engineering are employed. For example, glucometers are constructed by mixing the glucose oxidase (GO_x) and MnO_2 nanosheets. The GO_x molecules can easily entangle on the surface of MnO_2 nanosheets via hydrophobic effect and hydrogen bond interactions (Chen et al., 2017). The MnO_2 nanosheets functionalized with (3-aminopropyl) trimethoxy silane (APTMS) or NH_2-PEG2000−COOH (PEG) conjugate with the tumor-targeting group folic acid (FA) (Hao et al., 2015).

The organic molecules on the surface of LMOs form covalent or coordinate bonds with the surface atoms and modify the surface (Yang et al., 2018). It is of utmost important that the functional groups should get strongly attached to the LMO surface to tune the electronic/physical properties of the LMOs and remains unaffected by the physiological environment. The modifications of the physical/ electronic properties of LMOs depend upon the functional materials and reaction takes place via the specific chemical routes. The covalent or noncovalent bonds of polymers or small organic molecules with the surface atoms of LMOs improve the physiological stability and make it suitable for the biosensing applications and therapeutic purposes (Ren et al., 2020; Yu et al., 2017).

21.5 Toxicity

LMO nanosheets can easily accumulate in tumors for longer lifetime with high concentrations compared to the small molecules. This makes them suitable for cancer imaging and therapy (Chimene et al., 2015; Kurapati et al., 2016; Ren et al., 2020). Its rapid clearance from the body provides them with upper hand over large-sized inorganic nanoparticles. MoO_x−PEG nanosheets demonstrate the dramatic reduction of relative cell viabilities of murine breast cancer 4T1 cells under certain concentrations, signifying dose limits to avoid cell damage (Fig. 21.4A). They are not retained in the body organs for long time and can be easily excreted. This minimizes the tissue exposure to nanomaterial avoiding potential long-term toxicity as those in other inorganic nanoparticles (Fig. 21.4B and C) (Song et al., 2016). LMOs generally possess low toxicity. Still, some oxides are slightly genotoxic (V_2O_5) (Wang et al., 2016). However, the investigations on LMOs concerning the biological characteristics are somewhat limited. Initial studies suggest that these materials' toxicity is supposed to be influenced by the visible crystal facet at the edges, defect density, and chemical compositions (Kalantar-zadeh et al., 2016).

Layered MoO_3 platelets are toxic to bacteria. The in vivo toxicity study of TiO_2 nanosheets with exposure time of 30 days reveal that the accumulation of large-sized TiO_2 nanosheets in the liver and spleen leads to the slight abnormality of the liver (Song et al., 2014). There are reports on in vitro cell viability assays for assessing the toxicity of LMOs, including MoO_3 and MnO_2. It has been shown that by the caspase pathway activation, MoO_3 sheets induce toxicity in breast cancer cell line MCF-7, but not in HaCAT cells, which makes it suitable to be used for the treatment of breast cancer (Anh Tran, Krishnamoorthy, Song, Cho, & Kim, 2014). For layered MnO_2, a significant decrease in cell viability is reported in the breast cancer cell line MCF-7, suggesting that LMO does not need to be limited to avoid cell damage (Park et al., 2011).

Functionalized LMOs, such as MoO_x-PEG and MnO_2-PEG nanosheets disintegrate into the metal ions under physiological pH. The kidneys can easily metabolize

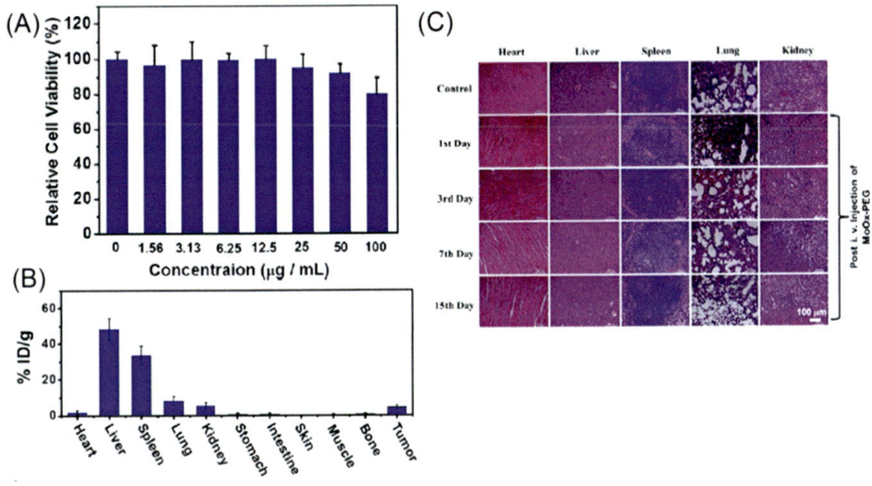

Figure 21.4 (A) Relative viabilities of 4T1 cells incubated with PEG-functionalized MoO_x with different concentrations. (B) Distribution of Mo in major organs in female Balb/c mice after 6 h injection of MoOx–PEG. (C) H&E-stained tissue sections of various organs in mice after intravenous injection of 20 mg/kg of MoOx–PEG for 15 days.
Source: Adapted with permission from Song, G., Hao, J., Liang, C., Liu, T., Gao, M., Cheng, L., et al. (2016). Degradable molybdenum oxide nanosheets with rapid clearance and efficient tumor homing capabilities as a therapeutic nanoplatform. *Angewandte Chemie International Edition*, *55*, 2122–2126, Copyright 2016 Wiley-VCH.

the metal ions without apparent in vivo toxicity (Chen et al., 2014; Song et al., 2016). However, several reports on in vitro relative cell viability reveal toxicity of LMOs, including MoO_3 and MnO_2. The cytotoxicity of MoO_3 nanoplates, induce apoptosis in breast cancer iMCF-7 cells, which makes MoO_3 nanoplates a suitable candidate for cancer cell treatment (Anh Tran et al., 2014). In case of ultra-thin MnO_2 nanoplates, their doses should be controlled to circumvent cytotoxicity toward the breast cancer cell line iMCF-7 (Park et al., 2011). The investigation of the cytotoxicity of MoO_3 nanosheets toward human embroynic kidney 293 T live cells reveals the applicability of B80% at high concentration with an exposure time of up to 48 h. However, we do not have sufficient information of the biocompatibility of MoO_3 nanosheets with long exposure time and higher concentration (Dhenadhayalan, Yadav, Sriram, Lee, & Lin, 2017).

21.6 Applications

21.6.1 Sensors

21.6.1.1 Electricity-based sensors

In general, LMOs have high isoelectric points (IEP), enabling charge transfer to biomolecules, and can possibly be used for restraining a series of biomolecules with relatively lower IEPs via electrostatic exchanges (Solanki et al., 2011). LMOs such as WO_3, CuO, TiO_2, and ZnO show considerably low detection limits for various biomolecules such as H_2O_2, nitrate, glucose, ascorbic acid, dopamine, and cortisol have been investigated for electrodes for electrochemical based sensing (Boroujerdi, Abdelkader, & Paul, 2020; Chimene et al., 2015; Kurapati et al., 2016; Li et al., 2012; Wen et al., 2018). Electrochemical sensing techniques, such as voltammetric/amperometric, electrochemical luminescence, chemiresistor/transistor-based, and photoelectrochemical sensors, have been explored on biomolecule detection in terms of the electrochemical alteration of electrode interfaces (Zhu, Yang, Li, Du, & Lin, 2015).

21.6.1.1.1 Voltammetry and amperometry-based biosensors

Metal oxides have large band gaps and are insulating which inhibit their catalytic activity and are used as electrochemical sensing materials. The most common electrochemical methods used for the detection of biomolecule are square wave, linear sweep, cyclic voltammetry, and differential pulse (Zhu et al., 2015). The electrocatalytic activity of LMOs can be improved by defect formation, doping and its hybridization with the functional materials (Luo et al., 2012; Zhang, Guo, & Li, 2014). The hybridization of functional materials with the surface atoms of MOs leads to the synergistic effect improving the biocompatibility, stability, electrocatalytic activity and conductivity by signal transduction and amplifying the biorecognition. Precisely designed identifiers produce highly sensitive biosensors while inducing vacancies/defects, and doping has various advantages in constructing electrochemical sensors. This is mainly because of the enhanced electrical conductivity via reduced conduction activation energy in various LMOs (Ren et al., 2018). Layered MnO_2 nanosheets have small limit of detection of 5 nM and a wide linear range and high sensitivity toward H_2O_2 (Fig. 21.5A and B) (Shu et al., 2017).

21.6.1.1.2 Chemiresistor/transistor-based biosensors

Owing to their large transconductance, LMOs are widely explored as a high carrier mobility doped regions in electronic field-effect transistors (FETs) to detect biomolecules (Balendhran, Walia, et al., 2013; Mu, Droujinine, Rajan, Sawtelle, & Reed, 2014). The band gaps of LMOs can be easily tuned by ion intercalation, doping, vacancy formation, and combining with metallic materials (Balendhran et al., 2013; Crowley, Ye, He, Abbasi, & Gao, 2018). The negatively charged BSA molecules are immobilized noncovalently on the nanoflakes of α-MoO_3 within 10 s through van der Waals forces and electrostatic interactions. This generates an n-type FET reducing the channel conductance and resulting in a detectable resistance change (Fig. 21.5C and D).

21.6.1.1.3 Electrochemiluminescence biosensors

LMO nanosheets have distinctive electrochemical and redox properties which make it suitable for its use as mediators for electrochemiluminescence (ECL) luminophores. The metal cations Mn^{2+} have been shown as an excellent quencher of the

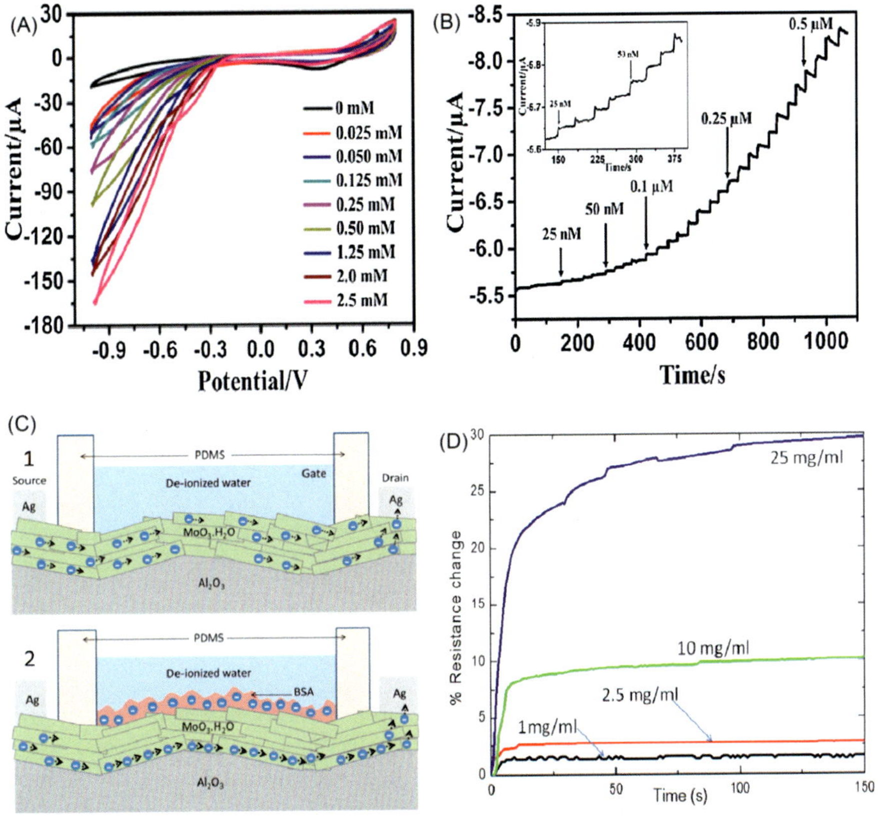

Figure 21.5 (A) The cyclic voltammetry (CV) of the glassy carbon electrode and MnO_2 nanosheet/GCE in N_2-saturated 0.1 M phosphate buffer solution (PBS) under various concentrations at a scan rate of 100 mV/s. (B) Amperometry responses of the MnO_2 nanosheet/GCE in N_2-saturated 0.1 M PBS at −0.6 V with different concentrations of H_2O_2. (C) Schematic of a 2D-MoO_3 field-effect transistor (FET)-based biosensor (1) without bovine serum albumin (BSA) and (2) with BSA. (D) The sensing performance of 2D-MoO_3 FET-based sensor.

Source: (B)Reprinted with permission from Shu, Y., Xu, J., Chen, J., Xu, Q., Xiao, X., Jin, D., et al. (2017). Ultrasensitive electrochemical detection of H_2O_2 in living cells based on ultrathin MnO_2 nanosheets. *Sensors and Actuators B: Chemical*, *252*, 72–78, Copyright 2017 Elsevier B.V.; (D) Reprinted with permission from Balendhran, S., Walia, S., Alsaif, M., Nguyen, E. P., Ou, J. Z., Zhuiykov, S., et al. (2013). Field effect biosensing platform based on 2D α-MoO3. ACS Nano, 7, 9753–9760, Copyright 2013 American Chemical Society.

ECL of luminophore lucigenin (He, Chai, Wang, Bai, & Yuan, 2014). The decomposition of MnO_2 nanosheets by GSH releases Mn^{2+} which drastically suppresses the ECL of lucigenin.

21.6.1.1.4 Photoelectrochemical biosensors

The bandgap tunability and exceptional photoelectrochemical (PEC) properties of LMOs makes it a suitable candidate for PEC biosensing applications. Photoelectrochemical sensing involves the combination of the advantage of optical and electrochemical methods and shows great potential in analytical applications. Owing to the strong *d-d* transition and high contact resistance, MnO_2 nanosheets show relatively small photocurrent (Sakai, Ebina, Takada, & Sasaki, 2005). Highly sensitive PEC biosensor based on MnO_2-CQDs hybrids has been demonstrated to be used for aflatoxin B1 (AFB1) detection (Lin, Zhou, Tang, Niessner, & Knopp, 2017).

21.6.2 Optical-based biosensors

LMOs show strong light absorption due to strong *d-d* transition. Due to Förster resonance energy transfer (FRET), LMOs can easily quench the luminescence of fluorophores once conjugated (Wang, Jiang, Zhu, Zhang, & Lin, 2015). When LMOs are exposed to an analyte, the metal oxide structure is changed, or the binding interaction between LMOs and the target luminescence material is lowered. Either mechanism results in the luminescence restoration.

21.6.2.1 Fluorescence assays

LMO nanosheets show excellent light absorption capacity and have fast electron transfer rates as compared to other layered materials (excluding graphene) which makes them an ideal candidate as fluorescence quenchers (Wang et al., 2015). LMO nanosheets exhibit broad absorption bands and large molar extinction coefficients in comparison to graphene and its derivatives. For example, MnO_2 nanosheets exhibit a broad absorption band from ~ 200 nm to 800 nm and has a molar extinction coefficient (ε) of $\varepsilon_{max} = 9.6 \times 10^3\ M^{-1}\ cm^{-1}$ at ~ 380 nm, whereas graphene and its derivative has a molar coefficient of $\varepsilon_{max} = \sim 10-10^2\ M^{-1}\ cm^{-1}$ at 660 nm (Lin, Cheng, Ouyang, & Wei, 2016; Tian et al., 2014). Also, it has been reported that the electron combination rate with donor in some MOs are faster than the commonly used graphene oxide. The luminescence properties of LMOs can be easily tuned by changing the concentration of oxygen vacancies and the oxidation states which makes them a suitable candidate for the fluorescence bioassays construction (Kumar, Babu, Karakoti, Schulte, & Seal, 2009).

LMOs have been widely used as acceptors to construct biosensing platforms via FRET (Chen, Meng, et al., 2019; Chimene et al., 2015; Deng et al., 2011; Kumar et al., 2009; Kurapati et al., 2016; Wang et al., 2015). Fluorescent donors and LMOs as a donor–acceptor pair are employed for the biomolecules detection in the FRET-based probes. After the addition of targeted biomolecules, there is a decrease in the binding affinity of donor–acceptor pair or MOs decomposition leads to the fluorescence recovery (Fig. 21.6A). MnO_2 nanosheets are the most used quencher for upconverted luminescence among other known layered MOs. The

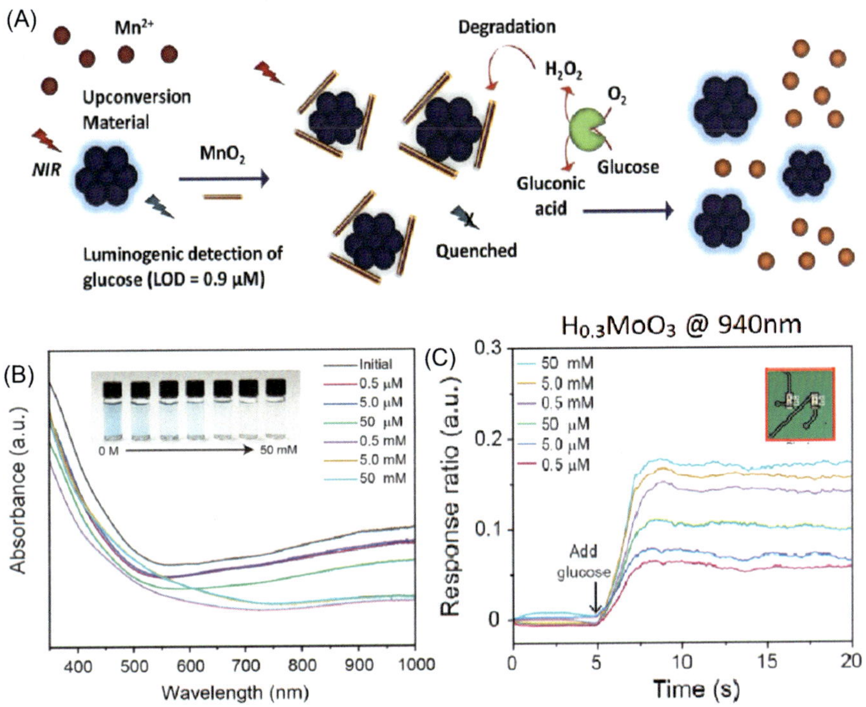

Figure 21.6 (A) Schematic of the operation principle of a representative layered MnO_2 upconversion particle FRET-based optical biosensor in the presence of glucose oxidase and glucose (He & Tian, 2016). (B) The UV-vis-NIR absorption spectra of $H_{0.3}MoO_3$ with varying concentrations of glucose in the presence of GO_x. The insets show corresponding optical images of the spectrum. (C) The real-time response of $H_{0.3}MoO_3{-}GO_x$ at 940 nm under varying concentrations of glucose using a sensitive LED-photodetector setup.
Source: (A) Reprinted with permission from He, X.-P., & Tian, H. (2016). Photoluminescence architectures for disease diagnosis: from graphene to thin-layer transition metal dichalcogenides and oxides. *Small (Weinheim an der Bergstrasse, Germany)*, *12*, 144–160, Copyright 2013 American Chemical Society; (B) Adapted with permission from Zhang, B. Y., Zavabeti, A., Chrimes, A. F., Haque, F., O'Dell, L. A., Khan, H., et al. (2018). (C) Degenerately hydrogen doped molybdenum oxide nanodisks for ultrasensitive plasmonic biosensing. *Advanced Functional Materials*, 28, 1706006, Copyright 2018 Wiley VCH.

MnO_2-induced extinction effect has been confirmed with several fluorescent labels, including organic fluorophores, carbon/semiconductor quantum dots, UPNPs, metal nanoclusters for recognition of GSH, glucose, H_2O_2, microRNAs, ascorbic acid (AA), and amantadine (Chen, Meng, et al., 2019; Ren et al., 2020; Solanki et al., 2011). MoO_3 nanosheets are used for the prostate-specific antigen (PSA) detection by acting as a sending platform (Dhenadhayalan et al., 2017). The adsorption of a dye-labeled ssDNA on MoO_3 nanosheets leads to the fluorescence quenching of the aptamer. The system exhibits fluorescence turn-on in the corresponding complementary DNA sequence due to dsDNA formation which prevents the FRET process (Dhenadhayalan et al., 2017).

The construction of glucose sensor is based on H_xMoO_3 with GO_x (Fig. 21.6B) (Zhang et al., 2018). It has been reported that the biomimic oxidase catalytic reaction between LMOs, such as MnO_2 and V_2O_5 nanosheets and a chromogenic substrate is hindered by the addition of targeted biomolecules which result in the significant color change of the coloring agents (Chen, Zhong, et al., 2019; Ganganboina & Doong, 2018; Yan et al., 2017). The decomposition of LMOs is triggered by the addition of an analyte which causes the change of solution absorbance directly (Ganganboina & Doong, 2018; Yan et al., 2017).

The electronic properties of MOs can be easily altered by the concurrent injection of electrons leading to an increase in the concentration of free electrons in ultra-doped MOs, with the ions intercalation (Fig. 21.6C) (Alsaif et al., 2016; Kalantar-zadeh et al., 2016). As a result of which MOs have high dielectric constant and shows the surface plasmon resonance in the NIR and visible regions (Lee, Nishi, & Tatsuma, 2017). Plasmonic biosensors based on ultra-doped MoO_3 have been shown useful for the detection of biomolecules such as BSA, enzymatic glucose, and H_2O_2 (Alsaif et al., 2014).

21.6.3 Bioimaging

21.6.3.1 Magnetic resonance imaging

Magnetic layered oxides such as iron oxides are broadly applied as magnetic resonance imaging contrast agents. Depending on lateral dimensions, it can be used as negative or positive contrast agents. The iron oxide with lateral dimensions larger than 10 nm improves the magnetic resonance imaging (MRI) performance significantly (Macher et al., 2015). Paramagnetic LMOs, such as MnO_2, exhibit low efficiency for MRI. The Mn ions in MnO_2 crystals are protected from water and hence they have no contribution to the spin−lattice or spin−spin relaxation of protons (Hao et al., 2015; Zhao et al., 2014). Layered MnO_2 undergo reductive dissolution and is reduced to Mn^{2+} initiating the activation of the T1 MRI in the presence of intracellular bioreductants (e.g., glutathione) after endocytosis (Hao et al., 2015; Meng et al., 2015; Zhao et al., 2014). However, paramagnetic Mn^{2+} ions in layered MnO exhibit high T1-weighted MRI contrast agents, permitting an efficient interaction with water molecules (Park et al., 2011).

21.6.3.2 Upconversion luminescence imaging

LMO nanosheets have exceptional photophysical and chemical stability, tunable emission, high fluorescence, and low biotoxicity with natural elimination and hence may be considered suitable for the upconversion luminescence (UCL) imaging (Lin et al., 2016; Tartaj et al., 2011). For example, MnO_2 nanosheets with attached upconversion nanoprobes (UCSMs) have been constructed (UCL imaging platforms). The MnO_2 nanosheets decompose into Mn^{2+} ions in the presence of endogenous H_2O_2 in solid tumors recovering the quenched UCL of UCSMs which provides the imaging guidance for the precise position in cancer theragnostic (Fan, Bu, et al., 2015).

21.6.4 Therapeutics

21.6.4.1 Therapeutics and diagnostics

Planar structures act as carriers for controlled drug delivery to targeted cells due to the large surface area for molecular binding. It is found that the drug and the carrier of polyethylene weight ratios are 90% in MnO_2 which is comparable to the other known layered materials (Hao et al., 2015). Multilayered MnO_2 nanosheets degrade in the low pH environment of cancer cells ensuring the apoptosis of target cancer cells induced by drug after nanosheets are endocytosed (drug delivery) (Chen et al., 2014; Meng et al., 2015). MnO_2 also function as a protective material for complexation with a DNAzyme (a therapeutic gene silencing agent) counter to enzymatic digestion proteins ensuing smooth delivery of the agent to target cells (Fan, Zhao, et al., 2015). Layered substoichiometric MoO_x is another example of LMOs which can be used for the therapeutic application. MoO_x possess the large number of adsorption sites and have excellent drug loading capacity, 230% for SN38 and 80% for Ce6 (Fig. 21.7) (Song et al., 2016). Based on its substoichiometry and the free charge carrier concentrations, MoO_x exhibit plasmonic absorption in the NIR region. This attribute makes MoO_x an ideal candidate for the efficient PTT when functionalized with PEG. As most LMOs have low absorbance in the NIR, the investigation of LMO-based PTT is still very limited.

21.6.4.2 Drug and gene delivery

The large surface area to volume ratio of LMOs make them suitable for drug/gene delivery, with loaded molecules to prevent gene expression and kill cancer cells (Chen et al., 2014; Wang et al., 2019). For example, MnO_2 nanosheets have a large number of binding sites for anticancer agents. The MnO_2 nanosheets decompose in the presence of the mild acidic microenvironments of tumor tissues reducing the harmful side effects to the normal cells (Chen et al., 2014). The loading efficiency of therapeutic molecules is quite high in case of MoO_x-PEG which makes it suitable for the use in the drug delivery and cancer therapy applications. The degradation of MoO_x nanosheets functionalized with PEG is dependent on the pH of microenvironments which enables their rapid excretion (Song et al., 2016). GSH-responsive layered-MnO_2 can be used for the DNAzyme transportation. This degrades EGR-1 mRNA which leads to gene silencing and the growth of cell is suppressed (Fig. 21.7) (Wang et al., 2019).

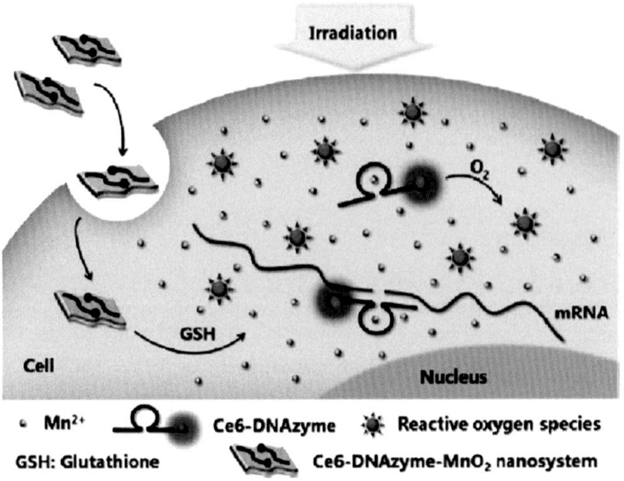

Figure 21.7 Schematic of the Ce6–DNAzyme–MnO_2 nanosystem mechanism for gene silencing and photodynamic therapy (PDT).
Source: Reprinted with permission from Song, G., Hao, J., Liang, C., Liu, T., Gao, M., Cheng, L., et al. (2016). Degradable molybdenum oxide nanosheets with rapid clearance and efficient tumor homing capabilities as a therapeutic nanoplatform. *Angewandte Chemie International Edition*, *55*, 2122–2126, Copyright 2016 Wiley-VCH.

21.6.4.3 Cancer treatment strategies

Noninvasive oxygen-dependent strategies such as radiation therapy (RT) and PDT are used for the cancer treatment which involve the targeting of tumor cells precisely by the radiation/light ionization to kill the tumor cells efficiently (Fan, Bu, et al., 2015). But solid tumor cells reveal hypoxic nature, which eventually leads to the strong resistance to the radiation/light ionization and ultimately leads to the cancer RT/PDT's failure (Wang, Huang, et al., 2018). The RT/PDT efficacy can be enhanced by directly producing the oxygen in solid tumors which will improve the oxygenation and in turn will help in overcoming the hypoxia (Fan, Bu, et al., 2015). It is a well-known fact that the lethal cancerous cells generate H_2O_2 in a large amount and hence tumor cells have an H_2O_2-rich microenvironment. MnO_2 nanosheets are reactive to the endogenous H_2O_2 in the tumor microenvironment and generate oxygen sustainably (Fig. 21.8A–C) (Gong et al., 2018; Zhu et al., 2016). Hence, for the promotion of oxygen supply inside tumors, MnO_2 nanosheets have gained tremendous attention in the recent years. PTT is another important method which employs photothermal conversion to slow down the growth of the cancer cells (Chen, Meng, et al., 2019). MnO_2 nanosheets have exceptional biocompatibility and have high efficient photothermal conversion capacity which makes them promising photothermal agents for highly efficient PTT against tumors (Wang, Song, Zhu, Zhang, & Liu, 2018). Also, under laser irradiation, the temperature of MnO_2 nanosheet solution is almost 12 times higher than the pure water (Liu et al., 2018). Both in vitro and in vivo PTT analyses show the high photothermal-conversion capability and efficient tumor eradication performance of MnO_2 nanosheets.

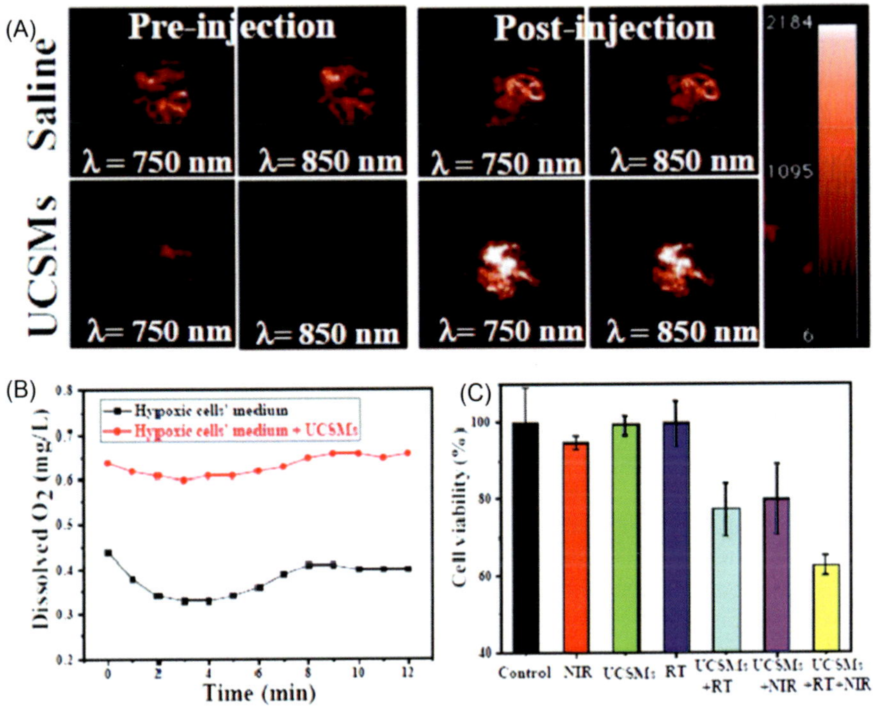

Figure 21.8 (A) Planar photoacoustic (PA) images within 4T1 solid tumors assess the saturated vascular O_2 by measuring deoxygenated and oxygenated hemoglobin at 750 and 850 nm, respectively, with/without injection of upconversion nanoprobes (UCSMs). Without the injection of UCSMs (injection of saline as a control), the signal intensity of both oxygenated and deoxygenated hemoglobin is much slighter than with UCSMs. (B) In vitro evaluation of generated O_2 in hc-4T1 cells incubated with UCSMs. (C) In vitro evaluation of the synergetic photodynamic therapy (PDT)/radiation therapy (RT) effects in hc-4T1 cells incubated with UCSMs upon NIR/X-ray irradiation.
Source: Reprinted with permission from Fan, W., Bu, W., Shen, B., He, Q., Cui, Z., Liu, Y., et al. (2015). Intelligent MnO2 nanosheets anchored with upconversion nanoprobes for concurrent pH-/H2O2-responsive UCL imaging and oxygen-elevated synergetic therapy. *Advanced Materials*, *27*, 4155–4161, Copyright 2015 Wiley-VCH.

However, the pristine LMO nanosheets exhibit the low extinction coefficients and inefficient photothermal-conversion in NIR region which suppress the PTT. The photothermal-conversion efficiency of MO nanosheets can be improved by creating vacancies or doping with another components (Yu et al., 2017).

21.7 Conclusion and outlook

The motivating properties and diverse applications of the LMOs have enthused broad exploration in biosystems. Manganese, molybdenum, and other layered oxides have shown impressive results in biochemical applications such as bioimaging, biosensors, drug/gene delivery, and cancer therapy. The investigations are still in the early stage as compared to other 2D materials such as graphene and its derivatives, and transition metal dichalcogenides. With significant advancement and possibilities in biomedical applications that LMOs in the future may offer, numerous unsolved critical topics and research perspectives need to be elucidated for biological/clinical translation.

The small lateral size and substantial stability with the particular concentration of LMOs are the prerequisite requirements for their transportation within the vasculature and effective penetration and accumulation in the tumors for the final therapeutic performances. LMOs have a promising future in the biomedical applications but we have still insufficient data of their biocompatibility both in the pristine and functionalized form. This calls out for the detailed evaluation of biocompatibility of LMOs which includes hemocompatibility, histocompatibility, acute/chronic toxicity, and metabolic routes. Also, further developments in the surface functionalization of LMOs are also needed to achieve excellent biostability and biocompatibility. The conventional functional groups or molecules employed for the functionalization of LMOs are PEG, lauric acid, and SDS. Other functional groups or molecules should be explored for the functionalization to improve their stability, biocompatibility, selectivity, and rapid elimination.

The functionalized LMO-based biosensing applications still require improvements in the sensitivities and specificities. One of the ways to improve the sensitivity of MOs for the target biomolecules detection is by building three-dimensional (3D) hierarchical structures from 2D basic MO building blocks. This has an enhanced active surface area as compared to the randomly organized 2D nanostructures, which eventually improve the sensitivity toward the target biomolecules. The functional groups/molecules employed for the surface functionalization regulate the charge transfer between the host MOs and target biomolecules, which plays a significant role to the specificity in the sensing performance. Limited number of functional groups have been explored for the functionalization of LMOs, we need to explore other functional groups to bring out the best properties in LMOs for their use in biomedical applications.

The plasmonic properties of 2D materials can be easily tuned by changing the free carrier density which can be easily achieved by changing the ionic intercalation. This has been achieved optically, electrically, and chemically. The 2D materials exhibit the surface plasmon resonance in the visible and NIR regions on sensing platforms and the effective immobilization of target molecules is facilitated by the large surface to volume ratio of 2D materials. The current plasmonic sensors are only based on MoO_x in the LMOs group. The focus of scientific exploration should be also shifted to the non-LMO candidates with 2D ultrathin morphology, for example, hematene, gallium oxides, indium oxides, and bismuth oxides.

References

Alsaif, M. M. Y. A., Field, M. R., Daeneke, T., Chrimes, A. F., Zhang, W., Zhen ou, J., Carey, B. J., ... Ou, J. Z. (2016). Exfoliation solvent dependent plasmon resonances in two-dimensional sub-stoichiometric molybdenum oxide nanoflakes. *ACS Applied Materials & Interfaces*, *8*, 3482–3493.

Alsaif, M. M. Y. A., Latham, K., Field, M. R., Yao, D. D., Medehkar, N. V., Beane, G. A., ... Kalantar-zadeh, K., et al. (2014). Tunable plasmon resonances in two-dimensional molybdenum oxide nanoflakes. *Advanced Materials*, *26*, 3931–3937.

Anh Tran, T., Krishnamoorthy, K., Song, Y. W., Cho, S. K., & Kim, S. J. (2014). Toxicity of nano molybdenum trioxide toward invasive breast cancer cells. *ACS Applied Materials & Interfaces*, *6*, 2980–2986.

Balendhran, S., Deng, J., Ou, J. Z., Walia, S., Scott, J., Kalantar-zadeh, K., Tang, J., et al. (2013). Enhanced charge carrier mobility in two-dimensional high dielectric molybdenum oxide. *Advanced Materials*, *25*, 109–114.

Balendhran, S., Walia, S., Alsaif, M., Nguyen, E. P., Ou, J. Z., Zhuiykov, S., ... Kalantar-zadeh, K. (2013). Field effect biosensing platform based on 2D α-MoO_3. *ACS Nano*, *7*, 9753–9760.

Boroujerdi, R., Abdelkader, A., & Paul, R. (2020). State of the art in alcohol sensing with 2D materials. *Nano-Micro Letters*, *12*, 33.

Breedon, M., Spizzirri, P., Taylor, M., du Plessis, J., McCulloch, D., Zhu, J., ... Kalantar-zadeh, K. (2010). Synthesis of nanostructured tungsten oxide thin films: A simple, controllable, inexpensive, aqueous Sol – Gel method. *Crystal Growth & Design*, *10*, 430–439.

Chen, J.-L., Li, L., Wang, S., Sun, X.-Y., Xiao, L., Ren, J.-S., & Gu, N. (2017). A glucose-activatable trimodal glucometer self-assembled from glucose oxidase and MnO_2 nanosheets for diabetes monitoring. *Journal of Materials Chemistry B.*, *5*, 5336–5344.

Chen, J., Meng, H., Tian, Y., Yang, R., Du, D., Li, Z., ... Lin, Y. (2019). Recent advances in functionalized MnO_2 nanosheets for biosensing and biomedicine applications. *Nanoscale Horizons*, *4*, 321–338.

Chen, P., Zhong, H., Wang, X., Shao, C., Zhi, S., Li, X.-R., & Wei., C. (2019). A label-free colorimetric strategy for facile and low-cost sensing of ascorbic acid using MnO_2 nanosheets. *Analytical Methods*, *11*, 1469–1474.

Chen, Y., Ye, D., Wu, M., Chen, H., Zhang, L., Shi, J., & Wang, L. (2014). Break-up of two-dimensional MnO_2 nanosheets promotes ultrasensitive pH-triggered theranostics of cancer. *Advanced Materials*, *26*, 7019–7026.

Chimene, D., Alge, D. L., & Gaharwar, A. K. (2015). Two-dimensional nanomaterials for biomedical applications: Emerging trends and future prospects. *Advanced Materials*, *27*, 7261–7284.

Crowley, K., Ye, G., He, R., Abbasi, K., & Gao, X. P. A. (2018). α-MoO_3 as a conductive 2D oxide: Tunable n-type electrical transport via oxygen vacancy and fluorine doping. *ACS Applied Nano Materials*, *1*, 6407–6413.

de Castro, I. A., Datta, R. S., Ou, J. Z., Castellanos-Gomez, A., Sriram, S., Daeneke, T., & Kalantar-zadeh, K. (2017). Molybdenum oxides – From fundamentals to functionality. *Advanced Materials*, *29*, 1701619.

Deng, R., Xie, X., Vendrell, M., Chang, Y.-T., & Liu, X. (2011). Intracellular glutathione detection using MnO_2-nanosheet-modified upconversion nanoparticles. *Journal of the American Chemical Society*, *133*, 20168–20171.

Dhenadhayalan, N., Yadav, K., Sriram, M. I., Lee, H.-L., & Lin, K.-C. (2017). Ultra-sensitive DNA sensing of a prostate-specific antigen based on 2D nanosheets in live cells. *Nanoscale*, *9*, 12087–12095.

Fan, H., Zhao, Z., Yan, G., Zhang, X., Yang, C., Meng, H., ... Tan, W. (2015). A smart DNAzyme–MnO_2 nanosystem for efficient gene silencing. *Angewandte Chemie International Edition*, *54*, 4801–4805.

Fan, W., Bu, W., Shen, B., He, Q., Cui, Z., Liu, Y., ... Shi, J. (2015). Intelligent MnO_2 nanosheets anchored with upconversion nanoprobes for concurrent pH-/H_2O_2-responsive UCL imaging and oxygen-elevated synergetic therapy. *Advanced Materials*, *27*, 4155–4161.

Ganganboina, A. B., & Doong, R.-A. (2018). The biomimic oxidase activity of layered V_2O_5 nanozyme for rapid and sensitive nanomolar detection of glutathione. *Sensors and Actuators B: Chemical*, *273*, 1179–1186.

Gong, F., Chen, J., Han, X., Zhao, J., Wang, M., Feng, L., ... Cheng, L. (2018). Core–shell TaOx@MnO_2 nanoparticles as a nano-radiosensitizer for effective cancer radiotherapy. *Journal of Materials Chemistry B*, *6*, 2250–2257.

Grisolia, M. N., Varignon, J., Sanchez-Santolino, G., Arora, A., Valencia, S., Varela, M., ... Bibes, M., et al. (2016). Hybridization-controlled charge transfer and induced magnetism at correlated oxide interfaces. *Nature Physics*, *12*, 484–492.

Gupta, U., & Rao, C. N. R. (2019). Graphene and other 2D materials. *Advances in the Chemistry and Physics of Materials: World Scientific*, 27–54.

Hao, Y., Wang, L., Zhang, B., Zhao, H., Niu, M., Hu, Y., & Zhang, Y. (2015). Multifunctional nanosheets based on folic acid modified manganese oxide for tumor-targeting theranostic application. *Nanotechnology*, *27*, 025101.

Haque, F., Zavabeti, A., Zhang, B. Y., Datta, R. S., Yin, Y., Yi, Z., ... Ou, J. Z. (2019). Ordered intracrystalline pores in planar molybdenum oxide for enhanced alkaline hydrogen evolution. *Journal of Materials Chemistry A*, *7*, 257–268.

Hayashi, E., Yamaguchi, Y., Kamata, K., Tsunoda, N., Kumagai, Y., Oba, F., & Hara, M. (2019). Effect of MnO_2 crystal structure on aerobic oxidation of 5-hydroxymethylfurfural to 2,5-furandicarboxylic acid. *Journal of the American Chemical Society*, *141*, 890–900.

He, Y., Chai, Y., Wang, H., Bai, L., & Yuan, R. (2014). A signal-on electrochemiluminescence aptasensor based on the quenching effect of manganese dioxide for sensitive detection of carcinoembryonic antigen. *RSC Advances*, *4*, 56756–56761.

He, X.-P., & Tian, H. (2016). Photoluminescence architectures for disease diagnosis: From graphene to thin-layer transition metal dichalcogenides and oxides. *Small (Weinheim an der Bergstrasse, Germany)*, *12*, 144–160.

Hu, C.-X., Shin, Y., Read, O., & Casiraghi, C. (2021). Dispersant-assisted liquid-phase exfoliation of 2D materials beyond graphene. *Nanoscale*, *13*, 460–484.

Jia, B., Yang, J., Hao, R., Li, L., & Guo, L. (2020). Confined synthesis of ultrathin amorphous metal-oxide nanosheets. *ACS Materials Letters*, *2*, 610–615.

Kalantar-zadeh, K., Ou, J. Z., Daeneke, T., Mitchell, A., Sasaki, T., & Fuhrer, M. S. (2016). Two dimensional and layered transition metal oxides. *Applied Materials Today*, *5*, 73–89.

Kijima, N., Yasuda, H., Sato, T., & Yoshimura, Y. (2001). Preparation and characterization of open tunnel oxide α-MnO_2 precipitated by ozone oxidation. *Journal of Solid State Chemistry*, *159*, 94–102.

Kumar, A., Babu, S., Karakoti, A. S., Schulte, A., & Seal, S. (2009). Luminescence properties of europium-doped cerium oxide nanoparticles: Role of vacancy and oxidation states. *Langmuir: The ACS Journal of Surfaces and Colloids*, *25*, 10998–11007.

Kurapati, R., Kostarelos, K., Prato, M., & Bianco, A. (2016). Biomedical uses for 2D materials beyond graphene: Current advances and challenges ahead. *Advanced Materials*, *28*, 6052–6074.

Lee, S. H., Nishi, H., & Tatsuma, T. (2017). Tunable plasmon resonance of molybdenum oxide nanoparticles synthesized in non-aqueous media. *Chemical Communications*, *53*, 12680–12683.

Li, Y., Cao, D., Liu, Y., Liu, R., Yang, F., Yin, J., & Wang, G. (2012). CuO nanosheets grown on cupper foil as the catalyst for H_2O_2 electroreduction in alkaline medium. *International Journal of Hydrogen Energy*, *37*, 13611–13615.

Li, Z., Zhang, X., Cheng, H., Liu, J., Shao, M., Wei, M., … Duan, X. (2020). Confined synthesis of 2D nanostructured materials toward electrocatalysis. *Advanced Energy Materials*, *10*, 1900486.

Lin, S., Cheng, H., Ouyang, Q., & Wei, H. (2016). Deciphering the quenching mechanism of 2D MnO_2 nanosheets towards Au nanocluster fluorescence to design effective glutathione biosensors. *Analytical Methods*, *8*, 3935–3940.

Lin, Y., Zhou, Q., Tang, D., Niessner, R., & Knopp, D. (2017). Signal-on photoelectrochemical immunoassay for aflatoxin B1 based on enzymatic product-etching MnO_2 nanosheets for dissociation of carbon dots. *Analytical Chemistry*, *89*, 5637–5645.

Liu, Z., Zhang, S., Lin, H., Zhao, M., Yao, H., Zhang, L., … Chen, Y., et al. (2018). Theranostic 2D ultrathin MnO_2 nanosheets with fast responsibility to endogenous tumor microenvironment and exogenous NIR irradiation. *Biomaterials*, *155*, 54–63.

Lou, X. W., & Zeng, H. C. (2002). Hydrothermal synthesis of α-MoO_3 nanorods via acidification of ammonium heptamolybdate tetrahydrate. *Chemistry of Materials*, *14*, 4781–4789.

Luo, L., Li, F., Zhu, L., Zhang, Z., Ding, Y., & Deng, D. (2012). Non-enzymatic hydrogen peroxide sensor based on MnO_2-ordered mesoporous carbon composite modified electrode. *Electrochimica Acta*, *77*, 179–183.

Macher, T., Totenhagen, J., Sherwood, J., Qin, Y., Gurler, D., Bolding, M. S., & Bao, Y. (2015). Ultrathin iron oxide nanowhiskers as positive contrast agents for magnetic resonance imaging. *Advanced Functional Materials*, *25*, 490–494.

Mannix, A. J., Kiraly, B., Hersam, M. C., & Guisinger, N. P. (2017). Synthesis and chemistry of elemental 2D materials. *Nature Reviews Chemistry*, *1*, 0014.

Meng, H.-M., Lu, L., Zhao, X.-H., Chen, Z., Zhao, Z., Yang, C., … Tan, W. (2015). Multiple functional nanoprobe for contrast-enhanced bimodal cellular imaging and targeted therapy. *Analytical Chemistry*, *87*, 4448–4454.

Miró, P., Audiffred, M., & Heine, T. (2014). An atlas of two-dimensional materials. *Chemical Society Reviews*, *43*, 6537–6554.

Mu, L., Droujinine, I. A., Rajan, N. K., Sawtelle, S. D., & Reed, M. A. (2014). Direct, rapid, and label-free detection of enzyme–substrate interactions in physiological buffers using CMOS-compatible nanoribbon sensors. *Nano Letters*, *14*, 5315–5322.

Nicolosi, V., Chhowalla, M., Kanatzidis, M. G., Strano, M. S., & Coleman, J. N. (2013). Liquid exfoliation of layered materials. *Science (New York, N.Y.)*, *340*, 1226419.

Nilius, N., & Freund, H.-J. (2015). Activating nonreducible oxides via doping. *Accounts of Chemical Research*, *48*, 1532–1539.

Novoselov, K. S., Geim, A. K., Morozov, S. V., Jiang, D., Zhang, Y., Dubonos, S. V., … Firsov, A. A. (2004). Electric field effect in atomically thin carbon films. *Science (New York, N.Y.)*, *306*, 666.

Park, M., Lee, N., Choi, S. H., An, K., Yu, S.-H., Kim, J. H., … Hyeon, T. (2011). Large-scale synthesis of ultrathin manganese oxide nanoplates and their applications to T1 MRI contrast agents. *Chemistry of Materials*, *23*, 3318–3324.

Ren, B., Sudarsanam, P., Kandjani, A. E., Hillary, B., Amin, M. H., Bhargava, S. K., & Jones, L. A. (2018). Electrochemical detection of As (III) on a manganese oxide-ceria (Mn_2O_3/CeO_2) nanocube modified Au electrode. *Electroanalysis*, *30*, 928–936.

Ren, B., Wang, Y., & Ou, J. Z. (2020). Engineering two-dimensional metal oxides via surface functionalization for biological applications. *Journal of Materials Chemistry B*, *8*, 1108–1127.

Sakai, N., Ebina, Y., Takada, K., & Sasaki, T. (2005). Photocurrent generation from semiconducting manganese oxide nanosheets in response to visible light. *The Journal of Physical Chemistry. B*, *109*, 9651–9655.

Shen, J., He, Y., Wu, J., Gao, C., Keyshar, K., Zhang, X., ... Ajayan, P. M., et al. (2015). Liquid phase exfoliation of two-dimensional materials by directly probing and matching surface tension components. *Nano Letters*, *15*, 5449–5454.

Shi, Y., Guo, B., Corr, S. A., Shi, Q., Hu, Y.-S., Heier, K. R., ... Stucky, G. D. (2009). Ordered mesoporous metallic MoO_2 materials with highly reversible lithium storage capacity. *Nano Letters*, *9*, 4215–4220.

Shu, Y., Xu, J., Chen, J., Xu, Q., Xiao, X., Jin, D., ... Hu, X. (2017). Ultrasensitive electrochemical detection of H_2O_2 in living cells based on ultrathin MnO_2 nanosheets. *Sensors and Actuators B: Chemical*, *252*, 72–78.

Solanki, P. R., Kaushik, A., Agrawal, V. V., & Malhotra, B. D. (2011). Nanostructured metal oxide-based biosensors. *NPG Asia Materials*, *3*, 17–24.

Song, G., Hao, J., Liang, C., Liu, T., Gao, M., Cheng, L., ... Liu, Z. (2016). Degradable molybdenum oxide nanosheets with rapid clearance and efficient tumor homing capabilities as a therapeutic nanoplatform. *Angewandte Chemie International Edition*, *55*, 2122–2126.

Song, S.-S., Xia, B.-Y., Chen, J., Yang, J., Shen, X., Fan, S.-J., ... Zhang, X. D. (2014). Two dimensional TiO_2 nanosheets: In vivo toxicity investigation. *RSC Advances*, *4*, 42598–42603.

Stamenkovic, V., Mun, B. S., Mayrhofer, K. J. J., Ross, P. N., Markovic, N. M., Rossmeisl, J., ... Nørskov, J. K. (2006). Changing the activity of electrocatalysts for oxygen reduction by tuning the surface electronic structure. *Angewandte Chemie International Edition*, *45*, 2897–2901.

Tartaj, P., Morales, M. P., Gonzalez-Carreño, T., Veintemillas-Verdaguer, S., & Serna, C. J. (2011). The iron oxides strike back: From biomedical applications to energy storage devices and photoelectrochemical water splitting. *Advanced Materials*, *23*, 5243–5249.

Tian, X., Sarkar, S., Pekker, A., Moser, M. L., Kalinina, I., Bekyarova, E., & Haddon, R. C. (2014). Optical and electronic properties of thin films and solutions of functionalized forms of graphene and related carbon materials. *Carbon*, *72*, 82–88.

Vila-Fungueiriño, J. M., Bachelet, R., Saint-Girons, G., Gendry, M., Gich, M., Gazquez, J., ... Genevrier, A. G. (2015). Integration of functional complex oxide nanomaterials on silicon. *Frontiers in Physics*, *3*, 38.

Wang, X., Dai, J., Wang, X., Hu, Q., Huang, K., Zhao, Z., ... Xia, F. (2019). MnO_2-DNAzyme-photosensitizer nanocomposite with AIE characteristic for cell imaging and photodynamic-gene therapy. *Talanta*, *202*, 591–599.

Wang, F., Di Valentin, C., & Pacchioni, G. (2011). Electronic and structural properties of WO_3: A systematic hybrid DFT study. *The Journal of Physical Chemistry C*, *115*, 8345–8353.

Wang, Y., Huang, X., Tang, Y., Zou, J., Wang, P., Zhang, Y., ... Dong, X. (2018). A light-induced nitric oxide controllable release nano-platform based on diketopyrrolopyrrole derivatives for pH-responsive photodynamic/photothermal synergistic cancer therapy. *Chemical Science*, *9*, 8103–8109.

Wang, Y., Jiang, K., Zhu, J., Zhang, L., & Lin, H. (2015). A FRET-based carbon dot–MnO_2 nanosheet architecture for glutathione sensing in human whole blood samples. *Chemical Communications*, *51*, 12748–12751.

Wang, Y., Song, Y., Zhu, G., Zhang, D., & Liu, X. (2018). Highly biocompatible BSA-MnO_2 nanoparticles as an efficient near-infrared photothermal agent for cancer therapy. *Chinese Chemical Letters*, *29*, 1685–1688.

Wang, Z., Zhu, W., Qiu, Y., Yi, X., von dem Bussche, A., Kane, A., ... Hurt, R. (2016). Biological and environmental interactions of emerging two-dimensional nanomaterials. *Chemical Society Reviews*, *45*, 1750–1780.

Wen, W., Song, Y., Yan, X., Zhu, C., Du, D., Wang, S., ... Lin, Y. (2018). Recent advances in emerging 2D nanomaterials for biosensing and bioimaging applications. *Materials Today*, *21*, 164–177.

Xu, K., Wang, Y., Zhao, Y., & Chai, Y. (2017). Modulation doping of transition metal dichalcogenide/oxide heterostructures. *Journal of Materials Chemistry C.*, *5*, 376–381.

Yan, X., Song, Y., Wu, X., Zhu, C., Su, X., Du, D., & Lin, Y. (2017). Oxidase-mimicking activity of ultrathin MnO_2 nanosheets in colorimetric assay of acetylcholinesterase activity. *Nanoscale*, *9*, 2317–2323.

Yang, B., Chen, Y., & Shi, J. (2018). Material chemistry of two-dimensional inorganic nanosheets in cancer theranostics. *Chemistry*, *4*, 1284–1313.

Yang, T., Song, T. T., Callsen, M., Zhou, J., Chai, J. W., Feng, Y. P., ... Yang, M. (2019). Atomically thin 2D transition metal oxides: Structural reconstruction, interaction with substrates, and potential applications. *Advanced Materials Interfaces*, *6*, 1801160.

Yao, D. D., Ou, J. Z., Latham, K., Zhuiykov, S., O'Mullane, A. P., & Kalantar-zadeh, K. (2012). Electrodeposited α- and β-phase MoO_3 films and investigation of their gasochromic properties. *Crystal Growth & Design*, *12*, 1865–1870.

Yu, N., Hu, Y., Wang, X., Liu, G., Wang, Z., Liu, Z., ... Chen, Z. (2017). Dynamically tuning near-infrared-induced photothermal performances of TiO_2 nanocrystals by Nb doping for imaging-guided photothermal therapy of tumors. *Nanoscale*, *9*, 9148–9159.

Zavabeti, A., Jannat, A., Zhong, L., Haidry, A. A., Yao, Z., & Ou, J. Z. (2020). Two-dimensional materials in large-areas: Synthesis, properties and applications. *Nano-Micro Letters*, *12*, 66.

Zavabeti, A., Ou, J. Z., Carey, B. J., Syed, N., Orrell-Trigg, R., Mayes, E. L. H., ... Daeneke, T. (2017). A liquid metal reaction environment for the room-temperature synthesis of atomically thin metal oxides. *Science (New York, N.Y.)*, *358*, 332.

Zhang, P., Guo, D., & Li, Q. (2014). Manganese oxide ultrathin nanosheets sensors for non-enzymatic detection of H_2O_2. *Materials Letters*, *125*, 202–205.

Zhang, Q., Mei, L., Cao, X., Tang, Y., & Zeng, Z. (2020). Intercalation and exfoliation chemistries of transition metal dichalcogenides. *Journal of Materials Chemistry A.*, *8*, 15417–15444.

Zhang, B. Y., Zavabeti, A., Chrimes, A. F., Haque, F., O'Dell, L. A., Khan, H., ... Ou, J. Z. (2018). Degenerately hydrogen doped molybdenum oxide nanodisks for ultrasensitive plasmonic biosensing. *Advanced Functional Materials*, *28*, 1706006.

Zhao, Z., Fan, H., Zhou, G., Bai, H., Liang, H., Wang, R., ... Tan, W. (2014). Activatable fluorescence/MRI bimodal platform for tumor cell imaging via MnO_2 nanosheet–aptamer nanoprobe. *Journal of the American Chemical Society*, *136*, 11220–11223.

Zhao, C., Zhang, H., Si, W., & Wu, H. (2016). Mass production of two-dimensional oxides by rapid heating of hydrous chlorides. *Nature Communications*, *7*, 12543.

Zhao, H., Zhu, Y., Li, F., Hao, R., Wang, S., & Guo, L. (2017). A generalized strategy for the synthesis of large-size ultrathin two-dimensional metal oxide nanosheets. *Angewandte Chemie International Edition*, *56*, 8766–8770.

Zhu, C., Yang, G., Li, H., Du, D., & Lin, Y. (2015). Electrochemical sensors and biosensors based on nanomaterials and nanostructures. *Analytical Chemistry*, *87*, 230–249.

Zhu, W., Dong, Z., Fu, T., Liu, J., Chen, Q., Li, Y., ... Liu, Z. Z. (2016). Modulation of hypoxia in solid tumor microenvironment with MnO_2 nanoparticles to enhance photodynamic therapy. *Advanced Functional Materials*, *26*, 5490–5498.

Metal oxide/graphene nanocomposites and their biomedical applications

22

Souravi Bardhan[1], Shubham Roy[1], Mousumi Mitra[2] and Sukhen Das[1]
[1]Department of Physics, Jadavpur University, Kolkata, India, [2]Department of Physics, University of Virginia, Charlottesville, VA, United States

22.1 Introduction

Graphene, a monolayer two-dimensional (2D) allotrope of carbon has emerged as a rising star since its discovery by Novoselov and his team (Novoselov et al., 2004). Novoselov and Geim received Noble Prize in Physics after discovery since during 2004 graphene was found to be the thinnest known material with tremendous scope for application in various fields. Graphene consists of benzene rings of sp^2-hybridized carbon atoms arranged in honeycomb network structure (Li & Kaner, 2008), which exhibit unique physicochemical properties (Brumfiel, 2009). The structural features of graphene impart high specific surface area (Dimiev & Eigler, 2016) and magnificent strength (breaking strength is $\sim$ 42 N/m and the tensile strength or Young modules is 1.0 TPa) (Huang et al., 2011). Since it is stronger than various other metals and has high flexibility, it is used as a substitute for metals in various sectors (Hu, Kulkarni, Choi, & Tsukruk, 2014). The mechanical, chemical, and thermal stability of graphene and its nanocomposites make them suitable for a wide range of applications. Graphene exhibits large theoretical surface area ($\sim$2630 m^2/g) which is nearly 260 times higher than graphite. Such high surface area provides greater efficiency for various catalytic activities (Chen, Wu, Jiang, Wang, & Chen, 2011). By virtue of the antibacterial nature, amphiphilicity, outstanding electrical conductivity, and easy surface functionalization capability (Balandin et al., 2008; Geim, 2009; Ranjbartoreh, Wang, Shen, & Wang, 2011; Wang, Zhang, Wu, & Wei, 2017), graphene is considered as a potential candidate in biological applications. Moreover biocompatible nature, surface-enhanced Raman scattering (SERS), fluorescence quenching ability, and low-energy requirement for electron movement due to $\pi-\pi^*$ transitions (Abbott's, 2007; Geim & Novoselov, 2010; Novoselov et al., 2005) are quite significant for bioimaging, biosensing, drug/gene delivery, photothermal and photodynamic therapies, and other biomedical studies. Based on the application purpose, several forms of graphene are used in biological applications, such as graphene oxide (GO), graphene quantum dots (GQDs), and reduced graphene oxide (rGO) (Muthurasu, Dhandapani, & Ganesh, 2016). In certain cases, graphene nanosheets (Chong et al., 2015), graphene nanocubes (Govindasamy et al., 2019), graphene nanocrystals (Zhang, Yuan, Zhang, Wang, & Liu, 2011), graphene nanotubes (Wu, Pei, & Zeng, 2009) are also used. Extensive studies regarding surface modification of addition of other

Metal Oxides for Biomedical and Biosensor Applications. DOI: https://doi.org/10.1016/B978-0-12-823033-6.00020-X

nanoparticles such as metals (Cu, Zn, Au), nonmetals (polymers, epoxy, chitosan or N_2, B, S) and metal oxides (TiO_2, ZnO) are widely performed (Giovannetti et al., 2008; Liu, Liu, & Zhu, 2011; Rajeshkumar & Veena, 2018; Stengl, Popelková, & Vlácil, 2011; Zheng, Jiao, Ge, Jaroniec, & Qiao, 2013). Graphene nanocomposites or graphene-based nanofillers are also popularizing in promoting specific tissue regeneration, providing binding sites for functionalization with biological molecules or in cell behavior regulation (Tadyszak, Wychowaniec, & Litowczenko, 2018). Graphene in combination with ZnO, Ag, chitosan, poly (lactic acid) (PLA) or poly (N-vinylcarbazole) (PVK) can also act as an antibacterial or sterilizing agent by imparting oxidative stress due to its electron transfer capability (Shi et al., 2016). Furthermore, deposition of various metal oxide nanoparticles such as zinc oxide, titanium dioxide, iron oxide, cobalt oxide, copper oxide, nickel oxide, and manganese dioxide on graphene sheets enhances the photophysical and electrochemical properties (Mohd Yazid et al., 2014). Thus graphene and its nanocomposites can be considered as "magic bullet" for the medical science (Fig. 22.1).

22.2 Synthesis of graphene and its derivatives

Graphene was discovered in 2004 during mechanical exfoliation of graphite to obtain single-layer samples (Novoselov et al., 2004). Since the discovery, vast

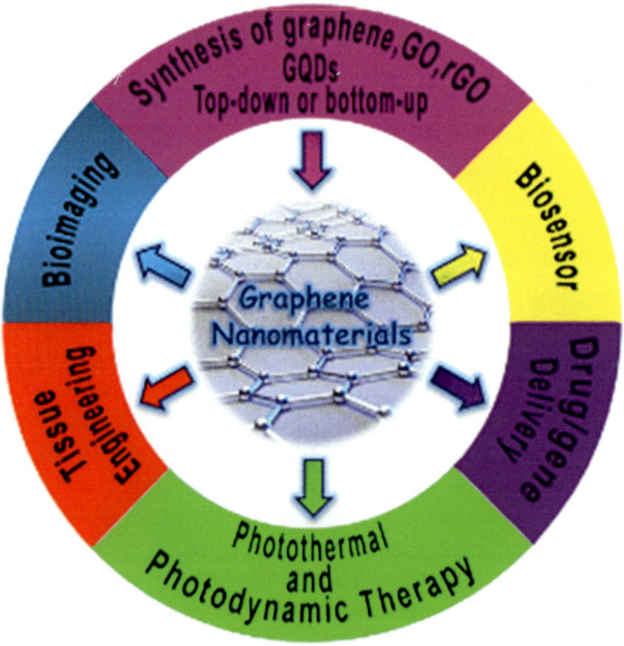

Figure 22.1 Different biomedical applications of graphene-based materials (Ghosal & Sarkar, 2018).

research was conducted because of its unique properties. Initially graphene was synthesized via epitaxial growth, bottom-up organic techniques, or top-down synthesis from graphite by splitting graphite into individual sheets through mechanical cleavage or chemical exfoliation (Li & Kaner, 2008). Graphene synthesis from graphite has gained widespread acceptance due to the cost-effectiveness and ease in synthesis along with large-scale production. Since 2006, graphene synthesis from graphite oxide by ultrasonication method (Stankovich et al., 2006) gained attention due to greater interlayer distance than graphite and hydrophilic nature that makes the exfoliation technique easier. The graphene oxide formed can be deoxygenated by chemical reduction to obtain graphene. A year later McAllister and coworkers (McAllister et al., 2007) introduced a new method to delaminate graphite into graphene sheets via oxidation and rapid thermal expansion. Since graphene sheets have a tendency to agglomerate spontaneously due to presence of strong Van der Waals interactions between the sheets and reform graphite, stabilizing individual sheets during synthesis became an important issue. Thus chemical functionalization and modification of graphene gained attention (Niyogi et al., 2006)(Mitra, Chatterjee, Kargupta, Ganguly, & Banerjee, 2013). This paved the path for graphene nanocomposite synthesis. Initially conversion of GO to graphene resulted in negatively charged surfaces of graphene on dispersion in water. Thus dispersing positively charged polycation into the solution helped in formation of graphene-based composites (Li, Müller, Gilje, Kaner, & Wallace, 2008). Subsequently modification of graphene shape and size also came into play. Graphene nanoribbons (>10 nm wide) were produced from thermally expanded graphite via controlled ultrasonication (Li, Wang, Zhang, Lee, & Dai, 2008). Various facile techniques such as drop-casting, dip or spin coating, or a simple filtration method was practiced to synthesize graphene film (Gilje, Han, Wang, Wang, & Kaner, 2007; Wu, Pisula, & Müllen, 2007). Graphene nanoplatelets were synthesized by covalently grafting polystyrene−polyacrylamide copolymer onto graphene sheets in order to improve graphene stability in both polar and nonpolar solvents (Shen, Hu, Li, Qin, & Ye, 2009). By 2010 several other techniques were adopted to produce graphene, which involved chemical vapor deposition (CVD) (Juang et al., 2010), arc discharge (Li et al., 2010), electrically assisted synthesis (Liu et al., 2008), solvothermal (Choucair, Thordarson, & Stride, 2009) unzipping CNTs and others methods. Furthermore, large quantity graphene production cost was found to be lower than carbon nanotubes, which fascinated various scientists to explore more dimensions of graphene. Thus various avenues in biomedical fields using graphene and its nanocomposites were thoroughly explored.

Metal or metal oxide-graphene/GO/rGO nanocomposites fabrication has gained special interest in the last decade. Metal or metal oxide nanoparticles provide stability by preventing restacking of individual graphene sheets. Their band gap has direct influence on conductivity and chemical reactivity of the nanocomposite, which can be enhanced according to the requirement (Sharma et al., 2014). Various synthesis routes are adopted for metal oxide-graphene-based nanocomposite formation (Sharma et al., 2014) such as hydrothermal/solvothermal method, sol-gel or precipitation technique, electrostatic or covalent interaction, microwave-assisted synthesis, atomic layer deposition, or layer-by-layer self- assembly, etc. Wang et al. (2010) had shown a novel strategy to fabricate flexible, free-standing metal oxide-graphene nanocomposite film by surfactant-based ternary self-assembly approach using low conductive metal oxide such as SnO_2. This method demonstrated a facile manner of enhancing electron transport of metal oxides of low conductivity and poor stability (MnO_2, NiO, SiO_2) by adding conductive phases such as graphene (Fig. 22.2).

The cytotoxicity of graphene-based materials was studied. Russier and team (Russier et al., 2013) conducted a detailed study on cellular functionality using graphene flakes. It was concluded that smaller flakes resulted in greater cytotoxicity due to higher cellular internalization. In another study, rGO was used which shows the relation between level of oxidation and cell viability (Shi et al., 2012). Cell viability using MTT ((3-(4,5-dimethylthiazol-2-yl)-2,5-diphenyltetrazolium bromide) assay or WST-based assay (water soluble tetrazolium) were performed using various cell lines such as human osteosarcoma (MG-63), human colon cancer (Caco2), human lung

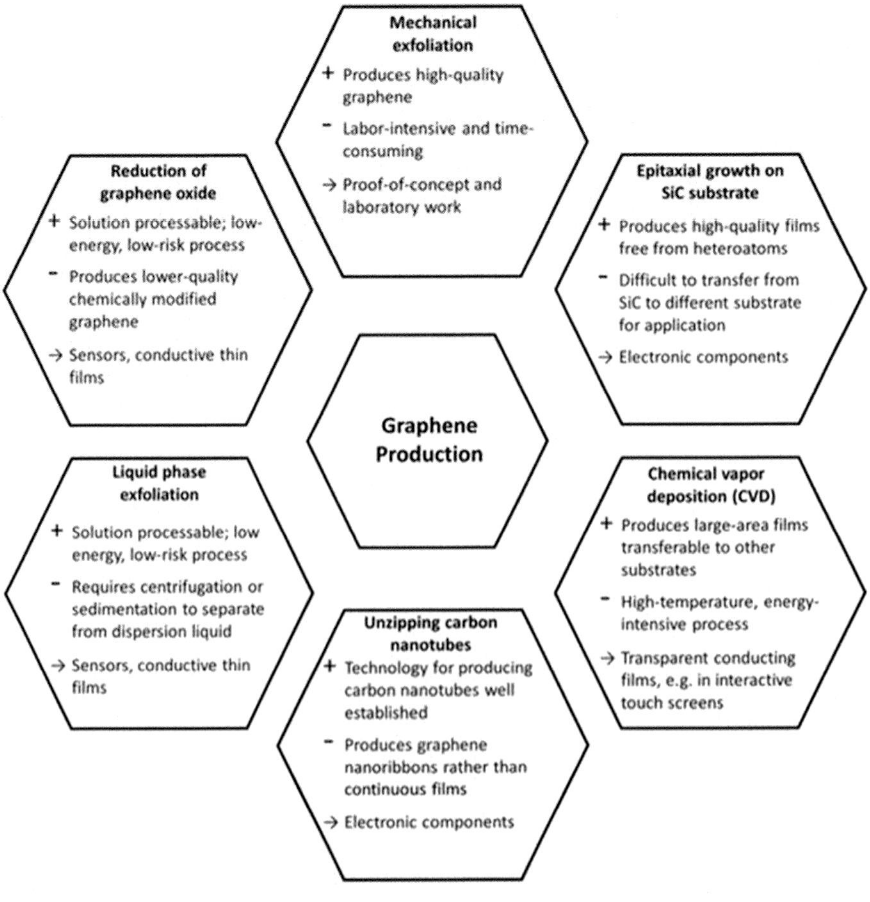

Figure 22.2 Summary of various graphene synthesis techniques (Hernaez et al., 2017).

adenocarcinoma cell line (A549), vero cell lines, etc. in order to quantify the biocompatibility of graphene-based materials (Coleman, Knight, Gies, Jakubek, & Zou, 2017; De Marzi et al., 2014; Ou et al., 2016). No significant reactive oxygen species (ROS) generation was observed in the cell viability assays (Tadyszak et al., 2018). Besides cytotoxicity assays, biodegradation of graphene-based materials studies were also conducted to determine whether the degradation products impart cell or DNA damage. A study by Kurapati et al. (2018) proved the nontoxic nature of biodegradable products of single-layer and few-layer graphene in human lung cells since neutrophils can successfully digest the biodegraded products.

22.3 Graphene-based biosensors

Among the various applications in biomedical field, such as drug delivery, cancer treatment and detection, tissue engineering, nanomedicine, etc., biosensing has emerged as a significant tool for detecting presence of specific biological component or biochemical reaction (Roy et al., 2020)(Bardhan et al., 2021). In general, biosensors are analytical devices that can interact with the target of interest and produce physicochemical signal which is measured by optical, thermal, electron transfer or by electrochemical technique (Chaubey & Malhotra, 2002). Various biosensors are developed throughout the world, but very few are efficient enough to detect biological analytes in picomolar range with high selectivity. Various graphene-based sensors have reported to overcome various drawbacks faced by other biosensors and have reported to be very sensitive due to high charge-carrier mobility and easiness in electron transfer (Ambrosi, Chua, Bonanni, & Pumera, 2014). The edge-plane-like defect sites of graphene provide active sites for electron transfer and also facilitate fluorescence resonance energy transfer (FRET) (Kim, Cote, Kim, & Huang, 2010). Very high surface area enhances electrochemical catalytic activity, thus exhibiting good potentiality in electrochemistry sensing (Shao et al., 2010). Moreover, high conductivity and high transparency of graphene monolayers make them suitable as electrical and optical sensor. The presence of hydrophobic domains or π-systems improves their ability to immobilize biomolecules and helps in target-specific interactions (Szunerits &Boukherroub, 2018). Graphene-based biosensors are often used for detecting DNA, glucose, dopamine, protein, etc. by immunosensing (Fig. 22.3). They are even used for pathogen detection (Jung, Cheon, Liu, Lee, & Seo, 2010). Few of the examples of graphene-based biosensors are elicited in Table 22.1.

Addition of metal oxides into graphene structure enhances sensitivity toward molecular detection with fast response and the biosensors exhibit better electrochemistry without the presence of any other electron mediator (Chen, Zhu, Wu, Han, & Wang, 2010). Graphene/metal oxide nanocomposites-based glucose sensors have gained immense popularity due to high stability, sensitivity, and selectivity with much lower limit of detection (Heller & Feldman, 2008; Toghill & Compton, 2010). Copper oxide was well known for its application in detection of glucose oxidation and presence of H_2O_2 because of its narrow band gap (1.2 eV), high-electron transfer capacity, great

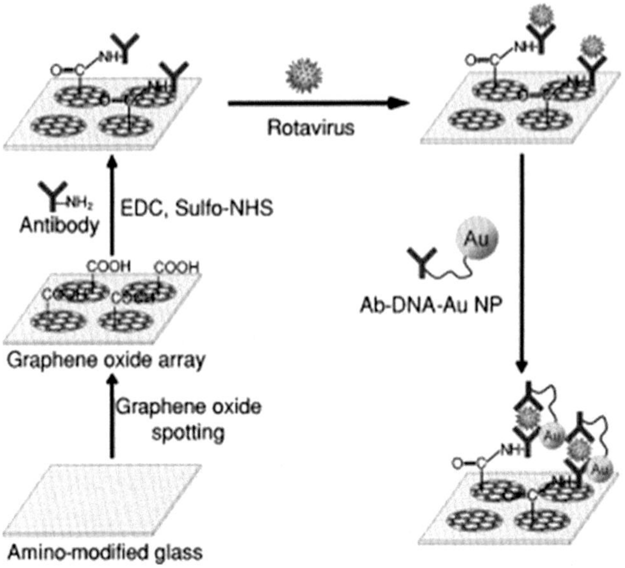

Figure 22.3 Schematic representation of graphene-based immuno-biosensor (Jung et al., 2010).

electrochemical activity, and high surface area (Zhuang et al., 2008). Incorporation of copper oxide into graphene results in a synergistic effects, thus improving sensing capacity (Luo, Zhu, & Wang, 2012). Similarly zinc oxide-graphene nanocomposite has been reported to exhibit high sensitivity with limit of detection for glucose as low as 0.02 mM (Norouzi et al., 2011). Incorporation of low cost metal oxides such as nickel oxide into graphene structure has also gained attention as cost-effective, sensitive electrochemical sensors (Yuan et al., 2013). Ecofriendly, biocompatible graphene/metal oxide biosensors such as titanium dioxide-graphene nanosensor are now gaining limelight. A report by Luo et al. (2013) shows the excellent performance of rGO-TiO_2 nanocluster in glucose detection with detection limit of 4.8 μM. Other than glucose detection, various other biomolecular sensing has been reported using metal oxide/graphene biosensors. Wu et al. (2014) have synthesized mesoporous Fe_3O_4-NH_2-graphene sheet-based electrochemical biosensor to detect the presence of smaller-sized biomolecules such as ascorbic acid, dopamine, and uric acid. The detection limit for ascorbic acid, dopamine, and uric acid was found to be 0.074 μmol/L, 0.126 μmol/L, and 0.056 μmol/L, respectively. Electrogenerated chemiluminescence-based cholesterol biosensor was prepared by Zhang, Yuan, Chai, Wang, and Wu (2013) by depositing cerium oxide on graphene. The sensor was sensitive enough to detect and quantify cholesterol in the linear range of 12.0 μM to 7.2 mM, with a limit of detection of 4.0 μM. Even graphene-metal oxide nanocomposite such as MnO_2-graphene-based nonenzymatic biosensor could successfully detect the presence of H_2O_2 with quite high sensitivity and stability (Li et al., 2010). The MnO_2 deposition enhanced the surface area, thus promoting better electrocatalytic activity.

Table 22.1 Few examples of graphene-based biosensors for different analytes (Szunerits & Boukherroub, 2018).

Analyte	Sensor design	Detection	Limit of detection (LOD)
Glucose	3D graphene foam–Co_3O_4 nanowires	Electrochemistry	20 nM
Glucose	Graphene + GO_x	Field effect transistor	0.1 mM
Glucose	GQDs–bipyridine boronic acid	Fluorescence	1 mM
Dopamine	rGO–polyvinylpyrrolidone	Electrochemistry	0.2 nM
DNA	DNA GO and GQD–ssDNA	Fluorescence resonance energy transfer	75 pM
DNA	Graphene–Au NPs–ssDNA	Surface plasmon resonance	500 aM
DNA	GO nanowalls	Differential pulse voltammetry	9.6 zM
DNA	Graphene	Field effect transistor	10 fM
Lysozyme	Au/PDDA–GO–*Micrococcus lysodeikticus*	Surface plasmon resonance	3.4 nM
Folic acid	Au–rGO	Differential pulse voltammetry	1 pM
Folic acid	Graphene	Surface plasmon resonance	5 fM
β-Amyloid	Magnetic/plasmonic GO	Surface-enhanced Raman substrate	100 fg/mL
Prostate-specific antigen	rGO	Field effect transistor	1 fM
Escherichia coli	Graphene–anti-*E. coli*	Field effect transistor	10 cfu/mL

22.4 Graphene-based nanocomposites for gene and drug delivery

Accurate designing and fabrication of efficient gene and drug delivery systems with ability for controlled release have always been a significant concern in biomedical science (Fig. 22.4). With the discovery of graphene, various studies were conducted regarding successful drug delivery using graphene-based nanocomposites (Su et al., 2015; Wang et al., 2015; Zhang, Wang, & Zhai, 2016). Graphene or graphene-based materials can act as a suitable carrier for drugs (Zhang, Xia, Zhao, Liu, & Zhang, 2010). TiO_2-ZnO nanohybrid was found to be a highly potential anticancer

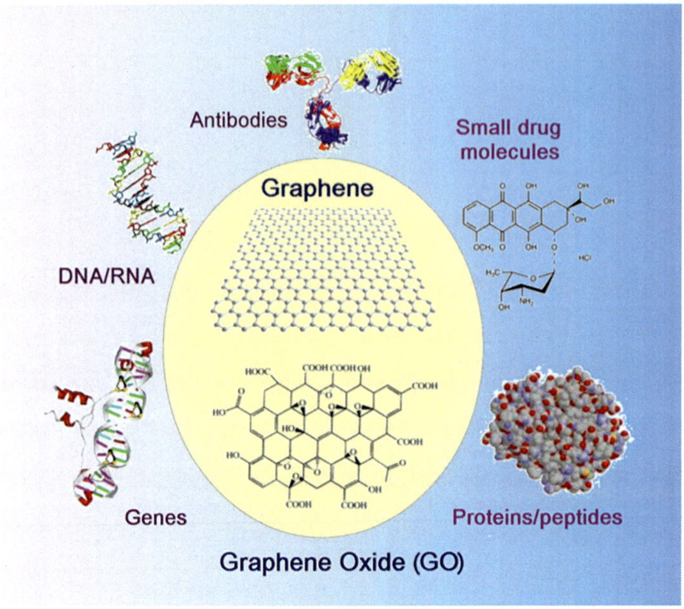

Figure 22.4 Schematic illustration of application of graphene-based materials for drug and various biomolecule delivery (Liu, Cui, & Losic, 2013).

agent (Rasmussen, Martinez, Louka, & Wingett, 2010). Hence Zamani et al. (2018) have designed an effective and efficient drug delivery system by incorporating TiO_2-ZnO into GO for slow and sustained colon-specific release. In another report by Fan, Jiao, Gao, Jin, and Li (2013), biocompatible hybrid of Fe_3O_4-graphene-CNT was prepared for controlled delivery of anticancer drugs with a high loading capacity (0.27 mg/mg).

Since graphene-based materials exhibit unique interactions with DNA and RNA, graphene nanocomposite-based DNA or RNA sensing and delivery have gathered tremendous attention. DNA has a tendency to interact with negative charge of graphene-based materials which enhances DNA adsorption, preferentially single-stranded (ss) DNA adsorption (Ren et al., 2010).

22.5 Metal oxide-modified graphene nanostructures for antibacterial applications

Antibacterial materials are gaining importance in various sectors such as prevention of diseases, prevention of biofilm growth, wastewater management, and especially in biomedical fields. Various antibacterial agents are explored throughout the world, but most of them are either complex or suffers drawback such as secondary toxicity. Hence graphene-based materials have gained prominence due to its unique morphological properties. The dimensions of graphene are suitable enough to provide extensive coverage to cell surfaces (Upadhyay, Soin, & Roy, 2014), thus exhibiting antibacterial effects. Graphene sheets (>0.4 mm^2) have the capability of covering the whole bacterial cells blocking all the available active sites, thus inhibiting cell proliferation. Certain metal oxide nanoparticles such as zinc oxide (ZnO), silver oxide (Ag_2O), titanium dioxide (TiO_2), and magnesium oxide (MgO) have been reported earlier to possess antimicrobial activity (Azam et al., 2012; Besinis, De Peralta, & Handy, 2014; Emami-Karvani & Chehrazi, 2011). The surface properties of graphene elevate the antibacterial efficacy of the metal oxides (Kumar, Shaikshavali, Srikanth, & Sankara Rao, 2013) and provide better possibility to penetrate the bacterial membrane to inhibit bacterial growth. Table 22.2 shows some of the examples of antibacterial activity of metal oxide-graphene-based nanocomposites.

22.6 Graphene-based wearable devices for biomedical applications

Wearable devices are a class of device that can be placed over human skin having extraordinary applicability. These new-age devices are flexible and advantageous over conventional rigid devices due to their flexibility and ease of access. Graphene is known to be a smart material for wearable electronics, flexible solar cells, and other energy-harvesting devices (Jang et al., 2016).

Previously, researchers have introduced pure graphene-based wearable biomedical devices such as electrogram monitoring system, heart activity monitoring device, body temperature meter, etc. (Huang et al., 2019; Liu, Pharr, & Salvatore, 2017; Liu et al., 2018). These wearable devices are designed to function alongside the human body or clothes (Mohan, Lau, Hui, & Bhattacharyya, 2018). Recent progress of modified graphene with various metallic and metal oxide nanoparticles is advantageous over pure graphene and enables us with numerous wearable biomedical applications (Kim & Ahn, 2017; Yang, Yang, Tan, & Yuan, 2017). Singh et al. (2020) reported a ZnO-doped graphene-epoxy thin film, which could be used in diverse biomedical applications such as electroactive sensing of glucose, cytochrome c, and cancer cells. Besides, this novel thin film is capable of high-frequency applications, such as ultrasound imaging or MRI. The high-electron transport property of graphene played a pivotal role to reduce the signal-to-noise ratio for better biosensing applications. Similarly, conductive polyoxometalate (POM) equipped with 2D graphene nanosheets was utilized by Xie, Zhang, Wang, Wang, and Wang (2019) to monitor pulse beats by using electrochemical method. This wearable device is super capacitive and insertion of POM into 2D graphene sheets enhanced its storage capacity remarkably, which in turn enriched the sensitivity of the device. Another excellent contribution has been made by Xiao et al. (2012) showing platinum nanoparticle and MnO_2 framework loaded

Table 22.2 Few examples of antibacterial activity of metal oxide/graphene-based nanocomposites.

Material	Hybrid component (metal oxide)	Type of interaction	Bacteria	References
Graphene	ZnO	Van der Waals	*Escherichia coli*	Kavitha, Gopalan, Lee, and Park (2012)
Graphene	ZnO	Van der Waals	*Streptococcus mutans*	Kulshrestha, Khan, Meena, Singh, and Khan (2014)
Graphene	ZnO	Van der Waals	*E. coli, Salmonella typhi*	Bykkam et al. (2015)
Graphene	TiO_2	Covalent	*E. coli*	Cao, Cao, Dong, Gao, and Wang (2013)
Graphene	TiO_2	Covalent	*Staphylococcus aureus, E. coli, Candida albicans* (fungi)	Karimi, Yazdanshenas, Khajavi, Rashidi, and Mirjalili (2014)
Graphene	TiO_2	Covalent	*S. aureus, E. coli, C. albicans* (fungi)	Karimi, Yazdanshenas, Khajavi, Rashidi, and Mirjalili (2015)
Graphene	Au-TiO_2	Covalent/ Van der Waals	*Rhodopseudanonas palustris, E. coli, Candida* (fungi)	He, Huang, Yan, and Zhu (2013)
Graphene	Fe_3O_4	Van der Waals	*E. coli*	Santhosh et al. (2014)
Graphene	SnO_2	Van der Waals	*Pseudomonas aeruginosa,*	Mohan and Manoj (2019)
Graphene oxide	SnO_2		*S. aureus*	Qiu, Liu, Zhu, and Liu (2018)
Graphene	Fe_3O_4-TiO_2		*E. coli*	Zhou, Zou, and Chen (2020)

free-standing film of rGO for H_2O_2 detection in living cells. This work is beneficial for detecting reactive oxygen species in cells to understand cellular functionalities. The large surface area of MnO_2 framework and high electrical transport property of graphene herein are responsible for excellent sensitivity and selectivity toward H_2O_2. In addition, this work provides an insight into the flexible electrodes for a wide-scale application in biosensing, bioelectronics, and lab-on-a-chip devices.

References

Abbott's, I. E. (2007). Graphene: Exploring carbon flatland. *Physics Today*, *60*(8), 35.

Ambrosi, A., Chua, C. K., Bonanni, A., & Pumera, M. (2014). Electrochemistry of graphene and related materials. *Chemical Reviews*, *114*(14), 7150–7188.

Azam, A., Ahmed, A. S., Oves, M., Khan, M. S., Habib, S. S., & Memic, A. (2012). Antimicrobial activity of metal oxide nanoparticles against gram-positive and gram-negative bacteria: A comparative study. *International Journal of Nanomedicine*, *7*, 6003.

Balandin, A. A., Ghosh, S., Bao, W., Calizo, I., Teweldebrhan, D., Miao, F., & Lau, C. N. (2008). Superior thermal conductivity of single-layer graphene. *Nano Letters*, *8*(3), 902–907.

Bardhan, S., Roy, S., Chanda, D. K., Mondal, D., Das, S., & Das, S. (2021). Flexible and reusable carbon dot decorated natural microcline membrane: a futuristic probe for multiple heavy metal induced carcinogen detection. *Microchimica Acta*, *188*(4). https://doi.org/10.1007/s00604-021-04787-0.

Besinis, A., De Peralta, T., & Handy, R. D. (2014). The antibacterial effects of silver, titanium dioxide and silica dioxide nanoparticles compared to the dental disinfectant chlorhexidine on *Streptococcus mutans* using a suite of bioassays. *Nanotoxicology*, *8*(1), 1–6.

Brumfiel, G. (2009). Graphene gets ready for the big time. *Nature*, *458*, 390–391.

Bykkam, S., Narsingam, S., Ahmadipour, M., Dayakar, T., Rao, K. V., Chakra, C. S., & Kalakotla, S. (2015). Few layered graphene sheet decorated by ZnO nanoparticles for anti-bacterial application. *Superlattices and Microstructures*, *83*, 776–784.

Cao, B., Cao, S., Dong, P., Gao, J., & Wang, J. (2013). High antibacterial activity of ultrafine TiO_2/graphene sheets nanocomposites under visible light irradiation. *Materials Letters*, *93*, 349–352.

Chaubey, A., & Malhotra, B. (2002). Mediated biosensors. *Biosensors and Bioelectronics*, *17* (6–7), 441–456.

Chen, S., Zhu, J., Wu, X., Han, Q., & Wang, X. (2010). Graphene oxide-MnO_2 nanocomposites for supercapacitors. *ACS Nano*, *4*(5), 2822–2830.

Chen, X. M., Wu, G. H., Jiang, Y. Q., Wang, Y. R., & Chen, X. (2011). Graphene and graphene-based nanomaterials: The promising materials for bright future of electroanalytical chemistry. *Analyst*, *136*(22), 4631–4640.

Chong, Y., Ge, C., Yang, Z., Garate, J. A., Gu, Z., Weber, J. K., ... Zhou, R. (2015). Reduced cytotoxicity of graphene nanosheets mediated by blood-protein coating. *ACS Nano*, *9*(6), 5713–5724.

Choucair, M., Thordarson, P., & Stride, J. A. (2009). Gram-scale production of graphene based on solvothermal synthesis and sonication. *Nature Nanotechnology*, *4*(1), 30.

Coleman, B. R., Knight, T., Gies, V., Jakubek, Z. J., & Zou, S. (2017). Manipulation and quantification of graphene oxide flake size: Photoluminescence and cytotoxicity. *ACS Applied Materials & Interfaces*, *9*(34), 28911–28921.

De Marzi, L., Ottaviano, L., Perrozzi, F., Nardone, M., Santucci, S., De Lapuente, J., ... Poma, A. (2014). Flake size-dependent cyto and genotoxic evaluation of graphene oxide on in vitro A549, $CaCo_2$ and vero cell lines. *Journal of Biological Regulators and Homeostatic Agents*, *28*(2), 281–289.

Dimiev, A. M., & Eigler, S. (2016). *Graphene oxide: Fundamentals and applications*. Hoboken: Wiley.

Emami-Karvani, Z., & Chehrazi, P. (2011). Antibacterial activity of ZnO nanoparticle on gram-positive and gram-negative bacteria. *African Journal of Microbiology Research.*, *5* (12), 1368–1373.

Fan, X., Jiao, G., Gao, L., Jin, P., & Li, X. (2013). The preparation and drug delivery of a graphene–carbon nanotube–Fe_3O_4 nanoparticle hybrid. *Journal of Materials Chemistry B*, *1*(20), 2658–2664.

Geim, A. K. (2009). Graphene: Status and prospects. *Science (New York, N.Y.)*, *324*(5934), 1530–1534.

Geim, A. K., & Novoselov, K. S. (2010). The rise of graphene. In *Nanoscience and technology: A collection of reviews from nature journals* (pp. 11–19). Singapore: World Scientific.

Ghosal, K., & Sarkar, K. (2018). Biomedical applications of graphene nanomaterials and beyond. *ACS Biomaterials Science & Engineering*, *4*(8), 2653–2703.

Gilje, S., Han, S., Wang, M., Wang, K. L., & Kaner, R. B. (2007). A chemical route to graphene for device applications. *Nano Letters*, *7*(11), 3394–3398.

Giovannetti, G. A., Khomyakov, P. A., Brocks, G., Karpan, V. V., van den Brink, J., & Kelly, P. J. (2008). Doping graphene with metal contacts. *Physical Review Letters*, *101* (2), 026803.

Govindasamy, M., Wang, S. F., Pan, W. C., Subramanian, B., Ramalingam, R. J., & Al-Lohedan, H. (2019). Facile sonochemical synthesis of perovskite-type $SrTiO_3$ nanocubes with reduced graphene oxide nanocatalyst for an enhanced electrochemical detection of α-amino acid (tryptophan). *Ultrasonics Sonochemistry*, *56*, 193–199.

He, W., Huang, H., Yan, J., & Zhu, J. (2013). Photocatalytic and antibacterial properties of Au-TiO_2 nanocomposite on monolayer graphene: From experiment to theory. *Journal of Applied Physics*, *114*(20), 204701.

Heller, A., & Feldman, B. (2008). Electrochemical glucose sensors and their applications in diabetes management. *Chemical Reviews*, *108*, 2482–2505.

Hernaez, M., Zamarreño, C. R., Melendi-Espina, S., Bird, L. R., Mayes, A. G., & Arregui, F. J. (2017). Optical fibre sensors using graphene-based materials: A review. *Sensors.*, *17*(1), 155.

Hu, K., Kulkarni, D. D., Choi, I., & Tsukruk, V. V. (2014). Graphene-polymer nanocomposites for structural and functional applications. *Progress in Polymer Science*, *39*(11), 1934–1972.

Huang, H., Su, S., Wu, N., Wan, H., Wan, S., Bi, H., & Sun, L. (2019). Graphene-based sensors for human health monitoring. *Frontiers in Chemistry*, *7*, 399.

Huang, X., Yin, Z., Wu, S., Qi, X., He, Q., Zhang, Q., ... Zhang, H. (2011). Graphene-based materials: Synthesis, characterization, properties, and applications. *Small (Weinheim an der Bergstrasse, Germany)*, *7*(14), 1876–1902.

Jang, H., Park, Y. J., Chen, X., Das, T., Kim, M. S., & Ahn, J. H. (2016). Graphene-based flexible and stretchable electronics. *Advanced Materials*, *28*(22), 4184–4202.

Juang, Z. Y., Wu, C. Y., Lu, A. Y., Su, C. Y., Leou, K. C., Chen, F. R., & Tsai, C. H. (2010). Graphene synthesis by chemical vapor deposition and transfer by a roll-to-roll process. *Carbon*, *48*(11), 3169–3174.

Jung, J. H., Cheon, D. S., Liu, F., Lee, K. B., & Seo, T. S. (2010). A graphene oxide based immuno-biosensor for pathogen detection. *Angewandte Chemie*, *122*(33), 5844–5847.

Karimi, L., Yazdanshenas, M. E., Khajavi, R., Rashidi, A., & Mirjalili, M. (2014). Using graphene/TiO_2 nanocomposite as a new route for preparation of electroconductive, self-cleaning, antibacterial and antifungal cotton fabric without toxicity. *Cellulose*, *21*(5), 3813–3827.

Karimi, L., Yazdanshenas, M. E., Khajavi, R., Rashidi, A., & Mirjalili, M. (2015). Optimizing the photocatalytic properties and the synergistic effects of graphene and nano titanium dioxide immobilized on cotton fabric. *Applied Surface Science*, *332*, 665–673.

Kavitha, T., Gopalan, A. I., Lee, K. P., & Park, S. Y. (2012). Glucose sensing, photocatalytic and antibacterial properties of graphene–ZnO nanoparticle hybrids. *Carbon, 50*(8), 2994–3000.

Kim, H., & Ahn, J. H. (2017). Graphene for flexible and wearable device applications. *Carbon, 120*, 244–257.

Kim, J., Cote, L. J., Kim, F., & Huang, J. (2010). Visualizing graphene based sheets by fluorescence quenching microscopy. *Journal of the American Chemical Society, 132*(1), 260–267.

Kulshrestha, S., Khan, S., Meena, R., Singh, B. R., & Khan, A. U. (2014). A graphene/zinc oxide nanocomposite film protects dental implant surfaces against cariogenic *Streptococcus mutans. Biofouling, 30*(10), 1281–1294.

Kumar, R. N., Shaikshavali, P., Srikanth, V. V., & Sankara Rao, K. B. (2013). Reducing agent free synthesis of graphene from graphene oxide. In *AIP Conference Proceedings* (Vol. 1538, No. 1, pp. 262–265). American Institute of Physics.

Kurapati, R., Mukherjee, S. P., Martín, C., Bepete, G., Vázquez, E., Pénicaud, A., ... Bianco, A. (2018). Degradation of single-layer and few-layer graphene by neutrophil myeloperoxidase. *Angewandte Chemie, 57*(36), 11722–11727.

Li, D., & Kaner, R. B. (2008). Graphene-based materials. *Science (New York, N.Y.), 320* (5880), 1170–1171.

Li, D., Müller, M. B., Gilje, S., Kaner, R. B., & Wallace, G. G. (2008). Processable aqueous dispersions of graphene nanosheets. *Nature Nanotechnology, 3*(2), 101–105.

Li, L., Du, Z., Liu, S., Hao, Q., Wang, Y., Li, Q., & Wang, T. (2010). A novel nonenzymatic hydrogen peroxide sensor based on MnO_2/graphene oxide nanocomposite. *Talanta, 82* (5), 1637–1641.

Li, N., Wang, Z., Zhao, K., Shi, Z., Gu, Z., & Xu, S. (2010). Large scale synthesis of N-doped multi-layered graphene sheets by simple arc-discharge method. *Carbon, 48*(1), 255–259.

Li, X., Wang, X., Zhang, L., Lee, S., & Dai, H. (2008). Chemically derived, ultrasmooth graphene nanoribbon semiconductors. *Science (New York, N.Y.), 319*(5867), 1229–1232.

Liu, H., Liu, Y., & Zhu, D. (2011). Chemical doping of graphene. *Journal of Materials Chemistry, 21*(10), 3335–3345.

Liu, J., Cui, L., & Losic, D. (2013). Graphene and graphene oxide as new nanocarriers for drug delivery applications. *Acta Biomaterialia, 9*(12), 9243–9257.

Liu, N., Luo, F., Wu, H., Liu, Y., Zhang, C., & Chen, J. (2008). One-step ionic-liquid-assisted electrochemical synthesis of ionic-liquid-functionalized graphene sheets directly from graphite. *Advanced Functional Materials, 18*(10), 1518–1525.

Liu, Y., Pharr, M., & Salvatore, G. A. (2017). Lab-on-skin: A review of flexible and stretchable electronics for wearable health monitoring. *ACS Nano, 11*(10), 9614–9635.

Liu, Y., Wang, H., Zhao, W., Zhang, M., Qin, H., & Xie, Y. (2018). Flexible, stretchable sensors for wearable health monitoring: Sensing mechanisms, materials, fabrication strategies and features. *Sensors, 18*(2), 645.

Luo, L., Zhu, L., & Wang, Z. (2012). Nonenzymatic amperometric determination of glucose by CuO nanocubes–graphene nanocomposite modified electrode. *Bioelectrochemistry (Amsterdam, Netherlands), 88*, 156–163.

Luo, Z., Ma, X., Yang, D., Yuwen, L., Zhu, X., Weng, L., & Wang, L. (2013). Synthesis of highly dispersed titanium dioxide nanoclusters on reduced graphene oxide for increased glucose sensing. *Carbon, 57*, 470–476.

McAllister, M. J., Li, J. L., Adamson, D. H., Schniepp, H. C., Abdala, A. A., Liu, J., ... Aksay, I. A. (2007). Single sheet functionalized graphene by oxidation and thermal expansion of graphite. *Chemistry of Materials: A Publication of the American Chemical Society*, *19*(18), 4396–4404.

Mitra, M., Chatterjee, K., Kargupta, K., Ganguly, S., & Banerjee, D. (2013). Reduction of graphene oxide through a green and metal-free approach using formic acid. *Diamond and Related Materials*, *37*, 74–79. https://doi.org/10.1016/j.diamond.2013.05.003.

Mohan, A. N., & Manoj, B. (2019). Surface modified graphene/SnO_2 nanocomposite from carbon black as an efficient disinfectant against *Pseudomonas aeruginosa*. *Materials Chemistry and Physics*, *232*, 137–144.

Mohan, V. B., Lau, K. T., Hui, D., & Bhattacharyya, D. (2018). Graphene-based materials and their composites: A review on production, applications and product limitations. *Composites Part B: Engineering*, *142*, 200–220.

Mohd Yazid, S. N., Md Isa, I., Abu Bakar, S., Hashim, N., Ab., & Ghani, S. (2014). A review of glucose biosensors based on graphene/metal oxide nanomaterials. *Analytical Letters*, *47*(11), 1821–1834.

Muthurasu, A., Dhandapani, P., & Ganesh, V. (2016). Facile and simultaneous synthesis of graphene quantum dots and reduced graphene oxide for bio-imaging and supercapacitor applications. *New Journal of Chemistry*, *40*(11), 9111–9124.

Niyogi, S., Bekyarova, E., Itkis, M. E., McWilliams, J. L., Hamon, M. A., & Haddon, R. C. (2006). Solution properties of graphite and graphene. *Journal of the American Chemical Society*, *128*(24), 7720–7721.

Norouzi, P., Ganjali, H., Larijani, B., Ganjali, M. R., Faridbod, F., & Zamani, H. A. (2011). A glucose biosensor based on nanographene and ZnO nanoparticles using FFT continuous cyclic voltammetry. *International Journal of Electrochemical Science*, *6*(11), 5189–5195.

Novoselov, K. S., Geim, A. K., Morozov, S. V., Jiang, D., Katsnelson, M. I., Grigorieva, I., ... Firsov, A. A. (2005). Two-dimensional gas of massless Dirac fermions in graphene. *Nature*, *438*(7065), 197–200.

Novoselov, K. S., Geim, A. K., Morozov, S. V., Jiang, D., Zhang, Y., & Dubonos, S. V. (2004). Grigorieva IV, Firsov AA. Electric field effect in atomically thin carbon films. *Science (New York, N.Y.)*, *306*(5696), 666–669.

Ou, L., Song, B., Liang, H., Liu, J., Feng, X., Deng, B., ... Shao, L. (2016). Toxicity of graphene-family nanoparticles: A general review of the origins and mechanisms. *Particle and Fibre Toxicology*, *13*(1), 1–24.

Qiu, J., Liu, L., Zhu, H., & Liu, X. (2018). Combination types between graphene oxide and substrate affect the antibacterial activity. *Bioactive Materials*, *3*(3), 341–346.

Rajeshkumar, S., & Veena, P. (2018). *Biomedical applications and characteristics of graphene nanoparticles and graphene-based nanocomposites. Exploring the realms of nature for nanosynthesis* (pp. 341–354). Cham: Springer.

Ranjbartoreh, A. R., Wang, B., Shen, X., & Wang, G. (2011). Advanced mechanical properties of graphene paper. *Journal of Applied Physics*, *109*(1), 014306.

Rasmussen, J. W., Martinez, E., Louka, P., & Wingett, D. G. (2010). Zinc oxide nanoparticles for selective destruction of tumor cells and potential for drug delivery applications. *Expert Opinion on Drug Delivery*, *7*(9), 1063–1077.

Ren, H., Wang, C., Zhang, J., Zhou, X., Xu, D., Zheng, J., ... Zhang, J. (2010). DNA cleavage system of nanosized graphene oxide sheets and copper ions. *ACS Nano*, *4*(12), 7169–7174.

Roy, S., Bardhan, S., Chanda, D. K., Roy, J., Mondal, D., & Das, S. (2020). In situ-grown Cdot-wrapped Boehmite nanoparticles for Cr (VI) sensing in wastewater and a theoretical probe for chromium-induced carcinogen detection. *ACS Applied Materials & Interfaces*, *12*(39), 43833–43843. https://doi.org/10.1021/acsami.0c13433.

Russier, J., Treossi, E., Scarsi, A., Perrozzi, F., Dumortier, H., Ottaviano, L., ... Bianco, A. (2013). Evidencing the mask effect of graphene oxide: A comparative study on primary human and murine phagocytic cells. *Nanoscale*, *5*(22), 11234–11247.

Santhosh, C., Kollu, P., Doshi, S., Sharma, M., Bahadur, D., Vanchinathan, M. T., ... Grace, A. N. (2014). Adsorption, photodegradation and antibacterial study of graphene-Fe_3O_4 nanocomposite for multipurpose water purification application. *RSC Advances*, *4*(54), 28300–28308.

Shao, Y., Wang, J., Wu, H., Liu, J., Aksay, I. A., & Lin, Y. (2010). Graphene based electrochemical sensors and biosensors: A review. *Electroanalysis*, *22*(10), 1027–1036.

Sharma, P., Hussain, N., Das, M. R., Deshmukh, A. B., Shelke, M. V., Szunerits, S., & Boukherroub, R. (2014). Metal oxide-graphene nanocomposites: Synthesis to applications. In Handbook of research on nanoscience. *nanotechnology, and advanced materials*, 196–225.

Shen, J., Hu, Y., Li, C., Qin, C., & Ye, M. (2009). Synthesis of amphiphilic graphene nanoplatelets. *Small (Weinheim an der Bergstrasse, Germany)*, *5*(1), 82–85.

Shi, L., Chen, J., Teng, L., Wang, L., Zhu, G., Liu, S., ... Ren, L. (2016). The antibacterial applications of graphene and its derivatives. *Small (Weinheim an der Bergstrasse, Germany)*, *12*(31), 4165–4184.

Shi, X., Chang, H., Chen, S., Lai, C., Khademhosseini, A., & Wu, H. (2012). Regulating cellular behavior on few-layer reduced graphene oxide films with well-controlled reduction states. *Advanced Functional Materials*, *22*(4), 751–759.

Singh, M., Kumar, S., Zoghi, S., Cervantes, Y., Sarkar, D., Ahmed, S., ... Banerjee, S. (2020). Fabrication and characterization of flexible three-phase ZnO-graphene-epoxy electro-active thin-film nanocomposites: Towards applications in wearable biomedical devices. *Journal of Composites Science*, *4*(3), 88.

Stankovich, S., Piner, R. D., Chen, X., Wu, N., Nguyen, S. T., & Ruoff, R. S. (2006). Stable aqueous dispersions of graphitic nanoplatelets via the reduction of exfoliated graphite oxide in the presence of poly (sodium 4-styrenesulfonate). *Journal of Materials Chemistry*, *16*(2), 155–158.

Stengl, V., Popelková, D., & Vlácil, P. (2011). TiO_2–graphene nanocomposite as high performace photocatalysts. *The JJournal of Physics Chemistry C*, *115*(51), 25209–25218.

Su, Z., Shen, H., Wang, H., Wang, J., Li, J., Nienhaus, G. U., ... Wei, G. (2015). Motif-designed peptide nanofibers decorated with graphene quantum dots for simultaneous targeting and imaging of tumor cells. *Advanced Functional Materials*, *25*(34), 5472–5478.

Szunerits, S., & Boukherroub, R. (2018). Graphene-based biosensors. *Interface Focus*, *8*(3), 20160132.

Tadyszak, K., Wychowaniec, J. K., & Litowczenko, J. (2018). Biomedical applications of graphene-based structures. *Nanomaterials*, *8*(11), 944.

Toghill, K. E., & Compton, R. G. (2010). Electrochemical non-enzymatic glucose sensors: A perspective and an evaluation. *International Journal of Electrochemical Science*, *5*(9), 1246–1301.

Upadhyay, R. K., Soin, N., & Roy, S. S. (2014). Role of graphene/metal oxide composites as photocatalysts, adsorbents and disinfectants in water treatment: A review. *RSC Advances*, *4*(8), 3823–3851.

Wang, D., Kou, R., Choi, D., Yang, Z., Nie, Z., Li, J., ... Liu, J. (2010). Ternary self-assembly of ordered metal oxide-graphene nanocomposites for electrochemical energy storage. *ACS Nano*, *4*(3), 1587–1595.

Wang, J., Ouyang, Z., Ren, Z., Li, J., Zhang, P., Wei, G., & Su, Z. (2015). Self-assembled peptide nanofibers on graphene oxide as a novel nanohybrid for biomimetic mineralization of hydroxyapatite. *Carbon*, *89*, 20–30.

Wang, L., Zhang, Y., Wu, A., & Wei, G. (2017). Designed graphene-peptide nanocomposites for biosensor applications: A review. *Analytica Chimica Acta*, *8*(985), 24–40.

Wu, D., Li, Y., Zhang, Y., Wang, P., Wei, Q., & Du, B. (2014). Sensitive electrochemical sensor for simultaneous determination of dopamine, ascorbic acid, and uric acid enhanced by amino-group functionalized mesoporous Fe_3O_4@ graphene sheets. *Electrochimica Acta*, *116*, 244–249.

Wu, J., Pisula, W., & Müllen, K. (2007). Graphenes as potential material for electronics. *Chemical Reviews*, *107*(3), 718–747.

Wu, X., Pei, Y., & Zeng, X. C. (2009). B2C graphene, nanotubes, and nanoribbons. *Nano Letters*, *9*(4), 1577–1582.

Xiao, F., Li, Y., Zan, X., Liao, K., Xu, R., & Duan, H. (2012). Growth of metal-metal oxide nanostructures on freestanding graphene paper for flexible biosensors. *Advanced Functional Materials*, *22*(12), 2487–2494.

Xie, T., Zhang, L., Wang, Y., Wang, Y., & Wang, X. (2019). Graphene-based supercapacitors as flexible wearable sensor for monitoring pulse-beat. *Ceramics International*, *45* (2), 2516–2520.

Yang, Y., Yang, X., Tan, Y., & Yuan, Q. (2017). Recent progress in flexible and wearable bio-electronics based on nanomaterials. *Nano Research*, *10*(5), 1560–1583.

Yuan, B., Xu, C., Deng, D., Xing, Y., Liu, L., Pang, H., & Zhang, D. (2013). Graphene oxide-nickel oxide modified glassy carbon electrode for supercapacitor and nonenzymatic glucose sensor. *Electrochimica Acta*, *88*, 708–712.

Zamani, M., Rostami, M., Aghajanzadeh, M., Manjili, H. K., Rostamizadeh, K., & Danafar, H. (2018). Mesoporous titanium dioxide@ zinc oxide–graphene oxide nanocarriers for colon-specific drug delivery. *Journal of Material Science*, *53*(3), 1634–1645.

Zhang, B., Wang, Y., & Zhai, G. (2016). Biomedical applications of the graphene-based materials. *Materials Science and Engineering C*, *61*, 953–964.

Zhang, C., Yuan, Y., Zhang, S., Wang, Y., & Liu, Z. (2011). Biosensing platform based on fluorescence resonance energy transfer from upconverting nanocrystals to graphene oxide. *Angewandte Chemie*, *50*(30), 6851–6854.

Zhang, L., Xia, J., Zhao, Q., Liu, L., & Zhang, Z. (2010). Functional graphene oxide as a nanocarrier for controlled loading and targeted delivery of mixed anticancer drugs. *Small (Weinheim an der Bergstrasse, Germany)*, *6*(4), 537–544.

Zhang, M., Yuan, R., Chai, Y., Wang, C., & Wu, X. (2013). Cerium oxide-graphene as the matrix for cholesterol sensor. *Analytical Biochemistry*, *436*(2), 69–74.

Zheng, Y., Jiao, Y., Ge, L., Jaroniec, M., & Qiao, S. Z. (2013). Two-step boron and nitrogen doping in graphene for enhanced synergistic catalysis. *Angewandte Chemie*, *125*(11), 3192–3198.

Zhou, X., Zou, T., & Chen, R. (2020). Sunlight-triggered dye degradation and antibacterial activity of graphene-iron oxide-titanium dioxide heterostructure nanocomposites. *Journal of Nanoscience and Nanotechnology*, *20*(7), 4158–4162.

Zhuang, Z., Su, X., Yuan, H., Sun, Q., Xiao, D., & Choi, M. M. (2008). An improved sensitivity non-enzymatic glucose sensor based on a CuO nanowire modified Cu electrode. *Analyst*, *133*(1), 126–132.

Liquid metal-based soft actuators and sensors for biomedical applications

23

Jun Shintake[1] *and Yegor Piskarev*[2]
[1]School of Informatics and Engineering, The University of Electro-Communications, Tokyo, Japan, [2]School of Engineering, Swiss Federal Institute of Technology Lausanne, Lausanne, Switzerland

23.1 Introduction

Actuators and sensors made of compliant materials are expected to find numerous applications in biomedical engineering as well as other fields such as soft robotics and stretchable electronics (Cianchetti, Laschi, Menciassi, & Dario, 2018; Gui, He, & Wang, 2021; Rus & Tolley, 2015; Shintake, Cacucciolo, Floreano, & Shea, 2018; Sun, Wang, Yuan, & Liu, 2020). In biomedical applications, the softness of the devices provided from their materials assures safety and adaptability to the human body. In most of the soft actuators and sensors, electrode materials are essential for powering them to perform physical movements and signal transmissions. As a conductive electrode material, liquid metals (LMs), especially gallium and its alloys, offer promising features such as high electrical/thermal conductivity and low toxicity (Daeneke, Khoshmanesh, & Mahmood, 2018; Dickey, 2014, 2017). Gallium has a melting temperature of 30°C, which can be lowered by adding other metals such as indium and tin. Eutectic gallium indium (EGaIn) and gallium indium tin (galinstan) have a melting temperature of 15.7°C and −19°C, respectively, keeping them in the form of liquid at room temperature. The fluidic nature of LMs exhibits negligible impact on the softness of actuators and sensors. In this chapter, soft actuators and sensors using LMs are discussed, which pave a way for the development of biomedical applications.

23.2 Soft actuators

Actuators are one of the key elements responsible for the physical movement of soft devices (El-Atab, Mishra, & Al-Modaf, 2020; Hines, Petersen, Lum, & Sitti, 2017; Polygerinos, Correll, & Morin, 2017). Researchers have developed LM-based soft actuators that work with the principle of electromagnetism. Guo et al. have demonstrated a stretchable electromagnet in which LM is formed in a planar coiled

Metal Oxides for Biomedical and Biosensor Applications. DOI: https://doi.org/10.1016/B978-0-12-823033-6.00021-1

shape and is encapsulated in an elastomeric structure (Guo, Sheng, Gong, & Liu, 2018). Applying an electric current to the coil creates a magnetic field that leads to deformation of the entire structure due to the presence of a permanent magnet nearby, which makes attractive force. Other examples exploit electrostatic actuation generated with a set of electrodes sandwiching a dielectric membrane. This type of soft actuator is classified as electroactive polymers, specifically dielectric elastomer actuators (DEAs) (Brochu & Pei, 2010; Pelrine, Kornbluh, Pei, & Joseph, 2000). Liu et al. have shown that a DEA with LM electrodes is capable of generating large planar deformations up to 300% of area strain (Liu, Gao, Mei, Han, & Liu, 2013). Interestingly, the LM electrodes have exhibited a self-healing capability allowing them to generate actuation even after getting electrical breakdown in the dielectric membrane. Wissman et al. have studied the behavior of a DEA in which LM electrodes are laminated with elastomeric substrates (Wissman, Finkenauer, Deseri, & Majidi, 2014). This configuration enables bending deformation upon application of the voltage, representing the high potential of LMs for achieving versatile movements.

The solid–liquid phase transition of LMs across the melting point can be an effective characteristic for soft actuators, namely variable stiffness (VS) (Manti, Cacucciolo, & Cianchetti, 2016). This functionality enables to control structural rigidity of the actuators and provides both adaptability through softness and force transmission through hardness. LM alloys of a melting temperature higher than room temperature are used for this purpose (typically, up to 62°C). Normally, those alloys are encapsulated in an elastomeric structure. Stiffness change factor achieved by LM alloys ranges from 25 to $10^4\times$ (Schubert & Floreano, 2013; Shan, Lu, & Majidi, 2013; Shintake, Schubert, Rosset, Shea, & Floreano, 2015; Tonazzini, Mintchev, & Schubert, 2016), which depends on the material composition and structural configuration. Those VS elements are integrated to various actuation technologies such as fluidic elastomer actuators (Hao et al., 2018; Tonazzini et al., 2016; Yoshida, Morimoto, Zheng, Onoe, & Takeuchi, 2018), DEAs (Piskarev et al., 2020; Shintake, Schubert, Rosset, Shea, & Floreano, 2015), shape memory alloys (Wang, Rodrigue, & Ahn, 2015), electromagnetic motors (Nakai, Kuniyoshi, Inaba, & Inoue, 2002), and external magnetic fields (Chautems, Tonazzini, & Boehler, 2020).

An example of VS actuator developed by the authors of this chapter is shown in Fig. 23.1A, consisting of LM embedded substrate and DEA (Shintake, Schubert, Rosset, Shea, & Floreano, 2015). The VS substrate is made of a polydimethylsiloxane (PDMS) in which a track of LM (Cerrolow 117, HiTech Alloys, melting temperature 47°C) is encapsulated. When an electric current is applied to the LM track, it changes the phase from solid to liquid by Joule-heating. This phase change softens the device structure which is initially rigid, as displayed in Fig. 23.1B and C. In the actuator, the DEA attached is initially prestretched so that voltage-controlled bending deformation is achieved (Fig. 23.2A). The working principle of the actuator is represented in Fig. 23.2B. As mentioned earlier, when an electric current is applied to the LM, the Joule heat causes the phase change of the alloy from solid to liquid, resulting in a soft state. In this state: (i) the entire structure bends with a

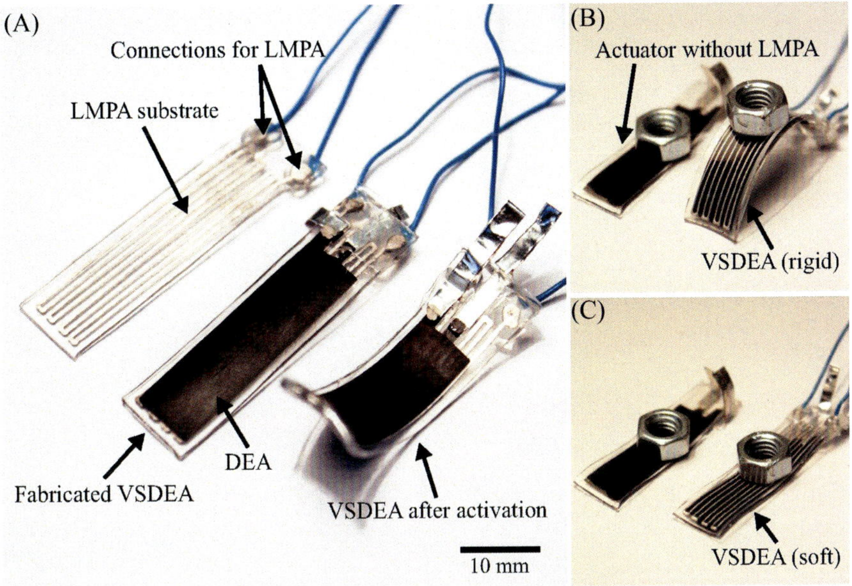

Figure 23.1 An example of variable stiffness (VS) actuator made with dielectric elastomer actuators (VSDEA). In this figure, LMPA refers to low-melting point alloy. (A) The actuator consists of a liquid metal (LM) embedded substrate and a DEA. (B) Compared to an actuator without the LM, the rigidity of VSDEA is visible. (C) The activation of the VS substrate makes the structure soft.

radius of curvature that minimizes the total energy of the device (strain energy of the DEA and bending energy of the substrate). (ii) When the voltage is applied to the DEA while keeping the LM substrate soft, the electrostatic energy of the DEA shifts the total energy and thus the device works to bend toward the flat state. The amount of bending can be controlled by the magnitude of voltage applied to the DEA. (iii) While the LM substrate deactivates, the DEA is kept active until the structure solidifies. (iv) After that, a rigid state for the desired bending (or planar) shape can be achieved. The rigid states (iv) and (v) mean that the device can rigidify all the shapes achievable in the soft state.

An application of variable stiffness dielectric elastomer actuator (VSDEA) discussed above is soft grippers from which high adaptability to the object and strong holding force are expected. Fig. 23.3A–I shows the sequence of operation of a gripper consisting of two VSDEAs (Shintake, 2016). In order to visualize the effect of VS, pick up of the object relying only on the soft state is firstly shown in the sequence (A)–(F). (A) In the first stage, the gripper is on top of the object, a plastic dish filled with a metal washer of mass $\sim$11 g ($\sim$108 mN). The fingers are softened by activating the LM substrate and (B) a voltage is applied to the DEA, making the fingers open. (C) Lower the gripper by the external motorized stage. (D) Deactivation of the DEA leads to closing the fingers to grasp the object with the

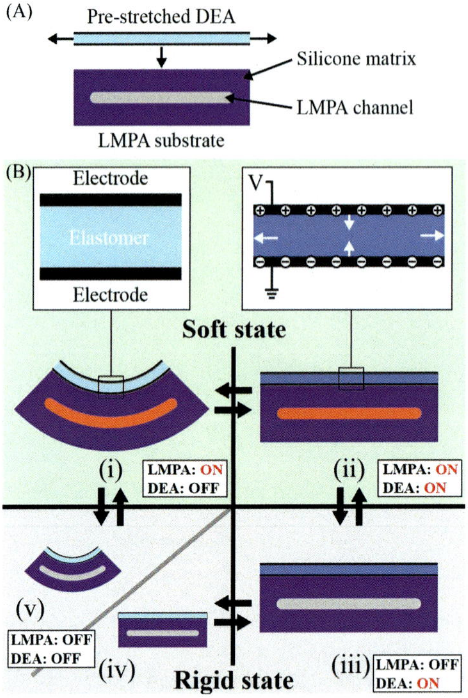

Figure 23.2 Working principle of variable stiffness dielectric elastomer actuator (VSDEA). In this figure, LMPA refers to low-melting point alloy. (A) The actuator is made of a prestretched DEA attached onto a VS substrate. (B) (i) The activation of the LMPA makes the structure soft, resulting in a bending shape. (ii) At this state, applying a voltage to the DEA leads to a bending actuation toward the flat shape. (iii and iv) Deactivation of the liquid metal (LM) substrate keeps a desired bending shape. (v) Re-activation of the LM substrate through state (i) allows changing of the rigid shape.

passive adaptation of the compliant structure. (E) The gripper is then raised to pick up the object. However, (F) the device eventually loses the object because of low holding force that results from the high compliance of the structure. Next, instead of manipulating the gripper only in a soft state, after the sequence of (A)–(D), (G) deactivates the LM substrate to rigidify the fingers. (H–I) Thanks to the improved rigidity, sufficient holding force is obtained and the object is successfully picked up.

23.3 Soft sensors

Highly stretchable/deformable sensors that can detect strains, pressures, proximity, etc. are an important element to assess the current status and perform control of

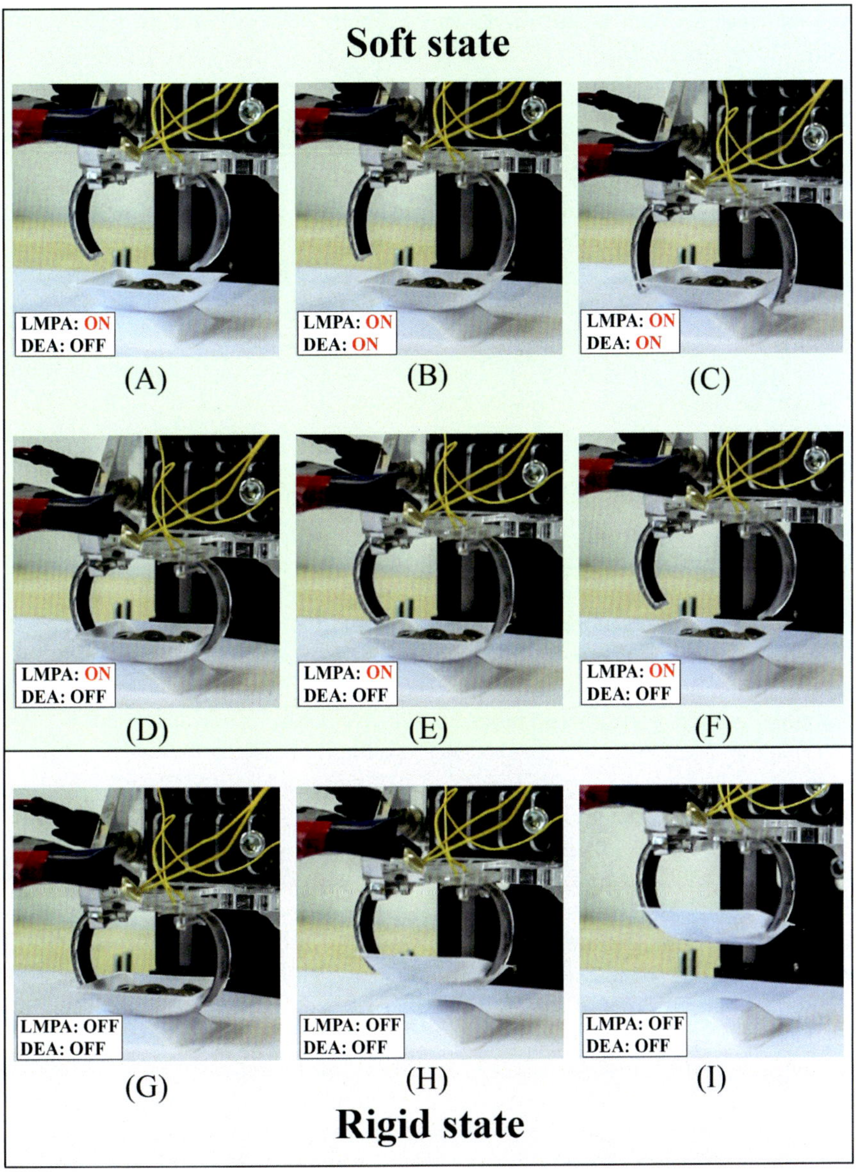

Figure 23.3 A gripper consisting of two variable stiffness dielectric elastomer actuators (VSDEAs). An external motorized stage is used to move the device up and down for picking up the object. Sequences (A)–(F) show picking up of the object based only on the soft state. Because of the high compline, it is not possible to pick up the object. When exploiting the rigid state in sequences (G)–(I), the gripper can pick up the object, thanks to sufficient holding force.

compliant systems such as soft robots and wearable devices (Amjadi, Kyung, Park, & Sitti, 2016; Souri, Banerjee, & Jusufi, 2020). LMs have a good matching to soft sensors thanks to their fluidic nature and electrical properties. So far, many different types of sensors have been developed (Kim, Kim, Lee, & Lee, 2019), in which LMs are used as electrodes. In some cases, LMs are mixed with elastomers to be used as dielectrics (Yang, Tang, & Ao, 2020). In those sensors discussed above, the change in resistance, capacitance, or inductance allows measuring the deformations. Chen et al. have demonstrated a resistive strain sensor capable of detecting strains up to 500% (Chen, Wang, & Guo, 2020). Their sensor has a fiber structure made of polyurethane and polymethacrylate, covered with EGaIn. A strong bonding between polymethacrylate and EGaIn keeps the alloy on the device surface. Lu et al. have instead employed a planar structure for a resistive sensor, which consists of a track of EGaIn encapsulated in an elastomer substrate (Lu, Chen, Hao, Luo, & Wang, 2020). With this sensor, they have shown that the curvature of fluidic elastomer actuators can be detected.

When LM is formed in a coiled shape, deformations can be read as the change in inductance. Zhou et al. have studied the sensing characteristics of coiled LM (galinstan) encapsulated in an elastomeric tube (Zhou, Gao, & Zhan, 2018). They have shown that the inductance of LM coil is proportional to the applied deformation, thus allowing to capture the amount of bending in both human fingers and fluidic elastomer actuators. In addition to the above methods, capacitive sensing is also one of the major ones. Cooper et al. have developed a capacitive sensor made of double twisted elastomeric tubes in which EGaIn is encapsulated (Cooper, Arutselvan, & Liu, 2017). By reading the capacitance between the tubes, sensing of strains up to 100% is achieved. Thanks to its unique configuration, the sensor also allows measuring torsion and touch.

Another example of LM-based capacitive sensor is shown in Fig. 23.4. The sensor, developed by the authors of this chapter (Shintake, Nagai, & Ogishima, 2019), has a structural configuration that aims to increase the sensitivity of strain detection. Normally, a capacitive strain sensor is composed of a dielectric laminated between two soft electrodes as represented in Fig. 23.4A. When strain is applied in the longitudinal direction, the area of the electrode increases, and the thickness of the dielectric decreases, resulting in an increase of the capacitance. The capacitance is linearly proportional to the strain, and sensitivity is theoretically 1, which is much lower than that of resistive type sensors (Shintake, Piskarev, Jeong, & Floreano, 2018). The idea of increasing the sensitivity is to integrate auxetic structures that enable to mechanically program the deformation of the sensor. Auxetics are a class of metamaterial having a negative Poisson's ratio (Ren, Das, Tran, Ngo, & Xie, 2018; Wang, Luan, & Liao, 2020), which increases the amount of change in the electrode area with applied strains, as illustrated in Fig. 23.4B. This results in an increase in the rate of capacitance change therefore the sensitivity. Fig. 23.4C details the structure of developed sensor. EGaIn and an acrylic elastomer (VHB4905, 3 M) are employed as the electrodes and the dielectric, respectively. The electrodes are encapsulated by laminating additional layers made of the acrylic elastomer. A couple of auxetic layers made of PDMS are attached to the top and

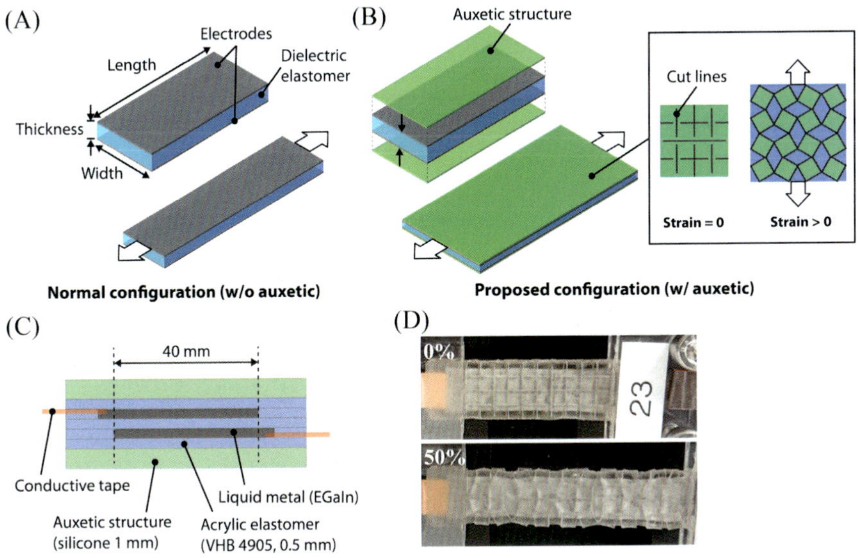

Figure 23.4 Capacitive strain sensors based on liquid metal (LM). (A) Schematics of a capacitive strain sensor. (B) Schematics of a capacitive strain sensor integrated with auxetic structures. (C) Cross-section view of the sensor (length direction). (D) Deformation of the sensor.

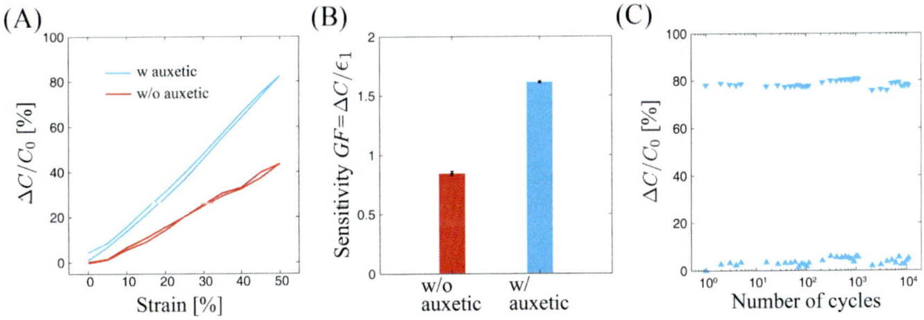

Figure 23.5 Characterization results of the liquid metal (LM)-based sensors. (A) Measured sensor response as a function of the applied strain. (B) Calculated sensitivity. (C) Measured sensor response for repeated strains of up to 10,000 cycles.

bottom of the structure. As displayed in Fig. 23.4D, the auxetic layers conform the sensor structure in such a way that the electrodes expand in the direction perpendicular to the applied strain. The measured sensor response is plotted in Fig. 23.5A. As a comparison, the response of a sensor without auxetic structure is also shown in this figure. It is obvious that the rate of change in the capacitance is increased by the auxetic structure. The sensitivity is calculated as 1.6, which is roughly two times higher than that of the normal one (Fig. 23.5B). Moreover, thanks to the excellent electrical and mechanical properties of the LM, the sensor is able to withstand the repeated strain of up to 10^4 cycles as shown in Fig. 23.5C.

23.4 Summary

In this chapter, existing LM-based soft actuators and sensors are discussed while the technical details of some examples are explained. The high electrical and thermal conductivity, as well as low toxicity of LM, makes it a promising conductive material. Moreover, its liquid phase is effective in minimizing the effect on the compliance of actuators and sensors, while tuning the melting temperature enables VS functionality. Future researches on the configuration of these devices and the composition of the materials are expected to advance the development and commercialization of biomedical applications.

References

Amjadi, M., Kyung, K.-U. U., Park, I., & Sitti, M. (2016). Stretchable, skin-mountable, and wearable strain sensors and their potential applications: A review. *Advanced Functional Materials*, *26*, 1678–1698. Available from https://doi.org/10.1002/adfm.201504755.

Brochu, P., & Pei, Q. (2010). Advances in dielectric elastomers for actuators and artificial muscles. *Macromolecular Rapid Communications*, *31*, 10–36. Available from https://doi.org/10.1002/marc.200900425.

Chautems, C., Tonazzini, A., Boehler, Q., Jeong, S. H., Floreano, D., & Nelson, B. J. (2020). Magnetic continuum device with variable stiffness for minimally invasive surgery. *Advances in Intelligent Systems and Computing*, *2*, 1900086. Available from https://doi.org/10.1002/aisy.201900086.

Chen, G., Wang, H., Guo, R., Duan, M., Zhang, Y., & Liu, J. (2020). Superelastic EGaIn composite fibers sustaining 500% tensile strain with superior electrical conductivity for wearable electronics. *ACS Applied Materials & Interfaces*, *12*, 6112–6118. Available from https://doi.org/10.1021/acsami.9b23083.

Cianchetti, M., Laschi, C., Menciassi, A., & Dario, P. (2018). Biomedical applications of soft robotics. *Nature Reviews Materials*, *3*, 143–153. Available from https://doi.org/10.1038/s41578-018-0022-y.

Cooper, C. B., Arutselvan, K., Liu, Y., Armstrong, D., Lin, Y., Khan, M. R., … Dickey, M. D. (2017). Stretchable capacitive sensors of torsion, strain, and touch using double helix liquid metal fibers. *Advanced Functional Materials*, *27*, 1605630. Available from https://doi.org/10.1002/adfm.201605630.

Daeneke, T., Khoshmanesh, K., Mahmood, N., de Castro, I. A., Esrafilzadeh, D., Barrow, S. J., … Kalantar-zadeh, K. (2018). Liquid metals: Fundamentals and applications in chemistry. *Chemical Society Reviews*, *47*, 4073–4111. Available from https://doi.org/10.1039/C7CS00043J.

Dickey, M. D. (2014). Emerging applications of liquid metals featuring surface oxides. *ACS Applied Materials & Interfaces*, *6*, 18369–18379. Available from https://doi.org/10.1021/am5043017.

Dickey, M. D. (2017). Stretchable and soft electronics using liquid metals. *Advanced Materials*, *29*, 1606425. Available from https://doi.org/10.1002/adma.201606425.

El-Atab, N., Mishra, R. B., Al-Modaf, F., Joharji, L., Alsharif, A. A., Alamoudi, H., … Hussain, M. M. (2020). Soft actuators for soft robotic applications: A review. *Advances in Intelligent Systems and Computing*, *2*, 2000128. Available from https://doi.org/10.1002/aisy.202000128.

Gui, Q., He, Y., & Wang, Y. (2021). Soft electronics based on liquid conductors. *Advanced Electronic Materials*, *7*, 2000780. Available from https://doi.org/10.1002/aelm.202000780.

Guo, R., Sheng, L., Gong, H. Y., & Liu, J. (2018). Liquid metal spiral coil enabled soft electromagnetic actuator. *Science China Technological Sciences*, *61*, 516–521. Available from https://doi.org/10.1007/s11431-017-9063-2.

Hao, Y., Wang, T., Xie, Z., Sun, W., Liu, Z., Fang, X., ... Wen, L. (2018). A eutectic-alloy-infused soft actuator with sensing, tunable degrees of freedom, and stiffness properties. *Journal of Micromechanics and Microengineering*, *28*, 024004. Available from https://doi.org/10.1088/1361-6439/aa9d0e.

Hines, L., Petersen, K., Lum, G. Z., & Sitti, M. (2017). Soft actuators for small-scale robotics. *Advanced Materials*, *29*, 1603483. Available from https://doi.org/10.1002/adma.201603483.

Kim, T., Kim, D., Lee, B. J., & Lee, J. (2019). Soft and deformable sensors based on liquid metals. *Sensors*, *19*, 4250. Available from https://doi.org/10.3390/s19194250.

Liu, Y., Gao, M., Mei, S., Han, Y., & Liu, J. (2013). Ultra-compliant liquid metal electrodes with in-plane self-healing capability for dielectric elastomer actuators. *Applied Physics Letters*, *103*, 064101. Available from https://doi.org/10.1063/1.4817977.

Lu, S., Chen, D., Hao, R., Luo, S., & Wang, M. (2020). Design, fabrication and characterization of soft sensors through EGaIn for soft pneumatic actuators. *Measurement*, *164*, 107996. Available from https://doi.org/10.1016/j.measurement.2020.107996.

Manti, M., Cacucciolo, V., & Cianchetti, M. (2016). Stiffening in soft robotics: A review of the state of the art. *IEEE Robotics and Automation Magazine*, *23*, 93–106. Available from https://doi.org/10.1109/MRA.2016.2582718.

Nakai, H., Kuniyoshi, Y., Inaba, M., & Inoue, H. (2002). Metamorphic robot made of low melting point alloy. In *IEEE/RSJ international conference on intelligent robots and system* (pp. 2025–2030). IEEE.

Pelrine, R., Kornbluh, R., Pei, Q., & Joseph, J. (2000). High-speed electrically actuated elastomers with strain greater than 100%. *Science (80-)*, *287*, 836–839. Available from https://doi.org/10.1126/science.287.5454.836.

Piskarev, E., Shintake, J., Ramachandran, V., Baugh, N., Dickey, M. D., & Floreano, D. (2020). Lighter and stronger: Cofabricated electrodes and variable stiffness elements in dielectric actuators. *Advanced Intelligent Systems*, *2*, 2000069. Available from https://doi.org/10.1002/aisy.202000069.

Polygerinos, P., Correll, N., Morin, S. A., Mosadegh, B., Onal, C. D., Petersen, K., ... Shepherd, R. F. (2017). Soft robotics: Review of fluid-driven intrinsically soft devices; manufacturing, sensing, control, and applications in human-robot interaction. *Advanced Engineering Materials*, *19*, 1700016. Available from https://doi.org/10.1002/adem.201700016.

Ren, X., Das, R., Tran, P., Ngo, T. D., & Xie, Y. M. (2018). Auxetic metamaterials and structures: A review. *Smart Materials and Structures*, *27*, 023001. Available from https://doi.org/10.1088/1361-665X/aaa61c.

Rus, D., & Tolley, M. T. (2015). Design, fabrication and control of soft robots. *Nature*, *521*, 467–475. Available from https://doi.org/10.1038/nature14543.

Schubert, B. E., & Floreano, D. (2013). Variable stiffness material based on rigid low-melting-point-alloy microstructures embedded in soft poly(dimethylsiloxane) (PDMS). *RSC Advances*, *3*, 24671. Available from https://doi.org/10.1039/c3ra44412k.

Shan, W., Lu, T., & Majidi, C. (2013). Soft-matter composites with electrically tunable elastic rigidity. *Smart Materials and Structures*, *22*, 085005. Available from https://doi.org/10.1088/0964-1726/22/8/085005.

Shintake, J. (2016). *Functional soft robotic actuators based on dielectric elastomers*. Thesis 6855:149. https://doi.org/10.5075/EPFL-THESIS-6855.

Shintake, J., Cacucciolo, V., Floreano, D., & Shea, H. (2018). Soft robotic grippers. *Advanced Materials*, *30*, 1707035. Available from https://doi.org/10.1002/adma.201707035.

Shintake, J., Nagai, T., & Ogishima, K. (2019). Sensitivity improvement of highly stretchable capacitive strain sensors by hierarchical auxetic structures. *Front Robot AI*, *6*, 1–7. Available from https://doi.org/10.3389/frobt.2019.00127.

Shintake, J., Piskarev, E., Jeong, S. H., & Floreano, D. (2018). Ultrastretchable strain sensors using carbon black-filled elastomer composites and comparison of capacitive vs resistive sensors. *Advanced Materials Technologies*, *3*, 1700284. Available from https://doi.org/10.1002/admt.201700284.

Shintake, J., Schubert, B., Rosset, S., Shea, H., & Floreano, D. (2015). *Variable stiffness actuator for soft robotics using dielectric elastomer and low-melting-point alloy* (pp. 1097–1102). IEEE.

Souri, H., Banerjee, H., Jusufi, A., Radacsi, N., Stokes, A. A., Park, I., ... Amjadi, M. (2020). Wearable and stretchable strain sensors: Materials, sensing mechanisms, and applications. *Advanced Intelligent Systems*, *2*, 2000039. Available from https://doi.org/10.1002/aisy.202000039.

Sun, X., Wang, X., Yuan, B., & Liu, J. (2020). Liquid metal–enabled cybernetic electronics. *Materials Today Physics*, *14*, 100245. Available from https://doi.org/10.1016/j.mtphys.2020.100245.

Tonazzini, A., Mintchev, S., Schubert, B., Mazzolai, B., Shintake, J., & Floreano, D. (2016). Variable stiffness fiber with self-healing capability. *Advanced Materials*, *28*, 10142–10148. Available from https://doi.org/10.1002/adma.201602580.

Wang, W., Rodrigue, H., & Ahn, S.-H. (2015). Smart soft composite actuator with shape retention capability using embedded fusible alloy structures. *Composites Part B: Engineering*, *78*, 507–514. Available from https://doi.org/10.1016/j.compositesb.2015.04.007.

Wang, Z., Luan, C., Liao, G., Liu, J., Yao, X., & Fu, J. (2020). Progress in auxetic mechanical metamaterials: Structures, characteristics, manufacturing methods, and applications. *Advanced Engineering Materials*, 1–23. Available from https://doi.org/10.1002/adem.202000312, 2000312.

Wissman, J., Finkenauer, L., Deseri, L., & Majidi, C. (2014). Saddle-like deformation in a dielectric elastomer actuator embedded with liquid-phase gallium-indium electrodes. *Journal of Applied Physics*, *116*, 144905. Available from https://doi.org/10.1063/1.4897551.

Yang, J., Tang, D., Ao, J., Ghosh, T., Neumann, T. V., Zhang, D., ... Dickey, M. D. (2020). Ultrasoft liquid metal elastomer foams with positive and negative piezopermittivity for tactile sensing. *Advanced Functional Materials*, *30*, 2002611. Available from https://doi.org/10.1002/adfm.202002611.

Yoshida, S., Morimoto, Y., Zheng, L., Onoe, H., & Takeuchi, S. (2018). Multipoint bending and shape retention of a pneumatic bending actuator by a variable stiffness endoskeleton. *Soft Robot*, *5*, 718–725. Available from https://doi.org/10.1089/soro.2017.0145.

Zhou, L.-Y., Gao, Q., Zhan, J.-F., Xie, C. Q., Fu, J. Z., & He, Y. (2018). Three-dimensional printed wearable sensors with liquid metals for detecting the pose of snakelike soft robots. *ACS Applied Materials & Interfaces*, *10*, 23208–23217. Available from https://doi.org/10.1021/acsami.8b06903.

Index

Note: Page numbers followed by "*f*" and "*t*" refer to figures and tables, respectively.

Printed in the United States
by Baker & Taylor Publisher Services